Fundamental Concepts of Molecular Spectroscopy

This practical and unique textbook explains the core areas of molecular spectroscopy as a classical teacher would, from the perspective of both theory and experimental practice. Comprehensive in scope, the author carefully explores and explains each concept, walking side by side with the student through carefully constructed text, pedagogy, and derivations to ensure comprehension of the basics before approaching higher level topics. The author incorporates both electric resonance and magnetic resonance in the textbook.

Fundamental Concepts of Molecular Spectroscopy

Abani K. Bhuyan

CRC Press
Taylor & Francis Group
Boca Raton London New York

CRC Press is an imprint of the
Taylor & Francis Group, an **informa** business

First edition published 2023
by CRC Press
6000 Broken Sound Parkway NW, Suite 300, Boca Raton, FL 33487-2742

and by CRC Press
4 Park Square, Milton Park, Abingdon, Oxon, OX14 4RN

CRC Press is an imprint of Taylor & Francis Group, LLC

ISBN: 9781032274850 (hbk)
ISBN: 9781032274959 (pbk)
ISBN: 9781003293064 (ebk)

DOI: 10.1201/9781003293064

Typeset in Times
by Newgen Publishing UK

Contents

Preface

Spectroscopy has an amazing history of development dating back some 350 years. The initial quest for the solar spectrum and light dispersion followed by flame spectroscopy, which is still performed in high school senior grades, already implied the importance of spectroscopy in physics, chemistry, and astronomy. The discovery of the hydrogen atom spectrum and the splitting of spectral lines in the presence of a magnetic field provided the first testing ground for quantum theory to be developed soon after. While small molecules became amenable to electric resonance measurements, the discovery of nuclear magnetic resonance (NMR) came as a giant leap, soon adding to the importance of spectroscopy in biology and medicine. Contemporary development of maser and lasers toward the second half of the twentieth century has been instrumental in the advancement of rotational and vibrational coherence spectroscopy.

Coverage of these later developments is scarcely found in the few available graduate-level texts on molecular spectroscopy. Coherence spectroscopy and NMR in particular, both of which have the footing of density matrix, only occasionally find their way into spectroscopy textbooks. One of the reasons for the marginal coverage could be the vastness of the areas as reflected by several excellent monographs and reference books in the respective fields. Even so, a fairly detailed introduction and discussion of these topics in graduate courses of molecular spectroscopy is imperative so that the whole subject is presented in a unified manner. Another opinion pertains to the use of supplements in textbooks. Inserts in the form of supplementary boxes often serve as tutorials and aids to the comprehension of student readers. Consideration of these aspects basically led to the development of this textbook, apart from the objective to present the overall subject more tangibly than before.

The material for the textbook grew from the general model of the two-level system taught by Professor Robin M. Hochstrasser in the fall of 1989. To it are added the course material, class notes, transcripts, and board scribbles of a core course in molecular spectroscopy I taught for some years intermittently at the University of Hyderabad. The concepts of rotational and vibrational coherence spectroscopy are prepared and updated by referring to several laser-based experiments published mainly from Professor Ahmed H. Zewail's laboratory. Material I have used already to teach an advanced one-semester course in magnetic resonance has been developed further to prepare the NMR chapter. Density matrix in spectroscopy is introduced along with the coherence phenomenon and is carried over to discuss the theoretical description of NMR experiments. I immensely profited from several monographs, in particular those by Born and Wolf, Bethe and Salpeter, Townes and Schawlow, and Ernst, Bodenhausen, and Wokaun. Images and figures, some of which represent actual calculations, simulations and experimental data, are used generously to aid in both explanation and comprehension. A few copyright figures are presented with permission courtesy of publishers or by purchasing reproduction licenses.

Considerable effort has been made to use standard symbols and abbreviations to the extent possible, although symbol usage may occasionally appear inconsistent from one chapter to another. For example, the kinetic energy of electronic motion is symbolized by V_N in Chapter 9 in contrast with T_r elsewhere. Whereas a hydrogen atom transition is depicted as $nlm \rightarrow n'l'm'$ where the primes refer to the excited state, vibrational transitions are denoted by $v' \rightarrow v$ or equivalently $v \leftarrow v'$, the prime now referring to the lower energy state. Small case letters **j**, **l**, **s** are generally used to denote the respective angular momentum vectors of one-electron system, reserving **J**, **L**, **S** for multi-electron cases. Even if the cases of the symbols appear non-uniform, they should be understood in the context of the system discussed. The inconsistency in the use of symbols at times arises from the lack of standard usage in the original research literature and monographs that have been consulted to prepare the material for the textbook.

Finding individuals to review this first edition proved to be difficult, perhaps due to the occupational business in academia. Nonetheless, I would like to thank Dr. David Tuschel for his critiques and comments on the chapter on Raman spectroscopy, Professor M. V. Rajasekharan for his valuable remarks and suggestions on the material concerning molecular eigenstates, and Professor Debasish Barik for his comments and criticisms on the chapter on rotational spectroscopy. I will be grateful to the readers if they would let me know of the remaining errors and conceptual difficulties, which will then be corrected in the next edition of the book.

It is hoped that the book provides a thorough basis for understanding molecular spectroscopy at graduate level and will serve as a reference for those who integrate theory and experiments in both electric resonance and nuclear magnetic resonance spectroscopy.

Abani K. Bhuyan
June 2022

About the Author

Abani K. Bhuyan has been in the Chemistry faculty at the University of Hyderabad since 2000 and is currently a Senior Professor of Physical Chemistry. He received his PhD in Molecular Biophysics from the University of Pennsylvania in 1995 and was a Visiting Fellow at Tata Institute of Fundamental Research from 1995 to 2000.

1 Electromagnetic Wave Nature of Light

This chapter summarizes the origin of the electromagnetic properties of light. It is mentioned that the laws of electrostatics and electromagnetic induction that emanated largely from experimental observations provide an unsophisticated approach to the derivation of the wave equation of light. The derivation of the wave equation from Maxwell's electromagnetic equations involves differential equations of the vector fields because the electric and magnetic field vectors are functions of space and time. The simplest form of the wave equation and the formation of a wave packet by superposition of plane monochromatic waves are described. The results explicate the principle of superposition of traveling plane waves and discuss light pulse compression as often done in laser spectroscopy.

1.1 GAUSS'S LAW OF ELECTROSTATICS

To determine electric and magnetic field vectors, a fundamental concept called flux is of great use. While the vector field implies both the magnitude and spatial location of electric or magnetic field, the quantity flux appears when the source of the field is enclosed by a two- or three-dimensional surfaces. Note that a three-dimensional surface can be constructed by summing up the two-dimensional surfaces just the way a sphere is obtained by appropriately juxtaposing loops of different circumferences. A three-dimensional enclosing surface is a closed one, often called a Gaussian surface. The interest here is the flux of vector fields unlike the general meaning of flux that refers to the rate of flow of energy or a substance through the surface. To define the flux of an electric or magnetic vector field, consider the electric lines of force due to a pair of charges in space (Figure 1.1). If a charge is enclosed by a Gaussian surface, the vector field **E** is, by definition, a function of its location at the surface, $\mathbf{E}(x,y,z)$ or $\mathbf{E}(\mathbf{r})$. For a small surface element Δs, there is a unit vector **s** pointing outward from Δs. The normal component of the electric field ϕ at Δs is then obtained from the scalar product of **E** and **s**:

$$\phi = \mathbf{E}\cdot\mathbf{s} = Es\cos\theta.$$

Because **E** is a function of **r**, its magnitude and direction will be different for different surface elements. With n number of elements making up the surface, the outward electric field flux through the surface is written as

$$\phi_\mathrm{E} = \sum_{i=1}^{n} \phi_i a_i,$$

where a_i is the surface area of the element ds_i, and the flux over the entire closed surface is obtained from the surface integral

$$\phi_\mathrm{E} = \oint \phi da.$$

The same treatment will work when a closed surface around the negative charge is considered; the only distinction made is inward electric field flux now. The outward and inward fluxes are also called positive and negative flux, respectively. Gauss's law of electrostatics provides a connection between ϕ_E for the surface and the total charge q enclosed

$$q = \varepsilon_\mathrm{o}\phi_\mathrm{E} = \varepsilon_\mathrm{o}\oint \phi da, \tag{1.1}$$

where ε_o, the electric permittivity in free space, is 8.85×10^{-12} Fm^{-1} (see Box 1.1). Permittivity is the ability of a substance to store electrical energy in an electric field.

1.2 GAUSS'S LAW OF MAGNETISM

This law states that the magnetic flux ϕ_B out of a Gaussian surface vanishes

$$\phi_\mathrm{B} = \oint \phi da = 0. \tag{1.2}$$

The equation does not imply that no magnetic vector field leaves and enters the surface; rather each field line leaving the surface is balanced out by a field line entering the surface. Equivalently, the flux into the surface is equal to the flux out of the surface (Figure 1.2), so net flux is zero. The balanced flux of field lines must mean that the two magnet poles cannot be distinguished as the source and the sink of the magnetic field lines, in contrast with the distinction of electric charges as the source and sink of electric field lines. The magnet poles are thus inseparable, which can be stated by the fact that magnetic monopoles have not been found to date.

1.3 FARADAY'S LAW OF INDUCED ELECTRIC FIELD

An induced current appears in a conducting loop when it is moved relatively near a magnetic field or another current-carrying loop. The meaning of relative motion is 'the movement of either of the two' with respect to each other. The induced emf or the electric current I in the wire loop is equal to the negative of the rate at which the magnetic flux through the loop changes

DOI: 10.1201/9781003293064-1

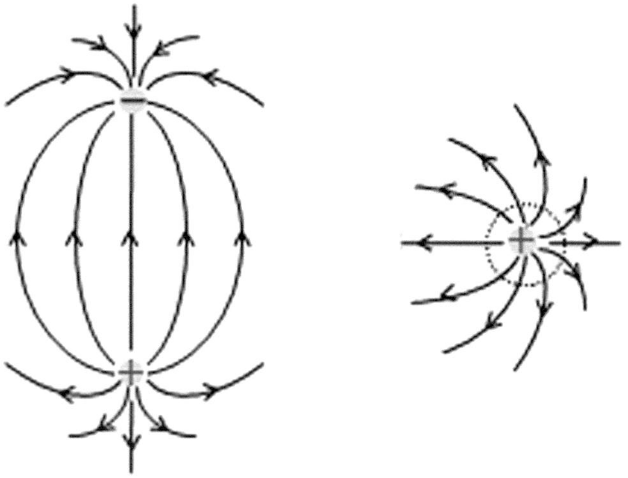

FIGURE 1.1 Electric lines of force for the cases of two opposite charges (*left*) and a single charge (*right*).

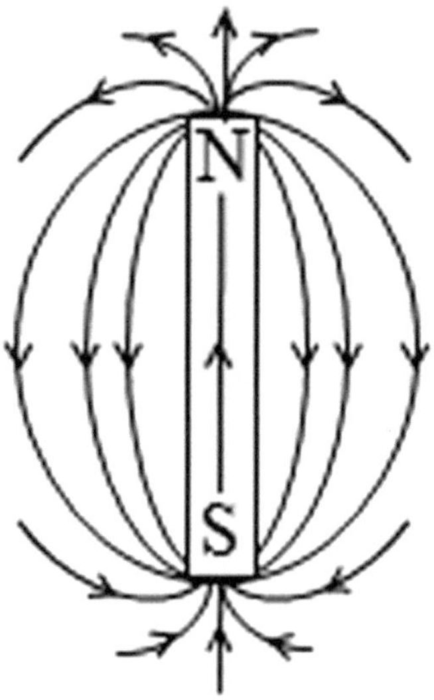

FIGURE 1.2 Magnetic flux equivalence for the case of a pair of charges.

$$I = -\frac{d\phi_B}{dt}.$$

The current is in Amp unit (1 Cs^{-1}). For a coil having n identical turns, each turn with the same ϕ_B,

$$I = -n\frac{d\phi_B}{dt}.$$

Faraday's law is often written in the form of the line integral of the electric field

$$I = \int_c \mathbf{E}\cdot d\mathbf{l} = -\frac{d\phi_B}{dt} \tag{1.3}$$

where C refers to a closed path. It is easy to see that no emf and hence no electric field will be produced in the wire if the magnet and the wire are held at a fixed distance. They must move relative to each other so as to provide a change in ϕ_B in real time.

1.4 AMPERE'S LAW OF INDUCED MAGNETIC FIELD

The quantitative relationship for the induction of magnetic effect by electric current due to Ampere is

$$\int_c \mathbf{B}\cdot d\mathbf{l} = \mu_o I_{enc}, \tag{1.4}$$

where μ_o is the permeability constant, also called permeability of free space ($4\pi\times10^{-7}$ Vs A^{-1} m^{-1} or $4\pi\times10^{-7}$ kg m s^{-2} A^{-2}), and I_{enc} is the electric current that passes through a surface bound by the closed path C (Figure 1.3). Note that μ_o is by definition an index of the degree of resistance to forming a magnetic field in vacuum.

Ampere's law initially did not consider the case of electric flux, the occurrence of which does not require the current passing through a surface. Maxwell added to the right side of Ampere's equation (1.4) a term containing the displacement current which is proportional to the time change in electric flux through a surface that is bounded by the closed path C, but no current passes through the surface. Maxwell-Ampere equation reads

$$\int_c \mathbf{B}\cdot d\mathbf{l} = \mu_o I_{enc} + \mu_o \varepsilon_o \frac{d}{dt}\int_c \phi da. \tag{1.5}$$

To note is the converse nature of Faraday and Maxwell-Ampere equations. If there were no charge carriers, then I_{enc} would be zero and the equation above reduces to

$$\int_c \mathbf{B}\cdot d\mathbf{l} = \mu_o \varepsilon_o \frac{d\mathrm{E}}{dt}, \tag{1.6}$$

which looks similar to the Faraday equation.

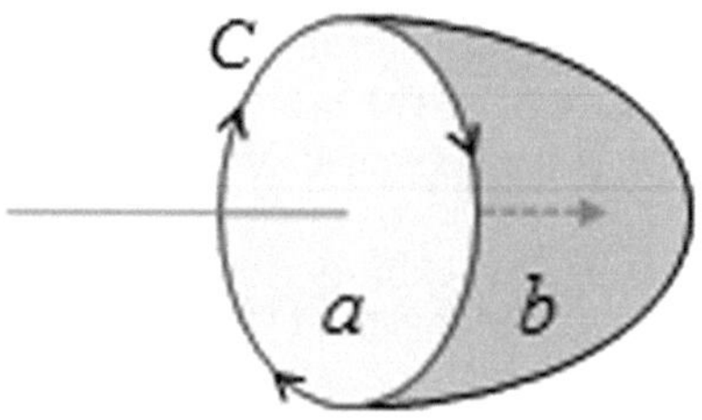

FIGURE 1.3 When an electric current passes through the surface a bounded by the closed path C (red loop), the current enclosed in this surface is I_{enc}, which induces a magnetic field looping upon itself – shown by arrowheads. The current enclosed by the surface b is zero, because no electric current passes through this surface. There can be an electric flux through surface b that would give rise to displacement current.

1.5 MAXWELL'S EQUATIONS

With the above supplement, Maxwell summed up the equations of electromagnetism with the ingenuity of connecting electromagnetism to optics so as to predict the electromagnetic nature of light. The starting point toward determining the possibility of existence of electromagnetic waves is to consider the interaction of an electromagnetic field, defined as a 'state of excitation' due to the presence of an electric charge somewhere in space, with material substances. The electromagnetic field is described by an electric field vector **E** and a magnetic field vector **B**, and to establish the effect of the fields on materials, a set of three other vectors called current density **j**, electric displacement **D**, and a magnetic vector **H** are introduced. Maxwell's equations read

$$\nabla\times\mathbf{H}-\frac{1}{c}\frac{d\mathbf{D}}{dt}=\frac{4\pi}{c}\mathbf{j} \tag{1.7}$$

$$\nabla\times\mathbf{E}+\frac{1}{c}\frac{d\mathbf{B}}{dt}=0 \tag{1.8}$$

$$\nabla\cdot\mathbf{D}=4\pi\rho \tag{1.9}$$

$$\nabla\cdot\mathbf{B}=0, \tag{1.10}$$

where c is the velocity of light in vacuum (~3×10^8 ms^{-1}), and ρ is electric charge density. The first two are vector relations entailing space and time derivatives **H** and **E**, respectively, and the latter two are scalar relations that require the divergence of vectors **D** and **B** (see Box 1.2).

Maxwell's equations are used in conjunction with a set of three material equations that describe the influence of the field on material behavior. The three equations are easy to write for isotropic material

$$\mathbf{j}=\sigma\mathbf{E} \tag{1.11}$$

$$\mathbf{D}=\varepsilon\mathbf{E} \tag{1.12}$$

$$\mathbf{B}=\mu\mathbf{H}, \tag{1.13}$$

where σ, the specific conductivity, determines electrical conductivity of a substance; ε, the dielectric constant or relative permittivity, which is the ratio of the absolute permittivity of the substance to the vacuum permittivity ε_o, determines the capacity of a material to store electric charge and hence generate electric flux; and the magnetic permeability μ determines whether a material is paramagnetic (μ>1) or diamagnetic (μ <1). Since the material referred to is isotropic, these three physical properties are constants and continuous throughout the material.

The material equations allow for expressing the **j D**, and **H** vectors in terms of **E** and **B**. The first of the two scalar equations (1.9) can be recast in cgs units (see Box 1.1)

$$\nabla\cdot\mathbf{E}=\frac{\rho}{\varepsilon_o}, \tag{1.14}$$

which is the differential form of Gauss's law for electric field in vacuum, and it shows that the electric field diverges from the positive charge and converges into the negative. The second scalar equation (1.10) straightforwardly provides the interpretation that magnetic density, unlike electric charge density, does not exist, because the two magnet poles are inseparable.

The two vector relations in the Maxwell's equations set above do not involve divergence. The second vector equation (1.8) can be recast as

$$\nabla\times\mathbf{E}=-\frac{1}{c}\frac{d\mathbf{B}}{dt}, \tag{1.15}$$

where c relates the unit of charge in electrostatic and electromagnetic units. If we took divergence of the equation, the rate of change of the magnetic field would be zero, because $\mathrm{div\,curl}=0$. We recognize that the induced electric field produced by electromagnetic induction is different from the electrostatic field involving a pair of opposite charges (Figure 1.4). The latter has a source (positive) and a sink (negative), and hence divergence near the source and sink exists, but the induced electric field lines in the case of electromagnetic induction have no source and sink – the field lines loop upon themselves, and therefore the divergence would be zero.

The first vector equation (1.7), written as

$$\nabla\times\mathbf{B}=\mu_o\mathbf{j}+\mu_o\varepsilon_o\frac{d\mathbf{E}}{dt}, \tag{1.16}$$

contains the conduction current density **j**, which is the amount of charge flowing through a unit cross-sectional area

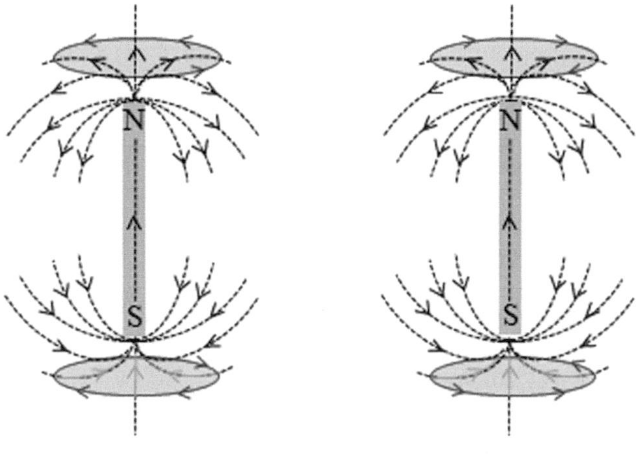

FIGURE 1.4 Direction of electric current induced in a loop is shown in red. *Left*, the magnet closes into the loop; *right*, the magnet pulls away from the current loop. The direction of current reverses according to the motion of the magnetic field. It is easy to see that magnetic flux lines will increase when the magnet closes in and this increase is offset by the magnetic flux produced by the induced electric current. The reverse will be the case when the magnet moves away.

perpendicular to the current direction. The displacement current density $\varepsilon_0 d\mathbf{E}/dt$ need not be taken as a physical flow of current; it rather implies that the time change of an electric field in space produces a magnetic field even when there is no physical passage of current. This must also mean that a time-varying magnetic field in space will produce an electric field. This is an underlying concept that conveys that electromagnetic waves propagate not only in space, but also through free space.

1.6 WAVE EQUATION

The derivation of the wave equation from Maxwell's electromagnetic equations involves differential equations of the vector fields because the electric and magnetic field vectors are functions of space and time. One considers only that region of the field which contains no charge and current, which means $\mathbf{j}$ and ρ are equal to zero. With $\mathbf{j}=0$, $\mathbf{D}=\varepsilon\mathbf{E}$, and $\mathbf{B}=\mu\mathbf{H}$, the first two of the Maxwell's equations are written as

$$\nabla\times\mathbf{H}-\frac{\varepsilon}{c}\frac{d\mathbf{E}}{dt}=0 \tag{1.17}$$

$$\nabla\times\mathbf{E}+\frac{\mu}{c}\frac{d\mathbf{H}}{dt}=0. \tag{1.18}$$

Differentiation of equation (1.17) with respect to time gives

$$\nabla\times\frac{d\mathbf{H}}{dt}=\frac{\varepsilon}{c}\frac{d^2\mathbf{E}}{dt^2},$$

and operating on equation (1.18) with curl yields

$$\nabla\times\nabla\times\mathbf{E}+\frac{\mu}{c}\nabla\times\frac{d\mathbf{H}}{dt}=0. \tag{1.19}$$

Comparison of the two equations (1.17) and (1.18) allows for writing

$$\nabla\times\left(\frac{1}{\mu}\nabla\times\mathbf{E}\right)+\frac{\varepsilon}{c^2}\frac{d^2\mathbf{E}}{dt^2}=0. \tag{1.20}$$

Using vector identities

$$\nabla\times(a\mathbf{A})=a\nabla\times\mathbf{A}+\nabla a\times\mathbf{A} \text{ and } \nabla\times(\nabla\times\mathbf{A})=\nabla(\nabla\cdot\mathbf{A})-\nabla^2,$$

equation (1.20) is expanded to

$$\frac{1}{\mu}\left[\nabla\times(\nabla\times\mathbf{E})\right]+\nabla\left(\frac{1}{\mu}\right)\times\nabla\times\mathbf{E}+\frac{\varepsilon}{c^2}\frac{d^2\mathbf{E}}{dt^2}=0$$

and then to

$$\frac{1}{\mu}\left[\nabla(\nabla\cdot\mathbf{E})-\nabla^2\mathbf{E}\right]+\nabla\left(\frac{1}{\mu}\right)\times\nabla\times\mathbf{E}+\frac{\varepsilon}{c^2}\frac{d^2\mathbf{E}}{dt^2}=0. \tag{1.21}$$

In a repeat of this procedure, equation (1.18) is differentiated with respect to time, and equation (1.17) is operated with curl to obtain

$$\nabla\times\frac{d\mathbf{E}}{dt}=-\frac{\mu}{c}\frac{d^2\mathbf{H}}{dt^2} \tag{1.22}$$

$$\nabla\times\nabla\times\mathbf{H}+\frac{\varepsilon}{c}\nabla\times\frac{d\mathbf{E}}{dt}=0. \tag{1.23}$$

Comparison of the two equations show that

$$\nabla\times\left(\frac{1}{\varepsilon}\nabla\times\mathbf{H}\right)+\frac{\mu}{c^2}\frac{d^2\mathbf{H}}{dt^2}=0. \tag{1.24}$$

Again, the vector identities invoked above can be used to expand equation (1.24) to

$$\frac{1}{\varepsilon}\left[\nabla(\nabla\cdot\mathbf{H})-\nabla^2\mathbf{H}\right]+\nabla\left(\frac{1}{\varepsilon}\right)\times\nabla\times\mathbf{H}+\frac{\mu}{c^2}\frac{d^2\mathbf{H}}{dt^2}=0. \tag{1.25}$$

For an isotropic homogeneous medium, gradients of μ and ε are zero so that equations (1.21) and (1.25) are simplified and reduced to

$$\nabla^2\mathbf{E}-\frac{\mu\varepsilon}{c^2}\frac{d^2\mathbf{E}}{dt^2}-\nabla(\nabla\cdot\mathbf{E})=0$$

$$\nabla^2\mathbf{H}-\frac{\mu\varepsilon}{c^2}\frac{d^2\mathbf{H}}{dt^2}-\nabla(\nabla\cdot\mathbf{H})=0.$$

By using equations (1.9), (1.10), (1.12), and (1.13), the gradients of divergence of $\mathbf{E}$ and $\mathbf{H}$ are found to be zero, which finally provides the equations of wave motion

$$\nabla^2\mathbf{E}=\frac{\mu\varepsilon}{c^2}\frac{d^2\mathbf{E}}{dt^2}$$

$$\nabla^2\mathbf{H}=\frac{\mu\varepsilon}{c^2}\frac{d^2\mathbf{H}}{dt^2}, \tag{1.26}$$

that is often also written in the form of a general one-dimensional wave equation

$$\frac{\partial^2\Psi(x,t)}{\partial x^2}=\frac{\mu\varepsilon}{c^2}\frac{\partial^2\Psi(x,t)}{\partial t^2}, \tag{1.27}$$

where $\Psi(x,t)$ is the wavefunction.

The derivation above involved the use of Maxwell's scalar equations and the material equations, resulting in the electromagnetic wave passing through a medium whose dielectric constant and magnetic permeability are ε and μ, respectively. Since electromagnetic waves propagate both in space and through free space, the wave equation is also derivable from Maxwell's equations in free space without including these auxiliary equations. One can simply operate with a curl on the Faraday's law equation and then on the Ampere-Maxwell's equation to check that a time-varying electric field is generated by a spatially varying magnetic field and the other way round. The one-dimensional wave equation in free space will appear as

$$\frac{\partial^2\Psi(x,t)}{\partial x^2} = \mu_o\varepsilon_o\frac{\partial^2\Psi(x,t)}{\partial t^2}$$

$$\frac{\partial^2\Psi(x,t)}{\partial x^2} = \frac{1}{c^2}\frac{\partial^2\Psi(x,t)}{\partial t^2}, \tag{1.28}$$

according to which an electromagnetic wave propagates at the velocity of light c in free space. For equation (1.27) the wave velocity is

$$v = \frac{c}{\sqrt{\varepsilon\mu}},$$

showing that the propagation velocity is reduced by the square root of the product of medium properties – dielectric constant and magnetic permeability.

The wave equation is a partial linear differential equation, so there is no unique solution. For example, both standing and traveling plane waves are solutions to the wave equation, and partial differential equations have the property that linear combinations of these two solutions would provide additional solutions. The solution multiplicity creates ambiguity in assigning a velocity to the wave. It is only the traveling plane wave for which the above velocity is defined straightforwardly. The standing wave would not have an associated velocity because the wave does not propagate but oscillates in time and space.

1.7 HOMOGENEOUS TRAVELING PLANE WAVE

The solution of the wave equation for the **E** field in free space provides a good physical insight into the orthogonality of directions of wave propagation and electric and magnetic field vectors. We rewrite equation (1.26) with the variables explicitly

$$\nabla^2\mathbf{E}(\mathbf{r},t) - \frac{\mu\varepsilon}{c^2}\frac{d^2\mathbf{E}(\mathbf{r},t)}{dt^2} = 0,$$

in which **r** is used for spatial coordinates. Variable separation yields

$$\mathbf{E}(\mathbf{r},t) = \mathbf{R}(\mathbf{r})T(t),$$

where both **E**-field and spatial coordinates $\mathbf{R}(\mathbf{r})$ are vectorial, but time $T(t)$ is scalar. Substitution gives

$$\frac{\nabla^2\mathbf{R}(\mathbf{r})}{\mathbf{R}(\mathbf{r})} - \frac{1}{c^2T(t)}\frac{d^2T(t)}{dt^2} = 0.$$

Since both terms are the same constant so as to produce zero on the right-hand side, one writes

$$\nabla^2\mathbf{R}(\mathbf{r}) + k^2\mathbf{R}(\mathbf{r}) = 0$$

$$\frac{d^2T(t)}{dt^2} + k^2c^2T(t) = 0, \tag{1.29}$$

in which $-k^2$ is the chosen constant. In wave phenomena, k is the angular wave number or simply the wave number

$$k = \frac{2\pi}{\lambda} = \frac{\omega}{c}.$$

The time equation above is a harmonic differential equation and the solution is obtained in both sine and cosine terms

$$\begin{aligned} T(t) &= \mathrm{Re}\left[ce^{-i\omega t}\right] \\ &= c_1\cos(\omega t) + c_2\sin(\omega t), \end{aligned} \tag{1.30}$$

in which c_1 and c_2 are real constants that define a complex constant $c = c_1 + ic_2$. Insertion of the $T(t)$ solution into the variable-separated equation shows

$$\mathbf{E}(\mathbf{r},t) = \mathrm{Re}[\{c_1 + ic_2\}\mathbf{R}(\mathbf{r})e^{-i\omega t}]. \tag{1.31}$$

Here, $\{c_1 + ic_2\}\mathbf{R}(\mathbf{r})$ is the complex electric field amplitude, **E**(**r**), so that

$$\mathbf{E}(\mathbf{r},t) = \mathrm{Re}[\mathbf{E}(\mathbf{r})e^{-i\omega t}]. \tag{1.32}$$

Note that $\mathbf{E}(\mathbf{r},t)$ is real and it contains the complex field **E**(**r**), meaning that the real time-dependent field contains the complex spatial part of the field. This equation can be cast in the more familiar trigonometric function

$$\begin{aligned} \mathbf{E}(\mathbf{r},t) &= \mathrm{Re}\{\mathbf{E}(\mathbf{r})\}\cos\omega t + \mathrm{Im}\{\mathbf{E}(\mathbf{r})\}\sin\omega t \\ &= \frac{1}{2}\left[\mathbf{E}(\mathbf{r})e^{-i\omega t} + \mathbf{E}(\mathbf{r})^{*}e^{i\omega t}\right] \\ &= \mathbf{E}(\mathbf{r})\left[\frac{e^{-i\omega t} + e^{i\omega t}}{2}\right] \\ &= \mathbf{E}(\mathbf{r})\cos\omega t, \end{aligned} \tag{1.33}$$

which is a widely used form of the electric field of radiation in semiclassical discussions of spectroscopic transitions. With the time-harmonic solution in the form of equation (1.32), the wave equation for the electric field in free space can be written as

$$\nabla^2\mathbf{E}(\mathbf{r}) + k^2\mathbf{E}(\mathbf{r}) = 0.$$

Assuming

$$\mathbf{E}(\mathbf{r}) = \mathbf{E}_o e^{\pm i\mathbf{kr}},$$

the expression for $\mathbf{E}(\mathbf{r},t)$ above appears as

$$\mathbf{E}(\mathbf{r},t) = \mathrm{Re}[\mathbf{E}_o e^{\pm i\mathbf{kr} - i\omega t}], \tag{1.34}$$

which is the plane wave solution. Notice that the wave vector $\mathbf{k}(=\boldsymbol{k}_x + \boldsymbol{k}_y + \boldsymbol{k}_z)$ can be positive or negative, meaning the wave can propagate along or against the direction of $\mathbf{k}$. If z is the direction of propagation, then an observer at the origin will see incoming waves if the solution is obtained with $-i\mathbf{kr}$ in the exponential, the wave is outgoing otherwise. However, according to the premise that considers only that region of the electric field which contains no charge ($\rho = 0$), the electric field cannot be divergent, $\nabla \cdot \mathbf{E}(\mathbf{r},t) = 0$. This consideration imposes orthogonality of $\mathbf{k}$ and $\mathbf{E}_o$ vectors

$$\mathbf{k} \cdot \mathbf{E}_o = |\mathbf{k}||\mathbf{E}_o|\cos\theta = 0.$$

As for the magnetic field $\mathbf{H}$, not only the zero-divergence condition holds, but Maxwell's equation for $\nabla \times \mathbf{E}$ also provides for

$$\mathbf{H}_o = \frac{1}{\omega\mu_o}(\mathbf{k} \times \mathbf{E}_o), \tag{1.35}$$

which imposes mutual orthogonality of $\mathbf{E}_o$, $\mathbf{H}_o$, and $\mathbf{k}$ vectors. Because $\mathbf{E}_o$ and $\mathbf{H}_o$ are always in phase

$$\frac{|\mathbf{E}_o|}{|\mathbf{H}_o|} = \frac{\omega}{k} = c, \tag{1.36}$$

and the ratio of the magnitudes of $\mathbf{E}$ and $\mathbf{H}$ fields will always be c at any point of time. With these restrictions, a plane electromagnetic wave traveling in the Cartesian x-direction is drawn in Figure 1.5. Since the $\mathbf{E}_o$ vector is restricted to orient only along y at any instant, the electric field is said to be linearly polarized. The time period of the wave is

$$\mathrm{T} = \frac{\lambda}{c} = \frac{1}{\nu} = \frac{2\pi}{\omega}. \tag{1.37}$$

The plane wave solution is valid only for an isotropic medium like the free space where permittivity and permeability do not change. The wave consists of an infinite number of planes, and the field is constant at every point in space defined by the xy plane. Although the magnitude of the field $|\mathbf{E}_o|$ is identical anywhere in a plane, it varies sinusoidally as one moves forward in space, goes to zero at the completion of a half cycle, and the direction is reversed for the next half cycle. The size of the plane in the plane-wave representation appears quite arbitrary, because one might as well draw planes of larger dimension without changing the features and characteristics of $\mathbf{E}_o$ and $\mathbf{H}_o$ fields. This arbitrariness in the extension of $\mathbf{E}_o$ and $\mathbf{H}_o$ makes plane waves unrealistic. Even if plane waves are unsophisticated forms of waves that do not really exist, they are useful to visualize local effects of sophisticated fields. They are helpful in at least two ways. One, when spherical electromagnetic waves originating from a point source in all directions have propagated sufficiently, the plane wave approximation can be made for a small region (Figure 1.6), and this approximation can be made for all complicated waves at a sufficient distance from the source. Two, superposition of plane waves produces other waveforms in space and time.

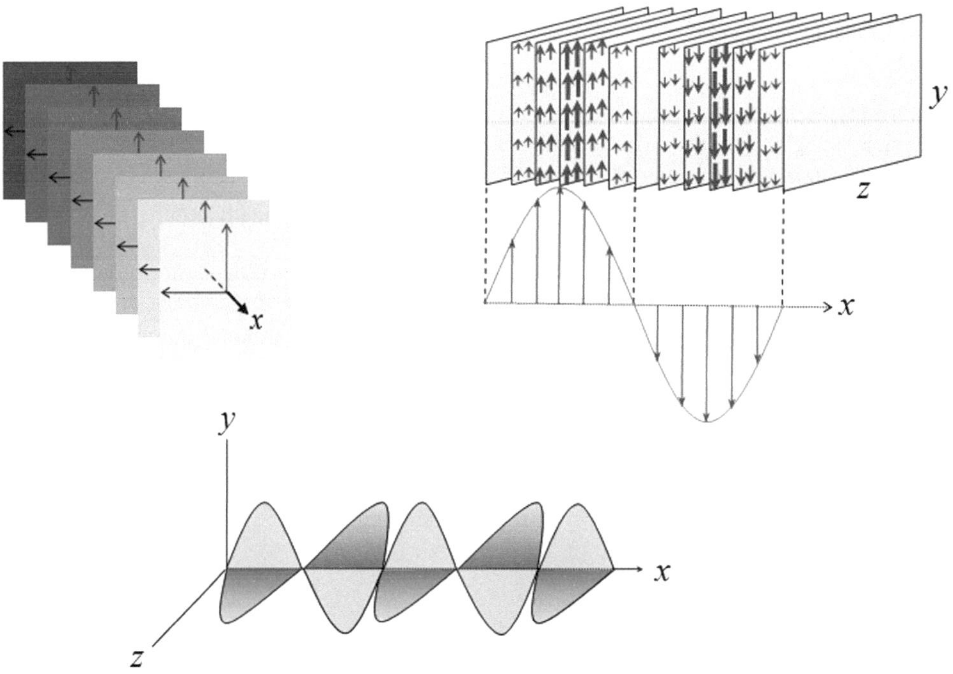

FIGURE 1.5 Visualization of a plane wave. *Top left*, the planar wavefronts with uniform separation means their propagation in phase. The mutually orthogonal electric (red) and magnetic (blue) field vectors are both orthogonal to the direction of propagation. *Top right*, sinusoidal variation of the amplitude of the electric field vector $\mathbf{E}_o$ in planar wavefronts. *Bottom*, plane-polarized traveling wave. It is customary to refer to the plane containing the $\mathbf{E}_o$ vector as the plane of polarization.

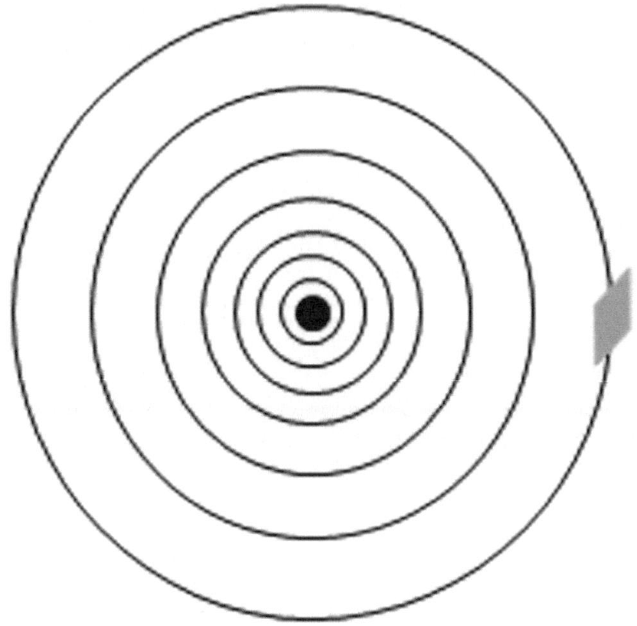

FIGURE 1.6 Plane wave approximation for a small region of a spherical circular wave at a large enough distance from the point source.

1.8 WAVE PACKET

The formation of a wave packet by superposition of plane monochromatic waves naturally follows from Fourier theorem. Consider the real part of n number of monochromatic plane waves of different frequencies. For simplicity we take only the electric field propagating along z-direction and write down the general solution (equation (1.34))

$$\mathbf{E}(z,t) = \mathrm{Re}\left[\mathbf{E}_{o} e^{ikz - i\omega t}\right]. \quad (1.38)$$

For a continuous range of frequencies of the plane waves considered, the summation of the waves according to Fourier transformation yields a superposed wave called a wave packet or wave group. Since the waves we have considered are propagating only along z with the $\mathbf{E}_o$ vector always parallel to the y-axis, the wave packet is one-dimensional. The real part of the complex wave is given by

$$\Psi(z,t) = \mathrm{Re}\int_0^{\infty} E_{o,\omega} e^{ik_{\omega}z - i\omega t} d\omega = \int_0^{\infty} E_{o,\omega} \cos\left(k_{\omega}z - \omega t\right) d\omega, \quad (1.39)$$

in which $E_{o,\omega}$ refers to the amplitude, and the subscript ω indicates dependence of the amplitude on the corresponding wave frequency. More generally, $E_{o,\omega}$ is the Fourier amplitude.

To provide a simple picture of combination of waves to produce a new waveform, let us consider just two waves of identical amplitude E_o, both propagating along z. To distinguish them, let the frequency and the wave vector be ω and k for one, and $\omega + \delta\omega$ and $k + \delta k$ for the other. The real part of the summation of the two waves reads

$$\Psi(z,t) = E_o\left[e^{ikz - i\omega t} + e^{i(k+\delta k)z - i(\omega + \delta\omega)t}\right], \quad (1.40)$$

which can be written as (see Box 1.3)

$$\Psi(z,t) = E_o\left[\left\{e^{-\frac{i}{2}(\delta kz - \delta\omega t)} + e^{\frac{i}{2}(\delta kz - \delta\omega t)}\right\} e^{iz\left(k + \frac{\delta k}{2}\right) - it\left(\omega + \frac{\delta\omega}{2}\right)}\right], \quad (1.41)$$

where

$$k + \frac{\delta k}{2} = \bar{k} = \frac{2\pi}{\bar{\lambda}} \quad \text{and} \quad \omega + \frac{\delta\omega}{2} = \bar{\omega}$$

are mean wavenumber and mean frequency, respectively. This shows that the wave packet $\Psi(z,t)$ formed by combining the two plane waves is another plane wave of wavelength $\bar{\lambda}$ and frequency $\bar{\omega}$, and the wave propagates along the same $+z$ direction that the initial two waves do. Using the formulas

$$\cos\theta = \frac{e^{i\theta} + e^{-i\theta}}{2} \quad \text{and} \quad e^{i\theta} = \cos\theta + i\sin\theta$$

equation (1.41) may be written in terms of cosine functions

$$\Psi(z,t) = 2E_o \cos\left[\frac{1}{2}(\delta\omega t - \delta kz)\right] e^{iz\left(k + \frac{\delta k}{2}\right) - it\left(\omega + \frac{\delta\omega}{2}\right)}$$

$$\Psi(z,t) = 2E_o \cos\left[\frac{1}{2}(\delta\omega t - \delta kz)\right]\cos\left(\bar{k}z - \bar{\omega}t\right). \quad (1.42)$$

This expression shows that the maximum amplitude of the wave packet is $2E_o$, whereas the amplitude of the two initial waves is E_o. The wave packet amplitude varies from 0 to $2E_o$ depending on the position of the wave with time. The half cycle of the wave during which the amplitude starts from and ends up at zero is a beat. While the first cosine term alone can be used to represent a beat, the entire expression including the second term represents the wave group (Figure 1.7).

The waveform is best described at a given instant of time. To illustrate, we set $t = 0$ in equation (1.42) to obtain

$$\Psi(z,0) = 2E_o \cos\left(\frac{-\delta kz}{2}\right)\cos\left(\bar{k}z\right) \quad (1.43)$$

and notice that $|\Psi(z)|$ is maximum for $z = 0$. Zero value for z is also significant for the fact that the two initial waves, which are in general out of phase due to their frequency difference, are in phase at $z = 0$, so they interfere constructively to produce the maximum of $|\Psi(z)|$. As the value of z changes on either side of $z = 0$, the two waves interfere destructively. When the phase shift between e^{ikz} and $e^{i(k+\delta k)z}$ becomes $\pm\pi$, the interference of the two waves becomes completely destructive, resulting in $|\Psi(z)| = 0$. The distance δz required for two successive $|\Psi(z)| = 0$ or $|\Psi(z)| = 2$ is given by

$$\delta k\,\delta z = 4\pi \quad (1.44)$$

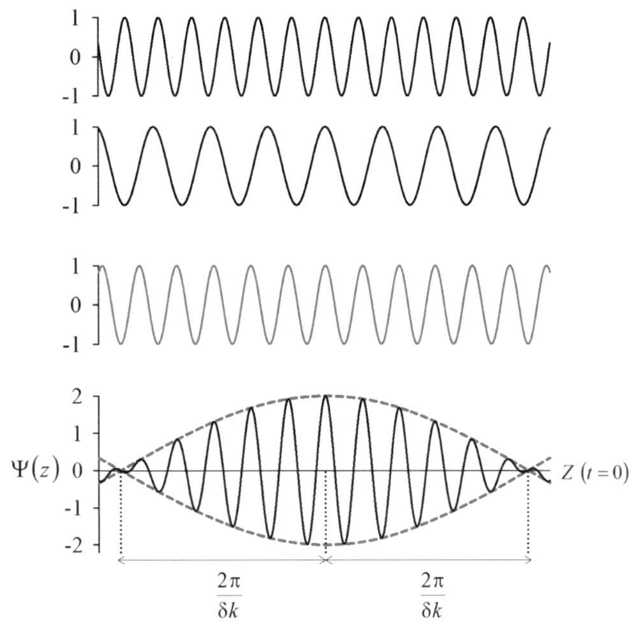

FIGURE 1.7 Two plane monochromatic waves of identical amplitude but different frequencies and wave vectors (top two panels). The $a\cos\left(\bar{k}z-\bar{\omega}t\right)$ part of the combined wave is shown in the third panel from top. In the last panel, the $2E_o\cos\left[\frac{1}{2}(\delta\omega t-\delta kz)\right]$ part of the combined wave is shown by the dashed-line waveform. The wave group inside this waveform is $2E_o\cos\left[\frac{1}{2}(\delta\omega t-\delta kz)\right]\cos\left(\bar{k}z-\bar{\omega}t\right)$. Implicitly, only the real part of each wave is shown.

and the velocity of propagation of the planes corresponding to $|\Psi(z)|=2$ is called the wave group (Figure 1.7). The wave group velocity, also called group velocity, is

$$v^{(g)}=\frac{\delta\omega}{\delta k}, \qquad (1.45)$$

where ω is the angular frequency of the wave packet and k is in angular wave number. Using $k=2\pi/\lambda$ and $\omega=2\pi v$, the group velocity can also be written as

$$v^{(g)}=\frac{\delta v}{\delta\left(\frac{1}{\lambda}\right)}. \qquad (1.46)$$

A compressed light pulse is essentially a wave packet because it is a superposition of many plane waves of different frequencies, not just two as illustrated. A short pulse of laser beam for example, which has a band width $\Delta\omega$ consists of a large number of plane waves of characteristic frequencies. The shorter the light pulse, the more is the bandwidth, resulting in the superposition of more plane waves of different frequency components. Both creation and evolution of wave packets, specific to the nature of molecular potential surfaces, play a significant role in the understanding of bond making and breaking phenomena.

Box 1.1

Electric field can be given in charge length^{-2} when the field is measured at a distance larger than the field source, say a dipole. To obtain the field in SI unit (NC^{-1} or force per unit charge or Vm^{-1}) the charge is divided by $4\pi\varepsilon_o=1.11\times10^{-10}$ J^{-1}C^{2}m^{-1}. If only cgs units are chosen for work the division of the charge by $4\pi\varepsilon_o$ is not needed. For example, the first scalar relation in Maxwell's set of equations is

$$\nabla\cdot\mathbf{D}=4\pi\rho,$$

which is in Gaussian electrostatic unit. Many textbooks show the equation in cgs units as

$$\nabla\cdot\mathbf{D}=\frac{\rho}{\varepsilon_o}.$$

Another point is that **B** and **H** in Maxwell's equations are given in electromagnetic unit of the Gaussian system of units. The constant c relates the units of charge in electrostatic and electromagnetic units.

Box 1.2

With the vector operator (call 'del') $\nabla=\mathbf{i}\frac{\partial}{\partial x}+\mathbf{j}\frac{\partial}{\partial y}+\mathbf{k}\frac{\partial}{\partial z}$, the divergence of a vector function **F** is a scalar product written as

$$\text{div }\mathbf{F}=\nabla\cdot\mathbf{F}=\left(\frac{\partial}{\partial x}+\frac{\partial}{\partial y}+\frac{\partial}{\partial z}\right)\cdot\left(\mathrm{i}F_x+\mathrm{j}F_y+\mathrm{k}F_z\right)$$
$$=\frac{\partial F_x}{\partial x}+\frac{\partial F_y}{\partial y}+\frac{\partial F_z}{\partial z}.$$

Curl of the **F** function is the vector product of ∇ and **F**

$$\text{curl }\mathbf{F}=\nabla\times\mathbf{F}=\begin{vmatrix}\mathrm{i} & \mathrm{j} & \mathrm{k}\\ \frac{\partial}{\partial x} & \frac{\partial}{\partial y} & \frac{\partial}{\partial z}\\ F_x & F_y & F_z\end{vmatrix}.$$

Box 1.3

Equations containing exponentials are widely encountered. An approach often used is shown below by providing the solution of equation (1.23) in the text.

$$\Psi(z,t) = E_o\left[e^{ikz-i\omega t} + e^{ikz+i\delta kz-i\omega t-i\delta\omega t}\right]$$
$$= E_o\left[e^{ikz-i\omega t} + e^{ikz-i\omega t}e^{i\delta kz-i\delta\omega t}\right]$$
$$= E_o\left[\left\{1+e^{i\delta kz-i\delta\omega t}\right\}e^{ikz-i\omega t}\right]$$
$$= E_o\left[\left\{1+e^{i(\delta kz-\delta\omega t)}\right\}e^{ikz-i\omega t}\right]$$

Continue by writing $e^{i\theta}$ as $e^{\frac{i}{2}\theta+\frac{i}{2}\theta}$

$$= E_o\left[\left\{1+e^{\frac{i}{2}(\delta kz-\delta\omega t)+\frac{i}{2}(\delta kz-\delta\omega t)}\right\}e^{ikz-i\omega t}\right]$$

$$= E_o\left[e^{\frac{i}{2}(\delta kz-\delta\omega t)}\left\{e^{-\frac{i}{2}(\delta kz-\delta\omega t)}+e^{\frac{i}{2}(\delta kz-\delta\omega t)}\right\}e^{ikz-i\omega t}\right]$$

$$= E_o\left[\left\{e^{-\frac{i}{2}(\delta kz-\delta\omega t)}+e^{\frac{i}{2}(\delta kz-\delta\omega t)}\right\}e^{\frac{i}{2}(\delta kz-\delta\omega t)+ikz-i\omega t}\right]$$

$$= E_o\left[\left\{e^{-\frac{i}{2}(\delta kz-\delta\omega t)}+e^{\frac{i}{2}(\delta kz-\delta\omega t)}\right\}e^{\frac{i}{2}\delta kz+ikz-\frac{i}{2}\delta\omega t-i\omega t}\right]$$

$$= E_o\left[\left\{e^{-\frac{i}{2}(\delta kz-\delta\omega t)}+e^{\frac{i}{2}(\delta kz-\delta\omega t)}\right\}e^{iz\left(k+\frac{\delta k}{2}\right)-it\left(\omega+\frac{\delta\omega}{2}\right)}\right]$$

PROBLEMS

1.1 Imagine a circular area of radius r in space through which an amount of charge Q passes along x direction. Show that the first of the Maxwell's scalar equations can be written in the form

$$\nabla\times\mathbf{H} - \frac{1}{c}\frac{d\mathbf{D}}{dt} = \frac{4}{cr^2}\frac{dQ}{dt} = -\sigma\frac{d\phi}{dx}$$

where ϕ is the electrical potential.

1.2 Using

$$\nabla\times\frac{d\mathbf{E}}{dt} = \frac{\varepsilon}{c}\frac{d^2\mathbf{H}}{dt^2}$$

verify that

$$\nabla\times\nabla\times\mathbf{H} + \frac{\varepsilon}{c}\nabla\times\frac{d\mathbf{E}}{dt} = 0.$$

1.3 Show that a time-varying magnetic field is produced by a spatially varying electric field.

1.4 Consider an electric charge in space producing an electromagnetic field. What could be the minimal spatial distance at which a plane wave approximation of the electromagnetic field can be made? Suppose we already have an answer regarding the length d, is this distance related to the wavelength of the field produced? Does this length depend on the amount of spatial charge Q we started with? Then, suppose one moves closer ($< d$) to the source. Is a wave packet still produced? How are these questions relevant to a real-life spectroscopy experiment?

1.5 Compare an electromagnetic wave vis-à-vis a vibrational wave of a diatomic molecule.

BIBLIOGRAPHY

Born, M. and E. Wolf (1970) *Principles of Optics*, Pergamon Press.

2 Postulates of Quantum Mechanics

The bond between quantum mechanics and spectroscopy is inseparable and complementary. Quantum mechanics provides theoretical ideas to design spectroscopic experiments and also a framework to interpret results. This process turns out to be a rigorous test for theoretical ideas themselves. One of the earliest and revealing examples of the marriage of spectroscopy and quantum mechanics comes from the discovery of spatial quantization of angular momentum of particles – electron and nucleus, for instance. We know that electron and nuclear spin angular momenta have no classical analog. To elucidate the close relationship of spectroscopy and quantum mechanics and to arouse interest in the subject, the collaborative evolution of the theoretical and experimental ideas of the quantization of magnetic moment of particle spin is briefly discussed before introducing the postulates of quantum mechanics.

2.1 STERN-GERLACH EXPERIMENT

Quantum mechanics itself had gone through a transformation from the earlier Bohr-Sommerfeld theory to the later new theory due to Schrödinger, Heisenberg, Born, and others. The Stern-Gerlach experiment, however, started out in 1922 with an idea of Sommerfeld's quantum theory, according to which the magnetic moment of an atom in a state of angular momentum $L = 1$ would have two spatially quantized components, $\pm eh / 4m_e$, where e and m_e are charge and mass of the electron, and h is Planck's constant. These two components should be observable if a force is applied on the magnetic moment μ by the use of an inhomogeneous magnetic field such that a field gradient, say along the z-direction, $\partial H/\partial z$ exists. The magnetic potential energy of the atom is the work done to rotate the magnetic dipole from $\theta = 90^\circ$ to a new value of q

$$U = -\mu_z H\cos\theta,$$

so that the force experienced by the dipole along the z-direction will be

$$F_z = \frac{\partial U}{\partial z}.$$

This should split the z-component of the magnetic moment μ_z into two components

$$\mu_z = \pm\frac{eh}{4m_e}.$$

The experiment that was initially based on passing a beam of silver atoms through a magnetic field gradient (Figure 2.1) relied on an estimate of the magnitude of atom magnetic moments provided by Sommerfeld's quantum theory, and used a $\partial H / \partial z$ value of about 1 T cm^{-1}. This value of the field gradient seemed good enough to deflect the atom beam into two, but only if the silver atoms were in a state of angular momentum $L = 1$. Stern and Gerlach assumed $L = 1$, although that was a wrong assumption because the atoms existed in a state of $L = 0$. Nevertheless, the silver atom beam did split into $\pm\mu_z$ components, which for assumed $L = 1$ appeared to provide a proof of Sommerfeld's prediction of two spatial quantizations. Later, the new quantum theory calculated that an atom in a state of $L = 1$ should have three quantized μ components +, −, and 0, not just + and − as Sommerfeld assumed earlier. Stern and Gerlach were puzzled, but stuck on to their two spots on the spectrograph and hence to the two spatial quantized states of μ_z for their assumed state of $L = 1$. Equally puzzling was the prediction of the new quantum theory that calculated three components for μ_z.

The puzzle was solved shortly later by the proposal of Uhlenbeck and Goudsmit in 1925 that the electron possesses intrinsic spin with spin angular momentum $\hbar / 2$, which they arrived at by analyses of atomic spectra existing at that time. Although this analysis was not oriented to explaining the Stern-Gerlach result, it became clear that the orbital angular momentum **L** and spin angular momentum **S** of an electron sum up to the total angular momentum $\mathbf{J} = \mathbf{L} + \mathbf{S}$, which states that the two spatial quantization of μ_Z observed by Stern and Gerlach is possible only when $L = 0$, so that $J = \hbar / 2$, and the two z-components of the electron spin magnetic moment would be $S_z = \pm\hbar / 2$. This exactly is the interpretation. Even though the silver atom was assumed to be in a state of $L = 1$, it is in fact in $L = 0$ state. There are 47 electrons in a silver atom, of which 46 form the closed inner core. The core electrons have no orbital angular momentum, and the electron spins pair up to produce zero spin angular momentum. The remaining electron is also devoid of orbital angular momentum, which leaves only its intrinsic spin ($s = \hbar / 2$) that produced the two spots in Stern-Gerlach's spectrograph (Figure 2.1). If there were n number of atoms in the silver atom beam, the magnetic field would split the beam into two along the direction of the field gradient to produce two equal intensity spots in the plate of the spectrograph.

It is also important to mention that Stern-Gerlach result did not contradict the prediction of the new quantum theory, simply because the silver atom exists in the angular momentum state of $L = 0$. Had an atom that existed in a state

DOI: 10.1201/9781003293064-2

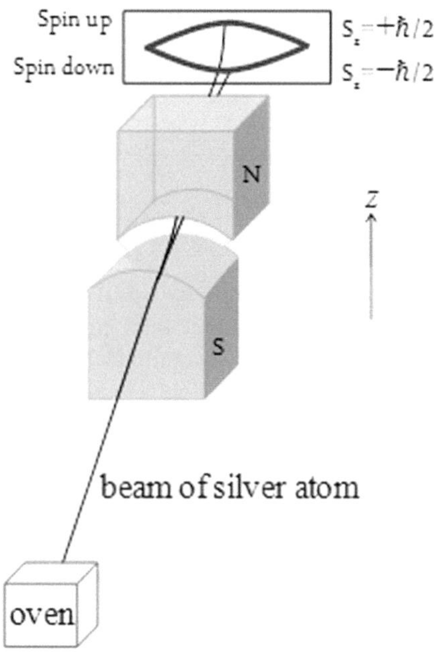

FIGURE 2.1 Sketch of the Stern-Gerlach experiment. Shown is the result expected for $L = 0$. Notice the splitting of the line in the spectrograph.

of $L = 1$ been chosen for their experiment, Stern and Gerlach would have observed the splitting of the atom beam into three spots corresponding to the three quantized μ components, +, –, and 0. Therefore, the prediction of the old quantum theory which is what Stern and Gerlach set out to test was wrong. Also wrong was the assumption that silver atoms existed in a state of $L = 1$. But their result was correct and consistent with the prediction of modern quantum theory.

The role of spectroscopy in the confirmation of quantum mechanical concepts need not be confined to analyses of atomic spectra alone. Since the principles of spectroscopy are similar, covering a wide range of radiation frequencies from x-ray to radio wave regions, a large body of spectroscopic approaches provides scope to deduce the quantum features.

2.2 POSTULATES OF QUANTUM MECHANICS

The postulates prescribe a formal mathematical basis to describe spectroscopic experiments and predict the results thereof. The postulates are, however, subject to verification, and spectroscopic experiments at times are required to be used intelligently to achieve this. One should also realize that knowing a set of postulates is just not enough. There is no unique set of postulates, the ideas developed are interconnected, and results and consequences of the postulates are introduced at different stages of learning. The following is only a summary of the postulates.

2.2.1 Postulate 1

The state of a physical system is defined by the state function $\Psi(\mathbf{r},t)$.

A system is a particle or a set of particles – electrons and nuclei, for example. State means a set of properties, the location of a particle is a property, for instance, and a property is associated with a value. However, the 'state' needs to be construed in a bit of abstract sense. Suppose the particle is in some state and has a corresponding value. The same particle at the same time can be in another state, and hence another value. Then the two states will combine to produce another state. In other words, different states or values of a property superpose to yield other states.

The state function $\Psi(\mathbf{r},t)$ is chosen as a function of a set of 3N coordinates ($\mathbf{r}$) for n-particles, one set at each instant of time t. But the state function need not be shown as a function of coordinate and time, but can be specified by a vector without referring to a coordinate system at a given instant (see Box 2.1). The state function is a complex function with both real and imaginary parts, which must mean that

$$\Psi^*(\mathbf{r},t)\Psi(\mathbf{r},t) = |\Psi(\mathbf{r},t)|^2, \tag{2.1}$$

where $|\Psi(\mathbf{r},t)|^2$, a real function, is postulated to be the probability density (Born rule) for finding the system at different locations in the $\mathbf{r}$ space. Then the probability of finding the system in a volume element $dxdydz = d^3r$ at instant t is given by

$$d\mathcal{P}(\mathbf{r},t) = c|\Psi(\mathbf{r},t)|^2 d^3r, \tag{2.2}$$

in which c is the normalization constant. Because the probability total over the whole space must be unity, the normalization condition is given by

$$c\int |\Psi(\mathbf{r},t)|^2 d^3r = 1, \tag{2.3}$$

where the integration limit covers the whole range of each coordinate.

2.2.2 Postulate 2

Every physical quantity that can be measured corresponds to a Hermitian operator such that the operator is the observable.

This is the operator postulate. As will be seen later, the Hermitian character of the operator yields an observable which is a real number. Finding an operator is not difficult. One starts with the classical expression for the observable defined in terms of space, mass, and velocity using a Cartesian coordinate system. Then the variables are transformed as given below for some commonly used properties.

Cartesian Coordinate Operator. Each coordinate x_i, y_i, z_i for i number of particles is replaced by the operator x_i, y_i, z_i, meaning the operator for a coordinate is just the coordinate. The operator for the x coordinate of a single particle is x.

Dipole Moment Operator. The classically written μ_i for the dipole moment of particle i is rewritten by adding a Cartesian coordinate component, $\mu_i x$, $\mu_i y$, and $\mu_i z$.

Kinetic Energy Operator. Consider the x component of the kinetic energy of a particle, for example. In the classical expression

$$\frac{1}{2}mv_x^2 = \frac{p_x^2}{2m} \quad (2.4)$$

the linear momentum p_x is replaced by

$$\frac{\hbar}{i}\frac{\partial}{\partial x} = -i\hbar\frac{\partial}{\partial x}, \quad (2.5)$$

where $i = \sqrt{-1}$, so the operator is

$$-\frac{\hbar^2}{2m}\frac{\partial^2}{\partial x^2}.$$

Potential Energy Operator. If the potential energy function in one dimension is $V(x) = bx^2$, where b is a constant, the x coordinate is replaced by x itself. The operator is the classical expression as it is. But the potential energy operator is generally written by just $V(x)$.

Angular Momentum Operator. When a particle at some instant can be described by a position vector r from the origin of a Cartesian axes system its angular momentum is classically given by

$$\mathbf{L} = \mathbf{r} \times \mathbf{P} = \begin{vmatrix} \mathrm{i} & \mathrm{j} & \mathrm{k} \\ x & y & z \\ p_x & p_y & p_z \end{vmatrix} \quad (2.6)$$

in which

$$\mathbf{r} = \mathrm{i}x + \mathrm{j}y + \mathrm{k}z$$

$$\mathbf{P} = p_x + p_y + p_z.$$

The three components of the angular momentum are

$$L_x = yp_z - zp_y$$

$$L_y = zp_x - xp_z$$

$$L_z = xp_y - yp_x. \quad (2.7)$$

Note that the $\mathbf{L} = \mathbf{r} \times \mathbf{P}$ expression is applicable in the discussion of orbital angular momentum only, not spin angular momentum, even though the spin angular momentum also has the Cartesian components. The linear momentum components are replaced as done before to obtain the following operators

$$L_x = -i\hbar\left(y\frac{\partial}{\partial z} - z\frac{\partial}{\partial y}\right)$$

$$L_y = -i\hbar\left(z\frac{\partial}{\partial x} - x\frac{\partial}{\partial z}\right)$$

$$L_z = -i\hbar\left(x\frac{\partial}{\partial y} - y\frac{\partial}{\partial x}\right). \quad (2.8)$$

2.2.3 Postulate 3

The expectation value or the mean value $\langle F\rangle$ of a physical observable associated with a Hermitian operator F at some instant t is

$$\langle F\rangle = \int \Psi^* F \Psi d\tau. \quad (2.9)$$

Suppose we have only one particle and the observable is its location at some instant of time t, then the integral above could be written explicitly as

$$\langle F\rangle = \iiint_{-\infty}^{+\infty} \Psi^*(x,y,z,t) F\, \Psi(x,y,z,t)\, dxdydz. \quad (2.10)$$

The integral is taken over the entire range of each coordinate. For a n-particle system, the integration will be carried out over the whole range of each coordinate for each particle. For convenience, the integral is shown simply as $\int d\tau$.

The expectation value is a real mathematical quantity, and for that reason it must satisfy the following condition

$$\langle F\rangle = \langle F\rangle^*$$

where $\langle F\rangle^*$ is the complex conjugate of $\langle F\rangle$. This allows for writing the postulate (equation (2.9)) as

$$\langle F\rangle = \int \Psi^* F \Psi d\tau = \int (F\Psi^*)\Psi d\tau. \quad (2.11)$$

Operators that satisfy this relation are called Hermitian operators. They also satisfy the relation

$$\int \Psi_A{}^* F \Psi_B d\tau = \int (F\Psi_A{}^*)\Psi_B d\tau, \quad (2.12)$$

where Ψ_A and Ψ_B are different state functions. Another property of a Hermitian operator is that it is equal to its adjoint operator or Hermitian conjugate, $F = F^\dagger$.

It is advantageous to work with Dirac notation to discuss the nature of the adjoint operator. Consider a Hermitian operator F and a ket vector $|\Psi\rangle$ along with the corresponding bra vector $\langle\Psi|$. Then the Hermitian operator associates another ket $|\Psi'\rangle$ when it operates on $|\Psi\rangle$, but the Hermitian conjugate operates on the bra vector $\langle\Psi|$ to yield another bra $\langle\Psi'|$ such that

$$|\Psi'\rangle = F|\Psi\rangle$$

$$\langle\Psi'| = \langle\Psi|F^\dagger. \quad (2.13)$$

From these relations another useful relation can be obtained by using the property of the scalar product. The scalar product of a ket $|\Psi\rangle$ by another ket $|\phi\rangle$ is given by

$$\langle\phi|\Psi\rangle = \langle\Psi|\phi\rangle^*, \quad (2.14)$$

which means the association of a complex number with a pair of kets $|\phi\rangle$ and $|\Psi\rangle$. Using this property equation (2.13) can be written as

$$\langle \Psi' | \phi \rangle = \langle \phi | \Psi' \rangle^*$$
$$\langle \Psi | F^\dagger | \phi \rangle = \langle \phi | F | \Psi \rangle^*. \tag{2.15}$$

Statistics. An important aspect of this postulate is that a particular value of an observable need not necessarily recur in multiple measurements of the same event. Quantum mechanics rather insists on the averaged nature of an observable, thus necessarily bringing statistics into the picture. Since it is important to understand the significance of the statistical nature of the expectation value $\langle F \rangle$, we briefly discuss the statistical features of observables and the mean value. The quantity $\langle F \rangle$ is a real number representing the mean of many measurements, and the mean need not necessarily correspond to any of the values measured in multiple measurements. For example, the result of a population census may declare a mean household size of 2.9. But the count of the number of persons in each household must always give an integral value. For n number of households the mean of the household size is

$$\langle F \rangle = \frac{\sum_{i=1}^{n} f_i}{n}, \tag{2.16}$$

where f_i is the count for the ith household. In our measurements of physical systems specified along one-dimension say, we consider multiple copies of the system, each in the same state $|\Psi(x,t)\rangle$, and measure the property F for each system. The average value $\langle F \rangle$ will be given by the same statistical average expression shown above.

In classical statistics, the above expression is cast in the form of an equivalent expression. If f is a set of j possible values that n number of measurements give, and m_{f_j} is the number of times the f_j value is obtained, then

$$\langle F \rangle = \frac{\sum_j m_{f_j} f_j}{n}. \tag{2.17}$$

Note that the i-index used in equation (2.16) referring to the number of measurements is now redundant. The summation now runs over the set of possible values. Equation (2.17) can be written as

$$\langle F \rangle = \sum_j \left(\frac{m_{f_j}}{n} \right) f_j = \sum_j p_j f_j, \tag{2.18}$$

where $p_j(f_j) = m_{f_j}/n$ is the probability function for obtaining the value f_j, and therefore p_j is a function of f_j. A simple analogy is this. At a cross road, one has the choice of taking any of the four directions, so the probability of taking one of the four directions at random is 1/4, and so $p_j = 1/4$. Since the probability of all four sample spaces is one, we have

$$\sum_{j=1}^{4} p_j = 1. \tag{2.19}$$

This leads to the probability distribution function $P(f)$ for the f variable

$$P(f) = \sum_j p_j f_j. \tag{2.20}$$

The probability function p_j and the probability distribution function $P(f)$ for the cross road example are shown in Figure 2.2.

In this example, the random variable, that is the road number, is discrete. For a continuous variable, the probability distribution function will be written as

$$P(f) = \int_{-\infty}^{\infty} p_j f_j \, df_j = 1, \tag{2.21}$$

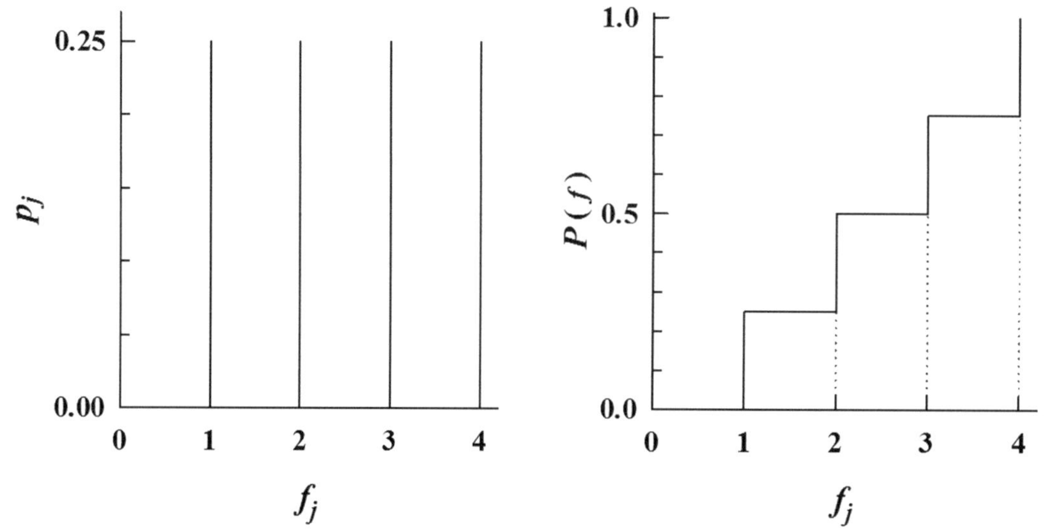

FIGURE 2.2 The probability function p_j and the probability distribution function $P(f)$ for taking a direction at a cross road.

where the integrand p_j is called probability density or distribution density or probability distribution. Clearly, the probability distribution is the derivative of the probability distribution function $P(f)$, which can be readily seen by differentiating equation (2.21). Up to now we have used letter symbols like p_j and f_j to suit the random example of passage from a cross road. In the following we will use standard symbols used in the discussion of probability distribution.

Gaussian Distribution. A continuous distribution with the probability distribution $p(f)$ given by

$$p(f)=\frac{1}{\sigma\sqrt{2\pi}}e^{-\frac{1}{2}\left(\frac{f-\langle f\rangle}{\sigma}\right)^2} \tag{2.22}$$

is called a normal distribution or Gaussian distribution, where $\langle f\rangle = \langle F\rangle$ and σ are the mean and standard deviation of the distribution. This distribution is of fundamental interest in the analysis of many problems, because the random variable commonly encountered can be expressed as normal random variables. The prefactor in equation (2.22) has been chosen so that the distribution is normalized

$$\int_{-\infty}^{\infty} p(f)\,df = 1. \tag{2.23}$$

The probability distribution given by equation (2.22) is drawn in Figure 2.3, which shows symmetry of the curves with respect to the mean $\langle f\rangle$. The value of $\langle f\rangle$ has been taken as zero to draw and show the significance of σ, but it need not be zero. The value of $\langle f\rangle$ can be positive or negative. In either case the curve will shift by $|\langle f\rangle|$ units to the right and left, respectively. An important characteristic of the curves is the magnitude of $p(f)$ at $f = 0$; as σ^2 decreases, the peak value of $p(f)$ increases, but the peak width decreases. Consider these trends keeping in mind the following relations

$$\sigma^2 = \text{variance}$$

$$\text{variance} = \langle (f-\langle f\rangle)^2\rangle = \langle f^2\rangle - \langle f\rangle^2. \tag{2.24}$$

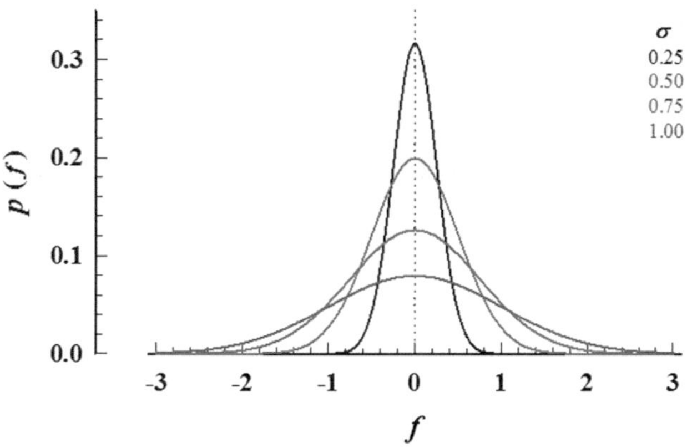

FIGURE 2.3 The probability distribution of equation (2.22) shown for zero mean, $\langle f\rangle = 0$, and varying value of σ.

Variance is precisely defined as the mean of the square of the deviation from the mean. Smaller value of variance means the measured f_i values are close to $\langle f\rangle$, and hence the dispersion or the spread of the peak is less. Note that the dispersion is often also called variance, and the mean value $\langle f\rangle$ of our measurement is called the expectation value.

To illustrate the use of a Gaussian distribution, let us consider the mean vibration energy of a classical harmonic oscillator. The energy is

$$\langle E\rangle = \frac{1}{2}kr^2, \tag{2.24}$$

in which k is the force constant and r refers to the amplitude a of undamped oscillation, implying the displacement from the equilibrium position (Figure 2.4).

For an ensemble of harmonic oscillators, the probability density for the distribution of energy is given by Boltzmann distribution

$$p(E)=\frac{1}{k_BT}e^{-\frac{E}{k_BT}}. \tag{2.25}$$

Here again, $p(E)$ is normalized. The mean value of the energy is

$$\langle E\rangle = \int_0^{\infty} p(E)E\,dE = k_BT \tag{2.26}$$

but the energy $E = \frac{1}{2}kr^2$ of the oscillator is a function of the variable r alone. To evaluate $\langle r^2\rangle$ we can use the following probability density

$$p(r)=\sqrt{\frac{k}{2k_BT}}e^{-\frac{kr^2}{2k_BT}}, \tag{2.27}$$

so that

$$\langle r^2\rangle = \int_{-\infty}^{\infty} p(r)r^2\,dr \sim \frac{2k_BT}{k} \tag{2.28}$$

$$\langle E\rangle = k_BT \sim \frac{1}{2}kr^2.$$

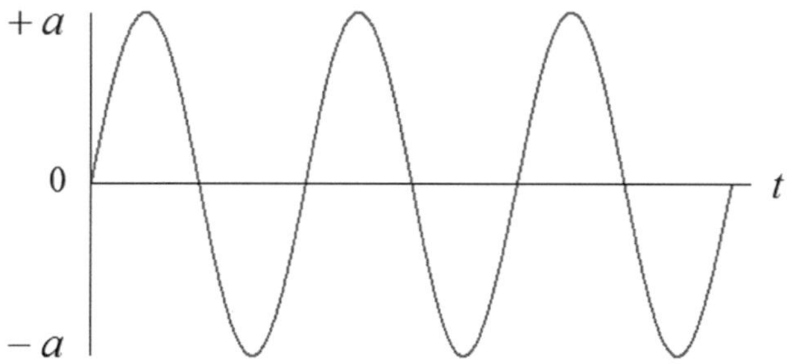

FIGURE 2.4 Undamped oscillations of a harmonic oscillator. The energy of the oscillator varies with the square of the amplitude only.

2.2.4 Postulate 4

When a measurement of an observable F is carried out, the only possible values of the observable are those that form a set of eigenvalues f of the equation

$$F\phi_k(\mathbf{r},t) = f_k\phi_k(\mathbf{r},t), \tag{2.29}$$

where F is the operator corresponding to the observable F, $\phi(\boldsymbol{r},t)$ is a state function or state vector commonly called eigenvector, and f is the eigenvalue.

This postulate is a general statement about stationary states, say eigenvectors as per the postulate, whose probability densities and mean values are time-independent. To obtain the eigenvalue equation containing the eigenvector, we will look into the solution of the time-dependent Schrödinger equation with the condition that the potential energy is time-independent. The time-dependent Schrödinger equation for a single particle in one dimension, say x-coordinate, is

$$i\hbar\frac{\partial}{\partial t}\Psi(x,t) = \frac{-\hbar}{2m}\frac{\partial^2\Psi(x,t)}{\partial x^2} + \mathrm{V}(x,t)\Psi(x,t), \tag{2.30}$$

and is solved by separating the variables

$$\Psi(x,t) = f(t)\,\psi(x). \tag{2.31}$$

Substitution in equation (2.30) yields

$$i\hbar\psi(x)\frac{\partial}{\partial t}f(t) = \frac{-\hbar}{2m}f(t)\frac{\partial^2\psi(x)}{\partial x^2} + \mathrm{V}(x)f(t)\psi(x)$$

$$i\hbar\frac{1}{f(t)}\frac{\partial}{\partial t}f(t) = \frac{-\hbar}{2m}\frac{1}{\psi(x)}\frac{\partial^2\psi(x)}{\partial x^2} + \mathrm{V}(x). \tag{2.32}$$

The left-hand side is a function of time and the right-hand side is a function of the coordinate x. But the equality of the two sides must mean that neither side has functional dependence on time and coordinate. The lack of variable dependence must make both functions $f(t)$ and $\psi(x)$ constant. Retaining the left-hand side, but substituting for the right-hand side with a constant $\hbar\omega$, we get

$$\frac{1}{f(t)}\frac{df(t)}{dt} = -i\omega$$

$$\frac{1}{f(t)}df(t) = -i\omega dt.$$

Integration of the two sides yields

$$\ln f(t) = -i\omega t + c$$

$$f(t) = e^{-i\omega t}e^{c}$$

$$= Ae^{-i\omega t}, \tag{2.33}$$

in which $A = e^c$ is a constant. Since equation (2.31) could have been written in the form

$$\Psi(x,t) = f(t)\{A\psi(x)\},$$

the constant A can be removed from equation (2.33) and written as

$$f(t) = e^{-i\omega t}. \tag{2.34}$$

Next, the right-hand side of equation (2.32) is taken and the left-hand side is equated by the constant $\hbar\omega$

$$\frac{\hbar^2}{2m}\frac{d^2}{dx^2}\psi(x) + \mathrm{V}(x)\psi(x) = \hbar\omega\psi(x). \tag{2.35}$$

This is the time-independent Schrödinger equation. Since $\psi(x)$ is a solution of this, together with equation (2.34) we write the variable-separated equation (2.31) as

$$\Psi(x,t) = \psi(x)e^{-i\omega t}. \tag{2.36}$$

States of this form, meaning $\psi(x)e^{-i\omega t}$, are called stationary states because both expectation values corresponding to these states and their probability densities are time-independent $|\Psi(x,t)|^2 = |\psi(x)|^2$.

An important property of stationary states is that each state is associated with only one frequency ω, so each stationary state has a constant energy $\hbar\omega = E$. Thus, equation (2.35) can be written as

$$\left[\frac{\hbar^2}{2m}\frac{d^2}{dx^2} + \mathrm{V}(x)\right]\psi(x) = E\psi(x).$$

The differential operator within the large brackets is the time-independent Hamiltonian $\mathcal{H}$, which is the total energy operator containing the kinetic energy and potential energy operators. The time-independent Schrödinger equation is generally written as

$$\mathcal{H}\psi(x) = E\psi(x), \tag{2.37}$$

which is the eigenvalue equation of the operator $\mathcal{H}$, and E is the energy eigenvalue.

The above was an illustrative discussion of stationary states and the eigenvalue equation. To maintain uniformity of using symbols, we will continue to write the eigenvalue equation as

$$F\phi = f\phi,$$

in which the coordinate dependence of the eigenfunction ϕ is implicit. An important aspect of the eigenvalue equation is its correspondence with 'zero variance'. In the discussion of the expectation value of a physical observable (Postulate 3), we invoked the statistical measure of variance $(=\sigma^2)$ associated with the expectation value (equation (2.24)). If the relation involving the operator F

$$\langle F \rangle = \int \phi^* F \phi d\tau \tag{2.38}$$

is squared, we get the square of the mean value

$$\langle F \rangle^2 = \left\{ \int \phi^* F \phi d\tau \right\}^2. \tag{2.39}$$

The relation for the square of the operator can also be written as

$$\langle F^2 \rangle = \int \phi^* F^2 \phi d\tau, \tag{2.40}$$

in which case one gets the mean of the square of the observable. Earlier we discussed the statistical significance of variance (see Figure 2.3), and stated that variance is the mean of the square of the deviation from the mean

$$\text{variance} = \langle (f - \langle f \rangle)^2 \rangle.$$

For convenience, we rewrite this as

$$\text{variance} = \langle (F - f)^2 \rangle.$$

The operator corresponding to variance in the distribution of an observable is $(F-f)^2$, so that

$$\langle (F - f)^2 \rangle = \int \phi^* (F - f)^2 \phi d\tau. \tag{2.41}$$

If the Hermitian property of the $(F-f)$ operator

$$(F - f) = (F - f)^\dagger$$

is used (see equation (2.13)), then

$$\langle (F - f)^2 \rangle = \int [(F - f)\phi]^* (F - f)\phi d\tau$$
$$= \int |(F - f)\phi|^2 d\tau. \tag{2.42}$$

If $\langle F \rangle^2 = f^2$, the variance in the distribution of the F observable vanishes (Figure 2.3), then

$$(F - f)\phi = 0$$
$$F\phi = f\phi. \tag{2.43}$$

Interestingly, we have recovered the eigenvalue equation that was also derived as the solution of the time-dependent Schrödinger equation with a time-independent potential energy operator (see equation (2.37)). The idea developed now states that the eigenvalues of the operator F has no variance

$$\langle (F - \langle f \rangle)^2 \rangle = 0. \tag{2.43}$$

The significance of zero variance for the eigenvalue is that one measurement of the observable F yields one of the f_k. Another measurement may yield the same eigenvalue or another eigenvalue. It should be clear that individually none of the f_k will have a variance, and hence no dispersion in the probability distribution (Figure 2.3). However, the mean of the distribution of eigenvalues can in principle be calculated from Postulate 3.

Eigenvalue Spectrum of a Property. The time-dependent Schrödinger equation that yielded the eigenvalue equation (2.37) is a linear equation, and therefore has a series of solutions represented by

$$\Psi(x,t) = \sum_k c_k \phi_k(x) e^{\frac{-iE_k t}{\hbar}}, \tag{2.44}$$

in which each of the ϕ_k is a stationary state

$$\phi_k(x,t) = \phi_k(x) e^{\frac{-iE_k t}{\hbar}}. \tag{2.45}$$

This means there exists a set of eigenfunctions $\{\phi_k\}$ of the operator F, and each eigenfunction of the set satisfies the eigenvalue equation

$$F\phi_k = f_k \phi_k, \tag{2.46}$$

which also implies the existence of a set of eigenvalues $\{f_k\}$. This set of eigenvalues is called the spectrum of the operator F. A discrete eigenvalue spectrum may have degenerate, non-degenerate, or both eigenvalues.

Eigenvalue Degeneracy and Non-degeneracy. Degeneracy of the eigenvalue f arises when the operator F has eigenfunctions ϕ_k such that

$$F\phi_k = f_k \phi_k \text{ and}$$
$$F\phi_k^j = f_k \phi_k^j, \tag{2.47}$$

where

$$\phi_k = \sum_{j=1}^{n} c_k \phi_k^j.$$

This derivation shows that an eigenvalue could be associated with a set of linearly independent eigenfunctions ϕ_k^j associated with the eigenfunction ϕ_k. The eigenvalue f_k is now said to be n-fold degenerate. Both ϕ_k and the set of ϕ_k^j are eigenfunctions of F having the same eigenvalue f_k. There may be some eigenvalues in the set of f_k whose corresponding eigenfunctions are unique. These eigenvalues are called non-degenerarate.

Discrete Eigenvalues and Quantization of a Physical Observable. When eigenvalues for a measurable property, such as the energy of a particle or a system, form a discrete but not continuous set of values, the property is said to be quantized. These states of the system are bound states. A property is not quantized if the corresponding eigenvalues are continuous. In all cases of measurement of physical properties, the basic idea is the probability of finding the system with a discrete – that is, not continuous – eigenvalue. The understanding that eigenvalues in a discrete eigenvalue spectrum may all be degenerate, non-degenerate, or a mixture of both is shown below in terms of the structure of the state function and the probability of finding the eigenvalue.

We are familiar by now that the eigenvalue equation in all cases is given by

$$F\phi_k = f_k\phi_k,$$

which says that ϕ_k forms a basis set $\{\phi_k, k = 1,2,3,\ldots,n\}$ that constitutes the state vector $|\Psi\rangle$, which is expressed as

$$|\Psi\rangle = \sum_k \sum_{j=1}^{n} c_k^j |\phi_k\rangle. \tag{2.48}$$

If the eigenvalue spectrum is discrete and contains only non-degenerate eigenvalues, this expression simplifies to

$$|\Psi\rangle = \sum_k c_k\phi_k, \tag{2.49}$$

and the probability distribution function $P(f_k)$ of finding the eigenvalue f_k is postulated to be

$$P(f_k) = |f_k|^2 = |\langle\phi_k|\Psi\rangle|^2 = \left|\int \phi_k^*\Psi d\tau\right|^2. \tag{2.50}$$

When the spectrum contains only degenerate eigenvalues or a mixture of degenerate and non-degenerate eigenvalues, the probability of finding the eigenvalue f_k in a measurement is

$$P(f_k) = \sum_{j=1}^{n} |c_k^j|^2 = \sum_{j=1}^{n} |\langle\phi_k^j|\Psi\rangle|^2 = \sum_{j=1}^{n} \left|\int \phi_k^{j*}\Psi d\tau\right|^2, \tag{2.51}$$

where the index j is the degree of degeneracy.

2.2.5 Postulate 5

The state function can be expressed as a linear combination of eigenstates of a Hermitian operator

$$\psi(x) = \sum_k c_k\phi_k(x),$$

in which c_k are probability coefficients, or simply numbers.

We have come across this postulate in some way in earlier discussions. However, what the postulate means is that the probability of finding the eigenvalue f_k according to

$$F\phi_k = f_k\phi_k$$

is $|c_k|^2$. To see that $|c_k|^2$ indeed is the probability of finding f_k in a measurement of the property F, we look at the equation for the expectation value of F

$$\begin{aligned}\langle F\rangle &= \int \Psi^* F \Psi d\tau \\ &= \int\left(\sum_k c_k\phi_k\right)^* F\left(\sum_k c_k\phi_k\right)d\tau \\ &= \int\left(\sum_k c_k\phi_k\right)^* \sum_k c_k f_k \phi_k d\tau \\ &= \sum_k |c_k|^2 f_k \int \phi_k^*\phi_k d\tau.\end{aligned} \tag{2.52}$$

Using the orthonormality property of eigenfunctions, by which

$$\int \phi_i^*\phi_j d\tau = \delta_{ij} \qquad \delta_{ij} = \begin{cases} 0\ for\ i \neq j \\ 1\ for\ i = j \end{cases} \tag{2.53}$$

we obtain

$$\langle F\rangle = \sum_k |c_k|^2 f_k. \tag{2.54}$$

A comparison of this expression with equation (2.18) for the expectation value shows directly that the classical probability function p_k corresponds to the quantum mechanical probability $|c_k|^2$.

An eigenfunction can be simultaneously associated with two eigenvalues corresponding to two operators. If F and G are the two operators with eigenvalues f and g, respectively, then an eigenfunction ϕ can satisfy

$$\begin{aligned} F\phi &= f\phi \\ G\phi &= g\phi \end{aligned} \tag{2.55}$$

provided that the products of the operators FG and GF also satisfy

$$(FG - GF) = 0. \tag{2.56}$$

The difference of the operator products, often also written as $[F,G] = -[G,F]$, is called the commutator of the two operators. It is then said that the operator F and G commute with each other, and the eigenvalues f and g are simultaneously observable. Here is an example from day-to-day life. In making a mug of tea, one can add sweetener first and cream later, or swap the two. It is immaterial as to which order we operate on the black tea – with sugar first and cream later or the reverse, the end result is the same.

A common example of commutator operators is the square of the angular momentum operator L^2 and any one of the operators L_x, L_y, and L_z for the Cartesian components of the angular momentum vector. Note that $L^2 = L_x^2 + L_y^2 + L_z^2$, but the commutators are

$$[L^2, L_x] = [L^2, L_y] = [L^2, L_z] = 0, \tag{2.57}$$

and the operators for the components of the angular momentum do not commute with each other. In fact,

$$[L_x, L_y] = i\hbar L_z$$
$$[L_y, L_z] = i\hbar L_x$$
$$[L_z, L_x] = i\hbar L_y. \quad (2.58)$$

Although the L^2 operator commutes with each of the component operators L_x, L_y, and L_z, traditionally L_z is specified with L^2. The eigenfunctions for angular momentum operators are spherical harmonics $Y_l^m(\theta,\varphi)$, given by

$$Y_l^m(\theta,\varphi) = \left[\frac{2l+1}{4\pi}\frac{l-|m|!}{l+|m|!}\right]^{\frac{1}{2}} P_l^{|m|}\cos\theta e^{im\varphi}, \quad (2.59)$$

where $P_l^{|m|}$ are associated Legendre functions, and l and m are quantum numbers such that $l = 0,1,2,\cdots$ and m can have values from $-l$, $-l+1$, $-l+2$, $\cdots$ to $\cdots l+2$, $l+1$, $+l$ through 0. The eigenvalue equations for a free particle orbital angular motion are

$$L^2 Y_l^m(\theta,\varphi) = l(l+1)\hbar^2 Y_l^m(\theta,\varphi)$$
$$L_z Y_l^m(\theta,\varphi) = m\hbar Y_l^m(\theta,\varphi). \quad (2.60)$$

The orientation of the angular momentum vector **L** in a Cartesian coordinate system is shown in Figure 2.5. The

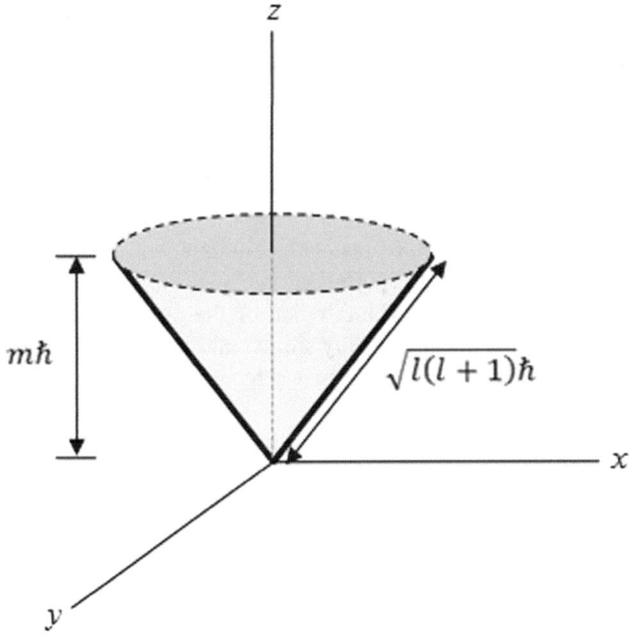

FIGURE 2.5 The orientation of **L** and the eigenvalues. The eigenvalue corresponding to the operator L^2 is $l(l+1)\hbar^2$, the square-root of which gives the length of **L**.

magnitude of the vector is specified by the quantum number l, and its z-projection, meaning the height of the L vector along the z-axis, is specified by the quantum number m. Notice how the eigenvalues $l(l+1)\hbar^2$ and $m\hbar$ appear in the figure. The L vector is shown to be everywhere on the surface of the cone. This depiction is based on the fact that the operators L_x and L_y do not commute, and therefore the x and y-components of the angular momentum cannot be specified simultaneously, which suggests that the **L** vector can be found anywhere on the cone surface.

2.2.6 Postulate 6

The time evolution of a state function is determined by the operator for the total energy of the system

$$i\hbar\frac{\partial}{\partial t}\Psi(x,t) = \mathcal{H}\Psi(x,t) \quad (2.61)$$

where $\mathcal{H}$ is the Hamiltonian operator for kinetic and potential energies. The energy eigenvalue is obtained by

$$\mathcal{H}\phi_k = E_k\phi_k.$$

This postulate essentially sets up the time-dependent Schrödinger equation, which in Dirac notation can be written as the superposition of eigenfunctions

$$i\hbar\frac{\partial}{\partial t}|\Psi(x,t)\rangle = \mathcal{H}|\Psi(x,t)\rangle$$
$$= \sum_k E_k c_k |\phi_k\rangle, \quad (2.62)$$

so that

$$\mathcal{H}|\phi_k\rangle = E_k|\phi_k\rangle.$$

If the Hamiltonian is composed of functions of coordinates and momenta but has no time dependence, the Schrödinger equation (2.61) can be integrated right away. This has been done earlier by the variable separation method in the discussion of the eigenvalue equation under Postulate 4. The result, which is

$$\Psi(x,t) = \sum_k c_k\phi_k(x)e^{-\frac{iE_kt}{\hbar}}, \quad (2.63)$$

can be used in the expansion of any state function if the Hamiltonian is time-independent. In this formalism, the state function can be viewed as a superposition of eigenstates corresponding to different energy levels, and $|c_k|^2$ is the probability of finding the system in state ϕ_k. Further, the energy that is an observable physical quantity associated with an eigenstate does not change with time. Hence, these eigenstates are called stationary states. Because the eigenstates are stationary, the probability density of the eigenstate

$$|\phi_k(x)|^2 = \phi_k^*(x)\phi_k \tag{2.64}$$

is also time-independent. But the complex amplitude

$$\phi_k(x)e^{-\frac{iE_kt}{\hbar}}$$

will change with time. By Euler's equation for exponential relation, the time variation of the complex amplitude becomes

$$\phi_k(x)\left[\cos\left(\frac{E_kt}{\hbar}\right) - i\sin\left(\frac{E_kt}{\hbar}\right)\right],$$

which is analogous to the variation of the electromagnetic radiation (equation (1.30)).

The principle of superposition of stationary states (see also Box 2.1) to describe a state function is centrally important in spectroscopy. To illustrate, let the state function at time $t = 0\,(t_0)$ be

$$\Psi(x,t) = \sum_k c_k\phi_k(x)e^{-\frac{iE_kt_0}{\hbar}}. \tag{2.65}$$

Then let the system interact with a time-varying field at time $t \geq 0$. The state function at time $t_0 + t$ is

$$\Psi(x,t) = \sum_k c_k(t)\phi_k(x)e^{-\frac{iE_k(t_0+t)}{\hbar}}. \tag{2.66}$$

Notice that the c_k coefficient is now given as a function of time. If $t_0 = 0$, the equation above is

$$\Psi(x,t) = \sum_k c_k(t)\phi_k(x)e^{-\frac{iE_kt}{\hbar}}. \tag{2.67}$$

Thus the stationary states still describe the system at all times, but the $c_k(t)$ coefficients will vary with time.

A thorough study of these principles of quantum mechanics provides a deeper understanding of spectroscopy. Spectroscopic experiments can be designed concurrently to verify the postulates of quantum mechanics. It is now obvious that spectroscopy and quantum mechanics are inseparable.

2.3 PERTURBATION THEORY

It is also necessary to point out some of the approximations that one does to explain observables in spectroscopy. In view of the quantum mechanical apparatus and the Schrödinger equation, it might appear that many properties of atomic and molecular systems can be predicted effortlessly. That simply is not true, the Schrödinger equation can be solved exactly only for a few selected simple systems. The hydrogen atom, for example, is an exactly solvable problem. In fact, the earliest experimental data on small atomic systems have greatly augmented the formulation of quantum mechanics. But the exact solution of the Schrödinger equation is hardly achieved beyond such simple systems. The most complex system that can be solved exactly is the molecular hydrogen ion H_2^+. This constraint forces adoption of approximate methods to evaluate properties of systems, and the perturbation theory is one such approach to calculate a property, say the energy of a system, if the property of a similar and close-by system is known. Not only that, perturbation calculation also estimates, at least in principle, the energy eigenvalue(s) of the system whose eigenstate(s) are unknown *a priori*. The calculations essentially endeavor to 'correct' the property and the associated eigenvector(s) over and above what is known. The perturbed property and eigenvectors are therefore often also said to introduce a correction to the eigenvalue(s) and eigenvector(s) of the system known already.

To justify the arguments above, consider an arbitrary harmonic oscillator whose eigenvalues and eigenvectors are known. This is an exactly solvable problem. In fact, the solution of the classical oscillator problem was known long before the advent of quantum mechanics, and Lorentz used the oscillator system to explain the Zeeman effect toward the very end of the nineteenth century. If the oscillator is perturbed by an electric field, both the energy eigenvalues and the eigenstates of the oscillator are going to change. The perturbed energy and eigenstates are assumed to be close to the unperturbed energy and eigenstates, so that a correction may be introduced to the unperturbed system to account for the electric field perturbation. It is important to keep in mind that the degree of perturbation should be reasonably small to carry out the calculations for correction. In summary, perturbation theory is applicable only to those systems that are externally perturbed to a small extent, else the perturbation expansion will not converge.

2.3.1 Perturbation of a Nondegenerate System

Let a system in the nth state of energy characterized by $\mathcal{H}_0$, $\psi_n^{(0)}$, and $E_n^{(0)}$ is somehow perturbed, by an external electric or magnetic field, for example. Assume that the external field is weak so that the perturbation expansion will converge suitably. Let the total Hamiltonian of the perturbed system be $\mathcal{H}$, which is mathematically similar and closer to $\mathcal{H}_0$. Then the eigenvalue equations can be written as

$$\mathcal{H}_0\psi_n^{(0)} = E_n^0\psi_n^{(0)} \quad \text{system properties known}$$

$$\mathcal{H}\psi_n = E_n\psi_n \quad \text{system properties unknown}$$

The unknown eigenstate and energy terms are now expanded in terms of the known ones

$$\mathcal{H} = \mathcal{H}_0 + \lambda\mathcal{H}^{(1)} + \lambda^2\mathcal{H}^{(2)} + \lambda^3\mathcal{H}^{(3)} + \cdots = \mathcal{H}_0 + \sum_{k=1}^{\infty}\lambda^k\mathcal{H}^{(k)}$$

$$\psi_n = \psi_n^{(0)} + \lambda\psi_n^{(1)} + \lambda^2\psi_n^{(2)} + \lambda^3\psi_n^{(3)} + \cdots = \psi_n^{(0)} + \sum_{k=1}^{\infty}\lambda^k\psi_n^{(k)}$$

$$E_n = E_n^{(0)} + \lambda E_n^{(1)} + \lambda^2 E_n^{(2)} + \lambda^3 E_n^{(3)} + \cdots = E_n^{(0)} + \sum_{k=1}^{\infty} \lambda^k E_n^{(k)}, \tag{2.68}$$

where $\lambda \ll 1$ is the perturbation scaling parameter whose smallness is necessary to have each consecutive term much smaller than the preceding term. The bracketed superscripts indicate the order of perturbation or the order of correction to $\mathcal{H}_0$, $\psi_n^{(0)}$, and $E_n^{(0)}$. The unknown eigenvalue equation can now be rewritten using these expansions, and we get the following

$$\left(\mathcal{H}_0 + \sum_{k=1}^{\infty} \lambda^k \mathcal{H}^{(k)}\right)\left(\psi_n^{(0)} + \sum_{k=1}^{\infty} \lambda^k \psi_n^{(k)}\right) = \left(E_n^{(0)} + \sum_{k=1}^{\infty} \lambda^k E_n^{(k)}\right)\left(\psi_n^{(0)} + \sum_{k=1}^{\infty} \lambda^k \psi_n^{(k)}\right)$$

$$\left(\mathcal{H}_0 - E_n^{(0)} + \sum_{k=1}^{\infty} \lambda^k \mathcal{H}^{(k)} - \sum_{k=1}^{\infty} \lambda^k E_n^{(k)}\right)\left(\psi_n^{(0)} + \sum_{k=1}^{\infty} \lambda^k \psi_n^{(k)}\right) = 0. \tag{2.69}$$

By keeping the energy terms of the same order together and multiplying them by the eigenstates of different order appropriately the following perturbation equations are obtained

$$\left(\mathcal{H}_0 - E_n^{(0)}\right)\psi_n^{(0)} = 0 \tag{2.70}$$

$$\left(\mathcal{H}_0 - E_n^{(0)}\right)\psi_n^{(1)} + \left(\mathcal{H}^{(1)} - E_n^{(1)}\right)\psi_n^{(0)} = 0 \tag{2.71}$$

$$\left(\mathcal{H}_0 - E_n^{(0)}\right)\psi_n^{(2)} + \left(\mathcal{H}^{(1)} - E_n^{(1)}\right)\psi_n^{(1)} + \left(\mathcal{H}^{(2)} - E_n^{(2)}\right)\psi_n^{(0)} = 0 \tag{2.72}$$

$$\left(\mathcal{H}_0 - E_n^{(0)}\right)\psi_n^{(3)} + \left(\mathcal{H}^{(1)} - E_n^{(1)}\right)\psi_n^{(2)} + \left(\mathcal{H}^{(2)} - E_n^{(2)}\right)\psi_n^{(1)} + \left(\mathcal{H}^{(3)} - E_n^{(3)}\right)\psi_n^{(0)} = 0. \tag{2.73}$$

Notice the systematic change of $\psi_n^{(i)}$ as the equations progress. The Hamiltonian $\mathcal{H}_0$ can be taken as Hermitian, and $\psi_n^{(0)}$ and $\psi_n^{(k)}$ are orthogonal to each other, $\langle \psi_n^{(0)} | \psi_n^{(k)} \rangle = 0$. Then, a standard procedure in quantum mechanics can be used, by which the equations above are multiplied through by $\psi_n^{(0)*}$ and then integrated. By invoking the orthogonal condition in the integrations, it is easy to see the emergence of the following integrals for energy corrections

$$E_n^{(1)} = \langle \psi_n^{(0)} | \mathcal{H}^{(1)} | \psi_n^{(0)} \rangle \tag{2.74}$$

$$E_n^{(2)} = \langle \psi_n^{(0)} | \mathcal{H}^{(2)} | \psi_n^{(0)} \rangle + \langle \psi_n^{(0)} | \mathcal{H}^{(1)} | \psi_n^{(1)} \rangle \tag{2.75}$$

$$E_n^{(3)} = \langle \psi_n^{(0)} | \mathcal{H}^{(3)} | \psi_n^{(0)} \rangle + \langle \psi_n^{(0)} | \mathcal{H}^{(2)} | \psi_n^{(1)} \rangle + \langle \psi_n^{(0)} | \mathcal{H}^{(1)} | \psi_n^{(2)} \rangle. \tag{2.76}$$

If one asks what are the forms of $\psi_n^{(0)}$, then the answer could just be that a general solution of eigenstates is hard to find because that demands the information about $\mathcal{H}^{(k)}$. An approach that can be used to evaluate $\psi_n^{(k)}$ is based on a property of an eigenstate, it is $\psi_n^{(k)}$ here, that any eigenstate can be expanded in a complete set of orthonormal functions $\{\psi_k\}$ (see Postulate 4), which include both discrete and continuum states. Accordingly,

$$\lambda \psi_n^{(1)} = \sum_{k \neq n} a_k \psi_n^{(0)}, \tag{2.77}$$

in which $\lambda \psi_n^{(1)}$ is orthogonal to $\psi_n^{(0)}$. Substitution for $\psi_n^{(1)}$ in equation (2.71) by this expansion will read

$$\sum_{k \neq n} a_k \left(\mathcal{H}_0 - E_n^{(0)}\right)\psi_k^{(0)} = -\left(\mathcal{H}^{(1)} - E_n^{(1)}\right)\psi_n^{(0)}. \tag{2.78}$$

Multiplication by $\psi_m^{(0)*}$ and integration gives

$$a_m\left(E_m^{(0)} - E_n^{(0)}\right) = -\langle \psi_m^{(0)} | \mathcal{H}^{(1)} | \psi_n^{(0)} \rangle$$

$$a_m = -\frac{\langle \psi_m^{(0)} | \mathcal{H}^{(1)} | \psi_n^{(0)} \rangle}{E_m^{(0)} - E_n^{(0)}} \tag{2.79}$$

so, one has

$$\psi_n^{(1)} = -\sum_{k \neq n} \frac{\langle \psi_m^{(0)} | \mathcal{H}^{(1)} | \psi_n^{(0)} \rangle}{E_k^{(0)} - E_n^{(0)}} \psi_k^{(0)}. \tag{2.80}$$

Substitution for this $\psi_n^{(1)}$ in the expression for $E_n^{(2)}$ in equation (2.75) yields

$$E_n^{(2)} = \langle \psi_n^{(0)} | \mathcal{H}^{(2)} | \psi_n^{(0)} \rangle - \sum_{k \neq n} \frac{\langle \psi_n^{(0)} | \mathcal{H}^{(1)} | \psi_k^{(0)} \rangle \langle \psi_k^{(0)} | \mathcal{H}^{(1)} | \psi_n^{(0)} \rangle}{E_k^{(0)} - E_n^{(0)}}. \tag{2.81}$$

This concludes the analytical solution to the second-order energy correction. Obtaining higher-order $E_n^{(k)}$ and $\psi_n^{(k)}$ becomes complicated. But higher than second-order correction is rarely necessary, because the scaling parameter λ is considered small enough so that perturbation expansion converges rapidly.

2.3.2 Perturbation of a Degenerate State

An energy state can at times contain a set of orthonormal eigenfunctions that have degenerate eigenvalues. The 2s state of hydrogen atoms, in which different l levels have the same energy, is an example. Degenerate state perturbation employs the same perturbation expansions considered above. Let $\mathcal{H}_0$, $E_n^{(0)}$, and $\psi_n^{(0,\mathrm{d})}$ be the unperturbed Hamiltonian, energy, and eigenfunction of the nth state. Note that the subscript 'n' in $E_n^{(0)}$ and $\psi_n^{(0,\mathrm{d})}$ has no special significance; it simply is a reminder that the perturbation is carried out with the initial unperturbed n^{th} state. The bracketed superscript '0' (zero) indicates unperturbed zeroth-order, and the 'd' that follows zero is used only to emphasize on the degeneracy of the state.

As done earlier, the perturbation expressions are written identically as

$$\mathcal{H} = \mathcal{H}_0 + \sum_{k=1}^{\infty} \lambda^k \mathcal{H}^{(k)} \tag{2.82}$$

$$E_n = E_n^{(0)} + \sum_{k=1}^{\infty} \lambda^k E_n^{(k)} \tag{2.83}$$

$$\psi_n = \psi_n^{(0,\mathrm{d})} + \sum_{k=1}^{\infty} \lambda^k \psi_n^{(k)}, \tag{2.84}$$

where $E_n^{(0)}$ is, say, j-fold degenerate, which also means $\psi_n^{(0,\mathrm{d})}$ is a complete set of j orthonormal functions $\phi_{n,i}$, such that

$$\psi_n^{(0,\mathrm{d})} = \sum_{i=1}^{j} c_i \phi_{n,i}. \tag{2.85}$$

The eigenvalue equations sorted out by the power of λ give

$$\left[\left(\mathcal{H}_0 - E_n^{(0)}\right) + \left(\mathcal{H}^{(1)} - E_0^{(1)}\right) + \left(\mathcal{H}^{(2)} - E_0^{(2)}\right) + \cdots\right]$$
$$\left[\psi_n^{(0,\mathrm{d})} + \psi_n^{(1)} + \psi_n^{(2)} \cdots\right] = 0,$$

from which the perturbations can be written down easily

$$\left(\mathcal{H}_0 - E_n^{(0)}\right)\psi_n^{(0,\mathrm{d})} = 0 \tag{2.86}$$

$$\left(\mathcal{H}_0 - E_n^{(0)}\right)\psi_n^{(1)} + \left(\mathcal{H}^{(1)} - E_n^{(1)}\right)\psi_n^{(0,\mathrm{d})} = 0 \tag{2.87}$$

$$\left(\mathcal{H}_0 - E_n^{(0)}\right)\psi_n^{(2)} + \left(\mathcal{H}^{(1)} - E_n^{(1)}\right)\psi_n^{(1)} + \left(\mathcal{H}^{(2)} - E_n^{(2)}\right)\psi_n^{(0,\mathrm{d})} = 0 \tag{2.88}$$

$\vdots$

The zeroth-order perturbation provides no new information, thus comes the first-order correction $E_n^{(1)}$. Multiplication of equation (2.87) by $\phi_{n,j}^*$ and integration gives

$$\langle \phi_{n,j} | \mathcal{H}^{(1)} | \psi_n^{(0,\mathrm{d})} \rangle - E_n^{(1)} \langle \phi_{n,j} | \psi_n^{(0,\mathrm{d})} \rangle = 0.$$

Substitution by the use of equation (2.85) yields

$$\sum_i c_i \langle \phi_{n,j} | \mathcal{H}^{(1)} | \psi_n^{(0,\mathrm{d})} \rangle - c_i E_n^{(1)} = 0, \tag{2.89}$$

in which j represents the 'fold-degeneracy' of $\psi_n^{(0,\mathrm{d})}$. This expression represents an eigenvalue problem involving a $j \times j$ operator matrix

$$\left(\langle \phi_{n,j} | \mathcal{H}^{(1)} | \phi_{n,i} \rangle\right)\left(c_i\right) = \left(E_{n,j}^{(1)}\right)\left(c_i\right). \tag{2.90}$$

The use of indices i and j may appear a bit confusing here. In fact, we used i to write down the complete set of $\psi_n^{(0,\mathrm{d})}$, but used j for fold-degeneracy. The square operator matrix must have diagonals $\mathcal{H}_{11}, \mathcal{H}_{22}, \mathcal{H}_{33}, \cdots, \mathcal{H}_{jj}$. From the matrix form of the eigenvalue equation, it is easy to find out the first-order correction of $\psi_n^{(0,\mathrm{d})}$ to a value close to $E_n^{(0)}$

$$E_{n,k}^{(1)} = E_n^{(0)} + \lambda E_{n,k}^{(1)}, \tag{2.91}$$

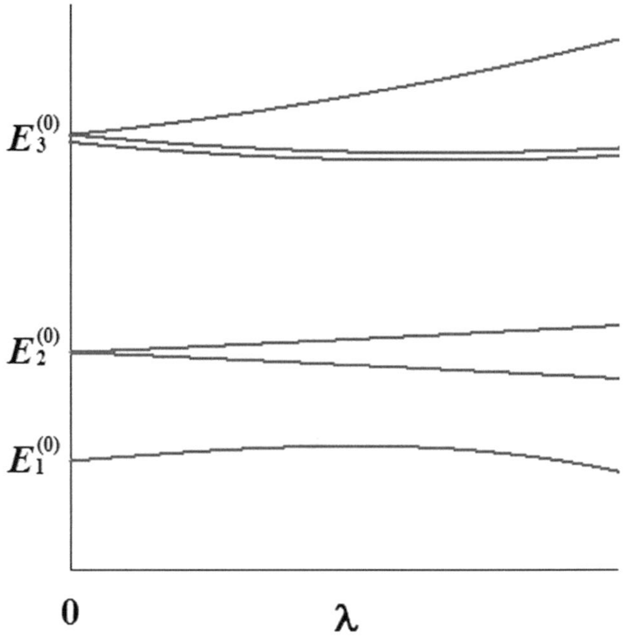

FIGURE 2.6 Changes in the energy eigenvalues corresponding to three eigenstates of a Hamiltonian perturbed to the extent of λ. The ordinate values labeled $E_1^{(0)}$, $E_2^{(0)}$, and $E_3^{(0)}$ correspond to the energies obtained with the unperturbed Hamiltonian ($\lambda = 0$). Notice that $E_1^{(0)}$ has no degeneracy, but $E_2^{(0)}$ and $E_3^{(0)}$ are doubly- and triply-degenerate, respectively. The degeneracy of $E_2^{(0)}$ is removed upon the application of perturbation, but $E_3^{(0)}$ splits into a non-degenerate and a doubly degenerate eigenvalue when $\lambda > 0$, implying only partial removal of degeneracy.

where $k = 1, 2, 3, \ldots, j$, implying that the energy correction is different for the j-fold unperturbed degenerate states. Different energy corrections for the initial degenerate states should also mean that the degeneracy is completely lifted. This is, however, not always true. The degeneracy in certain instances may be lifted only partially, in which case the same first-order energy correction will be applicable to those states which still remain degenerate (Figure 2.6).

The second-order correction to the energy of degenerate states can be derived from the second-order perturbation of the $\psi_{n,k}^{(0,\mathrm{d})}$ state. Here again, $\psi_{n,k}^{(0,\mathrm{d})}$ refers to the nth degenerate state which is still unperturbed or stays in zeroth-order. If the unperturbed state is j-fold degenerate, the index k will run from 1 to j. The second-order energy correction turns out to be

$$E_{n,k}^{(2)} = \langle \psi_{n,k}^{(0,d)} | \mathcal{H}^{(2)} | \psi_{n,k}^{(0,d)} \rangle - \sum_{m \neq n} \frac{\left\langle \psi_{n,k}^{(0,d)} | \mathcal{H}^{(1)} | \phi_m \right\rangle \left\langle \phi_m | \mathcal{H}^{(1)} | \psi_{n,k}^{(0,d)} \right\rangle}{E_m^{(0)} - E_n^{(0)}}. \tag{2.92}$$

Box 2.1

The principle of superposition of states and redundancy of coordinate system for the state function arise from the mathematical description of the state of the system (see Cohen-Tannoudji et al., 1977, for a detailed description). A wavefunction $\Psi(\mathbf{r},t)$ that defines the state of a particle is understood as the particle wave moving in three dimension $(\mathbf{r} = x,y,z)$. But the use of the wavefunction in quantum description of a system is limited for two reasons at least. First, for a n-particle system there are 3N coordinates and the time, and associating a physical wave to this kind of many-particle system is problematic. Second, even for a single particle the wavefunction does not always give a quantum description. For instance, the description of an electron by a wavefunction cannot be provided if the spin degrees of freedom of the electron are considered. These problems are overcome if the state of a system is described by a state vector, which belongs to a state space of the system. The state space, symbolized by ε_r, is an abstract space – a subspace of a Hilbert space.

The idea of using a state vector belonging to a vector space for the description of the state of a system is to gain the advantage of vector geometry and calculation, which does not require the reference to a set of coordinate systems. The vector calculation follows the superposition principle, which states that a linear combination of state vectors is a state vector. The superposition of state vectors belonging to a state space of a system is clearly described by

$$|\Psi(t)\rangle = \sum_{k=1}^{n} c_k(t)|k\rangle,$$

in which n is the dimension of the vector space of allowed state functions, and the time dependence of the state function is fulfilled by the time dependence of the coefficients. This is a direct implication of the first postulate. In many situations involving only two levels or two states, spatial orientation of spin in Stern-Gerlach experiment and polarization of photons, for example, the above equation can be written as

$$\begin{aligned} |\Psi(t)\rangle &= c_1(t)|1\rangle + c_2(t)|2\rangle \\ &= |1\rangle\langle 1||\Psi\rangle + |2\rangle\langle 2||\Psi\rangle, \end{aligned}$$

where we have used $\langle 1||\Psi\rangle = c_1$ and $|2||\Psi\rangle = c_2$.
For a many-level system, $1,2,\ldots n$,

$$\begin{aligned} |\Psi\rangle &= |1\rangle\langle 1||\Psi\rangle + |2\rangle\langle 2||\Psi\rangle + \ldots |n\rangle\langle n||\Psi\rangle \\ &= \sum_i |i\rangle\langle i||\Psi\rangle \\ &= \mathbb{1}|\Psi\rangle \end{aligned}$$

where $\mathbb{1}$ is the identity operator, but $\mathbb{1}$ has no effect on any ket. The relation

$$\sum_i |i\rangle\langle i| = \mathbb{1}$$

is called closure relation.

It is important to understand the vector signs and the way they are shown. The $|i\rangle$ vectors $(i = 1,2,\ldots n)$ are called kets. Each ket is associated with a bra vector, say $\langle\phi|$, that belongs to what is called the dual space, ε^*, of ε space. Consider $\langle x|$ in the vector space associated with the coordinates. We then have

$$\langle x|\psi(t)\rangle = |\Psi(x,t)\rangle.$$

Note that $|\psi(t)\rangle$ is a state vector, but $|\Psi(x,t)\rangle$ is a state function. In the bra and ket notation, a bra vector $\langle x|$ is the adjoint of the ket $|x\rangle$, meaning

$$\langle x| = |x\rangle^\dagger,$$

so that

$$\langle\psi(t)|x\rangle = \langle\Psi(x,t)| = \left(|\Psi(x,t)\rangle\right)^*.$$

The $\langle x|\Psi(t)\rangle$ notation is construed as the projection of the $|\Psi(t)\rangle$ ket onto a basis vector $\langle x|$ belonging to the coordinate space.

PROBLEMS

2.1 Operators that satisfy

$$\int \Psi^* F \Psi d\tau = \int (F\Psi^*)\Psi d\tau$$

are called Hermitian operators. Let $F = \dfrac{h}{2\pi i}\dfrac{\partial}{\partial x}$ be a linear operator.

(a) Show that F is Hermitian.
(b) Find the eigenfunctions explicitly.
(c) Show that F^{-1} is also Hermitian.

2.2 Consider a particle whose state function is given by

$$\Psi(x,t) = \sum_k c_k \phi_k(x) e^{-i\omega_k t}$$

where the energy eigenvalues E_k corresponding to ϕ_k eigenstates are $\hbar\omega_k$.

(a) What will be the outcome of a single measurement of the property F?
(b) How is the expectation value of the energy measured?
(c) If many measurements of energy are carried out, what will be the associated variance?
(d) Suppose a measurement of F was found to have the value f. What will be found for the energy immediately after?

2.3 Suppose a bond angle θ shows a Gaussian probability distribution given by

$$P(\theta) = \frac{1}{\sigma\sqrt{\pi}} e^{-(\theta_o - \theta)^2/\sigma^2}$$

where θ_o represents the most probable value, and σ is the standard deviation.

(a) Show that $\langle(\theta_o - \theta)^2\rangle = \frac{\sigma^2}{2}$.

(b) Also verify that $\langle \theta^2 - \langle\theta\rangle^2 \rangle = \frac{\sigma^2}{2}$.

(c) What will be the outcome if $\sigma = 0$?

(d) What is the physical significance of $\sigma = 0$?

2.4 The expectation value of an observable F has been postulated to be $F = \int \psi^* F \psi d\tau$. Let us make measurements of F, and suppose we obtained the same value for all the measurements we carried out. What is the physical significance of such an outcome?

2.5 Consider

$$A\phi_1 = a\phi_1$$

$$B\phi_2 = b\phi_2$$

where A and B are operators, and a and b are corresponding eigenvalues. Is or is not $\phi_1 = \phi_2$? The answer could be 'yes', but with a restriction related to whether A and B commute.

2.6 Given the expectation value of a property $\langle F\rangle$ corresponding to the operator F such that $[\mathcal{H}, F] = 0$, where $\mathcal{H}$ is the energy operator. Show that $\frac{d\langle F\rangle}{dt} = 0$.

2.7 In quantum mechanics, the time dependence of an observable is different from that in classical physics. In the former, the time dependence at certain instant may not have a definite value. The time derivative of an operator is equal to the time derivative of the expectation value

$$\frac{\partial F}{\partial t} = \frac{\partial\langle F\rangle}{\partial t}.$$

Using $\langle F\rangle = \int \psi^* F \psi d\tau$, where $\int \psi^* \psi d\tau = 1$, show that

$$\frac{\partial\langle F\rangle}{\partial t} = \int \psi^* \left[\frac{\partial F}{\partial t} + \frac{i}{\hbar}\{\mathcal{H}F - F\mathcal{H}\}\right] \psi d\tau.$$

What is the physical meaning of this result?

2.8 Let the Hamiltonian of an one-dimensional harmonic oscillator be

$$\mathcal{H}(x) = -\frac{\hbar^2}{2}\frac{\partial^2}{\partial x^2} + 2\pi^2 \nu^2 x^2,$$

where x ranges from $-\infty$ to $+\infty$. Let the oscillator be perturbed by an electric field **E** along $+z$ direction. Derive the effect of the perturbation on the oscillator.

BIBLIOGRAPHY

Cohen-Tannoudji, C., B. Diu, and F. Laloë (1977) *Quantum Mechanics* (Vol I and II), John Wiley & Sons.

Eyring, H., J. Walter, and G. E. Kimbell (1944) *Quantum Chemistry*, Wiley.

Fock, V. A. (1978) *Fundamentals of Quantum Mechanics* (English translation), Mir Publishers.

Kauzman, W. (1957) *Quantum Chemistry*, Academic Press.

Lande, A. (1955) *Foundations of Quantum Theory*, Yale University Press.

3 Semiclassical Theory of Spectroscopic Transition

It has been postulated that the state function can be expressed as a linear superposition of eigenstates of Hermitian operators. When a system is subjected to an external impulse in the form of radiation, the probability of finding the system in different eigenstates of the state function changes. The total probability is always unity, which means the change in probabilities for different states occurs at the expense of each other. This change in probability of finding the system in different eigenstates brought about by system-radiation interaction is called the spectroscopic transition. Here, we use a two-level system to develop the details of a general spectroscopic transition, and work further to examine the effects of interruption in the coherent evolution of the state function. We will derive the general forms of relaxation of the excited state, spectral line shape, and line broadening.

3.1 TWO-LEVEL SYSTEM

The time-development of a two-state system – a particle, a photon, an atom, or a molecule in isolation that can be described by two energy eigenstates, can be modeled using the same mathematical apparatus prescribed by quantum mechanics. Not only the mathematical development but also the implications are generally applicable to all two-state systems. With particular reference to spectroscopy of two energy levels, ground and excited, we call the model 'two-level spectroscopic transition', the propositions of which are experimentally demonstrable.

Since a spectroscopic transition involves time-dependent changes in the probability of finding the system in different states, it is necessary to know the time evolution of the state function, which is completely determined by the energy operator in the time-dependent Schrödinger equation

$$i\hbar\frac{\partial}{\partial t}\Psi(x,t)=\mathcal{H}\Psi(x,t), \tag{3.1}$$

where the total energy operator $\mathcal{H}$ consists of kinetic and potential energies. The solution for the state function in the time-dependent equation can be obtained in the time-independent form by assuming time-independence of the total energy operator, meaning the operator is a function of coordinate and momentum alone. Recall that the operators for coordinate and linear momentum are x and $-i\hbar\partial/\partial x$, respectively, so the Schrödinger equation can be written as

$$i\hbar\frac{\partial}{\partial t}\Psi(x,t)=-\frac{\hbar^2}{2m}\frac{\partial^2}{\partial x^2}\Psi(x,t)+V(x)\Psi(x,t). \tag{3.2}$$

The time-independent assumption for the energy operator allows immediate integration of equation (3.1) to obtain

$$\Psi(x,t)=\sum_k c_k\phi_k(x)e^{-i\omega_k t}, \tag{3.3}$$

where the absolute values of the complex coefficients $c_k e^{-i\omega_k t}$ yield $|c_k|^2$, which are probabilities of finding the system in ϕ_k eigenstates. The frequencies ω_1 and ω_2 correspond to energies E_1 and E_2, respectively; this follows from $E/\hbar=\omega$. Notice that c_k is still time-independent, but the complex amplitude would vary with time as

$$\phi_k(x)[\cos(\omega_k t)-i\sin(\omega_k t).$$

To note, the linear superposition relation (equation (3.3)) for the state function above can also be written to accommodate the lifetime of each eigenstate. If Γ (in rad s^{-1}) corresponds to the lifetime of a state through time-energy uncertainty ($\Delta E\Delta t\sim\hbar$) such that $2\Gamma=1/\tau$, then

$$\Psi(x,t)=\sum_k c_k\phi_k(x)e^{-i(\omega_k-i\Gamma_k)t}. \tag{3.4}$$

Usually the eigenstates are taken to have infinite lifetime ($\tau\to\infty$), hence the linear combinations are written without Γ.

3.2 SYSTEM-RADIATION INTERACTION

The physical description of a spectroscopic transition is greatly simplified when radiation-induced changes in the probability of finding the system are explored for two eigenstates only. In the model of a two-level system the state function is a linear combination of only two eigenstates, and is written explicitly in the absence of radiation as

$$\Psi(x,t)=c_1\phi_1(x)e^{-i\omega_1 t}+c_2\phi_2(x)e^{-i\omega_2 t}, \tag{3.5}$$

where c_1 and c_2 are coefficients such that $|c_1|^2$ and $|c_2|^2$ are probabilities of finding the system in states ϕ_1 and ϕ_2, respectively, when a measurement is done. It is realized immediately that the time-independence of c_1 and c_2 implies no change in probabilities of finding the system in the two eigenstates with time, meaning no occurrence of transitions, which is the consequence of employing a time-independent energy operator. In the model considered, spectroscopic transitions will occur if a time-varying part is included in the total energy operator to make the Hamiltonian time dependent

$$\mathcal{H}(t)=\mathcal{H}_0+V(t),$$

in which $\mathcal{H}_0$ is the time-independent unperturbed total energy operator (kinetic and potential energy), and $V(t)$ is the

DOI: 10.1201/9781003293064-3

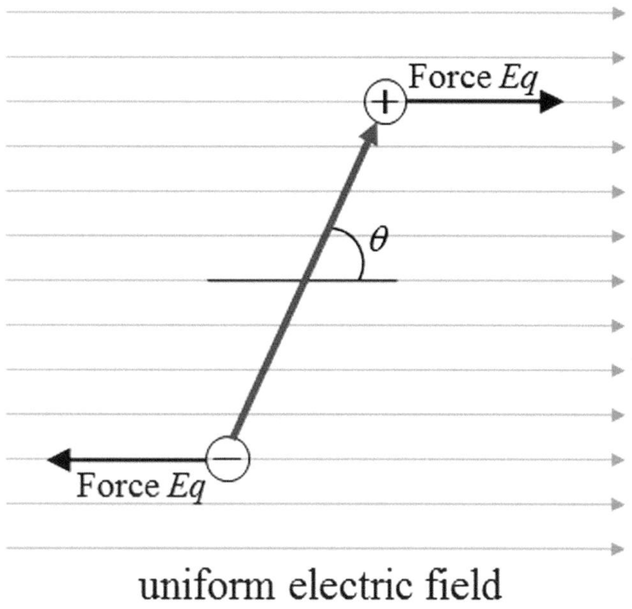

FIGURE 3.1 The rotation of a classical dipole through interaction with the electric field of radiation.

time-varying potential energy that comes into effect through the interaction of radiation with the molecule whose spectroscopic transitions are of interest.

To understand how $V(t)$ is generated, it is useful to consider an electric dipole in the molecule that interacts with the electric field $\mathbf{E}(t)$ of the electromagnetic wave. If the $\mathbf{E}(t)$ field is uniform along the dipole vector, the equal and opposite forces on the two ends of the dipole will be as depicted in Figure 3.1. The total torque τ acting around the center of the dipole so as to rotate it by the angle θ is the vector product of the dipole moment and electric field vectors

$$\tau = \mu E \sin\theta. \tag{3.6}$$

The work done to rotate the dipole by an angle $d\theta$ is $dW = \tau d\theta$, which is the potential energy V of the dipole. The change in the potential energy while the dipole is turned from θ_1 to θ_2 is

$$V_2 - V_1 = \int_{\theta_1}^{\theta_2} \tau d\theta = \mu E\left(\cos\theta_1 - \cos\theta_2\right). \tag{3.7}$$

To find the change in the potential energy during the course of interaction of a molecular dipole with the electric field of the radiation, one is at liberty to choose any initial value of θ_1. If the initial potential is given as $V_1 = 0$, the matching choice is $\theta_1 = \pi/2$, and thus the potential energy becomes the dot product of the vectors $\boldsymbol{\mu}$ and $\mathbf{E}(t)$

$$V(t) = -\boldsymbol{\mu}\cdot\mathbf{E}(t). \tag{3.8}$$

We should note the nature of the $V(t)$ function. It is nonlinear because of the cosine function in it, and unlike the Hamiltonian $\mathcal{H}_o$, it is not Hermitian.

3.3 TIME DEVELOPMENT OF EIGENSTATE PROBABILITIES

Let the time immediately before shining light on the molecule be $t = 0$ when the system is found in eigenstate $\phi_1(x)$ alone, even though it may also be found in $\phi_2(x)$ with a corresponding probability. The rationale of letting $\phi_1(x)$ alone is to simplify the development, because this supposition allows for writing the state function at $t = 0$ as

$$\Psi(x, t=0) = c_1\phi_1(x). \tag{3.9}$$

The time development of the system ensues as soon as the light in the form of a pulse or a continuous wave is allowed to interact with the molecules at $t > 0$. This time evolution will be described by the time-dependent Schrödinger equation (3.1), which after substitution for $\Psi(x,t)$ reads

$$i\hbar\frac{\partial}{\partial t}\left[c_1(t)\phi_1(x)e^{-i\omega_1 t} + c_2(t)\phi_2(x)e^{-i\omega_2 t}\right]$$
$$= \left[\mathcal{H}_o + V(t)\right]\left[c_1(t)\phi_1(x)e^{-i\omega_1 t} + c_2(t)\phi_2(x)e^{-i\omega_2 t}\right],$$

which can be written for the sake of convenience as

$$\begin{aligned} &i\hbar\left[\dot{c}_1(t)\frac{\partial}{\partial t}\phi_1(x)e^{-i\omega_1 t} + \dot{c}_2(t)\frac{\partial}{\partial t}\phi_2(x)e^{-i\omega_2 t}\right] \\ &= \left[\mathcal{H}_o + V(t)\right]\left[c_1(t)\phi_1(x)e^{-i\omega_1 t} + c_2(t)\phi_2(x)e^{-i\omega_2 t}\right]. \end{aligned} \tag{3.10}$$

Notice that the coefficients c_1 and c_2 are now shown to have time dependence, implying that the probability of finding the system in the two states will change with time, and hence we need to solve for $|c_1(t)|^2$ and $|c_2(t)|^2$. The solution is approached with the standard procedure of multiplying both sides with $\phi_1{}^*(x)$ and $\phi_2{}^*(x)$, one at a time, and integrate over all space. Because the exercise is fairly instructive, the method is explicitly worked out in Box 3.1. Multiplication with $\phi_1{}^*(x)$ followed by integration yields

$$i\hbar\dot{c}_1(t)e^{-i\omega_1 t} = V_{12}c_2(t)e^{-i\omega_2 t},$$

and that using $\phi_2{}^*(x)$ yields

$$i\hbar\dot{c}_2(t)e^{-i\omega_2 t} = V_{21}c_1(t)e^{-i\omega_1 t}, \tag{3.11}$$

in which V_{12} and V_{21} denote the integrals containing the $V(t)$ operator

$$\begin{aligned} V_{12} &= \int \phi_1{}^*(x)V(t)\phi_2(x)d\tau \\ V_{21} &= \int \phi_2{}^*(x)V(t)\phi_1(x)d\tau. \end{aligned} \tag{3.12}$$

The V_{12} integral is often expressed as

$$V_{12} = -\mathbf{E}(t)\int \phi_1{}^*(x)\boldsymbol{\mu}\phi_2(x)d\tau = -\mathbf{E}(t)\cdot\boldsymbol{\mu}_{12}, \tag{3.13}$$

and $\boldsymbol{\mu}_{12}$ is called the dipole transition moment, the magnitude and direction of which can be determined experimentally.

To indicate the resonance transition frequency ($\omega_{21}=\omega_2-\omega_1$) it is useful to rewrite equation (3.11) coupling $c_1(t)$ and $c_2(t)$ as

$$i\hbar\dot{c}_1(t)=V_{12}c_2(t)e^{-i\omega_{21}t}$$

$$i\hbar\dot{c}_2(t)=V_{21}c_1(t)e^{i\omega_{21}t}, \tag{3.14}$$

where $\omega_{21}=\omega_2-\omega_1$ is the resonance frequency and $V_{21}=V_{12}{}^*$.

3.4 PROBABILITY EXPRESSIONS

In the initial approach, the probability $|c_2(t)|^2$ can be calculated in the 'weak-field' limit where the system interacts with broad-band radiation, meaning light of a wide range of frequency. If a broad-band source is used, the absorption intensity at the resonance frequency will be small, because the perturbation due to the radiation field is small. This means that the transition probability to the upper state is small, which allows for assuming $|c_2(t)|^2 \ll |c_1(t)|^2$ at all times so that the coupled equations above (equation 3.14) can be reduced to one. Since the total probability is 1,

$$|c_1(t)|^2=1-|c_2(t)|^2\cong 1,$$

and hence the first of the two coupled equations is dropped out and the second one is written simply as

$$i\hbar\dot{c}_2(t)=V_{21}e^{i\omega_{21}t}. \tag{3.15}$$

This can be integrated in the time limit of 0 to t (Box 3.2) to obtain

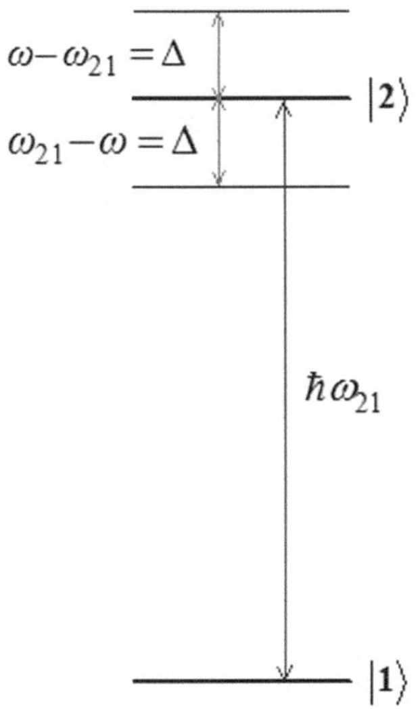

FIGURE 3.2 A two-level system with transition energy $\hbar\omega_{21}$. Exactly on-resonancce condition is achieved only when $\omega_{21}=\omega$, where ω is the frequency of light. Detuning is given by $|\Delta|=\omega_{21}-\omega$.

$$c_2(t)=\frac{\mathbf{E}_o(t)\cdot\boldsymbol{\mu}_{21}}{2\hbar}\left[\frac{e^{i(\omega_{21}-\omega)t}-1}{(\omega_{21}-\omega)}\right] \tag{3.16}$$

in the RWA approximation. The detuned frequency $|\Delta|=\omega_{21}-\omega$, where ω is the frequency of light (Figure 3.2).

The quantity

$$\frac{\mathbf{E}_o(t)\cdot\boldsymbol{\mu}_{21}}{\hbar}=\Omega \tag{3.17}$$

is called Rabi frequency.

The square of the two sides of equation (3.16) yields (see Box 3.3) the probability of finding the system in the eigen-state $\phi_2(x)$

$$|c_2(t)|^2=\frac{|\Omega|^2}{\Delta^2}\sin^2\Delta\frac{t}{2}, \tag{3.18}$$

which explicitly shows the inverse relationship between $|c_2(t)|^2$ and the detuned frequency Δ. But $|c_2(t)|^2$ is finite at the on-resonance condition. Sinusoidal oscillation of the $|c_2(t)|^2$ function with both detuned frequency and the time of the system-radiation interaction become clear at once (Figure 3.3). There are two ways of looking at the oscillation of $|c_2(t)|^2$. In one, the source radiation is detuned variably and each detuned frequency is allowed to interact with the system for a fixed time period, in which case the amplitude of $|c_2(t)|^2$ decreases and the oscillation disappears as Δ increases. The oscillations will die out because the radiation is prevented from interacting with the system after the fixed time. In the other, when Δ is held constant and the system is allowed to interact with the radiation continuously, the $|c_2(t)|^2$ oscillation continues with the same amplitude at all times. However, both oscillation period and amplitude are determined by the magnitude of the detuned frequency – both increase as Δ decreases (Figure 3.3).

The behavior of both $|c_1(t)|^2$ and $|c_2(t)|^2$ can be observed simultaneously under the so-called 'strong-field' condition where the source radiation frequency is tuned to achieve on-resonance excitation ($\omega_{21}-\omega=\Delta=0$). The $|c_2(t)|^2 \ll |c_1(t)|^2$ assumption is extraneous in the strong-field limit, because the perturbation field is strong enough to produce $|c_2(t)|^2$ substantially, and hence the expressions for $|c_1(t)|^2$ and $|c_2(t)|^2$ are obtained using the coupled equations (3.14). By expanding the V integrals with the $\mathbf{E}(t)$ field, and using the RWA approximation, the two equations can be written as

$$\dot{c}_1(t)=-\frac{i\Omega}{2}c_2(t)e^{-i(\omega_{21}-\omega)t}$$

$$\dot{c}_2(t)=-\frac{i\Omega}{2}c_1(t)e^{i(\omega_{21}-\omega)t}. \tag{3.19}$$

3.5 RABI OSCILLATIONS

A very useful concept emerges if the singled out two-level system is excited exactly on resonance ($\Delta=0$), which means

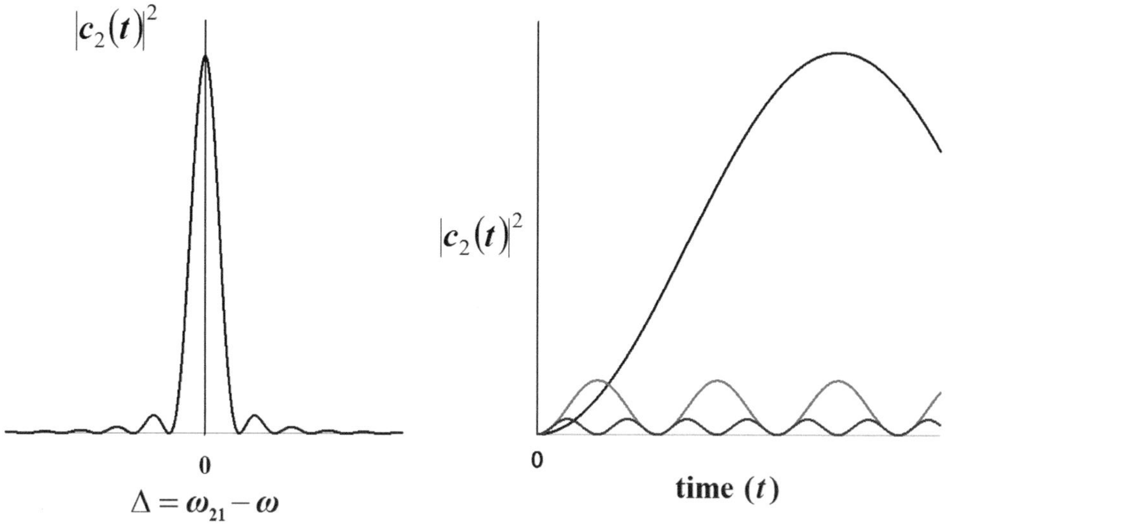

FIGURE 3.3 Oscillation of the $|c_2(t)|^2$ function. *Left*, decay of the oscillation with varying $\omega_{21} - \omega$ for a fixed time of system-radiation interaction. *Right*, amplitude and time period of oscillation for continuous system-radiation interaction are shown when $\omega_{21} - \omega$ is very large (blue), intermediate (red), and very small (black).

a perfectly monochromatic radiation whose frequency ω is tuned to the value of ω_{21}. The exponentials on the right-hand side of equation (3.19) then assume unity

$$\dot{c}_1(t) = -\frac{i\Omega}{2} c_2(t)$$

$$\dot{c}_2(t) = -\frac{i\Omega}{2} c_1(t). \tag{3.20}$$

Letting the transition begin from the lower state at time $t = 0$, the condition of $c_2(0) = 0$ can be set up to obtain the differential equation

$$\frac{d^2}{dt^2} c_2(t) + \frac{\Omega^2}{4} c_2(t) = 0,$$

which is solved by letting $d/dt = \pm i\omega$. After setting the values of the arbitrary constants appropriately, the solutions for the probability coefficients can be obtained as

$$c_2(t) = \sin\left(\frac{\Omega t}{2}\right)$$

$$c_1(t) = -i\cos\left(\frac{\Omega t}{2}\right), \tag{3.21}$$

We can square the coefficients to obtain the corresponding probabilities

$$|c_1(t)|^2 = \cos^2\left(\frac{\Omega t}{2}\right)$$

$$|c_2(t)|^2 = \sin^2\left(\frac{\Omega t}{2}\right). \tag{3.22}$$

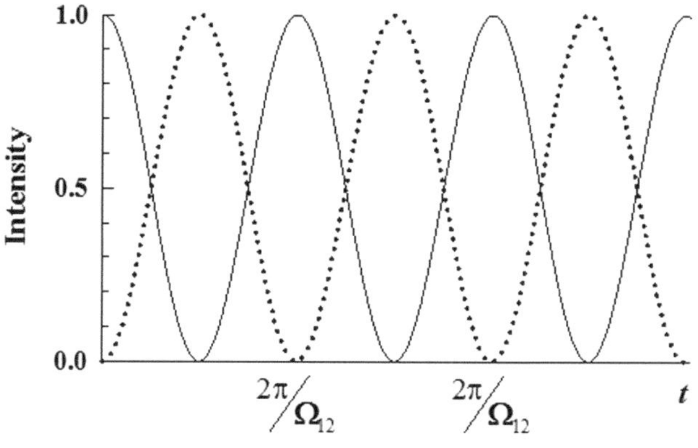

FIGURE 3.4 Rabi oscillations. Solid and dotted lines show $|c_1(t)|^2$ and $|c_2(t)|^2$ functions, respectively.

The $|c_1(t)|^2$ and $|c_2(t)|^2$ functions start out with values of 1 and 0, respectively, at time $t = 0$, and then oscillate between 0 and 1 with a time period $T = 1/\Omega_{12}$, where Rabi frequency Ω_{12} is in $s^{-1}(\nu)$. These are called Rabi oscillations (Figure 3.4) that will continue with time as long as the state function evolves coherently, meaning the electric field of radiation drives the state function to oscillate between eigenstates $\phi_1(x)$ and $\phi_2(x)$. This is an important concept because it pertains to coherence dephasing and system relaxation. If there were no relaxation, the system would continue to evolve coherently by oscillating between the two eigenstates. The decay of Rabi oscillations of a quantum state is a direct measure of coherence leakage.

3.6 TRANSITION PROBABILITY AND ABSORPTION COEFFICIENT

It is clear that light absorption due to a transition from the lower to the upper eigenstate and the resulting excited state

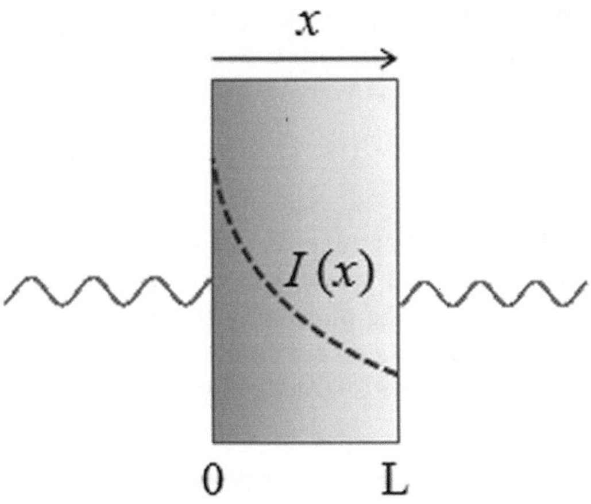

FIGURE 3.5 Exponential decrease of the intensity of light with the distance L from the front edge of the cuvette containing the molecules.

population will depend on both the number of molecules in the sample and $|c_2(t)|^2$. Since absorbance is often described referring to absorption or extinction coefficient, it will be useful to consider the behavior of the absorption coefficient predicted by the theory. With reference to the light-molecule interaction depicted rather ordinarily in Figure 3.5, where a light beam of intensity $I(0)$ enters into a cuvette of length L and exits with intensity $I(L)$, the intensity at some point along the path length of the cuvette is $I(x)$. The absorption coefficient α is defined by the x-derivative of the light intensity as

$$\frac{dI}{dx} = -I(x)\alpha,$$

from which $I(L)$ is obtained by integrating from $x = 0$ to $x = L$

$$\int_0^L \frac{1}{I(x)} dI = -\alpha \int_0^L dx$$

$$I(L) = I(0)e^{-\alpha L}. \tag{3.23}$$

Let the light pass through a volume V that contains N number of molecules. Since $|c_2(t)|^2$ is the probability of finding the system in the eigenstate $\phi_2(x)$, the number of molecules that have made a transition from $\phi_1(x)$ to $\phi_2(x)$ at time t will be $N|c_2(t)|^2$. Because the transition involves absorption of energy $\hbar\omega_{21}$, the energy absorbed per unit volume is

$$\frac{E}{V} = N\hbar\omega_{21}|c_2(t)|^2,$$

and the energy absorbed per unit volume per unit time is

$$\frac{1}{V}\frac{dE}{dt} = N\hbar\omega_{21}|c_2(t)|^2.$$

If the cross-sectional area of the cylindrical volume element of length L is A

$$\frac{1}{A}\frac{dE}{dt} = NL\hbar\omega_{21}|c_2(t)|^2, \tag{3.24}$$

in which the left-hand side defines the intensity of light, which is the energy or the number of photons per unit surface area per unit time. It is important to distinguish intensity from flux. As discussed earlier in Chapter 1, flux is the number of lines of force that cut through a surface. The left-hand side of equation (3.24) represents the change in the light intensity due to its traversing through the volume V. Because the relative change in the light intensity along the cuvette path x is

$$\frac{I(0) - I(L)}{I(0)} \approx \alpha L,$$

equation (3.24) can be written as

$$\alpha(t) = N\hbar\omega_{21}\frac{d}{dt}|c_2(t)|^2, \tag{3.25}$$

which shows that the absorption coefficient should be a function of time.

A simpler expression for time dependence of the absorption coefficient is obtained when $|c_2(t)|^2$ is differentiated. Substituting the right-hand side of equation (3.18) for $|c_2(t)|^2$ we have

$$\alpha(t) = N\hbar\omega_{21}\frac{|\Omega|^2}{\Delta^2}\frac{d}{dt}\left[\sin^2\Delta\frac{t}{2}\right],$$

and using the identity $2\sin\theta\cos\theta = \sin 2\theta$ while differentiating by chain rule yields

$$\alpha(t) = N\hbar\omega_{21}\frac{|\Omega|^2}{2}\left[\frac{\sin\Delta t}{\Delta}\right]. \tag{3.26}$$

A typical spectroscopic experiment often employs nearly 'on-resonance' condition, but a perfect on-resonance condition is really hard to achieve. So we use the limiting condition

$$\lim_{\Delta\to 0}\frac{\sin\Delta}{\Delta} = 1,$$

$$\alpha(t) = \text{constant}.t, \tag{3.27}$$

meaning the absorption coefficient will increase linearly with time. This demonstration, although acceptable with certain assumptions, is not what the reality is. Obviously, no single-frequency experiment with time will ever give us the result that the absorption keeps increasing with time. This must mean that some other process(es) curbs what the theory shows, and we know it is the loss of phase coherence and system relaxation that give rise to the reality.

3.7 LIMITATIONS OF THE THEORY

The theory developed above is based on a system that is somehow isolated so that the state function always evolves coherently, hence the system is measured to be found in the same two eigenstates with respective phase and frequency preserved at all the time. This behavior can be approached with small-molecule gas phase experiments carried out under on-resonance condition and at ultralow temperature and pressure where the molecule may not suffer a collision. However, the properties of the $|c_2(t)|^2$ function found above will not be observed in real experiments in condensed phase and solid states. For example, optical density at a given value of Δ does not oscillate with time, unlike what is depicted in Figure 3.3, nor does the absorption coefficient vary with time, unlike the result in equation (3.26). This happens because of two reasons. One, molecular collisions prevent the coherent evolution of the state function, meaning the two particular eigenstates that we have considered for the state function will not describe the state of the system any more once the system suffers a collision during its interaction with the radiation. Two, the light used is not absolutely monochromatic (not on resonance), but contains a band of frequencies whose distribution produces a frequency-averaged probability. The problem of broad-band frequency can be alleviated by taking the frequency-average of a probability function, $|c_2(t)|^2$ for example, as

$$\left\langle |c_2(t)|^2 \right\rangle = \int_{-\infty}^{+\infty} |c_2(t)|^2 f(\omega)d\omega,$$

in which $f(\omega)$ is the normalized frequency distribution function in the source light such that $f(\omega)d\omega$ is the probability that the light beam carries a frequency between ω and $\omega + d\omega$. But the problem of molecular collisions is not easy to overcome. Collisions that continuously modify the state function prevent oscillations of the probability functions. In general, the presence of the frequency distribution along with collision-interrupted coherent evolution of the state function produces a situation where different molecules interact with the radiation differently so as to yield the average value of $|c_2(t)|^2$.

3.8 COLLISIONAL LINE BROADENING

A simple sign of molecular collisions is the width of the spectral line. Consider an experiment that employs a broad band source to observe the absorption coefficient α of a sample of gas molecules at ordinary temperature where the molecules undergo collisions. Since the absorption coefficient has been shown to be time-dependent, the derivation considered below can be started with $\alpha(t)$. A gas molecule may be characterized by its free-flight time, which is the time it spends in linear motion before colliding with another molecule in motion. The free-flight time need not be the same for all molecules. Let $P(t)$ be the distribution function for free flight times such that $P(t)dt$ is the probability of finding a molecule in free-flight between times t and $t + dt$. Then the free-flight time-averaged absorption coefficient will be

$$\langle \alpha(t) \rangle = \int_0^{\infty} \alpha(t) P(t) dt. \tag{3.28}$$

In kinetic theory of gases, the probability distribution of free-flight time is given by the exponential

$$P(t) = \frac{1}{\tau_c} e^{\frac{-t}{\tau_c}}, \tag{3.29}$$

where τ_c is the collision time. In this distribution, the probability of free-flight is 1 at $t = 0$ when no molecule has undergone a collision, but the probability is 0 at $t \to \infty$ when all molecules have surely suffered collisions. A combination of equations (3.28) and (3.29), and substitution for $\alpha(t)$ from equation (3.26) gives

$$\langle \alpha(t) \rangle = \int_0^{\infty} \frac{1}{\tau_c} e^{\frac{-t}{\tau_c}} N\hbar\omega_{21} \frac{|\Omega|^2}{2} \left[\frac{\sin \Delta t}{\Delta} \right] dt. \tag{3.30}$$

Using the standard definite integral

$$\int_0^{\infty} e^{-ax} \left[\sin(mx) \right] dx = \frac{m}{a^2 + m^2},$$

the absorption coefficient is found to be

$$\langle \alpha(\omega) \rangle = \frac{\frac{B}{\tau_c}}{(\omega_{21} - \omega)^2 + \left(\frac{1}{\tau_c} \right)^2}, \tag{3.31}$$

where the constant B represents $N\hbar\omega_{21} \frac{|\Omega|^2}{2}$.

Notice that α is now a function of the frequency of light. The time dependence of α is lost because of collision-induced interruptions in the coherent evolution of the state function $\Psi(x,t)$. More specifically, $\langle \alpha(\omega) \rangle$ is a Lorentzian function of frequency shift from on-resonance ($\omega_{21} - \omega = 0$), and the width of the Lorentzian depends on the collision time. As Figure 3.6 shows, the maximum of $\langle \alpha(\omega) \rangle$ occurs under on-resonance condition, $\omega_{21} = \omega$, but different molecules absorb at different frequencies of the broad band source due to different values of ω_{21}. If the sample-radiation interaction time is sufficient, a molecule will absorb all frequencies of the source light, because constant collision that the molecule suffers will continuously modulate its ω_{21} value. The full width at half maximum (FWHM) of the spectral line is $2/\tau_c$.

3.9 LINE BROADENING FROM EXCITED STATE LIFETIME

In a real-situation experiment, an absorption transition is followed by spontaneous decay of the excited state population that emits radiation in the form of fluorescence and phosphorescence. This is a relaxation process, and each molecule relaxes independently because different molecules are in different evolution of the state function. If there are N number of molecules in the excited state, then the excited state population $N|c_2(t)|^2$ will

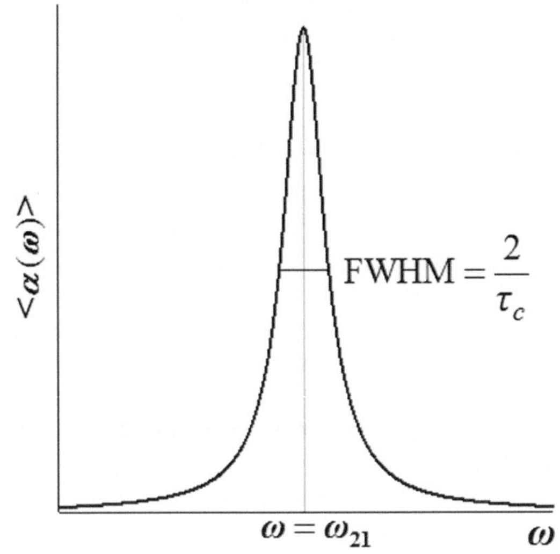

FIGURE 3.6 Collision broadening of the Lorentzian spectral line (flanking wings). The maximum of absorbance occurs at the on-resonance frequency of the light. The ordinate label is often also depicted as $\alpha(\omega)$.

decay exponentially with a time constant, say τ_f. Therefore, the excited state decays exponentially with a time constant $\tau_f/2$, and the normalized probability distribution is given by

$$P(t) = \frac{1}{2\tau_f} e^{\frac{-t}{2\tau_f}}, \tag{3.32}$$

so the time-averaged absorption coefficient is

$$\langle \alpha(t) \rangle = \int_0^\infty \frac{1}{2\tau_f} e^{\frac{-t}{2\tau_f}} N\hbar\omega_{21} \frac{|\Omega|^2}{2} \left[\frac{\sin \Delta t}{\Delta} \right] dt. \tag{3.33}$$

Arguing as we did in the case of collisional broadening (equations (3.30) and (3.31)), we obtain the frequency-averaged absorption coefficient

$$\langle \alpha(\omega) \rangle = \frac{\frac{B}{2\tau_f}}{(\omega_{21} - \omega)^2 + \left(\frac{1}{2\tau_f} \right)^2}, \tag{3.34}$$

where $B = N\hbar\omega_{21} \frac{|\Omega|^2}{2}$.

This again is a Lorentzian spectral band with FWHM = $1/\tau_f$.

The overall broadening of the Lorentzian line is calculated by combining the respective probability distributions for molecular collisions and spontaneous emission. The result will be

$$\langle \alpha(\omega) \rangle = \frac{\frac{B}{\tau_c + 2\tau_f}}{(\omega_{21} - \omega)^2 + \left(\frac{1}{\tau_c + 2\tau_f} \right)^2}, \tag{3.35}$$

showing that the half width at half maximum (HWHM) of the spectral band is

$$\frac{1}{T_2} \equiv \frac{1}{\tau_c} + \frac{1}{2\tau_f}. \tag{3.36}$$

Quite generally, if there are i processes that contribute to interruption of the evolution of the state function over a time t, and j processes contribute to spontaneous decay of the excited molecules,

$$\frac{1}{T_2} \equiv \frac{1}{\sum_i \tau_i} + \frac{1}{\sum_j \tau_j}. \tag{3.37}$$

The value of T_2, which is easily measurable from the spectral line, represents the overall dephasing time of the state function, and hence provides dynamical information about the molecular system.

3.10 SPECTRAL LINE SHAPE AND LINE WIDTH

3.10.1 Homogeneous or Lorentzian Line Shape

If the time-average response is the same for all molecules in the sample under long observation period, a homogeneous form of line broadening results. It is called homogeneous broadening, because each molecule in the sample absorbs all frequencies that span the spectral line. The basis for this is the continuous change in the state function. Suppose the band width of the source light has 100 frequencies, and the state function for the system has also changed 100 times over a finite time period t of the sample-radiation interaction so as to produce the corresponding ω_{21} frequencies that match the ω frequencies available in the source, then the molecule will absorb light of all of those 100 frequencies, and the spectral band spanning the 100 frequencies of the source will broaden homogeneously in the form of a Lorentzian. If the molecule shows transitions over the entire band width, then it must also emit over the whole band width. Essentially, each molecule in the sample absorbs and emits over the entire frequency width, which is why the line shape is homogeneous.

Considering the contribution of spontaneous decay to line width, homogeneous broadening is said to be a reflection of Heisenberg's time-energy uncertainty

$$\Delta E \Delta t \cong \hbar, \tag{3.38}$$

because there is an uncertainty in the energy of the excited state due to its finite lifetime Δt. Uncertainty in the energy of the excited state naturally broadens the spectral line with a FWHM $= 1/\Delta t$. The higher the energy of the excited state ($\hbar\omega_2$), the lesser is its lifetime and more is the uncertainty in its energy, and hence the broader the line would be. This can be compared from spectral line widths across the electromagnetic spectrum. NMR line widths are often tens of Hertz or lesser than 1 Hz for the protons of small hydrocarbons, which is exceedingly smaller than the observed line widths of several thousands of Hz in the visible-region optical spectra.

This is because NMR energy is about six orders of magnitude smaller than the resonance energy in the visible range, which means the excited state for the former is relatively far more stable, has a much longer lifetime, and hence a narrow spectral band results. In fact, the excited state of the nuclear spin has a lifetime which is ~ 10^9 times more than the lifetime of the excited state of an electron. Excited state lifetimes can also be shortened by more frequent molecular collisions. Based on Einstein's spontaneous decay by emission, the excited-state lifetime and angular transition frequency are related by

$$\tau \propto \frac{1}{\omega^3}.$$

3.10.2 Inhomogeneous or Gaussian Line Shape

When molecules in a sample are different because they are somehow in different environments or in different rotational energy states, for example, the same transition for different molecules occur at somewhat different frequencies of the radiation. This frequency shift in the transition gives rise to a spectral band that has a Gaussian shape, and is called inhomogeneous broadening. This kind of broadening occurs independently and apart from the processes that produce homogeneous broadening. If the molecules in the sample are divided into groups according to their environmental and motional properties, then the line shape of each group will be homogeneous and the maximum of each homogeneous transition will show a frequency shift. The composite of these two simultaneous occurrences produces an inhomogeneously broadened spectral band under which lies a large number of homogeneously broadened bands (Figure 3.7).

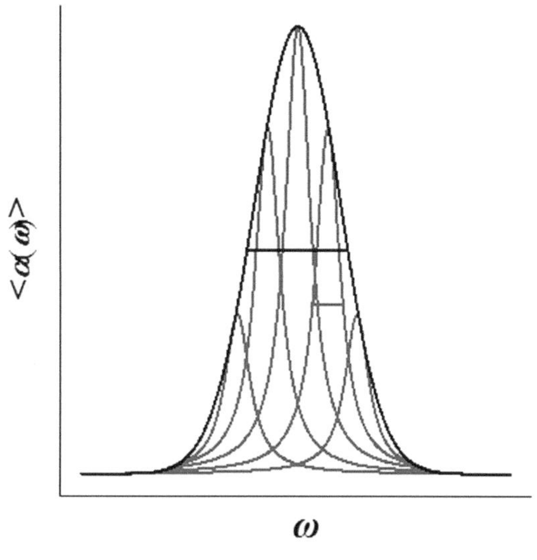

FIGURE 3.7 An inhomogeneous or Gaussian line shape (black), which is a superposition of a large number of homogeneously broadened Lorentzian bands (red). Very large broadening of the inhomogeneous band relative to that of the homogeneous ones is clearly seen.

3.10.3 Doppler Interpretation of Inhomogeneous Line Shape

Inhomogeneous line shape is often also called the Doppler line shape because the Doppler effect most commonly introduces Gaussian line broadening. The effect arises from Maxwell-Boltzmann velocity distribution of gas particles as understood from the kinetic theory of gases. To see how different velocities of the gas molecules in a sample would produce line broadening, it is useful to consider classically how the velocity components of a gas molecule are related to the frequency of light after radiation-molecule interaction. As a basic illustration, only one of the Cartesian components of the velocity may be looked into. Consider a gas molecule of mass m whose velocity vector at some instant is V so that the x-projection of the initial velocity $V_{x,i} = V\cos\theta$ (Figure 3.8). When a photon traveling along the x-direction interacts with the gas particle such that the x-projection of the final velocity is $V_{x,f}$, the total momentum of the system, that is the photon and the molecule together, is obtained from the momentum conservation principle

$$\frac{h\nu}{c} + mV_{x,i} = mV_{x,f}$$

$$\frac{h\nu}{mc} = V_{x,f} - V_{x,i}. \tag{3.39}$$

If light absorption by the molecule is described by a two-level transition, then the principle of energy conservation provides

$$\frac{1}{2}mV_{x,i}^2 + h\nu = \frac{1}{2}mV_{x,f}^2 + h\nu_{21},$$

where $h\nu_{21}$ is the absorbed energy which is equivalent to an increase in internal energy of the gas molecule,

$$h\nu - h\nu_{21} = \frac{1}{2}m\left(V_{x,f}^2 - V_{x,i}^2\right). \tag{3.40}$$

Assuming $V_{x,i} + \mathbf{V}_{x,f} \approx 2\mathbf{V}_{x,i}$,

$$h\left(\nu - \nu_{21}\right) = mV_{x,i}\left(V_{x,f} - V_{x,i}\right).$$

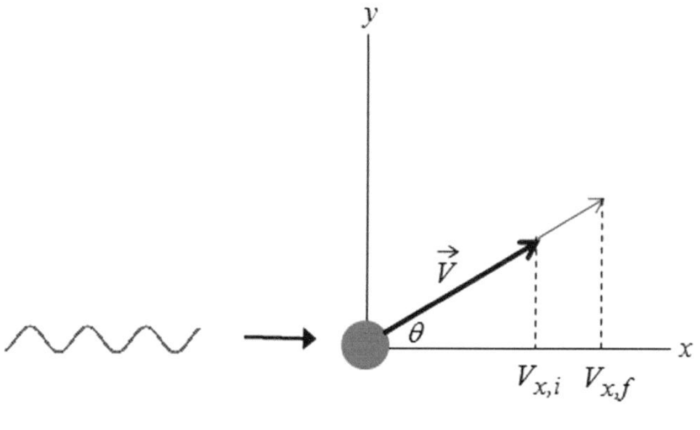

FIGURE 3.8 The source frequency ν and the excitation of a molecule with a velocity vector V, whose component along the direction of the light is $V_{x,i}$.

The use of equation (3.39) yields

$$V_{x,i} = c\left(\frac{\nu - \nu_{21}}{\nu}\right), \quad (3.41)$$

meaning the source radiation of frequency ν will excite only those molecules whose x-component of the velocity vector is $V_{x,i}$. If the above equation is written as

$$\nu = \frac{1}{\left(1 - \frac{V_{x,i}}{c}\right)}\nu_{21},$$

an approximate expression can be used

$$\nu \cong \nu_{21}\left(1 + \frac{V_{x,i}}{c}\right), \text{because } \frac{1}{1-x} \cong 1 + x \text{ for small values of } x. \quad (3.42)$$

We now come to notice that the light frequency ν absorbed exceeds the molecular frequency ν_{21} for all positive values of $V_{x,i}$ or positive $\cos\theta$ values. The value of ν will be less than ν_{21} unless $\cos\theta$ is positive. The situation $\nu = \nu_{21}$ will appear only when the velocity vector V is not in the xy plane.

The molecular velocity distribution irrespective of the direction of molecular motion is given by Maxwell-Boltzmann distribution

$$P(V) = 4\pi\left(\frac{m}{2\pi k_B T}\right)^{\frac{3}{2}} V^2\left(e^{-\frac{mV^2}{2k_B T}}\right), \quad (3.43)$$

from which the mean velocity can be found by

$$\langle V \rangle = \int_0^{\infty} P(V)VdV. \quad (3.44)$$

In a specified direction, say x,

$$P(V_x) = \left(\frac{m}{2\pi k_B T}\right)^{\frac{1}{2}} e^{-\frac{mV^2}{2k_B T}}. \quad (3.45)$$

Using equation (3.41), the above distribution is rewritten to obtain the probability per frequency unit $P(\nu)$ that a transition will occur between ν and $\nu + d\nu$. The velocity variable V_x is now replaced by the light frequency

$$P(\nu) = \sqrt{\frac{m}{2\pi k_B T}} e^{-\frac{(\nu-\nu_{21})^2}{\sigma^2}}, \quad (3.46)$$

in which $\sigma^2 = \frac{2\nu^2 k_B T}{mc^2}$ is the variance in the frequency shift from ν_{21}. For a normalized distribution given by

$$\int_{-\infty}^{+\infty} P(\nu)d\nu = 1,$$

the result is

$$P(\nu) = \frac{1}{\sigma\sqrt{\pi}} e^{-\frac{(\nu-\nu_{21})^2}{\sigma^2}}, \quad (3.47)$$

which indicates that the same nominal transition corresponding to ν_{21} is frequency-shifted for different molecules according to their velocity distribution. In the Doppler-broadened line, ν_{21} and σ correspond to the mean absorption frequency and the standard deviation, respectively, and σ approximates half-width of the Gaussian at $1/e$ of the full height corresponding to the frequency ν_{21} (Figure 3.9). A useful formula connecting σ and the molecular weight of a gas at temperature T is

$$\sigma = (0.428 \times 10^{-6})\nu_{21}\sqrt{\frac{T}{M}},$$

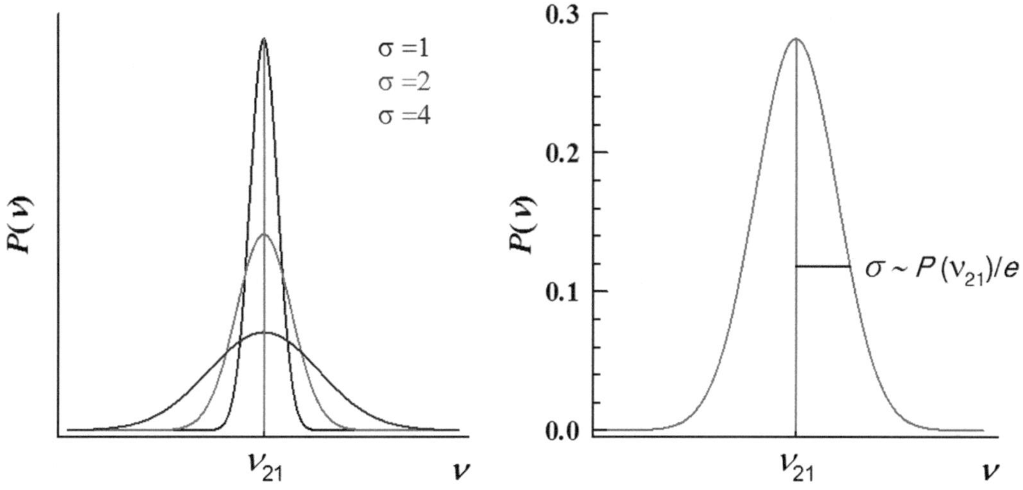

FIGURE 3.9 *Left*, dependence of the amplitude and width of a Gaussian line on the standard deviation σ. *Right*, the half width of a Doppler-broadened line at $1/e$ of the height at ν_{21} is approximated by σ.

which approximates half of the Doppler frequency width to the order of 10^{-6}.

Box 3.1

Using $\frac{d}{dt}\left[a(t)b(t)\right] = b(t)\frac{d}{dt}a(t) + a(t)\frac{d}{dt}b(t)$, the left-hand side of the equation

$$i\hbar\left[\dot{c}_1(t)\frac{\partial}{\partial t}\phi_1(x)e^{-i\omega_1 t} + \dot{c}_2(t)\frac{\partial}{\partial t}\phi_2(x)e^{-i\omega_2 t}\right]$$
$$= \left[\mathcal{H}_o + V(t)\right]\left[c_1(t)\phi_1(x)e^{-i\omega_1 t} + c_2(t)\phi_2(x)e^{-i\omega_2 t}\right] \quad (3.B1.1)$$

is written as

$$i\hbar\left[\dot{c}_1(t)\phi_1(x)e^{-i\omega_1 t} + c_1(t)\phi_1(x)e^{-i\omega_1 t}(-i\omega_1)\right.$$
$$\left.+\dot{c}_2(t)\phi_2(x)e^{-i\omega_2 t} + c_2(t)\phi_2(x)e^{-i\omega_2 t}(-i\omega_2)\right]$$

Multiplication by $\phi_1{}^*(x)$ and integration over all space $d\tau = dx\,dy\,dz$ yields

$$i\hbar\left[\dot{c}_1(t)e^{-i\omega_1 t}\int\phi_1{}^*(x)\phi_1(x)d\tau - i\omega_1 c_1(t)e^{-i\omega_1 t}\right.$$
$$\int\phi_1{}^*(x)\phi_1(x)d\tau + \dot{c}_2(t)e^{-i\omega_2 t}$$
$$\int\phi_1{}^*(x)\phi_2(x)d\tau - i\omega_2 c_2(t)e^{-i\omega_2 t}$$
$$\left.\int\phi_1{}^*(x)\phi_2(x)d\tau\right]$$
$$= i\hbar\dot{c}_1(t)e^{-i\omega_1 t} - E_1 c_1(t)e^{-i\omega_1 t}, \quad (3.B1.2)$$

where orthonormality property of the $\phi(x)$ functions is used.

Multiplication by $\phi_1{}^*(x)$ and integration of the right-hand side of equation (3.B1.1) proceed as

$$c_1(t)e^{-i\omega_1 t}\int\phi_1{}^*(x)\mathcal{H}_o\phi_1(x)d\tau + c_2(t)e^{-i\omega_2 t}$$
$$\int\phi_1{}^*(x)\mathcal{H}_o\phi_2(x)d\tau + c_1(t)e^{-i\omega_1 t}$$
$$\int\phi_1{}^*(x)V(t)\phi_1(x)d\tau + c_2(t)e^{-i\omega_2 t}$$
$$\int\phi_1{}^*(x)V(t)\phi_2(x)d\tau$$
$$= E_1 c_1(t)e^{-i\omega_1 t}\int\phi_1{}^*(x)\phi_1(x)d\tau + E_2 c_2(t)e^{-i\omega_2 t}$$
$$\int\phi_1{}^*(x)\phi_2(x)d\tau + c_1(t)e^{-i\omega_1 t}V_{11} + c_2(t)e^{-i\omega_2 t}V_{12}, \quad (3.B1.3)$$

where the integrals containing the $V(t)$ operator are represented by V_{11} and V_{12}, of which the first one can be set to zero.[1] Equating the expressions in (3.B1.2) and (3.B1.3) gives

$$i\hbar\dot{c}_1(t)e^{-i\omega_1 t} = V_{12}c_2(t)e^{-i\omega_2 t} \quad (3.B1.4)$$

Multiplication by $\phi_2{}^*(x)$ and integration performed throughout, and setting $V_{22} \cong 0$ yield

$$i\hbar\dot{c}_2(t)e^{-i\omega_2 t} = V_{21}c_1(t)e^{-i\omega_1 t} \quad (3.B1.5)$$

[1] The matrix representing the full Hamiltonian that couples the two states $\phi_1(x)$ and $\phi_2(x)$ is

$$(\mathcal{H}) = \begin{pmatrix} E_1 + V_{11} & V_{12} \\ V_{21} & E_2 + V_{22} \end{pmatrix}.$$

The existence of the off-diagonal elements is simply due to the coupling produced by $V(t)$ part of the total Hamiltonian, and both V_{12} and V_{21} which cannot go to zero. On the other hand, V_{11} and V_{22} do not couple the two eigenstates if E_1 and E_2 are retained. Therefore, either $V_{11} = V_{22} = 0$ or E_1 and E_2 are modified to, say, E_1' and E_2'.

For a simplified description of the coupling transition, V_{11} and V_{22} are assumed to vanish (see Cohen-Tannoudji et al., 1977, for a detailed discussion).

Box 3.2

Substitution by $-\mathbf{E}(t)\cdot\boldsymbol{\mu}_{21}(t)$ for $V_{21}(t)$ gives

$$c_2(t) = \frac{1}{i\hbar}\int_0^t -\mathbf{E}(t)\cdot\boldsymbol{\mu}_{21}(t)e^{i\omega_{21}t}dt$$

Substituting $\mathbf{E}(t)$ by $\mathbf{E}_o(t)\cos\omega t$, and using the identity

$$\cos\omega t = \frac{e^{i\omega t} + e^{-i\omega t}}{2}$$

$$c_2(t) = \frac{-\mathbf{E}_o(t)\cdot\boldsymbol{\mu}_{21}(t)}{2i\hbar}\int_0^t e^{i(\omega_{21}+\omega)t} + e^{i(\omega_{21}-\omega)t}dt$$

Using the standard integral $\int e^{i\omega t}dt = \frac{e^{i\omega t}}{i\omega}$, the following expression is obtained

$$c_2(t) = \frac{-\mathbf{E}_o(t)\cdot\boldsymbol{\mu}_{21}(t)}{2i\hbar}\left[\frac{e^{i(\omega_{21}+\omega)t} - 1}{i(\omega_{21}+\omega)} + \frac{e^{i(\omega_{21}-\omega)t} - 1}{i(\omega_{21}-\omega)}\right]$$

Since $(\omega_{21}+\omega) \gg (\omega_{21}-\omega)$, the first term within the parenthesis is dropped out

$$c_2(t) = \frac{\mathbf{E}_o(t)\cdot\boldsymbol{\mu}_{21}(t)}{2\hbar}\left[\frac{e^{i(\omega_{21}-\omega)t} - 1}{(\omega_{21}-\omega)}\right]$$

The exclusion of the $\frac{e^{i(\omega_{21}+\omega)t} - 1}{(\omega_{21}+\omega)}$ term corresponds to using $\mathbf{E}(t) = \mathbf{E}_o(t)e^{i\omega t}$ instead of $\mathbf{E}(t) = \mathbf{E}_o(t)\cos\omega t$, meaning a rotating perturbation instead of an oscillating perturbation of the electromagnetic field, and is called rotating wave approximation (RWA).

Box 3.3

Take the expression for $c_2(t)$ under RWA (see Box 3.2 above)

$$c_2(t) = \frac{\Omega_{12}}{2}\left[\frac{e^{i(\omega_{21}-\omega)t}-1}{(\omega_{21}-\omega)}\right]$$

in which the front factor has been written in the compact format of Rabi frequency. The idea here is to recognize that the exponential in the numerator inside the square bracket can be written as

$$e^{i(\omega_{21}-\omega)t} = e^{i(\omega_{21}-\omega)t/2} + e^{i(\omega_{21}-\omega)t/2}.$$

We also know that

$$\left|e^{i\theta}\right|^2 = 1, \text{ and } \sin\theta = 1/2i\left(e^{i\theta}-e^{-i\theta}\right).$$

Applying these two identities we can rewrite the $c_2(t)$ expression as

$$c_2(t) = \frac{\Omega_{12}}{2(\omega_{21}-\omega)}2i\left[\sin(\omega_{21}-\omega)t/2\right]e^{i(\omega_{21}-\omega)t/2}$$

$$= \frac{i\Omega_{12}}{(\omega_{21}-\omega)}\sin(\omega_{21}-\omega)t/2.e^{i(\omega_{21}-\omega)t/2}$$

Now square both sides and use the identity

$$\left[e^{i(\omega_{21}-\omega)t/2}\right]^2 = 1$$

to get the final expression

$$\left|c_2(t)\right|^2 = \frac{\left|\Omega_{12}\right|^2}{(\omega_{21}-\omega)^2}\sin^2(\omega_{21}-\omega)t/2.$$

We keep in mind that this expression is obtained with RWA strictly.

PROBLEMS

3.1 Consider a hydrogen atom whose electric dipole moment $\boldsymbol{\mu}$ originates from the electric dipole moment of the electron. Note that the atomic electric moment appears only under relativistic approximation. For the ground state of the hydrogen atom, we have $\boldsymbol{\mu} = -2Z^2\alpha^2 d_e$, where Z is the atomic number, $\alpha = e^2/\hbar c \approx 0.0073$ is called the fine-structure constant, and $d_e \sim 10^{-31} e$ nm $\approx 4.8\times10^{-30}$ D is the dipole moment of the electron. Let us place the atom in a laser field of intensity 1 Watt cm^{-2}.
  (a) Calculate the value of the electric dipole moment of hydrogen atom in the ground state.
  (b) Calculate the Rabi frequency within the precept of the two-level system.
  Note: When necessary, use 1 Watt $= 10^7$ erg s^{-1}, 1 D $= 10^{-18}$ esu cm.

3.2 Let there be a rectangular distribution of frequencies from 569.95×10^{12} to 561.409×10^{12} Hz in a laser beam of intensity 0.1 Watt cm^{-2}. Calculate the following:
  (a) the integrated energy density over the frequency bandwidth (in erg cm^{-3});
  (b) the energy density per unit frequency (in erg cm^{-3} s);
  (c) the transition rate the field would induce in an isotropic two-level system for which $\mu = 4.8\times10^{-30}$ D. Assume that the resonance frequency lays near 565 THz, and that the spontaneous relaxation time is $\leq 10^{-8}$ s.

3.3 Show that the absorption coefficient α can be written in the form

$$\alpha = \frac{4\pi^2\omega\left|\mu\right|^2 L(\Delta)}{3\hbar c}$$

where

$$L(\Delta) = \frac{1/(\pi T_2)}{\Delta^2 + 1/T_2^2},$$

Δ being the shift in frequency from the on-resonance condition, $\Delta = \omega_{12} - \omega$. Then consider a sample of molecules confined to 0.1 cm thickness, where the number density of molecules is 10^{10} cm^{-3}. Suppose, the dipole transition $\mu_{12} = 0.1$D at 350 nm, and the spontaneous emission $1/\tau_f = 8\times10^7$ s^{-1}.
  (a) Calculate α in cm^2.
  (b) How does the fraction of light absorbed, that is $1-e^{\alpha NL}$, vary with Δ? Draw a graph of the variation.

3.4 The graph below (Figure P.3.4) shows measured exponential decay time for spontaneous emission (τ_f) of a molecule at 505 nm as a function of solvent viscosity given in centiPoise.

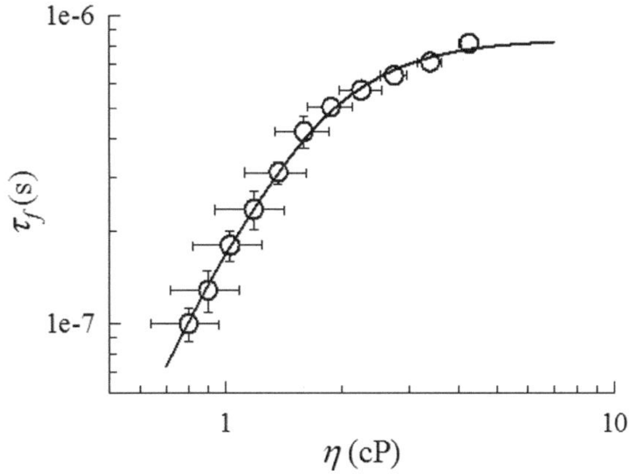

  (a) Determine the broadening (FWHM) of the Lorentzian absorption line for some of the viscosity conditions.

(b) Draw a graph to indicate the changes in FWHM with viscosity.
(c) Estimate the length of the transition dipole in nm and graph its variation with viscosity.
(d) Comment on the effect of the solvent viscosity on the transition dipole length.
For this problem, assume that spontaneous emission is the only source of line broadening.

3.5 The Doppler width is generally found, in atomic spectra at least, to be much larger than the radiative Lorentzian linewidth, although experimental techniques exist to narrow down the former. Give reason(s) why the Doppler width should be relatively larger.

FURTHER READING

Allen, L. and J. H. Eberly (1975) *Optical Resonance and Two-Level Atoms*, Wiley.

Cohen-Tannoudji, C., B. Diu, and F. Laloë (1977), *Quantum Mechanics* (Vol I and II), John Wiley & Sons.

Dyke, T. R., G. R. Tomasevich, W. Klemperer, and W. Falconer (1972) *J. Chem. Phys.* 57, 2277.

Portis, A. M. (1953) *Phys. Rev.* 91, 1071.

4 Hydrogen Atom Spectra

4.1 FREE HYDROGEN ATOM

By free we mean the hydrogen atom is not subjected to an external field. The hydrogen atom is described by an electron of charge $e(= -1.6 \times 10^{-19}\,\text{C})$ and mass $m_e(\approx 9.11 \times 10^{-31}\,kg)$ moving within a sphere relative to a central nucleus of charge Ze. Note that Z is the atomic number and e here is the charge of a proton $(= +1.6 \times 10^{-19}\,\text{C})$. The mass of the nucleus M is justifiably infinite (but only $\sim 2 \times 10^3$ times more) compared to the mass of the electron. At an instant, the problem is formulated as the motion of a particle like an electron with a reduced mass μ or effective mass m_e

$$\mu = m_e \cong \frac{m_e M}{m_e + M}$$

in the potential field of the nucleus. The Coulomb potential experienced by the electron at a distance r from the nucleus is $-\frac{Ze^2}{r}$, so the Schrödinger equation for the electron is

$$-\left(\frac{\hbar^2}{2m_e}\nabla^2 + \frac{Ze^2}{r}\right)\psi(\mathrm{r}) = E\psi(\mathrm{r}), \tag{4.1}$$

in which the Laplace operator is given by

$$\nabla^2 = \frac{\partial^2}{\partial_x^{\,2}} + \frac{\partial^2}{\partial_y^{\,2}} + \frac{\partial^2}{\partial_z^{\,2}}.$$

Now we use spherical polar coordinates to assign coordinates to the electron. As Figure 4.1 shows, the z-axis is chosen as the polar axis of the sphere and r, θ, φ are coordinates of the electron. In the Cartesian to spherical polar coordinate transformation, the r coordinates (x, y, z) of the electron transform as

$$x = r\sin\theta\cos\varphi$$

$$y = r\sin\theta\sin\varphi$$

$$z = r\cos\varphi.$$

Regarding the Schrödinger equation (equation 4.1), we recognize two properties. One, the Hamiltonian operator

$$\mathcal{H} = -\left(\frac{\hbar^2}{2m_e}\nabla^2 + \frac{Z_e^{\,2}}{r}\right) \tag{4.2}$$

commutes with the operator for the square of the electron angular momentum L^2 as well as the z-component of the angular momentum L_z (see the discussion under Postulate 5 in Chapter 2). The commutating operators are

$$\left[\mathcal{H}, L^2\right] = \left[\mathcal{H}, L_z\right] = \left[L^2, L_z\right] = 0. \tag{4.3}$$

We have also learned that

$$\begin{aligned} \left[L_x, L_y\right] &= i\hbar L_z \\ \left[L_y, L_z\right] &= i\hbar L_x \\ \left[L_z, L_x\right] &= i\hbar L_y. \end{aligned} \tag{4.4}$$

These commutation relations suggest that there is a basis which is a complete set of eigenstates for $\mathcal{H}, L^2$ and L_z so that

$$\mathcal{H}\psi = E_n\psi \tag{4.5}$$

$$L^2\psi = l(l+1)\hbar^2\psi \tag{4.6}$$

$$l_z\;\psi = m\hbar\psi, \tag{4.7}$$

where n, l, and m are principal quantum number, orbital angular momentum (azimuthal) quantum number, and magnetic quantum number, respectively. The magnetic quantum number m is often also referred to as angular momentum orientation quantum number. States with $l = 0, 1, 2, \cdots$ are labeled $s, p, d, \cdots$ in order. The second property refers to the mathematical nature of the wavefunction that occupies the central place in the description of a function in spherical polar coordinates. The wavefunction of a hydrogen-like particle in spherical polar coordinates is given by the product of radial functions $R(r)$, which are Laguerre polynomials, and angular functions $Y(\theta, \phi)$, which are spherical harmonics (see Box 4.1). The eigenfunctions with the variables take the form

$$\psi_{nlm}(r, \theta, \phi) = R_{nl}(r)Y_{lm}(\theta, \phi), \tag{4.8}$$

in which

$$Y_{lm}(\theta, \phi) = \Theta_{lm}(\theta)\Phi_m(\phi) \tag{4.9}$$

with

$$\Theta_{lm}(\theta) = (-1)^m\sqrt{\frac{(2l+1)(l-m)!}{2(l+m)!}}P_l^m(\cos\theta) \tag{4.10}$$

DOI: 10.1201/9781003293064-4

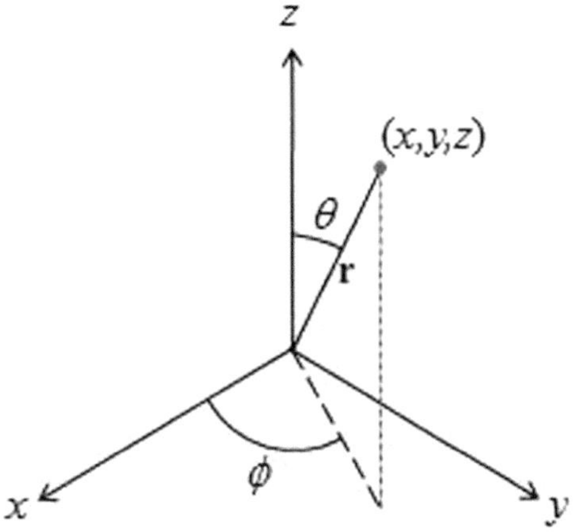

FIGURE 4.1 Coordinates of an electron trapped in the nuclear potential of the hydrogen atom represented in the spherical polar coordinate system. The electron is shown as a small circle in red and the nucleus is positioned at the origin of the Cartesian coordinates.

$$\Phi_m(\phi) = \frac{1}{\sqrt{2\pi}} e^{im\phi}. \tag{4.11}$$

Note that the Θ and Φ functions are shown here in the normalized form. Box 4.1 provides a short description of the derivation of these eigenfunctions from $\nabla^2 \psi(r,\theta,\phi)$.

4.2 EIGENVALUES, QUANTUM NUMBERS, SPECTRA, AND SELECTION RULES

We are continuing to discuss the hydrogen atom in the absence of an external potential. The eigenvalue E_n in equation (4.5) is the energy of the hydrogen atom. In the absence of an external field, only the radial part of the eigenfunction is relevant and the quantum number $n(=1,2,3,\ldots)$ solely determines the energy according to the formula

$$E_n = -\frac{1}{n^2}\left(\frac{\mu e^4}{2\hbar^2}\right) \cong -\frac{1}{n^2} \times 13.53 \text{ eV}, \tag{4.12}$$

in which 13.53 eV is the energy required for ionization of the hydrogen atom. The distance between the electron and the nucleus corresponding to an energy state E_n is given by

$$r_n = n^2 a_0,$$

where $a_0 \cong 0.52$ Å is the Bohr radius, often also used to quantify atomic dimensions. However, since the radial function also depends on the quantum number l (equation (4.8) and Box 4.1) and $l = 0,1,2,\ldots,n-1$, each E_n energy state should be n-fold degenerate. The involvement of l also brings in the quantum number m because $m = -l, -l+1, -l+2, \cdots l-2, l-1, l$. Therefore, the overall degeneracy of the radial energy is n^2.

The significance of energy levels corresponding to the quantum number n has been known for a long time due to the famous series of hydrogen spectral lines: the Lyman lines of ultraviolet absorption from $n = 1$ to $n = 2,3,4,\ldots$,; the Balmer

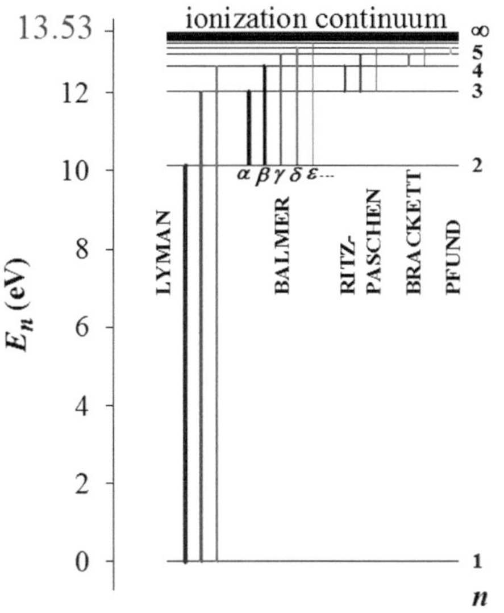

FIGURE 4.2 Hydrogen energy levels corresponding to states of different n. The transition energy is given in electron volts (eV) showing the spectral limit of 13.53 eV.

lines of spontaneous visible emission from all n greater than 1 to $n = 1$; the Ritz-Paschen lines arising from infrared transitions between $n = 3$ and higher n states; the Brackett series of infrared absorption or emission lines between $n = 4$ and higher n states; and Pfund series of infrared lines produced when the electron excited to higher energy states makes a transition down to $n = 5$. Figure 4.2 shows these transitions that are limited to the maximum transition energy of 13.53 eV (see equation (4.12)). Beyond this limit the atom ionizes, and hence the electron is no longer bound. From the transitions that produce hydrogen spectral lines it should be obvious that there is no selection rule for the change of states corresponding to different n.

The eigenvalues $l(l+1)$ and m, both in the unit of h (equations (4.6) and (4.7)), bring in the selectivity of optical transitions because the selection rules prescribe only certain changes in the quantum numbers l and m, and the allowed changes are determined by the angular functions $\Theta(\theta)$ and $\Phi(\phi)$, respectively. As discussed in Chapter 3, the probability of an electric dipole transition (equation (3.9), for example) between states nlm and $n'l'm'$ has the factor

$$\left|\mathbf{E} \cdot \boldsymbol{\mu}_{nlmn'l'm'}\right|,$$

in which $\mathbf{E}$ is a vector orthogonal to the direction of propagation of the electric field, and can be written as the sum of its components in the laboratory fixed axes

$$\mathbf{E} = E_x \mathbf{x} + E_y \mathbf{y} + E_z \mathbf{z}.$$

We will consider linearly polarized light with the $\mathbf{E}$ vector along the z-axis so that $E_x = E_y = 0$ and $E_z = 1$. The operator for the dipole moment of the hydrogen atom is simply the product of the electron charge e and the position vector $\mathbf{r}$.

$$-\mu = e\mathbf{r}.$$

By transforming the Cartesian coordinates of r to spherical polar coordinates, one writes

$$|\boldsymbol{\mu}| = e\left(r\ \sin\theta\cos\phi\mathbf{x} + r\ \sin\theta\sin\phi\mathbf{y} + r\ \cos\theta\mathbf{z}\right),$$

which gives

$$-\boldsymbol{\mu}\cdot\mathbf{E} = -er\left\{\sin\theta\left(E_x\cos\phi + E_y\sin\phi\right) + E_z\cos\theta\right\}. \quad (4.13)$$

When the electric field vector is along the z-axis $E_x = E_y = 0$, so the probability of a transition from the state nlm to $n'l'm'$ is

$$\left|-\mathbf{E}\cdot\boldsymbol{\mu}_{nlm,n'l'm'}\right|^2 = \left|\begin{matrix} -e\int_0^{\infty} R^*_{nl}R_{n'l'}r^3dr \\ \int_0^{\pi}\Theta^*_{lm}\cos\theta\Theta_{l'm'}\sin\theta d\theta\int_0^{2\pi}\Phi^*_m\Phi_{m'}d\phi \end{matrix}\right|^2, \quad (4.14)$$

in which the volume element integral is

$$\int d\tau = \int_0^{\infty} r^2 dr\int_0^{\pi}\sin\theta d\theta\int_0^{2\pi}d\phi. \quad (4.15)$$

Now consider the integral over ϕ. Substitution for Φ^*_m and $\Phi_{m'}$ (see equation (4.11)) in the last integral of equation (4.14) will read

$$\frac{1}{2\pi}\int_0^{2\pi}e^{-im\phi}e^{im'\phi}d\phi = \frac{1}{2\pi}\int_0^{2\pi}e^{-i\Delta m\phi}d\phi, \quad (4.16)$$

where $\Delta m = m - m'$. The integral goes to zero unless $\Delta m = 0$. For $m = m'$,

$$\frac{1}{2\pi}\int_0^{2\pi}d\phi = 1.$$

Note that we get the same result by expanding the exponential as

$$\frac{1}{2\pi}\int_0^{2\pi}e^{-i\Delta m\phi}d\phi = \frac{1}{2\pi}\int_0^{2\pi}\left(\cos\Delta m\phi + i\sin\Delta m\phi\right)d\phi. \quad (4.17)$$

This result is obtained when the electric field vector of the radiation is along the z-axis, i.e, z-polarization. When the light is x- or y-polarized the result will be different. For x-polarized light ($E_y = E_z = 0$), the ϕ-dependent part of equation (4.14) is

$$\frac{1}{2\pi}\int_0^{2\pi}e^{-i\Delta m\phi}\cos\phi d\phi,$$

which goes to zero if $\Delta m = 0$, but exists when $\Delta m = \pm 1$. An identical result will be obtained for y-polarized light. The selection rule for the change of m is thus

$$\begin{array}{ll} \Delta m = 0 & z\text{-polarized light} \\ \Delta m = \pm 1 & x\text{- or } y\text{-polarized light.} \end{array} \quad (4.18)$$

To find out the possible changes in the quantum number l during a transition, the integral over θ

$$\int_0^{\pi}\Theta^*_{lm}\Theta_{l'm'}\cos\theta\sin\theta d\theta$$

needs to be examined. For this purpose, we will apply the following three formulas of spherical harmonics given by Bethe and Salpeter (1977) that are useful for calculation of matrix elements involving Cartesian coordinates x, y, and z.

$$\sin\theta P_l^m(\cos\theta) = \sqrt{\frac{(l+m+1)(l+m+2)}{(2l+1)(2l+3)}}P_{l+1m+1} - \sqrt{\frac{(l-m)(l-m-1)}{(2l+1)(2l-1)}}P_{l-1\,m+1} \quad x\text{-coordinate}, \quad (4.19)$$

$$\sin\theta P_l^m(\cos\theta) = -\sqrt{\frac{(l-m+1)(l-m+2)}{(2l+1)(2l+3)}}P_{l+1m-1} + \sqrt{\frac{(l+m)(l+m-1)}{(2l+1)(2l-1)}}P_{l-1\,m-1} \quad y\text{-coordinate}, \quad (4.20)$$

$$\cos\theta P_l^m(\cos\theta) = \sqrt{\frac{(l+m+1)(l-m+1)}{(2l+1)(2l+3)}}P_{l+1m} + \sqrt{\frac{(l+m)(l-m)}{(2l+1)(2l-1)}}P_{l-1\,m} \quad z\text{-coordinate}. \quad (4.21)$$

Using the formula for the z-coordinate, the integral over θ is rewritten as

$$\int_0^{\pi}P^*_{lm}\left\{P_{l'+1m'}\sqrt{\frac{(l'+m'+1)(l'-m'+1)}{(2l'+1)(2l'+3)}} + P_{l'-1m'}\sqrt{\frac{(l'+m')(l'-m')}{(2l'+1)(2l'-1)}}\right\}\sin\theta d\theta. \quad (4.22)$$

Then one applies the orthogonality condition for associated Legendre functions

$$\int_0^{\pi}P^*_{lm}P_{l'm}\sin\theta d\theta = \delta_{ll'},$$

and argues that the integral over θ will exist only if the condition $l = l' \pm 1$ holds, else the integral vanishes. Thus, the selection rule

$$\Delta l = l' - l = \pm 1 \quad (4.23)$$

is obtained for the change of the quantum number for orbital angular momentum of the electron. This selection rule corresponds to not only the z-polarization of radiation, but also x- and y-polarization. To see this, the θ integral of equation (4.14) can be rewritten again by inserting equation (4.19) or

(4.20), and by using the orthogonality condition the selection rule $\Delta l = \pm 1$ can be recovered.

The radial integral in the transition probability expression (equation (4.14)) is not easy to evaluate. The value of this integral for transitions of the type $n,l \rightarrow n,l \pm 1$ where no change of the principal quantum number occurs is given by

$$\int_0^{\infty} R_{nl}^{*} R_{nl'} r^3 dr = \frac{3}{2} n\sqrt{n^2 - l^2}, \tag{4.24}$$

and the integral for transitions $n,l \rightarrow n',l \pm 1$ involving change in n is evaluated by the Gordon formula

$$\int_0^{\infty} R_{nl}^{*} R_{nl'} r^3 dr = \frac{(-1)^{n'-l}}{4(2l-1)!} \sqrt{\frac{(n+l)!(n'+l-1)!}{(n-l-1)!(n'-l)!}} \frac{(4nn')^{l+1}(n-n')^{n+n'-2l-2}}{(n+n')^{n+n'}}$$
$$\times \left[F\left\{-n+l+1, -n'+l, 2l, -\frac{4nn'}{(n-n')^2}\right\} - \left(\frac{n-n'}{n+n'}\right)^2 F\left\{-n+l+1, -2, +l, -n'+l, 2l, -\frac{4nn'}{(n-n')^2}\right\} \right] \tag{4.25}$$

where F is hypergeometric function.

4.3 HYDROGEN ATOM IN EXTERNAL MAGNETIC FIELD: ZEEMAN EFFECT AND SPECTRAL MULTIPLETS

We now look at the hydrogen atom made captive by an external potential. A physical captivity can be created by immersing the hydrogen atom in an external magnetic field. The effect of a steady external field on atomic spectra was observed by Zeeman in 1896, a year before the electron was discovered and much before quantum theory was known. Lorentz used classical electron theory to treat the electron motion as a three-dimensional harmonic oscillator and calculated the interaction of the components of the oscillator with the magnetic field. The calculations showed polarization of the oscillator components in the presence of a magnetic field. If the linear vibration of the oscillator is resolved into a linear component parallel to, and two opposite circular components perpendicular to the field direction, then the optical transition of the electron will show three lines.

That the Zeeman effect indeed produces three spectral lines for an electron possessing an angular momentum L is experimentally verifiable. The experiment involves the study of optical transitions of an atom having no spin. For example, two-electron systems such as beryllium ion that exists in two zero-spin states, namely $\mathrm{S}(L=0)$ and $\mathrm{P}(L=1)$, can be used to study transitions within the states. The atoms are placed in a magnetic field $\mathbf{B}_0$ oriented along the z-direction and optical transitions are studied with light polarized perpendicular or parallel to the direction of propagation. If the light is propagating along x-axis, then z-polarization of the electric vector yields a single line corresponding to the linear motion of the harmonic oscillator that Lorentz employed, but y-polarized light produces two transitions split by $2|\gamma_e|B_0$, where

$$\gamma_e = \frac{e\hbar}{2m_e c} \approx 9.724 \times 10^{-21} \text{ erg G}^{-1} \left(9.274 \times 10^{-24} \text{JT}^{-1}\right) \tag{4.26}$$

is Bohr magneton, which is the magnitude of the magnetic moment due to the orbital motion of an electron. If right or left circularly polarized light is allowed to propagate along the z-axis, which is collinear with the magnetic field direction, one transition for each circularly polarized light is obtained. These two transitions corresponding to the two oppositely circular components of the oscillatory motion of the electron in Lorentz theory are apart by $2|\gamma_e|B_0$ (Figure 4.3). The appearance of the Lorentz triplet of spectral line in an external magnetic field is known as normal Zeeman effect.

4.3.1 MAGNETIC MOMENT IN EXTERNAL MAGNETIC FIELD

Underlying the Zeeman effect is the existence of the magnetic moment of the electron which is quantized by the magnetic quantum number m. The magnetic moment of an electron in a fixed orbit oriented anyway in space may be thought as a degenerate quantity, and the degeneracy is removed in the presence of a magnetic field $\mathbf{B}_0$ that would produce Zeeman splitting of the spectral line. While the electron revolving in the space-fixed orbit has an orbital angular momentum $\mathbf{L}$ and the corresponding magnetic moment $\boldsymbol{\mu}_L$, each characterized by the vector perpendicular to the orbital plane, the external magnetic field turns $\boldsymbol{\mu}_L$ to align it along the field direction. The rotational inertia of the electron, however, resists this alignment, resulting in precession of $\boldsymbol{\mu}_L$ (Figure 4.4) analogous to the motion of a spinning top. The argument is easily seen from the equation of motion of the angular momentum vector in the presence of the external magnetic field

$$\frac{d\mathbf{L}}{dt} = \boldsymbol{\tau} = \boldsymbol{\mu}_L \times \mathbf{B}_0 = -\gamma_e \mathbf{B}_0 \times \mathbf{L} = \omega_L \times \mathbf{L} \tag{4.27}$$

in which $\boldsymbol{\tau}$ is torque, γ_e the electron gyromagnetic ratio, and ω_L the Larmor frequency.

Because angular momentum is quantized this motion should also be quantized, given by

$$l_\phi = m\hbar, \tag{4.28}$$

where l_ϕ is the angular momentum component in the field direction, and m the magnetic quantum number with values from $-l$ to $+l$ through zero. If the $\mathbf{B}_0$ field is along z, then m is the z-projection of l (Figure 4.4). The quantity l_ϕ is given here explicitly in $\hbar$ unit, which is the Bohr unit of angular momentum.

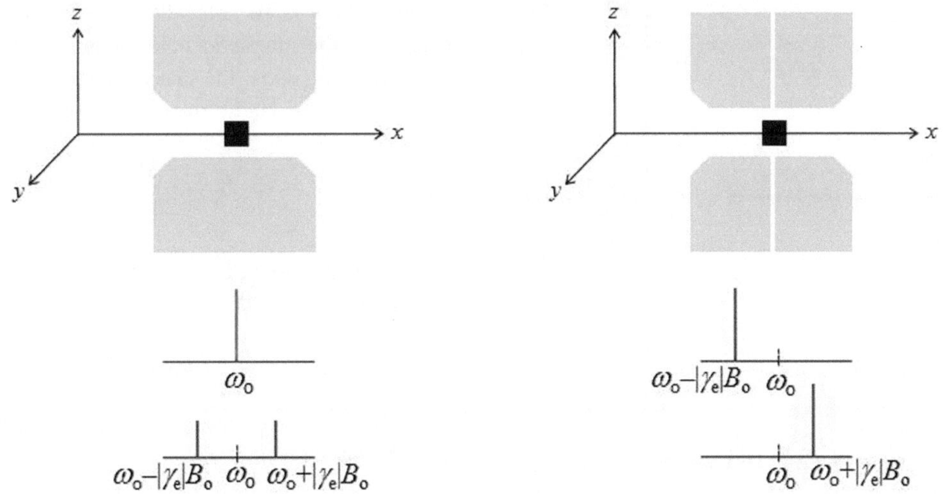

FIGURE 4.3 Lorentz splitting of spectral lines. *Left*, light polarized along the z-direction produces only one line ($m = 0$) at frequency ω_o, but y-polarization splits the line into two $(m = \pm 1)$ each with intensity half of that obtained with z-polarized light. The two lines are apart by $2|\gamma_e|B_o$. *Right*, excitation by left- and right-circularly polarized light propagating along z-direction each produces a line of equal intensity resonating at frequencies $\omega_o - |\gamma_e|B_o$ and $\omega_o + |\gamma_e|B_o$, respectively.

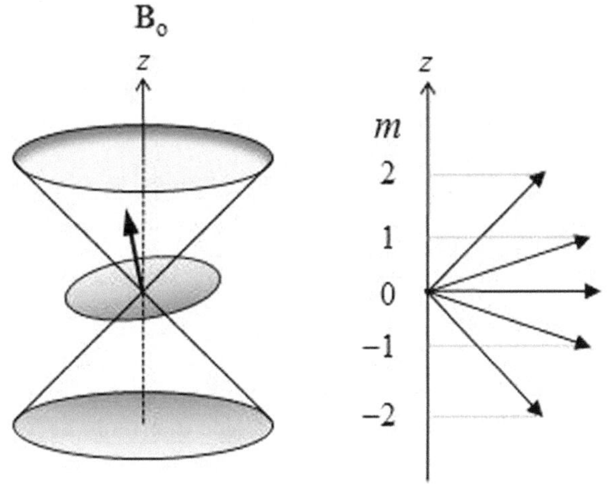

FIGURE 4.4 The motion of the orbital angular momentum vector in the presence of an external magnetic field along z (*left*), and spatially quantized components of the orbital angular momentum vector (quantum number $l = 2$) according to the angle ϕ and the z-projections of the components (*right*). There are $2l + 1$ allowed values of the quantum number m.

4.3.2 Larmor Precession

The precession of the angular momentum vector $\mathbf{L}$ can be derived from the classical potential energy of the magnetic dipole $\boldsymbol{\mu}$ in the presence of the external field $\mathbf{B}_o$. The work done by $\mathbf{B}_o$ to change the angle of $\boldsymbol{\mu}$ by dq with respect to the field direction is

$$dW = -\boldsymbol{\tau} d\theta = -\boldsymbol{\mu} \times \mathbf{B}_o d\theta = -\mu B_o \sin\theta d\theta, \quad (4.29)$$

where $\boldsymbol{\tau}$ is the torque. The change in the potential energy will be

$$\begin{aligned} dU &= \mu B_o \sin\theta d\theta \\ U &= \int \mu B_o \sin\theta d\theta \\ &= -\mu B_o \cos\theta + c. \end{aligned} \quad (4.30)$$

The constant $c = 0$ when $\theta = 90°$ and $U = 0$, so that

$$U = -\boldsymbol{\mu} \cdot \mathbf{B}_o = -\mu B_o \cos\theta. \quad (4.31)$$

Substituting with

$$\cos\theta = \frac{l_\phi}{l} = \frac{m}{l}$$

$$U = -\frac{m}{l} \mu B_o.$$

Since the electron motion in the circular orbit generates a circular current $I = e\omega / 2\pi$, where ω is the circular frequency of the electron revolution, the magnetic moment of the circulating electron is

$$\mu = \frac{1}{2} \omega r^2 \frac{e}{c}.$$

The orbital angular momentum of the electron of mass m_e in the unit of $\hbar$ is given by

$$l\hbar = m_e \omega r^2,$$

so that

$$\frac{\mu}{l} = \frac{e\hbar}{2m_e c}$$

$$U = -\frac{e\hbar}{2m_e c} m B_o. \quad (4.32)$$

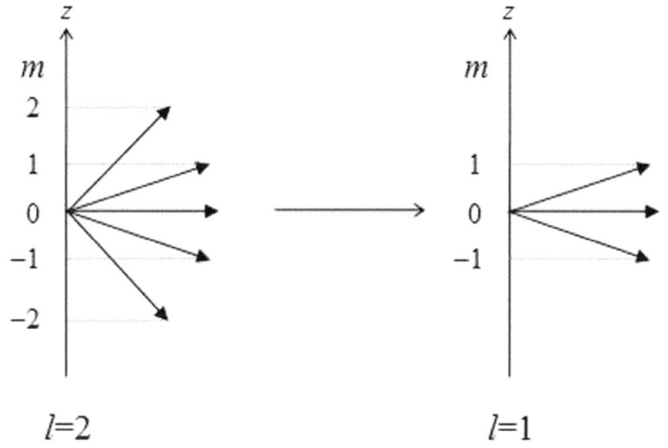

FIGURE 4.5 The transition $l = 2 \rightarrow l = 1$ causes a change in the magnetic quantum number m by one unit ($\Delta m = \pm 1$) as per the selection rule.

The use of Bohr magneton $e\hbar / 2m_e c$ as the unit of magnetic moment has been mentioned earlier.

If the change in the potential energy of the magnetic dipole in the presence of the external magnetic field $\mathbf{B}_o$ is equated with an increase in the energy of the atom, then

$$E = \hbar\omega = \left| -\frac{e\hbar}{2m_e c} \right| mB_o. \quad (4.33)$$

For $m = \pm 1$, the initial energy when the magnetic field was absent now splits by a separation

$$\omega = \frac{e}{m_e c} B_o \approx 8.79 \times 10^6 B_o \mathrm{s}^{-1} \quad (4.34)$$

where ω is Larmor precession frequency.

This analysis redirects one to the selection rules of atomic transitions discussed above (section 4.2). Atomic transitions involve a change in the angle between the z-direction and the orbital angular momentum vector, and this change is understood with respect to the $\cos\theta$ dependence of the energy. Figure 4.5 illustrates the $l = 2 \rightarrow l = 1$ transition entailing $m = 2 \rightarrow m = 1$. Since

$$\theta = \cos^{-1} \frac{m}{\sqrt{l(l+1)}},$$

the transition changes the angle from ~35.3 to ~45°.

4.3.3 Eigenstate, Operator, and Eigenvalue in External Magnetic Field

It is straightforward to put the classical description above into the Schrödinger equation. For example, consider the α-transition in the Balmer series, which is a $3d^1 \rightarrow 2p^1$ transition. For the initial state, $n = 3$, $l = 2$, and $m = -2, -1, 0, +1, +2$, the final excited-state quantum numbers are $n = 2$, $l = 1$, and $m = -1, 0, +1$. To note, the n-levels are degenerate in the absence of the magnetic field. When the field is turned on, say along the z-axis, the classical energy (equations (4.31)–(4.33)) enters the Hamiltonian as an additional potential

$$V = -\boldsymbol{\mu} \cdot \mathbf{B}_o = \frac{e\hbar}{2m_e c} B_o l_z = |\gamma_e| B_o l_z, \quad (4.35)$$

where l_z is the operator for the z-component of the angular momentum, such that

$$l_z \phi_m = m\phi_m. \quad (4.36)$$

If $\mathcal{H}_o$ was the energy operator in the absence of the magnetic field, the operator now is

$$\mathcal{H} = \mathcal{H}_o + V.$$

Because ψ_{nlm} is the eigenfunction for $\mathcal{H}_o$, L^2, and l_z (equations (4.5)–(4.7)) in the absence of the field, it must also be the eigenfunction in the presence of the field, so that

$$\begin{aligned} (\mathcal{H}_o + V)|\psi_{nlm}\rangle &= \left(\mathcal{H}_o + |\gamma_e| B_o l_z\right)|\psi_{nlm}\rangle \\ &= E_n + |\gamma_e| B_o m. \end{aligned} \quad (4.37)$$

For the Balmer α-transition, one has

$$(\mathcal{H}_o + V)|\psi_{32m}\rangle = \left[-13.6 + \hbar(\omega + m\omega)\right]|\psi_{32m}\rangle$$

where $\omega = \dfrac{E_3 - E_2}{\hbar} = \dfrac{1.8876}{\hbar} \mathrm{s}^{-1} = 1.8021 \times 10^6 \,\mathrm{rad\,s}^{-1}$ is the angular frequency corresponding to the 656.67 nm wavelength for the $3 \rightarrow 2$ transition (H_α line) of the Balmer series in the absence of the magnetic field. In the presence of thc ficld, thc sclcction rule allows nine transitions as listed in Table 4.1. It is however clear that the Larmor frequency corresponding to $\Delta m = 0$ is zero, and that corresponding to $\Delta m = \pm 1$ in an abnormally weak field of say 0.1 Gauss will

TABLE 4.1
Allowed Transitions for Balmer H_α-line in the Presence of a Magnetic Field

Selection rule	Transitions					
	from			to		
Δm	n	l	m	n	l	m
	3	2	1	2	1	1
0	3	2	0	2	1	0
	3	2	−1	2	1	−1
	3	2	2	2	1	1
+1	3	2	1	2	1	0
	3	2	0	2	1	−1
	3	2	0	2	1	1
−1	3	2	−1	2	1	0
	3	2	−2	2	1	−1

be ~1.4 GHz (214.2 nm), and the two lines will appear at 656.67±214.2 nm. Therefore, even though the probability of occurrence of all five ψ_{nlm} eigenstates corresponding to $3d^1$ is equal, the nine allowed transitions will appear only as three lines. In fact, the triplet of normal Zeeman effect is observable only at very high magnetic field, not at 0.1 Gauss considered for illustration here. Weak presence of Balmer α- and β-emissions, each of which splits into a widely separated triplet, is observable in astronomical spectra. Such splitting occurs due to very strong magnetic field, often in excess of 10^6 Gauss.

4.4 ANOMALOUS ZEEMAN EFFECT AND FURTHER SPLITTING OF SPECTRA

It is too little to say that Zeeman's observation in the pre-quantum years had a profound influence on the thinking that led to the discovery of electron spin, spin quantum number, and spin-orbit coupling. The Zeeman-Lorentz triplet discussed earlier which arises from $\Delta m = 0, \pm 1$ selection rule originates from the magnetic field-resolved z-projection of orbital angular momentum alone. There was no concern about the electron spin. The triplet of a spectral line observed in the presence of magnetic field as such is called normal Zeeman effect. But a year after Zeeman's detection of the sodium triplet, Preston observed further splitting of spectral lines into fine structure multiplets. For example, of the two sodium D-lines ($3^2P_{3/2} \leftrightarrow 3^2S_{1/2}$ and $3^2P_{1/2} \leftrightarrow 3^2S_{1/2}$) in the absence of magnetic field, the first splits into six and the second into four when an external field is turned on. This deviation from triplets to multiplets appeared as an anomaly, the interpretation of which was at the focus when quantum theory was being formulated.

4.4.1 Electron Spin and Spin Magnetic Moment

Based on Pauli's introduction of the spin quantum number s that can have only two possible values (see Box 4.2), Uhlenbeck and Goudsmit proposed that the electron is a charged particle, rotates on its own axis, possesses an angular momentum **S** ($s = \hbar/2$) around the spinning axis, and has a small magnetic moment $\boldsymbol{\mu}_s$ vectorially opposite to the direction of the angular momentum. The magnitude of the magnetic moment they suggested was

$$\mu_s = \frac{e\hbar}{2m_e c},$$

which is exactly equal to the Bohr magneton. The ratio of spin magnetic moment μ_s to spin angular momentum s, also called gyromagnetic ratio, is

$$\gamma_s = \frac{e\hbar}{m_e c}, \tag{4.38}$$

where $s = 1/2$ in the unit of $\hbar$ has been used. A simple and non-relativistic derivation of this result by assuming the electron as a classical particle has been obtained by Kramers. But this ratio of spin magnetic moment to angular momentum is two-times the ratio of orbital magnetic moment to angular momentum

$$\gamma_l = \frac{e\hbar}{2m_e c}. \tag{4.39}$$

Thus, the electron has two magnetic moments originating separately from orbital and spin motions

$$\mu_l = \frac{e\hbar}{2m_e c} l \qquad l = 0, 1, 2, \cdots \text{ orbital magnetic moment}$$

$$\mu_s = \frac{e\hbar}{2m_e c} s \qquad s = 1/2 \text{ spin magnetic moment.} \tag{4.40}$$

We notice that μ_s is a constant, but μ_l changes with the value of l. While normal Zeeman effect can be explained with μ_l alone, anomalous Zeeman effect also involves μ_s and the coupling of **L** and **S**. Note that angular momenta are usually given in capitals (L and S, both plain) instead of lower-case letters when they arise from more than one electron. Even if not indicated in capitals, they should be understood in the context of the system.

4.4.2 Lande *g*-Factor

Clearly, the magnetic dipole moment of the electron, or any particle or a molecule for that matter, can be given in terms of a number multiplied by Bohr magneton and the angular momentum – a vector. The appropriate number to multiply with is called the g-factor for the particle in consideration. From above we see that for an electron

$$\mu_s = g_s \frac{e\hbar}{2m_e c}|\mathbf{S}| \text{ and } \mu_l = g_l \frac{e\hbar}{2m_e c}|\mathbf{L}|, \tag{4.41}$$

where $g_s(\approx 2)$ and $g_l(=1)$ are g-factors for the spin and orbital magnetic moments. Put another way, the g-factor for a particle is the gyromagnetic ratio factor because

$$\gamma = |g| \frac{\boldsymbol{\mu}}{\mathbf{L}}.$$

The sign of the g-factor may be taken as positive when the dipole moment vector is parallel to the angular momentum vector, and negative when they are opposite in direction.

If several angular momentum vectors couple to produce a new angular momentum vector, then a new g-factor will characterize its magnetic moment. In the case of the electron, the total angular momentum J is the sum of orbital and spin angular momenta (**L** and **S**), so another g-factor, g_J, arises, which is seen at once from the following

$$\mathbf{J} = \mathbf{L} + \mathbf{S}$$

$$g_J\, \mathbf{J} = g_L \mathbf{L} + g_S \mathbf{S}.$$

Taking the dot product with **J** on both sides

$$g_J\ \mathbf{J}\cdot\mathbf{J} = g_L \mathbf{L}\cdot\mathbf{J} + g_S \mathbf{S}\cdot\mathbf{J}.$$

One can now use the properties of scalar multiplication and apply the principle that a vector projected onto itself is equal to the square of the magnitude of the vector, that is

$$\mathbf{L}\cdot\mathbf{J} = \mathbf{L}\cdot(\mathbf{L}+\mathbf{S}) \text{ and } \mathbf{S}\cdot\mathbf{J} = \mathbf{S}\cdot(\mathbf{L}+\mathbf{S})$$

$$\mathrm{J}^2 = \mathrm{J}^2 + \mathrm{S}^2 + 2\mathbf{L}\cdot\mathbf{S}$$

$$\mathrm{L}^2 = \mathrm{J}^2 + \mathrm{S}^2 - 2\mathbf{J}\cdot\mathbf{S}$$

$$\mathrm{S}^2 = \mathrm{J}^2 + \mathrm{L}^2 - 2\mathbf{J}\cdot\mathbf{L} \tag{4.42}$$

and substitute to obtain

$$g_J \mathrm{J}^2 = g_L\left(\mathrm{L}^2 + \mathbf{L}\cdot\mathbf{S}\right) + g_S\left(\mathrm{S}^2 + \mathbf{L}\cdot\mathbf{S}\right)$$

$$= g_L\left[\mathrm{L}^2 + \frac{1}{2}\left(\mathrm{J}^2 - \mathrm{L}^2 - \mathrm{S}^2\right)\right] + g_S\left[\mathrm{S}^2 + \frac{1}{2}\left(\mathrm{J}^2 - \mathrm{L}^2 - \mathrm{S}^2\right)\right].$$

Because $J^2 = \mathbf{J}^2$, $L^2 = \mathbf{L}^2$, and $S^2 = \mathbf{S}^2$, the eigenvalues for these operators are $J(J+1)$, $L(L+1)$, and $S(S+1)$, respectively (equation (4.6)). Then

$$g_J J(J+1) = \frac{1}{2} g_L\left[J(J+1)+L(L+1)-S(S+1)\right]$$
$$+\frac{1}{2} g_S\left[J(J+1)-L(L+1)+S(S+1)\right].$$

With $g_L = 1$ and $g_S = 2$

$$g_J = \frac{3}{2} + \frac{S(S+1) - L(L+1)}{2J(J+1)}. \tag{4.43}$$

The factor g_J is Lande g-factor for the electron.

4.4.3 Spin-Orbit Coupling

In the initial formulation of the hydrogen atom problem, the electron was taken as a spinless charged particle in the potential field of the nucleus, and as such Ze^2/r was included in the Hamiltonian. The Schrödinger equation in this form completely describes the energy levels for atomic transitions involving n, l, m quantum numbers. But atomic transitions split into closely spaced multiplets, the interpretation of which requires the fourth quantum number s. Therefore, the spinless Hamiltonian should be supplemented by energy terms due to the electron spin. Precisely, it is the spin-orbit interaction energy that arises from spin-orbit coupling.

Spin-orbit coupling is a relativistic effect that arises from the partial differential equations of Dirac. It is indeed hard to derive, but one should be familiar with the basis of the interaction. The relativistic effect sets in strongly for inner electrons of heavy atoms, but the velocity of electrons in lighter hydrogen-like atoms is about two orders of magnitude less than the velocity of light, rendering relativistic mass correction small. For a short discussion of spin-orbit coupling, let an observer also go on orbiting while riding on the electron. To the observer the nucleus will appear moving around the electron. Because the nucleus is a charged body its apparent rotating motion will produce an intrinsic magnetic field given by

$$\mathbf{B}_{\text{int}} = \frac{\mu_{\text{o}Ze}}{4\pi r^3}\mathbf{r}\times\mathbf{v}, \tag{4.44}$$

where Ze is the charge on the nucleus, **r** is the position vector directed from the nucleus to the electron, and **v** is the orbital velocity of the electron, which is ~10^8 cm s^{-1} for the electron of the hydrogen atom. The nucleus also produces an electric field given by

$$\mathbf{E}_{\text{int}} = \frac{Ze\mathbf{r}}{4\pi\varepsilon_{\text{o}} r^3}. \tag{4.45}$$

Combining equations (4.44) and (4.45), and using $c = \left(\varepsilon_{\text{o}}\mu_{\text{o}}\right)^{-0.5}$, where ε_{o} and μ_{o} are vacuum permittivity and free-space permeability, respectively, the magnetic field experienced by the electron in its rest frame is

$$\mathbf{B}_{\text{int,el}} = \frac{1}{c^2}\mathbf{E}_{\text{int}}\times\mathbf{v}. \tag{4.46}$$

Since the energy of a magnetic dipole in the presence of a magnetic field is

$$U = -\boldsymbol{\mu}\cdot\mathbf{B},$$

the spin-orbit interaction energy takes the form

$$E_{\text{SO}} = -\boldsymbol{\mu}\cdot\mathbf{B}_{\text{int,el}}$$

$$= -\frac{1}{c^2}\left(\mathbf{E}_{\text{int}}\times\mathbf{v}\right)\cdot\boldsymbol{\mu} \tag{4.47}$$

The electron orbital velocity $\mathbf{v} = \mathbf{P}/m$, and the force on the electron due to the electric field is

$$\mathbf{F} = -e\mathbf{E} = -\nabla V,$$

where $V = -e^2/r$ is the electrostatic energy of the electron. For hydrogen atom ($Z = 1$), this V also represents the Coulomb potential for the electron. As per the central-field approximation the electric field at the electron site is a function of its distance r from the nucleus. Accordingly,

$$\nabla V = \frac{1}{r}\frac{\partial V}{\partial r}\mathbf{r}.$$

Because the electron spin is a relativistic effect, the spin magnetic moment in SI unit is written as

$$\boldsymbol{\mu}_{\text{s}} = -g_e\frac{e}{2m_e}\mathbf{S} = -\frac{e}{m_e}\mathbf{S}, \tag{4.48}$$

in which the electron g-factor $g_e \approx 2$ as discussed earlier. Also a fact to be aware of is that a relativistic correction, called Thomas precession, is introduced to the precession frequency of the electron spin in the $\mathbf{B}_{\text{int,el}}$ field. To a stationary observer, the actual frequency of precession – Thomas precession – is reduced to half.

Combining these pieces of information, equation (4.47) is written as

$$E_{\text{SO}} = \frac{1}{2m_e c^2}\frac{1}{r}\frac{\partial V}{\partial r}(\mathbf{r}\times\mathbf{P})\cdot\mathbf{S}, \tag{4.49}$$

where S is the spin angular momentum vector. This expression is essentially the spin-orbit coupling Hamiltonian $\mathcal{H}_{\text{SO}}$ for a one-electron system like the hydrogen atom

$$\mathcal{H}_{\text{SO}} = \frac{1}{2m_e c^2}\frac{1}{r}\frac{\partial V}{\partial r}(\mathbf{L}\cdot\mathbf{S}). \tag{4.50}$$

For a many-electron system the Hamiltonian contains the sum of the spin-orbit interaction for each electron

$$\mathcal{H}_{\text{SO}} = \frac{1}{2m_e c^2}\sum_i \frac{1}{r_i}\frac{\partial V(r_i)}{\partial r_i}(\mathbf{L}_i\cdot\mathbf{S}_i).$$

To write the vector product $\mathbf{L}\cdot\mathbf{S}$ in terms of the total angular momentum $\mathbf{J}$, we take the square of the sum of $\mathbf{L}$ and $\mathbf{S}$, similar to that done earlier in equation (4.42),

$$\mathbf{L}\cdot\mathbf{S} = \frac{1}{2}\left(J^2 - L^2 - S^2\right)$$

where $J = L+S, L+S-1,\cdots,L-S$. By writing $\mathbf{L}\cdot\mathbf{S}$ in this form it is easier to see that m_L and m_S, which are the z-projections of $\mathbf{L}$ and $\mathbf{S}$, are mixed yielding $m_L + m_S = m_J$. The spin-orbit coupling thus involves five quantum numbers, n, l, s, J, and m_J. These quantum numbers in the context of a single-electron atom are n, l, s, j, and m_j.

4.4.4 Spin-Orbit Coupling Energy

The total Hamiltonian for the atomic problem at this stage can be written as

$$\mathcal{H} = \mathcal{H}^{\text{o}} + \mathcal{H}_{\text{SO}},$$

even though it is hard to solve the Schrödinger equation with the $\mathcal{H}_{\text{SO}}$ operator, which is

$$\mathcal{H}_{\text{SO}}\psi_{nlsjm_j(r,\theta,\phi)} = E_{\text{SO}}\psi_{nlsjm_j(r,\theta,\phi)}. \tag{4.51}$$

But it should be obvious that $\mathcal{H}_{\text{SO}} \ll \mathcal{H}^{\text{o}}$, which could allow for a first-order perturbation approximation to obtain the spin-orbit interaction energy

$$E_{\text{SO}}^{(1)} = \int \psi^*_{nlsjm_j}\mathcal{H}_{\text{SO}}\psi_{nlsjm_j}\,d\tau. \tag{4.52}$$

To calculate the radial part (n, l dependence of the radial function) of this integral we recall that the Coulomb potential for the electron at an average distance r from the nucleus is $-Ze^2/r$, so that

$$\frac{1}{r}\frac{\partial U}{\partial r} = \frac{Ze^2}{r^3}$$

$$Ze^2\int_0^\infty R^*_{nl}R_{nl}\frac{1}{r^3}dr = Ze^2\int_0^\infty R^2_{nl}\frac{1}{r^3}dr. \tag{4.53}$$

For the mean value of $1/r^3$ we use the formula

$$\left\langle\frac{1}{r^3}\right\rangle_{nl} = \left(\frac{m_e cZe^2}{n\hbar^2 c}\right)^3 \frac{1}{l\left(l+\frac{1}{2}\right)(l+1)} \text{ for } l>0.$$

Notice that the energy shift due to spin-orbit interaction is angular momentum-dependent.

To obtain the angular momentum part of the integral equation (4.52) we recognize that the angular functions are associated with orbital angular momentum l, spin angular momentum s, total angular momentum J due to $\mathbf{L}\cdot\mathbf{S}$ coupling, and the z-projection of J, which is m_J. For convenience, we can write the integral in Dirac notation

$$\left\langle Y_{lsjm_j}\left|\mathbf{L}\cdot\mathbf{S}\right|Y_{lsjm_j}\right\rangle = \left\langle Y_{lsjm_j}\left|\frac{1}{2}\left(J^2-L^2-S^2\right)\right|Y_{lsjm_j}\right\rangle$$
$$= \frac{\hbar^2}{2}\left[J(J+1) - L(L+1) - S(S+1)\right]. \tag{4.54}$$

Note again that capital symbols for quantum numbers J, L, S for multi-electron system corresponds to lower-case symbols j, l, s for a single-electron system. The spin-orbit interaction is

$$\begin{aligned} E_{\text{SO}}^{(1)} &= \frac{1}{2m_e{}^2c^2}Ze^2\left[\frac{1}{r^3}\right]_{nl}\langle\mathbf{L}\cdot\mathbf{S}\rangle_{lsjm_j} \\ &= \frac{1}{2m_e{}^2c^2}Ze^2\left(\frac{m_e ce^2 Z}{n\hbar^2 c}\right)^3\frac{\hbar^2}{2}\frac{1}{l\left(l+\frac{1}{2}\right)(l+1)} \\ &\quad \left[J(J+1)-L(L+1)-S(S+1)\right] \\ &= \frac{1}{2n^3}\frac{m_e c^2 Z^4 e^8}{\hbar^4 c^4}\frac{J(J+1)-L(L+1)-S(S+1)}{2L\left(L+\frac{1}{2}\right)(L+1)} \\ &= \frac{1}{2}\frac{m_e c^2 Z^4\alpha^4}{n^3}\frac{J(J+1)-L(L+1)-3/4}{2L\left(L+\frac{1}{2}\right)(L+1)}, \end{aligned} \tag{4.55}$$

where $\alpha = \frac{e^2}{\hbar c} \approx 0.0073$ is the fine-structure constant. The upfront appearance of the factor of $1/2$, which is omitted in some textbooks, reflects the fact that only half of the total internal magnetic field arising from the motion of the nucleus (equation (4.44)) is experienced by the electron. This correction, which is essentially due to relativistic precession of the electron with respect to the resting nucleus, leads to agreement of theoretical prediction and experimental observation of atomic spectra.

4.4.5 Spectroscopic Notation

The electron configuration of an atom can give rise to several energy levels, call them 'states', based on the *L*, *S*, and *J* values. The different energy states are named by the spectroscopic notation, a practice since the early days of atomic spectroscopy. The notation is generally given by

$$n^{2S+1}L_J^{\Pi}$$

where *n* (shell) is the principal quantum number. The $^{2S+1}L^{\Pi}$ part of the notation stands for the 'term' of the state of electron configuration. In the term, 'S' in the superscript is the total spin of the atomic system, and hence 2S + 1 provides the 'spin multiplicity' of the state. For example, the spin multiplicity is a singlet when $S = 0$, a doublet when $S = 1/2$, a triplet when $S = 1$, and so forth. The letter 'L' in the term stands for the value of the total orbital angular momentum, and L is labeled as S, P, D, F, ⋯, corresponding to 0, 1, 2, 3, ⋯. The superscript 'Π' following 'L' shows the parity of the state (see Box 4.3). For even parity, Π is shown as Π only, but Π is replaced by a small-case 'o' when the parity is odd. By adding the total angular momentum 'J' shown in the subscript one obtains 'levels' due to the splitting of a term. If there are *i* electrons in the configuration, then

$$J = L + S = \sum_i l_i + \sum_i s_i .$$

Even though the angular momenta in the spectroscopic notation refers to the total angular momentum of all electrons, it should be realized that completely filled shells and subshells in accordance with Pauli's exclusion principle yield L=0 and S=0. Therefore, electrons only from open or partly filled shells need to be considered. Also, excited states of atoms generally arise due to the excitation of the outermost electrons to higher orbitals.

4.4.6 Fine Structure of Atomic Spectra

The different terms and levels of energy associated with an atomic configuration will certainly produce many transitions in the atomic spectrum. These multiple transitions constitute the fine structure of the spectrum. Based on the discussions above, fine structure effects of simple atoms is easily understood. For illustration, we consider hydrogen and sodium – the two paradigm atoms whose spectral transitions have been studied very extensively since Zeeman's time.

Hydrogen Atom. For the lone electron of the hydrogen atom ($s = 1/2$) the total angular momentum $J = L + S$ for any *nl* configuration is always a half-integer. The atomic states, terms, and levels obtained by using the $n^{2S+1}L_J$ notation are listed in Table 4.2. The ground-state level is $1^2S_{\frac{1}{2}}$. The various levels can be explicitly written out

TABLE 4.2
Fine Structure Effect of Hydrogen Atom

nl configuration	*l*	*s*	*j*	Atom label	Term	Level
ns	0	$\frac{1}{2}$	$\frac{1}{2}$	$ns_{\frac{1}{2}}$	n^2S	$n^2S_{\frac{1}{2}}$
np	1	$\frac{1}{2}$	$\frac{1}{2}, \frac{3}{2}$	$np_{\frac{1}{2}}, np_{\frac{3}{2}}$	n^2P	$n^2P_{\frac{1}{2}}, n^2P_{\frac{3}{2}}$
nd	2	$\frac{1}{2}$	$\frac{3}{2}, \frac{5}{2}$	$nd_{\frac{3}{2}}, nd_{\frac{5}{2}}$	n^2D	$n^2D_{\frac{3}{2}}, n^2D_{\frac{5}{2}}$
nf	3	$\frac{1}{2}$	$\frac{5}{2}, \frac{7}{2}$	$nf_{\frac{5}{2}}, nf_{\frac{7}{2}}$	n^2F	$n^2F_{\frac{5}{2}}, n^2F_{\frac{7}{2}}$

Notes: Allowed values of *J* run from *L* – *S* to *L* + *S* in steps of integer

Because the hydrogen atom has only one electron, small letters (*l*, *s*, *j*) are used for atom labels, but to conform with the spectroscopic notation terms and levels are shown with capital S, P, D, F.

$$1^2S_{\frac{1}{2}}$$

$$2^2S_{\frac{1}{2}} \quad 2^2P_{\frac{1}{2}} \quad 2^2P_{\frac{3}{2}}$$

$$3^2S_{\frac{1}{2}} \quad 3^2P_{\frac{1}{2}} \quad 3^2P_{\frac{3}{2}} \quad 3^2D_{\frac{3}{2}} \quad 3^2D_{\frac{5}{2}}$$

$$4^2S_{\frac{1}{2}} \quad 4^2P_{\frac{1}{2}} \quad 4^2P_{\frac{3}{2}} \quad 4^2D_{\frac{3}{2}} \quad 4^2D_{\frac{5}{2}} \quad 4^2F_{\frac{5}{2}} \quad 4^2F_{\frac{7}{2}}$$

The energy level structures for the principal quantum number $n = 2$ are shown in Figure 4.6. To notice are the differences that are so small, of the order of meV, that the probability of spontaneous emission is very small. These transitions were detected by Lamb by subjecting the atom to oscillating magnetic field of appropriate frequency to induce the transitions. The splitting between $2^2P_{1/2}$ and $2^2P_{3/2}$ is the fine structure, and the smaller splitting between $2^2P_{1/2}$ and $2^2S_{1/2}$ produces lamb shift.

Sodium Atom. The sodium electrons are configured as $1s^2 2s^2 2p^6 3s$, so the ground state is $^2S_{1/2}$. The fine structure effect of sodium is the same as that of hydrogen (Table 4.2). The energy levels of sodium are drawn in Figure 4.7 and listed in Table 4.3 in the order of increasing energy. The compressed energy scale required to draw all energy levels does not show energy differences between various ^2L states. The spin-orbit coupling does split the states, even though the energy difference is very small. For example, the 3^2P term splits into $3^2P_{1/2}$ and $3^2P_{3/2}$ levels by only 2.1 meV (Figure 4.7), and the two transitions $3^2P_{1/2} \leftrightarrow 3^2S_{1/2}$ (589.59 nm) and $3^2P_{3/2} \leftrightarrow 3^2S_{1/2}$ (589 nm) are split by only 0.597 nm. These two transitions are the D_1 and D_2 single-line emission of sodium when no external magnetic field is present.

4.4.7 Splitting of m_j Degeneracy: Anomalous Zeeman Effect

Consider the spatial quantization of the total angular momentum J. The eigenvalue equations for the operators J^2 and J_z are

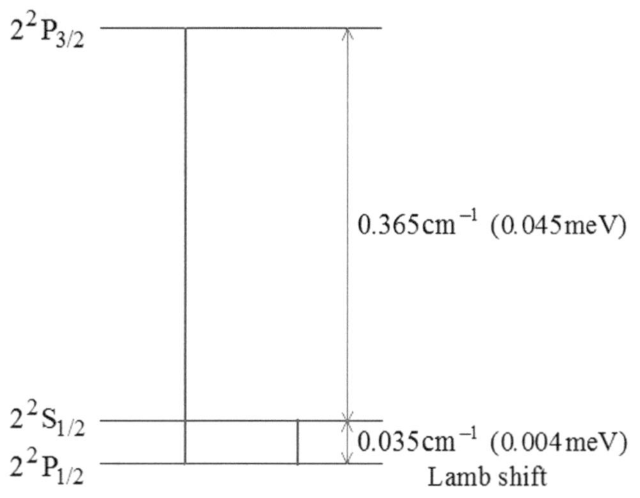

FIGURE 4.6 Fine structure splitting of the $n = 2$ level of atomic hydrogen in the absence of external magnetic field. The Lamb effect originates from quantum electrodynamics.

$$J^2\psi_{JM} = J(J+1)\hbar^2\psi_{JM}$$
$$J_z\,\psi_{JM} = m_j\hbar\psi_{JM}. \tag{4.56}$$

Since $\mathbf{J} = \mathbf{L}+\mathbf{S}$, the eigenfunction ψ_{JM} can be expressed in terms of the eigenfunctions of the operators L^2, L_z, S^2, and S_z (Box 4.4). The spatial quantization of J directly follows from the commutation relation between J^2 and J_z. The relation $\left[J^2, J_z\right] = 0$ allows for simultaneous specification of the magnitudes of $\mathbf{J}^2$ and $\mathbf{J}_z$. Since m_j is the z-projection of J, the m_j levels are degenerate as such, but can be split into sublevels by the application of an external magnetic field. Each value of J yields $2J+1$ levels of m_j, thus offering additional energy levels from and to which spectroscopic transitions can occur. This splitting of J into m_j levels in an external magnetic field is traditionally called anomalous Zeeman effect.

We note, however, that the directions of the total angular momentum and the corresponding magnetic moment vectors are not collinear. Unlike **l** and **s** angular momentum vectors, both of which are collinear to the respective magnetic moment vectors $\boldsymbol{\mu}_l$ and $\boldsymbol{\mu}_s$, only one component of the total angular momentum vector **j** is strictly in the direction same as that of the corresponding magnetic moment vectors $\boldsymbol{\mu}_j$. This difference arises from the fact that $\boldsymbol{\mu}_s$ is two-fold larger in magnitude than $\boldsymbol{\mu}_l$, which is obvious from magnitudes of $g_s(\approx 2)$ and $g_l(=1)$. Consider the vector model construction depicted in Figure 4.8, which shows $\mathbf{l}=\boldsymbol{\mu}_l$ but $\mathbf{s}=2\boldsymbol{\mu}_s$. The **j** and $\boldsymbol{\mu}_j$ vectors so obtained are not in the same direction. This directional difference causes the appearance of precession of the total magnetic moment vector $\boldsymbol{\mu}_j$ about the total angular momentum vector **j** (Figure 4.8). We also recognize that the precession of $\boldsymbol{\mu}_l$ and $\boldsymbol{\mu}_s$ about respective **l** and **s** is not apparent

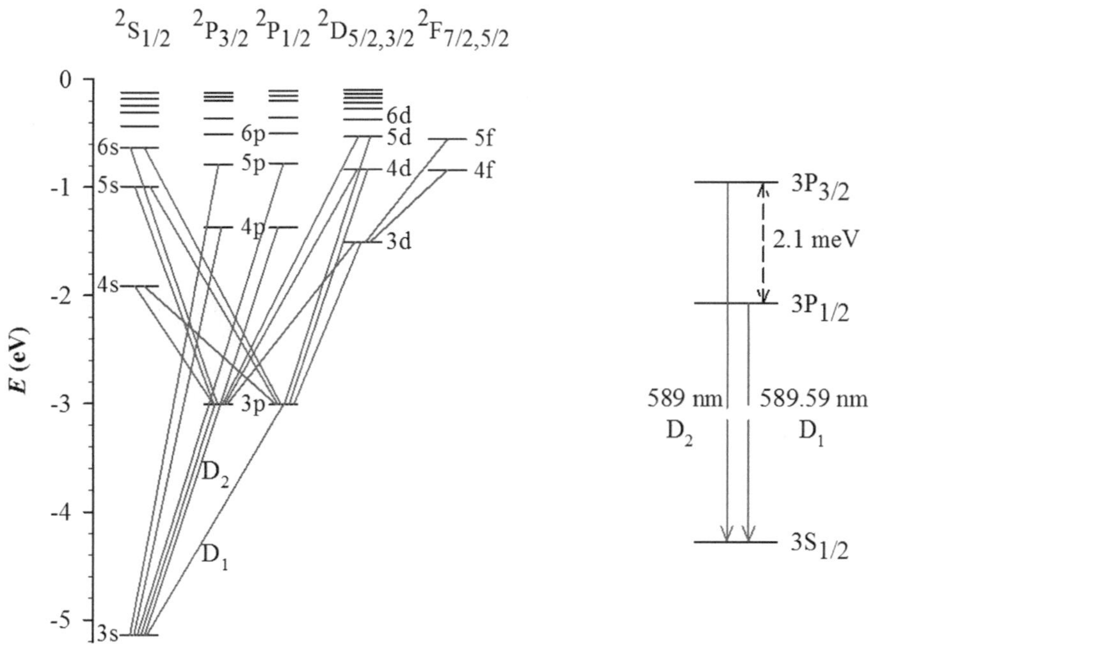

FIGURE 4.7 *Left*, Sodium atom energy level diagram in the absence of an external magnetic field. *Right*, expansion of 3^2P_J showing the D_1 and D_2 lines.

TABLE 4.3
Emission Lines of Sodium

From	to	Energy (eV)	Wavelength (nm)
$5^2P_{\frac{3}{2}}$	$3^2S_{\frac{1}{2}}$	4.34605	285.28
$5^2P_{\frac{1}{2}}$	$3^2S_{\frac{1}{2}}$	4.34575	285.30
$4^2P_{\frac{3}{2}}$	$3^2S_{\frac{1}{2}}$	3.75448	330.23
$4^2P_{\frac{1}{2}}$	$3^2S_{\frac{1}{2}}$	3.75380	330.29
$5^2D_{\frac{5}{2}}$	$3^2P_{\frac{1}{2}}$	2.49034	497.86
$5^2D_{\frac{5}{2}}$	$3^2P_{\frac{3}{2}}$	2.48819	498.29
$6^2S_{\frac{1}{2}}$	$3^2P_{\frac{1}{2}}$	2.40788	514.91
$6^2S_{\frac{1}{2}}$	$3^2P_{\frac{3}{2}}$	2.40787	514.912
$4^2D_{\frac{5}{2}\frac{3}{2}}$	$3^2P_{\frac{1}{2}}$	2.18178	568.27
$4^2D_{\frac{5}{2}\frac{3}{2}}$	$3^2P_{\frac{3}{2}}$	2.17967	568.82
$3^2P_{\frac{1}{2}}$	$3^2S_{\frac{1}{2}}$	2.10289	589.59
$3^2P_{\frac{3}{2}}$	$3^2S_{\frac{1}{2}}$	2.10146	589.99
$5^2S_{\frac{1}{2}}$	$3^2P_{\frac{1}{2}}$	2.01463	615.42
$3^2D_{\frac{5}{2}\frac{3}{2}}$	$3^2P_{\frac{1}{2}}$	1.51509	818.33
$3^2D_{\frac{5}{2}\frac{3}{2}}$	$3^2P_{\frac{3}{2}}$	1.52147	814.90
$4^2S_{\frac{1}{2}}$	$3^2P_{\frac{1}{2}}$	1.08930	1138.20
$5^2S_{\frac{1}{2}}$	$3^2P_{\frac{3}{2}}$	1.08922	1138.28
$4^2S_{\frac{1}{2}}$	$3^2P_{\frac{3}{2}}$	1.08720	1140.40
$5^2F_{\frac{7}{2}\frac{5}{2}}$	$3^2D_{\frac{5}{2}\frac{3}{2}}$	0.97795	1267.80
$4^2F_{\frac{7}{2}\frac{5}{2}}$	$3^2D_{\frac{5}{2}\frac{3}{2}}$	0.67167	1845.90

because of collinearity of μ_l and **l**, and μ_s and **s**. Remember, the lower-case letters j, l, and s have been used here to discuss the vector construction diagram for a single electron; the capital letters J, L, and S are used for multielectron system.

Now the frequency of precession of μ_j about **j** is easily estimated from the LS splitting of orbital energy. For example, the energy of splitting in the $n = 2$ level of the hydrogen atom into $2^2P_{3/2}$ and $2^2P_{1/2}$ is 0.365 cm^{-1}, which corresponds to a frequency of $\sim 1.1\times10^{10}$ s^{-1} for the precession of μ_j – that is really fast. This precession does not assume significance when the atom is placed in an external magnetic field where only a time-averaged vector $\langle \mu_j \rangle_t$ can be taken. This is equivalent to saying that the magnetic property of the atom does not change with time. The use of $\langle \mu_j \rangle_t$ lets one projecting μ_j onto **j** so as to obtain a component of the total magnetic moment parallel to **j**, that is $\langle \mu_{j\parallel} \rangle_t$. The time-averaged perpendicular component $\langle \mu_{j\perp} \rangle_t$ is zero because fast precession causes a continuous change of direction of $\mu_{j\perp}$ with time; the different directions cancel out. The existence of the $\mu_{j\parallel}$ dipole causes precession of **s** and **l** vectors about it (Figure 4.8, *right*). Henceforth we will call $\mu_{j\parallel}$ as μ_j only.

4.5 ZEEMAN EFFECT IN WEAK MAGNETIC FIELD

In the presence of a weak magnetic field $\mathbf{B}_o$ applied along the z direction the μ_j vector precesses about the field direction. Note that if $\mathbf{B}_o$ is weak then the weak coupling of μ_j and $\mathbf{B}_o$ still preserves **LS**-coupling so that **j** exists. The vector model then requires that the interaction of μ_j and $\mathbf{B}_o$ be described by the projection of **l** and **s**, first onto μ_j or **j** and then onto $\mathbf{B}_o$. Because $|j| = \hbar\sqrt{j(j+1)}$ where j runs from $l+s$ to $l-s$ in

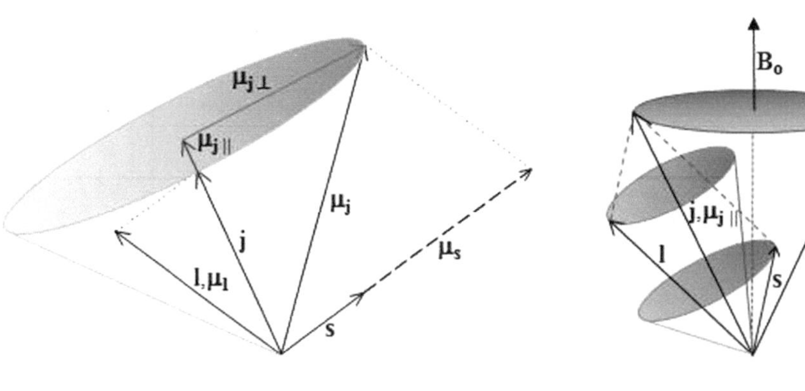

FIGURE 4.8 *Left*, construction of **j** and μ_j vectors by the addition of **l** and **s**, and μ_l and μ_s, respectively. Vectors **l** and μ_l, and **s** and μ_s are collinear, but **j** and μ_j are not. This directional difference gives rise to the precession of μ_j about **j**. *Right*, as mentioned in the text, the **l** and **s** vectors precess about $\mu_{j\parallel} = \mu_j =$ **j** when the atom is placed in an external magnetic field $\mathbf{B}_o$.

integer steps, the z-projection of $\mathbf{j}$, call it m_j, is quantized into $2j+1$ states with energies proportional to the magnitude of the $\mathbf{B}_o$ field

$$E=-(\mu_l+\mu_s)=\mu_j\cdot\mathbf{B}_o=-\mu_jB_o\cos\theta=-\frac{m_j}{j}\mu_jB_o, \qquad (4.57)$$

with

$$\mu_l=-\frac{e}{2m_ec}\mathbf{l}$$

$$\mu_s=-\frac{e}{2m_ec}\mathbf{s},$$

where

$$\cos\theta=\frac{m_j}{j} \text{ and } m_j=-j,-j+1,\cdots,j-1,j.$$

One needs to recognize that the effective total magnetic moment μ_j is determined by both orbital and spin quantum numbers as well as the cosine of the angle between **l** and **s** vectors. Also, because μ_s and μ_l are associated with the respective g-factors, μ_j too is expected to be identified with a corresponding g-factor, g_j. Thus, the energy formula above should be written by incorporating g_J

$$\mu_j=\frac{e\hbar}{2m_ec}g_jj.$$

Substitution of this into the energy formula provides

$$E=-\frac{e\hbar}{2m_ec}g_jm_JB_o. \qquad (4.58)$$

It is this g-factor in the energy formula which did not appear in the case of normal Zeeman energy (see equation (4.37)) that gives rise to the anomaly in the Zeeman effect.

To see the above argument more clearly consider the selection rules

$$\Delta m=0,\pm1 \quad l\rightarrow l' \quad \text{normal}$$

$$\Delta m_J=0,\pm1 \quad j\rightarrow j' \quad \text{anomalous.}$$

While the energy of l-levels do not depend on the g-factor, the energy of j-levels do, which means the energy splitting of j and j' levels will be different. To illustrate, consider Zeeman splitting of the sodium D-line into D_1 and D_2, which occurs due to transitions between different energy levels of the $3^2P_{3/2}$, $3^2P_{1/2}$, and $3^2S_{1/2}$ terms (Figure 4.9). The three terms differ by the value of J, and we saw earlier that g_J is determined by the value of J according to equation (4.43)

$$g_J=\frac{3}{2}+\frac{S(S+1)-L(L+1)}{2J(J+1)}.$$

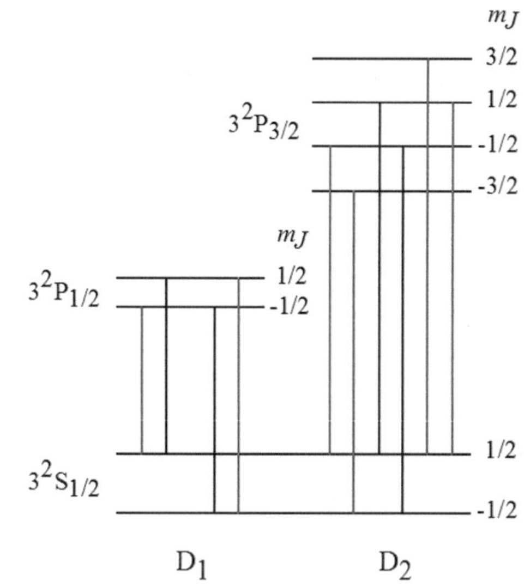

FIGURE 4.9 Anomalous Zeeman effect for a 3P→3S transition. Transitions corresponding to longitudinal observation by z-polarized light are in cyan, and those observed by transverse polarization of light are in black.

TABLE 4.4
Values of m_J and g in the Splitting of the Three Levels of $n=3$ of Sodium

Level	L	S	J	m_J	g_J	m_Jg_J
$3^2S_{\frac{1}{2}}$	0	$\frac{1}{2}$	$\frac{1}{2}$	$\pm\frac{1}{2}$	2	±1
$3^2P_{\frac{1}{2}}$	1	$\frac{1}{2}$	$\frac{1}{2}$	$\pm\frac{1}{2}$	$\frac{2}{3}$	$\pm\frac{1}{3}$
$3^2P_{\frac{3}{2}}$	1	$\frac{1}{2}$	$\frac{3}{2}$	$\pm\frac{1}{2},\pm\frac{3}{2}$	$\frac{4}{3}$	$\pm2,\pm\frac{2}{3}$

The splitting to all allowed m_J levels will vary according to the values of L, S, and J (Table 4.4). Since the selection rule for transitions involving J is $\Delta J=\pm1$, the energy splitting in the upper and lower J-levels will be different. This term-dependent splitting of a J-level to $2J+1$ levels gives rise to anomalous Zeeman effect. The transitions (Figure 4.9) are according to the polarization of the interrogating light discussed earlier.

In another example of Zeeman effect in weak magnetic field, we look at the splitting of the $1^2s_{1/2}$, $2^2s_{1/2}$, $2^2p_{1/2}$, and $2^2p_{3/2}$ term levels of the hydrogen atom in the presence of an external magnetic field of 1 T, which is indeed a weak field. The $1^2s_{1/2}$ and $2^2s_{1/2}$ levels have no fine structure, meaning no **LS**-coupling, so each of them responds to $\mathbf{B}_o$ by simply splitting into two energy levels corresponding to $m_s=\pm1/2$. The magnitude of splitting is $2\gamma_sB_o$ for the $1^2s_{1/2}$, and $(2/3)\gamma_sB_o$ for the $2^2s_{1/2}$ levels, where γ_s is Bohr magneton. But the 2^2p term already has two fine-structures or spin-orbit levels: $2^2p_{1/2}$ and $2^2p_{3/2}$. As per the $2j+1$ selection rule for splitting, the former splits into two ($m_j=\pm1/2$) and the latter into four ($m_j=\pm1/2,\pm3/2$) Zeeman levels. The spacing between the sublevels is $(2/3)\gamma_sB_o$ and $(4/3)\gamma_sB_o$,

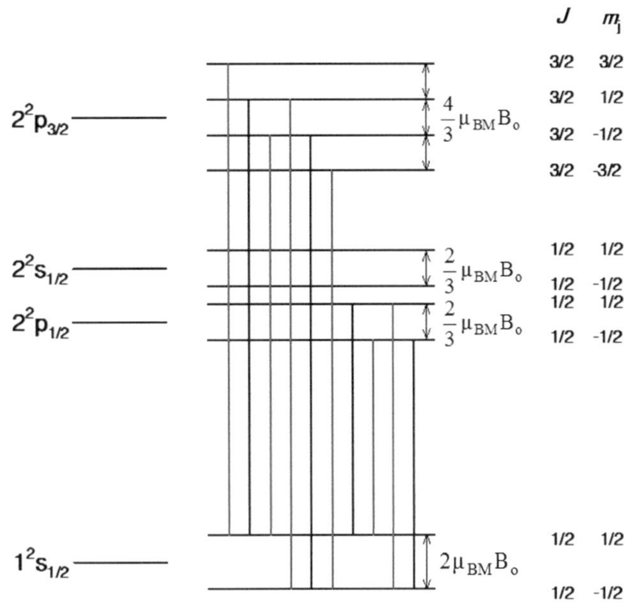

FIGURE 4.10 The J-split Zeeman energy levels for the fine-structure levels $1^2s_{1/2}$, $2^2s_{1/2}$, $2^2p_{1/2}$, and $2^2p_{3/2}$ of the hydrogen atom placed in a weak external magnetic field. The possible transitions between J-split levels are shown by the vertical lines distinguished according to transitions observable by z-polarized light (cyan) and transversely polarized light (black). The anomalous splitting values for these four levels of hydrogen in the presence of 1 T field cover the range from 38 μeV (9190 MHz) to 115 μeV (27810 MHz).

respectively (Figure 4.10). We pay attention to the transitions in the presence of light, keeping in mind that the electric dipole operator is an odd operator. The transition 1s↔2s is forbidden because both have even parity transformations. But the electron is in the ground state 1s, not in 2s. Therefore, the possible allowed transitions are between the 1s and 2p configurations. This situation should be contrasted with that already encountered in the case of the sodium electron (Figure 4.9) which is the outermost $n = 3$ electron that effectively is involved in the 3s ↔ 3p transitions. The 1s ↔ 2p Zeeman transitions of the hydrogen atom in the presence of the magnetic field are shown in Figure 4.10, where the $2^2s_{1/2}$ level is left out from the transitions because of the reason of parity transformation just mentioned.

4.6 ZEEMAN SPLITTING CHANGEOVER FROM WEAK TO STRONG MAGNETIC FIELD

It was mentioned earlier that normal Zeeman effect is observed in the presence of a strong magnetic field. The anomalous Zeeman effect that appears in a weak field swaps to show the normal Zeeman effect spectrum when the field strength turns considerably higher. This transition from the anomalous to the normal Zeeman effect with increasing magnetic field strength is due to the Paschen-Back effect, which implicates a zone of the B_0 field where the spin-orbit coupling and Zeeman splitting energies approach each other. This happens because **l** and **s** are individually decoupled in strong field, resulting in cessation of the **j** vector precession about the field and hence causing the reappearance of the degenerate fine structure.

In the discussion of the **LS** vector model earlier (Figure 4.8), it was mentioned that it is the **j** vector alone that precesses about the weak $\mathbf{B}_0$ field; the **l** and **s** vectors are coupled and hence do not precess individually. Under this condition, j is a good quantum number, but l and s are not. The Hamiltonian does not commute with l and s, so m_l and m_s are also not good quantum numbers. A preferred basis set of the eigenvectors here could be $\{n,l,j,m_j\}$. The eigenvalue calculations have been outlined already. In the absence of the magnetic field, the energy of the eigenstate is just the energy of the degenerate levels corresponding to a given value of j. Only different values of j determine different energies. Now let the $\mathbf{B}_0$ field along the z-direction be turned on and the field strength be increased gradually; at some value of B_0 the spin-orbit interaction energy will be equal to the Zeeman energy.

To appreciate the crossover of fine structure and Zeeman splitting we resort to the vector model introduced earlier (Figure 4.8). When B_0 is weak, the precession of $\boldsymbol{\mu}_j$ about the **j** vector is faster compared to the precession frequency of **j** about $\mathbf{B}_0$, and hence the component of $\boldsymbol{\mu}_j$ which is parallel to **j** (i.e, $\boldsymbol{\mu}_{j\|}$) is the effective time-averaged magnetic dipole. As the B_0 strength increases the precession frequency of **j** about the direction of $\mathbf{B}_0$ picks up, and when this precession frequency approaches the precession frequency of $\boldsymbol{\mu}_j$ about the direction of j the relevance of $\boldsymbol{\mu}_{j\|}$ ceases. Rather, the $\boldsymbol{\mu}_j$ vector itself is the effective magnetic moment. When the frequencies of the two precession motions are equal, the condition $E_{SO} = E_{Zeeman}$ is approached starting from $E_{SO} \gg E_{Zeeman}$. Because E_{SO} arises from the magnetic field intrinsic to the atom but E_{Zeeman} arises from an external field, one can say that the intrinsic precession energy is equal to the extrinsic precession energy at the Paschen-Back crossover. The B_0 strength at which this happens determines the saturation of **LS**-coupling, beyond which **l** and **s** are decoupled.

In the limit of strong B_0, the individual decoupling of **l** and **s** renders L and S good quantum numbers, but not J anymore. The **l** and **s** vectors will now precess individually about $\mathbf{B}_0$ (Figure 4.11) because the energy of both orbital and spin magnetic moments in the $\mathbf{B}_0$ field is larger than the spin-orbit interaction energy E_{SO}. The time-average of $\mathbf{L}\cdot\mathbf{S}$ is now given by the product of the time-average components of **l** and **s** in the direction of $\mathbf{B}_0$. The Zeeman Hamiltonian is then

$$\mathcal{H}_{Zeeman} \approx -\frac{e\hbar}{2m_e c}\mathbf{B}_0\left(l_z + 2s_z\right) + \frac{1}{2m_e^2c^2}\frac{1}{r}\frac{\partial V}{\partial r}\left(m_l m_s\right), \tag{4.59}$$

the second term of which can be compared with equation (4.50). The replacement of $\mathbf{L}\cdot\mathbf{S}$ with $m_l m_s$ here occurs due to the fact that the respective time-average components of **l** and **s** are now parallel to the field, even though the spin-orbit coupling term is independent of $\mathbf{B}_0$. The Zeeman energy is

$$E_{Zeeman} = E\left(m_l,m_s\right) = -\frac{e\hbar}{2m_e c}\left(m_l + 2m_s\right) + \frac{1}{2m_e^2c^2}\frac{1}{r}\frac{\partial V}{\partial r}\left(m_l m_s\right). \tag{4.60}$$

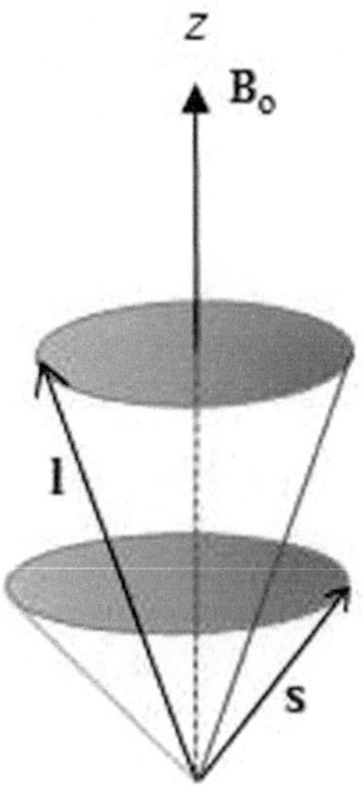

FIGURE 4.11 Individual precession of **l** and **s** vectors about a strong magnetic field. As discussed in the text, *j* is not a good quantum number under this decoupling condition, but *l* and *s* are.

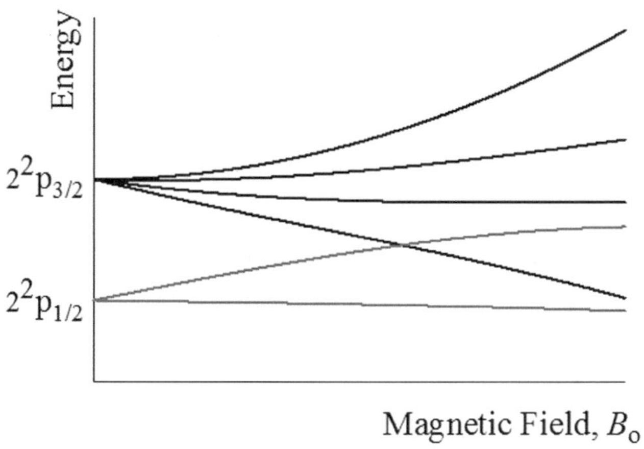

FIGURE 4.12 The pattern of Zeeman splitting for the 2p term.

An example of the splitting of $2^2P_{1/2}$ and $2^2P_{3/2}$ levels, as for sodium or hydrogen atom, is qualitatively shown in Figure 4.12.

4.7 ELECTRON-NUCLEAR HYPERFINE INTERACTION

Consider the hydrogen atom. By virtue of being a spinning charged particle the nucleus (proton) behaves as a tiny magnet that produces a tiny magnetic field. The field can be characterized by the vector potential **A** given by

$$\mathbf{A} = \frac{1}{r^3}\boldsymbol{\mu}_I \times \mathbf{r} \tag{4.61}$$

in which

$$\boldsymbol{\mu}_I = g_N \frac{e}{2m_N c}\mathbf{I}$$

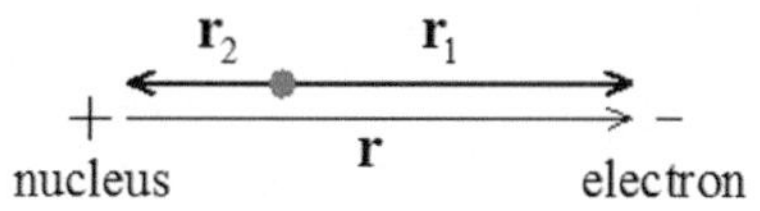

FIGURE 4.13 Relative coordinates of electron and nucleus.

is the nuclear magnetic dipole moment with g_N the nuclear *g*-factor, m_N the nuclear mass, and **I** the nuclear angular momentum vector. The static vector **r** defines relative coordinates of the electron and the nucleus for locating them with reference to a fixed point (Figure 4.13).

If m_e is the mass of the electron, then

$$m_e\mathbf{r}_1 + m_N\mathbf{r}_2 = 0$$

$$\mathbf{r}_1 = \frac{m_N}{m_e + m_N}\mathbf{r}$$

$$\mathbf{r}_2 = -\frac{m_e}{m_e + m_N}\mathbf{r}.$$

The length of the vector **r** is essentially the electron-nuclear distance in the hydrogen atom. The magnetic vector field produced by the nucleus is then the curl of **A**

$$\mathbf{B} = \nabla \times \mathbf{A}. \tag{4.62}$$

We will use the Maxwell vector relation (equation (1.7)) by replacing **H** with **B** for the magnetic vector to write

$$\nabla \times \mathbf{B} = \frac{1}{c}\frac{d\mathbf{D}}{dt} + \frac{4\pi}{c}\mathbf{j},$$

where **D** and **j** are electric displacement and current density, respectively. The equation of motion for the vector potential **A** can be written as

$$\nabla \times \mathbf{B} = \frac{1}{c}\frac{d\mathbf{D}}{dt} + \frac{4\pi}{c}\mathbf{j} - 4\pi\boldsymbol{\mu}_I \times \nabla\delta\mathbf{r}. \tag{4.63}$$

Assume there is no time-varying electric field so that the first two terms of this equation vanish

$$\nabla \times \mathbf{B} = -4\pi\boldsymbol{\mu}_I \times \nabla\delta\mathbf{r}.$$

To solve for **A** in equation (4.62), one requires Coulomb gauge (see Box 4.4) which enables writing

$$-\Delta\mathbf{A} = -4\pi\boldsymbol{\mu}_I \times \nabla\delta\mathbf{r}.$$

Using $\Delta\frac{1}{r^3} = -4\pi\delta\mathbf{r}$, the vector potential as a function of $\mathbf{r}_1$ is obtained as

$$A(\mathbf{r}_1) = -\boldsymbol{\mu}_I \times \nabla\frac{1}{r_1} = -\boldsymbol{\mu}_I \times \frac{\mathbf{r}}{|\mathbf{r}|^2}.$$

Since $\mathbf{B}(\mathbf{r}_1) = \nabla \times \mathbf{A}$

$$\mathbf{B}(\mathbf{r}_1) = -\boldsymbol{\mu}_I \frac{1}{r_1} = -\boldsymbol{\mu}_I \Delta \frac{1}{|\mathbf{r}|} + \nabla(\boldsymbol{\mu}_I \cdot \nabla)\frac{1}{|\mathbf{r}|}$$

$$= \frac{2}{3}\boldsymbol{\mu}_I \Delta = -\boldsymbol{\mu}_I \Delta \frac{1}{|\mathbf{r}|} + \left[\nabla(\boldsymbol{\mu}_I \cdot \nabla) - \frac{1}{3}\boldsymbol{\mu}_I \cdot \nabla\right]\frac{1}{|\mathbf{r}|}, \quad (4.64)$$

which has been written using the approximation of isotropic distribution of the magnetic dipole field outside the nucleus. Differentiation of equation (4.64) above yields

$$\mathbf{B}(\mathbf{r}_1) = -\frac{8\pi}{3}\boldsymbol{\mu}_I \delta\mathbf{r} + \frac{1}{r^3}\left[3\frac{\mathbf{r}}{r}\frac{\boldsymbol{\mu}_I \cdot \mathbf{r}}{r} - \boldsymbol{\mu}_I\right].$$

The Hamiltonian for the interaction energy of the electron and nuclear spin magnetic moments, $\boldsymbol{\mu}_s$ and $\boldsymbol{\mu}_I$, is given by

$$\mathcal{H} = -\boldsymbol{\mu}_I \cdot \mathbf{B}(\mathbf{r}_1) = \frac{8\pi}{3}\boldsymbol{\mu}_s \cdot \boldsymbol{\mu}_I \delta(\mathbf{r}) - \frac{1}{r^3}\left[3(\boldsymbol{\mu}_s \cdot \hat{\mathbf{r}})(\boldsymbol{\mu}_I \cdot \hat{\mathbf{r}}) - \boldsymbol{\mu}_s \cdot \boldsymbol{\mu}_I\right], \quad (4.65)$$

where $\hat{\mathbf{r}}$ should be read as a unit vector along the electron-nuclear separation. This Hamiltonian also needs to contain a term arising from the interaction of $\boldsymbol{\mu}_I$ with the magnetic field produced by the orbital motion of the electron. This field at the nuclear site is simply

$$\mathbf{B}(\mathbf{r}_2) = \frac{\mathbf{L}}{m_e r^3}, \quad (4.66)$$

and the energy of interaction of the nuclear magnetic dipole with this magnetic field is

$$E_{eN} = -\frac{1}{m_e r^3}\mathbf{L} \cdot \boldsymbol{\mu}_I. \quad (4.67)$$

When this term is added to equation (4.65), the total hyperfine Hamiltonian in SI unit becomes

$$\mathcal{H}_{hf} = -\frac{\mu_o}{4\pi}\left\{\frac{1}{m_e r^3}\mathbf{L} \cdot \boldsymbol{\mu}_I + \frac{8\pi}{3}\boldsymbol{\mu}_s \cdot \boldsymbol{\mu}_I \delta(\mathbf{r}) + \frac{1}{r^3}\left[3(\boldsymbol{\mu}_s \cdot \hat{\mathbf{r}})(\boldsymbol{\mu}_I \cdot \hat{\mathbf{r}}) - \boldsymbol{\mu}_s \cdot \boldsymbol{\mu}_I\right]\right\}, \quad (4.68)$$

where the nuclear and electron spin magnetic moments are

$$\boldsymbol{\mu}_I = -g_N \mu_N \frac{\mathrm{I}}{\hbar} \quad \text{and} \quad \boldsymbol{\mu}_s = -g_e \mu_e \frac{\mathrm{s}}{\hbar}. \quad (4.69)$$

Of the three terms in the electron-nuclear hyperfine Hamiltonian (equation (4.68)) the first one is the interaction of the magnetic field due to electron orbital motion with the nuclear magnetic moment. The second term arises from the interaction of the electron spin magnetic moment with the nuclear magnetic field in the interior of the nucleus which highlights the importance of the finite size of the nucleus. This interaction is known as Fermi contact interaction, which requires overlap of the electron and nuclear wavefunctions. The third corresponds to the dipole-dipole interaction between the electron spin magnetic moment and the nuclear magnetic field.

To look at the expectation value of the Hamiltonian we consider the example of hyperfine splitting in the 1s state of hydrogen atom. With the relevant quantum numbers n, l, s, and I, the vector basis could be $\{|u_i\rangle\}$ where $i = n, l, m_l, m_s$, and m_I, where m_s and $m_I = \pm\frac{1}{2}$.

But $l = 0$ for ns configurations, which causes the $\mathbf{L}$-containing first term in the Hamiltonian (equation (4.68)) to go to zero. The third term involving dipole-dipole interaction also vanishes, which is due to the fact that an integral containing the product of three spherical harmonics will have a non-zero value only when $l, l' \geq 1$. The Hamiltonian then reduces to contain the Fermi contact term alone

$$\mathcal{H}_{hf} = -\frac{2\mu_o}{3}\boldsymbol{\mu}_s \cdot \boldsymbol{\mu}_I \delta(\mathbf{r}) = -\frac{2}{3\mu_o c^2}\boldsymbol{\mu}_s \cdot \boldsymbol{\mu}_I \delta(\mathbf{r}), \quad (4.70)$$

where we have used $\varepsilon_o = \frac{1}{\mu_o c^2}$.

The expectation value is obtained as

$$E^{(1)} = \int \psi^*_{n,l,m_l,m_s,m_I}\left|-\frac{2}{3\mu_o c^2}\boldsymbol{\mu}_s \cdot \boldsymbol{\mu}_I \delta(\mathbf{r})\right|\psi_{n,l,m_l,m_s,m_I}\, d\tau. \quad (4.71)$$

By writing $\boldsymbol{\mu}_s$ and $\boldsymbol{\mu}_I$ with electron and nuclear Bohr magnetons in SI unit as

$$\mu_e = \frac{e\hbar}{2m_e} \quad \text{and} \quad \mu_N = \frac{q\hbar}{2m_N},$$

one obtains

$$\boldsymbol{\mu}_s \cong g_e \mu_e \frac{\mathbf{S}}{\hbar} \quad \text{and} \quad \boldsymbol{\mu}_I \cong g_N \mu_N \frac{\mathbf{I}}{\hbar}$$

Solving equation (4.71) further

$$E^{(1)} = \frac{2}{3}\frac{e^2}{\varepsilon_o}\frac{\hbar^2 g_e g_N}{4c^2 m_e m_N \hbar^2}\int \psi^*_{n,l,m_l}|\delta(\mathbf{r})|\psi_{n,l,m_l}\,d\tau \int \psi^*_{m_s,m_I}\mathbf{S}\cdot\mathbf{I}\,\psi_{m_s,m_I}\,d\tau$$
$$= \frac{2}{3}\frac{e^2}{\varepsilon_o}\frac{\hbar^2 g_e g_N}{4c^2 m_e m_N \hbar^2}\left|\psi_{1,0,0}(0)\right|^2 \int \psi^*_{m_s,m_I}\mathbf{S}\cdot\mathbf{I}\,\psi_{m_s,m_I}\,d\tau$$
$$= \frac{2}{3\hbar^2}g_e g_N \frac{m_e^2}{m_N}c^2\alpha^2\left(1 + \frac{m_e}{m_N}\right)^{-3}\int \psi^*_{m_s,m_I}\mathbf{S}\cdot\mathbf{I}\,\psi_{m_s,m_I}\,d\tau, \quad (4.72)$$

in which α is the fine structure constant and the factor $(1 + m_e/m_N)^{-3}$ appears because of the use of reduced mass (μ) in the radial function. The Bohr radius then contains μ, not m_e. The energy expression above is often written as

$$E = \mathcal{A}\int \psi^*_{m_s,m_I} \mathbf{S}\cdot\mathbf{I}\,\psi_{m_s,m_I}\, d\tau, \tag{4.73}$$

in which $\mathcal{A}$ represents the front factor

$$\mathcal{A} = \frac{2}{3\hbar^2} g_e g_N \frac{m_e^2}{m_N} c^2\alpha^2 \left(1+\frac{m_e}{m_N}\right)^{-3} \tag{4.74}$$

It is convenient to introduce the total angular momentum **F** arising from electron orbital, electron spin, and nuclear spin motions

$$\mathbf{F} = \mathbf{L}+\mathbf{S}+\mathbf{I}.$$

For the 1s state of the hydrogen atom $L = 0$, $S = I = 1/2$, so

$$\mathbf{F} = \mathbf{S}+\mathbf{I} = \begin{cases} 0 \\ 1 \end{cases}$$

$$\mathbf{F}^2 = \mathbf{S}^2 + \mathbf{I}^2 + 2\mathbf{S}\cdot\mathbf{I}$$

$$\mathbf{S}\cdot\mathbf{I} = \frac{1}{2}\left(\mathbf{F}^2 - \mathbf{I}^2 - \mathbf{S}^2\right).$$

The time dependent Schrödinger equation is now obtained as

$$\frac{\mathcal{A}}{2}\left(\mathbf{F}^2 - \mathbf{I}^2 - \mathbf{S}^2\right)\psi_F = \frac{\mathcal{A}\hbar^2}{2}\left[F(F+1) - I(I+1) - S(S+1)\right]\psi_F, \tag{4.75}$$

where we have used the fact that the eigenvalues for the square of the angular momenta $\mathbf{F}^2$, $\mathbf{I}^2$, and $\mathbf{S}^2$ are $F(F+1)$, $I(I+1)$, and $S(S+1)$, respectively. For $S = I = 1/2$ and $F = 0,1$ we get the eigenvalues

$$E = \begin{cases} \mathcal{A}\dfrac{\hbar^2}{4} & F = 1 \\ -\mathcal{A}\dfrac{3\hbar^2}{4} & F = 0 \end{cases}$$

The hyperfine interaction energy can also be given by the electron and nuclear spin matrices in the direct product space. Using the matrix form of the $\mathbf{S}\cdot\mathbf{I}$ operator (Box 4.5) we have

$$\mathcal{A}\mathbf{S}\cdot\mathbf{I} = \left(S_x I_x + S_y I_y + S_z I_z\right) = \mathcal{A}\frac{\hbar^2}{4}\begin{pmatrix} 1 & 0 & 0 & 0 \\ 0 & -1 & 2 & 0 \\ 0 & 2 & -1 & 0 \\ 0 & 0 & 0 & 1 \end{pmatrix}, \tag{4.76}$$

and the characteristic equation of the operator matrix is obtained as

$$\begin{vmatrix} 1-\lambda & 0 & 0 & 0 \\ 0 & -(1-\lambda) & 2 & 0 \\ 0 & 2 & -(1+\lambda) & 0 \\ 0 & 0 & 0 & 1-\lambda \end{vmatrix} = 0. \tag{4.77}$$

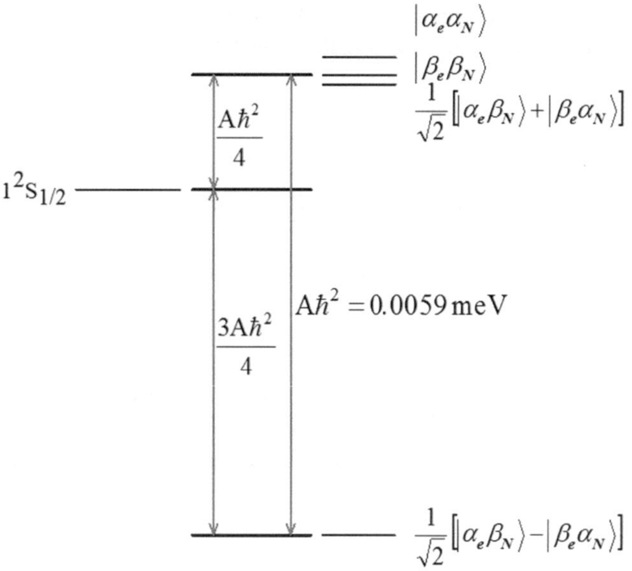

FIGURE 4.14 Hyperfine splitting of the 1s state of hydrogen atom.

The eigenvalues are $\lambda = +1$, which occurs three times, and $\lambda = -3$, occuring just once, implying three-fold degeneracy of the eigenstate of energy $\mathcal{A}\hbar^2/4$. We will see soon that this degeneracy can be broken by using an external magnetic field.

A simple evaluation of $\mathcal{A}$ that works out to ~5×10^{-2} in the unit of $B_o\gamma_e/2$, where $B_o = 0.5\,\text{T}$ and the gyromagnetic ratio of the electron $\gamma_e = -2.7\times10^4$ MHz T^{-1}, gives $\mathcal{A}\hbar^2 = \Delta E \cong 0.0059$ meV~1420.41 MHz (Figure 4.14). This is the $F = 0 \leftrightarrow F = 1$ emission line of hydrogen gas at 21 cm wavelength probed widely in radio astronomical studies.

4.8 ZEEMAN SPLITTING OF HYPERFINE ENERGY LEVELS

In the presence of an external field $\mathbf{B}_o$ the Hamiltonian takes the form

$$\mathcal{H} = \mathcal{H}_{hf} + \mathcal{H}_{\text{Zeeman}}, \tag{4.78}$$

in which the structure of the Zeeman Hamiltonian is already familiar. Since the hyperfine phenomenon involves **L**, **S**, and **I** vectors

$$\begin{aligned}\mathcal{H}_{\text{Zeeman}} &= -\left(\boldsymbol{\mu}_l + \boldsymbol{\mu}_s + \boldsymbol{\mu}_I\right)\cdot\mathbf{B}_o \\ &= -\left(\gamma_l\mathbf{L} + \gamma_s\mathbf{S} + \gamma_I\mathbf{I}\right)\cdot\mathbf{B}_o \\ &= -\left(\mathbf{L} + 2\mathbf{S} + \gamma_I\mathbf{I}\right)\cdot\mathbf{B}_o\end{aligned} \tag{4.79}$$

where

$$\gamma_e = g_l\frac{e\hbar}{2m_e c},\; \gamma_s = g_s\frac{e\hbar}{2m_e c},\; \text{and}\; \gamma_I = g_N\frac{e\hbar}{2m_N c}$$

are corresponding gyromagnetic ratios with g-factors $g_l = 1$, $g_s \approx 2$, and g_N ~ 5.58 (in nuclear magneton unit). The Bohr and nuclear magnetons are units for magnetic moments of electron orbital and spin, and nuclear spin, respectively. The Hamiltonian above can be written in terms of precession frequencies of **L+S** and **I** vectors. Using the general formulas of magnetic moments to angular momentum

$$\mu_l = \gamma_e \mathbf{L},\ \mu_s = \gamma_s \mathbf{S},\ \mu_I = \gamma_I \mathbf{I},\ \text{and}\ \omega_o = \gamma B_o,$$

where ω_o is the angular frequency of Larmor precession of the respective vectors about $\mathbf{B}_o$, the Hamiltonian is written as

$$\mathcal{H}_{\text{Zeeman}} = -\omega_{o,e}(L+2S) + \omega_{o,N} I. \quad (4.80)$$

By focusing on the 1s state of the hydrogen atom, the Hamiltonian can be considerably simplified without any loss of generality. In the 1s state, $l = 0$. The Larmor precession of the nuclear spin vector **I** is ~1836 times slower than that for the electron spin vector. These two arguments allow for reducing the Hamiltonian to

$$\mathcal{H}_{\text{Zeeman}} = -\omega_{o,e} 2S,$$

so that equation (4.78) can be written as

$$\mathcal{H} = \mathcal{A}\mathbf{S}\cdot\mathbf{I} - \omega_{o,e} 2S$$

where the operator form of $\mathcal{H}_{hf}$ is shown explicitly (see equations (4.73) and (4.76)). Retaining $\mathcal{H}_{hf}$ in the Hamiltonian is obvious because it is the magnetic field splitting of the hyperfine energy levels that we are interested in. But to note is the linear increase in Zeeman energy as the strength of B_o increases. When B_o is weak, say within ~500 G, the Zeeman splitting energy is less than the hyperfine splitting energy ($\mathcal{A}\hbar^2$), but in the strong-field limit the former exceeds the latter.

$$\hbar\omega_{o,e} \ll \mathcal{A}\hbar^2 \text{ weak field limit}$$

$$\hbar\omega_{o,e} \gg \mathcal{A}\hbar^2 \text{ strong field limit}$$

This distinction of magnetic field limit is necessary because F and m_F are good quantum numbers only in weak B_o where the total electronic angular momentum J and the nuclear spin angular momentum I remain coupled. But the coupling of J and I will break down as B_o turns stronger, and then it is m_J and m_I that are good quantum numbers, not F and m_F. This is analogous to the argument provided earlier regarding the goodness of J, L, and S quantum numbers with respect to **LS**-coupling status (see section 4.6). The vector precessions shown in Figure 4.15 distinguish the cases of persistence and breakdown of **JI** coupling.

4.8.1 Zeeman Splitting of Hyperfine States in Weak Magnetic Field

Weak-field Zeeman splitting of hyperfine states is treated as a small perturbation to the hyperfine energy for which the eigenstates of the hyperfine interaction need to be considered. In such a case, the eigenvalue of the Zeeman interaction can simply be added to the eigenvalue of hyperfine interaction.

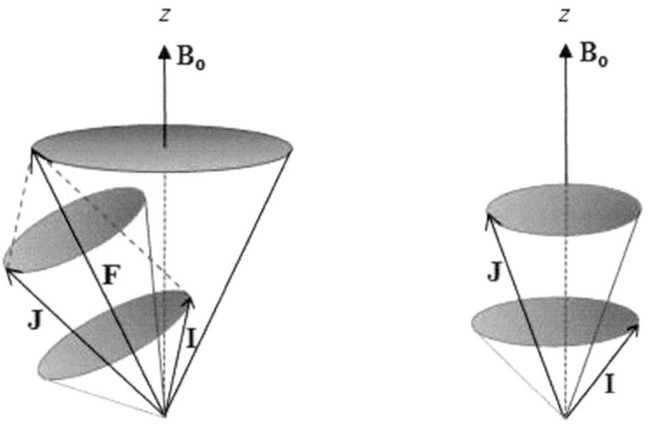

FIGURE 4.15 Vector diagram to show the breakdown of **J** and **I** coupling as the magnetic field becomes stronger. *Left*, in a weak B_o field **JI** coupling results in the precession of the **F** vector about $\mathbf{B}_o$, and **J** and **I** individually precess about **F**. *Right*, in a strong B_o field **F** does not exist due to decoupling, and **J** and **I** individually precess about $\mathbf{B}_o$.

Coming back to the 1s ground state of hydrogen atom we find no spin-orbit coupling because $l = 0$ and hence **J**=0. But the prevalence of $\mathbf{F} = \mathbf{S} + \mathbf{I}$ allows one to work with any of the two bases $\{|F, m_F\rangle\}$ and $\{|m_s, m_I\rangle\}$ in the weak limit of B_o. Let us use the $\{|F, m_F\rangle\}$ basis in which the hyperfine perturbation is diagonal. First, we find the matrix representation of the operator with the $\mathbf{B}_o$ field along the z-direction

$$\mathcal{H} = \mathcal{A}\hbar^2 + 2\omega_{o,e} 2S_z.$$

Because $\mathcal{A}\hbar^2$ is a constant, only the S_z operator needs to be considered. By reviewing the eigenstates corresponding to $F = 1$ and $F = 0$ in the hyperfine splitting (Figure 4.14; see also Box 4.6) the following eigenvector equations are obtained (see Cohen-Tannoudji et al., 1977):

$$\begin{aligned} S_z|F=1, m_F=1\rangle &= \frac{\hbar}{2}|F=1, m_F=1\rangle \\ S_z|F=1, m_F=0\rangle &= \frac{\hbar}{2}|F=0, m_F=0\rangle \\ S_z|F=1, m_F=-1\rangle &= -\frac{\hbar}{2}|F=1, m_F=-1\rangle \\ S_z|F=0, m_F=0\rangle &= \frac{\hbar}{2}|F=1, m_F=0\rangle, \end{aligned} \quad (4.81)$$

where the eigenstates have been arranged as they appear from top to bottom in the hyperfine splitting shown in Figure 4.14. From these equations we get the matrix of the S_z operator as

$$2\omega_o S_z = \hbar\omega_o \begin{pmatrix} 1 & 0 & 0 & 0 \\ 0 & 0 & 0 & 1 \\ 0 & 0 & -1 & 0 \\ 0 & 1 & 0 & 0 \end{pmatrix}. \quad (4.82)$$

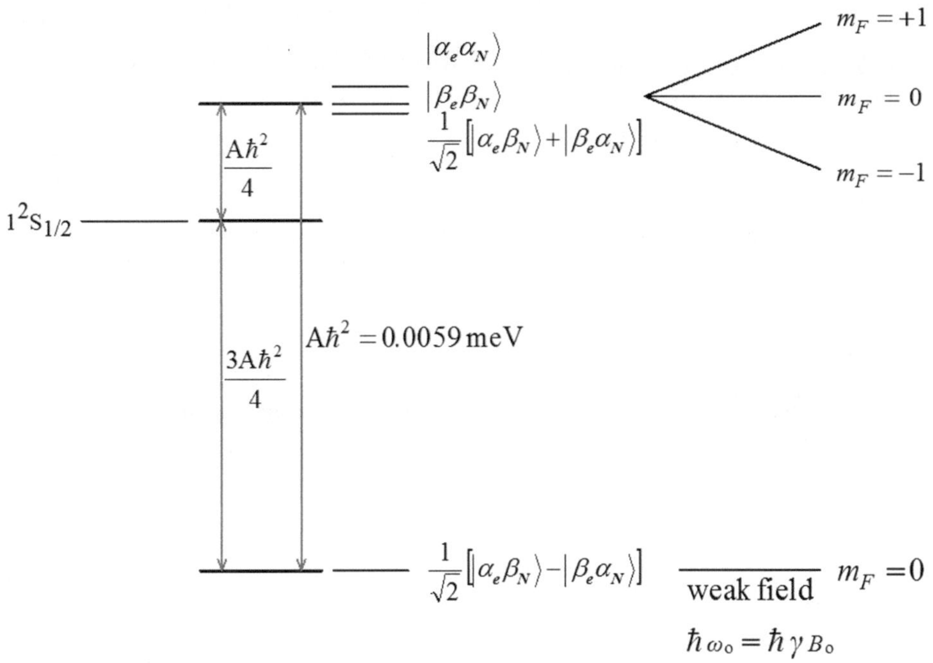

FIGURE 4.16 Isotropic hyperfine interaction and Zeeman splitting of hyperfine levels for the 1*s* ground state of hydrogen atom in weak B_o.

As a cross-check, we may multiply it out with the matrix for the basis vectors

$$\begin{pmatrix} 1 & 0 & 0 & 0 \\ 0 & 0 & 0 & 1 \\ 0 & 0 & -1 & 0 \\ 0 & 1 & 0 & 0 \end{pmatrix}\begin{pmatrix} 1 & 1 \\ 1 & 0 \\ 1 & -1 \\ 0 & 0 \end{pmatrix}=\begin{pmatrix} 1 & 1 \\ 0 & 0 \\ -1 & 1 \\ 1 & 0 \end{pmatrix}.$$

The diagonals of the 3×3 submatrix corresponding to the first three rows and columns in equation (4.82) correspond to the eigenvalues for the $F = 1$ level and the last 'zero' diagonal is that for $F = 0$ level. The ordering of the eigenstates is the same as in Figure 4.14. In the limit of weak B_o, the splitting of the hyperfine lines for the 1*s* state of hydrogen atom is shown in Figure 4.16.

4.8.2 Hyperfine States of Hydrogen Atom in Strong Magnetic Field

Here, hyperfine interaction is supposedly a small perturbation, and the idea is to use the eigenvectors and eigenvalues of $\mathcal{H}_{\text{Zeeman}}$ to calculate the expectation value of $\mathcal{H}_{\text{hf}}$. The eigenvectors and eigenvalues of $\mathcal{H}_{\text{Zeeman}}$ are given by the eigenvalue equation

$$\begin{aligned} 2\omega_o S_z|m_s,m_I\rangle &= 2m_s\hbar\omega_o|m_s,m_I\rangle \\ &= \pm\hbar\omega_o|m_s,m_I\rangle. \end{aligned} \tag{4.83}$$

Because $m_s = \pm 1/2$, the two eigenvalues are $+\hbar\omega_o$ and $-\hbar\omega_o$. Also, $m_I = \pm 1/2$ corresponding to $|\alpha\rangle$ and $|\beta\rangle$, so that

$$\begin{aligned} 2\omega_o S_z|\alpha_e\alpha_N\rangle &= \hbar\omega_o|\alpha_e\alpha_N\rangle \\ 2\omega_o S_z|\alpha_e\beta_N\rangle &= \hbar\omega_o|\alpha_e\beta_N\rangle \\ 2\omega_o S_z|\beta_e\alpha_N\rangle &= -\hbar\omega_o|\beta_e\alpha_N\rangle \\ 2\omega_o S_z|\beta_e\beta_N\rangle &= -\hbar\omega_o|\beta_e\beta_N\rangle. \end{aligned} \tag{4.84}$$

Having obtained the eigenvalues and eigenvectors for the Zeeman term we should find out the expectation values of $\mathcal{H}_{\text{hf}}$ by calculating the matrix elements

$$m_s,m_I|\mathcal{A}\mathbf{S}\cdot\mathbf{I}|m_s,m_I\rangle = \mathcal{A}m_s,m_I|\mathbf{S}\cdot\mathbf{I}|m_s,m_I\rangle.$$

We have already found the matrix form (see Box 4.5), which is

$$\mathbf{S}\cdot\mathbf{I}=\frac{\hbar^2}{4}\begin{pmatrix} 1 & 0 & 0 & 0 \\ 0 & -1 & 2 & 0 \\ 0 & 2 & -1 & 0 \\ 0 & 0 & 0 & 1 \end{pmatrix}, \tag{4.85}$$

and whose diagonals provide the eigenvalues listed in Table 4.5.

The eigenstates in the strong-field limit are shown in Figure 4.17 where $|\alpha_e\alpha_N\rangle$ and $|\alpha_e\beta_N\rangle$ are parallel with a slope of +1, and $|\beta_e\alpha_N\rangle$ and $|\beta_e\beta_N\rangle$ are parallel with a slope of −1.

4.9 STARK EFFECT

The splitting of electronic energy levels by external electric field was first studied by Johannes Stark in Germany and

TABLE 4.5
Eigenstates and Corresponding Eigenvalues of Hyperfine Interaction for the 1s Ground State of Hydrogen Atom in Strong Magnetic Field

Eigenstate	Eigenvalue
$\lvert\alpha_e\alpha_N\rangle$	$\hbar\omega_o + \mathcal{A}\frac{\hbar^2}{4}$
$\lvert\alpha_e\beta_N\rangle$	$\hbar\omega_o - \mathcal{A}\frac{\hbar^2}{4}$
$\lvert\beta_e\alpha_N\rangle$	$-\hbar\omega_o - \mathcal{A}\frac{\hbar^2}{4}$
$\lvert\beta_e\beta_N\rangle$	$-\hbar\omega_o + \mathcal{A}\frac{\hbar^2}{4}$

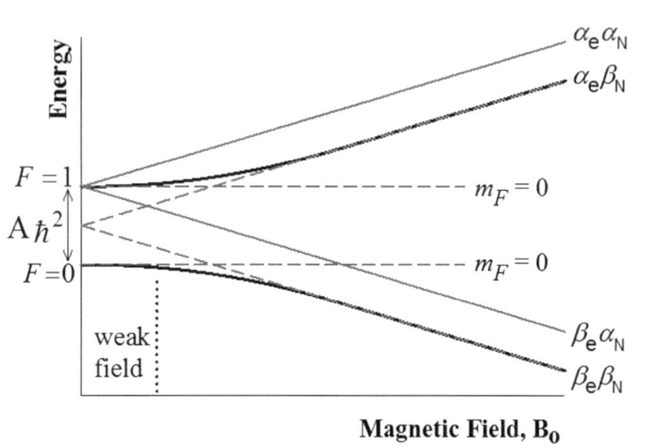

FIGURE 4.17 Zeeman splitting of the 1*s* ground state of hydrogen atom distinguishing the strong field effect (brown dashed lines). The solid lines represent calculated results assuming that the magnetic field much larger than ~500 Gauss falls in the strong field limit.

hence is called the Stark effect, which may also be called the electrochromic effect if one would not like to associate Stark's name with the effect because of his endorsement of Hitler's speeches. But we would like to call it after Stark because of the extraordinary influence his work has had. Quite similar to Stern-Gerlach's, the basic Stark experimental setup consists of an inhomogeneous electric field, say along *z*-direction, which deflects an atom beam passing through it. The electric field has no effect on electron and nuclear spins, so the energy levels of concern are those that are involved in electronic dipole transitions. Indeed, it is the electric dipole interaction that underlies Stark splitting. But the Coulomb electrostatic interaction energy which constitutes the basic atomic Hamiltonian is much stronger than electric dipole interaction, so the Stark effect is essentially treated as a perturbation to the Coulombic Hamiltonian. The order of perturbation applicable depends on the parity and degeneracy of eigenstates involved in transitions. In this discussion we will briefly consider 1*s* and 2*s* states of hydrogen atoms to examine the second-order quadratic and first-order linear perturbations.

4.9.1 Hydrogen Atom in External Electric Field

The Stark Hamiltonian (see Box 4.7) for a system of *i* electrons in an external electric field **E** is

$$\mathcal{H}_{\text{Stark}} = -\mathbf{E}\cdot\sum_i e\mathbf{r}_i, \tag{4.86}$$

where *e* is the charge of the electron and **r** is electron-nuclear position vector, so *e***r** is the electric dipole moment. If the field is along the *z*-axis, the Hamiltonian is

$$\mathcal{H}_{\text{Stark}} = -\mathcal{E}_z eZ, \tag{4.87}$$

which is now added to the unperturbed Coulomb potential Hamiltonian to obtain

$$\mathcal{H} = -\left(\frac{\hbar^2}{2m_e}\nabla^2 + \frac{Ze^2}{r}\right) - \mathcal{E}_z eZ \tag{4.88}$$

where $Z = \cos\theta$, and $\mathcal{E}_z eZ$ is the perturbation potential whose effect on the eigenstates and energies is to examine. Note that *Z* used here is not the atomic number, but is a spherical harmonic.

4.9.2 Effect on the *n* = 1 Level

The ground state in the absence of **E** is just the state with $n = 1$, $l = 0$, and $m = 0$ (ψ_{100}), which can be used to obtain the expectation value of the perturbation potential. The expectation value essentially corresponds to the first-order perturbation correction of the total energy of the atom

$$E^{(1)}_{100} = e\mathcal{E}_z\int Z\left|\psi_{100}\right|^2 r^2 dr\sin\theta d\theta. \tag{4.89}$$

This integral irrespective of the atom type and the field-free eigenstates involved goes to zero. For example,

$$E^{(1)}_{nlm} = e\mathcal{E}_z\int Z\psi^*_{nlm}\psi_{nlm}r^2 dr\sin\theta d\theta = 0. \tag{4.90}$$

The zero value of the integral is obviously due to required parity conditions (see Box 4.3). The state ψ_{100} or the 1*s* orbital is spherically symmetric that gives even parity. Regarding the parity of the operator $Z = \cos\theta$, which is a spherical harmonic, one may note that an inversion operation on spherical harmonics gives rise to a change of sign

$$Y_{lm}(\theta,\phi) \to (-1)^l\, Y_{lm}(\theta,\phi). \tag{4.91}$$

This change of even to odd parity causes the integral to go to zero, revealing that there is no first-order energy correction to the ground state of hydrogen. One then seeks for the second-order perturbation energy that involves the ground state ψ_{100} and other ψ_{nlm} states,

$$E^{(2)}_{nlm} = e^2\mathcal{E}_z^2\sum_{100\neq nlm}\frac{\left|\int Z\psi^*_{nlm}\psi_{nlm}r^2 dr\sin\theta d\theta\right|^2}{E^{(0)}_1 - E^{(0)}_n}, \tag{4.92}$$

in which energies $E_1^{(0)}$ and $E_n^{(0)}$ in the denominator are obtained from the eigenvalue equation in the absence of an external field (equation (4.5))

$$\mathcal{H}_o \psi_{nlm} = E_n^{(0)} \psi_{nlm},$$

where $E_n^{(0)} = -1/n^2 \times 13.53$ eV as discussed earlier. Since $E_{100}^{(2)}$ is the sum of expectation values of Z between the ground state and all other states, it is unlikely that $E_{100}^{(2)}$ will vanish because the required parity condition will be met with some of the connected states, if not all. Another apparent concern is the magnitude of the denominator. Smaller values of $E_1^{(0)} - E_n^{(0)}$ that arise if the states are very close in energy will yield very large perturbation energy. Calculations are then done using degenerate state perturbation theory, which we will use shortly in the discussion of linear Stark effect for the 2s state of hydrogen. But for the 1s state now $E_1^{(0)} - E_n^{(0)}$ (~31 eV for $n = 2$) is large.

The angular integral in equation (4.92) can be solved with some effort. Writing the integral as

$$\int R_{nl}^* r R_{10} r^2 dr Y_{lm}^* \cos\theta Y_{00} \sin\theta d\theta \tag{4.93}$$

with $\cos\theta Y_{00} = 1/\sqrt{4\pi}\,\cos\theta$, and using $Y_{10} = \sqrt{3/4\pi}\,\cos\theta$, we can write $Z = \cos\theta = \dfrac{Y_{10}}{\sqrt{3/4\pi}}$.

One can now invoke the orthogonality condition for associated Legendre functions, i.e.,

$$\int_0^\pi P_{lm}^* P_{l'm} \sin\theta d\theta = \delta_{ll'}$$

to reduce the integral (equation (4.93)) into a radial integral

$$\langle Z \rangle = \frac{\delta_{l1}\delta_{m0}}{\sqrt{3}} \int_0^\infty R_{nl}^* r R_{10} r^2 dr,$$

and then use the Gordon formula (equation (4.25)) to evaluate it. The numerator of equation (4.92) is obtained as

$$|\langle Z \rangle|^2 = \frac{\delta_{l0}\delta_{m0}}{3} \frac{2^8 n^7 (n-1)^{2n-5}}{(n+1)^{2n+5}} a_0^2 = f(n) a_0^2 \delta_{l0}\delta_{m0}, \tag{4.94}$$

where a_0 is Bohr radius. If the denominator in equation (4.92) is also expressed in terms of Bohr radius then

$$\begin{aligned} E_{100}^{(2)} &= e^2 \mathcal{E}_z^2 \sum_{n=2}^\infty \frac{f(n) a_0^2}{\dfrac{-e^2}{2a_0} + \dfrac{e^2}{2n^2 a_0}} \\ &= -2a_0^3 \mathcal{E}_z^2 \sum_{n=2}^\infty \frac{n^2 f(n)}{n^2 - 1} \\ &= -2a_0^3 \mathcal{E}_z^2 [0.74 + 0.10 + 0.03 + 0.01 + \cdots] \\ &\approx -1.8 a_0^3 \mathcal{E}_z^2 \end{aligned} \tag{4.95}$$

The negative sign that originates from the denominator $E_1^{(0)} - E_n^{(0)}$ indicates lowering of the ground state due to perturbation. From the second-order energy correction it should be obvious that the electric field first induces a dipole moment in the hydrogen, and the induced dipole then interacts with the same electric field to lower the ground state.

4.9.3 Effect on the $n = 2$ Level

The $n = 2$ level corresponds to the $2s2p$ configuration; $2s$ is matched with the state $|\psi_{200}\rangle$ where $l = 0$, and $2p$ are $|\psi_{210}\rangle$, $|\psi_{211}\rangle$, and $|\psi_{21-1}\rangle$, where $l = 1$ and $m = -l, 0, +l$. These four states are degenerate, which is lifted by perturbation. Unlike the result for $n = 1$ level, the energy correction for the $n = 2$ level is first order. For convenience, we will use Dirac notation to write the perturbation equation

$$\sum_{k=1}^4 \alpha_k \langle \psi^j | e\mathcal{E}Z | \psi^k \rangle = E^{(1)} \alpha_j \tag{4.96}$$

in which the index k runs from 1 to 4 due to the four-fold degeneracy of the unperturbed state. The column vector α_k is the eigenvector with eigenvalue $E^{(1)}$. Vectors $|\psi^j\rangle$ are the four degenerate eigenvectors of the unperturbed Hamiltonian $\mathcal{H}_o$ which yield the eigenvalue $E^{(0)}$

$$\mathcal{H}_o \langle \psi^j | = E^{(0)} | \psi^j \rangle. \tag{4.97}$$

Vectors $|\psi^k\rangle$ are uncorrected zeroth-order eigenvectors contained in the general form of the first-order perturbation equation

$$(\mathcal{H}_o - E^{(0)}) | \psi^k \rangle + (e\mathcal{E}Z - E^{(1)}) | \psi^j \rangle = 0.$$

This arises from the perturbation expansion of eigenstates $|\psi(\lambda)\rangle$ and eigenvalues $E(\lambda)$

$$|\psi(\lambda)\rangle = |\psi^j\rangle + \lambda |\psi^k\rangle + \cdots$$

$$E(\lambda) = E^{(0)} + \lambda E^{(1)} + \cdots$$

where the perturbation metric $\lambda \le 1$. The index j runs from 1 to 4 which can be seen by inspection as well. If α_k is a 1×4 column vector, so must α_j be. In the matrix form equation (4.96) reads

$$(\langle \psi^j | e\mathcal{E}Z | \psi^k \rangle) \begin{pmatrix} \alpha_1 \\ \alpha_2 \\ \alpha_3 \\ \alpha_4 \end{pmatrix} = E^{(1)} \begin{pmatrix} \alpha_1 \\ \alpha_2 \\ \alpha_3 \\ \alpha_4 \end{pmatrix} \tag{4.98}$$

There are sixteen matrix elements of the 4×4 matrix $(\langle \psi^j | e\mathcal{E}Z | \psi^k \rangle)$. The values of the integrals can be determined at once in view of parity conditions of eigenstates and operators. For example, the integral $\langle \psi_{200} | e\mathcal{E}Z | \psi_{200} \rangle$ corresponding to one of the matrix elements of the operator has a 'zero' value due to even parity of $|\psi_{200}\rangle$ and odd parity of $e\mathcal{E}Z$. Integrals containing basis states with opposite parity need not always yield zero. By working with the angular part of the integrals one finds

$$e\mathcal{E}\begin{pmatrix} 0 & \gamma & 0 & 0 \\ \gamma & 0 & 0 & 0 \\ 0 & 0 & 0 & 0 \\ 0 & 0 & 0 & 0 \end{pmatrix}\begin{pmatrix} \alpha_1 \\ \alpha_2 \\ \alpha_3 \\ \alpha_4 \end{pmatrix} = E^{(1)}\begin{pmatrix} \alpha_1 \\ \alpha_2 \\ \alpha_3 \\ \alpha_4 \end{pmatrix}. \tag{4.99}$$

The operator matrix is the Stark perturbation operator, and the eigenfunctions are ψ_{21-1}, ψ_{211}, $\frac{1}{\sqrt{2}}\left[\psi_{200} + \psi_{210}\right]$, and $\frac{1}{\sqrt{2}}\left[\psi_{200} - \psi_{210}\right]$, with respective eigenvalues $0, 0, +\gamma$ and $-\gamma$. To calculate γ, the corresponding integral

$$e\mathcal{E}\langle\psi_{200}|Z|\psi_{210}\rangle = \gamma$$

can be solved by using the appropriate radial eigenfunctions and spherical harmonics

$$\gamma = e\mathcal{E}\int\left(\frac{1}{2a_0}\right)^{\frac{3}{2}} 2\left(1 - \frac{r}{2a_0}\right)e^{-\frac{r}{2a_0}}\frac{1}{\sqrt{4\pi}} r\cos\theta\left(\frac{1}{2a_0}\right)^{\frac{3}{2}} \frac{1}{\sqrt{3}}\left(\frac{r}{a_0}\right)e^{-\frac{r}{2a_0}}\sqrt{\frac{3}{4\pi}}\cos\theta r^2 dr\sin\theta d\theta. \tag{4.100}$$

The result should be $\gamma = \pm 3e\mathcal{E}a_0$, where a_0 is Bohr radius.

These developments for the $n = 2$ level ($2s2p$ configuration) suggest that the first order perturbation does not completely lift the degeneracy of the four field-free states; the perturbed states also contain degeneracy. In the presence of $\mathcal{E}$, states ψ_{211} and ψ_{21-1} continue to remain degenerate without an energy shift, but the eigenstates $\frac{1}{\sqrt{2}}\left[\psi_{200} + \psi_{210}\right]$ and $\frac{1}{\sqrt{2}}\left[\psi_{200} - \psi_{210}\right]$, which are formed by linear combinations of ψ_{200} and ψ_{210}, show energy shift by $\pm 3ea_0$ with a constant slope of $\pm 3a_0$ (Figure 4.18).

From $2s$ and $2p$ wavefunctions

$$\langle\psi_{200}|Z|\psi_{210}\rangle = \langle\psi_{210}|Z|\psi_{200}\rangle = 3a_0$$

$$e\mathcal{E}\begin{pmatrix} 0 & \gamma \\ \gamma & 0 \end{pmatrix}\begin{pmatrix} \alpha_1 \\ \alpha_2 \end{pmatrix} = E^{(1)}\begin{pmatrix} \alpha_1 \\ \alpha_2 \end{pmatrix}$$

$$\gamma = \langle\psi_{200}|Z|\psi_{210}\rangle = \langle\psi_{210}|Z|\psi_{200}\rangle = 3a_0$$

$$E_1^{(1)} = +3ea_0$$

$$E_2^{(1)} = -3ea_0.$$

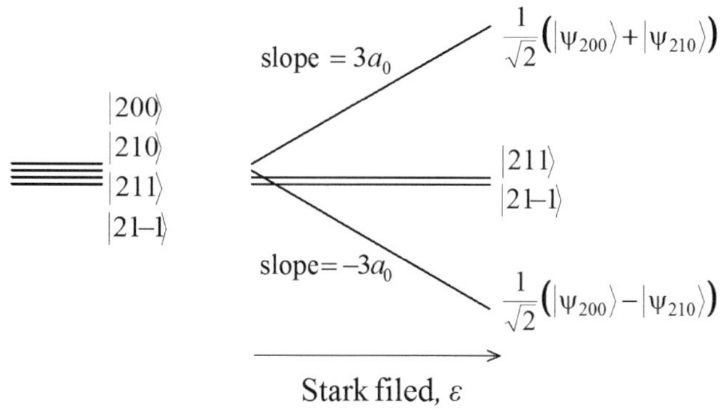

FIGURE 4.18 First-order Stark correction of the energy of the $n = 2$ level of hydrogen.

Box 4.1

Take only the Laplacian, and substitute with

$\psi(r,\theta,\phi) = R(r)\Theta(\theta)\Phi(\varphi)$, which yields

$$\frac{1}{r^2}\frac{\partial}{\partial r}\left(r^2\frac{\partial R(r)}{\partial r}\right) + \frac{1}{r^2\sin\theta}\frac{\partial}{\partial\theta}\left[\sin\theta\frac{\partial\Theta(\theta)}{\partial\theta}\right] + \frac{1}{r^2\sin^2\theta}\frac{\partial^2\Phi(\varphi)}{\partial\varphi^2} = 0.$$

Multiplying both sides by $\frac{r^2}{R(r)\Theta(\theta)\Phi(\varphi)}$ gives

$$\frac{1}{R(r)}\frac{\partial}{\partial r}\left(r^2\frac{\partial R(r)}{\partial r}\right) + \frac{1}{\Theta(\theta)}\frac{1}{\sin\theta}\frac{\partial}{\partial\theta}\left[\sin\theta\frac{\partial\Theta(\theta)}{\partial\theta}\right] + \frac{1}{\Phi(\varphi)\sin^2\theta}\frac{\partial^2\Phi(\varphi)}{\partial\varphi^2} = 0 \tag{4.B1.1}$$

We notice that the first term is a function of r and the second term is a function of θ, but the last term contains both θ and φ. To separate out this term from the former two so as to obtain it as a function of φ alone, we will multiply equation (4.B1.1) by $\sin^2\theta$. This yields the Φ function as

$$\frac{1}{\Phi(\varphi)}\frac{\partial^2\Phi(\varphi)}{\partial\varphi^2} = -m^2, \tag{4.B1.2}$$

where m is called the separation constant, and

$$\Phi = \begin{cases} \sin m\varphi \\ \cos m\varphi \end{cases}.$$

The separation constant m has to be –ve, because the function is periodic. The –ve sign will appear also for simple harmonic motion that we will see in the chapter on vibrational spectroscopy. One more condition is that the value of m has to be an integer, because the periodic functions sin $m\varphi$ and cos $m\varphi$ require m to be an integer, since a value at some angle φ will be repeated at $\varphi + 2m\pi$.

Now we consider the functions of r and θ variables. We use equation (4.B1.2) to rewrite equation (4.B1.1) as

$$\frac{1}{R(r)}\frac{\partial}{\partial r}\left(r^2\frac{\partial R(r)}{\partial r}\right)+\frac{1}{\Theta(\theta)}\frac{1}{\sin\theta}\frac{\partial}{\partial\theta}\left[\sin\theta\frac{\partial\Theta(\theta)}{\partial\theta}\right]-\frac{m^2}{\sin^2\theta}=0$$

and separate out the terms with respect to the variables r and θ. This gives

$$\frac{1}{R(r)}\frac{\partial}{\partial r}\left(r^2\frac{\partial R(r)}{\partial r}\right)=\lambda, \text{ and} \quad (4.B1.3)$$

$$\frac{1}{\sin\theta}\frac{\partial}{\partial\theta}\left[\sin\theta\frac{\partial\Theta(\theta)}{\partial\theta}\right]-\frac{m^2}{\sin^2\theta}\Theta=-\lambda\Theta. \quad (4.B1.4)$$

The equation with θ variable will turn out to be associated Legendre functions if $\lambda=l(l+1)$. In fact, equation (4.B1.4) can be solved only if $\lambda=l(l+1)$, because the solution will be finite only when $\cos\theta=\pm1$, which requires $\theta=0$ or $l\pi$. To solve equation (4.B1.4) we include $l(l+1)$ and set $x=\cos\theta$ to write

$$(1-x^2)\frac{\partial^2\Theta(\theta)}{\partial\theta^2}-2x\frac{\partial\Theta(\theta)}{\partial\theta}+\left\{l(l+1)-\frac{m^2}{1-x^2}\right\}\Theta(\theta)=0. \quad (4.B1.5)$$

To solve further (see Boas, 1983), substitute for

$$\Theta=(1-x^2)^{\frac{m}{2}}u$$

in equation (4.B1.5) to obtain

$$(1-x^2)\frac{\partial^2 u}{\partial\theta^2}-2(m+1)x\frac{\partial u}{\partial\theta}+\{l(l+1)-m(m+1)\}u=0.$$

This equation is called Legendre's equation when $m=0$. For $0 \le m \le l$, the general solution turns out to be

$$\Theta=P_l^m(\cos\theta)=(1-\cos^2\theta)^{\frac{m}{2}}\frac{d^m}{d(\cos\theta)^m}P_l(\cos\theta). \quad (4.B1.6)$$

The solutions for Θ above are associated Legendre functions.

Next, we take equation (4.B1.3) and substitute for $\lambda=l(l+1)$. The solution comes out as

$$R=\begin{cases} r^l \\ r^{-l-1} \end{cases}. \quad (4.B1.7)$$

Of these two solutions we should retain only r^l, because a hydrogen-like particle is confined to move within a sphere. The latter solution is relevant for motions outside the sphere.

Putting together the solutions for Φ (equation (4.B1.2)), Θ (equation (4.B1.6)) and R (equation (4.B1.7)) the wavefunction obtained is of the form

$$\psi(r,\theta,\varphi)=R(r)\Theta(\theta)\Phi(\varphi)=R(r)Y_l^m(\theta\varphi)=r^l P_l^m(\cos\theta)\begin{cases}\sin m\varphi\\ \cos m\varphi\end{cases}.$$

The radial functions $R(r)=r^l$ are Laguerre polynomials, and the angular function $Y_l^m(\theta\varphi)$ are called spherical harmonics, which can be written as the product of normalized functions of the variables θ and φ

Box 4.2

The spin refers to an angular momentum property that had to be invoked necessarily to explain fine structure of atomic spectra (see Section 2.1). Although the *spin* appears in Dirac's relativistic quantum mechanics, Pauli's theory allows for an easier comprehension and the use of electron and nuclear spins under non-relativistic condition. If the *z*-projection of the spin angular momentum has only two possible values $\pm\hbar/2$ such that

$$S_z|\alpha\rangle=\frac{\hbar}{2}|\alpha\rangle \text{ and } S_z|\beta\rangle=-\frac{\hbar}{2}|\beta \quad (4.B2.1)$$

then the spin eigenvectors $|\alpha\rangle$ and $|\beta\rangle$ are defined in a two-dimensional spin space

$$\alpha=\begin{pmatrix}1\\0\end{pmatrix} \text{ and } \beta=\begin{pmatrix}0\\1\end{pmatrix} \quad (4.B2.2)$$

A linear combination of these two vectors can generate any spin vector in the two-dimensional space. For example, the jth vector can be written as

$$|\phi_j\rangle=a_j|\alpha\rangle+b_j|\beta\rangle$$

and using the orthogonality condition

$$\sum_{-\frac{1}{2}}^{+\frac{1}{2}}\alpha^*\beta=0$$

one can obtain

$$\langle \phi_i | \phi_j \rangle = \begin{pmatrix} a_i^* & b_i^* \end{pmatrix} \begin{pmatrix} a_j \\ b_j \end{pmatrix}$$

From equations (4.B2.1) and (4.B2.2) it is straightforward to get the operator

$$S_z = \frac{\hbar}{2}\begin{pmatrix} 1 & 0 \\ 0 & -1 \end{pmatrix} \quad (4.B2.3)$$

Certain algebraic operations, including commutation relations and ladder operator formalism, applicable for orbital angular momentum are applicable to spin angular momentum as well. Thus

$$S^2 = S_x^2 + S_y^2 + S_z^2$$

$$[S_x, S_y] = i\hbar S_z \quad [S_y, S_z] = i\hbar S_x \quad [S_z, S_x] = i\hbar S_y$$

Having obtained the operator matrix for S_z in equation (4.B2.3), the commutation relations can be employed to obtain the matrices for S_x and S_y operators

$$S_x = \frac{\hbar}{2}\begin{pmatrix} 0 & 1 \\ 1 & 0 \end{pmatrix} \quad S_y = \frac{\hbar}{2}\begin{pmatrix} 0 & -i \\ i & 0 \end{pmatrix}$$

The S_x, S_y, and S_z matrices are Pauli matrices, and they appear identically for the nuclear spin. The three operator matrices give the matrix for the square of the total spin angular momentum

$$S^2 = \frac{3\hbar^2}{4}\begin{pmatrix} 1 & 0 \\ 0 & 1 \end{pmatrix}$$

which also implies

$$[S^2, S_z] = [S^2, S_y] = [S^2, S_x] = 0$$

$$S^2 |\alpha\rangle = \frac{3\hbar^2}{4} |\alpha\rangle$$

where the expectation value of S^2 for a spin-half particle, say the electron or the proton, is

$$\sqrt{\hbar^2 s(s+1)} = \frac{\sqrt{3}}{2}\hbar$$

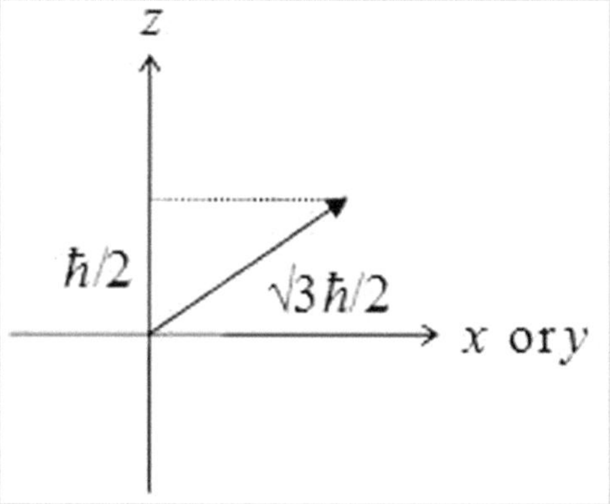

Box 4.3

An orbital, say ϕ in general, is by definition a one-electron stationary state the probability density of which is only coordinate dependent, not time (see Postulate 4, Chapter 2). The probability density is $|\phi(\mathbf{r})|^2$, in which $\mathbf{r}$ represents the total coordinate. Even though the probability surfaces of s and p orbitals, for example, are positive everywhere, '+' and '–' signs are often included just to indicate the wave nature. At the origin of the coordinate is the nucleus.

The parity of the stationary state wavefunction is obtained from the inversion operation, which entails inverting the wavefunction or the orbital. One can start with a point somewhere in the orbital and move to the origin of the coordinates – that is, the position of the nucleus – and continue to move the same distance across. This is an inversion operation which is equivalent to $\mathbf{r} \to -\mathbf{r}$, but the probability density is unchanged. The operation reads

$$\phi(r_1, r_2, r_3, \cdots, r_n) = \pm\phi(r_1, r_2, r_3, \cdots, r_n)$$

The parity of the state is even (П) when inversion leaves ϕ as $+\phi$, and odd (o) when ϕ transforms to $+\phi$. For many electrons in the atom, the orbital angular momentum of each electron is added up to determine the parity. More generally,

$$(-1)^{\sum_i l_i} = \begin{cases} 1 \text{ even parity} \\ -1 \text{ odd parity} \end{cases}$$

Parity is an important factor in the selection rule for dipole transition, because electric dipole transitions are allowed only between configurations of even and odd parity, often called Laporte selection rule.

Box 4.4

Maxwell's equations can be solved when scalar and vector potentials are available. While the fields **E** and **B** are physical observables, and hence are real functions of space and time variables (x,y,z,t), the meaning of potentials is restricted in the sense that their magnitudes are not direct observables. This allows for obtaining the same field vector from different sets of potentials. That is to say, the two sets of potentials can be transformed without any change in the physical observable of interest. For example, the potential **A** in our text is defined as

$$\mathbf{A} = \mathbf{A}_o - \nabla\chi(x,y,z,t) \tag{4.B4.1}$$

where $\nabla\chi$ is the gradient of the function χ. Then operating with curl on both sides

$$\nabla\times\mathbf{A} = \nabla\times\mathbf{A}_o - [\nabla\times(\nabla\chi)]$$

With the identity

$$\nabla\times(\nabla\chi) = 0$$

we get

$$\nabla\times\mathbf{A} = \nabla\times\mathbf{A}_o$$

which means the same field vector **B** is obtained from **A** and $\mathbf{A}_o$. Further, from the function ϕ defined as

$$\phi = \phi_o + \frac{1}{c}\frac{\partial\chi}{\partial t} \tag{4.B4.2}$$

for which

$$-\nabla\phi = -\nabla\phi_o$$

we write

$$-\nabla\phi = -\frac{1}{c}\frac{\partial\chi}{\partial t} = -\nabla\phi_o - \frac{1}{c}\frac{\partial\chi}{\partial t}$$

Thus, the same field vector **B** is obtained from **A** and ϕ, and $\mathbf{A}_o$ and ϕ_o. The transformation of the set of potentials $\mathbf{A}_o$ and ϕ_o according to equations (4.B4.1) and (4.B4.2) is called gauge transformation.

Lorentz gauze $\nabla\cdot\mathbf{A} + \frac{1}{c}\frac{\partial\phi}{\partial t} = 0$

Coulomb gauge $\nabla\cdot\mathbf{A} = 0$

Box 4.5

Earlier in Box 4.2 spin vectors in the two-dimensional space were written as

$$\alpha = \begin{pmatrix}1\\0\end{pmatrix} \text{and } \beta = \begin{pmatrix}0\\1\end{pmatrix}$$

Each of α and β states is the same for the electron and the hydrogen nucleus. For representation they may be distinguished by α_e, β_e, α_N, and β_N. For the two particles the entire space is obtained as the direct product space. The four base vectors, $|\alpha_e\alpha_N\rangle$, $|\alpha_e\beta_N\rangle$, $|\beta_e\beta_N\rangle$, and $|\beta_e\alpha_N\rangle$, are then

$$|\alpha_e\alpha_N\rangle = \begin{pmatrix}1\\0\end{pmatrix}\otimes\begin{pmatrix}1\\0\end{pmatrix} = \begin{pmatrix}1\\0\\0\\0\end{pmatrix}$$

$$|\alpha_e\beta_N\rangle = \begin{pmatrix}1\\0\end{pmatrix}\otimes\begin{pmatrix}0\\1\end{pmatrix} = \begin{pmatrix}0\\1\\0\\0\end{pmatrix}$$

$$|\beta_e\beta_N\rangle = \begin{pmatrix}0\\1\end{pmatrix}\otimes\begin{pmatrix}0\\1\end{pmatrix} = \begin{pmatrix}0\\0\\0\\1\end{pmatrix}$$

$$|\beta_e\alpha_N\rangle = \begin{pmatrix}0\\1\end{pmatrix}\otimes\begin{pmatrix}1\\0\end{pmatrix} = \begin{pmatrix}0\\0\\1\\0\end{pmatrix}$$

The matrix form of the $\mathbf{S}\cdot\mathbf{I}$ operator is obtained from the direct product of corresponding matrices.

$$S_xI_x = \frac{\hbar}{2}\begin{pmatrix}0 & 1\\1 & 0\end{pmatrix}\otimes\frac{\hbar}{2}\begin{pmatrix}0 & 1\\1 & 0\end{pmatrix} = \frac{\hbar^2}{4}\begin{pmatrix}0&0&0&1\\0&0&1&0\\0&1&0&0\\1&0&0&0\end{pmatrix}$$

$$S_yI_y = \frac{\hbar}{2}\begin{pmatrix}0 & -i\\i & 0\end{pmatrix}\otimes\frac{\hbar}{2}\begin{pmatrix}0 & -i\\i & 0\end{pmatrix} = \frac{\hbar^2}{4}\begin{pmatrix}0&0&0&-1\\0&0&1&0\\0&1&0&0\\-1&0&0&0\end{pmatrix}$$

$$S_zI_z = \frac{\hbar}{2}\begin{pmatrix}1 & 0\\0 & -1\end{pmatrix}\otimes\frac{\hbar}{2}\begin{pmatrix}1 & 0\\0 & -1\end{pmatrix} = \frac{\hbar^2}{4}\begin{pmatrix}1&0&0&0\\0&-1&0&0\\0&0&-1&0\\0&0&0&1\end{pmatrix}$$

$$\mathbf{S}\cdot\mathbf{I} = \left(S_xI_x + S_yI_y + S_zI_z\right) = \frac{\hbar^2}{4}\begin{pmatrix} 1 & 0 & 0 & 0 \\ 0 & -1 & 2 & 0 \\ 0 & 2 & -1 & 0 \\ 0 & 0 & 0 & 1 \end{pmatrix}$$

Box 4.6

Shown here is the representation of $|F,m_F\rangle$ states in terms of $|m_s,m_I\rangle$ states. In Box 4.5 above, the four $|m_s,m_I\rangle$ states are given as $|\alpha_e\alpha_N\rangle$, $|\alpha_e\beta_N\rangle$, $|\beta_e\beta_N\rangle$, and $|\beta_e\alpha_N\rangle$, which in some textbooks are shown using + and – for $|\alpha\rangle$ and $|\beta\rangle$ states, respectively. The $|F,m_F\rangle$ and $|m_s,m_I\rangle$ are connected as follows.

$$S_z|F=1,m_F=1\rangle = S_z|\alpha_e\alpha_N\rangle = |F=1,m_F=1\rangle$$

$$S_z|F=1,m_F=0\rangle = S_z\frac{1}{\sqrt{2}}\left[|\alpha_e\beta_N\rangle + |\beta_e\alpha_N\rangle\right] = \frac{1}{\sqrt{2}}\left[|\alpha_e\beta_N\rangle - |\beta_e\alpha_N\rangle\right] = |F=0,m_F=0\rangle$$

$$S_z|F=1,m_F=-1\rangle = S_z|\beta_e\beta_N\rangle = -|F=1,m_F=-1\rangle$$

$$S_z|F=0,m_F=0\rangle = S_z\frac{1}{\sqrt{2}}\left[|\alpha_e\beta_N\rangle - |\beta_e\alpha_N\rangle\right] = \frac{1}{\sqrt{2}}\left[|\alpha_e\beta_N\rangle + |\beta_e\alpha_N\rangle\right] = |F=1,m_F=0\rangle$$

Box 4.7

When a continuous charge distribution ρ (r), r=x,y,z, is allowed to interact with an applied potential, say an electrostatic potential $V(r)$, where $r = x,y,z$, the interaction can be expressed using the integral

$$\mathcal{H} = \int \rho(r)V(r)dr^3$$

Since the externally applied field is macroscopic and not limited to the volume of an atom or a molecule, we can confidently assume that the field is flat over the charge distribution. The flatness allows series expansion of the external potential about an origin ($r = 0$) which is some arbitrary point contained within ρ (r)

$$V(r)_{r=0} = V_0 + \sum_{i=x,y,z}\left(\frac{\partial V(r)}{\partial r_i}\right)_{r=0}(r - r_i) + \frac{1}{2!}\sum_{i,j=1,2,3}\left(\frac{\partial^2 V(r)}{\partial r_i \partial r_j}\right)_{r=0} r_i r_j + \cdots$$

As done generally, we take the first two terms of which the first one is $V_0 = 0$, that is zero energy. So we write

$$\mathcal{H}_{\text{Stark}} = -\int \rho(r)\sum_{i=x,y,z}\left(\frac{\partial V(r)}{\partial r_i}\right) r_i dr^3 = -\sum_{i=x,y,z} E_i\mu_i = -\mathbf{E}\cdot\boldsymbol{\mu}$$

where $E_i = -\left(\frac{\partial V}{\partial r_i}\right)_{r=0}$, and the dipole moment μ is the integral of the charge distribution.

PROBLEMS

4.1 The energy of hydrogen atom emission lines corresponding to the series of $n \to n'$ transitions is given by the formula

$$\Delta E\left(\text{cm}^{-1}\right) = Z^2 R_\infty\left(\frac{1}{n'^2} - \frac{1}{n^2}\right)$$

where

$R_\infty = \frac{me^4}{4\pi\hbar^3 c} = 109737.3\,\text{cm}^{-1}$ is the Rydberg constant for infinite nuclear mass. Consider the range of wavelength 10^2 (10^5 cm^{-1}) to 10^4 (10^3 cm^{-1}) nm. Of the Lyman, Balmer, … series of lines, which ones and how many lines of each type will be observed in this range of wavelength?

4.2 Consider hydrogen atom transitions $1s(n=1,l=0) \to 2p(n=2,l=1)$, $1s(n=1,l=0) \to 3p(n=3,l=1)$, and $1s(n=1,l=0) \to 4p(n=4,l=1)$.
  (a) Calculate the transition dipole moment strength $|\mu_{12}|$, $|\mu_{13}|$, and $|\mu_{14}|$.
  (b) Assuming that line broadening in these transitions arises only from spontaneous emission, calculate the ratio of maximum absorption coefficient for $1s \to 2p$ to $1s \to 3p$ and $1s \to 2p$ to $1s \to 4p$. How do the two ratios compare?

4.3 With respect to the quantum numbers n, l, and m of hydrogen in the absence of an external field, verify that the change of orbital angular momentum of the electron is $\Delta l = \pm 1$ irrespective of x, y, or z polarization of light.

4.4 Calculate the expectation value of the distance r between the electron and the proton of hydrogen atom in the state $|\psi_{200}\rangle$. How different is this r from that in the ground state $|\psi_{100}\rangle$?

4.5 Reconsider the two states $|\psi_{100}\rangle$ and $|\psi_{200}\rangle$, and calculate the potential energy corresponding to each state.

4.6 Suppose a hydrogen atom exists in states

$$|\Psi\rangle = \frac{1}{\sqrt{2}}\left[|\psi_{200}\rangle + |\psi_{210}\rangle\right]$$

$$|\Psi\rangle \frac{1}{\sqrt{2}}\left[|\psi_{200}\rangle - |\psi_{210}\rangle\right],$$

and is subjected to a Stark field along the $+z$ axis.

(a) Calculate the expectation value of the energy change for each of the two states.

(b) Consider a transition from $n = 1$ to $n' = 2$. Absorption should occur at 121.5 nm. How different are the absorption spectra in the absence and the presence of the electric field?

(c) Draw line diagrams of the spectra for different polarizations of light.

4.7 Consider the electron-nuclear hyperfine interaction in the state $|\psi_{220}\rangle$ of hydrogen atom. Write out the time-dependent Schrödinger equation, and calculate the eigenvalues in terms of the constant $\mathcal{A}$. How will the hyperfine lines split in the presence of a weak external magnetic field B_o? Graphing the Zeeman splitting at a few values, say 10, 50, 100, 250, and 500 Gauss, will help.

4.8 Take only the 1s ground state $|\psi_{100}\rangle$ of hydrogen and place it in an external magnetic field whose strength varies from 0.1 to 1 T. Graph the energy shifts using a few values of B_o.

BIBLIOGRAPHY

A large volume of work on hydrogen atom spectra has been published originally in German almost a century ago. For example, the contributions of W. Gordon have not been translated. However, Bethe and Salpeter describe the entire theoretical foundation, and provide essential information. This is a classic book the reader is encouraged to refer to. Additionally, Chapters VII and XII of Cohen-Tannoudji et al. discuss the hydrogen atom in detail.

Bethe, H. A. and E. E. Salpeter (1977) *Quantum Mechanics of One and Two-Electron Atoms*, Plenum Publishing Corporation.

Boas, M. L. (1983) *Mathematical Methods in the Physical Sciences*, Wiley.

Cohen-Tannoudji, C., B. Diu, and F. Laloë (1977) *Quantum Mechanics* (Vol I and II), John Wiley & Sons.

5 Molecular Eigenstates

A spectroscopic transition was defined in Chapter 3 as an allowed transition from one eigenstate to another. We wish to consider here what these eigenstates could be for a simple system, a diatomic molecule for example. Motions of both electrons and nuclei determine energy absorption, and calculation of energy eigenvalues to predict molecular spectra requires a detailed understanding of energy eigenstates, requiring consideration of motions of both of them. The effect of coupling of electronic and nuclear dynamics distinguishes molecular from atomic spectra. But knowing the form of the eigenstates and then solving the Schrödinger equation become increasingly difficult as molecular size increases. Most of the eigenstate and eigenvalue discussions therefore consider diatomic or small polyatomic molecules for which some of the solutions are achieved exactly. For instance, the use of Born-Oppenheimer (BO) approximation in the case of molecular hydrogen ion provides an exact solution to which vibronic effects can be added on. Discussed in this chapter is an approach that allows for such a treatment.

5.1 BORN-OPPENHEIMER APPROXIMATION

The eigenfunction of a molecule is a function of both electronic and nuclear coordinates. If there are n electrons and N nuclei in a molecule, then there will be $3n$ and $3N$ corresponding spatial coordinates. The complete set of electronic and nuclear coordinates, denoted by r and R, respectively, are referred to a laboratory fixed axes system, which means the changing internuclear coordinate of a diatomic molecule, for example, is given with reference to a XYZ Cartesian coordinate system. Let the time-independent energy eigenfunction of the molecule be $\psi_{\mathrm{BO}}(r,R)$. The Born-Oppenheimer approximation (Born and Oppenheimer, 1927) states that the total molecular eigenfunction $\psi_{\mathrm{BO}}(r,R)$ can be written as a factorized product of $\psi(r;R)$ at a fixed nuclear configuration R and $F(R)$ – a function that exclusively describes all nuclear motions. The rationale for the use of BO approximation is that the nuclear mass M is ~2000 times larger than the electron mass m_e, so that the latter can be considered moving with respect to a motionless nucleus. The molecular eigenfunction then is a product function of electronic and nuclear eigenstates

$$\psi_{\mathrm{BO}}(r,R) = \psi(r;R)F(R). \tag{5.1}$$

The total energy operator contains kinetic energy of all electrons T_r and nuclei T_R, and potential energy due to electron-electron, nuclei-nuclei, and electron-nuclear interactions, $V(r,R)$,

$$\mathcal{H}(r,R) = T_r + T_R + V(r,R).$$

If there are i electrons and j nuclei, then

$$T_r = \sum_{i=1}^{3n} -\frac{\hbar^2}{2m_e}\frac{\partial^2}{\partial r_i^2}$$

$$T_R = \sum_{j=1}^{3N} -\frac{\hbar^2}{2M_j}\frac{\partial^2}{\partial R_j^2}$$

$$V(r,R) = \sum_{i<i'}\frac{e^2}{r_{ii'}} + \sum_{j<j'}\frac{z^2e^2}{R_{jj'}} - \sum_{i<j}\frac{ze^2}{r_{ij}}. \tag{5.2}$$

In the potential energy operator $V(r,R)$, the first two terms describe electron-electron and nuclear-nuclear repulsion energies, and the last term represents the energy of electron-nuclear attractive interaction. Note that nuclear charges of a heteronuclear diatomic molecule may differ, in which case $z^2 = Z_jZ_{j'}$. The full Schrödinger equation reads

$$\left[T_r + T_R + V(r,R)\right]\psi(r;R)F(R) = E\psi(r;R)F(R). \tag{5.3}$$

If the $\psi_{\mathrm{BO}}(r;R)$ function is a product of electronic and nuclear eigenfunctions, how to separate the two? For simplicity let there be a diatomic molecule A – B, and the two nuclei of the molecule is somehow restricted to move so as to fix the internuclear distance $d_{\mathrm{A-B}}$ to some value. Because the nuclear motions are frozen the nuclear kinetic energy $T_R = 0$, due to which the function $F(R)$ that describes nuclear dynamics becomes unity. Under this situation, the molecular eigenfunction will be the electronic eigenfunction $\psi(r;R)$ alone at the specified $d_{\mathrm{A-B}}$. Specifying $d_{\mathrm{A-B}}$ is equivalent to parameterizing R. In the variables of the electronic eigenfunction, the parametric R is shown separated from r by a semicolon. The Schrödinger equation for the nth eigenfunction is

$$\mathcal{H}_{\mathrm{E}}\psi_n(r;R) = \varepsilon_n(R)\psi_n(r;R),$$

where the subscript 'E' is added to indicate that the Hamiltonian is specific for the electronic eigenfunction. Under this condition the nuclei are said to be clamped, and we have

$$\mathcal{H}_{\mathrm{E}} = T_r + V(r,R).$$

The eigenvalue equation for the nth electronic function is solved with the nuclei clamped to a specific value of $d_{\mathrm{A-B}}$ or the parameter R

DOI: 10.1201/9781003293064-5

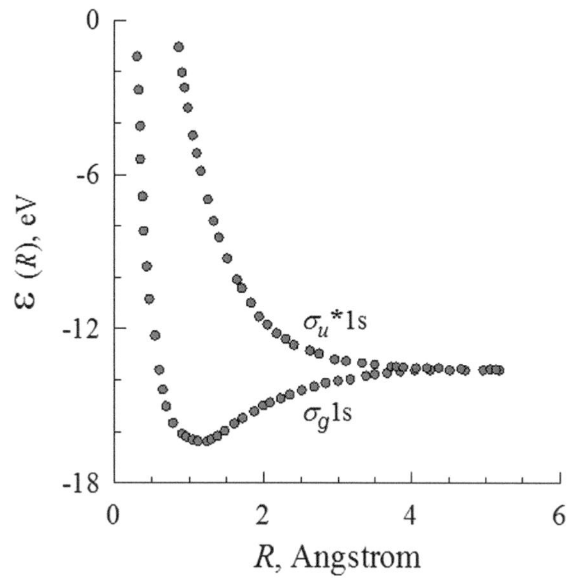

FIGURE 5.1 Calculated potential energy curves at several internuclear distances for the electronic states $\sigma_g 1s$ and $\sigma_u^* 1s$ of H_2^+ ion.

$$\int \psi_n^*(r;R)\mathcal{H}_E \psi_n(r;R)dr = \varepsilon_n(R) \quad (5.4)$$

The solution is then obtained for another value of R, and so on. One thus gets a set of eigenvalues $\varepsilon_n(R)$, and the graph of ε_n vs R yields a potential energy curve for the nth electronic state. Note that one calls the potential energy of an electronic state with R the potential energy curve, because diatomic molecules do not produce a surface. The potential curve should obviously be different for different values of n. In the case of H_2^+ for example, where the sole electron and the two nuclei form a stable ground state, the calculations for $\sigma_g 1s$ and $\sigma_u^* 1s$ states at different internuclear distances are carried out as shown in Figure 5.1. The electronic eigenvalues $\varepsilon(R)$ so obtained are then added to T_R to get the energy operator for nuclear motion $\mathcal{H}_N$, so the nuclear Schrödinger equation becomes

$$\mathcal{H}_N F(R) = EF(R) = [T_R + \varepsilon(R)]F(R) = EF(R). \quad (5.5)$$

5.2 SOLUTION OF THE TOTAL SCHRÖDINGER EQUATION

The description above relied on fixing the nuclei to solve the Schrödinger equation for an isolated electronic state first, and using the resultant eigenvalues to solve the nuclear equation. An isolated state that is not coupled to any other electronic state is often called a Born-Oppenheimer state. It is also clear that the energy operator $\mathcal{H}_E$ operates on the electronic eigenfunction alone. But is it not desirable to be able to use the full molecular eigenfunction so that motional degrees of freedom for both electrons and nuclei can be discussed simultaneously? If so, we need to consider the total Schrödinger equation

$$\mathcal{H}(r,R)\psi(r,R) = E\psi(r,R) \quad (5.6)$$

The full Hamiltonian $\mathcal{H}(r,R)$ is given in terms of all electronic and nuclear coordinates. The starting point is the recognition that the electronic Hamiltonian $\mathcal{H}_E$ is a Hermitian operator, which means the eigenfunctions of $\mathcal{H}_E$ form a complete set of orthonormal functions. Note also that the states corresponding to these eigenfunctions are in fact Born-Oppenheimer states. The idea now is that the total molecular eigenstate $\psi(r,R)$ can be linearly expanded in terms of the orthonormal set of Born-Oppenheimer states

$$\psi(r,R) = \sum_n \psi_n(r;R)F_n(R). \quad (5.7)$$

Inserting this expansion into the total Schrödinger equation we get

$$\mathcal{H}(r,R)\sum_n \psi_n(r;R)F_n(R) = E\sum_n \psi_n(r;R)F_n(R)$$

$$(\mathcal{H}_E + T_R - E)\sum_n \psi_n(r;R)F_n(R) = 0. \quad (5.8)$$

We can multiply this by ψ_l^* and integrate over the electronic coordinates. The integration is performed over electronic positions because the eigenfunctions used here are electronic eigenfunctions that correspond to Born-Oppenheimer states. Multiplication by ψ_l^* and integration of the $\mathcal{H}_E$ part of the Hamiltonian will give

$$\int \sum_n \psi_l^*(r;R)\mathcal{H}_E \psi_n(r;R)F_n(R)dr^3$$

$$= E_n(R)\sum_n \int \psi_l^*(r;R)\psi_n(r;R)dr^3 F_n(R)$$

$$= E_n(R)F_n(R). \quad (5.9)$$

Next, the eigenvalue equation with the nuclear kinetic energy operator can be expanded as

$$T_R \sum_n \psi_n(r;R)F_n(R) = \sum_{j=1}^{3N} -\frac{\hbar^2}{2M_j}\nabla_j^2 \sum_n \psi_n(r;R)F_n(R)$$

$$= \sum_n \Bigg[\int \psi_l^*(r;R)\psi_n(r;R)dr^3 \sum_{j=1}^{3N} -\frac{\hbar^2}{2M_j}\nabla_j^2 F_n(R) +$$

$$\int \psi_l^*(r;R)F_n(R)\sum_{j=1}^{3N} -\frac{\hbar^2}{2M_j}\nabla_j^2 \psi_n(r;R)dr^3 +$$

$$2\int \psi_l^*(r;R)\sum_{j=1}^{3N} -\frac{\hbar^2}{2M_j}\nabla_j \psi_n(r;R)\nabla_j F_n(R)dr^3 \Bigg]. \quad (5.10)$$

All of these terms can be collated in order to write down the solution of the full form of the Schrödinger equation

$$\left[E_n(R)\sum_{j=1}^{3N}-\frac{\hbar^2}{2M_j}\nabla_j^2-E\right]F_n(R)+\sum_n\int\psi_l^*(r;R)$$
$$\sum_{j=1}^{3N}-\frac{\hbar^2}{2M_j}\nabla_j^2\,\psi_n(r;R)dr^3F_n(R)+\sum_n\int\psi_l^*(r;R)$$
$$\sum_{j=1}^{3N}-\frac{\hbar^2}{M_j}\nabla_j\psi_n(r;R)dr^3\nabla_jF_n(R)=0 \tag{5.11}$$

These equations are referred to as coupled-channel equations which form a $n\times n$ matrix for the coupling of n electronic potential surfaces. The off-diagonal elements of the matrix come from the terms contained in the second and third parts of the solution above where $\delta_{ln}\neq 1$ if $l\neq n$. The off-diagonal terms are called nonadiabatic coupling terms whose magnitudes are very small and often ignored. The smallness of the nonadiabatic terms arises from the $1/M$ factor in the nuclear kinetic energy operator. Clearly, $1/m_e \gg 1/M$, so these terms can be conditionally ignored. We will revisit this point and discuss when and how the off-diagonals can be ignored.

5.3 STATES OF NUCLEAR MOTION

The BO approximation is particularly suitable for the description of molecular spectra because of the adiabaticity of electronic motion – faster motions of electrons render them following the slow-moving nuclei adiabatically. The electron distribution in a particular electronic state instantaneously adjusts to a change of nuclear configuration, which is why Born-Oppenheimer surfaces are called adiabatic states.

The BO approximation provides an approach to separate out the electronic states from the nuclear motional states. The method turns out to be exact when a simple system can be approximated to a harmonic oscillator or a rigid-body rotor or a center of mass particle represented in a Morse-like ideal potential. The slow motion of these overly idealized systems do not resolve the much faster electronic motion, thus facilitating derivation of nuclear motional states. Said another way, the nuclear motions are decoupled from electron motions which would mean that the nuclear rotations and vibrations must be contained within the electronic potential energy surface. Putting it still another way, ignoring the nonadiabatic coupling terms makes it possible to assume that the nuclei move on the electronic potential field. Considered below is an analytical discussion of the emergence of rotational and vibrational motional states of the nuclei, meaning how the rotational and vibrational energy levels appear when the nuclei move on an electronic potential energy surface.

The changing position of the rigid body nuclei, as it would when the orientation of the body changes, are described by three Euler coordinates, $\theta\phi\chi$, that determine the relative positions of the static laboratory frame and the rotating molecule-fixed frame. The procedure for determination of the rotation matrix is briefly outlined in Box 5.1. The positions of the nuclei in the molecule-fixed axes system are given by R, which represents the $3N-6$ independent coordinates of relative motion. The independent coordinates for a linear molecule is $3N-5$.

The problem of looking at the nuclear motional states is significantly reduced if the nuclear kinetic energy T_R can be expressed in terms of laboratory-frame coordinates (XYZ), Euler angles ($\theta\phi\chi$), and nuclear positional coordinates (R). If so, which actually works for small molecules at least, we are able to write for the nuclear kinetic energy

$$T_R=T(X,Y,Z)+T(R,\ \theta,\phi,\chi)+T(R), \tag{5.12}$$

where the three terms on the right-hand side stand for the translational kinetic energy of the center of mass, rotational kinetic energy due to rotation of the fixed nuclei, and internuclear vibrational motions relative to each other. Strictly, the independent representation of these three kinetic energy terms would work only in the absence of intermotional interferences. For example, the rotational centrifugal forces affect the nuclear vibrational motions, and the nuclear vibrations would modify the rotational moment of inertia. To account for such effects, the total nuclear kinetic energy is approximated by adding another term called rotational–vibrational kinetic energy or rovibrational coupling

$$T_R=T(X,Y,Z)+T(R,\theta,\phi,\chi)+T(R)+T_{RV}(R,\theta,\phi,\chi). \tag{5.13}$$

Nonetheless, the T_{RV} term can still be neglected assuming that rotational and vibrational motions may be decoupled because of smallness of the Coriolis effect.

The provision of separation of the nuclear kinetic energy into translational, rotational, and vibrational energy terms (equation (5.12)) also allows for expressing the nuclear motional wavefunction $F(R)$ as a product of translational, rotational, and vibrational wavefunctions

$$F(R)=\psi_{\text{trans}}(X,Y,Z)\,\psi_{\text{rot}}(\mathrm{R}_o;\theta,\phi,\chi)\,\psi_{\text{vib}}(R). \tag{5.14}$$

We mention that the rotational motion taken up here pertains to a rigid-body rotor at the equilibrium nuclear configuration R_o. The nuclear motional equation can then be written as

$$[T_R+\varepsilon(R)]F(R)=EF(R)$$

$$\left[T(X,Y,Z)+T(R_o;\theta,\phi,\chi)+T(R)+\varepsilon(R)\right]\psi_{\text{trans}}\psi_{\text{rot}}\psi_{\text{vib}}$$
$$=\left(E_{\text{trans}}+E_{\text{rot}}^{(R_o)}+E_{\text{vib}}\right)\psi_{\text{trans}}\psi_{\text{rot}}\psi_{\text{vib}}\Big), \tag{5.15}$$

which implies the following three equations, one each for the three types of nuclear motions

$$T(X,Y,Z)\,\psi_{\text{trans}}=E_{\text{trans}}\psi_{\text{trans}}$$
$$[T_R+\varepsilon(R)]\psi_{\text{vib}}=E_{\text{vib}}\psi_{\text{vib}}$$
$$T(R_o;\theta,\phi,\chi)\,\psi_{\text{rot}}=E_{\text{rot}}\psi_{\text{rot}}. \tag{5.16}$$

The rotational equation insists on choosing the equilibrium nuclear configuration R_o as the distance about which the nuclei oscillate.

Indeed, these analytical results are already seen in the total Schrödinger equation of the molecular eigenstate (equation 5.11) if the nonadiabatic coupling terms are ignored. Equation (5.11) then reduces to

$$\left[E_n(R)\sum_{j=1}^{3N}-\frac{\hbar^2}{2M_j}\nabla_j^2-E\right]F_n(R)=0, \tag{5.17}$$

which is essentially the Schrödinger equation involving the kinetic energy of the moving nuclei. The nuclei that move on the nth potential energy surface have the potential energy E_n, which we will write as $E_{n,\text{rot,vib}}$ indicating the presence of rotational and vibrational energy levels coded into the nth electronic state. Each vibrational and rotational state is referred with respect to a particular electronic state. Equation (5.17) is often also written as

$$\left[E_n(R)\sum_{j=1}^{3N}\frac{\hbar^2}{2M_j}\nabla_j^2-E_{n,\text{rot,vib}}\right]F_{n,\text{rot,vib}}(R)=0, \tag{5.18}$$

which is identical to the analytical expression (equation (5.15)) for the states of nuclear motion. Typical potential energy curves for $n = 0$ and 1, corresponding to the ground and first excited electronic states containing the vibrational and rotational energy levels of a diatomic molecule are illustrated in Figure 5.2.

5.4 ADIABATIC AND NONADIABATIC PROCESSES

It has been said frequently that Born-Oppenheimer surfaces are adiabatic surfaces on which the nuclear positions are instantaneously followed by the electrons. Processes where the BO approximation breaks down, meaning the surfaces are not ideally adiabatic, are called nonadiabatic. This could occur when two adiabatic surfaces come closer or cross each other. The nuclear motions in the crossover region will not be determined by just one, but both potential energy surfaces. Nonadiabatic coupling here could induce radiationless electronic transitions from one surface to the other. To illustrate the probabilistic choice of adiabatic vs nonadiabatic transition at the crossing region of two potential energy surfaces, we consider two surfaces corresponding to k^{th} and l^{th} electronic states as shown in Figure 5.3.

If the off-diagonal matrix element vanishes, then

$$\langle \psi_k | T_R | \psi_l \rangle = 0 \tag{5.19}$$

and the two states would not interact or they would not mix. If on the other hand, $\langle \psi_k | T_R | \psi_l \rangle \neq 0$, the nonadiabatic coupling can cause the two surfaces repel each other and they will mix conditionally, resulting in adiabatic and nonadiabatic transitions according to the probability of the occurrence

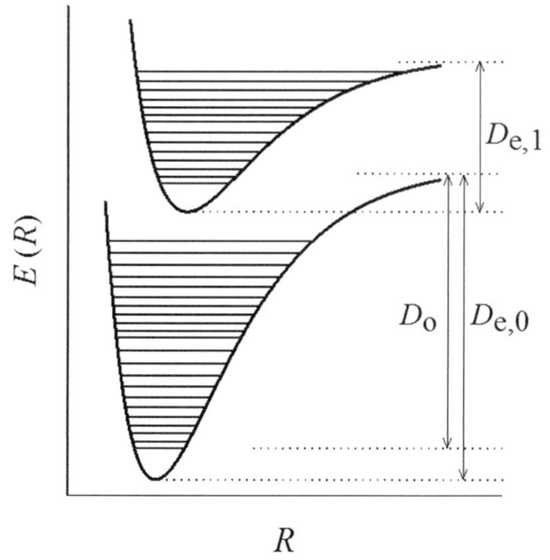

FIGURE 5.2 Schematic of potential energy curves for the ground and the first excited state of a simple diatomic system modeled by the analytical form of the Morse potential $E(R) = D_e\left\{1 - e^{[-\beta(R-R_e)]}\right\}^2$. The bond energy is designated by D_o. The symbol D_e represents dissociation energy, $D_{e,o}$ for the lower potential (ground state) and $D_{e,1}$ for the upper potential (excited state). The energy levels – vibrational (red horizontal lines) and rotational (black horizontal lines), are drawn to approximations only, not truly corrected for the anharmonicity of the potentials.

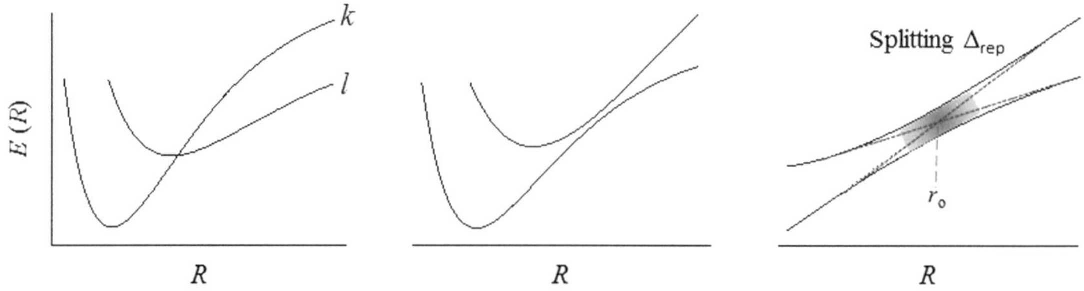

FIGURE 5.3 Schematic potential energy curves. *Left*, nonadiabatic energy levels. *Middle*, adiabatic energy levels showing crossing of electronic states. *Right*, magnification of the crossing region where the solid lines represent adiabatic eigenstates and the broken lines delineate the region of splitting.

of the two. The extent of repulsive separation of the two states (Landau and Lifshitz, 1958: 304–312), denoted by Δ_{rep} (Figure 5.3), is determined by the mean value of the coupling matrix element

$$\Delta_{\text{rep}} = 2\langle \psi_k | T_R | \psi_l \rangle. \tag{5.20}$$

The conditional mixing of the adiabatic states calls for parameterization of adiabaticity, the details of which have been worked out long ago by Landau (1932a, 1932b), Zener (1932), and Stuckelberg (1932). It is more useful to focus on the dynamics in the region of avoided crossing, centered about the internuclear distance r_0 at which the tips of the two cones meet. Since the $k \to l$ transition takes place in the region of r_0, the energy-time uncertainty of the system needs to be considered. If the energy uncertainty is

$$\Delta E \approx \frac{\hbar}{\Delta t_{\text{LZ}}}, \tag{5.21}$$

where t_{LZ} refers to the time the system spends in the mixing region, also called Landau-Zener (LZ) region. If the velocity of the system in the LZ region is v, then the time t_{LZ} is given by

$$t_{\text{LZ}} = \frac{l_{\text{LZ}}}{v}, \tag{5.22}$$

where l_{LZ} is the length of the LZ region along the nuclear coordinate in which adiabatic transition can occur. The l_{LZ} length can be approximated from the slopes of the conic slants in the vicinity of r_0,

$$\left[\frac{\partial E(R)}{\partial R}\right]^{(k)} = F_k \quad \text{and} \quad \left[\frac{\partial E(R)}{\partial R}\right]^{(l)} = F_l$$

$$l_{\text{LZ}} = \frac{\Delta_{\text{rep}}}{|F_k - F_l|}$$

From these relations the adiabaticity parameter γ_{LZ}, which is the ratio of Δ_{rep} to the energy uncertainty, is obtained as

$$\gamma_{\text{LZ}} = \frac{\Delta_{\text{rep}}}{\left(\dfrac{\hbar}{\Delta t_{\text{LZ}}}\right)} = \frac{\Delta_{\text{rep}}^{2}}{\hbar v |F_k - F_l|}. \tag{5.23}$$

A fully adiabatic transition is characterized by $\gamma_{\text{LZ}} \gg 1$. The system or the molecule then always stays in the lower potential but can undergo $k \rightleftharpoons l$ transition. If $\gamma_{\text{LZ}} \ll 1$, then the potentials can nonadiabatically cross each other, in which case the $k \to l$ transition will not occur, meaning the system will stay in the lower potential adiabatically. The probability of adiabatic transition in this case is given by

$$P = 1 - e^{\left(-\frac{\pi \gamma_{\text{LZ}}}{2}\right)}. \tag{5.24}$$

Clearly, $P = 1$ when $\gamma_{\text{LZ}} \gg 1$, which is the characteristic of an adiabatic transition or Franck-Condon transition. When $\gamma_{\text{LZ}} \ll 1$, the probability that the electronic structure will change is

$$P = \frac{\pi}{2}\gamma_{\text{LZ}} = \frac{\pi}{2}\frac{\Delta_{\text{rep}}^{2}}{\hbar v |F_k - F_l|} \tag{5.25}$$

showing that $P \propto \Delta_{\text{rep}}^{2}$ in this case. The proportionality of the probability of electronic structure change to the square of the mean value of the coupling matrix element is the characteristic of highly nonadiabatic transitions. In some situations, the coupling matrix element Δ_{rep} is treated as a perturbation that will be discussed later in the context of Golden rule.

5.5 MOLECULAR POTENTIAL ENERGY STATES

5.5.1 ONE-ELECTRON HYDROGEN-LIKE ATOM STATES

As studied in Chapter 4, wavefunctions of the form

$$\psi_{nlm}(r,\theta,\phi) = R_{nl}(r) Y_{lm}(\theta,\phi) \tag{5.26}$$

are called orbitals, where n, l, m are quantum numbers. With different values of n, l, m the one-electron hydrogen-like orbitals, such as

$$\psi_{100} = \psi_{1s} = \frac{1}{\sqrt{\pi}}\left(\frac{Z}{a}\right)^{\frac{3}{2}} e^{\frac{-Zr}{a}} \tag{5.27}$$

$$\psi_{210} = \psi_{2p_z} = \frac{1}{4\sqrt{2\pi}}\left(\frac{Z}{a}\right)^{\frac{5}{2}} r e^{\frac{-Zr}{2a}} \cos\theta, \tag{5.28}$$

and so on, are obtained. These one-electron orbitals are labeled with subscripts $s, p, d, f, \cdots$ according to the value of the orbital angular momentum quantum number l as

wavefunction label	$s\,p\,d\,f \cdots$
l	$0123\cdots$

If an one-electron wavefunction forms a part of the atomic wavefunction, then the one-electron wavefunction itself can be called an atomic orbital bearing the same s, p, d, f, ... labels. However, writing the exact form of the atomic wavefunction for a many-electron atom turns out to be a difficult problem because of existing repulsions among the electrons which do not render separation of the Schrödinger equation.

5.5.2 MOLECULAR ELECTRONIC STATES DERIVED FROM ATOM STATES

If s, p, d, f, are one-electron wavefunctions, what should the molecular wavefunctions be called? Stating otherwise, what will be the form and symmetry of the resultant stationary-state molecular wavefunction when two atomic orbitals are allowed to combine? The homonuclear diatomic orbitals are labeled by a set of symmetry symbols similar to the way atomic orbitals are labeled by quantum numbers l and m. We may recall that each value of l has $2l+1$ values of m, all having the same energy.

In a diatomic molecule, the two stationary nuclei can be seen to set up an internal Stark field ε_z, which may be considered as an uniform electric field along the z-direction due to the positive charges on the nuclei. The interaction of this electric field with the atoms removes the $2l+1$ fold degeneracy of m and induces a dipole μ_z. The field–dipole interaction energy is

$$E = \left|-\boldsymbol{\mu}_z \cdot \boldsymbol{\varepsilon}_z\right| \tag{5.29}$$

Because E would not change if $\boldsymbol{\varepsilon}_z$ was along $-z$, the atom states corresponding to $\pm m$ will have degenerate energy. This fact can be used as a basis to argue that molecular symmetry as such and atomic symmetry in the presence of an electric field are the same. To justify this analogy, it should be established that the total molecular Hamiltonian that would include both electronic and nuclear energies commutes with L_z, just the way the atomic Hamiltonian commutes with L_z (equation 4.4). The commutation of the molecular Hamiltonian and L_z has been formally established in literature by Bates et al (1953). In brief, the electronic orbital angular momentum vector is set into Larmor precession about the internuclear axis (Figure 5.4). The eigenvalue equation for the L_z operator is

$$L_z \psi_{\text{el}} = \frac{\hbar}{i}\frac{\partial \psi}{\partial \phi} = \pm \Lambda \hbar \psi_{\text{el}}, \tag{5.30}$$

where ϕ is the azimuthal angle about the molecular axis. Thus, the quantum number Λ represents the z-projection of the electronic orbital angular momentum, and the value of $\Lambda = 0,1,2,\ldots$. We have written the eigenvalue Λ for the total orbital angular momentum, but it should be understood that $\Lambda = \sum_i \lambda_i$, where λ_i is the orbital angular momentum of the ith electron. The normalized wavefunction should appear as

$$\psi_{\text{el}} = \frac{1}{\sqrt{2\pi}} e^{\pm i\Lambda\phi}, \tag{5.31}$$

but the dependence of ψ_{el} on ϕ is redundant because $\mathcal{H}_{\text{el}}$ is independent of ϕ, which is why the problem is dealt in the Cartesian coordinate system.

To state the development above clearly, the electronic Hamiltonian under BO approximation

$$\mathcal{H}_{\text{el}}\,\psi(r;R) = E_{\text{el}}(R)\,\psi(r;R)$$

commutes with L_z

$$\left[\mathcal{H}_{\text{el}}, L_z\right] = 0.$$

Interestingly, however, unlike the atomic context, L^2 does not commute with $\mathcal{H}_{el}$ because of very strong interaction of the Stark field with the orbital angular momentum vector to an extent that the angle between the z-axis and the vector is not defined. Under this condition, no spherical symmetry exists and hence the total orbital angular momentum of electrons in a diatomic molecule is not conserved, and is therefore not a constant of motion. However, the commuting relation

$$\left[\mathcal{H}_{\text{el}}, L_z\right] = 0$$

renders L_z a conserved quantity, meaning a constant of motion in the cylindrical symmetry of the diatomic.

It should be noted that the eigenvalues m and Λ in the atomic and molecular contexts, respectively, serve as symmetry symbols. Also to note is the removal of $2l+1$ degeneracy of m when the two atoms are united. Thus, molecular orbitals and the corresponding electron configuration in a diatomic molecule are labeled by the quantum number Λ which is the eigenvalue of the L_z operator. These results are listed in Table 5.1.

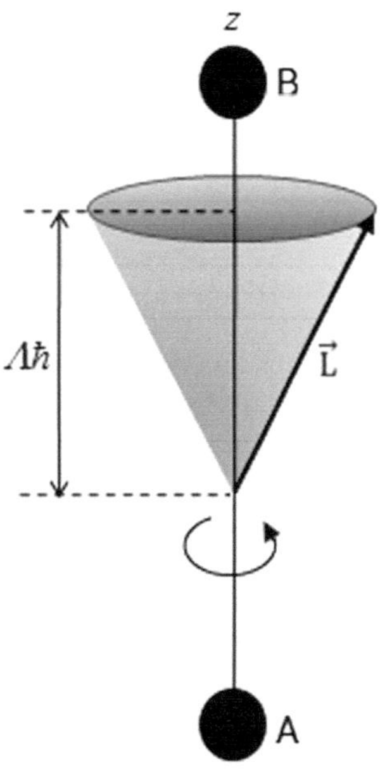

FIGURE 5.4 The precession of the total orbital angular momentum vector of electrons in the potential field of the nuclei of A and B. The z-projection of the vector **L** defines the azimuthal quantum number Λ. Molecular orbitals are labeled Σ, Π, Δ, Φ, and so on, with respective Λ values of 0, 1, 2, 3 … .

TABLE 5.1
Molecular States of Diatomic Molecules Derived from Separated Atoms

Atomic m	Atomic orbital	Molecular Λ	Molecular orbital	Molecular electronic state
0	s	0	σ	Σ
1	p	1	π	Π
2	d	2	δ	Δ
3	f	3	ϕ	Φ
⋮	⋮	⋮	⋮	⋮

TABLE 5.2
Molecular States of a Heteronuclear or a Homonuclear Molecule, the Two Atoms in the Latter Having Identical Electronic States

Molecular electronic states	States denoted with spin, spin-orbit coupling, and parity					
Δ	$^1\Delta_2$	$^3\Delta_3$	$^3\Delta_2$	$^3\Delta_1$		
Π	$^1\Pi_1$	$^3\Pi_2$	$^3\Pi_1$	$^3\Pi_0$		
Σ^+,Σ^-	$^1\Sigma^+$	$^3\Sigma_0^+$	$^3\Sigma_1^+$	$^1\Sigma^-$	$^3\Sigma_0^-$	$^3\Sigma_1^-$
Σ^+	$^1\Sigma^+$	$^3\Sigma_0^+$	$^3\Sigma_1^+$			

The molecular states can be written even more explicitly by accounting for electron spin, spin-orbit coupling, and the parity of the states in a manner similar to that adopted earlier for atomic states. The introduction of electron spin into the representation of molecular states is easy, but comes with a caution that the z-component of the total electron spins is traditionally denoted by 'Σ', which is not the same Σ that we have used in Table 5.1 to label the molecular state corresponding to $\Lambda = 0$. Indeed the z-component of the electron spin $\Sigma = 0, 1/2, 1, \ldots$ corresponds to singlet, doublet, triplet, ... spin states. The spin-orbit coupling is the same that we are familiar with, $\mathbf{J} = \mathbf{L} + \mathbf{S}$, but the z-projection of the $\mathbf{J}$ vector is denoted by Ω in the context of molecular states. In the limit of spin-orbit separation, the Stark field does not affect spin degeneracy. Using the electron spin Σ-values along with the possible values of Ω, we show the extension of the molecular electronic states in Table 5.2.

5.6 LCAO-MO

Molecular wavefunctions or molecular orbitals (MOs) are constructed by linear combination of atomic orbitals (LCAO)

$$\psi_{\text{MO}} = \sum_i^n c_i \phi_i, \tag{5.32}$$

where ϕ are atomic orbitals. The molecular wavefunctions ψ_{MO} are assumed to contain both the spatial and spin parts of the wavefunction, and they possess exchange antisymmetry to conform to the Pauli principle because of which the spatial part of ψ_{MO} should possess a definite symmetry or antisymmetry. Yet, these are assumptions only, and hence the choice of such wavefunctions to calculate the electronic energy levels of molecules under BO approximation may be considered somewhat arbitrary. The electronic potentials of a LCAO-constructed MO system is calculated to a first approximation using the variation principle

$$E = \frac{\left\langle \sum_i^n c_i\phi_i \middle| \mathcal{H}_{el} \middle| \sum_i^n c_i\phi_i \right\rangle}{\left\langle \sum_i^n c_i\phi_i \middle| \sum_i^n c_i\phi_i \right\rangle} = f\left(c_1, c_2, c_3, \ldots, c_n\right). \tag{5.33}$$

Because of the arbitrary choice of the initial ψ_{MO}, this energy needs to be minimized with regard to the parameters c_i. The energy minimization involves solving a set of n equations

$$\frac{\partial E}{\partial c_i} = 0 \text{ with } i = 1, 2, 3, \cdots, n,$$

that would yield a set of n simultaneous equations

$$\sum_i^n c_i \left(\mathcal{H}_{ij} - \left\langle \phi_i \middle| \phi_j \right\rangle E\right) = 0, \tag{5.34}$$

with $j = 1, 2, 3, \ldots, n$. The secular determinant so obtained provides n roots corresponding to E_n energies and ψ_n wavefunctions. Thus, the LCAO method yields as many molecular orbitals as the number of atomic orbitals combined.

5.7 MOLECULAR EIGENSTATES OF H_2^+

Molecular eigenstates (or wavefunctions) are conveniently discussed with the example of the hydrogen molecular ion H_2^+, because the problem is exactly solvable to derive the eigenstates and eigenvalues. Calling the two atoms (nuclei) A and B, the ground-state atomic orbitals are $1s_{\text{A}}$ and $1s_{\text{B}}$. The combinations of the atomic orbitals provide two molecular orbitals

$$\psi_1(R, r) = \frac{1}{\sqrt{2}}\left[\phi_{\text{A}}(r_{\text{A}}) + \phi_{\text{B}}(r_{\text{B}})\right]$$

$$\psi_2(R, r) = \frac{1}{\sqrt{2}}\left[\phi_{\text{A}}(r_{\text{A}}) - \phi_{\text{B}}(r_{\text{B}})\right] \tag{5.35}$$

where R and r variables refer to nuclear and electronic coordinates. The r_{A} and r_{B} vectors are defined in the context of the Hamiltonian discussed below. The factor $1/\sqrt{2}$ arises from the normalization condition $\int|\psi|^2 d\tau = 1$. The appearance of two molecular orbitals, ψ_1 and ψ_2, is the consequence of symmetry with respect to inversion through the center of the H_2^+ molecule. The molecular wavefunctions then must be either symmetric or antisymmetric. We notice that ψ_1 is symmetric due to even combination of ϕ_{A} and ϕ_{B}, but ψ_2 is antisymmetric. While ψ_1 is bonding, ψ_2 is the antibonding molecular orbital. Both are σ orbitals because they are cylindrically symmetric along the internuclear axis, but ψ_1 is called $\sigma_g 1s$ and ψ_2 is called $\sigma_u^* 1s$ in accordance with the traditional terminology gerade (even or symmetric) and ungerade (odd or antisymmetric).

To get the MO energy we write

$$\left(\mathcal{H}_{el} - E_{el}\right)\sum_i c_i\phi_i = 0, \tag{5.36}$$

in which the electronic Hamiltonian is given with reference to Figure 5.5

$$\mathcal{H}_{el} = \frac{\hbar^2}{2m_e}\nabla^2 - \frac{e^2}{r_{\text{A}}} - \frac{e^2}{r_{\text{B}}} - \frac{e^2}{r_{\text{AB}}}. \tag{5.37}$$

By multiplying the eigenvalue equation by ϕ_l^* and integration we obtain

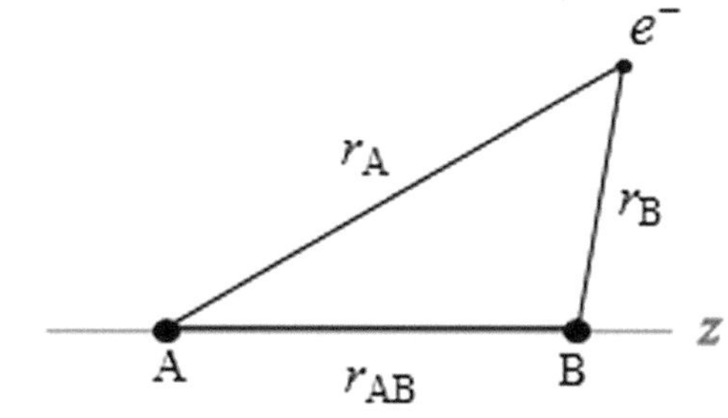

FIGURE 5.5 Electron and nuclear distance coordinates for H_2^+.

$$\sum_i c_i\left(\mathcal{H}_{li} - E\delta_{li}\right) = 0, \tag{5.38}$$

where i = A, B and l = A, B.

The resultant secular determinant

$$\begin{vmatrix} \mathcal{H}_{AA} - E & \mathcal{H}_{AB} - ES_{AB} \\ \mathcal{H}_{BA} - ES_{BA} & \mathcal{H}_{BB} - E \end{vmatrix} = 0 \tag{5.39}$$

contains three integrals

$$\text{Coulomb} \quad \mathcal{H}_{AA} = \int \psi_A^* \mathcal{H} \psi_A d\mathbf{r} = \mathcal{H}_{BB} = \int \psi_B^* \mathcal{H} \psi_B d\mathbf{r}$$

$$\text{Resonance} \quad \mathcal{H}_{AB} = \int \psi_A^* \mathcal{H} \psi_B d\mathbf{r} = \mathcal{H}_{BA} = \int \psi_B^* \mathcal{H} \psi_A d\mathbf{r}$$

$$\text{Overlap} \quad S = \int \psi_A \psi_B d\mathbf{r} = \int \psi_B \psi_A d\mathbf{r}. \tag{5.40}$$

Note that the integration variables $d\mathbf{r}$ are shown in this way to convey that they are invariant to swapping the coordinates of A and B, for example. They are often called dummy variables. The physical significance of these integrals becomes clear from the H_2^+ model in which the lone electron jumps from one to the other nuclear field. The off-diagonal resonance integrals signify this oscillation of the electron – a phenomenon called quantum resonance. The diagonal Coulomb integrals determine the electrostatic interaction between a nucleus and the electron charge distribution when the electron is closer in proximity with the other nucleus. The overlap integral provides the level of overlap of the atomic orbitals. If there were two orbitals only, then the overlap will be considered for the region where both orbitals are nonzero. The extent of overlap varies from 0 to ±1.

The solution of the determinant above (equation 5.39) has been provided by Cohen-Tannoudji et al. (1977: 1169–1180), as well as in earlier work of Eyring et al. (1944: 192–199) and King (1964: 133–137). The energy formula appears as

$$E_\pm = E(1s) - \frac{1}{1 \pm S}\frac{e^2}{a_o}\left[\frac{1}{\rho} - e^{-2\rho}\left(\frac{1}{\rho} + 1\right) \pm e^{-\rho}(1+\rho)\right] + \frac{e^2}{a_o}\frac{1}{\rho}, \tag{5.41}$$

where a_o is the Bohr radius, and $\rho = \frac{r_{AB}}{a_o}$. The energies E_+ and E_- correspond to electronic energies of ψ_1 and ψ_2 at a fixed r_{AB}. The plot of $E_\pm$ against r_{AB} has been found to qualitatively reproduce the energy calculated by solving the Schrödinger equation directly, but the dissociation energy D_e obtained by the LCAO method is found ~63% shorter than the value of 2.79 eV calculated directly. This suggests that LCAO need not be an imposing model for the description of MO. Nevertheless, LCAO does provide a conceptual framework for a qualitative description of molecular potential energy curves.

The eigenfunctions corresponding to bonding and antibonding states of the diatomic are

$$\psi_1(R,r) = \frac{1}{\sqrt{2(1-S)}}\left[\phi_A(r_A) + \phi_B(r_B)\right]$$

$$\psi_2(R,r) = \frac{1}{\sqrt{2(1+S)}}\left[\phi_A(r_A) - \phi_B(r_B)\right] \tag{5.42}$$

which, as mentioned earlier, are also labeled as $\sigma_g 1s$ and $\sigma_u^* 1s$, respectively. Note that S in the coefficients $1/\sqrt{2(1 \pm S)}$ is the overlap integral (see equation 5.40). The asterisk used in the symmetry symbol in $\sigma_u^* 1s$ indicates its antibonding character. Figure 5.6 depicts the contours of the orbitals, where the + sign indicates that the eigenfunction is positive everywhere – the case for the bonding orbital. For the otherwise antibonding case, the eigenfunction is positive on one side of the nodal plane and negative on the other. We notice that the electronic cloud is shared by both nuclei when it is the bonding scenario, but not so for the antibonding MO. Rather, there is no electron density in the nodal plane that bisects the two contour planes.

The construction of MO from p-orbitals is no different; symmetric (even) and antisymmetric (odd) combinations yield bonding and antibonding orbitals. In the case of the combination of p_z atomic orbitals, which we may choose to lie along the internuclear axis, the resultant MOs are designated bonding $\sigma_g 2p$ and antibonding $\sigma_u^* 2p$ orbitals (Figure 5.7). Both orbitals are cylindrically symmetric giving rise to $\lambda = 0$, and hence the use of the label σ. However, $\sigma_g 2p$ is symmetric with regard to both inversion and reflection operations, but $\sigma_u^* 2p$ is not. Thus, the two functions are even and odd, respectively, and hence the orbitals are of bonding and antibonding nature.

The result of combination of two $2p_x$ or two $2p_y$ orbitals is a bit different. Symmetric combination leads to a low-energy molecular orbital which lacks inversion symmetry, but the asymmetric combination results in the high-energy MO which is symmetric with respect to inversion operation. Because $\lambda = 1$ for the combined $2p_x$ or $2p_y$ orbitals, the low and high-energy molecular orbitals are designated $\pi_u 2p$ and $\pi_g^* 2p$, respectively (Figure 5.8). Notice the swapped appearance of 'g' and 'u' here compared to $\pi_g 2p$ and $\pi_u^* 2p$ labels for the $2p_z$ combinations. This is because the bonding and antibonding π orbitals are antisymmetric and symmetric, respectively, with respect to inversion.

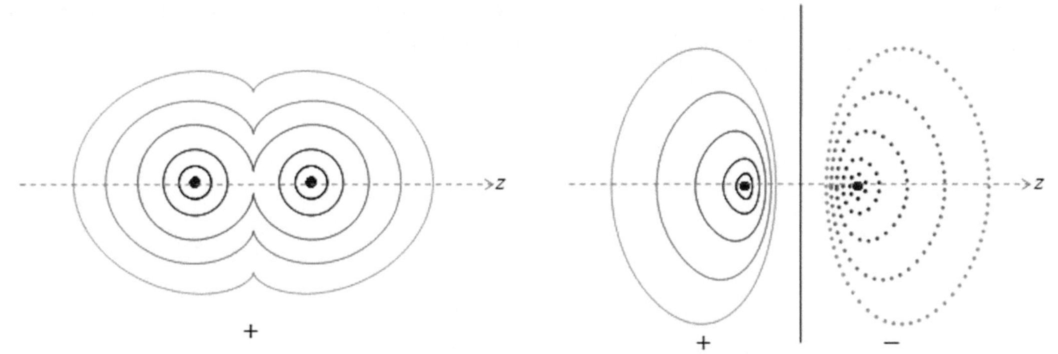

FIGURE 5.6 Contours of bonding ($\sigma_g 1s$) and antibonding ($\sigma_u^* 1s$) molecular orbitals. The bonding orbital is cylindrically symmetric, so the contour surfaces will appear identically when the orbital is rotated about the axis perpendicular to z. The antibonding orbital is positive and negative on the two sides of the nodal plane, so a 180° rotation about the axis perpendicular to z will swap the '+ and –' signs.

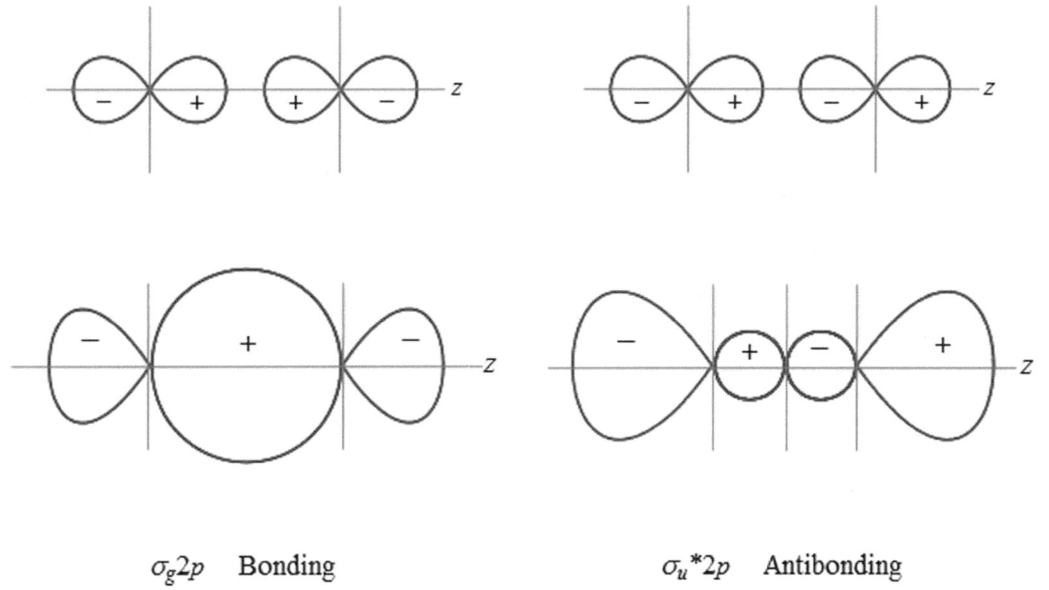

FIGURE 5.7 Linear combination of $2p_z$ atomic orbitals (top row) to produce bonding ($\sigma_g 2p$) and antibonding ($\sigma_u^* 2p$) molecular orbitals.

5.8 MOLECULAR EIGENSTATES OF H_2

The construction of molecular states of multielectron systems involves the same approach based on LCAO and BO approximation. For systems with more than one electron the total orbital angular momentum, $\Lambda = \Sigma_i \lambda_i$, and spin angular momentum, $S = \Sigma_i s_i$, are used to generate the total angular momentum $\Omega = \Lambda + \Sigma$, such that Σ represents the z-component of S. The electron-electron repulsions are approximated by perturbation, and the one-electron molecular orbitals are built up with regard to Pauli exclusion principle. Briefly discussed below is the construction of molecular states of H_2.

For the two electrons numbered 1 and 2, and the two nuclei labeled A and B, the required coordinate system is shown in Figure 5.9, with reference to which the electronic Hamiltonian can be written as

$$\mathcal{H}_{el} = -\frac{\hbar^2}{2m_e}\nabla_1^2 - \frac{\hbar^2}{2m_e}\nabla_2^2 - \frac{e^2}{r_{A1}} - \frac{e^2}{r_{B1}} - \frac{1}{r_{A2}} - \frac{1}{r_{B2}} + \frac{1}{r_{12}} + \frac{1}{r_{AB}}. \quad (5.43)$$

This is just an extension of the Hamiltonian that appeared for H_2^+ (equation 5.37). Leaving aside the nuclear repulsion term for the moment, which can be included later without a loss, we rewrite the Hamiltonian in a more convenient form

$$\mathcal{H}_{el} = \mathcal{H}_1 + \mathcal{H}_2 + \frac{1}{r_{12}}$$

$$\text{with } \mathcal{H}_1 = -\frac{\hbar^2}{2m_e}\nabla_1^2 - \frac{e^2}{r_{A1}} - \frac{e^2}{r_{B1}}$$

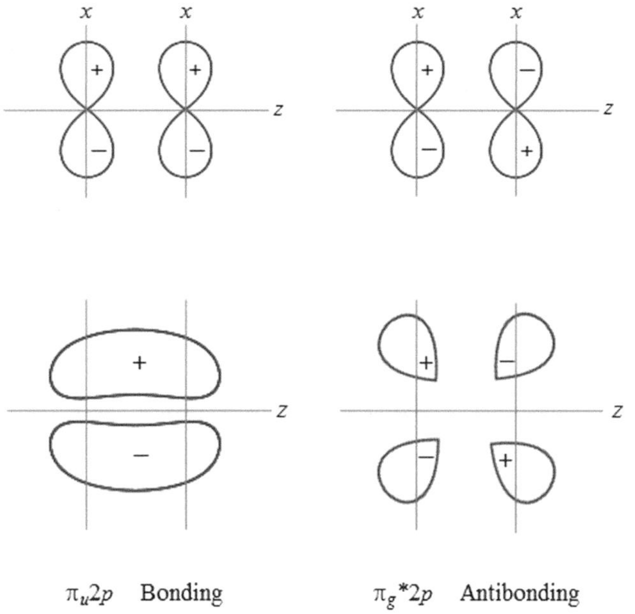

FIGURE 5.8 Molecular orbitals from linear combination of $2p_x$ atomic orbitals.

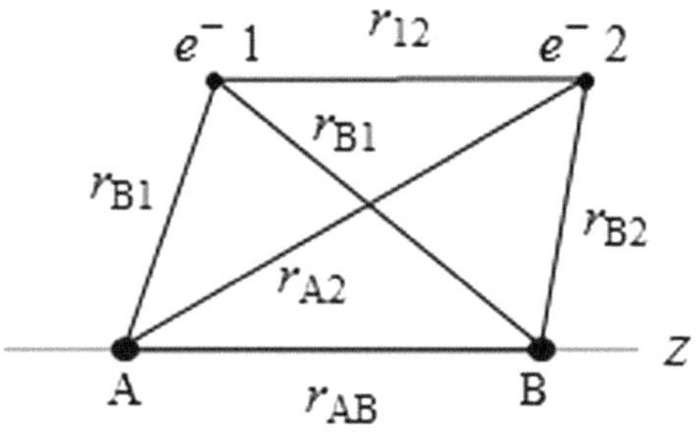

FIGURE 5.9 Electron and nuclear coordinate system for H_2.

$$\mathcal{H}_2 = -\frac{\hbar^2}{2m_e}\nabla_2^2 - \frac{e^2}{r_{A2}} - \frac{e^2}{r_{B2}}. \tag{5.44}$$

If the electron repulsions can be ignored as well, then the following two equations result

$$\mathcal{H}_1 \psi_{k,1} = E_k \psi_{k,1}$$

$$\mathcal{H}_2 \psi_{k,2} = E_k \psi_{k,2}, \tag{5.45}$$

whose solutions will provide the molecular orbitals. One can now include the nuclear repulsion term, so the total energy becomes

$$E = E_{el} + \frac{1}{r_{AB}}, \tag{5.46}$$

which can be written more generally including the nuclear charges as

$$E = E_{el} + \frac{Z_A Z_B}{r_{AB}}. \tag{5.47}$$

The normalized wavefunctions and corresponding energies can be found to be

$$\psi_1 = \frac{1}{\sqrt{2(1+S)}}\left(1s_A + 1s_B\right)$$

$$\psi_2 = \frac{1}{\sqrt{2(1+S)}}\left(1s_A - 1s_B\right)$$

$$E_1 = \mathcal{H}_{(1)AA} + \mathcal{H}_{(1)AB}\frac{1}{(1+S)}$$

$$E_2 = \mathcal{H}_{(2)AA} + \mathcal{H}_{(2)AB}\frac{1}{(1-S)}, \tag{5.48}$$

in which

$$\mathcal{H}_{(1\text{ or }2)AA} = E_{(1s)} - P$$

$$\mathcal{H}_{(1\text{ or }2)AB} = E_{(1s)}S - Q. \tag{5.49}$$

The $E_{(1s)}$ term denotes the energy of the $1s$ atomic orbital, which can be easily worked out and expressed in Dirac notation for convenience as

$$E_{(1s)} = \left\langle 1s_A \left| -\frac{1}{2}\nabla^2 - \frac{1}{r_A} \right| 1s_A \right\rangle. \tag{5.50}$$

The other integrals are

$$P = \left\langle 1s_A \left| \frac{1}{r_{B1}} \right| 1s_B \right\rangle$$

$$S - \left\langle 1s_A \middle| 1s_B \right\rangle$$

$$Q = \left\langle 1s_A \left| \frac{1}{r_{A1}} \right| 1s_B \right\rangle. \tag{5.51}$$

Of the two molecular orbitals, it is only the ψ_1 which is occupied by both electrons in the ground state of H_2. The ground state is generally given by

$$\psi_o = \frac{1}{\sqrt{2}}\left|\psi_1(1)\bar{\psi}_1(2)\right|. \tag{5.52}$$

The overbar on ψ_1 may seem new to the reader. This originates from Slater's (1963) work. Briefly, each orbital can have two electrons, one with α- and the other with β-spin. When we write an orbital, the overbar indicates that the corresponding electron indicated within small braces has the spin opposite to that of the first electron already assigned. Also to note in the above equation is the representation of only ψ_1 in the ground-state wavefunction, which should be obvious due to the bonding nature of the molecular orbital that contains both electrons 1 and 2. While $\psi_1(1)$ indicates electron 1 occupying the ψ_1 orbital with the electron spin in the α-state (+), $\bar{\psi}_1(2)$ indicates the occupation of electron 2 in the same molecular orbital but with the spin in the β-state. By identifying orbitals in this way both spin and space coordinates of the two electrons are ordered – a procedure used in the derivation of Slater determinant. The consequence of using both spin and space degrees of freedom is that the energy is derived by performing the integration of the eigenvalue equation over both spin and space coordinates of both electrons

$$E_{el} = \iint \psi_o^* \left| \mathcal{H}_1 + \mathcal{H}_2 + \frac{1}{r_{12}} \right| \psi_o d\tau_1 d\tau_2 + \frac{1}{r_{12}}$$

$$= \iint \psi_1^{2(*)}(1) \left| \mathcal{H}_1 + \mathcal{H}_2 \right| \psi_1^2 d\tau_1 d\tau_2 +$$

$$\iint \psi_1^{2(*)}(1) \left| \frac{1}{r_{12}} \right| \psi_1^2(2) d\tau_1 d\tau_2 + \frac{1}{r_{AB}}$$

$$= 2E_{(1 \text{ or} 2)} + \iint \psi_1^{2(*)}(1) \left| \frac{1}{r_{12}} \right| \psi_1^2 d\tau_1 d\tau_2 + \frac{1}{r_{AB}}$$

$$= 2E_{(1 \text{ or} 2)} + \left| \frac{1}{r_{12}} \right|_{11} + \frac{1}{r_{AB}}. \qquad (5.53)$$

The term $|1/r_{12}|_{11}$ is fairly involved. It contains two-centered Coulomb, exchange, and hybrid integrals, and a one-center integral that is similar to what is used in the derivation of the energy of the helium atom. The integrals are provided in detail by Bethe and Salpeter (1977: 127–139) and Kotani et al. (1955), and interested readers are encouraged to seek out such literature. For the present, we will carry along $|1/r_{12}|_{11}$ without expanding it. We note further that the $1/r_{12}$ term could be treated as a perturbation term. If so, the eigenfunctions at a fixed value of r_{AB} can be used to obtain the ground state energy (Davidson, 1961) as

$$E_o(r_{AB}) = 2E_1 + \frac{1}{r_{AB}}. \qquad (5.54)$$

Since we have chosen to use the $|1/r_{12}|_{11}$ integral here, we can write the ground state energy in the following way. Substituting for the $2E_1$ term above with equations (5.48) and (5.49), one obtains the ground state energy of H_2

$$E_o(r_{AB}) = 2E_{(1 \text{or} 2)} - \frac{2(P+Q)}{1+S} + \left| \frac{1}{r_{12}} \right|_{11} + \frac{1}{r_{AB}}. \qquad (5.55)$$

Both first and second terms on the right-hand side are energies of a single electron in the Stark field, whether electron 1 or 2 is immaterial. Recall that P, Q, and S are integrals as mentioned in equation (5.51). The third term obviously represents electron-electron repulsion. As cautioned earlier, LCAO is not a robust model. The bond energy calculated from equation (5.55) is 2.68 eV, which is largely different from the experimental value of 4.75 eV.

5.9 SINGLET AND TRIPLET EXCITED STATES OF H_2

To work with the excited state of H_2, we continue to label the two electrons of the molecule as 1 and 2. While both electrons occupy the ground-state bonding orbital ψ_1 where the spin multiplicity is singlet, $2S+1=1$, the promotion of one of the electrons to the ψ_2 molecular orbital can retain the same spin multiplicity or else produce a triplet state, $2S+1=3$. In the excited state an electron can be found in any of the two MOs, which yields two product functions given by $\psi_1(1)\psi_2(2)$ and $\psi_1(2)\psi_2(1)$. These two functions can be combined to construct two new normalized functions χ and χ'

$$\chi = \frac{1}{\sqrt{2}}\left[\psi_1(1)\psi_2(2)\right] + \psi_1(2)\psi_2(1)$$

$$\chi' = \frac{1}{\sqrt{2}}\left[\psi_1(1)\psi_2(2)\right] - \psi_1(2)\psi_2(1). \qquad (5.56)$$

It is noticed easily that χ and χ' are symmetric and antisymmetric, respectively, with respect to interchange of the two electrons between ψ_1 and ψ_2. Because electron spin has not been invoked yet, these functions provide only the space part of the electrons. Regarding the spin part, the combination of $\alpha(1)$, $\alpha(2)$, $\beta(1)$, and $\beta(2)$ yields four product spin functions

$$\sigma_1 = \alpha(1)\ \alpha(2)$$

$$\sigma_2 = \beta(1)\ \beta(2)$$

$$\sigma_3 = \alpha(1)\ \beta(2)$$

$$\sigma_4 = \beta(1)\alpha(2). \qquad (5.57)$$

The functions σ_3 and σ_4 are found to interchange when operated upon by an interchange operator P, such that

$$P\sigma_3 = P\left[\alpha(1)\beta(2)\right] = \alpha(2)\beta(1) = \sigma_4$$

$$P\sigma_4 = P\left[\beta(1)\alpha(2)\right] = \beta(2)\alpha(1) = \sigma_3. \qquad (5.58)$$

This result compels the use of a linear combination of σ_3 and σ_4, while σ_1 and σ_2 can be brought forth as they are. The spin functions are now categorized as

$$\left.\begin{array}{c} \alpha(1)\alpha(2) \\ \beta(1)\beta(2) \\ \frac{1}{\sqrt{2}}\left[\beta(1)\alpha(2) + \beta(2)\alpha(1)\right] \end{array}\right\} = \phi_T$$

$$\frac{1}{\sqrt{2}}\left[\alpha(1)\beta(2) - \beta(1)\alpha(2)\right] = \phi_s, \qquad (5.59)$$

where ϕ_T and ϕ_S are triplet- and singlet-state spin functions, respectively. Also, ϕ_T is symmetric but ϕ_S is antisymmetric with respect to electron interchange operation. The consideration of the symmetry of both spatial and spin functions provides two wavefunctions for the excited states

$$1_{\psi_1} = \chi\phi_s$$
$$3_{\psi_1} = \chi'\phi_T. \quad (5.60)$$

Substitution for χ, ϕ_s, and ϕ_T would show

$$1_{\psi_1} = \frac{1}{2}\left[\psi_1(1)\bar{\psi}_2(2) - \bar{\psi}_1(1)\psi_2(2)\right]$$

$$3_{\psi_1} = \begin{cases} \frac{1}{\sqrt{2}}\left[\psi_1(1)\psi_2(2)\right] \\ \frac{1}{2}\left[\psi_1(1)\bar{\psi}_2(2) + \bar{\psi}_1(1)\psi_2(2)\right]. \\ \frac{1}{\sqrt{2}}\left[\bar{\psi}_1(1)\psi_2(2)\right] \end{cases} \quad (5.61)$$

The eigenvalue equations for these functions use the same electronic Hamiltonian given in equation (5.43), so

$$E_{\text{el}}^{(s)} = \left\langle 1_{\psi_1}\middle|\mathcal{H}_{\text{el}}\middle|1_{\psi_1}\right\rangle$$
$$E_{\text{el}}^{(t)} = \left\langle 3_{\psi_1}\middle|\mathcal{H}_{\text{el}}\middle|3_{\psi_1}\right\rangle, \quad (5.62)$$

where (s) and (t) superscripts denote singlet and triplet states. The integrals can be solved in the way done earlier so as to obtain

$$E_{\text{el}}^{(s)} = E_1 + E_2 + \left|\frac{1}{r_{12}}\right|_{12} + \left|\frac{1}{r_{12}}\right|_{12}'$$

$$E_{\text{el}}^{(t)} = E_1 + E_2 + \left|\frac{1}{r_{12}}\right|_{12} - \left|\frac{1}{r_{12}}\right|_{12}', \quad (5.63)$$

in which E_1 and E_2 are energies of electron 1 and 2, $\left|1/r_{12}\right|_{12}$ represents the repulsion between electron 1 in ψ_1 and electron 2 in ψ_2, and $\left|1/r_{12}\right|_{12}^{/}$ describes repulsion between electrons 1 and 2 when both of them are associated with both ψ_1 and ψ_2.

5.10 ELECTRIC DIPOLE TRANSITION IN H_2

The general form of the transition moment appropriate for a two-level system has been worked out earlier (Chapter 3). In electric transitions of molecules, however, both space and spin parts of the molecular wavefunction need to be considered. This should be evident from the preceding section that has shown the forms of excited-state singlet- and triplet-state wavefunctions. We consider here the explicit form of the transition dipole moment for H_2 when an electron is promoted from the ground state to the excited triplet $\left(\psi_o \rightarrow 3_{\psi_1}\right)$ or singlet ($\psi_o \rightarrow 1_{\psi_1}$) states. The transition dipole moment for the $\psi_o \rightarrow 3_{\psi_1}$ excitation is

$$\mathbf{\mu}_{o3} = \left\langle\psi_o\middle|\mathbf{\mu}\middle|3_{\psi_1}\right\rangle = \left\langle\psi_o\middle|\Sigma_i e\mathbf{r}_i\middle|3_{\psi_1}\right\rangle. \quad (5.64)$$

The dipole is a function of the position vector of the electron $\mathbf{r}$; it is the product of the electron charge and the positions vector. Since the integral involves both spin and orbital space

$$\mathbf{\mu}_{o3} = \left\langle\chi_o\middle|\Sigma_i e\mathbf{r}_i\middle|\chi'\right\rangle\left\langle\phi_s|\phi_t\right\rangle, \quad (5.65)$$

where the form of the excited state wavefunction given in equation (5.60) has been used. But the spin functions are orthogonal, which means the integral $\left\langle\phi_s|\phi_t\right\rangle$ in the equation above is zero, and therefore the transition integral $\mathbf{\mu}_{o3}$ is zero. This is why the singlet-to-triplet transition is said to be forbidden. However, this result is appropriate only when spin and orbital motions of an electron are weakly coupled, which allows separating the orbital and spin functions. This is the case with light atoms where spin and orbital motions interact rather weakly, and singlet-to-triplet transitions are not frequent. The $\mathbf{\mu}_{o3}$ integral may have a finite value for molecules containing heavy atoms.

Regarding the ground to the singlet excited state transition ($\psi_o \rightarrow 1_{\psi_1}$), the dipole integral

$$\mathbf{\mu}_{o1} = \left\langle\chi_o\middle|\Sigma_i e\mathbf{r}_i\middle|\chi\right\rangle\left\langle\phi_s|\phi_s\right\rangle \quad (5.66)$$

reduces to

$$\mathbf{\mu}_{o1} = \left\langle\chi_o\middle|\Sigma_i e\mathbf{r}_i\middle|\chi\right\rangle, \quad (5.67)$$

which is due to the orthogonality of the spin functions. Substituting for χ_0 and χ from equation (5.56) in the integral above, we get

$$\mathbf{\mu}_{o1} = \frac{1}{\sqrt{2}}[\left\langle\psi_1(1)\psi_1(2)|\mathbf{r}_1|\psi_1(1)\psi_2(2) + \psi_1(2)\psi_2(1)\right\rangle$$
$$+\left\langle\psi_1(1)\psi_1(2)|\mathbf{r}_2|\psi_1(1)\psi_2(2) + \psi_1(2)\psi_2(1)\right\rangle]. \quad (5.68)$$

Since the molecular wavefunctions ψ_1 and ψ_2 belong to the same origin of the molecular system they must be orthogonal, and hence the integrals above reduce to

$$\mathbf{\mu}_{o1} = \frac{1}{\sqrt{2}}\left[\psi_1(1)|\mathbf{r}_1|\psi_2(1) + \psi_1(2)|\mathbf{r}_2|\psi_2(2)\right]. \quad (5.69)$$

However, the two integrals are equal because the two electrons can be swapped for each other, so

$$\mathbf{\mu}_{o1} = \sqrt{2}\left\langle\psi_1(1)|\mathbf{r}_1|\psi_2(1)\right\rangle$$

$$= \sqrt{2}\left[\frac{1}{\sqrt{2(1+S)}\sqrt{2(1-S)}}\right]\left\langle 1s_A(1) + 1s_B(1)|\mathbf{r}_1|1s_A(1) - 1s_B(1)\right\rangle$$

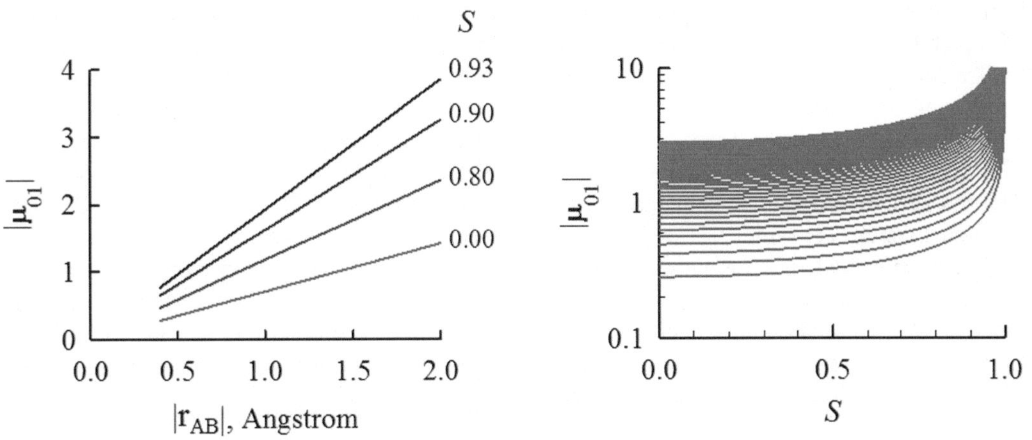

FIGURE 5.10 The dependence of the transition dipole on the internuclear distance (*left*), and on the magnitude of the overlap integral (*right*). In the calculation of the latter the value of r_{AB} was allowed to increase from 0.4 to 4 Å.

$$= \frac{\sqrt{2}}{\sqrt{4\left(1-S^2\right)}}\left[\left\langle 1s_A(1)|r_1|1s_A(1)\right\rangle - \left\langle 1s_B(1)|r_1|1s_B(1)\right\rangle\right], \quad (5.70)$$

where we have used the normalized wavefunctions for bonding and antibonding states. Note that the normalization coefficients contain the overlap integral S (see equation 5.42).

If the average location of an electron in a spherically symmetric $1s$ orbital is assumed to correspond to the location of the nucleus, meaning the center of the sphere, then the expectation value of the integrals will be the positive vector of the nuclei only when

$$\mu_{01} = \frac{\sqrt{2}}{\sqrt{4\left(1-S^2\right)}}\left[\mathbf{r}_A - \mathbf{r}_B\right]$$
$$= \frac{\left|\mathbf{r}_{AB}\right|}{\sqrt{2\left(1-S^2\right)}}. \quad (5.71)$$

This result is simple and seemingly obvious. The transition dipole moment is directly proportional to the internuclear separation for any value of the overlap integral $|S| < 1$, even when there is no overlap (Figure 5.10), although the gain in the magnitude of μ is infinitely large when $|S| \rightarrow 1$.

5.11 MOLECULAR ORBITAL ENERGY AND ELECTRONIC CONFIGURATION

The MO energy is generally determined by the atom orbital types that combine and the magnitude of orbital overlap as evaluated from the overlap integral. Although the knowledge of the relative energies of atomic orbitals can be used as the first principle, knowing the extent of orbital overlap greatly facilitates ordering MOs in terms of relative energies. The MOs formed from only $1s$ atomic orbitals will obviously be lower in energy compared with MOs formed of $2s$ atomic orbitals. Also, the energy separation between bonding and antibonding π-orbitals will be less than that in the case of σ orbitals because the overlap of two $2p_x$ or two $2p_y$ orbitals is less than the overlap between a pair of $2s$ or $2p_z$ orbitals. The energy order of MOs could be

$$\sigma_g 1s < \sigma_u 1s < \sigma_g 2s < \sigma_u 2s < \sigma_g 2_p < \pi_u 2p < \sigma_u 2p. \quad (5.72)$$

To describe the electronic state configurations of many-electron diatomic molecules, the total electron orbital angular momentum Λ is taken as the sum of the individual electron orbital angular momentum, $\Lambda = \Sigma_i \lambda_i$. Similarly, the total spin angular momentum, $S = \Sigma_i s_i$, so that the total angular momentum $\Omega = \Lambda + \Sigma$, where Σ represents the z-component of S. As mentioned earlier, the electron-electron repulsions are approximated by perturbation, and the one-electron molecular orbitals are filled according to the Pauli principle. Let us consider again the eigenvalue equation for the L_z operator. The capital L is being used only to indicate the multielectron context. The eigenvalue equation is

$$L_z\, \psi_{el} = \pm \Lambda \hbar \psi_{el}, \quad (5.73)$$

where the eigenvalue Λ, a quantum number used to indicate the quantized z-projection of the total orbital angular momentum, is used to label the molecular electronic states. We also need to consider the changes in the eigenfunction ψ_{el} with respect to a reflection in a plane containing the molecular axis along z-direction. If σ is the operator for reflection, then

$$\sigma\psi_{el} = c\psi_{el} \text{ and } \sigma^2\psi_{el} = c^2\psi_{el} = \psi_{el}, \quad (5.74)$$

because $c^2 = 1, c = \pm 1$. This is the case when $\Lambda = 0$ for which Σ is given as Σ^+ and Σ^- states. This conforms to the Pauli principle according to which the wavefunction that includes both space and spin parts must be antisymmetric. Thus, if Σ^+ exists, so does Σ^-, and vice versa. If $\sigma\psi_{el} = -\psi_{el}$, the sign of the eigenvalue is swapped, which is the case for $\Lambda = 1, 2, 3, \ldots,$

then each reflection operation changes the sign of Λ, so that $\Lambda = \pm1, \pm2, \pm3, \ldots$. This implies that the molecular electronic states $\Pi, \Delta, \Phi, \ldots$ are doubly degenerate, because swapping the sign of Λ will not change the energy.

If all molecular orbitals are doubly filled, then the electron configuration is called a 'closed shell' configuration. The importance of closed shell configurations is that all possible electronic states arising thereby are symmetric, and such filled molecular orbitals can be excluded in further determination of symmetry properties of electronic states by using higher partially filled shells. Conventionally, the closed part of the shell configuration, which is $(\sigma 1s)^2(\sigma^* 1s)^2$, is indicated by 'KK'. For example, configurations for three electronic states of O_2 are

Ground state	$\text{KK}(\sigma 2s)^2(\sigma 2s^*)^2(\sigma 2p_z)^2(\pi 2p_x)^2(\pi 2p_y)^2(\pi 2p_x^*)^1(\pi 2p_y^*)^1$	${}^3\Sigma_g^-$
First excited state	$\text{KK}(\sigma 2s)^2(\sigma 2s^*)^2(\sigma 2p_z)^2(\pi 2p_x)^2(\pi 2p_y)^2(\pi 2p_x^*)^2(\pi 2p_y^*)^0$	${}^1\Delta_g$
Second excited state	$\text{KK}(\sigma 2s)^2(\sigma 2s^*)^2(\sigma 2p_z)^2(\pi 2p_x)^2(\pi 2p_y)^2(\pi 2p_x^*)^1(\pi 2p_y^*)^1$	${}^1\Sigma_g^+$

Importantly, the configurations shown for the ground state and the second excited state appear identical. But it is easy to pick up that they could be different on the basis of spin multiplicity. Indeed, the ground state ${}^3\Sigma_g^-$ is more stable than the ${}^1\Sigma_g^+$ state due to Hund's rule of maximum multiplicity.

In a closed shell configuration, the number of electrons with α-spin must be equal to the number with β-spin, which implies that $2S+1=1$, meaning only one spin wavefunction is associated with the spatial wavefunction. The overall evenness of the inversion symmetry of states arising from closed shell configuration is denoted by the subscript 'g' as in Σ_g^+ and Σ_g^-. For example, the $(\sigma_g 1s)^2$ configuration of the ground state of H_2 is a closed shell configuration because the MO is filled by the two electrons. Hence, the ground state electronic configuration is Σ_g^+. The electronic states of homonuclear diatomic molecules should also show the spin multiplicity. As mentioned already, closed shell configurations always yield unit multiplicity $2S+1=1$, but open shell configurations like the excited electronic states can be singlet or triplet. The MO configuration for the excited state of H_2, for instance, is $(\sigma_g 1s)(\sigma_u^* 1s)$. Then the ground and excited electronic configurations of H_2 correspond to ${}^1\Sigma_g^+$ and ${}^3\Sigma_u^+$, respectively. The symmetry symbols along with the spin multiplicity indications are also called term symbols. The term symbols are sometimes also shown by a letter preceding, which serves to identify the ground state X, but has no special significance otherwise. The ground state electronic configuration is thus shown by $X^1\Sigma_g^+$.

5.12 MOLECULAR ORBITALS OF HETERONUCLEAR DIATOMIC MOLECULE

The LCAO method is applicable also for construction of heteronuclear diatomic molecular orbitals, and the one-electron molecular orbitals are identically denoted by $\sigma, \pi, \delta, \ldots$, as done for the homonuclear case. Even the electronic states are given by the same set of term symbols $\Sigma^+, \Sigma^-, \Pi, \Delta, \ldots$. However, heteronuclear diatomic molecules do not possess center of symmetry, because of which associating the 'g' and 'u' symbols with these molecular orbitals is not permitted. Nevertheless, the symmetric and antisymmetric combinations here are distinguished by the letters A and B, respectively.

The atomic orbitals that combine to yield a heteronuclear molecular orbital need not both originate from an isoelectronic shell. Unlike the homonuclear case, where only combinations of the type $1_{s_A} + 1_{s_B}$, $2_{s_A} + 2_{s_B}$, and $2_{p_{zA}} + 2_{p_{zB}}$ occur, a heteronuclear orbital may form by the combination of atomic orbitals originating from different shells. Thus, the $1s$ orbital of one atom can combine with the $2p_z$ orbital of the other atom. But such combinations must yield orbital overlap, which leads to the condition that the two combining atomic orbitals must be characterized by the same m value. If we let the molecular axis lie along the z-direction, then the combination of the $1s$ orbital ($m=0$) of one atom with the $2p_x$ orbital ($m=+1$ or -1) of the other does not lead to orbital overlap (Figure 5.11), because the regions of positive and negative overlap cancel

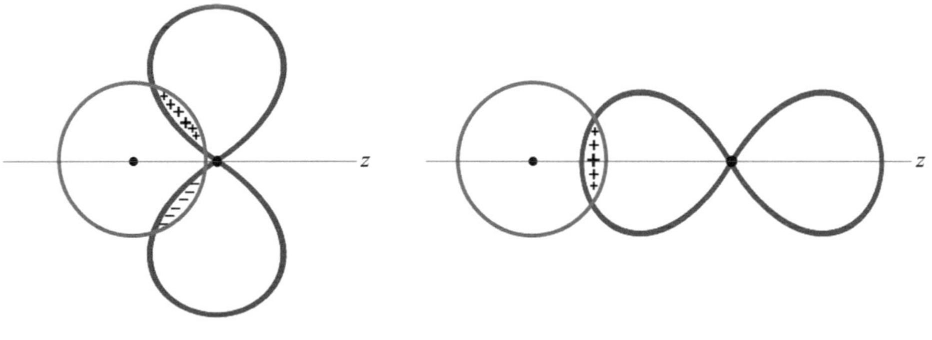

FIGURE 5.11 Linear combination of $1s$ and $2p$ orbitals. *Left*, the P-orbital is $2p_x$ or $2p_y$ type, and the positive and negative overlaps cancel out to produce no net overlap. *Right*, the $1s$ and $2p_z$ combination yields a net orbital overlap.

each other out, which leads to a net overlap of zero. The combination of $1s$ orbital of one atom and the $2p_z$ orbital of the other, the m value for both orbitals being 0, will yield a nonzero value of the overlap integral.

The LCAO-constructed MO wavefunction for the A–B heteronuclear diatomic should be

$$\psi = c_1 1s_{\mathrm{A}} + c_2 2s_{\mathrm{B}}, \tag{5.75}$$

but the determination of the atomic orbital coefficients c_1 and c_2 here is difficult due to the lack of symmetry. Approaches based on the variation theorem are adopted to carry out MO calculations of heteronuclear systems, including di- and conjugated polyatomic systems (Salem, 1966: 1–158). Let us take a brief look at an approximate calculation of molecular orbitals of lithium hydride that Mulliken (1936) did a little less than a century ago. The ground state molecular orbital electron configuration is $(\sigma 1s)^2(\sigma 2s)^2$, where $(\sigma 1s)^2$ is the lithium closed shell, because the orbital is localized almost entirely to the $1s$ orbital of the lithium atom. This allows performing the calculation considering the $\sigma 2s$ orbital alone. From the wavefunction

$$\psi = c_1 1s_{\mathrm{H}} + c_2 2s_{\mathrm{Li}}$$

one gets the following secular determinant

$$\begin{vmatrix} \mathcal{H}_{11} - E & \mathcal{H}_{12} - ES_{\mathrm{AB}} \\ \mathcal{H}_{21} - ES_{\mathrm{BA}} & \mathcal{H}_{22} - E \end{vmatrix} = 0,$$

in which the matrix elements have the usual meaning. The diagonal elements are estimated from the ionization potentials of $1s_{\mathrm{H}}$ and $2s_{\mathrm{Li}}$ orbitals that are known from experiments. The overlap integral S estimated from the Li–H bond length turns out to be ~ 1.6Å, and the off diagonals $\mathcal{H}_{12} = \mathcal{H}_{21}$ is given by the Mulliken formula

$$\mathcal{H}_{12} \approx \kappa S\left(\frac{\mathcal{H}_{11} + \mathcal{H}_{22}}{2}\right). \tag{5.76}$$

Without noting the respective numerical values of these quantities, we simply write down the roots obtained from the determinant. The number of roots should be equal to the number of molecular orbitals. For the present

$$E = \begin{cases} -0.52 \\ +0.06 \end{cases}.$$

The value of –0.52 eV represents the energy of the lowest molecular orbital, and by using the value of –0.52 in a secular equation the ratio c_1/c_2 can be determined. The higher-value root (0.06 eV) represents the energy of the lowest excited state $(\sigma 1s)^2(\sigma 2s)(\sigma 2p)$.

5.13 MOLECULAR ORBITALS OF LARGE SYSTEMS

The LCAO is a general approach and can be applied to construct molecular orbitals of large molecules, including hydrocarbons, and linear and cyclic conjugated systems. The procedure for molecular orbital calculation is the same – orthogonal atomic orbitals are linearly combined to generate molecular orbitals whose energies and atomic orbital coefficients are determined using an effective electronic Hamiltonian in conjunction with the variational method. Molecular orbitals can also be constructed using one-electron theories based on π-electrons confined to a region of constant potential. Results obtained by the LCAO approach are consistent with those from one-electron theories of absorption spectra. As an example of molecular orbital calculations of large molecules, we will provide literature results briefly for porphin, which is discussed widely to establish the structure-spectroscopy relationship for large molecules. In fact, the MO theory of porphyrins has played a crucial role in the discussion of structure and spectra of various hemes and heme-containing systems (Eaton and Hofrichter, 1981).

5.13.1 LCAO-MO of Porphyrins

The problem of molecular orbitals of heme porphyrin considers the $2p$ orbitals of all 24 atoms in the localized neutral porphine backbone (Figure 5.12). It is useful to consider first the symmetry group of porphin because the molecular orbitals or the orbital wavefunctions are bases for the irreducible representations of the symmetry groups considered. This follows from a basic principle that the solutions of the Schrödinger equation

$$\mathcal{H}_{\mathrm{el}}\psi_{\mathrm{el}} = E_{el}\psi_{\mathrm{el}} \tag{5.77}$$

belong to a set of irreducible representations of the symmetry group, but the Hamiltonian remains invariant to such symmetry transformations. The molecular point group for the planar porphin is D_{4h} for the delocalized dianion, and since the π-molecular orbitals are antisymmetric to inversion and reflection with respect to the plane of the molecule, a set of five irreducible representations – A_{1u}, A_{2u}, B_{1u}, B_{2u}, E_g – results (see the following chapter). Porphin has 20 carbon atoms, each contributing one π-electron, and four nitrogen atoms that contribute six π-electrons. These 26 π-electrons should generate 13 MOs of lowest energy that belong to the irreducible representations of D_{4h} symmetry. This level of information is obtained from the π-electron counting and the consideration of the group theory alone. According to the procedure already outlined, each MO should correspond to one of the roots of the secular equations. Original calculations of Longuet-Higgins et al. (1950) have listed a total of 24 molecular orbitals, 13 of which are lowest-energy occupied and 11 are highest-energy unoccupied orbitals. Further, the molecular orbitals of E_g representation occur in doubly degenerate pairs, generating three pairs each of lowest occupied and highest unoccupied molecular orbitals (Figure 5.12).

5.13.2 Free-Electron Orbitals of Porphyrins

A particularly simple approach for constructing molecular orbitals of conjugated systems, both linear and circular, insists

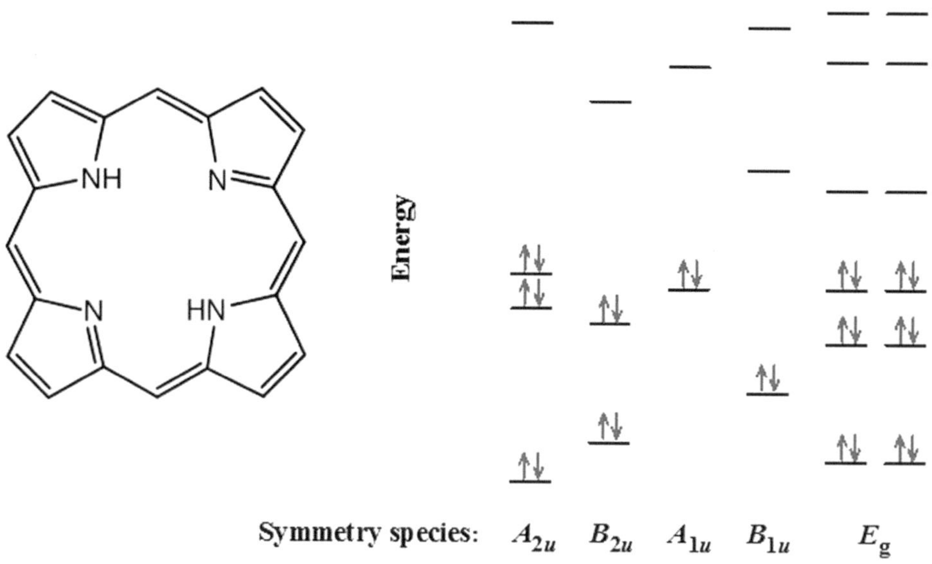

FIGURE 5.12 *Left*, porphin backbone. *Right*, LCAO–MO of porphin (porphyrin) drawn on the basis of data of Longuet-Higgins et al. (1950).

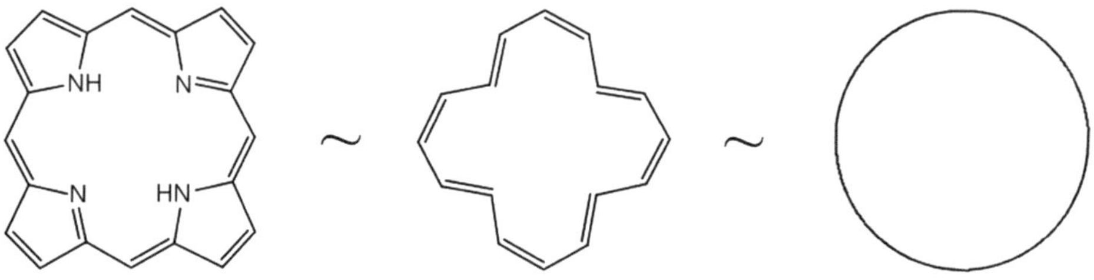

FIGURE 5.13 The perimeter approximation in which the carbon and nitrogen atoms of porphin lie in a ring.

on obtaining free-electron orbitals for electrons contained in a one-dimensional potential well. The idea is to consider the entire length of the chain skeleton as a box containing the π-electrons, in analogy with particles in a one-dimensional potential well. From the 'particle in a box' problem solved in elementary quantum chemistry, the wavefunction and energy are known to be

$$\psi = \sqrt{\frac{2}{L}}\sin\frac{j\pi x}{L} \text{ with } j = 1,2,3,\cdots$$

$$E = \frac{j^2h^2}{8mL^2}, \tag{5.78}$$

in which L is the length of the one-dimensional box along the x-axis, m is the particle (electron) mass, and j is the quantum number for the energy states. This is the same quantum number denoted by n in many textbooks, but we will reserve n here to serve for another purpose of the model explanation. Also to note is that unlike the 'particle in a box' where $j = 0$ is not allowed, the zero-quantum number is allowed in the 'particle in a ring' analogy. The two formulas above serve for the wavefunction and the energy of the molecular orbitals projected in the free-electron molecular orbital approach. The value of L for molecular orbital calculations may be taken as the total length of the circularized molecular skeleton. When applied to heme porphyrins the free-electron model takes the form of Platt's (1949) perimeter model that considers the porphyrin ring as a one-dimensional loop or ring of constant potential (Figure 5.13).

Quite generally, the one-electron molecular orbitals in the perimeter model are hydrogen-like atomic orbitals that have a dependence on the azimuthal angle for the orientation of the electron orbital angular momentum about the ring axis (Figure 5.14), which is equivalent to the molecular symmetry axis. If θ is the azimuthal angle, then the molecular wavefunction

$$\psi_{\text{MO}} \propto e^{im\theta}, \tag{5.79}$$

where the azimuthal quantum number is indicated by m, which will vary from one to another atomic orbital. Henceforth the quantum number will be given by l specific to each atomic orbital.

If the normalized atomic coefficient a for the rth atomic orbital is

$$c_r = ae^{ir\theta}, \tag{5.80}$$

where c_r is the amplitude coefficient, then as described earlier, the secular equation can be written for all atomic

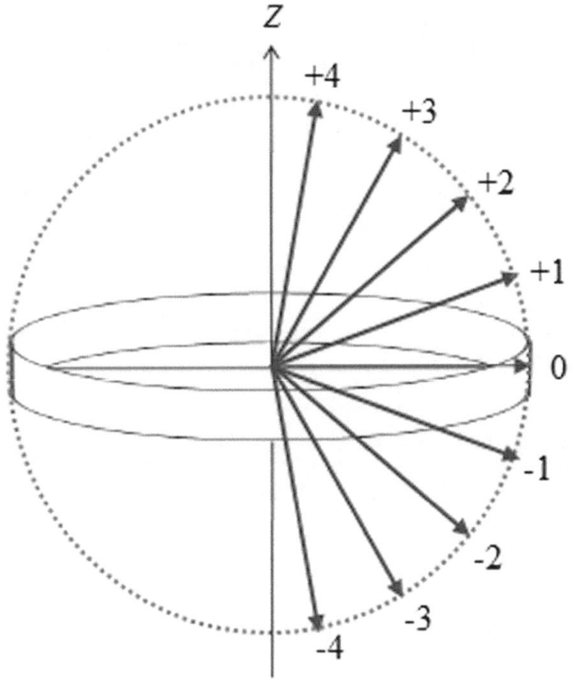

FIGURE 5.14 Electron orbital angular momenta about the ring axis-z.

orbitals. If there are p atomic orbitals, then there are p roots corresponding to energies and wavefunctions of the molecular orbitals. Taking up the porphyrin example, the ring is assumed to form from the 16 inner atoms (Figure 5.13), such that each of the 12 carbon atoms contribute one π-electron, and the four nitrogen atoms contribute six π-electrons in view of the two hydrogens attached to two of these. There are thus 18 π-electrons $(=4n+2)$. In molecular orbital calculations of cyclic systems like polyene and porphyrin, the number of atoms in the ring is given by $4n$ so that the number of π-electrons available is $4n+2$. By a linear combination of the $2p_\pi$ atomic orbitals (ϕ) we expect nine molecular orbitals (ψ) for the porphyrin, which are given by

$$\psi_k = \frac{1}{\sqrt{4n}}\sum_{p=1}^{4n} e^{\left(\frac{\sqrt{-1}\pi}{2n}jp\right)}\phi_r, \tag{5.81}$$

where $j = 0, \pm1, \pm2, \cdots, \pm n$, $n=4$ in the present case

The form of the molecular orbital energy can be shown to be

$$E = 2\beta\cos\theta, \tag{5.82}$$

where β is resonance energy and θ is the azimuthal angle. Obtaining θ needs boundary conditions, but this can be achieved by using the argument that in counting the atoms in a ring the last atom-count is always followed by the first-atom count. If there are p atoms in the ring, then after counting the $(p-1)^{\text{th}}$ atom one arrives at the pth atom. This means

$$e^{ir\theta} = e^{i(r+p)\theta}$$
$$e^{ip\theta} = 1, \tag{5.83}$$

which can be solved easily to obtain

$$\theta = \frac{2l\pi}{p},$$

where $l(=0,1,2,\cdots,p-1)$ is the electron orbital angular momentum.

The explicit form of the molecular wavefunctions can also be written down. Since the amplitude coefficient c of an atom depends on the quantum number l, the coefficient for the rth atom is

$$c_{(l)r} = a\, e^{\left[i\left(\frac{2l\pi}{p}\right)r\right]}. \tag{5.84}$$

Using $\sum_r a^2 = 1$ for normalization, we can write

$$\psi_l = \frac{1}{\sqrt{p}}\sum_{r=1}^{p} e^{\left[i\left(\frac{2l\pi}{p}\right)r\right]}\phi_r. \tag{5.85}$$

To specify the $(4n+2)$ π-electrons for porphyrin-like cyclic systems we should write

$$\psi_l = \frac{1}{\sqrt{4n}}\sum_{r=1}^{4n} e^{\left[\frac{\sqrt{-1}\pi}{2n}lr\right]}\phi_r, \tag{5.86}$$

where $l = 0, \pm1, \pm2, \cdots, (2n-1), 2n$. The 18 π-electrons can then be filled up to $l = \pm4$, two orbitals of degenerate energy arising from each non-zero value of l.

The physical meaning of ψ_l molecular wavefunctions can be understood by rewriting the azimuthal angle as

$$\theta = \frac{2l\pi}{p} = \frac{2\pi}{\lambda}r = \mathbf{k}r \tag{5.87}$$

where $\mathbf{k}$ is the wave vector (see Chapter 1), and r is approximated to an average bond length. Now, $\mathbf{k}$ will be in the $+ve$ or clockwise direction if l is $+ve$, and the anticlockwise direction will correspond to the negative l. This means each π-molecular orbital can be characterized by a wavelength which is determined by the total number of π-electrons that we started with, and the wavelength is

$$\lambda = \frac{pr}{|l|}, \tag{5.88}$$

which will be infinitely long for $l = 0$. Using $p = 18$ and r~1.5Å, the wavelengths are estimated to be 27, 14, 9, and 7 Å for $|l|$ values of 1, 2, 3, and 4, respectively. Further, the wavefunctions corresponding to a degenerate pair of molecular orbitals ($\pm l$) are traveling waves in the opposite direction, one clockwise (+) and the other anticlockwise (–). We will encounter later the analogous situation when we analyze the motion of a rigid diatomic molecule rotating in a two-dimensional plane. For the present, the two senses of the wave are

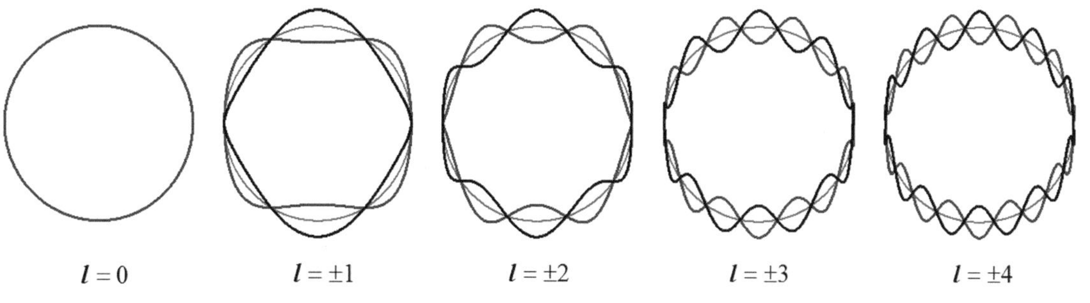

FIGURE 5.15 The wavefunctions for π-molecular orbitals of porphyrin, $l = 0, \pm 1, \pm 2, \pm 3, \pm 4$. The stationary-state wave of the $l = 0$ function coincides with the ring, shown in red. The waves in yellow and black are oppositely traveling.

$$\psi_{+l} \propto e^{\frac{i2\pi l}{p}}$$

$$\psi_{-l} \propto e^{\frac{-i2\pi l}{p}}. \tag{5.89}$$

The waves drawn according to the orbital quantum number are depicted in Figure 5.15.

The quantum number l that we have called electron orbital quantum number in fact gives the magnitude of the orbital angular momentum. As the value of l increases the electron orbital angular momentum increases too, leading to higher energy. The wavefunctions and the angular momentum discussed here are similar to those of a freely-rotating rigid rotor whose motion is confined to a two-dimensional plane.

Box 5.1

Consider two coordinate systems: a molecule-fixed frame (a,b,c) and a spatial or laboratory-fixed frame (X,Y,Z) which serves as the reference axes system. In fact, XYZ represents the center of mass coordinates. Let the position of a nucleus of the rigid rotor be labeled R, such that the molecule-fixed coordinate of the nucleus r is R(x,y,z). Strictly, the nuclear positions R can be given by a vector **R**, and the coordinates of this vector in the molecule-fixed frame is R(x,y,z) – this **R** is referred to as the coordinate vector. Since the body is rigid, the coordinate vector is a constant. In other words, the coordinate of a point, say the nucleus, are constant in the molecule-fixed frame. The molecule-fixed frame is now placed on the laboratory frame so that the two are aligned exactly, and then the former is rotated by an angle φ about the Z-axis. The rotations breaks the aX and bY alignment displacing a and b anticlockwise to a' and b' by angle φ each. The resultant molecule-fixed frame is next rotated about a' by an angle θ, which will displace b' to b'' and c to c'. This resultant molecule-fixed frame is finally rotated about c' by an angle χ. The final relative orientations of the molecule-fixed and laboratory frame are shown in the Figure below. The rotational transformation of the molecule-fixed coordinate vector $R(x,y,z)$ to the laboratory coordinate vector $R'(X,Y,Z)$ involves the orthogonal rotation matrix $\tilde{R}$

$$R'(X,Y,Z) = \tilde{R}R(x,y,z)$$

The final orientation of the molecule-fixed frame relative to the laboratory-fixed axes, and the rotation matrix consisting of the Euler angles is shown below. In fact, any initial Cartesian axis system can be related to any general rotated Cartesian system by using the Euler angles.

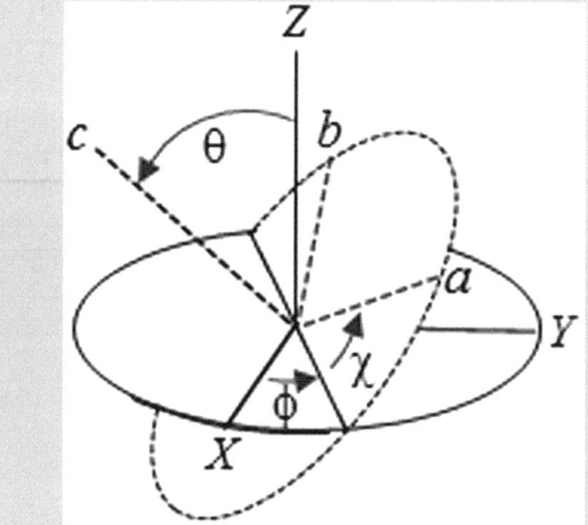

$$\begin{array}{c} \\ a \\ b \\ c \end{array}\left[\begin{array}{ccc} X & Y & Z \\ \cos\theta\cos\phi\cos\chi - \sin\phi\sin\chi & \cos\theta\sin\phi\cos\chi + \cos\theta\sin\chi & -\sin\theta\cos\chi \\ -\cos\theta\cos\phi\sin\chi - \sin\phi\cos\chi & -\cos\theta\sin\phi\sin\chi + \cos\phi\cos\chi & \sin\theta\sin\chi \\ \sin\theta\cos\phi & \sin\theta\sin\phi & \cos\theta \end{array}\right]$$

PROBLEMS

5.1 Consider the surfaces S_o, S_1, and S_2 such that the latter two define a mixing region (see the figure). In the $S_o \rightarrow S_1$ electronic absorption spectrum under ideal conditions (no mixing) a number of discrete vibrational lines appear due to vertical transitions to different vibrational levels in S_1.

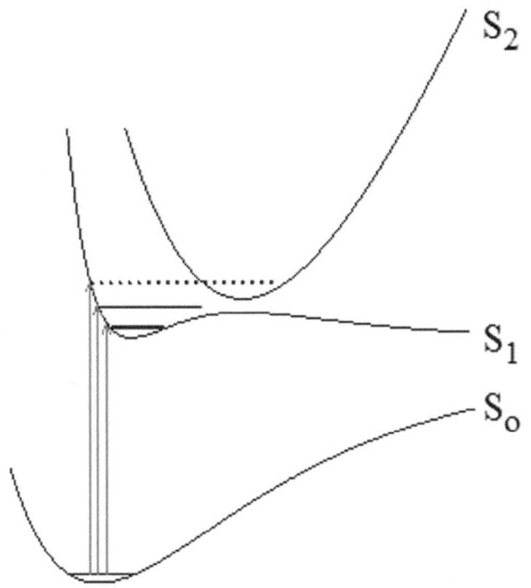

Sketch the possible changes in the absorption spectrum when the excited vibrational level lie within the splitting region. Assume a finite probability that the electronic structure will change in the mixing region.

5.2 Suppose the length of the Landau-Zener (LZ) region for the mixing of two surfaces k and l is 2 pm and the splitting energy $\Delta = 1 \times 10^{-3}$ eV. If the system is taken from temperature T to T' K such that the nuclear vibration changes from 10^{13} to 10^{14} Hz without a change in the length of the LZ region, how will the adiabaticity probability P change with thermal energy? Draw a P vs T graph. The question is based on a hypothetical case. Assume the molecule is diatomic.

5.3 Suppose somehow one brings about a significant compression or extension in the bond length of a diatomic molecule. How will such changes influence the applicability of the LCAO method? Extend the explanation to cases of polyatomic molecules. Also cite reasons why the LCAO method for the description of molecular orbitals is not satisfactory.

5.4 For a simple homonuclear diatomic like H_2 we could express the electric transition dipole moment associated with an excitation from the ground to the excited singlet state as

$$\mu = \frac{\sqrt{2}}{\sqrt{4(1-S^2)}}(r_{\text{H-H}})$$

where S is the ovelap integral and $r_{\text{H-H}}$ is the internuclear separation. Derive a similar expression for a simple heteronuclear diatomic.

5.5 How many π electrons are available in the ring of benzene? How many molecular orbitals can be constructed by a linear combination of the $2p$ atomic orbitals?

5.6 Describe the π molecular orbitals of a cyclic system.

5.7 A 'wavelength' λ determined by the total number of π -electrons for each π-molecular eigenstate of a cyclic molecule has been discussed. What is the physical significance of this wavelength? Why is not a wavelength associated with an eigenstate of a linear conjugated molecule?

BIBLIOGRAPHY

Bates, D. R., K. Ledsham, and A. L. Stewart (1953) *Phil. Trans. R. Soc.* 246, 215.

Bethe, H. and E. E. Salpeter (1977) *Quantum Mechanics of One- and Two-Electron Atoms*, Plenum.

Born, M. and J. R. Oppenheimer (1927) *Ann. Phys.* 84, 457.

Cohen-Tannoudji, C., B. Diu, and F. Laloë (1977) *Quantum Mechanics* Vol II, John Wiley & Sons.

Davidson, E. R. (1961) *J. Chem. Phys.* 35, 1189.

Eaton, W. A. and J. Hofrichter (1981) *Methods Enzymol.* 76, 175.

Eyring, H., J. Walter, and G. W. Kimball (1944) *Quantum Chemistry*, Wiley, pp 192–199.

King, G. W. (1964) *Spectroscopy and Molecular Structure*, Holt, Rinehart and Winston.

Kotani, M., A. Amemiya, E. Ishiguro, and T. Kimura (1955) *Table of Molecular Integrals*, Maruzen & Co, Tokyo.

Landau, L. (1932a) *Sov. Phys.* 1, 89.

Landau, L. (1932b) *Z. Phys. Sov.* 2, 1932.

Landau, L. D. and E. M. Lifshitz (1958) *Quantum Mechanics*, Pergamon Press.

Longuet-Higgins, H. C., C. W. Rector, and J. R. Platt (1950) *J. Chem. Phys.* 18, 1174.

Mulliken, R. S. (1936) *Phys. Rev.* 50, 1028.

Platt, J. R. (1949) *J. Chem. Phys.* 17, 484.

Salem, L. (1966) *Molecular Orbital Theory of Conjugated Systems*, W. A. Benjamin. Inc.

Slater, J. C. (1963) *Quantum Theory of Molecules and Solids*, Vol 1, McGraw-Hill.

Stuckelberg, E. G. C. (1932) *Helv. Phys. Acta* 5, 369.

Zener, C. (1932) *Proc. R. Soc. Ser* A, 137, 696.

6 Elementary Group Theory

At the heart of molecular spectroscopy is the task of comprehending spectra due to transitions from one to another eigenstate. We have learned about the $\psi_1^* \mu \psi_2 d\tau$ integral to a certain extent. But, how do we know that the integral will indeed give an expectation value that we are interested in? The integral will actually give a 'zero' expectation value if the symmetries of $\psi_1^* \psi_2$ and μ are not the same. This warning is just a basic but ordinary way of bringing out the importance of symmetry and group theory in molecular spectroscopy. Symmetry is a common terminology in the macroscopic world too. Interestingly, the symmetry principles of the macroscopic world hold also for microscopic entities like atoms and molecules. Even more, mathematical functions like the wavefunctions and operators are constrained to work under symmetry principles. For that matter, a vector too can be subjected to symmetry transformations. It is then obvious that the structural symmetry of macroscopic bodies like rotation and reflection could be applied equally well to atoms and molecules, wavefunctions, and operators. This transition of macroscopic to microscopic symmetry is what is called a 'point group', and the symmetry properties and transformations under various symmetry operations constitute the vast area of group theory that has numerous applications in studies of photons, subatomic particles, behavior of orbitals and wavefunctions, molecular transitions, spectra, and many more. Having said so, the learning of all of the principles and applications of symmetry and group theory is beyond the scope of the present study. Authorities, including Wilson et al. (1955), Cotton (1963), Hochstrasser (1966), and Bunker and Jensen (1998), have described group theoretic principles and their applications in molecular spectroscopy. Here we will restrictively describe the meaning of symmetry, operations of symmetry, the products of symmetry operations, and the applications of such products to appreciate the symmetric principles for the determination of electronic transitions from one to another molecular orbital. Before leaving this introductory note, we should say that symmetry is thought-provoking with respect to the creation and philosophy. It is certain that the universe and all of its constituents are symmetric or at least nearly symmetric. So are atoms and molecules in their stable ground state, which should also imply that symmetry considerations would be less interesting if the molecules are allowed to exist perpetually in their ground states. That is not to be the case, for molecules do undergo transitions from symmetric ground to frequently distorted excited states, and to know the very nature of such transitions the ground-state symmetry operations need to be known.

6.1 SYMMETRY OPERATIONS

All symmetry operations involve only two things – rotation of the molecule about appropriate axes (operation c), and reflection of the molecule in a mirror plane (operation σ) which would be σ_v if the plane contains the major rotation axis, and σ_h if the plane is perpendicular to that axis. Occasionally, additional operations are invoked – improper rotation $S = c\sigma_h$, which means c and σ_h operations in succession, and inversion $i = c\sigma_h$ that refers specifically to a π-rotation about a rotation axis followed by σ_h.

6.1.1 Rotation

Rotation is given by c_n such that a rotation by an angle $2\pi / n$ leaves the object identical. To exemplify, a $2\pi / n$ rotation about each of axes a, b, c of the planar triangle in Figure 6 .1 regenerates the triangle identically. These are c_2 rotational symmetry axes, the subscript '2' obviously referring to a π-rotation. The axes are sometimes written explicitly as c_{2a}, c_{2b}, and c_{2c}. But the d-axis which passes vertically through the center of the triangle is a c_3 axis because every 120°-rotation about d generates the triangle exactly of the c_2 and c_3 axes, the latter is called the principal axis of rotation which in general corresponds to the axis with the largest value of n. The c_3 rotation could be given further values by $c_3^{\,1}$, $c_3^{\,2}$, and $c_3^{\,3}$, where the superscript numbers indicate the times the c_3-rotation is carried out in succession. For example,

$$c_3 \times c_3 = c_3^2.$$

There is one more operation called the identity operation E, which implies doing nothing to the system. Yet, the operation E is always indicated in all symmetry groups and the tables that follow. In fact, E is an unit element analogous to the unit matrix in matrix algebra. We will appreciate the usefulness of having E as we proceed.

With these operations the symmetry group or rotation group of the triangle (Figure 6.1) can be written as

$$D_3 = \left\{E,\ c_{2a},\ c_{2b,}c_{2c},\ c_{3d}^{\ 1}, c_{3d}^{\ 2},\ c_{3d}^{\ 3}\right\}. \qquad (6.1)$$

For the square drawn, the rotational symmetry axes a, b, c, d, e can be used to perform clockwise rotation to generate the square identically. The operations are indicated by a clockwise rotation by π about the a-axis. Using the other two-fold rotation axis we get c_{2a}, c_{2b}, c_{2c}, c_{2d}. Continuing with the same

DOI: 10.1201/9781003293064-6

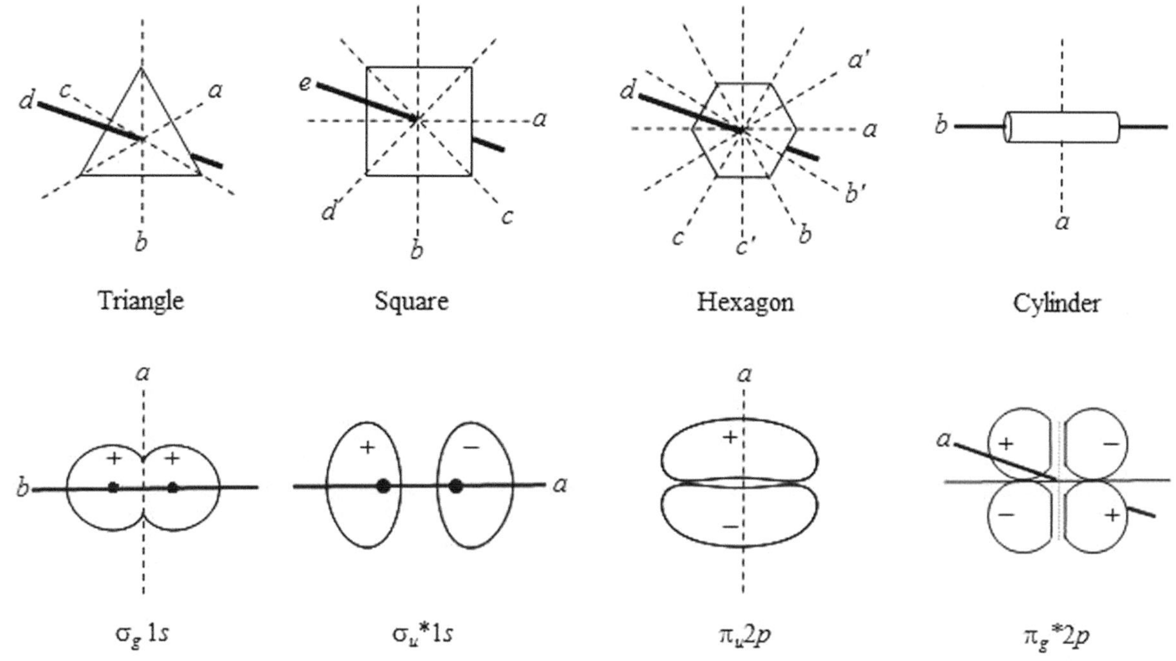

FIGURE 6.1 Some geometries and molecular orbitals, and the results of their rotation and reflection (see text).

spirit of operations we can write down the symmetry group or the rotation group of the square

$$D_4 = \{E,\ c_{2a},\ c_{2b},\ c_{2c},\ c_{2d}, c^1_{4e}, c^2_{4e},\ c^3_{4e},\ c^4_{4e}\} \quad (6.2)$$

A hexagon (cyclohexane or benzene, for example) has a six-fold axis of rotational symmetry c_6, and the rotation group is

$$D_6 = \{E,\ c_{2a},\ c_{2b}, c_{2c}, c_{2a'},\ c_{2b'}, c_{2c'}, c^1_{6d}, c^2_{6d},\ c^3_{6d},\ c^4_{6d}, c^5_{6d}\} \quad (6.3)$$

Each of the D_3, D_4, and D_6 rotation groups contains more than one rotation axis. For the hexagon geometry, the axes are C_n where $n = 2$ and 6. Conventionally, the axis with the largest value of n, called the principal axis, is used to represent the rotation group. Thus, the rotation group for benzene is D_6, and not D_2. The cylinder that typifies a homonuclear diatomic molecule has a two-fold rotation axis c_{2a} and an infinity-fold axis of symmetry c_∞, so that

$$D_\infty = \{E,\ c_{2a},\ c_{\infty b}\} \quad (6.4)$$

Considering the σ and π bonding and antibonding orbitals next in Figure 6.1, we notice that both $\sigma_g 1s$ and $\sigma_u{}^* 1s$ have c_∞ symmetry with respect to rotation about the internuclear axis, even though the former also possesses a c_2 axis between the two atoms. The π-molecular orbitals lack the symmetry of rotation with respect to the internuclear axis in spite of the fact that both have a two-fold rotational symmetry axis c_{2a}. The rotation groups for these molecular orbitals are as follows

$$\sigma_g 1\text{s} \quad D_\infty = \{E, c_{2a}, c_{\infty b}\}$$

$$\sigma^*_u 1\text{s} \quad D_\infty = \{E, c_{\infty a}\}$$

$$\pi_u 2\text{p} \quad D_\infty = \{E, c_{2a}\}$$

$$\pi^*_g 2\text{p} \quad D_\infty = \{E, c_{2a}\}. \quad (6.5)$$

It should be clear that it is the presence of the c_∞ symmetry axis in σ and σ^* orbitals that distinguishes them from the π and π^* orbitals which lack c_∞.

6.1.2 Reflection

A plane of reflection is generally symbolized by σ, and if a molecule does not change in a reflection operation then the reflection plane is called a plane of symmetry. If the plane of reflection contains the principal axis, then the reflection plane σ is said to be vertical σ_v, and if the reflection plane is perpendicular to the principal axis then σ is horizontal σ_h. Relooking at the geometries in Figure 6.1 we find the following reflection groups

square	$\{\sigma_{va}, \sigma_{vb}, \sigma_{vc}, \sigma_{vd}, \sigma_h\}$
triangle	$\{\sigma_{va}, \sigma_{vb}, \sigma_{vc}, \sigma_h\}$
hexagon	$\{\sigma_{va}, \sigma_{vb}, \sigma_{vc}, \sigma_{va'}, \sigma_{vb'}, \sigma_{vc'}, \sigma_h\}$
cylinder	$\{\sigma_{v\infty}, \sigma_h\}$
$\sigma_g 1\text{s}$	$\{\sigma_{v\infty}, \sigma_h\}$
$\sigma^*_u 1\text{s}$	$\{\sigma_{v\infty}\}$

$\pi_u 2p \quad \{c_{h\infty}\}$

$$\pi_g^* 2p \quad \text{no mirror plane symmetry.} \tag{6.6}$$

For a planar geometry, the plane of the body itself is the σ_h plane.

6.1.3 Improper Rotation

This symmetry element consists of a rotation and a reflection operation in tandem. For the cube shown below (Figure 6.2), a c_2 rotation about an axis followed by a reflection in a plane perpendicular to the rotation axis restores the cube identically, and the symmetry element is generally given by S_n, which is S_2 for the body we have considered

$$S_n = c_n \sigma_h. \tag{6.7}$$

It is also noticed that if the rotation axis is along z, then the improper operation changes the axes system as the following

$$xyz \xrightarrow{c_{2z}} -x,-y,z \xrightarrow{\sigma_{h(x,y)}} -x,-y,-z, \tag{6.8}$$

which means S_2 is essentially an inversion operation, $S_2 = i$.

6.1.4 Inversion

This operation denoted i entails inversion of any point of the body through the center of mass of the body. If the inverted point is identical to the initial in sign, the body is said to have an inversion symmetry element. For example, in the $\sigma_g 1s$ orbital we can take a point anywhere in the orbital and carry it straight through the center of the orbital i to a distance equal to its original position and the center of the orbital. In its new position the inverted point will have the same sign (+). This means the orbital is symmetric with respect to inversion. Inspection of the four molecular orbitals shown in Figure 6.1 will reveal that inversion is a symmetry element for $\sigma_g 1s$ and $\pi_g^* 2p$, but not for $\pi_u^* 1s$ and $\pi_u 1s$. This is the reason why the subscript g (*gerade*), meaning even or symmetric with respect to inversion, is used to indicate the bonding orbitals $\sigma_g 1s$ and $\pi_g^* 2p$. The subscript u (*ungerade*), meaning uneven or antisymmetric with respect to inversion operation, is used to indicate the antibonding orbitals $\sigma_u^* 1s$ and $\pi_u 1s$.

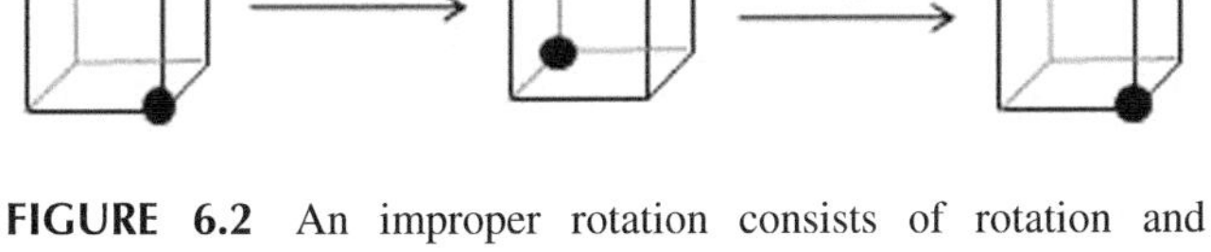

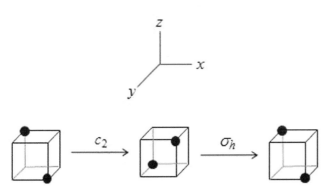

FIGURE 6.2 An improper rotation consists of rotation and reflection operations in tandem.

The following are the common changes when the symmetry elements E, c_n, σ, and i are applied to a Cartesian coordinate system

$$(x,y,z) \xrightarrow{E} (x,y,z)$$

$$(x,y,z) \xrightarrow{c_{2z}} (-x,-y,z)$$

$$(x,y,z) \xrightarrow{\sigma_{xy} \text{ or } \sigma_h} (x,y,-z)$$

$$(x,y,z) \xrightarrow{\sigma_{xz} \text{ or } \sigma_v} (x,-y,z)$$

$$(x,y,z) \xrightarrow{i} (-x,-y,-z). \tag{6.9}$$

6.2 POINT GROUP

A symmetry point group is a collection of symmetry elements. More generally, a point group is defined for a body if at least one point of the body stays fixed while the symmetry operations are performed. An object or a molecule that can be operated by all symmetry elements contained in a point group belongs to that point group. Some point groups and examples are listed below.

- c_n — only one n-fold rotation axis; H_2O_2 (c_2)
- c_{nv} — a c_n group (n-fold rotation axis) that also has n reflection planes each containing the rotation axis; H_2O (c_{2v}), HCN ($c_{\infty v}$)
- c_{nh} — a c_n group that also has one reflection plane perpendicular to the rotation axis; trans-$H_2C_2Cl_2$ (c_{2h})
- D_n — a n-fold principal rotation axis and n-number of two-fold rotation axes each of which is perpendicular to the principal axis; staggered ethane (D_3)
- D_{nd} — a D_n group having n-number of reflection planes each containing the n-fold principal rotation axis in such a way that the reflection planes bisect the angle between the successive two-fold rotation axes; C_3H_3 (D_{3d})
- D_{nh} — a D_{nd} point group that also has a reflection plane perpendicular to the n-fold principal rotation axis; naphthalene (D_{2h}), BF_3 (D_{3h}); C_6H_6 (D_{6h})
- $D_{\infty h}$ — a c_∞ rotation axis, an infinite number of c_2 axes perpendicular to the c_∞ axis, an infinite number of reflection planes σ_v containing the c_∞ axis, one reflection plane perpendicular to the c_∞ axis, and one center of symmetry i; O_2, N_2, and homonuclear diatomic molecules are some examples.

6.2.1 Properties of Point Groups

A few guidelines about the symmetry elements of a point group are the following.

1. Every point group invariably contains the identity operation E which serves as the unit element.
2. Symmetry elements of a point group can be applied in succession without a specific order of application, and the resultant is another symmetry element of the same point group. Figure 6.3 shows that successive operations $c_3\sigma_{va}$ does not produce the same result as $\sigma_{va}c_3$ does. The results of successive operations are given by a group multiplication table which also indicates the order in which the operations are carried out – usually, the elements arranged in the row are applied first followed by those in the left column. Table 6.1 shows a group multiplication table for the c_{3v} group as an example.
3. The multiplication of symmetry elements of a point group is associative. For example,

$$c_2\left(\sigma_v\sigma_{v'}\right)=\left(c_2\sigma_v\right)\sigma_{v'}. \tag{6.10}$$

4. The product of every symmetry element and its inverse is equivalent to an identity transformation E

$$\sigma_v\sigma_v^{-1}=\sigma_v^{-1}\sigma_v=E. \tag{6.11}$$

5. When a set of symmetry elements, not necessarily all, within a point group obey the above rules then the respective set of elements constitute a subgroup. The product of each element of the subgroup with another element outside this subgroup will not belong to this same subgroup, but surely is an element of this point group. Note that E belongs to each possible subgroup.
6. A pair of symmetry elements within a point group is a conjugate pair if they show similarity transformation.

TABLE 6.1
Group Multiplication Table Obtained for the c_{3v} Group

	First operation					
Second operation	E	c_3	c_3^2	σ_{va}	σ_{vb}	σ_{vc}
E	E	c_3	c_3^2	σ_{va}	σ_{vb}	σ_{vc}
c_3	c_3	c_3^2	E	σ_{vc}	σ_{va}	σ_{vb}
c_3^2	c_3^2	E	c_3	σ_{vb}	σ_{vc}	σ_{va}
σ_{va}	σ_{va}	σ_{vb}	σ_{vc}	E	c_3	c_3^2
σ_{vb}	σ_{vb}	σ_{vc}	σ_{va}	c_3^2	E	c_3
σ_{vc}	σ_{vc}	σ_{va}	σ_{vb}	c_3	c_3^2	E

For example, a conjugate pair consisting of c_n and σ_v shows the following

$$c_n=\sigma_{v'}^{-1}\sigma_v\sigma_{v'}$$

$$\sigma_{v'}^{-1}c_n\sigma_{v'}=\sigma_v, \tag{6.12}$$

in which $\sigma_{v'}$ is the third element. Each of these two relations is a similarity transformation. A set of symmetry elements where all elements are mutually conjugate forms a class of the point group. The element E is a class of its own in which no other element needs to be entered. Also, a class need not contain E as an element. We will see later in the discussion that the number of classes in a group is equal to the number of irreducible representations.

6.2.2 Representation of Symmetry Operators of a Group

The results of different symmetry operations that we have considered can also be derived by considering the changes of a

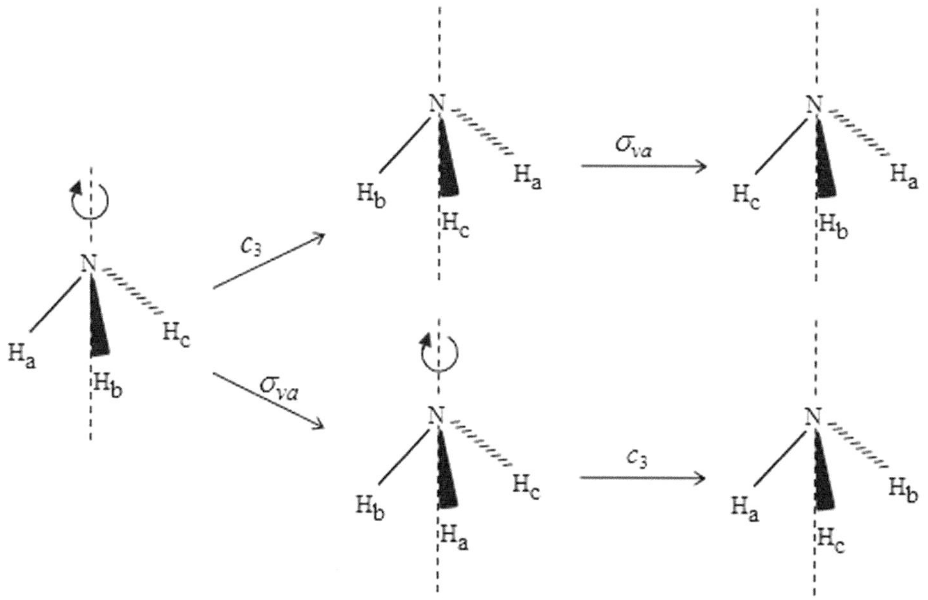

FIGURE 6.3 Multiplication of elements to show that $c_3\sigma_{va} \neq \sigma_{va}c_3$.

molecule in a basis of representation. The basis for a representation could be the xyz coordinates, the 3N coordinates for the nuclei, a set of atomic orbitals, the bond angles, and the bond vectors. This was briefly mentioned earlier for the xyz basis. We can take it up here to generate a set of matrices that represent the symmetry as numbers. Below is illustrated the changes in the sign of coordinates of a vector as a result of transformation by a few symmetry operators, each result of the operations can be shown by a 3 × 3 diagonal matrix (Figure 6.4). The diagonal nature of the 3 × 3 matrices containing +1 and −1 suggests

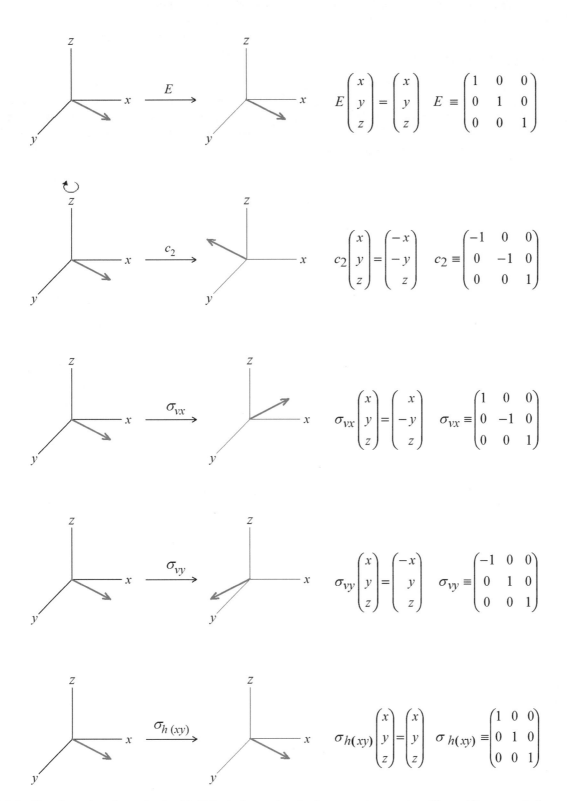

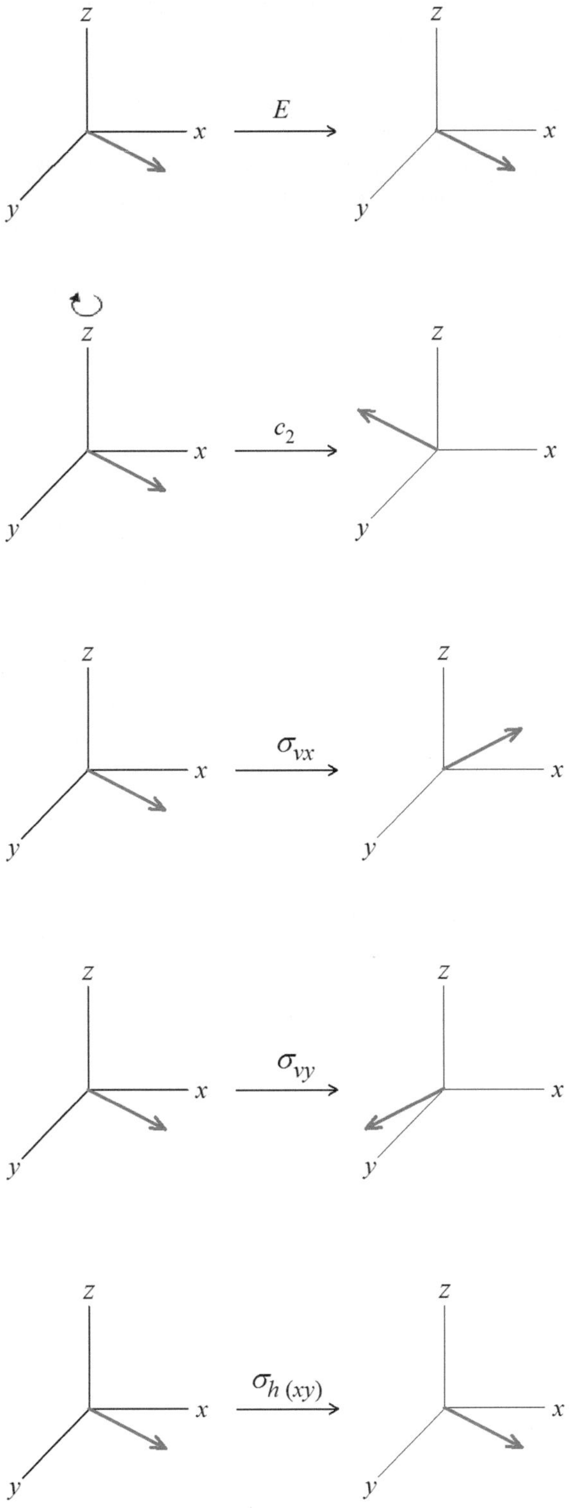

$$E\begin{pmatrix}x\\y\\z\end{pmatrix}=\begin{pmatrix}x\\y\\z\end{pmatrix}\quad E\equiv\begin{pmatrix}1&0&0\\0&1&0\\0&0&1\end{pmatrix}$$

$$c_2\begin{pmatrix}x\\y\\z\end{pmatrix}=\begin{pmatrix}-x\\-y\\z\end{pmatrix}\quad c_2\equiv\begin{pmatrix}-1&0&0\\0&-1&0\\0&0&1\end{pmatrix}$$

$$\sigma_{vx}\begin{pmatrix}x\\y\\z\end{pmatrix}=\begin{pmatrix}x\\-y\\z\end{pmatrix}\quad \sigma_{vx}\equiv\begin{pmatrix}1&0&0\\0&-1&0\\0&0&1\end{pmatrix}$$

$$\sigma_{vy}\begin{pmatrix}x\\y\\z\end{pmatrix}=\begin{pmatrix}-x\\y\\z\end{pmatrix}\quad \sigma_{vy}\equiv\begin{pmatrix}-1&0&0\\0&1&0\\0&0&1\end{pmatrix}$$

$$\sigma_{h(xy)}\begin{pmatrix}x\\y\\z\end{pmatrix}=\begin{pmatrix}x\\y\\z\end{pmatrix}\quad \sigma\,h(xy)\equiv\begin{pmatrix}1&0&0\\0&1&0\\0&0&1\end{pmatrix}$$

FIGURE 6.4 Transformation of a vector (xyz) initially in the xy plane by the symmetry operators. The matrices generate the numbers +1 (symmetric) or −1 (antisymmetric).

that the representations can be given by 1×1 matrices as well. In fact, the way the symmetry species or objects transform under the influence of symmetry operations is given by the set of numbers $\{0, +1, -1\}$. All operators are assigned the value of 0, but +1 or −1 are assigned subject to whether the function changes to itself (+1) or to its negative (−1) under the influence of the symmetry operator.

6.3 GROUP REPRESENTATIONS

Let us look at the procedures for obtaining symmetry operator representations for the D_{2h} point group which may be taken as a case study found in many textbooks, but the reader can try out more using other point groups. The basis set for the representation of the D_{2h} point group can be the ten π-atomic orbitals $\phi_1, \phi_2, \phi_3, \cdots, \phi_{10}$, of naphthalene for example (Salem, 1966: 334–342). The symmetry elements for D_{2h} are E, c_{2x}, c_{2y}, c_{2z}, i, σ_{xy}, σ_{xz}, and σ_{yz} (Figure 6.5), and each of these symmetry operators is applied to each of the ten atomic orbitals. The result of the operations is given in a character table (Table 6.2). Say we take the orbital ϕ_1, let it transform under each of the eight symmetry operators, and record the outcomes sequentially in the first row of the character table. The outcome will be +1 or −1 depending on whether the orbital is untransformed or transformed. Then orbital ϕ_2 is taken and subjected to the same set of operations one by one, and the outcomes of +1 or −1 are noted in the second row of the table, and so on. Thus, we generate ten rows for ten orbitals in the character table. The table however shows only eight rows; each of these eight rows represents what is called an irreducible representation. The ten rows together stand for reducible representation.

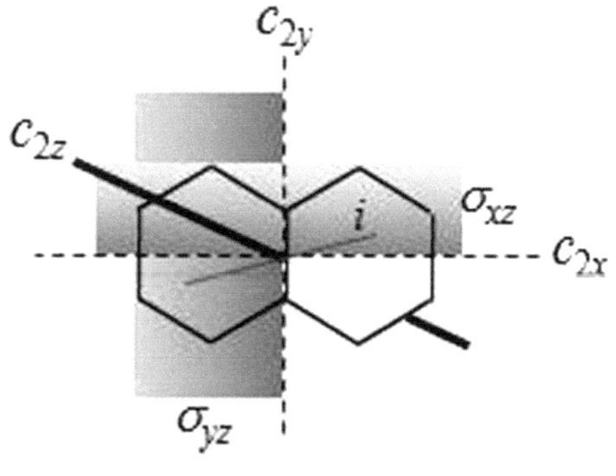

FIGURE 6.5 Symmetry elements of the D_{2h} point group.

Let us look at what reducible and irreducible representations are. All of those ten rows that could have been generated by operating on each of the ten orbitals constitute a set of reducible representations, and are indicated by $\Gamma_i (i = 1, 2, 3, ...)$ according to Bethe notation. The minimum number of reducible representations corresponds to the number of functions in the basis for the representation of the group. But many more reducible representations can be generated by using the idea that two reducible representations can be combined to obtain additional reducible representations. For example, the following matrices generated by combining Γ_1 and Γ_2 serve for additional reducible representations (Γ_9), and a combination of Γ_4 and Γ_5 will provide Γ_{10}.

$$\Gamma_9 \quad \begin{pmatrix} 1 & 0 \\ 0 & 1 \end{pmatrix}, \begin{pmatrix} 1 & 0 \\ 0 & 1 \end{pmatrix}, \begin{pmatrix} 1 & 0 \\ 0 & 1 \end{pmatrix}, \begin{pmatrix} 1 & 1 \\ 0 & 1 \end{pmatrix}, \begin{pmatrix} 1 & 0 \\ 0 & -1 \end{pmatrix}, \begin{pmatrix} 1 & 0 \\ 0 & -1 \end{pmatrix}, \begin{pmatrix} 1 & 0 \\ 0 & -1 \end{pmatrix}, \begin{pmatrix} 1 & 0 \\ 0 & -1 \end{pmatrix}$$

$$\Gamma_{10} \quad \begin{pmatrix} 1 & 0 \\ 0 & 1 \end{pmatrix}, \begin{pmatrix} -1 & 0 \\ 0 & -1 \end{pmatrix}, \begin{pmatrix} -1 & 0 \\ 0 & 1 \end{pmatrix}, \begin{pmatrix} 1 & 0 \\ 0 & -1 \end{pmatrix}, \begin{pmatrix} -1 & 0 \\ 0 & 1 \end{pmatrix}, \begin{pmatrix} -1 & 0 \\ 0 & -1 \end{pmatrix}, \begin{pmatrix} 1 & 0 \\ 0 & 1 \end{pmatrix}, \begin{pmatrix} 1 & 0 \\ 0 & -1 \end{pmatrix}$$

One can generate more reducible representations by using this procedure. But all of the reducible representations, no matter how many are generated in the form of numbers and matrices, must always satisfy the multiplication rule of the group. This means that the additional reducible representations that are generated are duplicates of the group multiplications. These representations $\Gamma_i (i = 1, 2, 3, ...)$ are said to be reducible because the total number of such representations can be reduced to a set of representations that cannot be reduced any further. This set is called irreducible representations.

TABLE 6.2
Character Table for the D_{2h} Point Group

Irreducible representation	E	σ_{xy}	σ_{xz}	σ_{yz}	i	c_{2z}	c_{2y}	c_{2x}	Rotational and linear functions	Quadratic functions
A_g	1	1	1	1	1	1	1	1		x^2, y^2, z^2
A_u	1	−1	−1	−1	−1	1	1	1		
B_{1g}	1	1	−1	−1	1	1	−1	−1	R_z	xy
B_{1u}	1	−1	1	1	−1	1	−1	−1	z	
B_{2g}	1	−1	1	−1	1	−1	1	−1	R_y	xz
B_{2u}	1	1	−1	1	−1	−1	1	−1	y	
B_{3g}	1	−1	−1	1	1	−1	−1	1	R_x	yz
B_{3u}	1	1	1	−1	−1	−1	−1	1	x	
Γ_{red}	10	−10	0	2	0	0	−2	0		

The row at the end of the table shows the characters for reducible representation. The representations to which x, y, z coordinate axes, the rotations R_x, R_y, R_z, and the products of the coordinates belong are shown in the last two columns.

6.4 LABELS OF IRREDUCIBLE REPRESENTATIONS

It was mentioned earlier that the number of classes of symmetry elements in a group equals the number of irreducible representations. From the set of symmetry elements of the D_{2h} group eight classes can be formed, so there are eight irreducible representations appearing in the first column of the character table. Listed below are the conventions of labeling the irreducible representations following the Placzek notation that can also be found in other textbooks.

1. Irreducible representations of dimension 1 or 1×1 matrix are indicated by A and B. The letter E, not to be confused with the identity operator E, is used to denote two-dimensional representations, meaning having a character of 2. The letter T is used for irreducible representation of character 3. The A representation can be distinguished by +1 for all of the symmetry operations of the group in context. Remember, character means the sum of the diagonal elements.
2. The representation A is used if the character corresponding to the rotation operation about the principal axis is +1, which is consistent with the rule that A representation is totally symmetric. The label B is used when the character for rotation about the principal axis is −1, which means asymmetric rotation

$$\chi_A(c_n)=+1 \quad \chi_B(c_n)=-1. \tag{6.13}$$

3. The representations A and B carry with them the subscript 1 or 2 depending on whether the representation is symmetric (1) or anstisymmetric (2) with respect to a reflection on a vertical mirror (σ_v)

$$\chi_{A_1}(\sigma_v)=+1 \quad \chi_{A_2}(\sigma_v)=-1. \tag{6.14}$$

4. The representations that are symmetric and antisymmetric to horizontal plane of reflection (σ_h) are distinguished by a single prime (′) and double primes (″), respectively,

$$\chi_{A'}(\sigma_h)=+1 \quad \chi_{A''}(\sigma_h)=-1. \tag{6.15}$$

5. With regard to the center of symmetry (inversion symmetry i) the A and B representations are indicated further by g or u subject to symmetric or antisymmetric outcomes

$$\chi_u(i)=+1 \quad \chi_g(i)=-1. \tag{6.16}$$

6.5 REDUCTION OF REPRESENTATIONS TO IRREDUCIBLE REPRESENTATIONS

The large numbers of reducible representations that can be generated are reducible to a limited set of irreducible representations. The group theory basically claims that irreducible representations are among the reducible representations, Γ_{red}. To justify this contention, we need to consider the concept of the trace of a symmetry transformation. Shown below by way of illustration is what will happen if the σ_h element is used to transform the ten π-orbitals of naphthalene as an example.

$$\sigma_h \begin{bmatrix}\phi_1\\ \phi_2\\ \phi_3\\ \phi_4\\ \phi_5\\ \phi_6\\ \phi_7\\ \phi_8\\ \phi_9\\ \phi_{10}\end{bmatrix} = \begin{bmatrix} 0&0&0&1&0&0&0&0&0&0\\ 0&0&1&0&0&0&0&0&0&0\\ 0&1&0&0&0&0&0&0&0&0\\ 1&0&0&0&0&0&0&0&0&0\\ 0&0&0&0&0&0&0&1&0&0\\ 0&0&0&0&0&0&1&0&0&0\\ 0&0&0&0&0&1&0&0&0&0\\ 0&0&0&0&1&0&0&0&0&0\\ 0&0&0&0&0&0&0&0&0&1\\ 0&0&0&0&0&0&0&0&1&0 \end{bmatrix} \begin{bmatrix}\phi_1\\ \phi_2\\ \phi_3\\ \phi_4\\ \phi_5\\ \phi_6\\ \phi_7\\ \phi_8\\ \phi_9\\ \phi_{10}\end{bmatrix} \tag{6.17}$$

Clearly, the trace (sum of diagonal elements) of the eigenvalue matrix is zero, $Tr(\sigma_h)=0$. We can also use σ_v, in which case the mirror plane may contain either xz or yz plane. For either of these two vertical plane reflections we need to be convinced that $Tr(\sigma_v)=2$. Continuing in the manner of the transformations above, the following traces are obtained for naphthalene under the relevant symmetry transformations

$$Tr(E)=10 \quad Tr(\sigma_h)=-10 \quad Tr(i)=0$$

$$Tr(c_{2z})=0 \quad Tr(c_{2x})=0 \quad Tr(c_{2y})=-2. \tag{6.18}$$

The illustrations up to this point set up the basis for connecting reducible and irreducible representations. We state without providing a proof that any of the reducible representations, called Γ collectively, can be reduced to the corresponding irreducible representation, Γ_j, by

$$\Gamma=\sum_j a_j\Gamma_j, \tag{6.19}$$

where the coefficient a_j represents the number of times Γ_j appears in the reducible representation. The reduction will not change the character. A symmetry operation, say O_i in which i is the index for the symmetry elements of the group, on the set of functions Γ yields a transformation matrix characterized by a trace

$$Tr\ \Gamma(O_i)=\chi_\Gamma(O_i),$$

where χ is the character. One says that the reduction of a reducible representation does not change the trace (or character) of any of the transformation matrices

$$Tr\ \Gamma(O_i)=\sum_j a_j\chi_{\Gamma_j}(O_i)$$

$$\Gamma = a_1 A_g + a_2 A_u + a_3 B_{1g} + a_4 B_{1u} + a_5 B_{2g} + a_6 B_{2u} + a_7 B_{3g} + a_8 B_{3u}. \tag{6.20}$$

To determine the coefficients a_j, the equation above relating the traces of reducible and irreducible representation can be multiplied on both sides by $\chi_{\Gamma_k}(O_i)$

$$\chi_\Gamma(O_i)\chi_{\Gamma_k}(O_i) = \sum_j a_j \chi_{\Gamma_j}(O_i)\chi_{\Gamma_k}(O_i). \tag{6.21}$$

Then sum is taken over all the operations O_i

$$\sum_{O_i}\chi_\Gamma(O_i)\chi_{\Gamma_k}(O_i) = \sum_j\sum_{O_i} a_j \chi_{\Gamma_j}(O_i)\chi_{\Gamma_k}(O_i). \tag{6.22}$$

Now we will make use of a theorem of irreducible representation which states that 'the matrix elements of two different irreducible representations are orthogonal'. For Γ_j and Γ_k in the context, if the dimensions of the respective matrices are l_j and l_k, then $\delta_{jk} = 1$ only if $j = k$, otherwise zero. We also need to know about a parameter h, called the order or the total number of operations in the group,

$$h = \sum_j l_j^2 = \sum_{O_i}\left|\chi_{\Gamma_j}(O_i)\right|^2. \tag{6.23}$$

The coefficient a_j is obtained as

$$a_j = \frac{1}{h}\sum_{O_i}\chi_{\Gamma_j}(O_i)\chi_\Gamma(O_i). \tag{6.24}$$

By taking the values of the set of Γ appearing in the last line of Table 6.2, the coefficient a_1 is determined as

$$a_1 = \frac{1}{8}(1\times 10 - 1\times 10 + 1\times 0 + 1\times 2 + 1\times 10 + 1\times 0 - 1\times 2 + 1\times 0) = 0. \tag{6.25}$$

The other coefficients calculated in the same way are $a_2 = 2$, $a_3 = 0$, $a_4 = 3$, $a_5 = 2$, $a_6 = 0$, $a_7 = 3$, and $a_8 = 0$, suggesting that there are 2, 3, 2, and 3 molecular orbitals of symmetry species A_u, B_{1u}, B_{2g}, and B_{3g}, respectively. These ten molecular orbitals of naphthalene are shown in Figure 6.6. Note that these are one-electron orbitals generated by linear combination of the ten π-atomic orbitals, and as discussed earlier one-electron orbitals are built up by the *Auf bau* principle.

6.6 DIRECT PRODUCT OF IRREDUCIBLE REPRESENTATIONS

The representation obtained from the direct product of two irreducible representations of molecular wavefunctions plays an important role in determining the symmetry species of ground and excited states, and also in the prediction of whether a transition integral exists or not. We briefly show the matrix for the direct product of two 3x3 matrices.

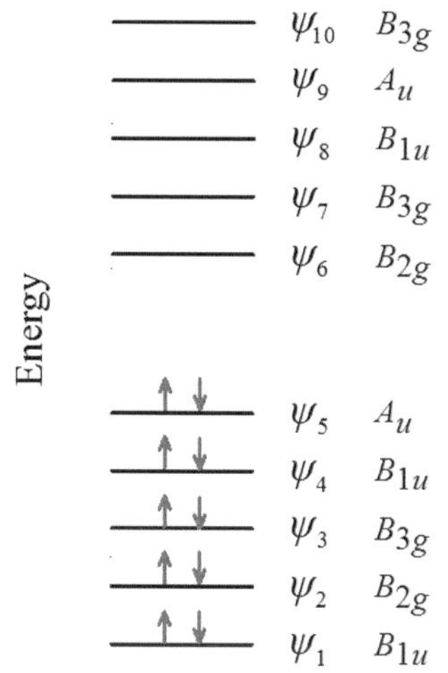

FIGURE 6.6 Symmetry of the one-electron orbitals of naphthalene formed by linear combination.

$$\begin{pmatrix} a_{11} & a_{12} & a_{13} \\ a_{21} & a_{22} & a_{23} \\ a_{31} & a_{32} & a_{33} \end{pmatrix} \otimes \begin{pmatrix} b_{11} & b_{12} & b_{13} \\ b_{21} & b_{22} & b_{23} \\ b_{31} & b_{32} & b_{33} \end{pmatrix} =$$

$$\begin{pmatrix} a_{11}b_{11} & a_{12}b_{11} & a_{13}b_{11} & & a_{11}b_{13} & a_{12}b_{13} & a_{13}b_{13} \\ a_{21}b_{11} & a_{22}b_{11} & a_{23}b_{11} & \cdots & a_{21}b_{13} & a_{22}b_{13} & a_{23}b_{13} \\ a_{31}b_{11} & a_{32}b_{11} & a_{33}b_{11} & & a_{31}b_{13} & a_{32}b_{13} & a_{33}b_{13} \\ & \vdots & & \ddots & & \vdots & \\ a_{11}b_{31} & a_{12}b_{31} & a_{13}b_{31} & & a_{13}b_{33} & a_{13}b_{33} & a_{13}b_{33} \\ a_{21}b_{31} & a_{22}b_{31} & a_{23}b_{31} & \cdots & a_{23}b_{33} & a_{23}b_{33} & a_{23}b_{33} \\ a_{31}b_{31} & a_{32}b_{31} & a_{33}b_{31} & & a_{33}b_{33} & a_{33}b_{33} & a_{33}b_{33} \end{pmatrix}. \tag{6.26}$$

The direct product of two 3x3 matrices is a 9x9 matrix as above. The two multiplying matrices each need not have the same number of rows and columns, one could be of $n\times n$ size and the other $n\times m$. In general,

$$(n\times n)\otimes(n\times m) = (mn\times nm). \tag{6.27}$$

This direct product definition can be used to show that the character of the direct product is equal to the product of the character of the two multiplying matrices. To see this result, one may like to consider two functions f and g, each doubly degenerate due to the component functions f_1 and f_2, and g_1 and g_2. Expressing f and g as linear combinations of their individual component functions, and then multiplying them, we get

$$f = a_{11}f_1 + a_{12}f_2$$

$$g = b_{11}g_1 + b_{12}g_2$$

$$fg = (a_{11}f_1 \times b_{11}g_1) + (a_{11}f_1 \times b_{12}g_2) + (a_{12}f_2 \times b_{11}g_1) + (a_{12}f_2 \times b_{12}g_2). \tag{6.28}$$

In the matrix form, one writes

$$\begin{pmatrix} a_{11} & a_{12} \\ a_{21} & a_{22} \end{pmatrix} \otimes \begin{pmatrix} b_{11} & b_{12} \\ b_{21} & b_{22} \end{pmatrix}, \quad (6.29)$$

the direct product of which will be a 4×4 matrix

$$\begin{pmatrix} a_{11}b_{11} & a_{11}b_{12} & a_{12}b_{11} & a_{12}b_{12} \\ a_{11}b_{21} & a_{11}b_{22} & a_{12}b_{11} & a_{12}b_{22} \\ a_{21}b_{11} & a_{21}b_{12} & a_{22}b_{11} & a_{22}b_{12} \\ a_{21}b_{21} & a_{21}b_{22} & a_{22}b_{21} & a_{22}b_{22} \end{pmatrix}. \quad (6.30)$$

The traces of the two matrices above can be equated to obtain

$$(a_{11} + a_{22}) \times (b_{11} + b_{22}) = a_{11}b_{11} + a_{11}b_{22} + a_{22}b_{11} + a_{22}b_{22}. \quad (6.31)$$

This shows that the trace of the direct product is equal to the product of the traces of the multiplying matrices. So the character of the direct product representation will be equal to the product of the characters of the two irreducible presentations Γ_a and Γ_b for all the operations i of the group

$$\chi_{\Gamma_{ab}}(O_i) = \chi_{\Gamma_a}(O_i)\chi_{\Gamma_b}(O_i). \quad (6.32)$$

In general, Γ_{ab} is a reducible representation, although Γ_a and Γ_b are irreducible.

The character table for naphthalene can be taken up to verify the characters of a direct product representation. The characters of any of the two representations can be multiplied to check if the product of the characters corresponds to any of the other representations. By taking A_u and B_{1u}, one gets the following

	E	σ_{xy}	σ_{xz}	σ_{yz}	i	c_{2z}	c_{2y}	c_{2x}
$A_u \otimes B_{1u}$	1	1	−1	−1	1	1	−1	−1

By inspection of the direct product table (Table 6.3) we get at once that

$$A_u \otimes B_{1u} = B_{1g}.$$

The other direct products can be made out identically. It should also be kept in mind that $g \otimes u = u$, $u \otimes u = g$, and $g \otimes g = g$. This is a general procedure for construction of direct product tables, and the one for our example of naphthalene is shown in Table 6.3. All representations and characters in the table are one-dimensional so the characters of a direct product representation will also be one-dimensional.

In another example, the c_{3v} symmetry point group to which CH_3F belongs may be considered. The character for the irreducible representation E is 2 (Table 6.4).

Because the representation for $E \otimes E$ does not correspond to any of the irreducible representations of this group, it can be decomposed by using the formula introduced earlier

TABLE 6.3
Direct Product Table for Irreducible Representations of Naphthalene

D_{2h}	A_g	A_u	B_{1g}	B_{1u}	B_{2g}	B_{2u}	B_{3g}	B_{3u}
A_g	A_g	A_u	B_{1g}	B_{1u}	B_{2g}	B_{2u}	B_{3g}	B_{3u}
A_u	A_u	A_g	B_{1u}	B_{1g}	B_{2u}	B_{2g}	B_{3u}	B_{3g}
B_{1g}	B_{1g}	B_{1u}	A_g	A_u	B_{3g}	B_{3u}	B_{2g}	B_{2u}
B_{1u}	B_{1u}	B_{1g}	A_u	A_g	B_{3u}	B_{3g}	B_{2u}	B_{2g}
B_{2g}	B_{2g}	B_{2u}	B_{3g}	B_{3u}	A_g	A_u	B_{1g}	B_{1u}
B_{2u}	B_{2u}	B_{2g}	B_{3u}	B_{3g}	A_u	A_g	B_{1u}	B_{1g}
B_{3g}	B_{3g}	B_{3u}	B_{2g}	B_{2u}	B_{1g}	B_{1u}	A_g	A_u
B_{3u}	B_{3u}	B_{3g}	B_{2u}	B_{2g}	B_{1u}	B_{1g}	A_u	A_g

TABLE 6.4
Character Table for c_{3v} Point Group

c_{3v}	E	$2c_3$	$3\sigma_v$
A_1	1	1	1
A_2	1	1	−1
E	2	−1	0
$E \otimes E$	4	1	0

$$a_j = \frac{1}{h}\sum_{O_i}\chi_{\gamma_j}(O_i)\chi_{\gamma}(O_i). \quad (6.33)$$

The coefficients work out to

$$a_1 = \frac{1}{6}\left[4\times1 + 2(1\times1) + 3(0\times1)\right] = 1$$

$$a_2 = \frac{1}{6}\left[4\times1 + 2(1\times1) + 3(0\times(-1))\right] = 1$$

$$a_3 = \frac{1}{6}\left[4\times1 + 2(1\times1) + 3(0\times0)\right] = 1, \quad (6.34)$$

from which one obtains

$$E \otimes E = A_1 \oplus A_2 \oplus E = 1 + 1 + 2.$$

The last line of the character table can now be written as

$A_1 \oplus A_2 \oplus E$	4	1	0

6.7 APPLICATIONS

There are numerous applications of group theory in all branches of science and philosophy. For specific applications to spectroscopy, there are several authoritative books and articles that we would recommend the reader to seek out, such as Hamermesh (1962), Tinkham (1964), Ferraro and Ziomek (1969), and Bunker (1979). Mentioned below are just a few applications that are directly relevant to the subject of the present study.

6.7.1 Energy Eigenvalues of Molecular Orbitals

The group theoretic basis for MO energy calculations is based on the principle that the molecular Hamiltonian of the Schrödinger equation commutes with all the elements of the symmetry that the molecule belongs to, and therefore the Hamiltonian itself is a symmetry element of the group. As a corollary, any element of the symmetry group will transform the eigenfunction ψ to a new eigenfunction ψ_n. The proof of this claim is omitted here, but can be provided readily by looking at the effect of permutation of a set of identical nuclei of the molecule on the solution of the Schrödinger equation $\mathcal{H}\psi = E\psi$. It turns out that a new eigenfunction, call it $\psi_n^{(\text{symmetry operated})}$, emerges each time a different nuclear-permuted Hamiltonian is allowed to operate upon, which is

$$\begin{aligned} \mathcal{H}^{(\text{permutation 1})}\psi &= E\psi_1^{(1)} \\ \mathcal{H}^{(\text{permutation 2})}\psi &= E\psi_2^{(2)} \\ &\vdots \\ \mathcal{H}^{(\text{permutation n})}\psi &= E\psi_n^{(n)} \end{aligned} \quad . \tag{6.35}$$

The energy E in all of these solutions has to be degenerate provided that the result of a symmetry group of transformation of any of these eigenfunctions is a linear combination of all of the nondegenerate ψ_n functions. We see from this argument that the wavefunctions are bases of irreducible representations of the molecular point group. These results are often summarized by the statement that 'the molecular Hamiltonian is invariant to a symmetry operation by any of the elements of the group to which it belongs'.

The procedure for determination of the irreducible representations to which molecular orbitals belong was discussed earlier. With regard to the calculation of molecular orbital energies, a constraint is imposed by another property of group theoretic principles that states that the element $\mathcal{H}_{ij}$ of the matrix $\langle\psi_i|\mathcal{H}|\psi_j\rangle$ does not exist if $|\psi_i\rangle$ and $|\psi_j\rangle$ belong to two different irreducible representations (Wigner, 1959: ch. 11–12). Therefore, the determination of the symmetry species to which the molecular orbitals belong, which can be decided by symmetry considerations alone, can provide information about orbital energy eigenvalues. For n-number of molecular orbitals the nth order secular determinant will yield n roots corresponding to energy eigenvalues $E_1, E_2, \ldots, E_n$.

6.7.2 Removal of Energy Degeneracy by Perturbation

If a function $f(r)$ is the sum of two or more functions $f_1^{(r)} + f_2^{(r)} + \cdots$, then the lowest symmetry across the functions determines the symmetry of $f(r)$. If the Hamiltonian function is $\mathcal{H} = \mathcal{H}_o + V$, where V is an external potential that may or may not be time-varying, the lower of the two symmetries – one each for $\mathcal{H}_o$ and V, determines the symmetry of $\mathcal{H}$. The consequence of perturbation of the $\mathcal{H}$ function by V is that the eigenfunctions of $\mathcal{H}_o$ which were once the basis of the group to which $\mathcal{H}_o$ belonged, will now be the representations of the symmetry group for V. These new representations of the perturbed Hamiltonian are expected to be irreducible because of the lowering of the symmetry, but the representations can be decomposed. As a result, the energy eigenvalues that were degenerate in the unperturbed state will now split.

TABLE 6.5
Character Table for the O_h Point Group

O_h		E	$8c_3$	$6c_2$	$6c_4$	$3c_3$	i	$6s_4$	$8s_6$	$6\sigma_h$	$6\sigma_d$
d_{xy}, d_{yz}, d_{zx}	Γ_π	12	0	0	0	−4	0	0	0	0	0
$d_{x^2-y^2}$, d_{z^2}	Γ_σ	6	0	0	2	2	0	0	0	4	2

$$\Gamma_\sigma = A_{1g} + E_g + T_{1u}$$
$$\Gamma_\pi = T_{1g} + T_{2g} + T_{1u} + T_{2u}$$

A very well studied example of perturbation splitting is the energy of the five d-orbitals of the first-series transition metal atoms placed in an octahedral field (O_h symmetry). The atomic Hamiltonian $\mathcal{H}$ is perturbed by the additional crystal field potential $V(\mathcal{H} = \mathcal{H}_o + V)$ when the atom is liganded to six atoms. This crystal field potential arises from either point charges of ligands or the induced dipole of neutral ligands. The free atom approximated to a sphere has the highest symmetry known – an infinite-fold rotation axis and mirror planes containing these axes. The d-orbital wavefunctions of the atomic Hamiltonian have degenerate energy. In the crystal field of octahedral geometry, the symmetry is lowered to O_h, so that the d-orbitals that transformed according to the spherical symmetry earlier will now transform under the O_h symmetry group. For a ready reference, the character table for O_h is produced in Table 6.5.

While making Table 6.5 we have used six vectors colinear with $\pm x, \pm y, \pm z$, whose origin is coincident with the metal atom, to obtain the characters for σ-bonding representations (Γ_σ). Twelve vectors are used to obtain the characters for representations of π-bonding (Γ_π). If the electric field in the octahedron is due to the six point charge ligands, then the electrons in the t_{2g} orbitals will have energy lower than those of the e_g orbitals. Note that capital 'T' and 'E' are reserved for denoting character representations, and lower-case 't' and 'e' are used to label the orbitals. The splitting of the orbitals is shown in Figure 6.7.

6.7.3 General Selection Rules for Electronic Transitions

Earlier in Chapter 3 we came across the general form of the dipole transition integral for a spectroscopic transition from state $|\phi_1\rangle$ to $|\phi_2\rangle$. The operator $\boldsymbol{\mu}$ for a many-electron polyatomic system can be written explicitly as

$$\boldsymbol{\mu} = \sum_i e\mathbf{r}_i - e\sum_j z_j\mathbf{R}_j \tag{6.36}$$

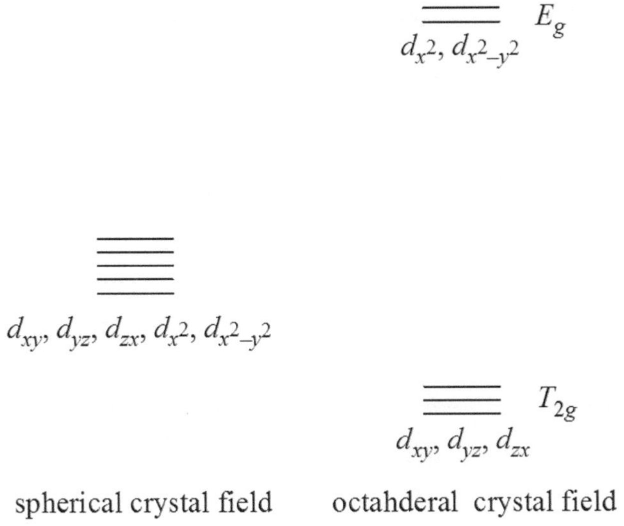

FIGURE 6.7 The d-orbitals in a spherical crystal field (*left*) and the splitting in an octahedral crystal field (*right*).

where e and z are electron and nuclear charges, and **r** and **R** are position vectors for electrons and nuclei. The inclusion of nuclear dipole requires that the integration be carried out over both electronic and nuclear coordinates. But if the integration is performed over electronic coordinates alone the nuclear dipole moment operator can be dropped out, and the integral is simply called electronic transition moment μ_{12} for a transition from ϕ_1 to ϕ_2. The quantity $|\mu_{12}|^2$ is proportional to the intensity of the absorption band.

Because the operator $\boldsymbol{\mu}$ is a vector

$$\boldsymbol{\mu} = \boldsymbol{\mu}_x + \boldsymbol{\mu}_y + \boldsymbol{\mu}_z, \tag{6.37}$$

the dipole transition integral for a transition from state 1 to 2 can be written separately for each of the Cartesian components

$$\mu_{12,x} = \int \psi_1^* \sum_i e x_i \psi_2 d\tau$$

$$\mu_{12,y} = \int \psi_1^* \sum_i e y_i \psi_2 d\tau$$

$$\mu_{12,z} = \int \psi_1^* \sum_i e z_i \psi_2 d\tau. \tag{6.38}$$

An electronic transition will not be allowed (forbidden) if all of these dipole component integrals have a zero value. But a transition will be allowed even if one of them has a nonzero value. To yield a value different from zero the integrand must be invariant to any of the possible transformations under the symmetry group. As discussed earlier, a transition will occur even when one of the above three integrands is symmetrical (even) to all symmetry operations permitted under the molecular point group. This requires knowing the symmetry species of $\boldsymbol{\mu}_x$, $\boldsymbol{\mu}_y$, $\boldsymbol{\mu}_z$, and ϕ_1 and ϕ_2.

Representation for the Dipole Moment Operator. Whether the transformations are symmetric or asymmetric can be determined simply by operating on the x, y, and z coordinates of $\boldsymbol{\mu}$, one by one, using the symmetry elements of the group. Since we have been working with naphthalene all through let us consider the operations under the D_{2h} group. Table 6.6 shows symmetric (+1) and antisymmetric (−1) transformations of the x, y, z basis. The last column introduces 'equivalent' representation, which means generating irreducible representations by mapping each matrix of the already existing representations through similarity transformation. A representation Γ consisting of $\boldsymbol{n}_i$ matrices can generate a mapped set of matrices $\boldsymbol{n}_i'$ by the similarity transformation

$$\mathbf{n}_i' = \mathbf{A}^{-1}\mathbf{n}_i\mathbf{A}, \tag{6.39}$$

where **A** is another matrix. The representation containing these mapped matrices $\boldsymbol{n}_i'$ is called an equivalent representation Γ_{eq}.

TABLE 6.6
Transformation of the Cartesian Coordinates under the Operation of the Symmetry Elements of the D_{2h} Point Group

D_{2h}	E	σ_{xy}	σ_{xz}	σ_{yz}	i	C_{2x}	C_{2y}	C_{2z}	Γ_{eq}
x or $\sum_i x_i$	1	1	1	−1	−1	1	−1	−1	B_{3u}
y or $\sum_i y_i$	1	1	−1	1	−1	−1	1	−1	B_{2u}
z or $\sum_i z_i$	1	−1	1	1	−1	−1	−1	1	B_{1u}

Representations for Ground and Excited States. Finding representations of the ground and excited states, ψ_g and ψ_e, are particularly simple when a direct product table for the relevant symmetry group is made. For the ground state function first, we write down the configuration keeping in mind that the one-electron molecular orbitals are filled up in accord with the *Auf bau* principle. For instance, consider the five ground-state molecular orbitals of naphthalene (Figure 6.8), the configuration for which is

$$B_{1u}^2, B_{2g}^2, B_{3g}^2, B_{1u}^2, A_u^2.$$

The representation for the ground state wavefunction ψ_g can be obtained simply by taking the direct product

$$B_{1u} \otimes B_{1u}, B_{2g} \otimes B_{2g}, B_{3g} \otimes B_{3g}, B_{1u} \otimes B_{1u}, A_u \otimes A_u.$$

So the symmetry species are

$$(A_g)(A_g)(A_g)(A_g)(A_g).$$

They all are totally symmetric, and the ground state symmetry is given as A_g (^1A_g). We are using italic capitals for all irreducible representations as well as to indicate the overall state. Some literature distinguishes them by using small letters for representations, reserving the corresponding capital letters

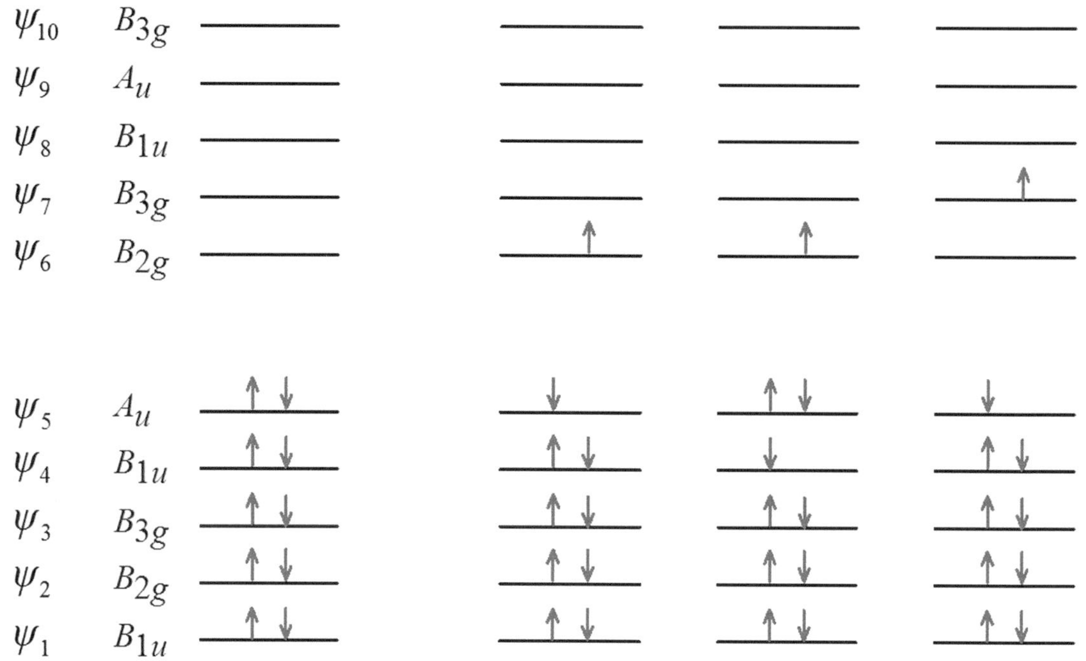

FIGURE 6.8 Ground and excited states of naphthalene.

for the state. The direct product of representations for finding out the representation for the ground state wavefunction ($\psi^2 = \psi\psi'$) works when both ψ and ψ' are nondegenerate. In the present case, the individual molecular wavefunctions do not have the same representation, so the direct product can be evaluated as usual – multiplying the characters of the two representations for each permitted operation and matching this new row of characters with one among those existing already.

To find out the representation for the excited state wavefunction, a simple rule can be used: the symmetry of the excited state is equivalent to the symmetry of the ground state multiplied by the symmetry of the orbital from which the electron is already extracted and the symmetry of the orbital which is already occupied by the electron. For example, if ϕ_j provides the electron then the symmetry of ϕ_j without that electron, and if ϕ_k accepts the electron, then the symmetry of ϕ_k with the electron in it have to be considered. For instance, consider the $A_u \leftrightarrow B_{3g}$ transition in naphthalene (Figure 6.8) for which the symmetry of the excited state is

$$\begin{aligned}\Gamma(\psi_2) &= \Gamma(\psi_1)\times\Gamma(\phi_j)\times\Gamma(\phi_k) \\ &= A_g \times A_u \times B_{3g} \\ &= A_g \times B_{3u} \\ &= B_{3u}.\end{aligned} \tag{6.40}$$

The multiplication of the representations is associative, so the order of multiplication does not affect the result. Using the same procedure, the excited state (ψ_2) for the $A_u \leftrightarrow B_{2g}$ and $B_{1u} \rightarrow B_{2g}$ transitions are found to be B_{2u} and B_{3u}, respectively. We notice that the excited state symmetry may vary with the transition studied.

Symmetry-Allowed Transitions. To find out if the transitions $A_u \leftrightarrow B_{2g}$ and $B_{1u} \rightarrow B_{2g}$, and $A_u \leftrightarrow B_{3g}$ are allowed, one needs to inspect the transition dipole integrals with the respective symmetry representation of the integrand. The requirement for a transition to be allowed is that the product of all three species in the integrand be totally symmetric, which means the direct product representation of the integrand must belong to an irreducible representation for which all characters across the relevant group symmetry operations is +1. This is the A_g representation for D_{2h}. Therefore, the product of the representations in the integrand must yield A_g in order for a transition to be allowed. Let us check if this condition holds for the transition $A_u \leftrightarrow B_{2g}$, for example,

$$\int_{-\infty}^{+\infty} \psi_g^* \sum x_i \psi_e \, d\tau.$$

The representation $\Gamma(\psi_g) \otimes \Gamma(\psi_e) \otimes \Gamma(x)$ can be analyzed as follows

$$\begin{aligned}&\text{for } x\text{-polarized light} : A_g \otimes B_{2u} \otimes B_{3u} = B_{2u} \otimes B_{3u} \neq A_g, \\ &\text{for } y\text{-polarized light} : A_g \otimes B_{2u} \otimes B_{2u} = B_{2u} \otimes B_{2u} = A_g, \\ &\text{for } z\text{-polarized light} : A_g \otimes B_{2u} \otimes B_{1u} = B_{2u} \otimes B_{1u} \neq A_g.\end{aligned} \tag{6.41}$$

Clearly, a transition would occur only when a y-polarized light beam is incident upon. Similarly, the selection rules for the other two transitions are found

$B_{1u} \rightarrow B_{2g}$ transition allowed only for y-polarized light
$A_u \rightarrow B_{3g}$ transition allowed only for x-polarized light.

A basic review of the results above will tell that a transition is allowed if the product of the orbitals – one vacated and one occupied in order, involved in the electronic transition has a symmetry representation identical to that of the electric dipole moment operator. It should also occur to the reader that the use of a two-headed arrow to indicate a transition implies the occurrence of both absorption and emission, meaning the selection rule(s) applicable to an absorption transition automatically applies to emission.

6.7.4 Specific Transition Rules

For a moment we merely mention three specific selection rules imposed by group theoretic principles. One, the transitions involving molecular orbitals allow transition of only one electron – from a ground to an excited state orbital, provided the integrand products satisfy the requirements discussed above. The reason for this restriction is that the μ_{12} matrix element will go to zero if the determinants involve more than one ground or excited state orbitals. Two, because the x, y, z components of a vector are antisymmetric with respect to the inversion operation, a transition will be allowed only when $g \rightarrow u$ or $u \rightarrow g$ (even-to-odd or odd-to even) transformation occurs. Three, the electronic transition integral does not exist when the spin coordinates are included. The spin selection rule is given by $\Delta s = 0$, which must imply that only a singlet-to-singlet is allowed, not a singlet-to-triplet. The triplet-to-triplet transition is forbidden in terms of group theoretic principles, even if it is spin-allowed.

PROBLEMS

6.1 Classify molecular hydrogen, formaldehyde, carbon tetrachloride, methanol, cyclohexane, phenol, pyrene anion, indole, and porphin according to the symmetry point group.

6.2 Find the group multiplication tables for the point groups to which carbon tetrachloride, cyclohexane, and pyrene anion belong.

6.3 Consider a net macroscopic nuclear spin magnetization vector $\mathbf{M}_z$ in the laboratory frame xyz as shown

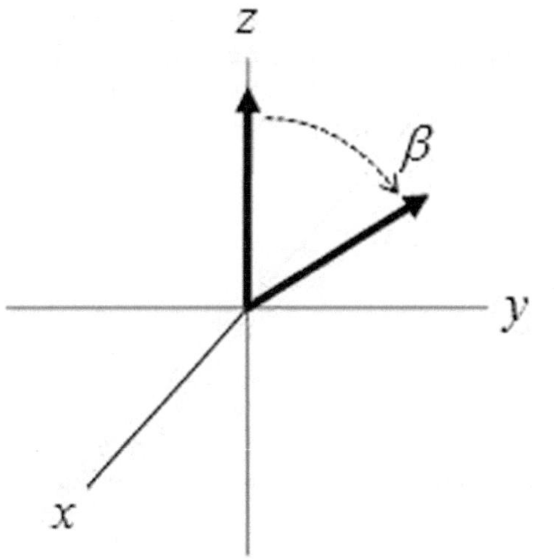

The vector is rotated by a pulsed radiofrequency (RF) field applied along $+x$. The angle β by which $\mathbf{M}_z$ is rotated depends on the nuclear gyromagnetic ratio, and the duration and strength of the RF field. Suppose the $\mathbf{M}_z$ vector is rotated by 45°, 60°, 90°, and 180° in turn. Show the 3×3 matrices generated by the rotation operator c_n.

6.4 Show the character tables for the point groups to which carbon tetrachloride, cyclohexane, and indole belong. The first row of a table should identify the reducible representations.

6.5 An important foundation for group theoretic application to molecular eigenstates is that the eigenstates ϕ_i form a complete basis set of irreducible representations of the point group to which the molecule belongs. But there are several irreducible representations; if the matrix element of two eigenstates is zero, $\langle \phi_k | \mathcal{H} | \phi_l \rangle = 0$, then ϕ_k and ϕ_l belong to different irreducible representation, and if $\langle \phi_k | \mathcal{H} | \phi_l \rangle \neq 0$, then the two eigenstates belong to the same irreducible representation. This property has been used in the text to study naphthalene. Use cyclohexane for a repeat of the exercise to show the possible number of irreducible representations. Also identify if there is a degeneracy within.

6.6 Construct a direct product table for irreducible representations of cyclohexane.

6.7 Show the ground and excited state representations for cyclohexane. Are the ground state symmetry species all symmetric?

6.8 List the equivalent representations Γ_{eq} for the Cartesian components of the electric diple moment μ of cyclohexane. Mention the allowed electronic transitions along with the polarization of the incident light.

BIBLIOGRAPHY

Bunker, P. R. (1979) *Molecular Symmetry and Spectroscopy*, Academic Press.

Bunker, P. R. and P. Jensen (1998) *Molecular Symmetry and Spectroscopy*, NRC Research Press.

Cotton, F. A. (1963) *Chemical Applications of Group Theory*, Interscience.

Ferraro, J. R. and J. S. Ziomek (1969) *Introductory Group Theory and Its Application to Molecular Structure*, Plenum.

Hamermesh, M. (1962) *Group Theory and Its Applications to Physical Problems*, Addison-Wesley.

Hochstrasser, R. M. (1966) *Molecular Aspects of Symmetry*, W. A. Benjamin, Inc.

Salem, L. (1966) *Molecular Orbital Theory of Conjugated Systems*, W. A. Benjamin, Inc.

Tinkham, M. (1964) *Group Theory and Quantum Mechanics*, McGraw Hill.

Wigner, E. P. (1959) *Group Theory*, Academic Press.

Wilson, E. B., J. C. Decius, and P. C. Cross (1955) *Molecular Vibrations*, McGraw-Hill.

7 Rotational Spectra

The total energy of a molecule may be thought to arise from translational, rotational, and vibrational degrees of freedom, the electronic motional degrees of freedom, and electron and nuclear spin variables. The three rotational degrees of freedom of a molecule rotating in space are described by the angular coordinates while placing the center of mass of the molecule at the origin of an orthogonal molecule-fixed axes system. In the simplest case of a non-vibrating diatomic molecule the frequency of rotation is in MHz range, corresponding to rotational absorption in nm (microwave) region. This is true when molecular moments of inertia are involved in end-over-end rotation. In rotational motion about the internuclear axis of the diatomic, the frequency of rotation can be higher by at least 3-orders of magnitude because the rotational frequency for this rotation is determined not by the nuclear (molecular) mass, but by the much lighter electrons. Such rotational frequencies correspond to rotational absorption in the UV-visible optical region. However, molecular rotational resonance is generally measured in the microwave region, and hence is also called microwave spectroscopy.

7.1 ROTATIONAL SPECTRA OF DIATOMIC MOLECULES

Suppose two atoms, A and B, of masses m_A and m_B constitute a rigid rotor. It is called rigid because the equilibrium interatomic distance R_e is fixed to a constant value even when the molecule is allowed to rotate. This clamped assumption (constant R_e) allows for separation of the rotational motion from all other degrees of freedom. We also position the center of mass of the rotor at the origin of the coordinate system, and let the center of mass move along with the laboratory-fixed coordinate system (X,Y,Z) at a constant velocity corresponding to the translational energy and momentum of the molecule (Figure 7.1). In some textbooks, this coordinate system that moves along with the center of mass of the rotor is called the 'space-fixed axes' system. Here, we call this system the lab-fixed coordinate system, which actually moves with the velocity of the center of mass. The molecule is allowed to rotate in the lab-fixed frame, and the rotational motion is described in terms of the angular coordinates θ, ϕ, χ, which bring in the molecule-fixed frame. Of these three rotations, the χ-rotation contributes little to the rotational energy of the molecule, because χ-rotation corresponds to the rotation of the molecular mass equivalent only to a spinning motion. Rotational energy must be due to rotational motion of the molecule about an axis, not its spinning motion.

Note that the center of mass or the center of gravity of the diatomic rigid rotor satisfies

$$m_1 r_1 = m_2 r_2. \tag{7.1}$$

The reduced mass, also called the single mass, is expressed as

$$\mu = \frac{m_1 m_2}{m_1 + m_2}, \tag{7.2}$$

and the moment of inertia about the ith axis of rotation is given by

$$I = \mu R_e^2. \tag{7.3}$$

In the principal inertial axes (see Box 7.1) the rotor may be visualized as shown in Figure 7.2 in which the moment of inertia $I_a = 0$, $I_b = I_c = 1$, and $\boldsymbol{\omega}$ and $\mathbf{J}$ are rotational velocity and rotational angular momentum vectors.

7.1.1 Schrödinger Equation for Diatomic Rotation

Just the way kinetic energy of a particle of mass m having a linear velocity v is $\text{KE} = \frac{1}{2}mv^2$, the kinetic energy of the rotor, often given the symbol T, is

$$\text{KE} = T = \frac{1}{2}\mu R_e^2 \omega^2, \tag{7.4}$$

where μR_e^2 is the single mass (reduced mass) moment of inertia of the rotor and ω is the angular velocity. Note that $\omega_b = \omega_c = \omega$. This kinetic energy expression can be written as

$$T = \frac{P^2}{2I}, \tag{7.5}$$

in which $P = \mu\omega$ is the angular momentum, and $I = \mu R_e^2$. The same expression is arrived at using the following matrix multiplication. In general,

$$\begin{aligned} T &= \frac{1}{2} I \omega^2 \\ &= \frac{1}{2}\omega^{\mathrm{T}} I \omega \\ &= \frac{1}{2}\begin{pmatrix}\omega_a & \omega_b & \omega_c\end{pmatrix}\begin{pmatrix} I_a\omega_a \\ I_b\omega_b \\ I_c\omega_c \end{pmatrix} \\ &= \frac{P_a^2}{2I_a} + \frac{P_b^2}{2I_b} + \frac{P_c^2}{2I_c}, \end{aligned} \tag{7.6}$$

DOI: 10.1201/9781003293064-7

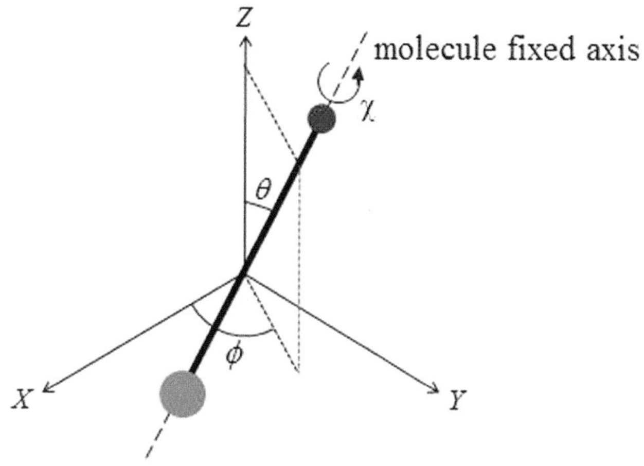

FIGURE 7.1 Rotation of a diatomic molecule in the laboratory-fixed Cartesian frame. The center of mass of the molecule is placed at the origin of the lab frame.

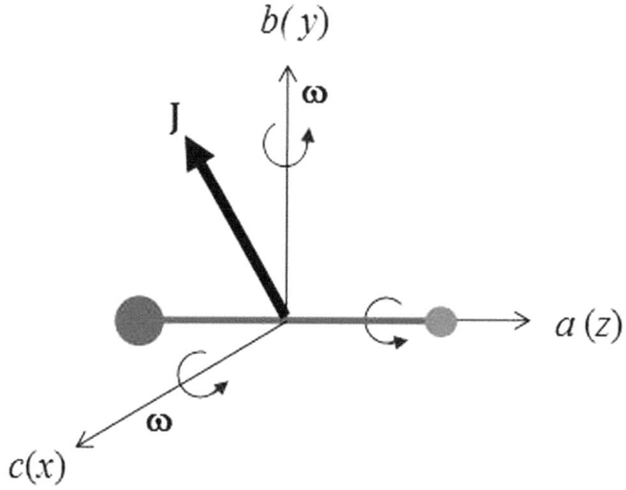

FIGURE 7.2 Disposition of a rigid rotor in the principal axes system.

where a, b, c are principal inertial axes (Box 7.1). But I_a does not exist for the rigid rotor, which means

$$I_b = I_c,\ P^2 = P_b^2 + P_c^2,\ \text{and } T = \frac{P^2}{2I}. \tag{7.7}$$

This kinetic energy, taken to be time-independent for now, is the energy operator for finding out the quantized rotational energy levels of the diatomic molecule. We also say that the rotation is free, meaning no work is involved in the rotation of the molecule, and because work is equivalent to the potential energy of the rotor, there is no potential energy. The total energy operator is then simply

$$\mathcal{H} = \frac{P^2}{2I}, \tag{7.8}$$

and the Schrödinger equation is

$$\mathcal{H}\psi_{JM} = E_{\text{rot}}\psi_{JM}$$

$$\frac{P^2}{2\mu R_e^2}\psi_{JM} = E_{\text{rot}}\psi_{JM}. \tag{7.9}$$

The quantum mechanical operator for the total angular momentum is constructed easily. We get

$$P^2 = -\hbar^2\left[\frac{1}{\sin\theta}\left(\frac{\partial}{\partial\theta}\sin\theta\frac{\partial}{\partial\theta}\right) + \frac{1}{\sin^2\theta}\frac{\partial^2}{\partial\phi^2}\right] \tag{7.10}$$

$$\frac{P^2}{2I}\psi_{JM} = \frac{\hbar^2}{2I}J(J+1)\psi_{JM}, \tag{7.11}$$

indicating that the total angular momentum – spin plus orbital, is quantized as $\hbar^2 J(J+1)$. Hence, the rotational eigenvalue is

$$E_{\text{rot}} = \frac{\hbar^2}{2I}J(J+1). \tag{7.12}$$

7.1.2 Rotational Energy of Rigid Rotor

It is convenient to express the rotational energy by a rotational energy unit called rotational constant B_e. Using $E_{\text{rot}} = h\nu$

$$h\nu = \frac{\hbar^2}{2I}J(J+1),$$

$$\frac{1}{\lambda} = \frac{h}{8\pi^2 Ic}J(J+1),$$

where

$$\frac{h}{8\pi^2 Ic} = B_e. \tag{7.13}$$

This also suggests that the rotational energy levels are given in wavenumber $\text{cm}^{-1} = 1/\lambda$

$$\frac{E_{\text{rot}}}{hc} = B_e J(J+1) = F_{\text{rot}}(J), \tag{7.14}$$

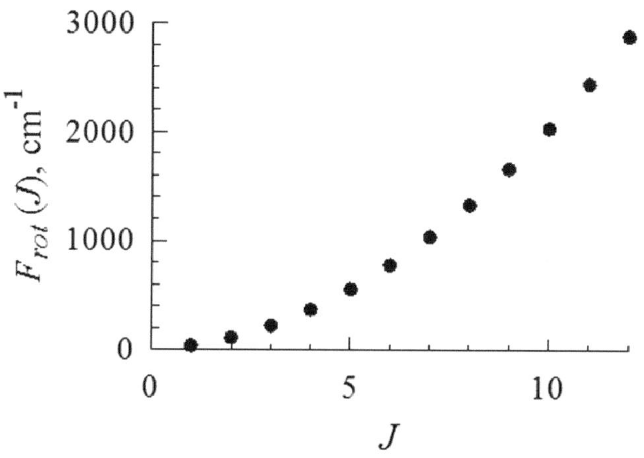

FIGURE 7.3 The variation of the rotational term value with J for the rigid molecular hydrogen. The value taken for R_e is 135.13 pm. The frequencies given in wavenumber correspond to the microwave region.

where $F_{rot}(J)$ is referred to as rotational term value. The unit will be linear frequency rather than wavenumber if E_{rot} is divided by h alone. Clearly, the F_{rot} shown for rigid molecular hydrogen with J varying from 1 to 12 (Figure 7.3) corresponds to the GHz region.

7.1.3 Rotational Energy of Non-Rigid Rotor

The rotational energy expressed by equations (7.12) or (7.14) applies to a rigid rotor, but a rotating diatomic cannot be considered rigid because the centrifugal force due to rotation will elongate the internuclear separation from the initial value. This effect of centrifugal distortion can be accounted for by adding to the E_{rot} formula of equation (7.12) an energy term proportional to the restoring force that counteracts the centrifugal force. The centrifugal force F_c is a function of the internuclear separation R

$$F_c = IR = \mu R^2 R. \tag{7.15}$$

The restoring force F_r whose magnitude depends on the steepness of the potential energy surface $E(R)$ is

$$F_r = \frac{dE(R)}{dR}. \tag{7.16}$$

If the potential surface is assumed to be parabolic (harmonic) in the proximity of R_e, then

$$F_r = k(R - R_e)R. \tag{7.17}$$

For $R > R_e$, we will have $E(R) > E(R_e)$, and the energy due to the displacement E_d will appear as

$$E_d = \frac{1}{2}k(R - R_e)^2,$$

where $R - R_e$ represents the amplitude of the displacement from R_e. We will add E_d to the rigid rotor energy (equation (7.12)) to obtain the rotational energy of the nonrigid rotor,

$$E_{rot,nonrigid} = \frac{\hbar^2}{2I}J(J+1) + \frac{1}{2}k(R - R_e)^2. \tag{7.18}$$

If this formula is cast in terms of the rotational constant

$$B_e = \frac{h}{8\pi^2 Ic} = \frac{h}{8\pi^2 \mu c}\frac{1}{R^2},$$

and $1/R^2$ is expanded about R_e in a power series

$$\frac{1}{R^2} = \frac{1}{R_e^{\,2}} - \left(\frac{2}{R_e^3}\right)R + \left(\frac{3}{R_e^4}\right)R^2 - \cdots,$$

the energy can be expressed as

$$F_{rot,nonrigid}(J) = \frac{E_{rot,nonrigid}}{hc} = B_e J(J+1) - D_e J^2 (J+1)^2 + H_e J^3 (J+1)^3 - \cdots. \tag{7.19}$$

The two additional rotational constants, called centrifugal distortion constants, appearing now are

$$D_e = B_e \frac{h^2}{4\pi^2 c I R_e^2 k}$$

$$H_e = D_e \frac{3\hbar^2}{\mu R_e^4 k}. \tag{7.20}$$

7.1.4 Stationary State Eigenfunctions and Rotational Transitions

Since the rotational angular momentum operator involves angles θ and ϕ in spherical polar coordinates, the eigenfunction ψ_{JM} is also a function of these two angles. This is reflected by the fact that $\psi_{JM}(\theta,\phi)$ are just normalized spherical harmonics

$$\psi_{JM}(\theta,\phi) = \Theta_{JM_J}(\theta)\Phi_{M_J}(\phi), \tag{7.21}$$

where Θ_{JM} and Φ_{M_J} are associated Legendre functions $P_J^M(\cos\theta)$. The normalization constant is obtained easily

$$\int_0^{2\pi} \Phi^*_{m_j}(\phi)\Phi_{m_j}(\phi)d\phi = 1$$

$$\int_0^{2\pi} Ae^{\mp i m_j \phi} Ae^{\pm i m_j \phi} d\phi = 1$$

$$A_\pm^2 \int_0^{2\pi} d\phi = 1$$

$$A_\pm = \sqrt{\frac{1}{2\pi}}. \tag{7.22}$$

The stationary-state rotational wavefunction (Figure 7.4) is then

$$\Phi_{m_j}(\phi) = \sqrt{\frac{1}{2\pi}}e^{\pm i m_j \phi} = \frac{1}{\sqrt{2\pi}}\left(\cos m_j\phi + i\sin m_j\phi\right). \tag{7.23}$$

The subscripts J and M in ψ_{JM} stand for the total angular momentum quantum number 0, 1, 2, ⋯, and the z-projection

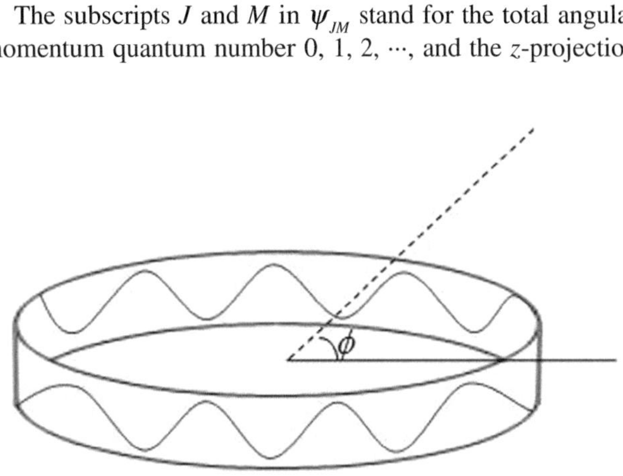

FIGURE 7.4 The real part of the rotational wavefunction.

quantum numbers $-J \le 0 \le +J$, respectively. Simply, the function $\psi_{JM}(\theta, \phi)$ describes the rotation of the molecule in terms of $J(J+1)$. The number of permitted values of M specifies the degeneracy of the Jth rotational energy level $(2J+1)$. For example, the degenerate quintet energy levels for $J = 2$ consist of energy levels $M = -2, -1, 0, 1, 2$ (Figure 7.5).

Having obtained the rotational energy levels from the Schrödinger equation one should enquire about the permitted transitions between the energy levels. To find the selection rules for rotational transitions, one considers the time-dependent potential $V(t)$, which will appear only when microwave radiation is shined. As before

$$V(t) = |-\boldsymbol{\mu}| \cdot \mathbf{E}(t) = \mu_x E_x + \mu_y E_y + \mu_z E_z. \qquad (7.24)$$

If the rotor has a permanent dipole moment $\boldsymbol{\mu}_o$ that orients along the internuclear axis (Figure 7.6) and rotates as the diatomic rotor does, the lab-axes components of the dipole moment are

$$\mu_x = \mu_o \sin\theta \cos\phi$$

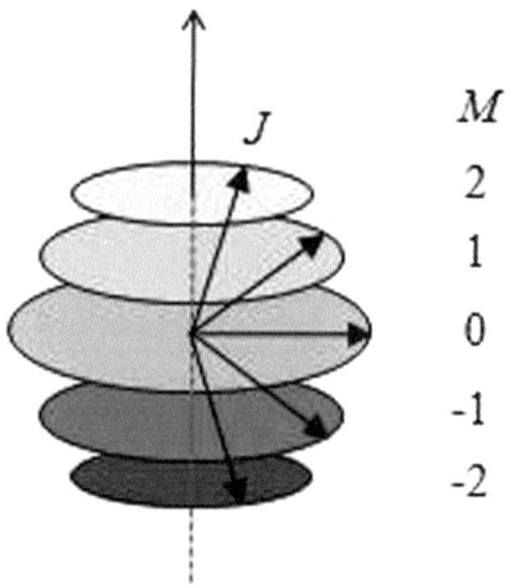

FIGURE 7.5 The degeneracy of the angular momentum J. The z-projections of J reflect the extent of degeneracy.

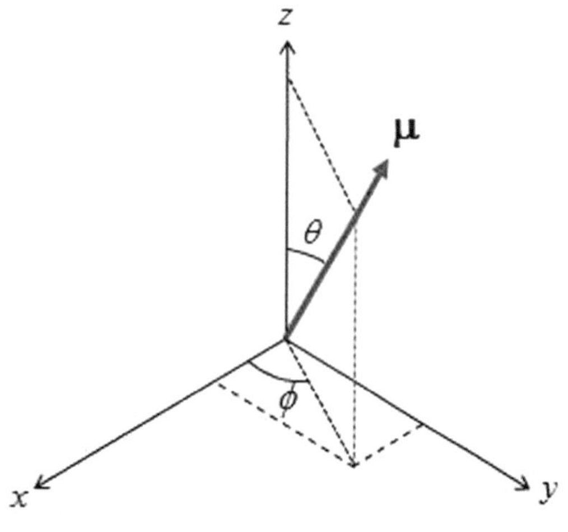

FIGURE 7.6 The orientation of the permanent dipole moment vector $\boldsymbol{\mu}$ of a diatomic.

$$\mu_y = \mu_o \sin\theta \sin\phi$$

$$\mu_z = \mu_o \cos\theta.$$

Notice the systematic emergence of these terms across the c-row of the matrix given in Box 5.1.

Writing the dipole transition integral should be easy. We have

$$< \psi_{JM} \left| \boldsymbol{\mu}_o \cdot \mathbf{E}(t) \right| \psi_{J'M'} >= \int_0^{2\pi}\int_0^{\pi} P_J^M(\cos\theta) e^{-iM\phi} \boldsymbol{\mu}_o \cdot \mathbf{E}(t) P_{J'}^{M'} (\cos\theta) e^{iM'\phi} \sin\theta d\theta d\phi = |\mu_o| |E(t)| \int_0^{2\pi}\int_0^{\pi} P_J^M(\cos\theta) e^{-iM\phi} \begin{pmatrix} \sin\theta\cos\phi \\ \sin\theta\sin\phi \\ \cos\theta \end{pmatrix} P_{J'}^{M'} (\cos\theta) e^{iM'\phi} \sin\theta d\theta d\phi, \qquad (7.25)$$

Clearly, the integral goes to zero if $\mu_o = 0$ or $J = J'$ (see Box 7.2), even though M and M' can have any permitted value $(-J \le 0 \le +J)$. Thus, we get the following selection rules for rotational transitions

1. The molecule must have a permanent dipole moment, $\mu_o \ne 0$. Symmetric molecules including H_2, N_2, and Cl_2 whose center of mass and charge coincide do not have pure rotational spectra.
2. $J = J' \pm 1$.
3. $M = M' - 1, M', M' + 1$.

Permitted changes in J and M values are $\Delta J = \pm 1$ and $\Delta M = 0, \pm 1$.

7.1.5 Energy Levels and Representation of Pure Rotational Spectra

The rotational energy levels in an electronic potential energy surface are not equally spaced. The energy levels corresponding to the rotational angular momentum quantum number J are according to $J(J+1)$, and are consistent with equation (7.14). Figure 7.7 shows the levels for a hypothetical diatomic molecule. Two parameters should be paid attention to, the reduced mass μ and the internuclear distance R, which we can take as R_e for the time being. These two considerations will tell that rotational energy spectra must appear in the microwave region.

The energy of rotational transitions expressed by

$$\Delta E = E_{J+1} - E_J = B_e (J+1)(J+2) - B_e J(J+1) = 2B_e (J+1)$$

is given in a general way. If there are q rotational energy levels, then

$$\Delta E = E_{J+k+1} - E_{J+k} = 2B_e (J+k+1) \qquad (7.26)$$

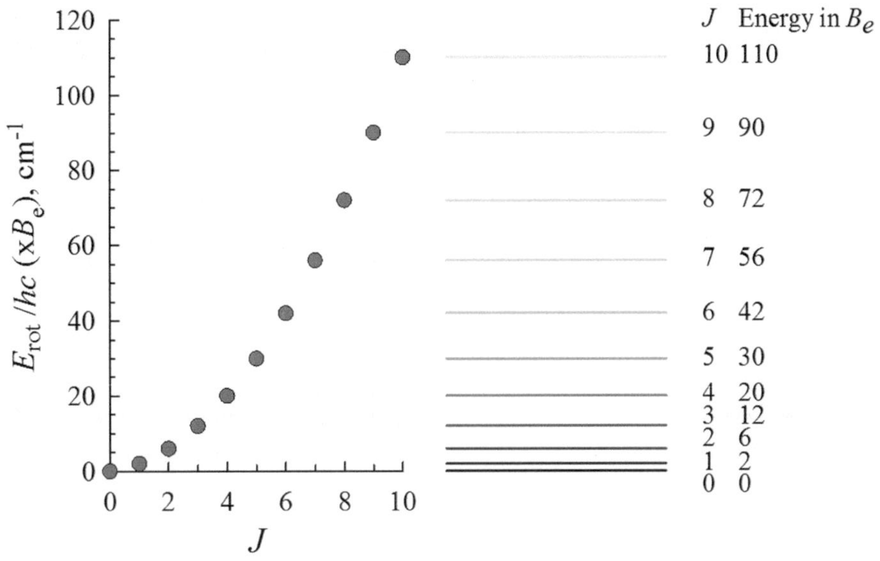

FIGURE 7.7 Energy levels according to the J-value of a rigid rotor.

where $k = 0,1,2,\cdots,q$, which should mean that different rotational absorption bands are spaced by the constant $2B_e$ even though the absorption energy increases as the value of k increases. Yet, to examine the intensity of a rotational absorption band one may consider the population of absorbing molecules according to their value of J. Of course, the absorption intensity will not be uniform across the absorption bands, which should be obvious from the fact that absorption intensity is proportional to the population in the lower-energy absorbing state. The population ratio in the two levels, say J'' (lower energy) and J' (higher energy) is

$$\frac{N_{J''}}{N_{J'}} = \exp\left(\frac{\Delta E_{\mathrm{rot}}}{k_{\mathrm{B}}T}\right), \tag{7.27}$$

where ΔE_{rot} is the energy difference and k_{B} is the Boltzmann constant. From statistical thermodynamics, the population distribution for rotational states according to the quantum number J is given by

$$P_J = (2J+1)\frac{e^{\frac{-B_eJ(J+1)}{k_{\mathrm{B}}T}}}{q_r}, \tag{7.28}$$

where the rotational partition function is

$$q_r = \sum_J (2J+1)e^{\frac{-B_eJ(J+1)}{k_{\mathrm{B}}T}}. \tag{7.29}$$

If the total population size is N, then the population in a rotational state corresponding to the quantum number J is

$$N_J = NP_J.$$

The P_J distribution determines the intensity of $J'' \rightarrow J'$ transitions, but to obtain P_J the value of q_r should be known. The evaluation of q_r involves integrating the function over J from 0 to ∞,

$$q_r = \int_0^\infty q_r dJ = \frac{k_{\mathrm{B}}T}{B_e} \sim \frac{k_{\mathrm{B}}T}{\sigma B_e}, \tag{7.30}$$

which is valid for $\Delta E_{\mathrm{rot}} < k_{\mathrm{B}}T$, particularly when the rotational states are contained in most of the low-lying vibrational energy levels. The σ-factor, termed the symmetry number, is introduced to correct for the partition function according to the rotational symmetry of the rotor. For example, a symmetric homodiatomic when rotated by π about an axis perpendicular to the internuclear axis will generate an indistinguishable configuration. But the partition function already contains the factor needed for counting distinguishable configurations. This means that if the same partition function is used for the indistinguishable case, then the microstate is overcounted, meaning the same configuration of the homodiatomic will be counted twice. To avoid this, the partition function is divided by a symmetry number σ, which is 2 for a homodiatomic, but 1 for a heterodiatomic like HCl, and 3 for a heterotriatomic molecules like NH_3.

The rotational partition function can be easily evaluated to a number, say a two-digit number for ordinary diatomics, with which the population distribution for different J-states (equation 7.28) is plotted in Figure 7.8. The frequency at which the absorption is maximum is given by $\nu_{\max} = 1.18\sqrt{(B_eT)}$.

A pure rotational spectrum of CO in argon vapor recorded at 129 K is reproduced in Figure 7.9. The spectrum shows far-infrared absorption of different rotational molecular states, and the distribution of absorption intensity follows the population distribution of the rotational states determined by equation (7.28). Another interesting observation is a gradual vanishing of the rotational band structure as the CO-argon gas pressure increases from 8.3 to 17.4

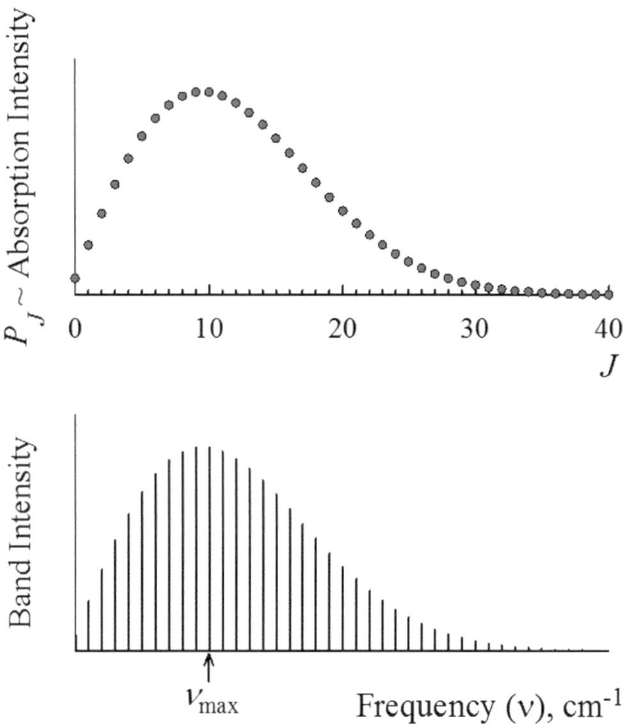

FIGURE 7.8 Appearance of pure rotational absorption bands. The envelope covering the bands is set by the P_J function that contains q_r, B_e, and T.

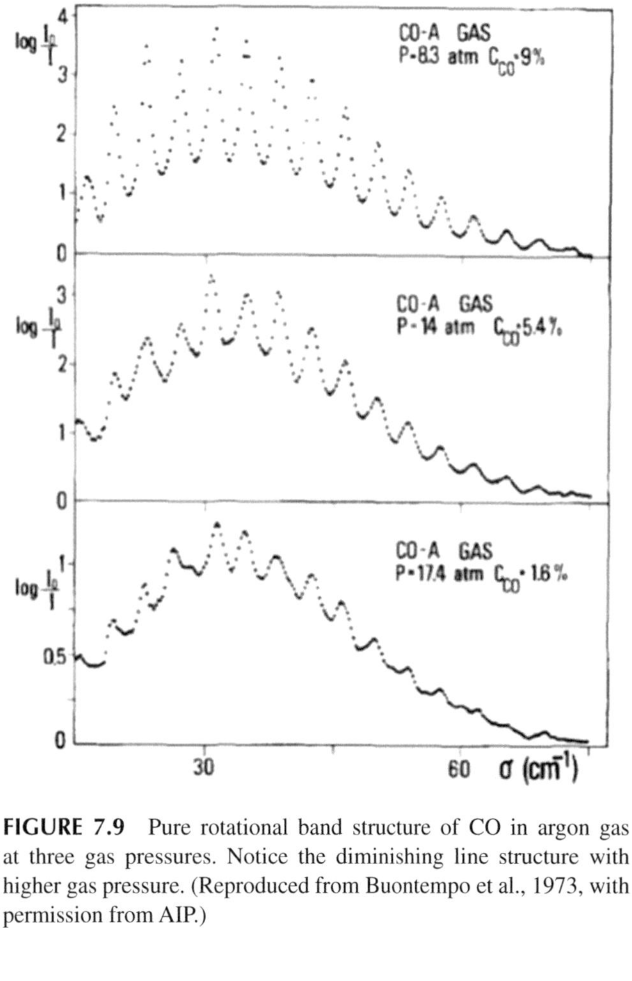

FIGURE 7.9 Pure rotational band structure of CO in argon gas at three gas pressures. Notice the diminishing line structure with higher gas pressure. (Reproduced from Buontempo et al., 1973, with permission from AIP.)

atm. This happens due to collisional suffering of rotational motion. At sufficiently high gas pressure, collision can overwhelmingly broaden the rotational bands to yield only an envelope of absorption.

A rotational spectrum can provide information about the moment of inertia I of the rotating molecule. The total area of an enveloped rotational band, which includes all observed rotational transitions, is given by the rigid rotor sum rule

$$\int \sigma d\lambda^{-1} = \frac{2\pi}{3\times 10^7}\frac{\mu_o^2}{I}, \tag{7.31}$$

where σ is the absorption cross section, λ^{-1} is inverse wavelength, and μ_o is the permanent dipole of the molecule. This sum rule relation also tells that a molecule without a permanent dipole moment does not produce a rotational spectrum (see the selection rules). The calculation of I from the total integrated area, however, requires the knowledge of μ_o. If information about μ_o is not available I can be calculated after obtaining the value of the equilibrium bond length under rigid rotor assumption.

Extracting the value of the bond length from the rotational spectrum is fairly easy. For example, in the CO spectrum (Figure 7.9) the rotational band positions from the extreme left to the right with increasing frequency can be assigned to $J \to J+1$ transitions, $0 \to 1$, $1 \to 2$, $2 \to 3$, $\cdots$. One can choose any of these absorption bands and read out the corresponding absorption frequency in order to know the rotational absorption energy. If the leftmost transition $(0 \to 1)$ is chosen that occurs at ~3.842 cm^{-1}, the corresponding transition energy is equal to $2B_e$ (see Figure 7.7). Since B_e can be equated to the bond length R_e through

$$B_e = \frac{h}{8\pi^2 Ic} = \frac{h}{8\pi^2 \mu R_e^2 c}, \tag{7.32}$$

the values of I and R_e can be immediately determined. The value of μ_o so obtained should be ~0.12 D in the present case.

We have discussed pure rotational spectra of rigid diatomic molecules here, but real molecules are not rigid because the nuclei vibrate. Rotational energy levels are contained in each vibrational energy level that affords detection of rotational–vibrational combination spectra when the transitions are excited by infrared radiation of higher frequency, say toward mid- and near-infrared. We will return to this aspect of rotational spectroscopy after we have learned about molecular vibrations.

7.2 ROTATIONAL SPECTRA OF POLYATOMIC MOLECULES

7.2.1 Rotational Inertia

Polyatomic rotation is substantially simplified by the assumption of rigid body rotation in the molecule-fixed axes superposed on the Cartesian axes system. By convention, the molecule is visualized in the axes system in a way that $I_c > I_b > I_a$. The determination of these rotational moment of inertia, I_a, I_b, and I_c is straightforward, and is shown for the planar formaldehyde molecule in Figure 7.10.

Using the rotational inertia formula

$$I = \sum_i m_i R_i^2 \quad (7.33)$$

one finds

$$I_b = 2\left[m_{\text{H}}(r_{\text{C-H}} \sin 30^\circ)^2\right] + m_{\text{O}} r_{\text{C-H}}{}^2 = \frac{1}{2} m_{\text{H}} r_{\text{C-H}}{}^2 + m_{\text{O}} r_{\text{C=O}}^2$$

$$I_a = 2\left[m_{\text{H}}(r_{\text{C-H}} \sin 60^\circ)^2\right] = \frac{3}{2} m_{\text{H}} r_{\text{C-H}}{}^2$$

$$I_c = 2\left[m_{\text{H}}\left(r_{\text{C-H}}\right)^2\right] + m_{\text{O}} r_{\text{C=O}}^2,$$

where $r_{\text{C-H}} \sin 30^\circ$ and $r_{\text{C-H}} \sin 60^\circ$ are vector distances of the hydrogen atoms from *b*- and *c*-axes, respectively. With atomic masses of 1.008 and 15.999 amu for hydrogen and oxygen, respectively, and bond lengths $r_{\text{C-H}} = 1.113\text{Å}$ and $r_{\text{C=O}} = 1.208\text{Å}$, we obtain $I_a = 1.8728$, $I_b = 23.9715$, and $I_c = 25.8444$, all three in amu Å^2. The equalities and inequalities within values of I_a, I_b, and I_c arise from symmetry properties of rigid rotors as shown in Table 7.1 and Figure 7.11.

But the equality of two or all three principal rotational inertia is found only for a small set of molecules; a large majority of molecules exhibit $I_a \neq I_b \neq I_c$, as seen for HCHO. Some of these strictly asymmetric tops can still be classified by approximating the equality of two of the inertial components. For example, although I_a, I_b, and I_c of formaldehyde are significantly different from each other, the approximation $I_a \neq I_b \sim I_c$ would allow its classification as a 'near-prolate' symmetric top.

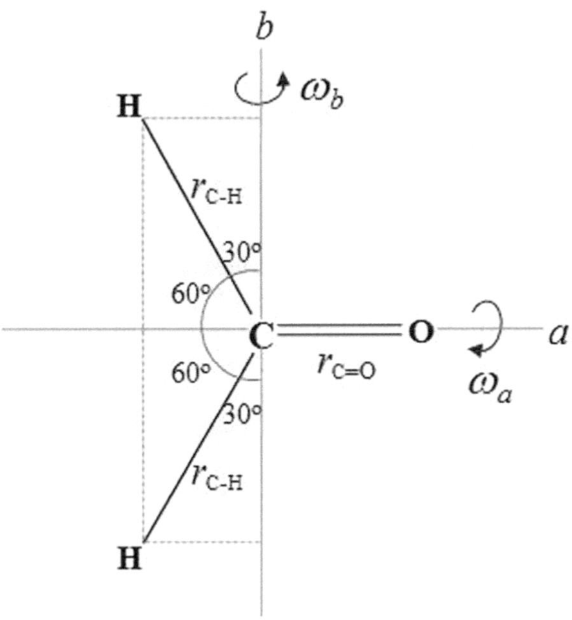

FIGURE 7.10 The geometry and angular velocities for the rotation of formaldehyde. The axis *c*, which is perpendicular to the plane of the molecule (paper), is not shown.

TABLE 7.1
Inertia and Symmetry of Rigid Rotors

Equality of rotational inertia	Symmetry classification	Example
$I_a = I_b = I_c$	Spherical top	CH_4, CCl_4
$I_a = I_b \neq I_c$	Oblate symmetric top	C_6H_6
$I_a \neq I_b = I_c$	Prolate symmetric top	NH_3, CH_3I
$I_a \neq I_b \neq I_c$	Asymmetric top	HCHO, C_2H_4

7.2.2 Energy of Rigid Rotors

Rotational energy of a rigid body is purely kinetic, provided the body rotates freely in space. When the center of mass of the body is placed at the origin of the principal axes system the rotational energy is the energy specified by the rotational moment of inertia, and is given by

$$E_{\text{rot}} = \frac{1}{2}\left(\sum_i m_i R_i a^2\right)\omega_a^2 + \frac{1}{2}\left(\sum_i m_i R_i b^2\right)\omega_b^2 + \frac{1}{2}\left(\sum_i m_i R_i c^2\right)\omega_c^2, \quad (7.34)$$

where ω is the angular velocity of rotation about the axis indicated by the subscript, m_i is the mass of the ith atom, and $R_i a$, $R_i b$, and $R_i c$ are vector distances of the ith atom from the rotation axes *a*, *b*, and *c*, respectively. This kinetic energy forms the basis for the time-independent Hamiltonian

$$\mathcal{H}_{\text{rot}} = \frac{P_a^2}{2I_a} + \frac{P_b^2}{2I_b} + \frac{P_c^2}{2I_c}, \quad (7.35)$$

which is the same operator that was encountered earlier (equation 7.6) in the discussion of diatomic molecules. The square of the total rotational angular momentum P^2 is the sum of the square of the individual components representing the principal axes system. As was done earlier for diatomic rotation, we can associate a rotational constant with each component of rotation

$$\left(\frac{\hbar^2}{hc}\right)\mathcal{H}_{\text{rot}} = A_e P_a^2 + B_e P_b^2 + C_e P_c^2, \quad (7.36)$$

where

$$A_e = \frac{h}{8\pi^2 I_a c}, \; B_e = \frac{h}{8\pi^2 I_b c}, \text{ and } C_e = \frac{h}{8\pi^2 I_c c}. \quad (7.37)$$

The energy expression above is appropriate for a spherical top, $I_a = I_b = I_c$, but can be extended easily for the other two symmetric tops by subtracting the relevant *P*-components from P^2. For the oblate top, for which $I_a = I_b \neq I_c$

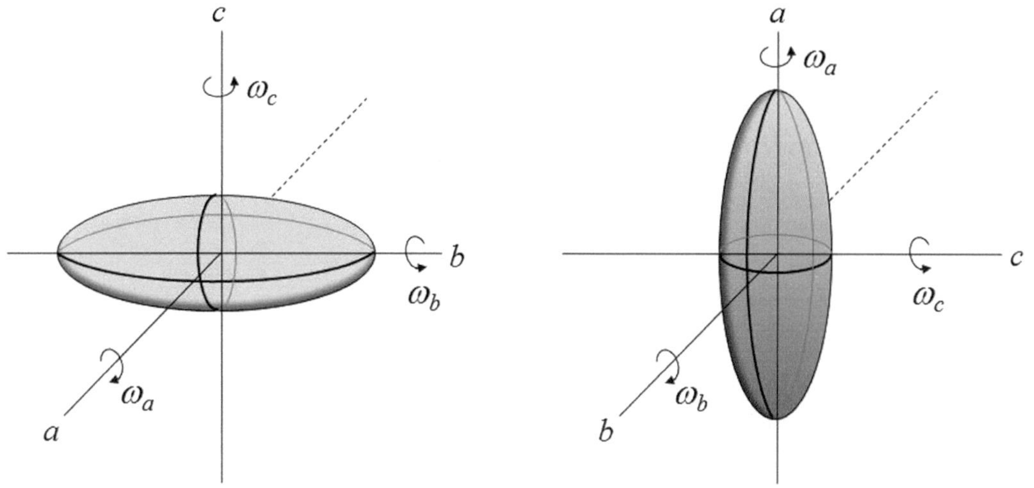

FIGURE 7.11 Rotation of oblate (*left*) and prolate (*right*) rigid rotors.

$$\mathcal{H}_{rot} = \frac{P_a^2}{2I_a} + \frac{P^2 - P_a^2 - P_c^2}{2I_b} + \frac{P_c^2}{2I_c} = \frac{P^2}{2I_b} - \frac{P_c^2}{2I_b} + \frac{P_c^2}{2I_c}, \quad (7.38)$$

and for the prolate top where $I_a \neq I_b = I_c$

$$\mathcal{H}_{rot} = \frac{P_a^2}{2I_a} + \frac{P^2 - P_a^2 - P_c^2}{2I_b} + \frac{P_c^2}{2I_c} = \frac{P^2}{2I_b} - \frac{P_a^2}{2I_b} + \frac{P_c^2}{2I_a}. \quad (7.39)$$

One should recognize the physical significance of the P_a, P_b, and P_c components of the angular momentum operator. They may be written in terms of the Euler angles θ, ϕ, and χ, which describe the motion of the principal axes with reference to a lab-fixed frame,

$$\begin{pmatrix} P_a \\ P_b \\ P_c \end{pmatrix} = \begin{pmatrix} \cos\chi\cot\theta & \sin\chi & -\cos\chi\operatorname{cosec}\theta \\ -\sin\chi\cot\theta & \cos\chi & \sin\chi\operatorname{cosec}\theta \\ 1 & 0 & 0 \end{pmatrix} \begin{pmatrix} -i\hbar\frac{\partial}{\partial\chi} \\ -i\hbar\frac{\partial}{\partial\theta} \\ -i\hbar\frac{\partial}{\partial\phi} \end{pmatrix}. \quad (7.40)$$

The components of the angular momentum operator P_X, P_Y, and P_Z in the lab-fixed axes XYZ are then obtained by multiplying out the P_a, P_b, P_c column vectors by the inverse of the rotation matrix $S(\theta,\phi,\chi)$

$$\begin{pmatrix} P_X \\ P_Y \\ P_Z \end{pmatrix} = \left(S^{-1}(\theta,\phi,\chi)\right) \begin{pmatrix} P_a \\ P_b \\ P_c \end{pmatrix}. \quad (7.41)$$

7.2.3 Wavefunctions of Symmetric Tops

Rotational motion of rotors is described by the motion of the molecule-fixed principal axes system a,b,c with reference to a laboratory-fixed axes system XYZ, such that cosines of angles between the two axes systems can be specified by Euler angles θ, ϕ, and χ. One often finds the motion difficult to visualize. At the start of rotation, abc and XYZ axes systems overlap. The body is then rotated about the Z-axis by an angle ϕ, followed by rotation about the a-axis by an angle θ, and then a rotation about the c-axis by an angle χ. The eventual disposition of the axes is shown in the figure contained in Box 5.1, which also shows the cosines of angles between the lab frame and the molecule-fixed frame in respect of the Euler angles as the 3×3 rotation matrix $S(\theta,\phi,\chi)$.

These developments make it clear that the rotational wavefunction of symmetric tops is a function of the three Euler angles, and the function is of the form

$$\psi_{rot} = \Theta_{JKM} e^{im\phi} e^{iK\chi}, \quad (7.42)$$

and the Schrödinger equation is

$$\mathcal{H}_{rot}\psi_{JKM}(\theta,\phi,\chi) = E_{rot}\psi_{JKM}(\theta,\phi,x). \quad (7.43)$$

7.2.4 Commutation of Rotational Angular Momentum Operators

For simultaneous assignment of definite energy values corresponding to the energy operators P^2 and the component operators in the principal and the laboratory-fixed axes systems, it is necessary to consider the commutation relations of these operators. From elementary quantum chemistry we know that angular momentum operators follow the commutation relations

$$\left[P^2, P_a\right] = \left[P^2, P_b\right] = \left[P^2, P_c\right] = 0, \quad (7.44)$$

which means a complete basis set of eigenstates would serve for the two commuting operators. It is also easy to show that

$$\left[\mathcal{H}_{rot}, P^2\right] = 0$$
$$\left[P^2, P_Z\right] = 0. \quad (7.45)$$

The Schrödinger equations with these operators are

$$P^2\psi_{JKM} = \hbar^2 J(J+1)\ \psi_{JKM}$$
$$P_c\psi_{JKM} = \hbar K\psi_{JKM}$$
$$P_Z\psi_{JKM} = \hbar M\psi_{JKM}, \tag{7.46}$$

and the allowed values of J, K, M quantum numbers are given as

$$J = 0, 1, 2, \cdots$$
$$K = -J \le 0 \le +J$$
$$M = -J \le 0 \le +J. \tag{7.47}$$

In equation (7.46), $\hbar^2 J(J+1)$ is the magnitude of the square of the total angular momentum vector, $\hbar K$ is its projection onto the c-axis of the principal axes system, and $\hbar M$ is its projection onto the Z-axis of the laboratory frame. Because K can take up $2J+1$ values, the number of K-states in a JK level is $2J+1$. But each JK level also has $2J+1$ values corresponding to M energy levels. Therefore, the degeneracy of each Jth energy level is $(2J+1)^2$.

7.2.5 Eigenvalues for Tops

Spherical top. Because $I_a = I_b = c$ and $P_a^2 = P_b^2 = P_c^2$,

$$E_{\text{rot}} = \frac{\hbar P^2}{2I_a} = \frac{\hbar^2 J(J+1)}{2I_a} = A_e J(J+1). \tag{7.48}$$

We may choose to use I_b or I_c instead of I_a, and would get

$$E_{\text{rot}} = B_e J(J+1)$$
$$E_{\text{rot}} = C_e J(J+1), \tag{7.49}$$

which means $A_e = B_e = C_e$ for spherical top.

Oblate Top. Here, $I_a = I_b \neq I_c$ and $P_a^2 = P_b^2 \neq P_c^2$, which yield

$$E_{\text{rot}} = \frac{\hbar^2 J(J+1)}{2I_b} + \frac{\hbar^2 K^2}{2}\left(\frac{1}{I_c} - \frac{1}{I_b}\right) = B_e J(J+1) + (C_e - B_e)K^2. \tag{7.50}$$

Prolate Top. The components of rotational inertia in this case are related as $I_a \neq I_b = I_c$; also, $P_a^2 \neq P_b^2 = P_c^2$. The rotational energy is then

$$E_{\text{rot}} = \frac{\hbar^2 J(J+1)}{2I_b} + \frac{\hbar^2 K^2}{2}\left(\frac{1}{I_a} - \frac{1}{I_b}\right) = B_e J(J+1) + (A_e - B_e)K^2. \tag{7.51}$$

Asymmetric top. For this class of molecules, $I_a \neq I_b \neq I_c$ (Table 7.1) and $P_a^2 \neq P_b^2 \neq P_c^2$. Therefore, rotational energies of these molecules cannot be obtained by straightforward algebraic manipulations that were possible for symmetric tops. Approximate expressions for J-energy levels can be found by a literature search and using computational tools. Rotational constants of asymmetric tops are often found empirically to follow

$$\kappa = \frac{2B_e - A_e - C_e}{A_e - C_e}, \tag{7.52}$$

which tells if such rotors can be approximated by $I_a \sim I_b \neq I_c$ or $I_a \neq I_b \sim I_c$. Values of the parameter κ lie in the +1 to −1 range. When κ approaches +1 the top is taken as a near-oblate one, and for κ nearing −1 a near-prolate approximation is made. The quantum number K for near-oblate and near-prolate is written as K_o and K_p, respectively, and corresponding rotational levels are labeled J_{K_o} and J_{K_p}.

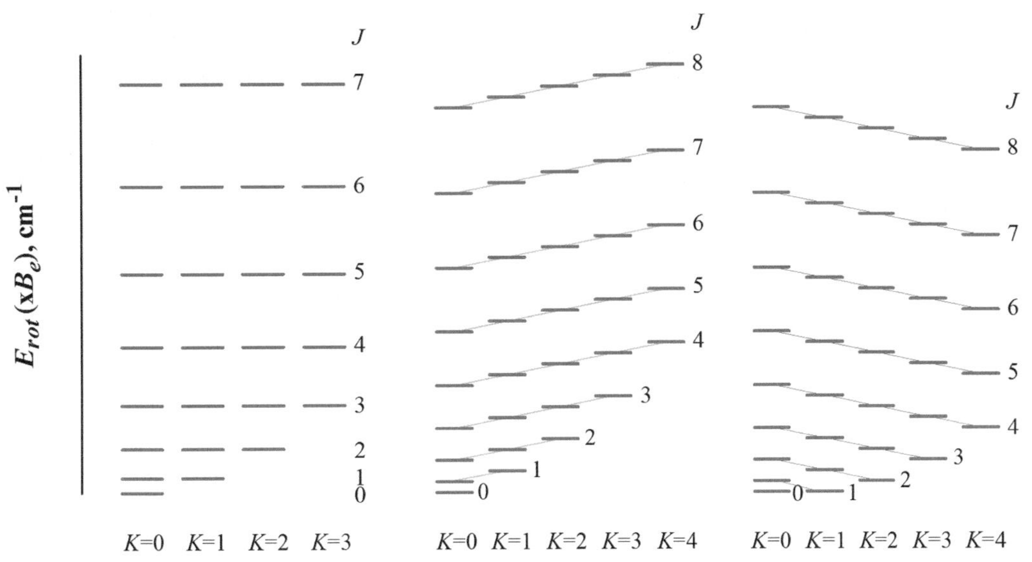

FIGURE 7.12 Energy levels for spherical, prolate, and oblate tops (from *left* to *right*) according to values of quantum numbers J and K. Note that rotational spectra will not be observed if the transition dipole moment is zero.

The energy levels for different tops are shown in Figure 7.12, but a spherical rotor whose transition dipole moment is zero obviously does not show rotational spectra. For that matter, any situation of zero-transition dipole will be spectroscopically silent. The figure shows that the energy corresponding to a J-value of a prolate increases with higher values of K, because the $(A_e - B_e)$ value in the energy expression $E_{rot} = B_e J(J+1) + (A_e - B_e)K^2$ is positive due to the $I_c > I_b > I_c$ condition. In the case of the oblate top, the energy level corresponding to a J-value decreases with increasing K -value, because the value of $(C_e - B_e)$ in the energy expression $E_{rot} = B_e J(J+1) + (A_e - B_e)K^2$ is negative.

7.2.6 Selection Rules for Polyatomic Rotational Transition

Similar to that for diatomics, a polyatomic molecule must have a permanent electric dipole moment for absorption and emission of radiation in near-IR and microwave regions. The interaction of the permanent dipole with the electric field of the radiation gives rise to the time-dependent potential, $V(t) = -\mu \cdot \mathbf{E}(t)$, and as derived earlier for the case of diatomic rotation, the selection rule for the change of J is $\Delta J = \pm 1$. If μ lies along the rotation symmetry axis of a prolate symmetric top, then $\Delta J = 0, \pm 1$, $\Delta K = 0$, and $\Delta M = 0$. As regard to the change of K, the determining factor is the orientation of the molecular electric dipole moment vector in the principal axes system. If the dipole is oriented along the rotation axis that has the highest symmetry, then $\Delta K = 0$. When the dipole is oriented perpendicular to one of the principal axes, $\Delta K = \pm 1$.

Box 7.1

Tensor and principal axes system (PAS). Suppose we observe an effect A on a body in response to an applied impulse or force or something B. In a Cartesian coordinate system *XYZ* defined with respect to the molecule-fixed coordinates *abc* (Figure 7.13) the effect A will have components A_a, A_b, and A_c when the impulse is applied along any of the axes *X*, *Y*, and *Z*.

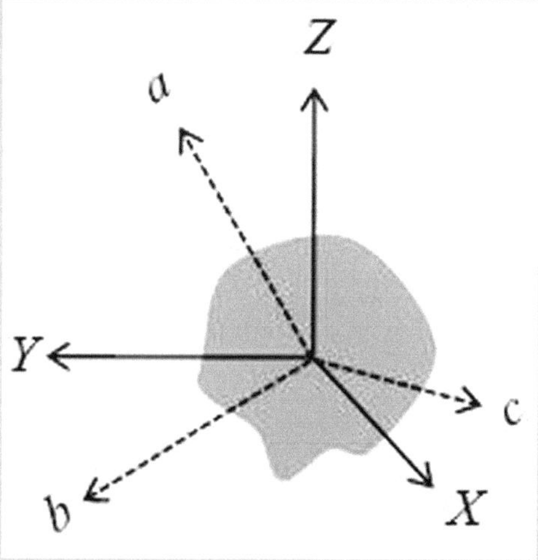

FIGURE 7.13 Relative positioning of the laboratory (*XYZ*) and molecular fixed axes (*abc*) systems.

The A_a, A_b, and A_c components are given by

$$A_a = N_{ax}B_x + N_{ay}B_y + N_{az}B_z$$

$$A_b = N_{bx}B_x + N_{by}B_y + N_{bz}B_z$$

$$A_c = N_{cx}B_x + N_{cy}B_y + N_{cz}B_z$$

Small letters *x*, *y*, *z* will be used to write the equations. The equations above can be represented by

$$\begin{pmatrix} A_a \\ A_b \\ A_c \end{pmatrix} = \begin{pmatrix} N_{ax} & N_{ay} & N_{az} \\ N_{bx} & N_{by} & N_{bz} \\ N_{cx} & N_{cy} & N_{cz} \end{pmatrix} \begin{pmatrix} B_x \\ B_y \\ B_z \end{pmatrix}$$

The nine N_{ij} numbers, which are constants, constitute a second-order or second rank tensor. The indices *i* and *j* of N_{ij} refer to the ith component of the observable A for the impulse along the *j*-direction. Suppose, the impulse was applied specifically along the *y*-direction and not along *x*- or *z* direction. It is easy to see that

$$\begin{pmatrix} A_a \\ A_b \\ A_c \end{pmatrix} = \begin{pmatrix} N_{ax} & N_{ay} & N_{az} \\ N_{bx} & N_{by} & N_{bz} \\ N_{cx} & N_{cy} & N_{cz} \end{pmatrix} \begin{pmatrix} 0 \\ B_y \\ 0 \end{pmatrix} = \begin{pmatrix} N_{ay} \\ N_{by} \\ N_{cy} \end{pmatrix} B_y$$

which yields

$$A_a = N_{ay}B_y$$

$$A_b = N_{by}B_y$$

$$A_c = N_{cy}B_y$$

The $\begin{pmatrix} N_{ay} \\ N_{by} \\ N_{cy} \end{pmatrix}$ array of numbers is a vector now.

It is possible to choose a condition where the *XYZ* and *abc* axes systems are parallel and collinear

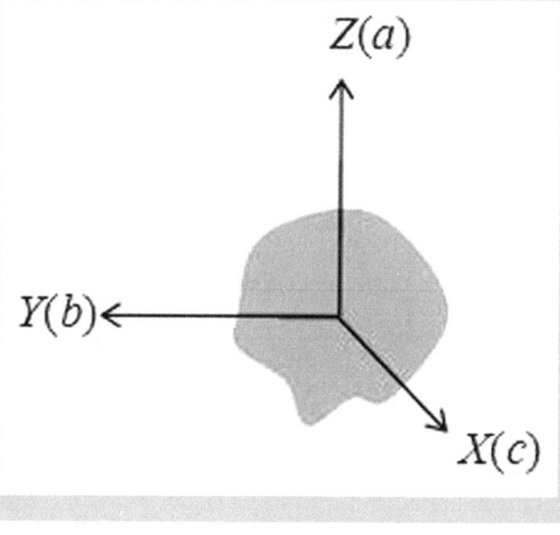

Now, if the impulse B is applied along *X*-direction alone, the effect A will be produced along the *X*-direction only, and we simply write

$$A_x = N_{xx} B_x$$

$$A_y = A_z = 0$$

Note that $N_{xx}B_x$ is a scalar. Similarly, the impulse B applied specifically along *Y* or *Z* will produce the effect A along the corresponding axis only. If the impulse is applied along all three directions simultaneously, then A will be related to B by the diagonal tensor

$$\begin{pmatrix} A_a \\ A_b \\ A_c \end{pmatrix} = \begin{pmatrix} N_{ax} & 0 & 0 \\ 0 & N_{by} & 0 \\ 0 & 0 & N_{cz} \end{pmatrix} \begin{pmatrix} B_x \\ B_y \\ B_z \end{pmatrix}$$

This relation can also be written as

$$\begin{pmatrix} A_a \\ A_b \\ A_c \end{pmatrix} = \begin{pmatrix} 1 & 0 & 0 \\ 0 & 1 & 0 \\ 0 & 0 & 1 \end{pmatrix} \begin{pmatrix} N_{ax} & 0 & 0 \\ 0 & N_{by} & 0 \\ 0 & 0 & N_{cz} \end{pmatrix} \begin{pmatrix} B_x \\ B_y \\ B_z \end{pmatrix}$$

$$= \delta_{ij} \begin{pmatrix} N_{ax} & 0 & 0 \\ 0 & N_{by} & 0 \\ 0 & 0 & N_{cz} \end{pmatrix} \begin{pmatrix} B_x \\ B_y \\ B_z \end{pmatrix}$$

The unit tensor δ_{ij} refers to the fact that $\delta_{ij} = 1$ if $i = j$, and $\delta_{ij} = 0$ if $i \neq j$. The unit tensor is often called Kronecker delta analogous to the description of orthogonality and orthonormality of wavefunctions

$$\int_{-\infty}^{+\infty} \psi_i^* \psi_j d\tau = \delta_{ij}$$

The *abc* axes system which diagonalizes the tensor as shown above is called principal axes for the observable A. In other words, *abc* axes are eigenvectors of the nine-component N_{ij} tensor.

How to find the principal axes *abc*? This is important because many descriptions in spectroscopy rely on the principal axes system (PAS). Suppose a two-dimensional frame *xy* is rotated by an angle θ. This is an orthogonal transformation represented by

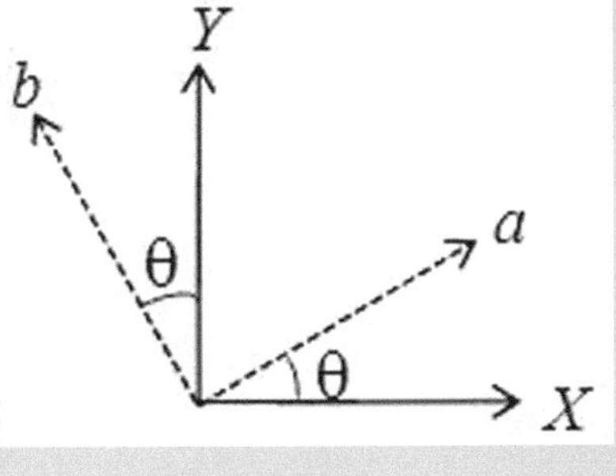

$$a = x\cos\theta + y\sin\theta$$

$$b = -x\sin\theta + y\cos\theta$$

$$\begin{pmatrix} a & b \end{pmatrix} = \begin{pmatrix} x & y \end{pmatrix} \begin{pmatrix} \cos\theta & \sin\theta \\ -\sin\theta & \cos\theta \end{pmatrix}$$

$$\begin{pmatrix} a & b \end{pmatrix} = \begin{pmatrix} x & y \end{pmatrix} \tilde{R}$$

where $\tilde{R}$ is the rotation matrix, and the inverse of the matrix is equal to its transpose $\frac{1}{\tilde{R}} = \tilde{R}^T$, so

$$\begin{pmatrix} a & b \end{pmatrix} \tilde{R}^T = \begin{pmatrix} x & y \end{pmatrix}$$

This transformation can be extended to three dimensions, where the two orthogonal systems will be related as

$$\begin{pmatrix} a & b & c \end{pmatrix} = \begin{pmatrix} x & y & z \end{pmatrix} \begin{pmatrix} \cos\theta_{xa} & \cos\theta_{xb} & \cos\theta_{xc} \\ \cos\theta_{ya} & \cos\theta_{yb} & \cos\theta_{yc} \\ \cos\theta_{za} & \cos\theta_{zb} & \cos\theta_{zc} \end{pmatrix}$$

so that the observable property A, which was read along *XYZ* axes system will now be given in the transformed *abc* system by

$$A^T = A\tilde{R}$$

Similarly, the impulse B in the newly transformed system will be given by

$$B^T = B\tilde{R}.$$

Therefore,

$$A^T = N^T B^T \text{ where } N^T = \frac{1}{\tilde{R}} N\tilde{R}$$

The rotation matrix $\tilde{R}$ can be chosen in a way that N^T contains only the diagonal elements. Then the transformation $\frac{1}{\tilde{R}} N\tilde{R}$ is called principal axis transformation.

In general, the correspondence of N_{ij} components in the *XYZ* frame to those in the *abc* frame is given by

$$N_{ij} = \sum_{k,l} \cos\theta_{il} \cos\theta_{jl} N_{kl}$$

where $i,j = x,y,z$ and $k,l = a,b,c$.

Many physical properties are related to the inducing sources of these observable properties by tensors. A few simple examples are the following.

Polarizability in atoms and molecules in response to an applied electric field **E**

P = $\tilde{\alpha}$**E**, where $\tilde{\alpha}$ is a polarizability tensor in general.

Current density or the amount of current flowing through a conductor in response to an applied electric field **E**

$\mathbf{J} = \kappa\mathbf{E}$, where κ is the electrical conductivity tensor

Angular momentum due to the rotation of a molecule about the center of mass with an angular velocity $\boldsymbol{\omega}$

$$\mathbf{P} = I\boldsymbol{\omega}.$$

Box 7.2

The dipole transition integral

$$|\mu| = |\mu_o||E|\int_0^{2\pi}\int_0^{\pi} P_J^M(\cos\theta)e^{-iM\phi}\sin\theta\cos\phi P_{J'}^{M'}(\cos\theta)e^{-iM'\phi}\sin\theta d\theta d\phi + \int_0^{2\pi}\int_0^{\pi} P_J^M(\cos\theta)e^{-iM\phi}\sin\theta\sin\phi P_{J'}^{M'}(\cos\theta)e^{-iM'\phi}\sin\theta d\theta d\phi + \int_0^{2\pi}\int_0^{\pi} P_J^M(\cos\theta)e^{-iM\phi}\cos\theta P_{J'}^{M'}(\cos\theta)e^{-iM'\phi}\sin\theta d\theta d\phi$$

can be solved by substituting

$$\cos\phi = \frac{1}{2}\left(e^{i\phi} + e^{-i\phi}\right) \text{ and } \sin\phi = \frac{1}{2i}\left(e^{i\phi} - e^{-i\phi}\right),$$

and using the recursion relation for the associated Legendre functions

$$(2l+1)\cos\theta P_l^M(\cos\theta) = (l+m)P_{l-1}^m(\cos\theta) + (l-m+1)P_{l+1}^m(\cos\theta).$$

The integrals go to zero when $J = J'$, even though M can be M' or $M' \pm 1$.

PROBLEMS

7.1 Rotational lines are distinguished by J values of ground and excited states restricted by the selection rule $\Delta J = \pm 1$. A rotational band envelope contains a spread of lines, each corresponding to a different pair of J levels in the ground and excited states, and for each rotational line the formula for E_{rot}/hc is applicable. To what accuracy one can estimate the bond length of a diatomic molecule considering each rotational line separately?

7.2 Continuing with the above question, it has been mentioned that with higher pressure of CO gas (see Figure 7.9) increasing intermolecular collisions set in which should broaden the individual rotational lines. Let one argue that higher pressure also increases friction, which should counter intermolecular collisions. How does one reconcile with these two prospects, and how the trade-off will be reflected?

7.3 It is stated that the total area defined by the envelope containing a large number of rotational absorption bands can provide information about the moment of inertia I of a diatomic rotation. Estimate the accuracy to which I can be determined by this approach.

7.4 Consider rotational bands contained within a vibrational transition. How will the line shape vary from the head to the tail of the rotational band structure?

7.5 The kinetic energy of a rigid rotor having N atoms is

$$\mathcal{H}_{rot} = \frac{1}{2}\sum_{\alpha=1}^{N} m_\alpha |\omega r_\alpha|^2$$

where m_α is the mass of the αth atom, and r_α is the value of its position vector. The angular velocity of rotation of the molecule is ω. Let ω and r_α both be described by a molecule-fixed coordinate system (x,y,z) whose origin coincides with the center of mass of the molecule. Write down the inertial tensor $\tilde{I}$ and show the energy in the principal axes system (a,b,c). (Note: A rotation about a principal axis produces an angular momentum along that axis alone.)

7.6 Use the properties of the angular momentum operators to show that

$$[P^2, P_z] = 0, [P^2, P_c] = 0, [P^2, \mathcal{H}_{rot}] = 0.$$

7.7 For both diatomic and polyatomic molecules the selection rule for the change of the quantum number J reads $\Delta J = \pm 1$ or $J = J' \pm 1$. Evaluate the integral

$$\int_0^{\infty} P_J^M(\cos\theta) P_{J'}^{M'}(\cos\theta)\sin\theta\, d\theta.$$

BIBLIOGRAPHY

Buontempo, U., S. Cunsolo, and G. Jacucci (1973) *J. Chem. Phys.* 59, 3751.

Steinfeld, J. I. (1985) *Molecules and Radiation*, MIT Press.

Townes, C. H. and A. L. Schawlow (1975) *Microwave Spectroscopy*, Dover Publications, Inc.

8 Diatomic Vibrations, Energy, and Spectra

Two vibrational levels in an electronic potential energy curve of a diatomic molecule sandwich many rotational energy levels, suggesting that vibrational resonance frequencies should be much higher than those for rotational. Indeed so, vibrational transitions occur in ~10^3–10^5 nm (~10^4–10^2 cm^{-1}) range compared to the microwave region where rotational absorption occurs. Unlike thermal accessibility of excited rotational states, all molecules always populate the ground state vibrational level at ambient or moderately elevated temperature. A vibrational absorption is normally fundamental which is a single quantum transition from the ground state, although accessibility of molecules to excited vibrational levels at higher temperatures can give rise to overtone absorptions. A diatomic molecule is the simplest system to analyze molecular vibration because stretch and compression of the internuclear distance give rise to just one mode of vibration.

8.1 CLASSICAL DESCRIPTION OF AN OSCILLATOR

We first discuss vibrations of strongly bound diatomic molecules for which a consideration of harmonic approximation of an oscillator is a good starting point. The basic concepts of the model and the motion of a classical oscillator are briefly given here. A one-dimensional harmonic oscillator for a classical body of mass m can be visualized as a spring, one end of which is fixed to an immovable surface and the other end to the body itself (Figure 8.1). When the body is displaced either away from or toward the surface along the line of the spring, which is the x-axis here, two restoring forces act to restore the equilibrium position of the body. One restoring force is

$$F = -kx$$

that can be written as

$$m\frac{d^2x}{dt^2} = -kx$$

$$\frac{d^2x}{dt^2} + \omega^2 x = 0$$

$$\left(\frac{d^2}{dt^2} + \omega^2\right)x = 0, \tag{8.1}$$

where k is the spring constant that will vary from one to another spring depending on the mechanical properties of the spring used, and $\omega^2 = k/m$. Denoting d/dt by D for convenience, we write the characteristic displacement equation as

$$\left(D^2 + \omega^2\right)x = 0, \tag{8.2}$$

which is known to yield two roots, $D = \pm i\omega$, so that the solution of the differential equation is

$$x(t) = Ae^{i\omega t} + Ae^{-i\omega t} = A\sin(\omega t + \gamma), \tag{8.3}$$

where A is the amplitude, γ is the initial phase angle that can be set to zero, and the oscillation frequency $\omega = \sqrt{k/m}$.

This is a measurable mechanics problem and we have used only one mass attached to the spring. But the concern should be as to what will happen to the internuclear oscillation scenario. The model still works if the equilibrium position of the body is taken to coincide with the equilibrium distance of separation of the two nuclei of a diatomic molecule. But there also is another restoring force proportional to the velocity of the spring-attached body. This is the second restoring force that one might like to call 'frictional force', which will be proportional to the velocity of the body. This is the second force that acts to restore the body to its resting position. For a displacement of the body by a distance x, the retarding frictional force is just the negative of the velocity of the body multiplied by a friction constant, say l. So the full equation of motion for the body is

$$m\frac{d^2x}{dt^2} = -kx - l\frac{dx}{dt}. \tag{8.4}$$

The roots of the characteristic equation corresponding to this differential equation are known to be

$$-\frac{l}{2m} \pm \sqrt{\frac{l^2}{4m^2} - \frac{k}{m}}, \tag{8.5}$$

which means the relative values of $l^2/4m^2$ and k/m will determine the fate of the motion of the body, irrespective of whether the periodic motion will sustain or disappear with time. The oscillation will damp with time t for $l^2/4m^2 > k/m$, but will continue to be periodic when $l^2/4m^2 < k/m$ (Figure 8.2). The total energy of displacement of a classical oscillator (vibrational energy) is given by

$$E_{\text{vib}} = \frac{1}{2}kA^2 = \frac{1}{2}m\omega^2A^2. \tag{8.6}$$

8.2 SCHRÖDINGER EQUATION FOR NUCLEAR VIBRATION

For the quantum mechanical description of diatomic vibrational oscillations, it is presumed that the Schrödinger

DOI: 10.1201/9781003293064-8

equation for electronic motion with fixed nuclei is solved using the Born-Oppenheimer approximation. But the nuclear repulsion that has been ignored so far may be taken on board. When the equation

$$(\mathcal{H}_{el} + V_{\mathrm{NN'}})\psi_{el} = E(R)\psi_{el}, \tag{8.7}$$

in which $V_{\mathrm{NN'}}$ is the nuclear repulsion term, is solved the variation of the electronic energy E with internuclear separation R will be known. The argument simply means solving for the electronic energy with an effective Hamiltonian that includes the electronic Hamiltonian and the nuclear repulsion potential. In a typical curve depicting the $E(R)$ function (Figure 8.3), the internuclear distance at which E is minimum is called equilibrium internuclear distance R_e for the specified electronic state. Clearly, R_e will shift, generally toward larger R, as the energy of the electronic state increases.

When $E(R)$ is obtained, the Schrödinger equation for nuclear vibration can be solved by using the $E(R)$ function as the potential energy function for nuclear motion with the additional assumption that the molecule is not undergoing rotation. This potential energy can be thought of as the work done to increase or decrease the internuclear separation relative to R_e. It must be clarified again that the potential energy depends only on the internuclear separation, meaning that the

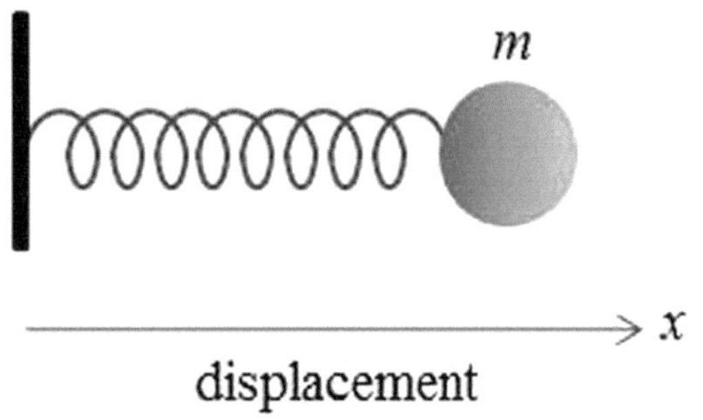

FIGURE 8.1 A body attached to a spring and subjected to a one-dimensional force.

total energy of a diatomic molecule is the sum of this potential energy and the kinetic energy of the single mass (reduced mass) whose coordinates are the coordinates of one of the atoms relative to the other atom. It is also important to recognize that diatomic vibration is an internal motion arising from a change of the internuclear distance, and therefore the vibrational energy of the molecule corresponds to the potential energy. But the kinetic energy of the system arises from the motion of the hypothetical reduced mass. Therefore, the total energy of the diatomic oscillator consists of the kinetic energy of the reduced mass and the potential energy due to internuclear separation.

However, the potential energy function $E(R)$ of a real molecule is not exactly a harmonic potential all through the internuclear separation R. It is harmonic at best for R values not too far from R_e. In order to describe the vibrational energy levels in a harmonic potential, the lowest parabolic part of $E(R)$ can be expanded in a Taylor series about $R_e (R = R_e)$

$$E(R) = E(R_e) + (R - R_e)\left[\frac{\partial E(R)}{\partial R}\right]_{R=R_e} + \frac{(R-R_e)^2}{2!}\left[\frac{\partial^2 E(R)}{\partial R^2}\right]_{R=R_e} + \cdots. \tag{8.8}$$

The first term $E(R_e) = 0$ (Figure 8.1), and the second term vanishes because the first derivative will go to zero for $R = R_e$. Thus the $E(R)$ potential function is approximated with the quadratic term

$$E(R) \sim \frac{1}{2}(R-R_e)^2\left[\frac{\partial^2 E(R)}{\partial R^2}\right]_{R=R_e}$$

$$\sim \frac{1}{2}k(R-R_e)^2, \tag{8.9}$$

in which k is the bond force constant, which represents the curvature of the potential. The $E(R)$ function is now a parabolic potential whose curvature corresponds to k. Although

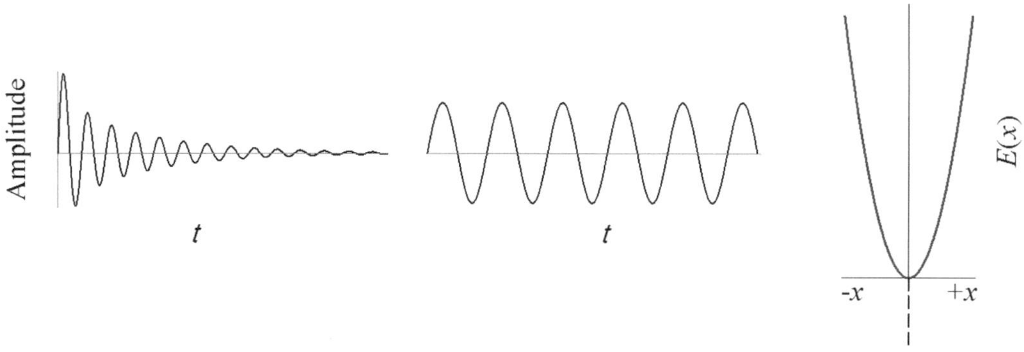

FIGURE 8.2 Overdamped (*left*) and sustained (*middle*) periodic motions. *Right*, if the rest position of the body is coincided with the equilibrium bond separation, then the sustained periodic oscillation (vibration) of the body between the bounds of $+x$ and $-x$ reproduces the periodic motion of the two nuclei between the highest and lowest internuclear separations permissible. The dependence of the total energy of the classical vibrator on the displacement (x) from the resting position is parabolic.

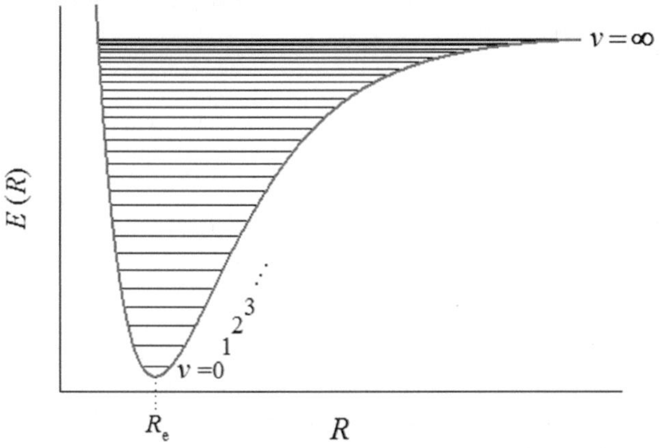

FIGURE 8.3 Potential energy inclusive of internuclear repulsion for an arbitrary diatomic molecule as a function of R.

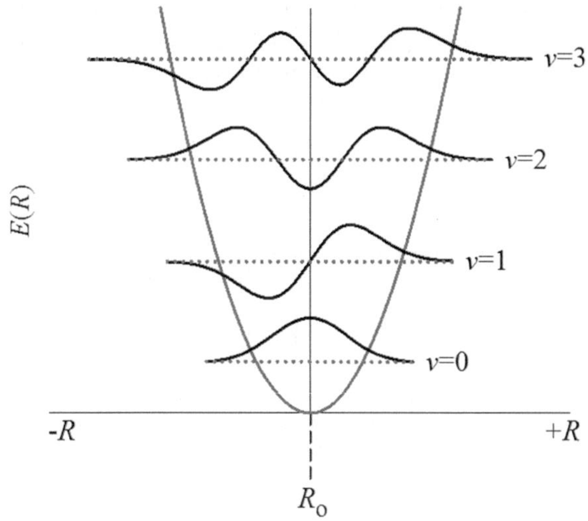

FIGURE 8.4 The Hermite function as vibrational wavefunctions for $\upsilon = 0,1,2,3$ levels. Notice the finite probability of the wavefunctions even outside the potential well.

the use of a parabolic potential in the description of harmonic motion is a good beginning, the anharmonicity of the real molecular potential brings in overtone vibrational transitions, as we will see below.

Now we use the potential energy of electronic motion obtained above by series expansion to write down the total energy operator for the nuclear motion

$$\mathcal{H} = \left[-\frac{\hbar^2}{2\mu}\frac{\partial^2}{\partial x^2} + \frac{1}{2}k\left(R - R_e\right)^2 \right], \quad (8.10)$$

and the Schrödinger equation is

$$\mathcal{H}\psi_{n,\text{vib}} = E_{\text{vib}}\psi_{n,\text{vib}}$$

$$-\frac{\hbar^2}{2\mu}\frac{\partial^2\psi_{n,\text{vib}}}{\partial x^2} + \frac{1}{2}k\left(R - R_e\right)^2\psi_{n,\text{vib}} = E_{\text{vib}}\psi_{n,\text{vib}}, \quad (8.11)$$

in which the subscript n attached to the wavefunction refers to the nth electronic state. The eigenenergy from this equation could be found as

$$E_{\text{vib}} = \hbar\sqrt{\frac{k}{\mu}}\left(n+1\right) = \hbar\omega\left(\upsilon + \frac{1}{2}\right). \quad (8.12)$$

We just got the bond vibrational frequency $\omega = \sqrt{k/\mu}$ and the vibrational quantum number $v = 0,1,2,3,\cdots$. It should be clear at once from the energy value that the vibrational energy levels are equally spaced in a parabolic potential, in contrast with the rotational energy levels that are squarely placed. The vibrational energy corresponding to $v = 0$ is called the 'zero-point energy ($E_o = h\nu/2$)'.

At this point we wish to introduce the appropriate wavefunctions without proof. The normalized wavefunctions for the harmonic oscillator are Hermite functions given as

$$\psi_\upsilon = A_\upsilon H_\upsilon\left(\alpha, R\right)e^{\frac{-\alpha R^2}{2}}$$

$$= \frac{1}{\sqrt{2^\upsilon \upsilon!}}\left(\frac{\alpha}{\pi}\right)^{\frac{1}{4}} H_\upsilon\left(\gamma, R\right)e^{\frac{-\alpha R^2}{2}}, \quad (8.13)$$

in which $\alpha = \mu\omega/\hbar$, and R is the deviation from equilibrium internuclear separation R_e. Importantly, $H_\upsilon\left(\alpha, R\right)$ are Hermite polynomials, which assume extra significance because these functions spill out of the classical boundary (see below). The vibrational wavefunctions for $v = 0,1,2,$ and 3 are reproduced below from literature:

$$\psi_0 = \left(\frac{\alpha}{\pi}\right)^{\frac{1}{4}} e^{\frac{-\alpha R^2}{2}}$$

$$\psi_1 = \left(\frac{\alpha}{\pi}\right)^{\frac{1}{4}} \sqrt{2\alpha}Re^{\frac{-\alpha R^2}{2}}$$

$$\psi_2 = \frac{1}{\sqrt{2}}\left(\frac{\alpha}{\pi}\right)^{\frac{1}{4}} (2\alpha R^2 - 1)e^{\frac{-\alpha R^2}{2}}$$

$$\psi_3 = \sqrt{3}\left(\frac{\alpha}{\pi}\right)^{\frac{1}{4}} \left(2\alpha^{\frac{3}{2}} - \alpha^{\frac{1}{2}}R\right)e^{\frac{-\alpha R^2}{2}}. \quad (8.14)$$

A few vibrational energy levels and the associated wavefunctions within the harmonic potential are graphed in Figure 8.4, where we notice that each wavefunction has a finite probability even outside the classical potential boundary.

8.3 SELECTION RULES FOR VIBRATIONAL TRANSITIONS

Derivation of the selection rules that specify allowed transitions between vibrational levels requires consideration of the electric dipole moment of the molecule. We have learned that the time-varying potential in the presence of radiation is $|-\boldsymbol{\mu}\cdot\mathbf{E}(t)|$. The dipole would oscillate as the diatomic oscillator vibrates about the equilibrium internuclear separation R_e. Because the dipole is a function of R, it can be approximated by Taylor expansion about $R = R_e$

$$\mu(R) = \mu(R_e) + (R - R_e)\left[\frac{\partial \mu(R)}{\partial R}\right]_{R=R_e} + \frac{(R-R_e)^2}{2!}\left[\frac{\partial^2 \mu(R)}{\partial R^2}\right]_{R=R_e} + \cdots. \tag{8.15}$$

The dipole transition integrals are then

$$\mu_{\upsilon\upsilon'} = \mu(R_e)\int_{-\infty}^{+\infty}\psi_\upsilon^*\psi_{\upsilon'}dr + \left[\frac{\partial \mu(R)}{\partial R}\right]_{R=R_e}\int_{-\infty}^{+\infty}\psi_\upsilon^*(R-R_e)\psi_{\upsilon'}dr + \cdots. \tag{8.16}$$

The front factor outside the second integral is a constant. If electrical anharmonicity is taken into account, only the first integral could be retained for vibrational transitions involving two different electronic states. Transitions between vibrational levels within the same electronic state can be approximated by the first two integrals. However, if we allow vibrational transitions within the same electronic state, the first integral will go to zero simply because vibrational wavefunctions of the same electronic state are orthogonal

$$\int \psi_{n\upsilon}^*\psi_{n\upsilon'}dr = 0. \tag{8.17}$$

The intensity of vibrational transitions within the same electronic state, given by,

$$\mu_{\upsilon\upsilon'} = \left(\frac{\partial \mu(R)}{\partial R}\right)_{R=R_e}\int_{-\infty}^{+\infty}\psi_\upsilon^*(R-R_e)\psi_{\upsilon'}dr \tag{8.18}$$

will then depend on both the first derivative of the electric dipole with respect to the internuclear separation and the value of the integral

$$\langle \upsilon|R-R_e|\upsilon'\rangle = \int_{-\infty}^{\infty}\psi_\upsilon^*(R-R_e)\psi_{\upsilon'}dr. \tag{8.19}$$

The integral can be evaluated by employing the Hermite polynomial recursion formula, which can be written as

$$\sqrt{\alpha}|R-R_e|H_\upsilon(R-R_e) = \frac{1}{2}H_{\upsilon+1}(R-R_e) + \upsilon H_{\upsilon-1}(R-R_e). \tag{8.20}$$

But one can also rewrite the relation using Dirac notation in an abbreviated form

$$|R-R_e|\upsilon'\rangle = \frac{1}{2}\frac{N_{\upsilon'}}{N_{\upsilon'}+1}|\upsilon'+1\rangle + \upsilon'\frac{N_{\upsilon'}}{N_{\upsilon'+1}}|\upsilon'-1\rangle \tag{8.21}$$

where $|\upsilon'\rangle$ and $|\upsilon'\pm 1\rangle$ are vibrational eigenstates, and

$$N_{\upsilon'} = \frac{1}{\sqrt{2^{\upsilon'}\upsilon'!}}\left(\frac{\alpha}{\pi}\right)^{\frac{1}{4}} \text{ and } \frac{N_{\upsilon'}}{N_{\upsilon'-1}} = \frac{1}{\sqrt{2\upsilon'}}. \tag{8.22}$$

We see that we have obtained above both $|\upsilon'+1\rangle$ and $|\upsilon'-1\rangle$ as eigenfunctions on the right, although the eigenvalue equation entailed operation of $|R-R_e|$ on the $|\upsilon'\rangle$ function. This implies that vibrational transitions can occur from $|\upsilon'\rangle$ to both $|\upsilon'+1\rangle$ and $|\upsilon'-1\rangle$. But the vibrational eigenstates arising from the same electronic state have to be orthogonal $\langle \upsilon|\upsilon'\rangle = 1$, which allows us to write $\upsilon - \upsilon' = \Delta\upsilon = \pm 1$. This conclusion implies a single fundamental absorption line for pure vibrational transition. Thus, the rules for the occurrence of vibrational transitions are:

1. The molecule must have a permanent electric dipole moment, and
2. Pure fundamental transitions must maintain $\Delta\upsilon = \pm 1$.

Another way to think about the $\Delta\upsilon = \pm 1$ result rests on the fact that a symmetric integral yields a zero value when the integrand consists of an even and an odd function. For an even function of x, one can write $f(-x) = f(x)$ so that

$$\int_{-\infty}^{+\infty} f(x)dx = 2\int_0^{\infty} f(x)dx. \tag{8.23}$$

For an odd function, $g(-x) = -g(x)$,

$$\int_{-\infty}^{+\infty} g(x)dx = 0,$$

which implies

$$\int_{-\infty}^{+\infty} f(x)g(x)dx = 0, \tag{8.24}$$

because the product of an even function and an odd function is an odd function. Consider the harmonic oscillator wavefunctions corresponding to $\upsilon = 0$ and $\upsilon = 1$. The function $\psi_{v=0}$ is an odd function of the internuclear distance, R. So, if $\psi_{v=2}$ is the other integrand in the transition dipole integral, a zero value of the integral results. This means a $\upsilon' = 0 \rightarrow \upsilon = 2$ has a zero transition value, but a $\upsilon' = 0 \rightarrow \upsilon = 1$ transition that satisfies $\Delta\upsilon = \pm 1$ has a definite value, and hence the transition is allowed.

Even if a permanent dipole is required to observe vibrational transitions, the strength of a transition need not be proportional to the magnitude of the dipole. What matters is the value of the derivative of the dipole with respect to the internuclear separation. A molecule may have a small electric dipole moment, and yet can show a strong vibrational absorption if the dipole derivative is large. The converse, that is small dipole derivative in spite of a large permanent dipole, produces weak vibrational absorption.

If the second derivative term in the Taylor expansion of the dipole is used in the transition dipole integral

$$\mu_{\upsilon\upsilon'} = \left[\frac{\partial^2 \mu(R)}{\partial R^2}\right]_{R=R_e} \int_{-\infty}^{\infty} \psi_\upsilon^* \left(R - R_e\right) \psi_{\upsilon'} dr, \qquad (8.25)$$

the $\langle \mu_{\upsilon\upsilon'} \rangle$ value would turn out to be relatively small. This is due to the fact that the second derivative itself is small, meaning the oscillator-well or the parabola is wide. The smaller the value of the second derivative of the dipole, the wider is the parabola, implying increasing anharmonicity. Therefore, vibrational transitions from a lower to a higher level involving $\Delta\upsilon = \pm 2, \pm 3, \cdots$ are increasingly weaker. They are called overtone absorptions – first overtone ($\Delta\upsilon = \pm 2$), second overtone ($\Delta\upsilon = \pm 3$), and so on. Overtone transitions are forbidden in a perfect harmonic potential, but such weak transitions do result with increasing anharmonicity. To account for the anharmonicity of the potential surface the vibrational energy of a harmonic oscillator (equation 8.12) is modified as

$$E_{\text{vib}} = \hbar\omega\left(\upsilon + \frac{1}{2}\right) - \frac{(\hbar\omega)^2}{4D_e}\left(\upsilon + \frac{1}{2}\right)^2, \qquad (8.26)$$

where $\upsilon = n = 0, 1, 2, \cdots$.

8.4 ROTATIONAL–VIBRATIONAL COMBINED STRUCTURE

Molecular rotational transitions are observed when pure rotational spectra are measured under less colliding conditions in the far-infrared (~50 μm – 1 mm) approaching the microwave region. We mention collision here because rotational lines can be smeared or blurred by molecular collisions. But if measured in the mid- or near-infrared regions (~800 nm – 50 μm) the rotational structure is invariably contained within the vibrational band. This is to be expected, because the radiation energy in these regions resonates with the energy of the allowed vibrational transitions and each vibrational level also contains the manifold of rotational levels. So an allowed vibrational excitation ($\Delta\upsilon = \pm 1$) can involve rotational levels contained within each of the connected vibrational levels, provided the rotational transition selection rule ($\Delta J = 0, \pm 1$) is not violated. At ordinary temperatures, ambient and below, most molecules are in the ground vibrational state ($\upsilon = 0$), but the population of the molecules in different rotational states will be distributed according to the population distribution

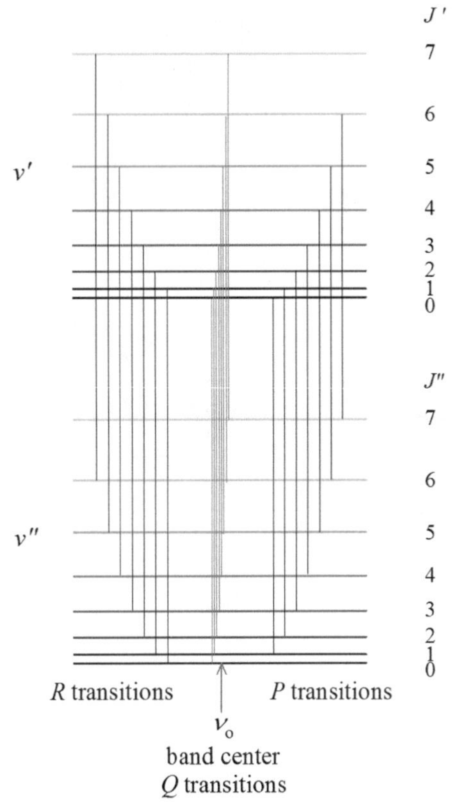

FIGURE 8.5 Pictorial rotational branches of transitions according to the selection rule $\Delta J = +1$ (R branch) and $\Delta J = -1$ (P branch). The $\Delta J = 0$ transition (Q branch), when occurs, is positioned at the center frequency, ν_o.

relation discussed earlier (equation 7.28). This means a vibrational transition $\upsilon_0 \rightarrow \upsilon_1$ can be substructured into the manifold of initial rotational states corresponding to $\upsilon = 0$ and the manifold of final rotational levels contained within $\upsilon = 1$. Also, the rotational transition accompanying the $\upsilon_0 \rightarrow \upsilon_1$ transition follows the $\Delta J = 0, \pm 1$ selection rule. Accordingly, a $\Delta J = +1$ occurrence $\left(J''_n \rightarrow J'_{n+1}\right)$ appears in the R-branch, and a $\Delta J = -1$ occurrence $\left(J''_n \rightarrow J'_{n-1}\right)$ appears in the P-branch of the vibrational transition, each branch involving a unit change in the total rotational quantum number J (Figure 8.5).

The rotational transition $\Delta J = 0$ is observed only in certain special cases where the z-projection of the electronic angular momentum of the molecular state has a finite value ($\Lambda \neq 0$). The molecular state corresponding to $\Lambda = 0$ is the Σ-state in which vibrational transition can occur, but $\Delta J = 0$ is not allowed. Instances where molecules are not in the Σ-state, and hence a vibrational transition without a change in the relevant rotational levels ($\Delta J = 0$) can occur, produce a large-intensity vibrational band called the Q-band. For example, rotational–vibrational spectrum of $^{14}N^{16}O$ shows P, Q, and R branches.

The rotational–vibrational spectrum of CO, on the other hand, does not show the Q-band structure because the ground state of CO is a singlet designated by the term symbol $^1\Sigma^+$ ($S = 0$ singlet, and $\Lambda = 0$, Σ-state). Thus, the rotational–vibrational bands of CO are only P- and R-type,

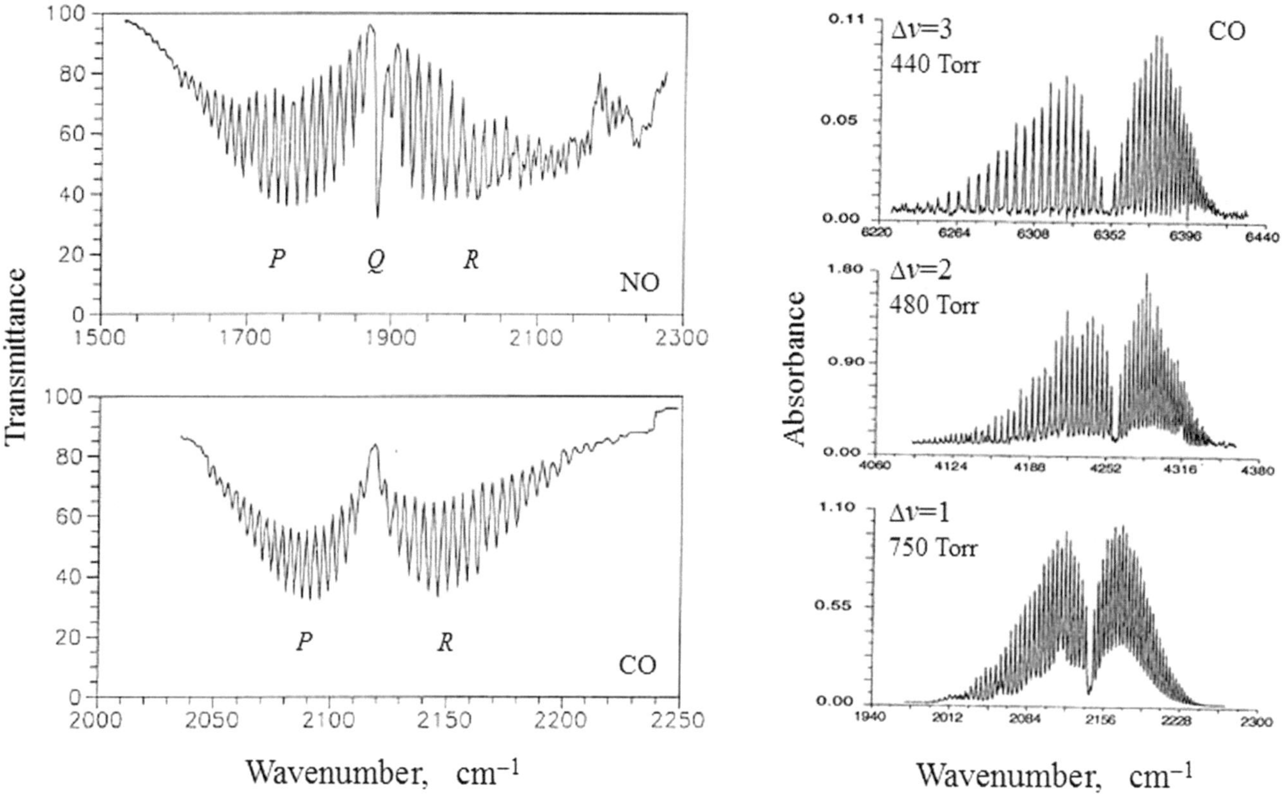

FIGURE 8.6 *Left*, rotational branch structures in the fundamental ($\Delta v = 1$) vibrational transition of NO and CO. Notice the absence of the Q line at the band center (v_0) of the CO spectrum. *Right*, rotational band structures (P and R) in the vibration-rotation spectrum of CO associated with the fundamental ($\Delta v = 1$), first overtone ($\Delta v = 2$), and second overtone ($\Delta v = 3$) vibrational transitions. (Reproduced with permission from H. H. R. Schor and E. L. Teixeira (1994) *J. Chem. Edu.* 71, 771, Copyright © 1994 American Chemical Society; N. Mina-Camilde, C. Manzanares, and J. Caballero (1996) *J. Chem. Edu.* 73, 804, Copyright © 1996 American Chemical Society).

the Q-band is missing. For a moment we will look at a nice set of rotational–vibrational combination spectra of CO and NO that were obtained in a laboratory teaching class experiment (Figure 8.6). The spectra for CO show the fundamental ($\upsilon_0 \rightarrow \upsilon_1$), first overtone ($\upsilon_0 \rightarrow \upsilon_2$), and second overtone ($\upsilon_0 \rightarrow \upsilon_3$) vibrational transitions, each containing the rotational structure. The overtone transitions occur due to anharmonicity of the potential energy surface. Since almost all molecules exist in the lowest vibrational level under ambient temperature conditions, the overtone transitions almost always involve the ground vibrational level υ_0. Relative to the frequency of the fundamental transition, the overtones increasingly absorb at higher frequency approaching mid- to near-infrared regions.

PROBLEMS

8.1 Show that

$$\int \psi_{nv}^{*} \psi_{nv'} dr = 0$$

where n refers to the nth th electronic state, and v and v' are vibrational levels.

8.2 Consider a vibrational transition from $v' = 0$ to $v = 1$. Evaluate the dipole transition integral

$$\mu_{vv'} = \int_{-\infty}^{+\infty} \psi_v^{*} \left\{ \mu(R_e) + (R - R_e) \left[\frac{\partial \mu_R}{\partial R} \right]_{R=R_e} \right\} \psi_{v'} d(R - R_e).$$

Use $\psi_{v'} = \left(\frac{\alpha}{\pi} \right)^{1/4} e^{-\alpha}$

$$\psi_v = \left(\frac{\alpha}{\pi} \right)^{1/4} \sqrt{2\alpha} R e^{-\alpha R^2/2}$$

in which $\alpha \equiv 2\pi\omega/\hbar$, and R is the deviation from the equilibrium distance R_e.

8.3 What is the physical significance of having a finite probability of a wavefunction, like the Hermite functions, outside the classical potential boundary?

8.4 The FWHM of both rotational structure lines and vibrational envelope of an overtone band should be more than those in a fundamental band. Why?

8.5 Why are overtone transitions forbidden in a harmonic potential? Consider looking for such transitions in the anharmonic part of the potential. What experimental strategies could be adopted to compensate for the weakness of overtone transitions?

BIBLIOGRAPHY

Eyring, H., J. Walter, and G. W. Kimball (1944) *Quantum Chemistry*, Wiley.

Herzberg, G. (1950) *Molecular Spectra and Molecular Structures*, Vol. I, Van Nostrand Reinhold.

Huber, K. P. and G. Herzberg (1979) *Constants of Diatomic Molecules*, Van Nostrand Reinhold.

9 Polyatomic Vibrations and Spectra

Polyatomic vibrations can be almost exclusively described using classical mechanics. The only apparent problem lies in the unification of the coordinate systems, one of which is Cartesian displacement of atoms and another is atom-centric polar angle-dependent bending of adjacent atoms. These two sets of coordinates are unified to a common set of internal coordinates by the technique of coordinate transformation which is extended further to evolve the normal coordinates that form the basis of normal modes of polyatomic vibrations. The coordinate transformations en route the normal modes require the use of Lagrangian mechanics instead of solving the Newton's second law equation.

9.1 A SIMPLE CLASSICAL MODEL TO DEFINE A NORMAL MODE

The rudimentary form of many-body vibrations in the Cartesian coordinates can be demonstrated by simply solving for the Newton's equation of motion $F = ma$. An atom or a group of atoms can be coarse-grained to a sphere, and many of these spheres are assumed to be connected by mechanical springs. The simplest model would consist of a system of two spring-connected spheres each of which is connected to two immovable supports by springs (Figure 9.1).

We assume that both spheres are of the same mass m, the spring constant for the middle spring is k, and that for the terminal ones is identically k_1. Under these assumptions, both uniform stretching and compression of the middle spring can be given by the change in the Cartesian coordinate $(x_2 - x_1)$ by $\pm d(x_2 - x_1)$. The one-dimensional equations of motion can be written down immediately to obtain the following coupled differential equations

$$m\frac{dx_1^2}{dt^2} = -k_1x_1 - k(x_1 - x_2)$$

$$m\frac{dx_2^2}{dt^2} = -k_1x_2 - k(x_2 - x_1). \tag{9.1}$$

Solving these equations to obtain the eigenvalues and eigenvectors is easy. We assume solutions

$$x_1(t) = Ae^{\lambda t} = Ae^{i\omega t}$$

$$x_2(t) = Be^{\lambda t} = Be^{i\omega t}, \tag{9.2}$$

where A and B are eigenvectors and $\lambda = i\omega$ is the eigenvalue. After differentiating, dividing by $e^{i\omega t}$, and then rearranging one gets

$$A(-m\omega^2 + k_1 + k) - Bk = 0$$

$$-Ak + B(-m\omega^2 + k_1 + k) = 0. \tag{9.3}$$

To avoid zero value for the eigenvectors A and B, the determinant is set to zero

$$\begin{vmatrix} -m\omega^2 + k_1 + k & -k \\ -k & -m\omega^2 + k_1 + k \end{vmatrix} = 0, \tag{9.4}$$

and the eigenvalue solutions are obtained. As shown by the ± signs below, they are four in number

$$\omega_s = \pm\sqrt{\frac{k_1}{m}} \quad \text{or} \quad \omega_f = \pm\sqrt{\frac{k_1 + 2k}{m}}, \tag{9.5}$$

and they specify frequencies of intersphere vibrations, ω_s for 'slow' and ω_f for 'fast' frequencies of motion. Each of these four possible vibrational frequencies specifies the two eigenvectors, meaning the amplitudes of displacements,

$$x_1(t) = A_se^{i\omega_st} + A_se^{-i\omega_st} + A_fe^{i\omega_ft} + A_fe^{-i\omega_ft}$$

$$x_2(t) = A_se^{i\omega_st} + A_se^{-i\omega_st} - A_fe^{i\omega_ft} - A_fe^{-i\omega_ft}. \tag{9.6}$$

Since the eigenvectors must be real, the imaginary terms are left out and the two equations are written with the real components only

$$x_1(t) = A_s\cos(\omega_st + \phi_s) + A_f\cos(\omega_ft + \phi_f)$$

$$x_2(t) = A_s\cos(\omega_st + \phi_s) - A_f\cos(\omega_ft + \phi_f), \tag{9.7}$$

where we have used the identity

$$e^{i\omega t} = \cos\omega t + i\sin\omega t. \tag{9.8}$$

Further, ϕ_s and ϕ_f are just the phase factors. Now we use the easier approach. If the vibrational amplitude corresponding to the 'fast frequency oscillation' is zero, then

$$x_1(t) = A_s\cos(\omega_st + \phi_s)$$

$$x_2(t) = A_s\cos(\omega_st + \phi_s). \tag{9.9}$$

By doing so we notice that the two eigenvectors are the same, implying that $x_1(t)$ and $x_2(t)$ displacements are identical,

DOI: 10.1201/9781003293064-9

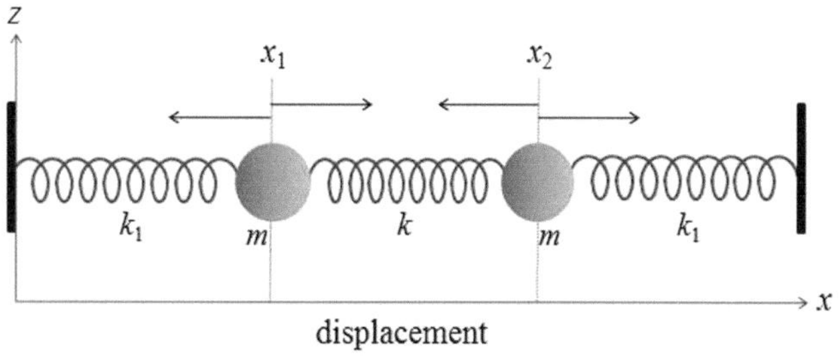

FIGURE 9.1 A model of ball and spring system to explain a normal mode. The spheres have identical mass m and are connected to each other and also to immovable surfaces by springs. Spring constants are denoted by k.

which means the motions of both spheres are identical in terms of modes. If sphere 1 moved to the right, so did sphere 2, and to the left in an identical manner – same frequency of motion. This motion where both masses 1 and 2 move with an identical to-and-fro frequency – either to the right or to the left together, is called a 'normal mode', and the eigenvector A_s is called normal mode of vibration. Application of this model to a diatomic molecule would tell us that these molecules have only one normal mode of vibration, which is just the stretching of the bond.

9.2 VIBRATIONAL ENERGY FROM CLASSICAL MECHANICS

As discussed earlier, polyatomic vibrations can be treated adequately in the classical perspective. In the classical description of kinetic and potential energies of vibrations below we will arrive at normal coordinates of vibration starting from Cartesian displacement coordinates. Consider for simplicity a resting linear triatomic molecule consisting of nuclei 1, 2, and 3 ($N = 3$) whose Cartesian coordinates are $x_1, y_1, z_1, x_2, y_2, z_2$, and x_3, y_3, z_3 respectively, defined with reference to a molecule-fixed axes system a, b, c (Figure 9.2).

If the atoms are subjected to an isotropic force their displacements are given by Cartesian displacement coordinates

$$\Delta x_i = a_i - a_{i,0},\ \Delta y_i = b_i - b_{i,0},\ \Delta z_i = c_i - c_{i,0} \quad (9.10)$$

in which i runs from 1 to N, and the subscript 'zero' refers to the equilibrium positions of the atoms. The square of each time derivatives of Δx_i, Δy_i and Δz_i gives the vibrational kinetic energy T_{vib} in a straightforward manner

$$T_{\text{vib}} = \frac{1}{2}\sum_{i=1}^{N} m_i \left[\left(\frac{d\Delta x_i}{dt}\right)^2 + \left(\frac{d\Delta y_i}{dt}\right)^2 + \left(\frac{d\Delta z_i}{dt}\right)^2\right]. \quad (9.11)$$

But the mass of each of these atoms can be accommodated if the Cartesian coordinates are expressed as mass-weighted coordinates denoted by q_i, such that

$$q_i = \sqrt{m_i}\, x_i. \quad (9.12)$$

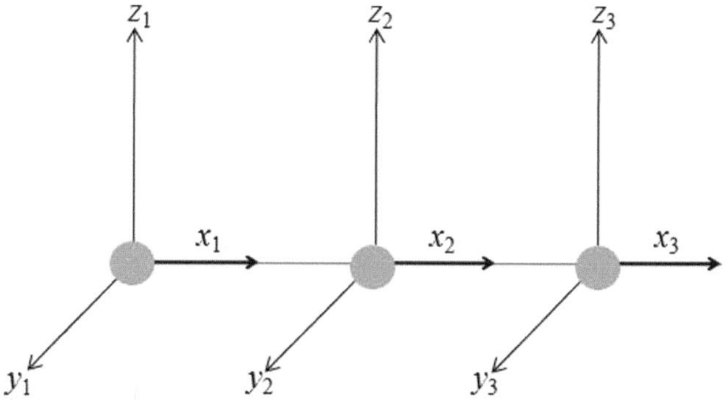

FIGURE 9.2 Cartesian displacements of the atoms in a linear triatomic molecule.

There will be $3N$ mass-weighted coordinates corresponding to the $3N$ Cartesian coordinates. The mass-weighting of the coordinates accounts for different response of the atoms of different mass with respect to the uniform force experienced by all of them. Thus, the vibrational kinetic energy expressed in mass-weighted coordinates is

$$T_{\text{vib}} = \frac{1}{2}\sum_{i=1}^{3N} \Sigma\left(\frac{dq_i}{dt}\right)^2$$

$$= \frac{1}{2}\left(\dot{q}_1^{\,2} + \dot{q}_2^{\,2} + \dot{q}_3^{\,2} + \cdots \dot{q}_{3N}^{\,2}\right)$$

$$= \frac{1}{2}\left(\dot{q}_1\ \dot{q}_2\ \dot{q}_3 \cdots \dot{q}_{3N}\right)\begin{pmatrix} \dot{q}_1 \\ \dot{q}_2 \\ \dot{q}_3 \\ \vdots \\ \dot{q}_{3N} \end{pmatrix} = \frac{1}{2}\dot{\mathbf{q}}^{\mathrm{T}}\dot{\mathbf{q}}, \quad (9.13)$$

in which the sum of $\dot{q}_i^2$ terms has been written as the matrix product of a row vector of $\dot{q}_i$ and a column vector of $\dot{q}_i$.

Regarding the nuclear potential energy V_N, we state that V_N conforms to the Born-Oppenheimer approximation

$$V_N = V_n + V_{NN'} - E_n, \quad (9.14)$$

where V_n is the energy of electronic motion, $V_{NN'}$ is the nuclear repulsion energy, and E_n is the electronic energy (equation (5.2)). Note that equation (9.14) conveys the same meaning that does equation (5.2); the change of symbols and indices here was necessary to conform to the setting of the problem. Now $V_N(\boldsymbol{x}) = 0$ if the nuclei are not displaced, i.e, when they exist in their equilibrium configuration. Note that the variable $\boldsymbol{x}$ (in bold) implies a single-index coordinate, meaning all xyz coordinates written collectively. Therefore, the potential energy in the neighborhood of the equilibrium nuclear configuration can be expressed by Taylor series expansion of the $V_N(\boldsymbol{x})$ function about the equilibrium value of $\boldsymbol{x}$. Due to its dependence on multiple Cartesian coordinates the potential energy is a multivariable function whose

Taylor expansion about the equilibrium position is somewhat complicated. To make it a bit simpler we have written the set of Cartesian coordinates $\{x_1y_1z_1, x_2y_2z_2, \cdots x_Ny_Nz_N\}$ as a one-index coordinate, which means $\boldsymbol{x}$ would symbolize $\{x_1, x_2, x_3, \cdots x_{3N-2}, x_{3N-1}, x_{3N}\}$. In this notation of variables, read $x_1 = x_1y_1z_1$, $x_2 = x_2y_2z_2$, $x_3 = x_3y_3z_3$, and so on. The Taylor expansion of the $V_N(\boldsymbol{x})$ function about the equilibrium, $V_N(\boldsymbol{x}) = 0$, will yield

$$\begin{aligned} V_N(\boldsymbol{x}) = V_o + \sum_{i=1}^{3N}\left(\frac{\partial V_N}{\partial x_i}\right)_0 x_i + \frac{1}{2!}\sum_{i,j=1}^{3N}\left(\frac{\partial^2 V_N}{\partial x_i \partial x_j}\right)_0 x_i x_j + \\ \frac{1}{3!}\sum_{i,j,k=1}^{3N}\left(\frac{\partial^3 V_N}{\partial x_i \partial x_j \partial x_k}\right)_0 x_i x_j x_k + \\ \frac{1}{4!}\sum_{i,j,k,l=1}^{3N}\left(\frac{\partial^4 V_N}{\partial x_i \partial x_j \partial x_k \partial x_l}\right)_0 x_i x_j x_k x_l + \cdots, \end{aligned} \tag{9.15}$$

in which the first term $V_o = 0$. The second term also vanishes because the derivative $\frac{\partial V_N}{\partial x_i}$ will be zero for the equilibrium value of x_i, i.e., $x_i = x_i(0)$. We then rewrite the above expansion as

$$\begin{aligned} V_N(\boldsymbol{x}) = \frac{1}{2!}\sum_{i,j=1}^{3N} f_{ij}x_i x_j + \frac{1}{3!}\sum_{i,j,k=1}^{3N} f_{ijk}x_i x_j x_k + \\ \frac{1}{4!}\sum_{i,j,k,l=1}^{3N} f_{ijkl}x_i x_j x_k x_l + \cdots \end{aligned} \tag{9.16}$$

where we now denote the second, third, and fourth derivatives of $V_N(\boldsymbol{x})$ by f_{ij}, f_{ijk}, and f_{ijkl}, respectively, all evaluated at the equilibrium of each x. These derivatives are called force constants.

The expansion above could be considered assuming the ball and spring model of bonded atoms, which is another way of saying that the molecular potential energy can be approximated by a harmonic potential. This consideration allows us to work with the multidimensional harmonic approximation alone so that only the quadratic term in the expansion is retained, leaving out the cubic, quartic, and higher order terms. In fact, multivariate Taylor expansion rarely considers a term beyond the quartic one. It should however be borne in mind that analyses of cubic, quartic, and higher order force constants do provide information about the Born-Oppenheimer surfaces, which we know are anharmonic. But the objective here is to conceptualize the normal modes of vibration, and the simplified potential energy function within the harmonic approximation is just

$$V_N(\boldsymbol{x}) = \frac{1}{2!}\sum_{i,j=1}^{3N} f_{ij}x_i x_j. \tag{9.17}$$

To make things even simpler we can write the $V_N(\boldsymbol{x})$ function in a matrix format

$$V_N(\boldsymbol{x}) = \frac{1}{2!}\begin{pmatrix} x_1 & x_2 & x_3 & \cdots & x_{3N} \end{pmatrix} \begin{pmatrix} f_{11} & f_{12} & f_{13} & & f_{1,3N} \\ f_{21} & f_{22} & f_{23} & \cdots & f_{2,3N} \\ f_{31} & f_{32} & f_{33} & & f_{3,3N} \\ \vdots & & & \ddots & \vdots \\ f_{3N,1} & f_{3N,2} & f_{3N,3} & \cdots & f_{3N,3N} \end{pmatrix} \begin{pmatrix} x_1 \\ x_2 \\ x_3 \\ \vdots \\ x_{3N} \end{pmatrix}. \tag{9.18}$$

Because the row vector $\begin{pmatrix} x_1 & x_2 & x_3 & \cdots & x_{3N} \end{pmatrix}$ and the corresponding column vector are transposes of each other, the $V_N(\boldsymbol{x})$ function can be casted as

$$V_N(\boldsymbol{x}) = \frac{1}{2}\mathbf{x}^{\mathrm{T}}\mathbf{f}\mathbf{x}$$

so as to obtain the f_{ij} derivatives in the form of a symmetric matrix called Cartesian force constant matrix, sometimes also called Hessian matrix of second-order partial derivative. Certain literature and textbooks show the $\mathbf{f}$ matrix more explicitly

$$\begin{pmatrix} \frac{\partial^2 V_N}{\partial x_1^2} & \frac{\partial^2 V_N}{\partial x_1 \partial x_2} & & \frac{\partial^2 V_N}{\partial x_1 \partial x_N} \\ \frac{\partial^2 V_N}{\partial x_2 \partial x_1} & \frac{\partial^2 V_N}{\partial x_2^2} & \cdots & \frac{\partial^2 V_N}{\partial x_2 \partial x_N} \\ \vdots & & \ddots & \vdots \\ \frac{\partial^2 V_N}{\partial x_N \partial x_1} & \frac{\partial^2 V_N}{\partial x_N \partial x_2} & \cdots & \frac{\partial^2 V_N}{\partial x_N^2} \end{pmatrix}. \tag{9.19}$$

As was done for the kinetic energy T_{vib}, we should write the potential energy function V_N also in mass-weighted Cartesian coordinates in a compact form

$$V_N(\boldsymbol{q}) = \frac{1}{2}\mathbf{q}^{\mathrm{T}}\frac{f_{ij}}{\sqrt{m_i m_j}}\mathbf{q} = \frac{1}{2}\mathbf{q}^{\mathrm{T}}\mathbf{W}\mathbf{q}. \tag{9.20}$$

9.3 SOLUTION OF LAGRANGE'S EQUATION

Having obtained the kinetic and potential energies of vibration in the generalized coordinates $\boldsymbol{q}$ we look for a solution of the equation of motion starting from Lagrange's equation (see Box 9.1). To suit the variables used we write Lagrange's equation as

$$L(q,\dot{q},t) = T - V$$

$$\frac{d}{dt}\left(\frac{\partial L}{\partial \dot{q}_i}\right) - \left(\frac{\partial L}{\partial q_i}\right) = 0. \tag{9.21}$$

By writing the kinetic and potential energy terms separately

$$\frac{\partial L}{\partial \dot{q}_i} = \frac{\partial T}{\partial \dot{q}_i} = \dot{q}_i \text{ and } \frac{\partial L}{\partial q_i} = -\frac{\partial V}{\partial q_i} = -\frac{1}{2}\sum_{j=1}^{3N} W_{ij}q_j, \tag{9.22}$$

we see that the generalized coordinate q_i $(i = 1,2,3,\cdots,3N)$ and the sum in the potential energy part of the equation involve $i, j = 3N - 6$, but the sum over i in the case of derivatives is eliminated. The equation of motion now takes the form

$$\frac{d}{dt}\dot{q}_i + \frac{1}{2}\sum_{j=1}^{3N} W_{ij}q_j = 0$$

$$\ddot{q}_i + \frac{1}{2}\sum_{j=1}^{3N} W_{ij}q_j = 0$$

$$\underset{\ddot{\mathbf{q}}}{\begin{pmatrix} \ddot{q}_1 \\ \ddot{q}_2 \\ \ddot{q}_3 \\ \vdots \\ \ddot{q}_{3N} \end{pmatrix}} + \frac{1}{2}\underset{\mathbf{W}}{\begin{pmatrix} W_{11} & W_{12} & W_{13} & & W_{1,3N} \\ W_{21} & W_{22} & W_{23} & \cdots & W_{2,3N} \\ W_{31} & W_{32} & W_{33} & & W_{3,3N} \\ & \vdots & & \ddots & \vdots \\ W_{3N,1} & W_{3N,2} & W_{3N,3} & \cdots & W_{3N,3N} \end{pmatrix}}\underset{\mathbf{q}}{\begin{pmatrix} q_1 \\ q_2 \\ q_3 \\ \vdots \\ q_{3N} \end{pmatrix}} = 0. \tag{9.23}$$

We then make change of variables

$$\mathbf{q} = \mathbf{cQ}$$

where $\mathbf{c}$ is an orthogonal matrix, $\mathbf{cc}^{\mathrm{T}} = \mathbf{c}^{\mathrm{T}}\mathbf{c} = \mathbf{I}$, with $\mathbf{I}$ a unitary matrix. The unitary transformation changes variables, which is necessary to transform to a new set of coordinates where they are uncoupled. The change of variables allows for writing

$$\mathbf{Q} = \mathbf{c}^{\mathrm{T}}\mathbf{q},$$

where

$$Q_k = \sum_{i=1}^{3N} U_{ik}q_i \text{ and } \mathbf{q{-}cQ}. \tag{9.24}$$

This change of variables affords expressing both kinetic and potential energy terms in the transformed coordinates. Because $\mathbf{c}$ is just a matrix of some constant numbers, we can write

$$\mathbf{q} = \mathbf{cQ} \rightarrow \ddot{\mathbf{q}} = \mathbf{c}\ddot{\mathbf{Q}}. \tag{9.25}$$

The Lagrange's equation can now be written as

$$\mathbf{c}\ddot{\mathbf{Q}} + \mathbf{WcQ} = 0$$

$$\mathbf{c}\left(\frac{d^2Q_k}{dt^2}\right) + \mathbf{WcQ} = 0. \tag{9.26}$$

Since the orthogonal matrix $\mathbf{c}$ satisfies $\mathbf{c}^{\mathrm{T}} = \mathbf{c}^{-1}$, the equation above can be divided on the left by $\mathbf{c}$ to rewrite

$$\frac{d^2Q_k}{dt^2} + \left(\mathbf{c}^{-1}\mathbf{Wc}\right)\mathbf{Q} = 0. \tag{9.27}$$

Notice that the choice of the orthogonal matrix $\mathbf{c}$ has already yielded a similarity transformation

$$\mathbf{c}^{-1}\mathbf{Wc} = \mathbf{\Lambda}, \tag{9.28}$$

which means the matrix $\mathbf{W}$ is diagonalized to yield the diagonal matrix $\mathbf{\Lambda}$

$$\mathbf{c}^{-1}\mathbf{Wc} = \begin{pmatrix} \lambda_1 & 0 & & 0 \\ 0 & \lambda_2 & \cdots & 0 \\ & \vdots & \ddots & \vdots \\ 0 & 0 & \cdots & \lambda_{3N} \end{pmatrix}. \tag{9.29}$$

With these solutions one can write

$$\ddot{\mathbf{Q}} + \mathbf{\Lambda Q} = 0$$

$$\ddot{\mathbf{Q}} + \begin{pmatrix} \lambda_1 & 0 & & 0 \\ 0 & \lambda_2 & \cdots & 0 \\ & \vdots & \ddots & \vdots \\ 0 & 0 & \cdots & \lambda_{3N} \end{pmatrix}\begin{pmatrix} q_1 \\ q_2 \\ q_3 \\ \vdots \\ q_{3N} \end{pmatrix} = 0. \tag{9.30}$$

Each of the uncoupled equations contains a λ_k and a Q_k, where $k = 1,2,3,\ldots,3N$

$$\frac{d^2Q_k}{dt^2} + \lambda_k Q_k = 0. \tag{9.31}$$

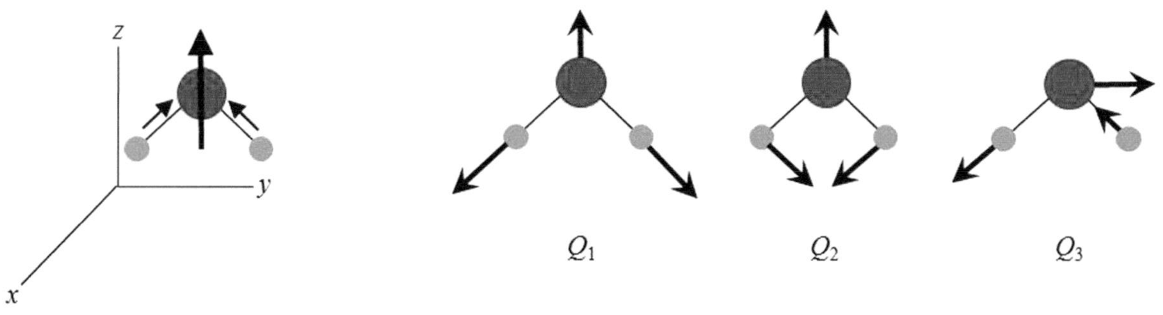

FIGURE 9.3 The normal modes of vibration in water. *Left*, the permanent dipole moment. *Right*, the normal modes of vibration. The modes are described by Q_1: symmetric stretch, A_1 symmetry, $\nu_1 = 3657$ cm^{-1}; Q_2: scissoring motion, A_1 symmetry, $\nu_2 = 1595$ cm^{-1}; Q_3: asymmetric stretch, B_1 symmetry, $\nu_3 = 3756$ cm^{-1}.

The solution of this type of differential equation is known to be

$$Q_k = A_k \sin\left(\omega_k t + \phi_k\right), \tag{9.32}$$

in which A_k and ϕ_k are constants and the frequency of the Q_k vibration is

$$\omega_k = \sqrt{\lambda_k}.$$

But the index k should now run from 1 to $3N-6$ and not to $3N$, because the form of the solution indicates harmonic motion. Since translation and rotation are not harmonic oscillations, we leave out the six degrees of freedom so that $k = 1,2,3,\ldots,3N-6$. The Q_k coordinates are called normal coordinates of vibration, so there will be $3N-6$ normal modes of vibration. The characteristic of a normal mode of vibration k is such that all the nuclei exhibit harmonic oscillation in which they all have the same phase ϕ_k and the same frequency ω_k. By convention, normal mode diagrams are shown by drawing from each atom a small vector, the length of which is determined by mass-weighting, and which indicates displacement of normal coordinates resulting in small displacement of the atom. The three normal modes of vibration of water as given in many textbooks and literature are shown in Figure 9.3. Each Q_k is a linear combination of the mass-weighted atomic displacements

$$Q_k = \sum_i c_{ki} q_i, \tag{9.33}$$

the sum being over all atoms in the molecule.

9.4 VIBRATIONAL HAMILTONIAN AND WAVEFUNCTION

The Schrödinger equation for nuclear vibrations becomes simple when the cubic and higher-order terms in the expansion of $V_N(\boldsymbol{x})$ are neglected (equation 9.15). This was discussed earlier, leading to the retention of the quadratic term alone (equation 9.17). Because there are $3N-6$ modes of vibration, the Schrödinger equation must also be $3N-6$ dimensional. Accordingly, we write the Hamiltonian as a sum of uncoupled harmonic oscillator Hamiltonians

$$\mathcal{H}_{\text{vib}} = \sum_k \left\{ T_{Q_k} + \frac{1}{2}\omega_k^2 Q_k^2 \right\}, \tag{9.34}$$

where $k = 1,2,3,\ldots,3N-6$, and Q_k specifies the kth mode of vibration (see equations (9.11) and (9.33)). By writing the form of the kinetic energy operator explicitly, we obtain

$$\mathcal{H}_{\text{vib}} = \sum_k \left\{ -\hbar^2 \frac{\partial}{\partial Q^2} + \frac{1}{2}\omega_k^2 Q_k^2 \right\} = \sum_k \mathcal{H}_k. \tag{9.35}$$

The appearance of the summation over the index k implies that the vibrational Hamiltonian is obtained by summing up the $3N-6$ harmonic oscillator Hamiltonians, each oscillator characterized by the normal coordinate Q_k. Each individual operator of the total operator $\mathcal{H}_{\text{vib}}$ corresponds to an energy eigenvalue, so the total energy is obtained as the sum of energies of the individual oscillators. Thus

$$E_{\text{vib}} = \sum_k \hbar\omega_k \left(v_k + \frac{1}{2} \right), \tag{9.36}$$

where v is the vibrational quantum number. The vibrational wavefunction is the product of individual one-dimensional oscillator wavefunctions

$$\psi_{\text{vib}} = \psi_{\text{v1}}\left(Q_1\right)\psi_{\text{v2}}\left(Q_2\right)\psi_{\text{v3}}\left(Q_3\right)\cdots\psi_{\text{v}k}\left(Q_k\right). \tag{9.37}$$

The wavefunction for a one-dimensional harmonic oscillator ψ_{v} is the Hermite function that we came across earlier in the discussion of diatomic vibrations (see equation 8.13). The function ψ_{vib} written in this form specifies that the vibrational quantum number for the kth oscillator is associated with the normal mode Q_k. Often the vibrational state is shown by $|\text{v}_1\text{v}_2\cdots\text{v}_k>$. A level is occasionally shown as $2_1 4_3$ only to indicate that the normal modes 2 and 4 have 1 and 3 quanta, respectively.

9.5 SYMMETRY OF NORMAL MODES

Normal coordinates form a basis for irreducible representations of a symmetry group, which implies that each normal mode has a symmetry corresponding to one of the irreducible representations. It does not mean one-on-one correspondence that would require the same number of normal modes and irreducible representations. Rather, one or more normal modes of vibration can belong to the same irreducible representation of the molecular point group. To outline the procedure for determination of symmetry of normal modes we consider the widely studied molecule ethylene for illustration. Ethylene is, of course, abundant on the Earth's surface, and is found in trace in the atmospheres of Jupiter, Neptune, Saturn, and Titan. It is also often observed in interstellar clouds, and hence is of interest to astrophysicists and astronomical spectroscopy. Indeed, numerous efforts have been made to study the infrared and Raman activity of ethylene.

The molecular point group of ethylene is D_{2h} with symmetry elements $\left\{E, c_{2x}, c_{2y}, c_{2z}, i, \sigma_{xy}, \sigma_{xz}, \sigma_{yz}\right\}$. As the procedure, each of these operations is applied one by one. We note down the number of atoms that do not move during an operation, and call them stationary atoms for that symmetry operation. The character contribution is the trace of the matrix which represents the operation in the Cartesian coordinate system. Instead of laboriously determining the character contribution for each operation in xyz coordinate system, one might like to consult the following chart for general applications

Operation	E	c_2	c_2	i	σ
Character contribution	3	−1	0	−3	1

TABLE 9.1
Determination of Reducible Representation from Atomic Motions Using Cartesian Displacement Coordinates as the Basis

D_{2h}	E	C_{2x}	C_{2y}	C_{2z}	i	σ_{xy}	σ_{xz}	σ_{yz}
Stationary atoms	6	2	0	0	0	2	6	0
Character contribution	3	−1	−1	−1	−3	1	1	1
Γ_{red}	18	−2	0	0	0	2	6	0

The symmetry analysis of ethylene at this stage yields Table 9.1, the last row of which gives the characters of the symmetry operations in the reducible representation, Γ_{red}.

In the next step, the array of Γ_{red} is decomposed by using the number h, which is the number of irreducible representations. The D_{2h} point group has eight symmetry elements – each a class by itself, so there will be eight irreducible representations. That the number of classes in a point group is equal to the number of irreducible representations is a theorem in group theory. The decomposition of Γ_{red} is done as described earlier in Chapter 6. For this we will need the character table of D_{2h} shown in Table 6.2. The last two columns of that table also provide the representations to which x, y, z coordinate axes, the rotations R_x, R_y, R_z, and the products of the coordinates belong. Now to get the number of normal modes corresponding to each of the irreducible representations we use the formula given by equation (6.24). The coefficient for the A_g representation, for example, is

$$a_1 = \frac{1}{8}\begin{bmatrix} 1\times 8 - 2\times 0 + 1\times 0 + 1\times 0 + 1\times 0 + \\ 2\times 1 + 1\times 6 + 1\times 0 \end{bmatrix} = 3.$$

The other coefficients are $a_2 = 1, a_3 = 2, a_4 = 3, a_5 = 3, a_6 = 2, a_7 = 1$, and $a_8 = 3$. With these coefficients the decomposition of Γ_{red} is written as

$$\Gamma_{red} = 3A_g + A_u + 2B_{1g} + 3B_{1u} + 3B_{2g} + 2B_{2u} + B_{3g} + 3B_{3u}. \tag{9.38}$$

Notice that the coefficients, which represent the number of modes that belong to the associated irreducible representations, sum up to $3N = 18$ for ethylene. Since only $3N - 6 = 12$ normal modes are genuine vibrational modes, one may dispense with six of the 18 possible normal modes. These six modes correspond to three degrees each of translational and rotational motions. These six non-genuine modes are culled out easily by inspecting representations for x, y, and z coordinates (translational motion) and R_x, R_y, and R_z axes (rotational motion). Thus, the coefficients for $B_{1g}, B_{1u}, B_{2g}, B_{2u}, B_{3g}$, and B_{3u} are reduced by 1 to rewrite equation (9.38) as

$$\Gamma_{red} = 3A_g + A_u + B_{1g} + 2B_{1u} + 2B_{2g} + B_{2u} + 2B_{3u}. \tag{9.39}$$

This expression for Γ_{red} shows that no normal mode of vibration belongs to B_{3g} representation, one mode belongs to each of A_u, B_{1g}, and B_{2u}, two modes belong to each of B_{1u}, B_{2g}, and B_{3u}, and three modes belong to the A_g representation.

We wish to make a comment here – the number of vibrational modes belonging to different irreducible representations is found inconsistent among numerous literature on ethylene vibration. The decomposition given in the equation above is what was found in this work, and the reader is encouraged to verify the decomposition. Although examining which irreducible representation a normal mode belongs to is instructive, the exercise per se provides no clue to specific properties and frequencies of normal modes.

9.6 FINDING THE VIBRATIONAL FREQUENCIES

These frequencies are calculated by the general procedure of solving the characteristic equation for eigenvalues λ. But setting up the characteristic equation for vibrational frequencies requires extra attention, mainly due to the involved manner in which the coordinates enter the total energy operator. Without discussing the difficulties associated with manipulation of coordinates, we will describe the approach comprehensively using ethylene again as an example.

Let us first introduce the internal coordinates of the molecule. Internal coordinates are defined in terms of atom displacements in bond stretching and bending vibrations. In the case of planar vibrations of ethylene for example, the construction of internal coordinates involve changes in bond lengths of the four C–H and one C=C bonds, the four bond angles of the C=C–H group, and the two bond angles of the H–C–H part. In this way we have 11 internal coordinates as shown in Figure 9.4.

These internal coordinates can be linearly combined to produce what is known as symmetry coordinates, so called because of complete utilization of the molecular symmetry to obtain the set of coordinates. The linear combination of the 11 internal coordinates will give a set of 11 symmetry

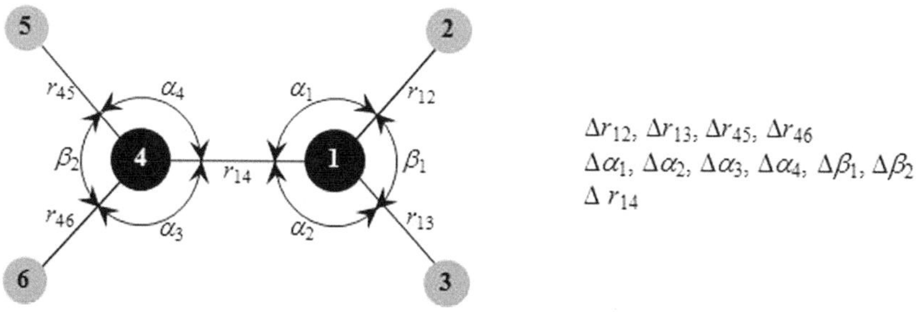

FIGURE 9.4 *Left*, atom numbering and the internal coordinates for ethylene. *Right*, the sets of basis functions.

coordinates. In the symmetry-adapted linear combinations (SALC) introduced by Cotton (1963), a SALC function is obtained by taking the sum over all basis functions. For example, in the expression for the ith SALC function

$$\phi_i = \sum_j c_{ij} b_j, \tag{9.40}$$

the coefficient c_{ij} determines the contribution b_j of the jth basis function. The normalized form of the function is

$$\phi_i = N \sum_j c_{ij} b_j,$$

with

$$N = \frac{1}{\sqrt{\sum_{j=1}^{n} c_{ij}^2}}. \tag{9.41}$$

In the case of ethylene, the sets of basis functions are $\{\Delta r_{12}, \Delta r_{13}, \Delta r_{45}, \Delta r_{46}\}$, $\{\Delta r_{14}\}$, and $\{\Delta\alpha_1, \Delta\alpha_2, \Delta\alpha_3, \Delta\alpha_4, \Delta\beta_1, \Delta\beta_2\}$. In the construction of SALC functions, we will find that two of them formed by combining the basis functions of the angles vanish. There are then nine SALC functions, each of which can be identified with one of the irreducible representations of the ethylene point group (D_{2h}). The normalized symmetry coordinates are listed below

$$\begin{aligned}
A_g\ \phi_1 &= \frac{1}{2}\left(\Delta r_{12} + \Delta r_{13} + \Delta r_{45} + \Delta r_{46}\right) \\
B_{3u}\ \phi_2 &= \frac{1}{2}\left(\Delta r_{12} + \Delta r_{13} - \Delta r_{45} - \Delta r_{46}\right) \\
B_{1g}\ \phi_3 &= \frac{1}{2}\left(\Delta r_{12} - \Delta r_{13} + \Delta r_{45} - \Delta r_{46}\right) \\
B_{2u}\ \phi_4 &= \frac{1}{2}\left(\Delta r_{12} - \Delta r_{13} - \Delta r_{45} + \Delta r_{46}\right) \\
A_g\ \phi_7 &= \frac{1}{6}\left(-2\Delta\beta_1 + \Delta\alpha_1 + \Delta\alpha_2 - 2\Delta\beta_2 + \Delta\alpha_3 + \Delta\alpha_4\right) \\
&= \frac{1}{2}\left(\Delta\alpha_1 + \Delta\alpha_2 + \Delta\alpha_3 + \Delta\alpha_4\right) \\
B_{3u}\ \phi_8 &= \frac{1}{6}\left(-2\Delta\beta_1 + \Delta\alpha_1 + \Delta\alpha_2 + 2\Delta\beta_2 - \Delta\alpha_3 - \Delta\alpha_4\right) \\
&= \frac{1}{2}\left(\Delta\alpha_1 + \Delta\alpha_2 - \Delta\alpha_3 - \Delta\alpha_4\right) \\
B_{1g}\ \phi_9 &= \frac{1}{2}\left(\Delta\alpha_1 - \Delta\alpha_2 + \Delta\alpha_3 - \Delta\alpha_4\right) \\
B_{2u}\ \phi_{10} &= \frac{1}{2}\left(\Delta\alpha_1 - \Delta\alpha_2 - \Delta\alpha_3 + \Delta\alpha_4\right) \\
A_g\ \phi_{11} &= \Delta r_{14}.
\end{aligned} \tag{9.42}$$

The two functions formed of the angle bases that tend to vanish are

$$\begin{aligned}
\phi_5 &= \frac{1}{6}\sqrt{6}\left(\Delta\beta_1 + \Delta\alpha_1 + \Delta\alpha_2 + \Delta\beta_2 + \Delta\alpha_3 + \Delta\alpha_4\right) \equiv 0 \\
\phi_6 &= \frac{1}{6}\sqrt{6}\left(\Delta\beta_1 + \Delta\alpha_1 + \Delta\alpha_2 - \Delta\beta_2 - \Delta\alpha_3 - \Delta\alpha_4\right) \equiv 0.
\end{aligned} \tag{9.43}$$

In addition to these nine in-plane stretching and bending vibrations, there are three out-of-plane wagging vibrations – trans-bending, cis-bending, and twisting of the two CH_2 groups. Let the internal coordinates be $\Delta\gamma_1$ and $\Delta\gamma_2$ for bending, and $\Delta\phi$ for twisting vibrations. From these bending vibrations we can construct the following three additional symmetry coordinates

$$\begin{aligned}
B_{1u}\ \phi_{12} &= \frac{1}{\sqrt{2}}\left(\Delta\gamma_1 + \Delta\gamma_2\right) \\
B_{2g}\ \phi_{13} &= \frac{1}{\sqrt{2}}\left(\Delta\gamma_1 - \Delta\gamma_2\right) \\
A_u\ \phi_{14} &= \Delta\phi.
\end{aligned} \tag{9.44}$$

There are thus 12 normal vibrations for ethylene (Figure 9.5).

The coefficients of the transformation equations for symmetry coordinates can be arranged in a 14×14 matrix shown below.

$$\begin{pmatrix}
\frac{1}{2} & \frac{1}{2} & \frac{1}{2} & \frac{1}{2} & 0 & 0 & 0 & 0 & 0 & 0 & 0 & 0 & 0 & 0 \\
\frac{1}{2} & \frac{1}{2} & -\frac{1}{2} & -\frac{1}{2} & 0 & 0 & 0 & 0 & 0 & 0 & 0 & 0 & 0 & 0 \\
\frac{1}{2} & -\frac{1}{2} & \frac{1}{2} & -\frac{1}{2} & 0 & 0 & 0 & 0 & 0 & 0 & 0 & 0 & 0 & 0 \\
\frac{1}{2} & -\frac{1}{2} & -\frac{1}{2} & \frac{1}{2} & 0 & 0 & 0 & 0 & 0 & 0 & 0 & 0 & 0 & 0 \\
0 & 0 & 0 & 0 & 0 & 0 & 0 & 0 & 0 & 0 & 0 & 0 & 0 & 0 \\
0 & 0 & 0 & 0 & 0 & 0 & 0 & 0 & 0 & 0 & 0 & 0 & 0 & 0 \\
0 & 0 & 0 & 0 & \frac{1}{2} & \frac{1}{2} & \frac{1}{2} & \frac{1}{2} & 0 & 0 & 0 & 0 & 0 & 0 \\
0 & 0 & 0 & 0 & \frac{1}{2} & \frac{1}{2} & -\frac{1}{2} & -\frac{1}{2} & 0 & 0 & 0 & 0 & 0 & 0 \\
0 & 0 & 0 & 0 & \frac{1}{2} & -\frac{1}{2} & \frac{1}{2} & -\frac{1}{2} & 0 & 0 & 0 & 0 & 0 & 0 \\
0 & 0 & 0 & 0 & \frac{1}{2} & -\frac{1}{2} & -\frac{1}{2} & \frac{1}{2} & 0 & 0 & 0 & 0 & 0 & 0 \\
0 & 0 & 0 & 0 & 0 & 0 & 0 & 0 & 0 & 0 & 1 & 0 & 0 & 0 \\
0 & 0 & 0 & 0 & 0 & 0 & 0 & 0 & 0 & 0 & 0 & \frac{1}{\sqrt{2}} & \frac{1}{\sqrt{2}} & 0 \\
0 & 0 & 0 & 0 & 0 & 0 & 0 & 0 & 0 & 0 & 0 & \frac{1}{\sqrt{2}} & -\frac{1}{\sqrt{2}} & 0 \\
0 & 0 & 0 & 0 & 0 & 0 & 0 & 0 & 0 & 0 & 0 & 0 & 0 & 1
\end{pmatrix} \tag{9.45}$$

This is called a $\mathbf{U}$-matrix such that

$$\phi = \mathrm{U}\Gamma, \tag{9.46}$$

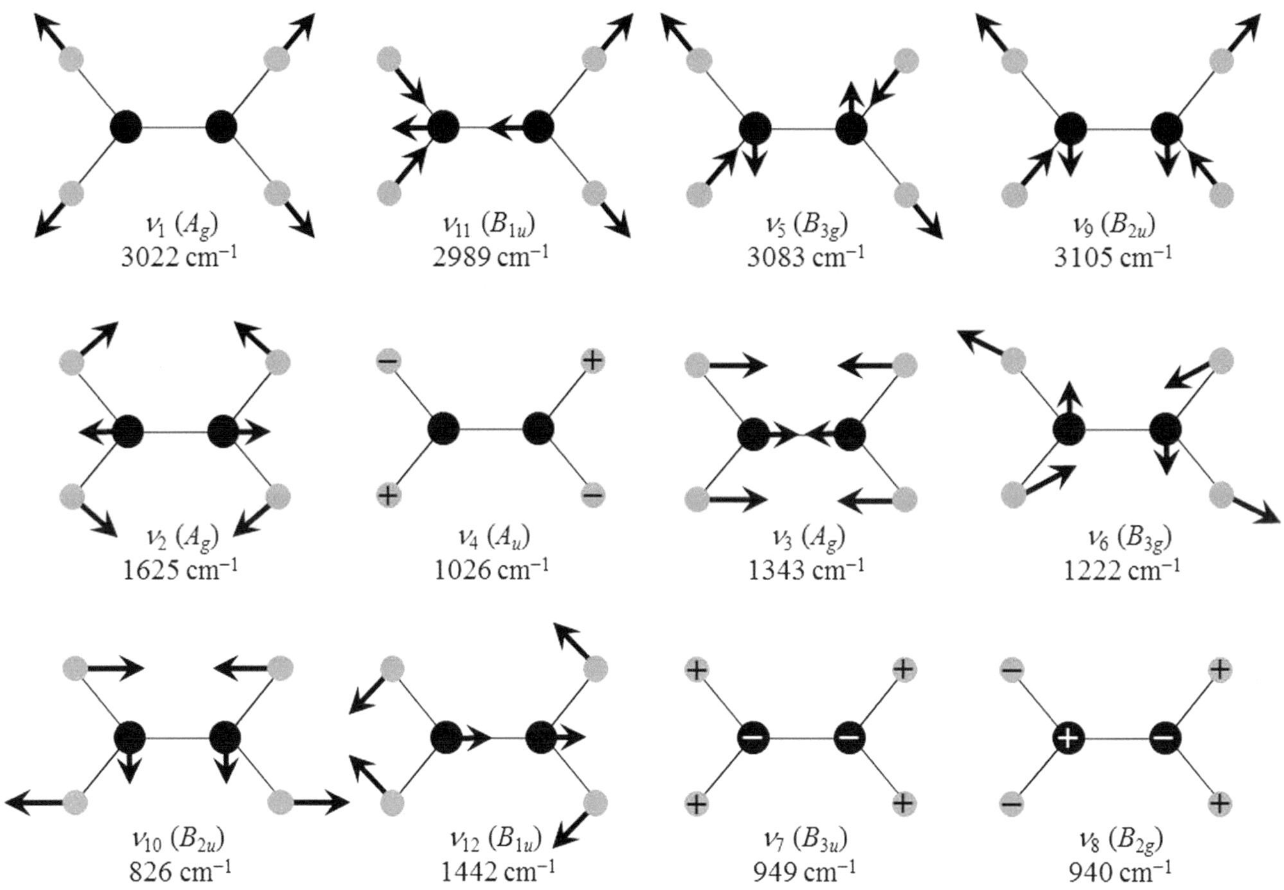

FIGURE 9.5 Normal modes of vibration of ethylene with symmetry labels and frequencies.

where $\mathbf{\Gamma}$ is the array of vectors for internal coordinates. The matrix $\mathbf{U}$ plays a central role here because it is used to define two important matrices called the potential energy matrix $\mathcal{F}$ (for force constant) and the inverse kinetic-energy matrix $\mathcal{G}$ (for geometry)

$$\mathcal{F} = \frac{1}{\mathbf{U}^{\mathrm{T}}}\mathbf{F}\frac{1}{\mathbf{U}}$$

$$\mathcal{G} = \mathbf{UG}\frac{1}{\mathbf{U}^{\mathrm{T}}}. \tag{9.47}$$

The $\mathbf{F}$-matrix is the $(3N-6)\times(3N-6)$ matrix of the force constants for the oscillators from normal coordinate analysis, and has the form

$$\begin{pmatrix} f_{11} & f_{12} & f_{13} & \cdots \\ f_{21} & f_{22} & f_{23} & \cdots \\ f_{31} & f_{32} & f_{33} & \cdots \\ \vdots & \vdots & \vdots & \ddots \end{pmatrix}. \tag{9.48}$$

The $\mathbf{G}$-matrix is essentially the inverse of the reduced mass matrix. But finding the $\mathbf{G}$-matrix even for a small molecule involves laborious and often complicated calculations. It is recommended that the $\mathbf{G}$-matrix elements are obtained from compilations found in Wilson, Decius, and Cross (1955).

It should be understood that each symmetry type of vibration will have corresponding $\mathcal{F}$ and $\mathcal{G}$ matrices. Thus, each vibrational mode is associated with a secular determinant

$$|\mathcal{GF} - \lambda E| = 0. \tag{9.49}$$

The roots λ for different vibrational modes yield the respective vibrational frequencies

$$\sqrt{\lambda} = \omega = 2\pi\nu. \tag{9.50}$$

As said, the $\mathcal{FG}$ matrix calculations become increasingly difficult with the size of the molecule. Even for computer calculations, the matrices are considerably simplified by assigning frequencies based on semi-empirical rules. Then further corrections to the input frequencies are introduced by accounting for neighboring groups. In spite of such efforts, vibrational modes having lower frequencies are rarely computed. This is one of the reasons why understanding vibrational energy patterns is still an active area of research (Georges et al., 1999; Tan et al., 2015). The IR absorption spectrum of ethylene has been studied extensively, and portions of a somewhat earlier spectrum is shown in Figure 9.6 for the sake of an example, and a compiled list of the normal modes of vibration, symmetry species, and fundamental frequencies is given in Table 9.2.

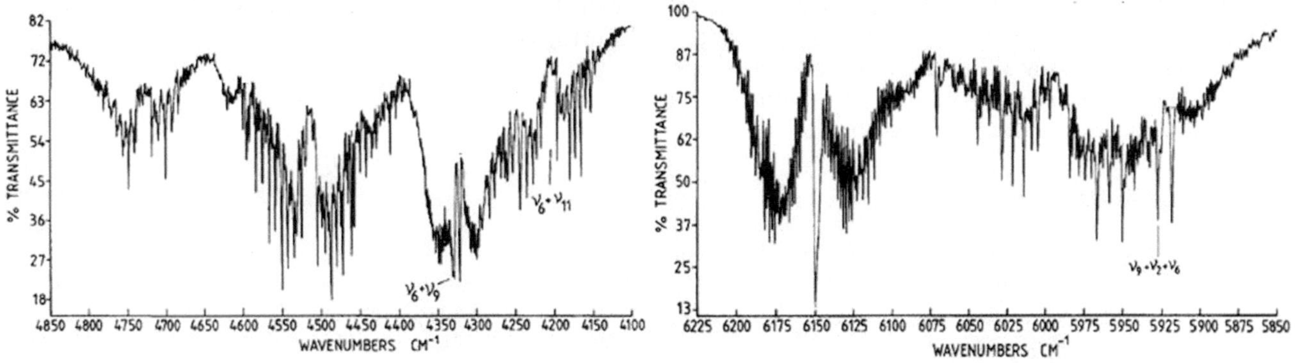

FIGURE 9.6 Two regions of a low-resolution FTIR spectrum of C_2H_4 showing a few normal mode band frequencies. Each normal mode band may be studied separately under high resolution conditions to extract rotation-vibration combined band structures. (Reproduced from Duncan and Ferguson, 1988, with permission from AIP.)

TABLE 9.2
Normal Modes and Corresponding Vibrational Modes, Symmetry Species, and Fundamental Frequencies of C_2H_4

Normal mode	Vibrational mode	Symmetry species	Fundamental frequency cm^{-1}	
Q_1	C–H stretch	A_g	ν_1	3022
Q_2	C=C stretch	A_g	ν_2	1625
Q_3	H–C–H bend	A_g	ν_3	1343
Q_4	$H_2C{=}CH_2$ twist	A_u	ν_4	1026
Q_5	trans-C–H stretch	B_{3g}	ν_5	3083
Q_6	H–C–H wag	B_{3g}	ν_6	1222
Q_7	out-of-plane	B_{3u}	ν_7	949
Q_8	out-of-plane	B_{2g}	ν_8	940
Q_9	C–H stretch	B_{2u}	ν_9	3105
Q_{10}	H–C–H wag	B_{2u}	ν_{10}	826
Q_{11}	C–H stretch	B_{1u}	ν_{11}	2989
Q_{12}	H–C–H bend	B_{1u}	ν_{12}	1422

One may consult Figure 9.5 to deduce the symmetric or antisymmetric attribute of each vibrational motional mode.

9.7 ACTIVITY OF NORMAL MODES OF VIBRATION

All normal modes may not be infrared active. Whether a mode will be observed to change the vibrational quantum number when irradiated in the infrared region depends on whether the dipole moment changes with the displacement of nuclei in that vibrational mode. As was the condition for diatomic vibrational absorption in the harmonic approximation, a polyatomic vibrational transition can occur from v_k to $v_{k\pm1}$ only when the dipole derivative with respect to a normal coordinate taken at the equilibrium internuclear configuration does not vanish

$$\left[\frac{\partial\mu(Q)}{\partial Q_k}\right]_{\mathrm{o}} \neq 0. \tag{9.51}$$

The subscript 'o' indicates the derivative taken about the equilibrium configuration. This condition must hold because the transition moment integral for infrared absorption

$$\mu_{v\to v'} = \int_{-\infty}^{+\infty} \psi_{v'}^{*}\,\boldsymbol{\mu}\,\psi_{v}\,d\mathbf{r} \tag{9.52}$$

will be zero if $\boldsymbol{\mu}$ is a constant, because $\psi_{v'}$ and ψ_v are orthogonal. In order to be infrared active, the vibrational mode must generate an oscillating dipole moment. The larger the oscillation of the dipole, the stronger is the absorption of infrared light. Consider, for example, water, which has three normal modes (Figure 9.3). It is easy to determine by inspection alone that both Q_1 and Q_2 change dipole moments along z

$$\left(\frac{\partial\mu_z}{\partial Q_{1,2}}\right)_{\mathrm{o}} \neq 0, \tag{9.53}$$

but Q_3 changes dipole moment along y

$$\left(\frac{\partial\mu_y}{\partial Q_3}\right)_0 \neq 0. \tag{9.54}$$

All three modes of water are therefore infrared active.

As another example, consider the 12 normal modes of ethylene (Table 9.2). A close inspection would tell that all but the five modes – Q_7, Q_9, Q_{10}, Q_{11}, and Q_{12}, do not involve a change in dipole moment. Only these five modes are infrared active. This is also evident from the character table for D_{2h} (Table 6.2) in which the second column from the right registers linear functions x, y, and z for B_{1u}, B_{2u}, and B_{3u} symmetries only. The five infrared active fundamental modes of ethylene belong to these three symmetry species. Whether or not a fundamental mode is infrared active can be determined at once by inspecting the character table. The normal modes are conventionally labeled with $Q_{k(or\,i)}$, Q_i, $i = 1, 2, 3, \ldots$, and the associated absorption frequencies are indicated by ν_i.

9.8 SECONDARY BAND MANIFOLD IN INFRARED SPECTRA

In addition to the infrared-active fundamental transitions involving a single-quantum transition ($v_0 \to v_{\pm1}$) a large number of low-intensity transitions crowd the IR spectra of polyatomic molecules, and indeed a great deal of effort is made to understand and assign these bands – an area of basic spectroscopy. But the problem compounds with the size of the molecules. We discuss below the origin and nature of these bands.

9.8.1 Overtone Band

Unlike the case for diatomic molecules in harmonic approximation where transitions can occur from v_k to $v_{k\pm1}$, transitions in polyatomics often also involve $v_0 \rightarrow v_2$ (first overtone, two quanta), $v_0 \rightarrow v_3$ (second overtone, three quanta), ..., $v_0 \rightarrow v_n$ ($n-1$ overtone, n quanta) changes. Since almost all molecules populate the ground vibrational state at ambient temperature or even higher (~400 K), the overtone transitions occur from the ground v_0 state and are excited in the near-infrared region (~1400–4000 cm^{-1}). It should not be construed that all overtones occur in a molecule. For example, none of the 12 fundamental modes of ethylene show the first overtone, but all infrared active modes belonging to B_{1u}, B_{2u}, and B_{3u} symmetries show the second overtone. Overtone transitions are forbidden in strictly harmonic approximation, which means they arise due to mechanical anharmonicity of the potential energy surface. The association of overtone bands with anharmonicity bears on three characteristic results. One, the transition frequency does not increase linearly with successive higher overtones. In other words, $\nu_{v_0\rightarrow v_n} \neq n\nu_{v_0\rightarrow v_1}$. The observed overtone frequencies are found to be rather consistent with what we calculate from the energy of a vibrational level v in a Morse-type anharmonic potential

$$E_v = h\nu\left(v+\frac{1}{2}\right) - \frac{(h\nu)^2}{4D_e}\left(v+\frac{1}{2}\right)^2. \tag{9.55}$$

Two, as evident from the form of the equation itself, the measurement of overtone transition frequency yields the value of bond dissociation energy D_e. Three, the intensity of infrared absorption exponentially decays as the index of overtone transition increases. The basis for this claim is that a vibrational state sufficiently higher in the anharmonic potential is a quasiclassical state. Let us elaborate on this concept a little bit. The infrared absorption intensity for the kth mode is proportional to the magnitude of the change in dipole moment

$$I_{\text{IR}} \propto \left(\frac{\partial\mu}{\partial Q_k}\right)^2, \tag{9.56}$$

so that the transition moment integral involving v_0 and v_n is

$$I_{\text{IR}} \propto \left\langle \psi_0 \left| \left(\frac{\partial\mu}{\partial Q_k}\right)^2 \right| \psi_n \right\rangle \sim \mu_{0n}. \tag{9.57}$$

For $n = 2,3,4,...$, the corresponding dipole moment matrix elements would be μ_{02}, μ_{03}, μ_{04}, ... But the dipole moment matrix for a transition from v_0 to a quasiclassical state is given by (Medvedev, 1985)

$$\mu_{0n} \propto \rho e^{-\sigma_{0n}}$$

$$\sigma_{0n} = \frac{1}{\hbar}\,\text{Im}\left(\int_{a_n}^{q_0} P_n dq - \int_{a_0}^{q_0} P_0 dq\right), \tag{9.58}$$

where ρ is a function of the two vibrational states involved, P_n and P_0 are momenta for the energies of the vibrational states v_n and v_0, respectively, a_n and a_0 may be taken as internuclear distances to the left of the equilibrium internulear configuration, and the upper integration limit q_0 is a real or complex point where the potential energy tends to infinity.

The quantity σ_{0n} is of special significance because the matrix μ_{0n} will decay exponentially as σ_{0n} exceeds 1. Path integration of equation (9.58) carried out by Medvedev (1985) gives the following normal intensity distribution

$$\log\left|\mu_{0n}\right|^2 \propto f_{0n} = \text{constant} - a\sqrt{v_n}, \tag{9.59}$$

where f_{0n} is oscillator strength (proportional to the mean transition dipole moment), a is another constant containing the molecular mass, and $v_n (= E_n / \hbar\omega)$ is the reduced energy. In Figure 9.7 we show the exponential decay of the oscillator strength with the excitation energy of benzene C–H stretching involving 1 to 9 quanta.

9.8.2 Hot Band

An infrared-active fundamental transition may occasionally split to give rise to a low-intensity side band positioned a few cm^{-1} away from the main band. Such side bands arise from infrared absorption by a small thermal population of molecules in low-lying energy levels above the ground-state vibrational energy level, and are called hot bands. A hot band represents the same fundamental transition that gives rise to the main band; while the main band is due to the transition

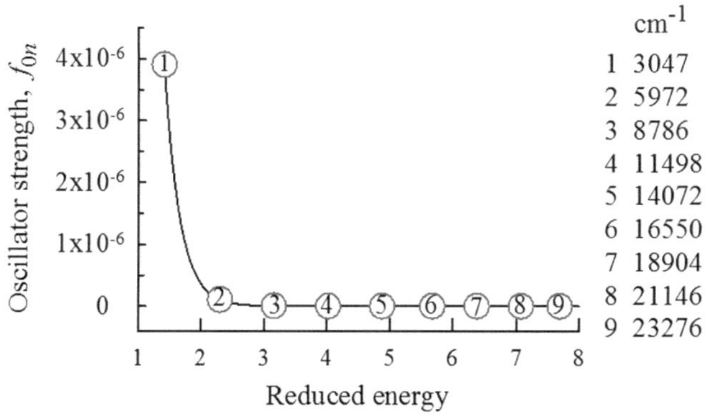

FIGURE 9.7 Exponentially decreasing oscillator strength of overtone bands involving 1 to 9 quanta excitations of C–H stretching vibration in benzene vapor.

from the ground vibrational state, the hot band absorption occurs from the low-lying excited state. A characteristic of hot bands is temperature dependence of their intensities which is easily understood from the temperature dependence of the Boltzmann population distribution between the ground and excited states

$$\frac{n}{n_0} = \frac{g}{g_0} e^{-\frac{E}{k_B T}} = \frac{g}{g_0} e^{-\frac{h\nu_f c}{k_B T}}, \qquad (9.60)$$

where n_0 is the population in the ground state, n is the relative population in the excited state having energy E, and g_0 and g are corresponding degeneracies. If the probabilitites of transition from the ground and excited states are equal, then

$$\frac{n}{n_0} \sim \frac{\text{OD}_{\text{fundamental}}}{\text{OD}_{\text{hot}}}. \qquad (9.61)$$

The Boltzmann relation is invoked here only to recall the distribution of thermal population in the excited state relative to the ground state. The ground vibrational state is not involved in hot band transition; the transition is rather between two excited states. The appearance of hot band in infrared spectra was first detected for acetylene derivatives by studying the ratio of intensities of the main and side bands as a function of temperature (Kraihanzel and West, 1962). The side band arises from transition of the population in the excited state energy level corresponding to the bending or deformation vibration of acetylenic $\equiv$C-H.

9.8.3 Combination Band

Two or more normal modes that have similar transition frequencies can couple with each other so that an infrared photon can simultaneously excite all coupled vibrations. To see how such a coupling may occur we consider two coupled normal mode coordinates Q_1 and Q_2. The total Hamiltonian can be written as

$$\mathcal{H} = \mathcal{H}_0 + V(Q) + \mathcal{H}' \qquad (9.62)$$

where $V(Q)$ is the vibrational potential energy, and $\mathcal{H}'$ is the coupling Hamiltonian. The vibrational potential energy function for the two normal modes here, $V(Q_1,Q_2)$, contains the anharmonicity terms, and the $\mathcal{H}'$ function consists of oscillating dipoles of the individual and coupled normal modes

$$V(Q_1,Q_2) = AQ_1^2Q_2 + BQ_1Q_2^2 + CQ_1^2Q_2^2, \qquad (9.63)$$

$$\mathcal{H}' = -\left(\frac{\partial\mu_1}{\partial Q_1}\right)Q_1\mathbf{E} - \left(\frac{\partial\mu_2}{\partial Q_2}\right)Q_2\mathbf{E} - \left(\frac{\partial^2\mu_{12}}{\partial Q_1\partial Q_2}\right)Q_1Q_2\mathbf{E}. \qquad (9.64)$$

The effect of the coupled Hamiltonian can occur in two ways. One, an infrared photon can simultaneously excite the two vibrational modes via the third coupled term. Two, the photon can excite a vibration via one of the two terms temporally followed by the influence of the appropriate anharmonic term $AQ_1^2Q_2$ or $BQ_1Q_2^2$ in the $V(Q_1,Q_2)$ function. In other words, the direct coupling effect of the Hamiltonian produces simultaneous excitation, and indirect coupling through anharmonicity leads to a follow-through excitation of the second mode.

All coupled modes need not necessarily be infrared active, but at least one has to be. For example, combination absorption will occur even if one of two initial terms is vanishing. This means an infrared active mode can absorb a photon first which is then shared by a mode that may otherwise be infrared inactive. Clearly, combination bands are forbidden in harmonic approximation, and it is the extent of anharmonicity of the vibrational potential that determines the intensity of a combination band – more intense with larger anharmonicity.

The number of quanta absorbed, the number of modes involved, the number of combination bands, and the frequencies of these bands may all present a confusing picture to some. It is necessary therefore to explain these aspects in some detail. The combination phenomenon may involve two or more normal modes. For a combination of two normal modes, we may like to show the transitions as

$$00 \rightarrow 01,\ 00 \rightarrow 10,\ 01 \rightarrow 12,\ 01 \rightarrow 22,\ \ldots$$

Each transition is distinct. In the above, the first and second are single-quantum transitions from the ground state of the respective normal mode, the third one involves both fundamental and hot band transitions, and the fourth is a overtone–hot band combination. Possible transitions involving the combination of three normal modes can be shown as

$$000 \rightarrow 100,\ 001 \rightarrow 012,\ 100 \rightarrow 010,\ 101 \rightarrow 121,\ \ldots$$

We realize by now that the combination manifolds connecting fundamental, overtone, and hot band transitions make infrared band assignment challenging, and one has to work through a wide range of experimental results in conjunction with group theoretic principles to assign them with confidence. For example, analyses of near infrared absorption bandshape, maximum frequency of absorption, and absorption intensity at varying sample density (pressure) and temperature form the list of groundwork to distinguish combination bands from single-mode overtones. Combination bands are conventionally represented by summation of the coupled modes involved. For example, a few ethylene combination bands are given in Table 9.3, the coupled modes in which can be identified by inspecting Figure 9.5.

The integrated absorption band intensity is

$$A = \int \alpha(\nu)\,d\nu = \frac{1}{pl}\int ln\frac{I_0}{I}\,d\nu$$

where α is molar absorption coefficient, p is pressure that changes the sample density or effective molar concentration,

TABLE 9.3
Combination Bands in Ethylene

Assignment	Band center (cm^{-1})	Integrated intensity, A [cm^{-1}/(cm atm)]
ν_9 and $\nu_2 + \nu_{12}$ *	3105 and 3079	112.20
$\nu_6 + \nu_{10}$	2048	2.56
$\nu_7 + \nu_8$	1889	21.70

Note: * The ν_9 band is an independent fundamental absorption and is not a part of the combination of ν_2+ν_{12}. Since the ν_9 band overlaps with the ν_2+ν_{12} combination band, the intensity together is measured.

l is the path length, and I_0 and I are intensities of incident and transmitted light, respectively.

In rare occasions, a combination band may appear as what is called a 'difference band'. Unlike a true combination transition, a difference transition involves two excited states, and hence a difference band is essentially a hot band, which is the reason why difference transition rarely occurs at ordinary temperature. The transitions $001 \rightarrow 110$, $100 \rightarrow 001$, and $100 \rightarrow 002$, in which both initial and final states are excited states, represent difference transitions. The first two are single-quantum transfers, and the third one involves the transfer of two quanta.

9.8.4 Fermi Resonance Band

A fundamental transition can couple with an overtone or a combination band or both when all these transitions are closer in energy and belong to the same symmetry of normal modes. For example, a normal mode of, say, A_{1u} symmetry in which the fundamental vibrational level lies close to an overtone level of another normal mode belonging to A_{1u} symmetry can strongly couple and mix to modify the initial energy levels of the transitions. The coupling referred to here occurs through the anharmonicity constants, which arise from the cubic and higher order force constants in the V_N function described earlier (equation 9.16), and because the energy levels of the coupled transitions are very close to each other, the coupling is called resonance coupling.

To illustrate the phenomenon, we consider the coupling of a fundamental ν_1 with an overtone $2\nu_2$. The fundamental transition generally produces a strong absorption relative to the much weaker overtone absorption. Let ϕ_1 and ϕ_2 be the wavefunctions of the states that give rise to the fundamental and the overtone, and let E_1 and E_2 be the corresponding unperturbed energies. We can invoke first-order perturbation to produce the mixed states

$$\psi_1 = c_{11}\phi_1 + c_{12}\phi_2$$
$$\psi_2 = c_{21}\phi_1 + c_{22}\phi_2. \tag{9.65}$$

Application of degenerate perturbation theory yields the following secular determinant for mixing of the two states

$$\begin{vmatrix} E_1 - \lambda & \frac{1}{2}k_{122} \\ \frac{1}{2}k_{122} & E_2 - \lambda \end{vmatrix} = 0, \tag{9.66}$$

in which the root λ will give the modified or new energy levels for the interacting transitions, and the off-diagonal $\frac{1}{2}k_{122}$ represents the anharmonic coupling constant. Solution of the determinant yields

$$\lambda = \frac{(E_1 + E_2) \pm \sqrt{(E_1 + E_2)^2 + k_{122}{}^2}}{2}. \tag{9.67}$$

Because $\lambda_\pm$ are modified energies of transition for the Fermi resonance arising from the coupling of ν_1 and $2\nu_2$, the original or unperturbed transitions will shift – the initial low-frequency transition will be lowered further, and the high-frequency one will be shifted higher. The shifts follow a symmetric pattern. This is called Fermi shift, depicted in Figure 9.8. The relative peak intensities are obtained as

$$\frac{\ln I_2}{\ln I_1} = \frac{(\lambda_1 - \lambda_2) - (E_1 - E_2)}{(\lambda_1 - \lambda_2) + (E_1 - E_2)}. \tag{9.68}$$

The calculation can be easily extended to Fermi interaction of three modes, $j = 1,2,3$. If the fundamental ν_1 and two overtones $2\nu_2$ and $2\nu_3$ were to interact, we would have the wavefunctions

$$\psi_i = \sum_{j=1,2,3} c_{ij}\phi_j, \tag{9.69}$$

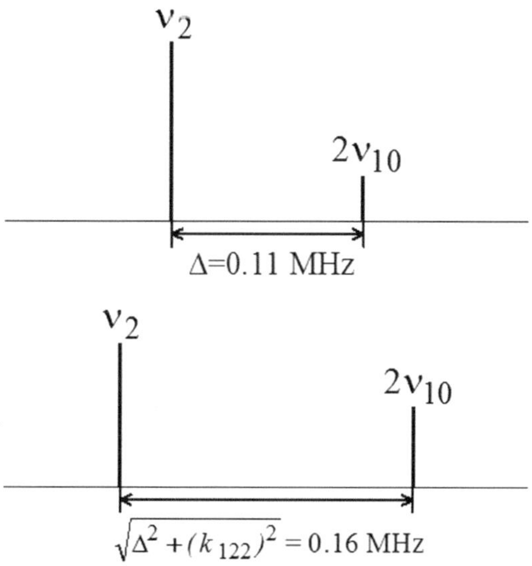

FIGURE 9.8 Depiction of Fermi shift of ν_2 and $2\nu_{10}$ of ethylene due to mixing of the eigenstates. *Top*, unperturbed (pure); *bottom*, perturbed (mixed). Notice that the low-frequency line is shifted to lower frequency and the high-frequency one to higher. For this calculation we took k_{122} ~0.01 MHz, which is a very crude estimate.

where c_{ij} are mixing coefficients. The secular determinant reads

$$\begin{vmatrix} E_1-\lambda & \frac{1}{2}k_{122} & \frac{1}{2}k_{133} \\ \frac{1}{2}k_{122} & E_2-\lambda & 0 \\ \frac{1}{2}k_{133} & 0 & E_3-\lambda \end{vmatrix} = 0, \quad (9.70)$$

in which E_1, E_2, and E_3 are unperturbed energies of the fundamental (ν_1), and second ($2\nu_2$) and third ($2\nu_3$) overtones, respectively.

9.8.5 Vibrational Angular Momentum and Coriolis-Perturbed Band Structure

Vibrational band structure due to Coriolis interaction is seldom discussed in detail because of the apparent complexity involved. The basic idea, however, is the emergence of a vibrational angular momentum due to Coriolis force in the rotating molecule. Coriolis force is a fictitious force operative only in a noninertial frame of reference, which rotates with acceleration, in contrast with an inertial frame of reference, which is either stationary or rotating with a constant velocity. A physical property appears different; the weight of an object, for example, is more in a noninertial frame relative to that in an inertial frame. We reason that the Newtonian force $F = ma$ experienced in the noninertial frame is a little larger because of accelerated rotation of the frame, even though Newton's law of motion does not hold in this frame, which is why the apparently larger force is a fictitious force.

In the discussion of molecular vibration, we know that the nuclei vibrate while the molecule is rotating. Consider a rotating polyatomic molecule that as a whole will expand and contract as the bonds stretch out and compress in. The molecular expansion due to bond stretching will slow down the rotation, but when the molecule contracts, the rotation will be accelerated due to a decrease in the moment of inertia. This aspect was briefly discussed in Chapter 5 under the states of nuclear motion. The reader should be convinced that molecular vibrations force the molecule to rotate at variable acceleration, and hence the vibrations should be treated in a noninertial frame of reference that decelerates and accelerates periodically. The emergence of the noninertial frame in the picture necessitates the consideration of Coriolis interaction of vibrational and rotational motions. An observer standing in the molecule-fixed rotational coordinate system will say that a vibrational angular momentum arises due to Coriolis force. A vibrating atom in a rotating molecule curves out from one of the vector components of vibration due to Coriolis force, but since the components of vibration are mutually orthogonal, the trajectory of the atom is now described by a circle or ellipse instead of a straight line as it would have been in an inertial frame of reference. The circular or elliptic trajectory of atomic motion then is associated with an angular momentum called vibrational angular momentum **p**, which interacts with the rotational angular momentum **P**. As a consequence of this interaction, the value of the vibrational angular momentum enters into the rotational energy levels as the Coriolis zeta coefficient ξ. The vibrational angular momentum turns out to be a sum of ξ, one ξ for each pair of normal coordinates. If there were j normal coordinates (not normal modes), then

$$p = \sum_j \xi_j. \quad (9.71)$$

The Coriolis corrections to rotational energy levels and intensity of transitions are generally small, and hence can be introduced to the rotational–vibrational problem by adding a perturbation Hamiltonian. We will briefly consider the theory here. In the absence of the Coriolis effect, the rotational–vibrational Hamiltonian is

$$\mathcal{H}^0 = \mathcal{H}_{\text{rot}} + \mathcal{H}_{\text{vib}}, \quad (9.72)$$

and as learned already, the eigenfunctions of $\mathcal{H}^0$ are

$$\psi = \Theta_{JKM} e^{im\phi} e^{iK\chi} \prod_k \psi_{v_k}(Q_k), \quad (9.73)$$

where J, K, M are rotational quantum numbers, θ, ϕ, χ are Eulerian angles, and v_k are vibrational quantum numbers corresponding to the normal coordinate Q_k. Recalling the earlier discussions of polyatomic rotational spectra, we can easily write down the energy eigenvalue. For a prolate top, for example,

$$E^0 = B_e J(J+1) + (A_e - B_e)K^2 + \sum_k \hbar\omega_k \left(v_k + \frac{1}{2}\right). \quad (9.74)$$

The rotational constants A_e and B_e are as described earlier.

Then one brings in the Coriolis perturbation and writes the total Hamiltonian as

$$\begin{aligned} \mathcal{H} &= \mathcal{H}^0 + \mathcal{H}^{\text{Coriolis}} \\ &= \mathcal{H}^0 + \left(\frac{-p_a P_a}{I_a} - \frac{-p_b P_b}{I_b} - \frac{-p_c P_c}{I_c} + \frac{p_a}{2I_a} + \frac{p_b^2}{2I_b} + \frac{p_c^2}{2I_c} \right). \end{aligned} \quad (9.75)$$

The terms representing $\mathcal{H}^{\text{Coriolis}}$ need some explanation. The components of the rotational angular momentum about the molecule fixed axes system a, b, c are shown by P_a, P_b, and P_c, the corresponding principal moment of inertia are I_a, I_b, and I_c, and the components of the vibrational angular momentum due to the Coriolis effect are denoted by p_a, p_b, and p_c. We notice coupling of the two kinds of angular momenta P and p in the first three terms, and these three terms give rise to rotational perturbation. The last three terms do not contain the rotational operator. In fact, $p \ll P$, implying that the vibrational momentum alone would not be involved in the calculation of matrix elements of $\mathcal{H}^{\text{Coriolis}}$. The form of p is interesting. For example, the c-component of p is

$$p_c = -\sum_i i\hbar\left[\left(\frac{\partial q_{(x)i}}{\partial q_{(y)i}}\right) - \left(\frac{\partial q_{(y)i}}{\partial q_{(x)i}}\right)\right], \qquad (9.76)$$

in which $q_{(x)i}$ and $q_{(y)i}$ are mass-weighted Cartesian displacement coordinates for the ith atom in the molecule. Having the expression for p_c, the reader may write down the expressions for p_a and p_b using the principle of cyclic permutation.

We should look at the matrix elements of the perturbation Hamiltonian alone. Using the vibration-rotation product function $|\psi_{\text{vib}}\psi_{\text{rot}}\rangle = |vr\rangle$ and the corresponding operators, we can write

$$\left\langle vr\left| -\frac{p_c P_c}{I_c}\right| v'r'\right\rangle = -\frac{1}{I_c}\langle v|p_c|v'\rangle\langle r|P_c|r'\rangle. \qquad (9.77)$$

The matrix elements for P_c, as discussed earlier under polyatomic rotations, are

$$\langle JKM|P_c|J'K'M'\rangle = \hbar^2 K. \qquad (9.78)$$

Regarding the matrix elements for the vibrational angular momentum, an interesting feature emerges. It has been shown that the vibrational angular momentum can be expressed in terms of normal coordinates (Smith and Mills, 1964; Boyd and Longuet-Higgins, 1952). The c-component of the vibrational angular momentum is

$$p_c = \sum_{r,s}\zeta^c_{r,s}[Q_r P_s - Q_s P_r], \qquad (9.79)$$

in which P_s and P_r are conjugate momenta (see Box 9.2) corresponding to coordinates Q_s and Q_r. The coefficient $\zeta^{(c)}_{r,s}$, defined along the molecule-fixed axis c, is called the Coriolis zeta constant, which, when expanded, appears as

$$\zeta^{(c)}_{r,s} = \sum_i\left(\frac{\partial q_{(x)i}}{\partial Q_r}\frac{\partial q_{(y)i}}{\partial Q_s} - \frac{\partial q_{(x)i}}{\partial Q_s}\frac{\partial q_{(y)i}}{\partial Q_r}\right). \qquad (9.80)$$

It is important to understand the meaning this equation conveys. Each zeta-constant ξ couples two mutually interacting vibrations through molecular rotation about a molecule-fixed axis. A value of $\zeta^{(\alpha)}_{r,s}$, where $\alpha = a, b, c$ denotes a molecule-fixed axis, will exist only when the product of the symmetry species of the normal coordinates Q_r and Q_s contains the rotational function R_α; otherwise $\zeta^{(\alpha)}_{r,s} = 0$. For example, the modes v_7 and v_{10}, and v_7 and v_{12} of ethylene are Coriolis-coupled (Nakanaga et al., 1979). By inspection of the direct product table for the D_{2h} point group (Table 6.3) we find

$$B_{3u}\otimes B_{2u} = B_{1g}, \text{ which contains } R_z \text{ or } R_c, \text{ implying } \zeta^{(c)}_{7,10} \neq 0$$

$$B_{3u}\otimes B_{1u} = B_{2g}, \text{ which contains } R_y \text{ or } R_b, \text{ implying } \zeta^{(b)}_{7,12} \neq 0. \qquad (9.81)$$

If the vibrational modes are taken as simple harmonic oscillators, then the product functions of the two Coriolis-coupled normal coordinates are

$$|v_r v_s\rangle, |v_r v_{s+1}\rangle, |v_s v_{r+1}\rangle, \text{ and } |v_{r+1}v_{s+1}\rangle, \qquad (9.82)$$

with which the matrix elements of the vibrational angular momentum p_c (equation 9.79) are obtained. The details of the form of the matrix elements can be somewhat complicated. For simplicity, we will take two fundamental transitions to write the matrix element as

$$\langle rs|p_c|sr\rangle = -i\hbar\zeta^{(c)}_{rs}\Omega_{rs}, \qquad (9.83)$$

where

$$\Omega_{rs} = \frac{1}{2}\left[\sqrt{\frac{v_r}{v_s}} + \sqrt{\frac{v_s}{v_r}}\right]. \qquad (9.84)$$

The quantity Ω_{rs} as a part of the eigenvalue determines the variation in the intensities of the coupled normal coordinates. We will look at it in more detail later.

Confusion often arises concerning the excitation frequencies of two Coriolis-coupled vibrations. Coriolis interaction can take place between two vibrational levels, both of which are degenerate or nondegenerate. The discussion presented above is a generalized one, regardless of the degeneracy of the vibrational levels. When it comes to symmetric top molecules (c_{3v} symmetry), for example, one or more degenerate modes of a state may be excited. The wavefunction of such a doubly degenerate normal vibration is a function of not just one but both coordinates of the degenerate modes, say Q_{r1} and Q_{r2}. Each degenerate mode has two quantum numbers that are traditionally denoted by v_r (vibrational quantum number) and l_r (vibrational angular momentum quantum number) such that

$$v = v_{r1} + v_{r2}. \qquad (9.85)$$

The operator p_c (equation 9.79) can now be written in terms of these two degenerate modes alone

$$p_c = Q_{r1}P_{r2} - Q_{r2}P_{r1}, \qquad (9.86)$$

so that the diagonal matrix elements of p_c is obtained from

$$\langle v_r^{lr}|Q_{r1}P_{r2} - Q_{r2}P_{r1}|v_r^{lr}\rangle = l_r\hbar. \qquad (9.87)$$

Delving into the form of the off-diagonal elements would not be worthwhile because they are not only complicated, but also the results are specific to a molecular point group.

Let us look at the rotational–vibrational line intensities due to Coriolis interaction. Consider a vibrational transition $v''r'' \rightarrow v'r'$ belonging to a normal coordinate, and let this be Coriolis-coupled to another normal coordinate. Then the excited-state wavefunction is not just $v'r'$, but is a combination of basis functions

$$|v'r'\rangle = a|v_1 r_1\rangle + b|v_2 r_2\rangle, \tag{9.88}$$

in which a and b are mixing coefficients. The transition dipole strength is given by

$$\left[\langle v''r''|\boldsymbol{\mu}_{\text{lab}}|v'r'\rangle\right]^2 = \left[\langle v''r''|\boldsymbol{\mu}_{\text{lab}}|\,av_1r_1\rangle + \langle v''r''|\boldsymbol{\mu}_{\text{lab}}|bv_2r_2\rangle\right]^2$$

$$= a^2\mu_1{}^2 + b^2\mu_2{}^2 + 2ab\mu_1\mu_2. \tag{9.89}$$

Here, $\boldsymbol{\mu}_{\text{lab}}$ is the dipole moment operator defined along the laboratory axes X,Y,Z, and can be related to the molecule-fixed axes a,b,c by the direction cosines given in Box 5.1. The transition moments μ_1 and μ_2 are unperturbed moments. The first two terms are always positive irrespective of the signs of the coefficients a and b and moments μ_1 and μ_2, but the third term can be either positive or negative. It is the magnitude and the sign of the third term that can enhance (+ sign) or deplete (− sign) the strength of the rotational lines in the P and R branches of the rotational spectrum. If the rotational bands are enhanced in the P branch, then those in the R branch will be weakened, and vice versa. The result will follow the same pattern if the fundamental transition of the other Coriolis-coupled normal mode is excited.

An IR spectrum of a fundamental band as such will not reveal if rotational–vibrational line intensities contained within are due to Coriolis coupling of the vibrational mode with another vibrational mode. This requires a detailed analysis based on the dipole moment derivative with respect to the normal coordinates aided by computation. Figure 9.9 shows IR spectra of the ν_{10} band region of ethylene where ν_{10} and ν_7 are Coriolis-coupled. The experimental spectrum is reproduced by a spectrum calculated assuming Coriolis enhancement by ν_7 (note that the third term in equation (9.89) is +ve), and not by the one that considered Coriolis-depletion of intensities. The spectrum calculated without assuming Coriolis interaction does not reproduce the experimental spectrum. It is thus clear that the intensity of the rotational–vibrational structures of ν_{10} is due to Coriolis borrowing from ν_7.

Suppose bands ν_r and ν_s are Coriolis-coupled, then the change in the intensity of the ν_r band is given by a very useful expression

$$\Delta I_{\nu_r} = \frac{4A_e\zeta_{rs}\Omega_{rs}}{\nu_r - \nu_s}\left(\frac{\dfrac{\partial\mu}{\partial Q_s}}{\dfrac{\partial\mu}{\partial Q_r}}\right), \tag{9.90}$$

in which the rotational constant A_e is specified by that axis about which Coriolis interaction occurs, ζ_{rs} is the zeta constant, Ω_{rs} is the eigenvalue shown in equation (9.83), and $\partial\mu/\partial Q_s$ and $\partial\mu/\partial Q_r$ are dipole moment derivatives with respect to the normal coordinates Q_s and Q_r. These dipole derivatives can be determined from the absolute intensity of the corresponding fundamental transitions. If I_i is the intensity of the ith fundamental, then by Beer's law we have

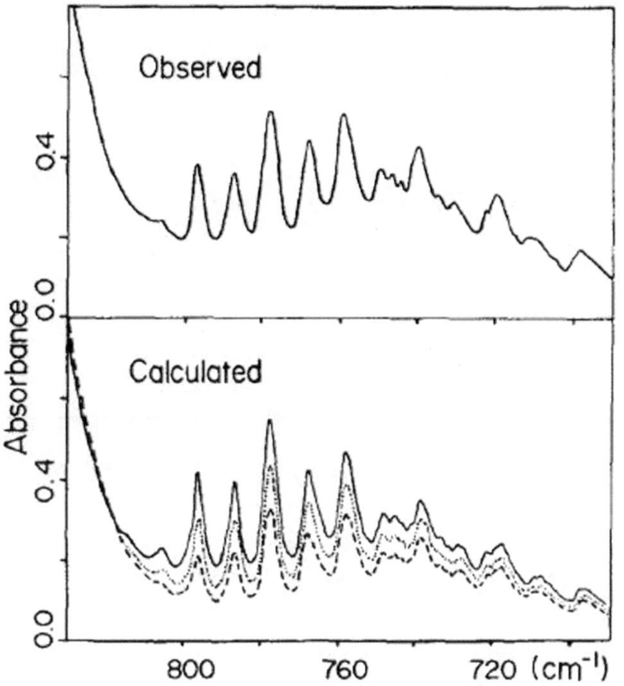

FIGURE 9.9 *Top*, observed IR spectrum of the ν_{10} band of C_2H_4. *Bottom*, calculated spectrum of the ν_{10} band. The solid-lined spectrum was calculated assuming Coriolis-enhancement, the broken-lined spectrum assumed Coriolis-depletion, and the dotted-lined spectrum considered no coupling with the ν_7 transition. (Reproduced from Nakanaga et al., 1979, with permission from AIP.)

$$I_i = \int \alpha_\nu d\nu = \int \ln\frac{I_0}{I}d\nu = \frac{\pi N}{3c^2}\left(\frac{\partial\mu}{\partial Q_i}\right)^2, \tag{9.91}$$

where α_ν is molar absorption coefficient at frequency ν, n is the concentration in moles L^{-1}, l is pathlength, I_0 and I are intensities of incident and transmitted light, N is Avogadro's number, and c is the velocity of light. Beer's law plots are commonly prepared for each vibrational mode measured (Figure 9.10). In this way the numerical value and the sign of the Coriolis constant ζ_{rs} determined from the individual absolute intensities of the coupled bands ν_r and ν_s will tell whether the Coriolis perturbation is positive or negative.

9.9 ROTATIONAL BAND STRUCTURE IN VIBRATIONAL BANDS

The details of rotational structures appearing in vibrational spectra depend on the type of symmetry of the top and the direction of the transition dipole moment – whether parallel or perpendicular to the principal axis of rotation. The rotational spectral features for linear molecules are simple, but those for tops are complicated in most cases, requiring a detailed study for assignment. Nonetheless, a few general patterns identified are listed in Table 9.4.

We identify that the principal axis for a symmetric top is the one used for quantization of the quantum number K. Rotational transitions involving $\Delta K = \pm 2, \pm 3, \pm 4, \cdots$, albeit allowed are too weak for detection. As regard to the frequencies of the P and Q branches and those of the sub-bands therein, the rotational and vibrational term values are used. For a transition $|v''r''\rangle \rightarrow |v'r'\rangle$, the frequency is generally

$$\tilde{\nu} = F'_{v'}(J'K') - F''_{v''}(J''K'') + (G_{v'} - G_{v''}), \quad (9.92)$$

in which

$$F_v(JK) = B_v J(J+1) + (A_v - B_v)K^2. \quad (9.93)$$

This is the same expression for E_{rot} of oblate and prolate symmetric tops given by equation (7.51). The constants A_e and B_e used there for rotational constants are equivalent to A_v and B_v used here for vibrational. By convention, term values are written with the subscript 'v'. Using the term value expression in frequency unit, meaning E_{rot}/h rather than E_{rot}/hc, the frequencies of the sub-band ν_0 and the respective branches $\nu[P(J)]$, $\nu[Q(J)]$, and $\nu[R(J)]$ are obtained as

$$\nu_0^{(P)} = \nu_{\text{origin}} + (A'_v - B'_v) - 2(A'_v - B'_v)K + [(A'_v - B'_v) - (A''_v - B''_v)K^2$$

$$\nu_0^{(R)} = \nu_{\text{origin}} + (A'_v - B'_v) + 2(A'_v - B'_v)K + [(A'_v - B'_v) - (A''_v - B''_v)K^2$$

$$\nu[P(J)] = \nu_0 - (B'_v + B''_v)J + (B'_v + B''_v)J^2$$

$$\nu[Q(J)] = \nu_0 + (B'_v - B''_v)J + (B'_v + B''_v)J^2$$

$$\nu[R(J)] = \nu_0 + (B'_v + B''_v)(J+1) + (B'_v - B''_v)(J+1)^2. \quad (9.94)$$

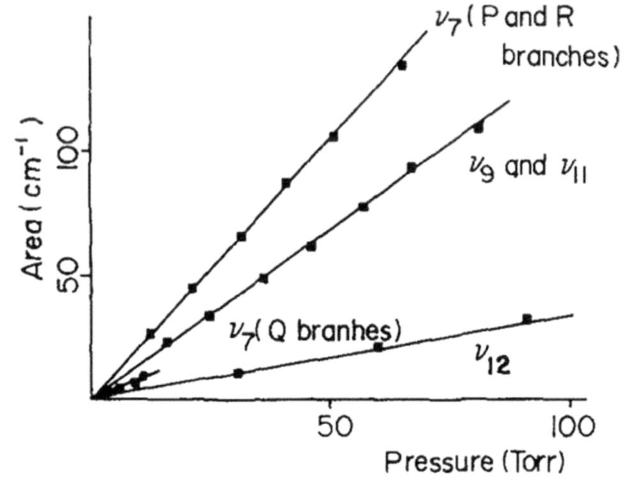

FIGURE 9.10 IR absorption of some bands vs ethylene pressure (concentration) showing the validity of Beer's law. (Reproduced from Nakanaga et al., 1979, with permission from AIP.)

Since the A_v and B_v terms are related to the rotational inertia, a variety of rotational structures, both apparent and hidden, are possible. Coriolis interactions may render the structures more complex. These are interesting structures, however, and they are amenable to investigation by ultrahigh-resolution spectrometers (resolution of 0.001 cm^{-1} or better) that use synchrotron beamline (Figure 9.11).

TABLE 9.4
General Characteristics of Rotational Structures in Vibrational Bands

Direction of μ	Band type	Selection rule	Rotational branches
Diatomic and linear polyatomic molecules			
along linear axis	∥	$\Delta J = \pm 1$	P, R
perpendicular to the linear axis	⊥	$\Delta J = 0, \pm 1$	P, Q, R
Symmetric top molecules			
along principal axis of rotation	∥	$\Delta K = 0,\ \Delta J = 0, \pm 1$	${}^QP_k(J),\ {}^QQ_k(J),\ {}^QR_k(J)$
perpendicular to principal axis of rotation	⊥	$\Delta K = \pm 1,\ \Delta J = 0, \pm 1$	${}^PP_k(J),\ {}^PQ_k(J),\ {}^PR_k(J)$ ${}^RP_k(J),\ {}^RQ_k(J),\ {}^RR_k(J)$
Asymmetric rotor-type molecules			
along principal axis a	A	$\Delta K^{(a)} = 0, \pm 2^{\dagger}$ $\Delta K^{(c)} = \pm 1, \pm 3$ $\Delta J = 0, \pm 1$	P, Q, R
along principal axis b	B	$\Delta K^{(a)} = \pm 1, \pm 3, \pm 5, \cdots$ $\Delta K^{(c)} = \pm 1, \pm 3, \pm 5, \cdots$ $\Delta J = 0, \pm 1$	P, Q, Q, R^*
along principal axis c	C	$\Delta K^{(a)} = \pm 1, \pm 3, \pm 5, \cdots$ $\Delta K^{(c)} = 0, \pm 2, \pm 4, \cdots$ $\Delta J = 0, \pm 1$	P, Q, R

Notes:
† The transitions corresponding to $\Delta K = \pm 2, \pm 3, \pm 4, \cdots$, although allowed, are too weak for detection
* The Q-band appears split due to a dip at the center frequency

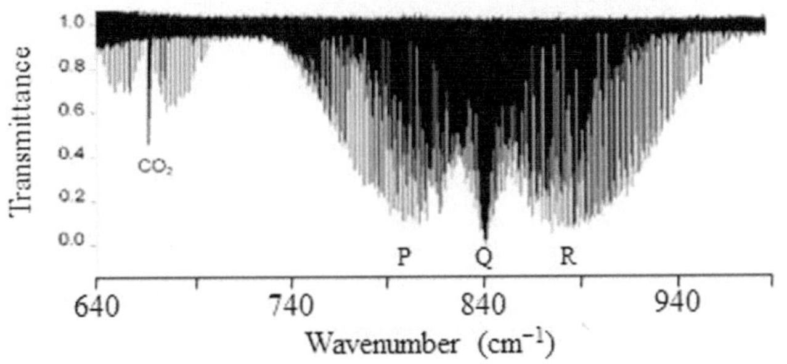

FIGURE 9.11 Synchrotron FTIR spectrum of the ν_7 band of cis-$C_2H_2D_2$ showing the rotational structures. The lines appearing between 640 and ~720 cm^{-1} are due to CO_2 impurity in the gas phase sample. (Adapted from Tan et al., 2016, with permission.)

9.10 SELECTION RULES FOR VIBRATIONAL TRANSITION

A vibrational transition would, but not necessarily, occur if the dipole moment changes with nuclear displacement during vibration because the intensity of infrared absorption is proportional to the dipole derivative (see equation 9.91). For the kth normal mode

$$I \propto \left| \left\langle \psi'' \left| \frac{\partial \mu}{\partial Q_k} \left(Q_k - Q_k^{\text{eq}} \right) \right| \psi' \right\rangle \right|^2 = \left(\frac{\partial \mu}{\partial Q_k} \right)^2 \left| \left\langle \psi'' \left| \left(Q_k - Q_k^{\text{eq}} \right) \right| \psi' \right\rangle \right|^2 . \tag{9.95}$$

Accordingly, the following two criteria should be fulfilled for observation of an infrared transition.

1. The dipole derivative should not vanish. Since $\mu = \mu_x + \mu_y + \mu_z$, at least one of the dipole components must have a non-zero value.
2. The matrix elements of $Q_k - Q_k^{\text{eq}}$ (normal coordinate displacement) must exist, for then only the integral does not vanish.

To check whether a $\psi'' \rightarrow \psi'$ transition is allowed, the integral needs to be evaluated on the basis of symmetry and group theoretic principles. For small linear molecules it is convenient to work with the symmetry of electronic states. A widely used textbook illustration is the infrared activity of CO_2 (D_h) whose vibrational modes (Figure 9.12) are symmetric –CO stretch (ν_1), asymmetric –CO stretch (ν_2), and in-plane and out-of-plane O–C–O bending (ν_3). The vibrational quantum number for all fundamentals is zero, and the ground-state molecular state is Σ_g^+. The symmetries associated with the ν_1, ν_2, and ν_3 vibrations are respectively Σ_g^+, Σ_u^+, and Π_u. The *x*, *y*, and *z* coordinates with respect to the molecular axes have Π_u, Π_u, and Σ_u^+ symmetry, respectively. We then determine the direct product of the symmetry species in the integral for each mode

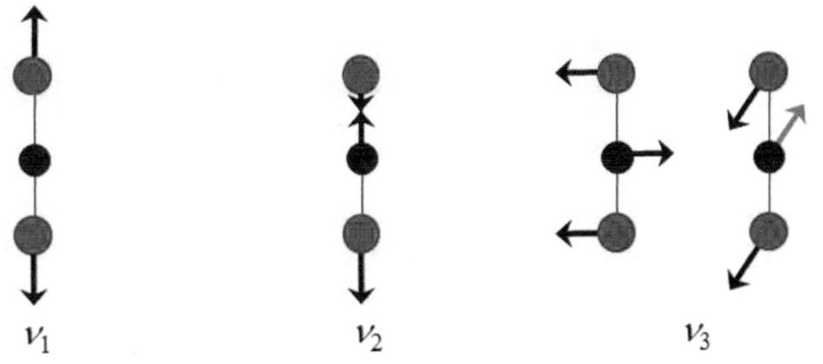

FIGURE 9.12 Vibrational modes of CO_2.

$$\nu_1 \quad \text{z-polarized}, \mu_z \quad \Sigma_g^+ \otimes \Sigma_u^+ \otimes \Sigma_g^+ = \Sigma_u^+ \otimes \Sigma_g^+ = \Sigma_u^+$$

$$\nu_2 \quad \text{z-polarized}, \mu_z \quad \Sigma_g^+ \otimes \Sigma_u^+ \otimes \Sigma_u^+ = \Sigma_u^+ \otimes \Sigma_u^+ = \Sigma_g^+$$

$$\nu_3 \quad \text{z-polarized}, \mu_z \quad \Sigma_g^+ \otimes \Sigma_u^+ \otimes \Pi_u = \Sigma_u^+ \otimes \Pi_u = \Sigma_g . \tag{9.96}$$

Because Σ_g^+ representation is obtained only for the ν_2 mode, this is the only IR-active transition for *z*-polarization. The same procedure can be applied to check for the activity of the modes for μ_x and μ_y components of the dipole. In this case, the Q_3 mode (ν_3) will also be IR-active because the bending vibrations occur along *x* and *y* coordinates, not along *z*. We will revisit the IR activity of CO_2 in the study of Raman activity. The use of a direct product table is particularly recommended for examining such integrals. For a glance, the table for the $D_{\infty h}$ point group relevant for the present discussion is reproduced in Table 9.5.

For relatively larger nonlinear molecules it is more convenient to use the character-based symmetry species Γ. If the two vibrational states are nondegenerate then the integrand should satisfy

$$\Gamma_{\psi''} \otimes \Gamma_{\mu} \otimes \Gamma_{\psi'} = A, \tag{9.97}$$

where *A* is a totally symmetric representation. This aspect is already discussed at length under 'Symmetry-Allowed Transitions' in section 6.7.3 in Chapter 6. If one or both vibrational states involved in the transition are degenerate then the direct product representation of the integral will involve more than one symmetry species. For example, representations of the kind

$$\begin{aligned} \Gamma_{\psi''} \Gamma_{\mu} \Gamma_{\psi'} &= A_1 \oplus A_2 \oplus E \\ &= A_1 \oplus A_2 \oplus B_1 \oplus B_2 \end{aligned} \tag{9.98}$$

are written as

$$\Gamma_{\psi''} \Gamma_{\mu} \Gamma_{\psi'} \subset A, \tag{9.99}$$

only to indicate that the integral representation contains a totally symmetric representation. The symmetry species of μ_x, μ_y, and μ_z correspond to the symmetry species of the *x*, *y*, and *z* translation operators, which can be read out at once from the

TABLE 9.5
Direct Product of Symmetries of Molecular States Based on Electron Angular Momentum

$D^{\infty}{}_h$	Σ_g^+	Σ_u^+	Σ_g^-	Σ_u^-	Π_g	Π_u	...	Operators
Σ_g^+	Σ_g^+	Σ_u^+	Σ_g^-	Σ_u^-	Π_g	Π_u	...	z^2, x^2+y^2
Σ_u^+	Σ_u^+	Σ_g^+	Σ_u^-	Σ_g^-	Π_u	Π_g	...	z
Σ_g^-	Σ_g^-	Σ_u^-	Σ_g^+	Σ_u^+	Π_g	Π_u	...	R_z
Σ_u^-	Σ_u^-	Σ_g^-	Σ_u^+	Σ_g^+	Π_u	Π_g	...	R_x, R_y, xz, yz
Π_g	Π_g	Π_u	Π_g	Π_u	$\Sigma_g^+ \oplus \Sigma_g^- \oplus \Delta_g$	$\Sigma_u^+ \oplus \Sigma_u^- \oplus \Delta_u$	...	xy
Π_u	Π_u	Π_g	Π_u	Π_g	$\Sigma_u^+ \oplus \Sigma_u^- \oplus \Delta_u$	$\Sigma_g^+ \oplus \Sigma_g^- \oplus \Delta_g$	...	x^2-y^2, xy
⋮								

This table is limited to $\Lambda = 0,1(\Sigma,\Pi)$; extension to higher molecular state symmetries can be done, and are also available in the literature.

appropriate character table. If the vibrational quantum number in the ground state wavefunction is zero ($\psi_{v''}=0$) then $\Gamma_{\psi''}=A$ already, so it is redundant to include this representation in the integrand. The integrand is shown simply as

$$\Gamma_{\mu} \otimes \Gamma_{\psi'} = A. \tag{9.100}$$

Box 9.1

Newton' second law of motion provides a fundamental equation to solve for dynamics in Cartesian coordinate system alone. Motion of bodies in other coordinate systems is solved by assumptions that are different from Newton's, although the assumptions in both cases are equivalent. An important postulate for studying dynamics by non-Newtonian mechanics states that

$$I = \int_{t_1}^{t_2} L dt$$

where $L = T - V$ is the Lagrangian. The Legrange's equation uses generalized coordinates q_i. Since the derivation of the vibrational energy involves transformation of Cartesian coordinates in which the problem is initially set up, it is the Lagrange's equation which is solved to find out the normal coordinates of vibration. In fact, all motional problems – be it particle motion in gravity or vibrations of polyatomic molecules – that involve coordinate transformations are analyzed by Lagrangian mechanics. A sufficient indication of when to apply Lagrangian mechanics is the necessity of using matrix methods to make changes of variables, which we have been doing to arrive at the classical vibrational energy.

Box 9.2

A conjugate momentum can be associated with each of the vibrational coordinates $Q_k\,(k = 1,2,3,,3N-6)$ because the molecular system possesses kinetic and potential energies. Denoting them by T and V, the momentum p_k conjugate to the normal coordinate Q_k is given by

$$p_k = \frac{\partial(T-V)}{\partial \dot{Q}_k},$$

in which $\dot{Q}_k$ is the velocity of the kth coordinate. It depends whether V is dependent or independent of $\dot{Q}_k$. If independent, then

$$p_k = \frac{\partial T}{\partial \dot{Q}_k}.$$

For simple systems like a particle moving in a constant potential, the conjugate momentum in the spherical coordinate (r,θ,ϕ) can be written as

$$p_r = \frac{\partial T}{\partial \dot{r}}$$

$$p_\theta = \frac{\partial T}{\partial \dot{\theta}}$$

$$p_\phi = \frac{\partial T}{\partial \dot{\phi}}$$

PROBLEMS

9.1 Consider the ball and spring model of a linear triatomic molecule A_3 of the atom A. Attach the terminal atoms to fixed surfaces connected by springs, and assume no bending motions. This description is an extension of what has been described with reference to Figure 9.1 in the main text. Repeat the two-ball exercise to derive the eigenvectors (normal modes) of vibration for A_3.

9.2 Draw the symmetric and antisymmetric stretching vibrations of cyclohexane. Also indicate the associated group representations (symmetry species). How would these vibrational modes change in the case of a ring distortion?

9.3 Make an independent effort to draw the normal modes of ethane indicating the associated symmetry species. Specify by inspection the infrared active modes, and explain why so.

9.4 Acetylene can exist in two different geometries as shown below

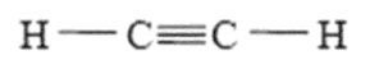
X-state

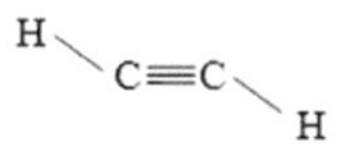
A-state

How do the normal modes of the A-state will differ from those of the X-state? Indicate the IR-active modes.

9.5 It has been said that the oscillator strength of vibrational overtone bands exponentially decrease with increasing quanta of excitation (see Figure 9.7). Show that this analysis can also be reproduced approximately by considering the anharmonicity-corrected vibrational energy levels in a Morse potential. The energy correction may be done up to the first anharmonicity constant according to the following

$$E_v = \omega_e\left(v+\frac{1}{2}\right) - x_e\omega_e\left(v+\frac{1}{2}\right)^2 + y_e\omega_e\left(v+\frac{1}{2}\right)^3 + \cdots,$$

where $x_e\omega_e$ and $y_e\omega_e$ are the first and second anharmonicity constants, respectively. The harmonic frequency ω_e is given by

$$\omega_e = \frac{\beta}{c}\sqrt{\frac{D_e}{2\pi^2\mu}}$$

in which μ is the reduced mass of the molecule, D_e is dissociation energy, and the parameter β is related to the force constant k by

$$\beta = \sqrt{k/2D_e}.$$

9.6 Could the measurement of overtone bands of different quanta of excitation be used to approximate the bond dissociation energy of a diatomic molecule?

9.7 Consider Fermi resonance interaction due to coupling of a fundamental transition at frequency v_1 with an overtone $2v_i$, where $i = 2,3,4,\cdots$. Show that Fermi shift increases as $\Delta = |2v_i - v_1|$ increases. An analytical or graphical presentation may be used to answer.

9.8 With reference to Coriolis-coupled modes Q_1 and Q_2, it has been mentioned that the excited state wavefunction of a vibrational transition belonging to one of the normal modes will be a combination of the functions of the basis state, i.e.,

$$|v_r^{'}\rangle = a|v_1r_1\rangle + b|v_2r_2\rangle$$

where r is vibrational quantum number, and a and b are mixing coefficients. Whether the magnitude of the transition dipole moment is positive (Coriolis-enhanced) or negative (Coriolis-depleted) depends on whether both a and b are positive or one of them is negative. What is the physical significance of a negative coefficient and what is the practical utility of the linear combination?

BIBLIOGRAPHY

Boyd, D. R. J. and H. C. Longuet-Higgins (1952) *Proc. Royal Soc. Ser A*. 213, 55.

Cotton, F. A. (1963) *Chemical Applications of Group Theory*, Interscience.

Duncan, J. L. and A. M. Ferguson (1988) *J. Chem. Phys.* 89, 4216.

Georges, R., M. Bach, and M. Herman (1999) *Mol. Phys.* 97, 279.

Kraihanzel, C. S. and R. West (1962) *J. Am. Chem. Soc.* 84, 3670.

Medvedev, E. S. (1985) *Chem. Phys. Lett.* 120, 173.

Nakanaga, T., S. Kondo, and S. Saëki (1979) *J. Chem. Phys.* 70, 2471.

Smith, W. L. and I. M. Mills (1964) *J. Chem. Phys.* 40, 2095.

Tan, T. L., L. L. Ng, and M. G. Gabona (2015) *J. Mol. Spectrosc.* 312, 6.

Tan, T. L., L. L. Ng, M. G. Gabona, G. Aruchunan, A. Wong, and D. R. T. Appadoo (2016) *J. Mol. Spectrosc.* 331, 23.

Wilson, E. B., J. C. Decius, and P. C. Cross (1955) *Molecular Vibrations*, McGraw-Hill.

10 Raman Spectroscopy

The Raman effect, which is essentially rooted in molecular vibrations, relies on monitoring the frequency of photons inelastically scattered rather than photons absorbed by molecules when irradiated by light in the ultraviolet to near-infrared regions, and since it is not a light absorption phenomenon there is no restriction that the molecules possess a permanent dipole moment. The different frequencies of incident and scattered photons imply the involvement of some molecular properties that are changed by the radiation, as a result of which the molecule excited from one of the ground vibrational states to a so-called virtual excited state emits by scattering to another vibrational state. It is the emissive decay of this virtual excited state to another ground vibrational state that causes a shift in the frequency of the scattered photons. Raman called this scattered light 'a new type of secondary radiation' only to emphasize on the light scattered by an oscillating dipole, the frequency of which is different from that of the incident radiation. A classical description of scattering is a good starting point in the discussion of Raman spectroscopy, although the quantum mechanical description, especially the perturbation approach, is inevitable when it comes to the resonance Raman effect. Several monographs and research reviews on the Raman effect are available; Long (2002), for example, has monographed Raman scattering in an excellent manner, describing all the principles and derivations thoroughly.

10.1 LIGHT SCATTERING

One of the material equations studied in Chapter 1 (equation 1.12) relates Maxwell's displacement vector **D** to the electric field vector **E** through ε, the dielectric constant or the permittivity of the medium

$$\mathbf{D} = \varepsilon \mathbf{E}. \tag{10.1}$$

But electrodynamics also connects the vector **D** to a molecular property called total polarization P of N molecules per unit volume

$$\mathbf{D} = \mathbf{E} + 4\pi P, \tag{10.2}$$

so that

$$\varepsilon = \frac{\mathbf{E} + 4\pi P}{\mathbf{E}}. \tag{10.3}$$

Polarizability is a measure of deformation of an isolated molecule when it is placed in an electric field, and the deformation occurs due to the attraction of the nuclei in the direction of the electric lines of force and the concomitant attraction of the electrons in the direction opposite. In the discussion of scattering here, we assume that the scattering molecules are much smaller than the wavelength of the incident light. The charge separation induces a temporal dipole moment in the molecule

$$\boldsymbol{\mu}(t) = \alpha \mathbf{E}(t), \tag{10.4}$$

where $\alpha = P / N$ is the molecular polarizability or deformability of the molecule that has the dimension of a volume. It is important to note the difference between the polarizability symbols P and α because both are used to describe light-induce dipole in a medium. The property P is the macroscopic polarization such that

$$P(t) = \chi E(t),$$

where χ is the electric or electronic susceptibility of molecules. The value of χ depends on internuclear distance. But each molecule is characterized by an induced dipole $\boldsymbol{\mu}(t)$, as equation (10.4) shows, and in which the polarizability α is the molecular polarizability. If the molecule already has a permanent dipole moment, then the induced dipole will be added up to or subtracted from the permanent dipole to give the total effective dipole. The creation of an induced dipole by the electric field of the incident radiation is just a generic phenomenon; it happens to all molecules. Thus, in principle, all molecules are amenable to Raman spectroscopy unlike infrared spectroscopy, which requires an existing permanent dipole of the molecule. The polarizability α is a scalar quantity as far as the molecule can be approximated to a sphere, but generally is a tensor ($\tilde{\boldsymbol{\alpha}}$) taken as an ellipsoid of polarization.

The origin of the tensor nature of polarizability is briefly described in Box 10.1. It is the matrix elements of $\tilde{\boldsymbol{\alpha}}$ that are of interest in Raman spectroscopy. The overhead squirrel denotes that it is generally a second-rank tensor. We will look at the tensor $\tilde{\boldsymbol{\alpha}}$ in detail later, but the classical picture of scattering should be developed first. The total molecular polarizability is related to the dielectric constant ε by equation (10.3). In a real isotropic sample, ε of different volume elements is different in terms of the position $\mathbf{r}$ and the time t, $\varepsilon(\mathbf{r},t)$, and the steady-state dielectric constant averaged over all volume elements of the sample will be

$$\langle \varepsilon \rangle \mathbf{I} = \varepsilon(\mathbf{r},t) - \delta\varepsilon(\mathbf{r},t), \tag{10.5}$$

DOI: 10.1201/9781003293064-10

where $\delta\varepsilon(\mathbf{r},t)$ is the dielectric fluctuation tensor, and $\mathbf{I}$ is a unit tensor of rank 2. The vector $\mathbf{r}$ that has three Cartesian components $x_1 = x$, $x_2 = y$, and $x_3 = z$, defines a displacement vector $u = \mathbf{r} - \mathbf{r}'$ where $\mathbf{r}'$ is the new position of the particle.

The importance of $\langle\varepsilon\rangle$ is that it is inversely related to the scattered electric field. To see this, consider the electric field of an incident beam $\mathbf{E}(\mathbf{r},t)$ as in equation (1.34),

$$\mathbf{E}(\mathbf{r},t) = \mathbf{E}_{0,i} e^{i(\mathbf{k}_i \mathbf{r} - \omega_i t)},$$

where $\mathbf{k}_i$ is the propagation vector, and let the field illuminate a volume element dv of the sample. We might even like to polarize the incident electric field by placing an x, y , or z polarizer appropriately in front of the sample, and analyze the scattered light by placing another appropriate polarizer called a Raman polarization analyzer in the optical path of the light to the detector (Figure 10.1). In fact, the isotropy and uncorrelated fluctuations of the small volume element dv will cause scattering in all directions, but for simplicity the figure shows scattering in one direction alone. The meaning of different symbols and vectors used is given in the legend to Figure 10.1.

scattered electric field $\mathbf{E}_s(R,t)$, which is a function of time t and the distance R between the scattering volume and the detector, is obtained by solving the Maxwell equation (equation 10.1), which relates the electric displacement to the electric field through ε of the isotropic material that scatters. The solution to the Maxwell equation starts with

$$\mathbf{D} = \mathbf{D}_i + \mathbf{D}_s = (\langle\varepsilon\rangle\mathbf{I} + \delta\varepsilon)(\mathbf{E}_i + \mathbf{E}_s)$$

$$\mathbf{D}_s = \langle\varepsilon\rangle\mathbf{E}_s + (\delta\varepsilon)\mathbf{E}_i, \qquad (10.6)$$

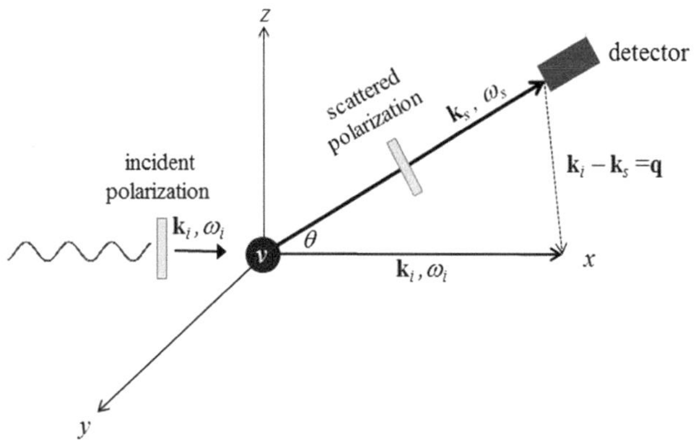

FIGURE 10.1 Scattering of an incident polarized light beam characterized by the propagation vector $\mathbf{k}_i$ and frequency ω_i. The propagation vector of the polarized scattered light is $\mathbf{k}_s$, and has a frequency ω_s, which may or may not be equal to ω_i. Other symbols used are v for a small sample volume, θ for a particular angle at which scattering is being monitored, and $\mathbf{q}$ for the scattering wave vector. Since $\mathbf{q} = \mathbf{k}_i - \mathbf{k}_s$, the amplitude of scattering will be $|\mathbf{q}|^2 = q^2 = (\mathbf{k}_i - \mathbf{k}_s)^2 = k_i^2 + k_s^2 + 2\mathbf{k}_i\mathbf{k}_s$.

Source: This figure is a modified version of the figure found in Berne and Pecora (2000).

where $\mathbf{D}_i$ and $\mathbf{D}_s$ are displacement vectors for the incident and scattered electric fields $\mathbf{E}_i$ and $\mathbf{E}_s$, respectively, and the average dielectric constant of the scattering volume is as given by equation (10.5). The solution for the scattered field has been worked out by Berne and Pecora (2000), and appears as

$$\mathbf{E}_s(R,t) = \frac{-k_s^2 E_{0,i}}{4\pi r\langle\varepsilon\rangle} e^{i(\mathbf{k}_s R - \omega_i t)\delta\varepsilon_{is}(\mathbf{q},t)}, \qquad (10.7)$$

where $\mathbf{k}_s$ is the scattered wave vector. The exponential also contains a component of the dielectric constant fluctuation tensor along the incident and the scattering polarization directions

$$\delta\varepsilon_{is}(\mathbf{q},t) \rightarrow n_s \delta\varepsilon(\mathbf{q},t) n_i.$$

One notices that the component $\delta\varepsilon_{is}(\mathbf{q},t)$ is a function of the difference of the incident and the scattered wave vector, the wave vector being $\mathbf{q} = \mathbf{k}_i - \mathbf{k}_s$ (Figure 10.1).

The temporal fluctuations are analyzed by the autocorrelation function

$$G(\tau) = \langle E_s(R,0) E_s(R,t)\rangle = \frac{k_s^4 |E_{i,0}|^2}{16\pi^2 r^2 \langle\varepsilon\rangle^2} \langle\delta\varepsilon_{is}(\mathbf{q},t)\rangle e^{-i\omega_i t}. \qquad (10.8)$$

Since the autocorrelation function is the time-domain function corresponding to the spectral density in the frequency domain, and the two functions are connected by Fourier transformation, one can write

$$I_E(\omega) = \frac{1}{2\pi}\int_{-\infty}^{+\infty} G(\tau) e^{-i\omega t} d\tau$$

$$I_{is}(\mathbf{q},\omega_s,\mathbf{R}) = \frac{|E_{I,0}|^2}{16\pi^2 R^2 \langle\varepsilon\rangle^2 \lambda^4} \qquad (10.9)$$

$$\frac{1}{2\pi}\int_{-\infty}^{+\infty} \langle\delta\varepsilon_{is}(\mathbf{q},0)\,\delta\varepsilon_{is}(\mathbf{q},0)\rangle e^{-i\Delta\omega t} dt,$$

in which $\Delta\omega(= \omega_i - \omega_s)$ is the detuning of incident and scattered frequencies. Remember, $\mathbf{R}$ in all of these expressions determines the distance between the scattering volume element and the detector. We also made the substitution of λ^{-4} for k_s^4 in the expression above. The integrated part of this equation implies that the spectral density is a function of $\Delta\omega$. More importantly, the front factor shows that the molecules scatter with an intensity proportional to λ^{-4}, implying that shorter-wavelength radiation is scattered more (Raman's question as to why the color of the sky is blue). The R^{-2} dependence of scattered intensity arises from the fact that scattering occurs in all directions, so the scattered waves can be considered spherical electromagnetic waves, but we make a plane wave approximation at the position of the detector or an observer standing at a distance $\mathbf{R}$ away from the scatterer. This explanation is already given in Chapter 1 (Figure 1.6).

10.2 FREQUENCIES OF RAYLEIGH AND RAMAN-SCATTERED LIGHT

Based on the energy of the scattered photons, scattering could be elastic where the incident and scattered lights have the same frequency, called Rayleigh scattering, or inelastic, in which the frequency of scattered light is different from the incident. The latter type occurs in Raman and Brillouin scattering. While Rayleigh and Raman processes arise from oscillation of the light induced dipole moment that acts as the secondary source of radiation, the Brillouin process is quite different and arises not from induced dipole, but from 'liquid motions', which are low-frequency motions of intermolecular distance fluctuations that can create or annihilate a photon, thereby shifting the frequency of the scattered light relative to that of the incident.

Rayleigh and Raman scattering occurs because the molecule acquires an induced dipole that oscillates at the frequency of the incident light. In the Raman process, however, the amplitude of the oscillating dipole is modulated by rotational and vibrational motions of the molecule. Vibrational frequencies are roughly four–six-fold smaller than the frequency of the incident light, which implies that the vibration of a diatomic molecule, for example, will modulate the oscillating wave at the vibrational frequency ω_v as shown in Figure 10.2. Fourier transformation of the time-domain function will produce three frequencies – a center Rayleigh line (ω) flanked by anti-Stokes ($\omega + \omega_v$) and Stokes ($\omega - \omega_v$) Raman lines.

One may like to look at the scattering frequencies by expanding equation (10.4) so as to account for both incident radiation frequency and allowed molecular vibrational frequencies. The induced dipole oscillates with time because the radiation field is time-dependent

$$\begin{aligned}\boldsymbol{\mu}(t) &= \tilde{\boldsymbol{\alpha}}\,\mathbf{E}(t)\\ &= \tilde{\boldsymbol{\alpha}}\,\mathbf{E}_o(t)\cos\omega t.\end{aligned} \qquad (10.10)$$

The molecular polarizability $\tilde{\boldsymbol{\alpha}}$ oscillates at the frequency of the incident light α_o as well as at all frequencies of periodic

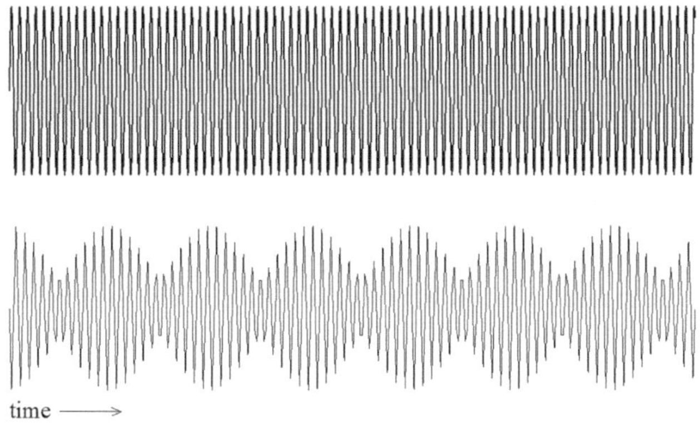

FIGURE 10.2 *Top*, oscillation of the dipole at the frequency of the incident light which alone gives rise to Rayleigh scattering. *Bottom*, vibrational frequency-modulated oscillation of the dipole which gives rise to scattering in addition to Rayleigh's.

molecular vibrations α_k, k being the index for the number of fundamental vibrations of the molecule. Expansion of the polarizability in a Taylor series about the equilibrium internuclear geometry is

$$\alpha = \alpha_o + \Sigma_k \alpha_k \cos\left(\omega_k t + \phi_k\right), \qquad (10.11)$$

where ϕ_k is a phase constant for the kth vibrational oscillation, even though ϕ_k is undetermined. With this expression for α, equation (10.10) is written as

$$\begin{aligned}\boldsymbol{\mu}(t) &= \alpha_o \mathbf{E}_o(t)\cos\omega t + \boldsymbol{\Sigma}_k \alpha_k \cos(\omega_k t + \phi_k)\mathbf{E}_o(t)\cos\omega t\\ &= \alpha_o \mathbf{E}_o(t)\cos\omega t + \frac{\mathbf{E}_o(t)}{2}\Sigma_k \alpha_k\\ &\quad \left[\cos\left\{\left(\omega + \omega_k\right)t + \phi_k\right\} + \cos\left\{\left(\omega - \omega_k\right)t + \phi_k\right\}\right]\end{aligned} \qquad (10.12)$$

showing that $\boldsymbol{\mu}(t)$ has several oscillatory components in addition to the component oscillating at the frequency of the incident light. The meaning of 'several' should be understood in terms of the variable k. If there are n (=3) fundamental vibrations ($k = 1, 2, 3$) then the induced dipole will oscillate at seven different frequencies, $2n + 1 (= 7)$, one at the frequency of the incident light ω and others at $\omega \pm \omega_k$. An oscillating electric dipole acts as a source of secondary radiation and emits light. This emitted radiation is what we have been calling the scattered wave. The scattered radiation may overlap with fluorescence, but it is not fluorescence. We see that scattering is related to emission of radiation, but is distinct from the conventional fluorescence preceded by vibrational relaxation.

The intensity of the scattered radiation into 4π steradians is

$$I = \frac{2\left|\mu(t)\right|^2}{3c^2} \equiv \frac{\omega^4}{12\pi\varepsilon_0 c^3}\left|\mu(t)\right|^2 \qquad (10.13)$$

where ε_o is free-space permittivity. Substitution for $\boldsymbol{\mu}(t)$ from equation (10.12) yields

$$\begin{aligned} I \equiv{}& \frac{\omega^4}{12\pi\varepsilon_o c^3}\alpha_o^2\left|\mathbf{E}_o(t)\right|^2\cos^2\omega t + \frac{1}{24\pi\varepsilon_0 c^3}\sum\nolimits_k \alpha_k^2\left|\mathbf{E}_o(t)\right|^2\\ &\left(\omega + \omega_k\right)^4\left[\cos^2\left\{\left(\omega + \omega_k\right)t + \phi_k\right\}\right] + \frac{1}{24\pi\varepsilon_o c^3}\sum\nolimits_k \alpha_k^2\left|\mathbf{E}_o(t)\right|^2\\ &\left(\omega - \omega_k\right)^4\left[\cos^2\left\{\left(\omega - \omega_k\right)t + \phi_k\right\}\right] + \cdots\end{aligned} \qquad (10.14)$$

The first term corresponds to Rayleigh scattering, and the second and third terms give rise to anti-Stokes and Stokes Raman scattering, respectively. The appearance of ω^4, $(\omega + \omega_k)^4$, and $(\omega - \omega_k)^4$ indicates λ^4 dependence of scattering intensity. For a typical incident wavelength of 500 nm (~2×10^4 cm^{-1}) and a C–H stretching resonance at 3000 nm (3.334×10^4 cm^{-1}), the intensity of Rayleigh scattering will be ~3 orders of magnitude stronger than Raman's, and the frequencies of anti-Stokes ($\omega + \omega_k$) and Stokes ($\omega - \omega_k$) Raman scattering will be respectively higher and lower than Rayleigh (ω) scattering. In real experimental procedures, the high intensity Rayleigh-scattered light is forced to damp out by using a filter device, typically a holographic filter, placed before the detector. Strong attenuation of the Rayleigh line helps detecting the low intensity Raman-scattered emission. It is also important that the damped Rayleigh band is carefully narrowed down such that the rotational Raman lines that closely flank the Rayleigh line can be observed.

Having mentioned about rotational Raman scattering above we should discuss the origin of this. The molecular rotational effect in the Raman process is best described by an experimental setup in which a z-polarized incident light (laboratory Cartesian frame) is used to excite a rigid rotating molecule. The incident and scattered lights are analyzed in the same laboratory frame of x,y,z axes system. Let a molecule with a set of internal coordinates ζ,η,ξ be placed at the origin of the lab frame such that the angle between z and ζ axes is given by θ (Figure 10.3). To a first approximation, we will consider θ to describe the rotation of the molecule. When the molecule is excited by a z-polarized light propagating from the $+x$ direction, the amplitudes of the ζ- and ξ components of the electric field are given by (see Box 7.1)

$$\mathbf{E}_\zeta(t) = -\mathbf{E}_z(t)\sin\theta$$

$$\mathbf{E}_\xi(t) = \mathbf{E}_z(t)\cos\theta, \tag{10.15}$$

with which one can write for the induced dipole

$$\boldsymbol{\mu}_\zeta(t) = -\alpha_\perp \mathbf{E}_z(t)\sin\theta$$

$$\boldsymbol{\mu}_\xi(t) = \alpha_\parallel \mathbf{E}_z(t)\sin\theta\cos\theta, \tag{10.16}$$

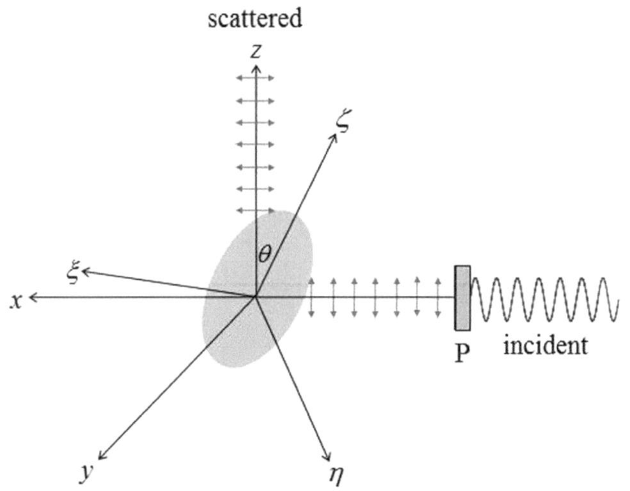

FIGURE 10.3 A typical experimental setup where a z-polarized beam excites the molecule and the induced dipole is analyzed as a function of the molecular rotation angle θ. The x,y,z coordinates represent the lab-fixed frame and ζ,η,ξ represent molecular internal coordinates.

where the tensor property of the molecular polarizability $\tilde{\alpha}$ is invoked. The polarizability ellipsoid is approximated with axial symmetry so that $\alpha_\parallel = \alpha_{zz}$ and $\alpha_\perp = \alpha_{xx} = \alpha_{yy}$. Then the z-component of the dipole in terms of ζ- and ξ components of the induced dipole can be written as

$$\begin{aligned}\boldsymbol{\mu}_z(t) &= -\boldsymbol{\mu}_\zeta(t)\sin\theta + \boldsymbol{\mu}_\xi(t)\cos\theta \\ &= \alpha_\perp \mathbf{E}_z(t)\sin^2\theta + \alpha_\parallel \mathbf{E}_z(t)\cos^2\theta\end{aligned} \tag{10.17}$$

Since the description of the molecular rotation involves θ alone, it can be expressed in terms of the rotational frequency and time

$$\theta(t) = \omega_{\text{rot}} t. \tag{10.18}$$

Using the power relations of the following identities

$$\sin^2\theta = \frac{1}{2}(1-\cos 2\theta)$$

$$\cos^2\theta = \frac{1}{2}(1+\cos 2\theta),$$

the expression for $\boldsymbol{\mu}_z(t)$ can be expanded. For convenience we will use below linear frequency ν_{rot} instead of the angular frequency ω_{rot},

$$\begin{aligned}\boldsymbol{\mu}_z(t) &= \mathbf{E}_z(t)\left[\frac{\alpha_\parallel}{2}\{1+\cos(2\nu_{\text{rot}})2\pi t\} + \frac{\alpha_\perp}{2}\{1+\cos(2\nu_{\text{rot}})2\pi t\}\right] \\ &= \mathbf{E}_z(t)\left(\frac{\alpha_\perp + \alpha_\parallel}{2}\right) + \mathbf{E}_z(t)\left(\frac{\alpha_\parallel - \alpha_\perp}{2}\right)\cos(2\nu_{\text{rot}})2\pi t.\end{aligned} \tag{10.19}$$

Inserting $\mathbf{E}_z(t) = \mathbf{E}_o(t)\cos 2\pi\nu t$, and using the identity

$$\cos A\cos B = \frac{1}{2}\left[\cos(A+B) + \cos(A-B)\right],$$

one obtains

$$\boldsymbol{\mu}_z(t) = \left(\frac{\alpha_\parallel + \alpha_\perp}{2}\right)\mathbf{E}_o(t)\cos 2\pi\nu t + \left(\frac{\alpha_\parallel - \alpha_\perp}{2}\right)$$

$$\left[\frac{\mathbf{E}_o(t)}{2}\cos 2\pi t(\nu + 2\nu_{\text{rot}}) + \frac{\mathbf{E}_o(t)}{2}\cos 2\pi t(\nu - 2\nu_{\text{rot}})\right]. \tag{10.20}$$

This is the z-component of the oscillating dipole, which will produce scattered light in the z-direction in the form of Rayleigh scattering that has the same frequency ν as that of the exciting radiation, and Raman-scattered light at frequencies $\nu + 2\nu_{\text{rot}}$ (anti-Stokes) and $\nu - 2\nu_{\text{rot}}$ (Stokes). Note also that the Raman-scattered frequency is shifted on either side of the excitation frequency by two-times the rotational frequency here, but only by a unit vibrational frequency as seen earlier (equations (10.12) and (10.14)). The reason for twice the rotational frequency shift is that the molecular polarizability changes only in the direction of the excitation field, not in the opposite direction, hence the polarizability changes at twice the rotational frequency.

10.3 LIMITATION OF THE CLASSICAL THEORY OF RAMAN SCATTERING

Although the classical model of induced dipole moment and the polarizability oscillation provides a qualitative description of Raman scattering, it does not explain the pattern of scattered intensity distribution and the appearance of rotational Raman bands closer to the frequency of excitation light. In fact, the pure rotational Raman bands are difficult to observe in a low-resolution spectrum because of the appearance of the rotational-Raman lines closer to the Rayleigh's, albeit the latter can be filtered out efficiently. The classical theory also does not describe the absence of anti-Stokes lines in diatomic molecules. Further, the correlation of Stokes–anti-Stokes intensity ratio to the frequency ratio, given by

$$\frac{I(\text{Stokes})}{I(\text{anti}-\text{Stokes})}=\frac{\left(\nu-\nu_k\right)^4}{\left(\nu+\nu_k\right)^4} \tag{10.21}$$

is not explained by the classical theory.

10.4 BRILLOUIN SCATTERING

Before we leave the classical description of scattering, we wish to consider the basic concept of Brillouin scattering, an inelastic process that monitors reflection of the incident light rather than the incident light causing oscillation of the induced dipole. We may like to think about the scattering medium as stacked layers, each layer having a periodic variation in the local dielectric constant (ε). The rippled layer is akin to a slow-moving wave with a periodicity Λ (Figure 10.4) like an acoustic wave. This model reproduces a Bragg-like condition

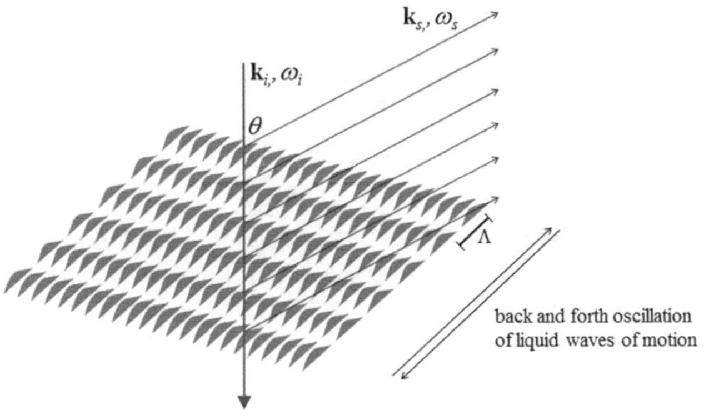

FIGURE 10.4 A schematic representation of Brillouin scattering. The incident beam is reflectively scattered by each layer of the medium. The layers are different in terms of the local dielectric constant and vary with a periodicity Λ. As indicated, the same periodicity is maintained as the wave oscillates.

in the sense that the incident light with the propagation vector $\mathbf{k}_i$ is multiply reflected by the layers of periodic ε. We also recognize that the typical velocity of 'liquid motion' or low-frequency fluctuations of the medium is of the order of $\sim 10^3$ m s^{-1}, which is much slower than the velocity of the incident light ($c \approx 3\times10^8$ m s^1), so the periodic wave motion of the medium is essentially static. We assume that the reflective scattering is quasi-elastic. Then

$$\mathbf{k}_i \sim \mathbf{k}_s = \frac{2\pi\sqrt{<\varepsilon>}}{\lambda}, \tag{10.22}$$

where λ is the wavelength of the incident light that allows for expression of the length of the propagation vector in terms of $\langle\varepsilon\rangle$ and λ. Since the scattering wave vector $\mathbf{q}=\mathbf{k}_i-\mathbf{k}_s$ and $\mathbf{k}_i \sim \mathbf{k}_s$ as assumed, we have

$$\mathbf{q}=2k_i\sin\frac{\theta}{2}=\frac{4\pi\sqrt{\langle\varepsilon\rangle}}{\lambda}\sin\frac{\theta}{2}, \tag{10.23}$$

where we have used the cosine rule of scattering, which is admissible if $\mathbf{k}_i \sim \mathbf{k}_s$.

Now the multiply reflected (scattered) light, all with the same scattering angle θ, will interfere constructively or destructively. If constructive, then

$$2\sqrt{\langle\varepsilon\rangle}\Lambda\sin\frac{\theta}{2}=\lambda. \tag{10.24}$$

Combining equations (10.23) and (10.24) we get

$$\mathbf{q}=\mathbf{k}=\frac{2\pi}{\Lambda}=\frac{4\pi\sqrt{\langle\varepsilon\rangle}}{\lambda}\sin\frac{\theta}{2}, \tag{10.25}$$

where we have used the assumption $\mathbf{q}=\mathbf{k}=\mathbf{k}_i \sim \mathbf{k}_s$ made already. Keep in mind that $\mathbf{q}$ is the scattering wave vector and $\mathbf{k}$ is the wave vector of the wave traveling in the medium.

The identity $\mathbf{q}=\mathbf{k}$ obtained above is central to the understanding of Brillouin scattering, because the condition of conservation of momentum is obtained only if $\mathbf{q}=\mathbf{k}$. The argument is similar to what we discussed earlier in Chapter 3, section 3.10.3 on the Doppler effect. In the present context, the total of the momentum – that of the traveling wave of the medium and the probing photon, is obtained from the principle of the conservation of momentum. The total momentum of the traveling wave of the medium is derivable starting with Maxwell's equations. If the wave of the medium is akin to small-amplitude disturbances, often called an acoustic limit, then the total momentum of the wave is

$$P=\frac{E_0}{c_0}, \tag{10.26}$$

in which E_0 is the energy of the traveling wave of the medium and c_0 is the velocity of the wave. This energy is associated with the density of the undisturbed medium, but we notice that the total momentum of the wave originating in the medium has the same form that an electromagnetic radiation has ($h\nu/c$, where c is the velocity of light). The momentum P points to the direction in which the wave travels.

If the scattered light has a different frequency than that of the incident, then the momentum conservation condition tells us that the interaction of the incident light with the medium is inelastic, and because the frequency shift of the scattered light is not due to electronic dipole oscillation, it must be the result of creation or annihilation of a phonon of the scattering medium. It should be taken to mean that the formation and the decay of phonons are due to light-medium interactions. In fact, collective excitation and deexcitation of array of atoms (particles) in the medium create and annihilate phonons perpetually. A phonon is created when the wave travels in one direction, say $+x$, such that the total angular momentum has a positive sign, and a phonon is annihilated when the wave travels in the reverse direction, in which case the sign of the angular momentum is reversed too. When a phonon is created, momentum is transferred from the incident light to the medium – this is Stokes Brillouin scattering – and when phonon is annihilated the momentum of the scattered photon is higher, which is anti-Stokes Brillouin scattering. In terms of the exchange of energy, it is the phonon frequency that increases (Stokes Brillouin) or decreases (anti-Stokes Brillouin). Since the acoustic wave is traveling toward and then away from the scattering detector, the detected frequency undergoes a Doppler shift given by

$$\omega_s = \omega_i \pm c_o 4\pi\sqrt{\langle\varepsilon\rangle}\sin\frac{\theta}{2}, \tag{10.27}$$

in which c_o denotes the velocity of the acoustic wave (the symbol c reserved for the velocity of light) and θ is the scattering angle (Figure 10.4). The '+' and '–' signs in the formula above represent respectively the wave traveling toward and away from the detector. Much of the Brillouin scattering studies in recent times has focused on temperature and density, and thermodynamic properties such as adiabatic compressibility of condensed phase systems, and interested readers may consult advanced literature (e.g., Damzen et al., 2003).

The frequency shifts in Brillouin and Raman scattering, both with reference to the Rayleigh frequency (incident light frequency) are indicated in Figure 10.5. Just a passing note is that elastic scattering of light by spherical particles of any size, although limited to geometric optics, is also treated by Mie scattering theory, the formulation of whose functions is rather complex and beyond the scope of this study.

10.5 RAMAN TENSOR

At the heart of the Raman scattering process is the effect of normal vibrations on the molecular polarizability tensor. The frequencies of nuclear vibrations are coded in the tensor and so in the oscillating dipole, and these frequencies become evident in the light scattered by the dipole. A good understanding of the role of the tensor is crucial to the discussion of Raman scattering.

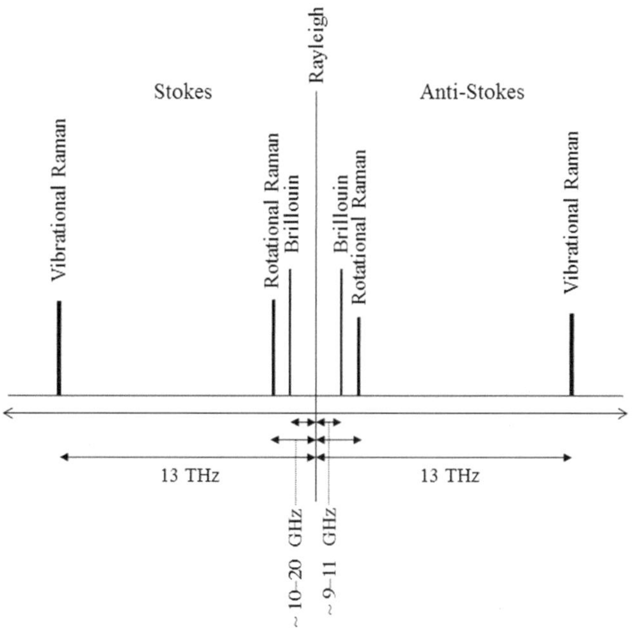

FIGURE 10.5 Approximate frequency shifts of Brillouin- and Raman-scattered lights with reference to the frequency of the incident light (Rayleigh frequency).

We consider a space-fixed molecule that is prevented from rotating, but vibrates freely. The idea of choosing a rotationally frozen molecule is the lack of a clear understanding of molecular rotational frequencies that could modulate the polarizability tensor. Further, as will be seen below, neglecting molecular rotation allows for working with a diagonal tensor. We start with the basic equation for the induced dipole in a molecule

$$\boldsymbol{\mu}(t) = \tilde{\alpha}\,\mathbf{E}(t)$$

where $\tilde{\alpha}$ is the polarizability tensor of rank 2. Each of the nine components of $\tilde{\alpha}$ is individually modulated by molecular vibrations. For example, in the case of a diatomic molecule that has only one normal mode of vibration we expand each component α_{is} $(i,s = x,y,z)$ in a Taylor series about the equilibrium nuclear configuration R_e

$$\alpha_{is} = (\alpha_{is})_o + (R - R_e)\left[\frac{\partial \alpha_{is}}{\partial R}\right]_o + \frac{(R-R_e)^2}{2!}\left[\frac{\partial^2 \alpha_{is}}{\partial R^2}\right]_o + \cdots, \tag{10.28}$$

in which the subscript 'o' refers to $R = R_e$.

For a polyatomic molecule having normal coordinate of vibration $Q_j, Q_k, Q_l, \cdots$, the expansion of a component of the molecular polarizability tensor is

$$\alpha_{is}(Q) = (\alpha_{is})_0 + \sum_j \left[\frac{\partial \alpha_{is}}{\partial Q_j}\right]_0 Q_j + \frac{1}{2!}\sum_{j,k}\left[\frac{\partial^2 \alpha_{is}}{\partial Q_j \partial Q_k}\right]_0 Q_j Q_k + \frac{1}{3!}\sum_{j,k,l}\left[\frac{\partial^3 \alpha_{is}}{\partial Q_j \partial Q_k \partial Q_l}\right]_0 Q_j Q_k Q_l + \cdots. \tag{10.29}$$

One should be clear about the form of this expansion. The tensor component being expanded is α_{is}. The first term is a Rayleigh tensor component, which does not relate to nuclear coordinates. The second term sums up the derivative of α_{is} with respect to all normal coordinates Q_j. This term represents the derived α_{is} for all modes ν_j. The third term represents the sum of the dipole derivatives of α_{is} with respect to normal coordinates Q_j and Q_k. The expansion can be carried out for each component of the α_{is} tensor, and each expansion could be approximated by retaining only the first two terms. The neglect of terms containing the second and higher powers of Q is permissible in the electrical harmonic approximation which assumes that the change in the polarizability is proportional to the first power of vibrational modes. This is the counterpart of the mechanical harmonic approximation in molecular vibrations. When electrical anharmonicity is not assumed or invalid, then the third term in equation (10.29) has to be considered. We learned earlier that when electrical anharmonicity is allowed, then the overtone and combination bands can be accounted for. Indeed, first overtones and combination overtones are sometimes observed in Raman spectra, but not frequently, and are weak when they do appear. This is one reason why the Raman peaks appear narrower and generally fewer in number than the counterpart infrared absorption spectrum which does consist of many overtone and combination modes. The exception is when resonance occurs in Raman excitation and one can observe many overtone and combination modes coupled to the electronic transition. This aspect will be discussed in some detail later. In the electrical harmonic limit, each vibrational mode will generate nine expansion equations, one for each of the nine components of the second rank polarizability tensor $\tilde{\boldsymbol{\alpha}}$. Thus, the expansion in equation (10.29) repeated nine times will give a new tensor for the polarizability of the molecule in the jth vibrational mode, call it $\tilde{\boldsymbol{\alpha}}^{(j)}$. This new tensor is the sum of two tensors – the equilibrium polarizability tensor, called the Rayleigh tensor $\tilde{\boldsymbol{\alpha}}_0$, and the derived tensor $\tilde{\boldsymbol{\alpha}}'^{(j)}$, called the Raman tensor for the jth mode of vibration

$$\tilde{\boldsymbol{\alpha}}^{(j)} = \tilde{\boldsymbol{\alpha}}_0 + \tilde{\boldsymbol{\alpha}}'^{(j)} Q_j. \tag{10.30}$$

Using the same procedure, the polarizability tensors for the other modes can be found

$$\begin{aligned} \tilde{\boldsymbol{\alpha}}^{(k)} &= \tilde{\boldsymbol{\alpha}}_0 + \tilde{\boldsymbol{\alpha}}'^{(k)} Q_k \\ \tilde{\boldsymbol{\alpha}}^{(l)} &= \tilde{\boldsymbol{\alpha}}_0 + \tilde{\boldsymbol{\alpha}}'^{(l)} Q_l \\ &\vdots \end{aligned} \tag{10.31}$$

It is essential to recognize that the polarizability tensor for each vibrational mode is different because the Raman tensor for each mode must be distinct. Since the Rayleigh tensor $\tilde{\boldsymbol{\alpha}}_0$ is common to polarizabilities of all modes, the Rayleigh line will always appear at the frequency of the incident light ω irrespective of which vibrational mode one is analyzing, but Raman lines will appear at frequencies $\omega \pm \omega_j$, $\omega \pm \omega_k$, $\omega \pm \omega_l$, and so on, where ω_j, ω_k, $\omega_l \cdots$ are vibrational frequencies associated with Q_j, Q_k, $Q_l \cdots$ normal coordinates of vibration.

If the Raman tensor for a mode vanishes, then that mode will not be Raman active. Whether a mode is Raman active or not is driven by selection rule, which will appear in some way in later discussions.

The Raman tensor can be analyzed as a second- and higher-rank tensors, although the latter are discussed in advanced treatise only. The idea for a general treatment of the Raman tensor at the second-rank level is based on rapid series convergence of the induced dipole. The induced dipole can be expanded in a series

$$\begin{aligned} \boldsymbol{\mu} &= \boldsymbol{\mu}_1 + \boldsymbol{\mu}_2 + \boldsymbol{\mu}_3 + \cdots \\ &= c_1 \tilde{\boldsymbol{\alpha}} \mathbf{E} + c_2 \tilde{\tilde{\boldsymbol{\alpha}}} \mathbf{E}^2 + c_3 \tilde{\tilde{\tilde{\boldsymbol{\alpha}}}} \mathbf{E}^3 + \cdots \end{aligned} \tag{10.32}$$

The $\boldsymbol{\alpha}$s are polarizability $\tilde{\boldsymbol{\alpha}}$, hyperpolarizability $\tilde{\tilde{\boldsymbol{\alpha}}}$, and second-hyperpolarizability $\tilde{\tilde{\tilde{\boldsymbol{\alpha}}}}$ tensors of rank 2, 3, and 4, respectively. Because the series for $\boldsymbol{\mu}$ converges rapidly, only the first term can be retained. Hence the Rayleigh and Raman tensors are generic second-rank tensors (or just polarizability tensors).

10.5.1 Polarizability Tensor Ellipsoid

For spherically symmetric or isotropic systems like an atom, the polarizability α is a number – a scalar quantity, because the electron cloud in such systems is isotropic, which produces polarization in all directions. These systems are mostly diamagnetic because the electron cloud belonging to the atoms is not delocalized. These also include gas molecules, the rapid rotation of which averages out the polarization, and the result is a mean polarizability $\langle \alpha \rangle$. For non-spherical systems, however, where the electron cloud is anisotropic, the polarizability is directional and hence is a tensor. Let the tensor be symmetric

$$\begin{pmatrix} \alpha_{xx} & \alpha_{xy} & \alpha_{xz} \\ \alpha_{yx} & \alpha_{yy} & \alpha_{yz} \\ \alpha_{zx} & \alpha_{zy} & \alpha_{zz} \end{pmatrix} \tag{10.33}$$

such that the tensor components straddling the diagonal are equal, $\alpha_{xy} = \alpha_{yx}$, $\alpha_{xz} = \alpha_{zx}$, and $\alpha_{yz} = \alpha_{zy}$. Thus, we have six independent components, and they enter into the equation of the ellipsoid given by

$$\alpha_{xx}x^2 + \alpha_{yy}y^2 + \alpha_{zz}z^2 + 2\alpha_{xy}xy + 2\alpha_{yz}yz + 2\alpha_{zx}zx = 1. \tag{10.34}$$

Since the coefficients belong to the polarizability tensor, the ellipsoid is called the polarizability ellipsoid, which represents the tensor. The principal axes of the polarizability ellipsoid are a, b, c, which need not necessarily be aligned with the space-fixed frame x,y,z. In fact, equation (10.34) is applicable for the situation in which the principal axes of the polarizability ellipsoid and the laboratory (space) frame axes are not aligned, even though the center of the ellipsoid is at the origin of the x ,y,z frame. The significance of the polarizability ellipsoid lies in the fact that if an electric field $\mathbf{E}$ is applied along the direction of one of the principal axes, a, b, or c, the polarizability will also be along the direction of $\mathbf{E}$. This means, if the electric vector of the incident light is along, say, the *a* axis then the induced dipole moment will also vibrate along the *a*-axis. The same stipulation works with respect to *b* and *c* axes,

$$\mu_a = \alpha_a \mathbf{E}_a$$

$$\mu_b = \alpha_b \mathbf{E}_b$$

$$\mu_c = \alpha_c \mathbf{E}_c.$$

If, on the other hand, the incident light acts on the ellipsoid not along the direction of a principal axis but along a space-frame axis, say *x*, then to determine the polarizability of the ellipsoid we will have to first resolve the incident electric field vector in terms of the components along *a*, *b*, and *c* axes, and then find the polarizability along these three axes separately. Not only that, the induced dipole will not be in the direction in which the electric vector of the incident light is. How nice it would be if the *x*,*y*,*z* frame could be rotated to *X*,*Y*,*Z* such that the rotated axes now coincide with the principal axes (see Figure 10.6). If granted and the rotation is indeed carried out, then the equation of the polarizability ellipsoid transforms to

$$\alpha_{XX}X^2 + \alpha_{YY}Y^2 + \alpha_{ZZ}Z^2 = 1. \tag{10.35}$$

Since the polarizability tensor is represented by the polarizability ellipsoid, the off-diagonal components α_{XY}, α_{YZ}, and α_{ZX} of the tensor go to zero, and we obtain a diagonal polarizability tensor, called Raman tensor

$$\begin{pmatrix} \alpha_{XX} & 0 & 0 \\ 0 & \alpha_{YY} & 0 \\ 0 & 0 & \alpha_{ZZ} \end{pmatrix}$$

in which the diagonal components represent the principal values of the polarizability. The values corresponding to the semi-axes of the polarizability ellipsoid are $(\alpha_{XX})^{-1/2}$, $(\alpha_{YY})^{-1/2}$, and $(\alpha_{ZZ})^{-1/2}$, using which equation (10.35) can be cast as the Raman polarizability ellipsoid

$$\frac{X^2}{\left(\frac{1}{\sqrt{\alpha_{XX}}}\right)^2} + \frac{Y^2}{\left(\frac{1}{\sqrt{\alpha_{YY}}}\right)^2} + \frac{Z^2}{\left(\frac{1}{\sqrt{\alpha_{ZZ}}}\right)^2} = 1. \tag{10.36}$$

In this form of the polarizability ellipsoid, the smallest of the principal values of polarizability arises from the largest of the ellipsoid axes. Also, larger values of $\sqrt{\alpha_{XX}}$, $\sqrt{\alpha_{YY}}$, and $\sqrt{\alpha_{ZZ}}$ correspond to smaller ellipsoid volume. Hence, the ellipsoid volume representing the polarizability of a symmetrically stretching linear molecule is inversely proportional to $\sqrt{\alpha}$. Note that some literature and textbooks incorrectly state that the ellipsoid volume is directly proportional to $\sqrt{\alpha}$. Since the polarizability ellipsoid graphs the tensor, the tensor components will vary as the ellipsoid makes a transition with varying internuclear separation. To a first approximation, one forms an idea about the changes of the ellipsoid for a triatomic molecule, CO_2 for a common example, as depicted in Figure 10.7.

10.5.2 Nomenclature of the Polarizability Tensor

Light scattering tensors are of three kinds. (1) The equilibrium polarizability tensor $\tilde{\alpha}_o$ or $\tilde{\alpha}^{Ray}$ is a Rayleigh tensor which is always there, and which alone will be effective in motionally frozen molecules where no vibrational and rotational motions occur ($v_f = v_i$, $\Delta J = 0$). All irradiated molecules will have the Rayleigh tensor, and hence will scatter light at the frequency of the incident light. (2) The derived polarizability is the Raman tensor, often shown by $\tilde{\alpha}$ only, which will have at least one non-zero component if the derivative of the polarizability tensor with respect to a normal coordinate taken at the equilibrium nuclear configuration does not vanish.

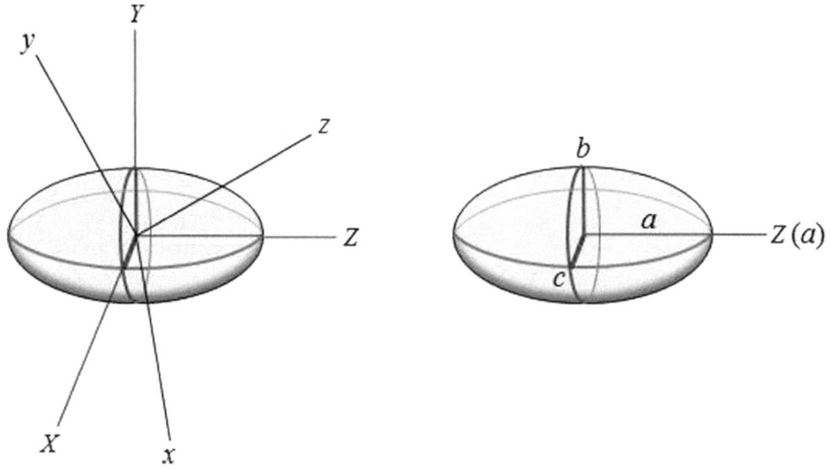

FIGURE 10.6 *Left*, rotation of the *x*,*y*,*z* frame to *X*,*Y*,*Z*. *Right*, the polarizability ellipsoid shown with reference to principal and molecular axes system.

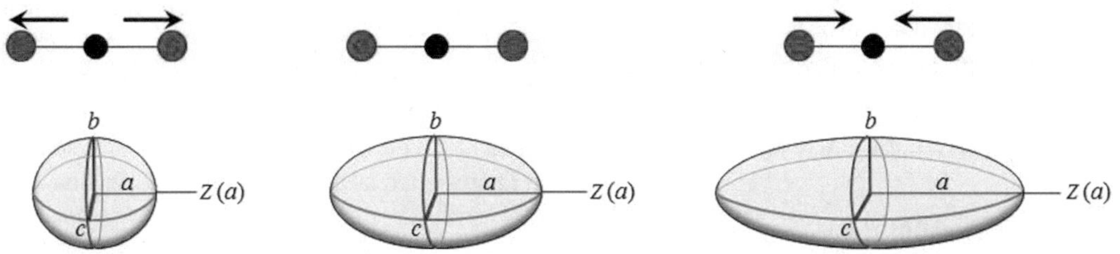

FIGURE 10.7 Changes in the polarizability ellipsoid representation with symmetric stretching vibrations of CO_2. *Left*, symmetric stretching; *middle*, equilibrium nuclear configuration; *right*, symmetric compression. Contraction of the bonds leads to a larger ellipsoid whereas bond stretching corresponds to a smaller ellipsoid.

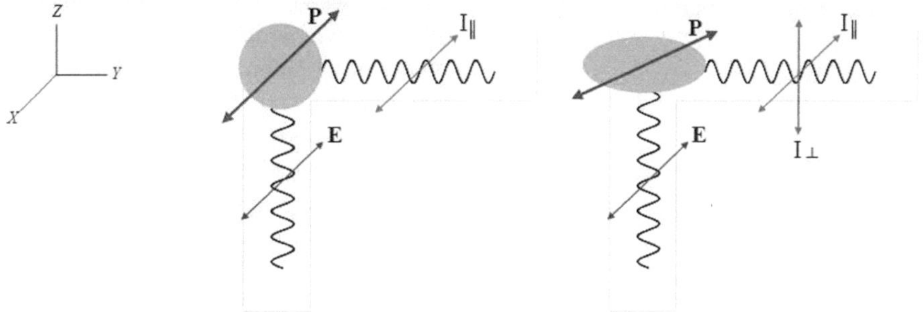

FIGURE 10.8 The induced dipole μ in an isotropic system (*left*) oscillates in the direction of **E** of the incident field, and **E** of the scattered light has the same polarization plane. In an anisotropic system (*right*), the direction of oscillation of μ is not the same as that of the incident **E**. The light scattered by anisotropic systems will only have partial polarization. There will be two components of scattered **E**, one parallel and the other perpendicular to the incident polarization.

The Raman tensor appears only when $v_f \neq v_i$ and $\Delta J \neq 0$. (3) The transition polarizability tensor, $\tilde{\alpha}_{ii}$ for Rayleigh and $\tilde{\alpha}_{fi}$ for Raman, is used in conjunction with the induced transition dipole moment in quantum mechanical perturbation treatment of Raman transitions. The importance of the transition polarizability tensor is deeply associated in the quantum treatment of rotational Raman transitions for which classical description is still inadequate.

10.5.3 Anisotropy of Polarizability

A number of factors, including molecular shape and symmetry, and medium dielectric in the case of freely rotating gas molecules at low pressure, are mapped onto the polarizability tensor via the polarizability ellipsoid. To see quantitatively as to how anisotropy enters into depolarization of the incident light in the scattered emission, we consider two polarizability ellipsoids – one due to an isotropic system like an atom or a spherical molecule and another due to an anisotropic molecule. The observation required is to determine the polarization of the scattered light in a direction perpendicular to the polarization of the incoming incident light. Let a Z-polarized (lab axis) light traveling along X fall on the system, and the scattered light is observed along Y. For an isotropic system the scattered light is completely polarized, which means the electric field vectors of incident and scattered light are parallel (Figure 10.8). This is due to the oscillation of the induced dipole in the direction of the incident electric vector. We note that spherical systems yield rotationally scattered Raman emission, which will be discussed in a following section. For the anisotropic system, on the other hand, the scattered light is only partly polarized. The extent of depolarization in this case depends on the direction of the induced dipole moment, which is not the same direction along which the incident electric vector oscillates. The observed partial polarization arises from the polarizability component obtained by resolving the initial incident polarization into the two perpendicular components.

Clearly, the degree of depolarization of the scattered electric vector is a measure of anisotropy of polarizability. The analysis is based on two simple relations. One, the depolarization parameter for scattered light is

$$\rho = \frac{I_\perp}{I_\parallel} \tag{10.37}$$

where $I_\parallel$ and $I_\perp$ are intensities of scattered light polarized parallel and perpendicular to the direction of the electric vector of the incident light. The parameter ρ is not uniform across bands, but is different for different vibrations depending on the symmetry of a vibration. A vibrational mode belonging to a totally symmetric irreducible representation yields $\rho = 0$, because the plane of polarization is not rotated. But an antisymmetric vibration will give a definite value of ρ because antisymmetric vibrations are depolarized. We thus recognize that ρ can provide information about the symmetry of vibrations. Two, the scattering intensities are determined by the square of the components of the diagonal tensor, $|\alpha_{xx}|^2$, $|\alpha_{yy}|^2$, $|\alpha_{zz}|^2$, or $|\alpha_{xx}\alpha_{yy}|$ etc., which at times are called quadratic products of tensor components. For an axisymmetric molecule, where $a = b \neq c$, the following two relations may be used

$$\text{mean polarizability } a = \frac{1}{3},$$

$$\text{mean anisotropy } \gamma = \left(\alpha_{\parallel} - \alpha_{\perp}\right). \tag{10.38}$$

Another parameter, called anisotropic invariant $\gamma_a^{\ 2}$ can be used to distinguish isotropic cases from anisotropic. For the tensor component written in the molecule-fixed frame

$$\begin{pmatrix} \alpha_{aa} & 0 & 0 \\ 0 & \alpha_{bb} & 0 \\ 0 & 0 & \alpha_{cc} \end{pmatrix}$$

the anisotropic invariant is

$$\gamma_a^{\ 2} = \frac{1}{2}\left[\left(\alpha_{aa} - \alpha_{bb}\right)^2 + \left(\alpha_{aa} - \alpha_{cc}\right)^2 + \left(\alpha_{bb} - \alpha_{cc}\right)^2\right]. \tag{10.39}$$

For an isotropic system like a spherical top, however,

$$\gamma_a^{\ 2} = 0. \tag{10.40}$$

The mean polarizability parameter a is determined by other methods.

10.5.4 Isotropic Average of Scattered Intensity

For freely rotating molecules – gas at sufficiently low pressure, for example – where intermolecular cohesive forces can be neglected, the intensity of the scattered light is obtained from the isotropic average of the polarizability tensor components. Four things are important here. One, because the molecules are rotating, the molecular polarizability tensor component at any instant is given relative to the space-fixed axes system X,Y,Z. Two, the orientation of a molecule at all instants must be averaged out. Three, the orientations of all molecules in the ensemble at all instants need to be averaged out. And, four, the square of the averaged tensor components relative to the space-fixed axes are used to determine the scattering intensity, which is the isotropically averaged intensity. The isotropic average can be calculated from the polarizability tensor obtained in any of the coordinate bases, Cartesian and spherical. It is not necessary to delve into the details of this aspect here, but we just mention that the method in brief involves expressing the polarizability tensor initially in the spherical basis, reducing the tensor in this basis to a set of three irreducible polarizability tensors, transforming the components of each irreducible tensor by rotation so as to generate angular momentum states called $\alpha_m^{(j)}$ with $j = 0,1,2,\cdots$ and $m = 2j+1$, and finally expressing the nine $\alpha_m^{(j)}$ components in terms of products of components of Cartesian or spherical-based polarizability tensor. The products of components so obtained yield the isotropic average of scattering intensity. Although this is an elegant method in the realm of computational spectroscopy to analyze transition probabilities and transition intensities, the procedure is lengthy and somewhat messy, so it is presented as additional material in Box 10.1.

10.6 SEMI-CLASSICAL THEORY OF RAMAN SCATTERING

A simple theory to analyze light scattering by a molecule can be developed using the same semiclassical approach that was used earlier to formulate the two-level transition (Chapter 3). We start with a rotating molecule having k number of normal modes of vibration. Let the initial vibrational level of the molecule correspond to the eigenstate $|\phi_1\rangle$, which can possibly make transitions to a final vibrational state $|\phi_k\rangle$, $k = 2,3,4\ldots$. It is implicit that each vibrational level has rotational levels that can be occupied by the molecule. In the Raman process, the molecule in state $|\phi_1\rangle$ is excited to a less understood virtual state, say $|\phi_V\rangle$, from which emission occurs en route to the final state $|\phi_k\rangle$. We do not include $|\phi_V\rangle$, but rather consider the initial and all the connected final states. The energy differences of the initial and final states are observed in the Raman spectrum.

The state function $\Psi(x,t)$ is a linear superposition of these states, and we write

$$\Psi(x,t) \cong c_1(t)\phi_1(x)e^{-i\omega_1 t} + \sum_k c_k(t)\phi_k(x)e^{-i\omega_k t}. \tag{10.41}$$

For convenience we will not write the space variable x, but take it implicitly. Let $c_1(t) = 1$ initially. When the incident radiation field is turned on, an electric dipole is induced

$$\boldsymbol{\mu}(t) = \langle\mu\rangle = \int_{-\infty}^{t} \Psi^*(t)\boldsymbol{\mu}\Psi(t)dx. \tag{10.42}$$

The rationale for integrating from $t = \infty$ instead of $t = 0$ is to avoid spurious high-frequency oscillations in $\langle\mu\rangle$

$$\langle\mu\rangle = \sum_k \left[c_k^*(t)\boldsymbol{\mu}_{k1}e^{i\omega_{k1}t} + c_k(t)\boldsymbol{\mu}_{1k}e^{-i\omega_{1k}t}\right]. \tag{10.43}$$

As was done earlier for the two-state model, substitution for c_k and c_k^*, and simplification yields

$$\boldsymbol{\mu}(t) = \left[\sum_k \frac{2\omega_{k1}\boldsymbol{\mu}_{1k}\boldsymbol{\mu}_{k1}}{\hbar^2\left(\omega_{k1}^{\ 2} - \omega_1^{\ 2}\right)}\right]\mathbf{E}_o\cos\omega t$$

$$= \left[\sum_k \frac{2\omega_{k1}\boldsymbol{\mu}_{1k}\boldsymbol{\mu}_{k1}}{\hbar^2(E_{k1}^{\ 2} - E_1^{\ 2})}\right]\mathbf{E}_o\cos\omega t = \tilde{\boldsymbol{\alpha}}\,\mathbf{E}_i(t)$$

This is the familiar molecular polarizability equation in which $\mathbf{E}_i(t)$ is the incident electric field vector. The expression for the polarizability obtained this way indicates that the

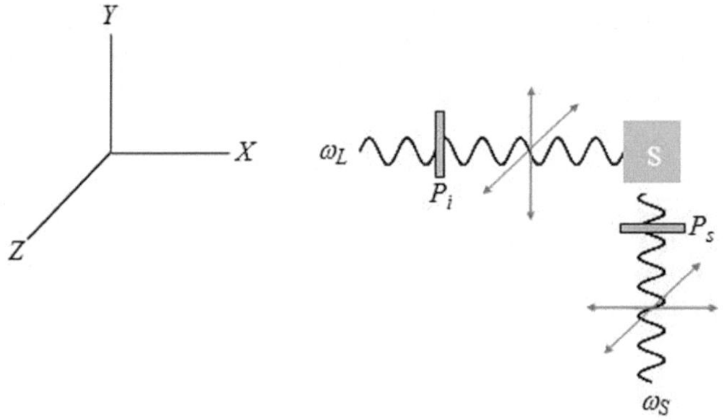

FIGURE 10.9 The incident light of frequency ω_L propagating along X is polarized by the polarizer P_i and the scattered light along Y is polarized by P_s. A monochromator placed before the detector is used to select a desired frequency of the scattered light.

polarizability is inversely proportional to the difference of the energy of the initial state $|\phi_1\rangle$ and the final state $|\phi_k\rangle$.

Light scattering experiments are generally carried out by specified experimental geometry in which the polarization state of the electric vector in the incident and scattered light is manipulated. The polarization states are identified in the lab- or space-fixed axes system X,Y,Z. Since the molecules are rotating in general, the molecular polarizability tensor components are defined in the molecule-fixed axes system a,b,c whose orientation relative to the space-fixed axes is determined by the appropriate direction cosines for the rotation of the axes. First we define a polarizability tensor component with the space-fixed polarization index $\alpha_{is}\ (i,s = X,Y,Z)$. The geometry and the axes system to label light polarization are shown in Figure 10.9. The incident polarizer may be Z or Y, and the scattering polarizer could be Z or X. If the scattered light is found to have a Z-polarization, then the molecule should have an electric dipole oscillating along Z, and similarly for X. With these polarizer configurations we have the following components of the induced dipole with the description given against each

$$\begin{aligned}
\mu_Z &= \alpha_{ZZ}E_Z \text{ excite with } Z\text{-polarization, detect } Z\text{-polarization}\\
\mu_Y &= \alpha_{ZY}E_Y \text{ excite with } Y\text{-polarization, detect } Z\text{-polarization}\\
\mu_X &= \alpha_{ZX}E_Z \text{ excite with } Z\text{-polarization, detect } X\text{-polarization}\\
\mu_X &= \alpha_{XY}E_Y \text{ excite with } Y\text{-polarization, detect } X\text{-polarization.}
\end{aligned}$$

These α_{is} components are indexed in the space-fixed (laboratory) frame. To examine the spectroscopic transitions, we consider the tensor components derived with respect to nuclear coordinates (equation 10.29). The general transition integral will read

$$\langle \alpha_{is} \rangle_o = \langle \psi_v \psi_{JKM} | \alpha_{is}(Q_o) | \psi_{v'} \psi_{J'K'M'} \rangle. \tag{10.44}$$

10.6.1 Rotational Raman Spectra

We imagine a rotating molecule with clamped nuclei so that $v = v'$ for which only the first term in the Taylor expansion of α_{is} needs to be considered

$$\langle \alpha_{is} \rangle_o = \langle \psi_{JKM} | \alpha_{is}(Q_o) | \psi_{J'K'M'} \rangle. \tag{10.45}$$

The use of $\langle \alpha_{is} \rangle_o$ implies that the integration is performed at equilibrium internuclear separation. The operator $\alpha_{is}(Q_o)$ whose indices denote the space-fixed axes X,Y,Z can be transformed to the molecule-fixed axes system a,b,c by using the direction cosines which relate the orientations of the two axes systems (see Box 5.1). One way of writing this transformation is

$$\alpha_{is} = \sum_{l,m} \alpha_{lm} \cos\theta_{il} \cos\theta_{sm} \tag{10.46}$$

where $l,m = a,b,c$, and the summation involves nine individual equations, one for each of the nine α_{is} components. The pure rotational Raman transition integral is

$$\langle \alpha_{is} \rangle_o = \left\langle \psi_{JKM} \left| \sum_{a,b,c}^{\text{(molecule fixed axes)}} \alpha_{ab} \cos\theta_{ai} \cos\theta_{bs} \right| \psi_{J'K'M'} \right\rangle. \tag{10.47}$$

The expectation value $\langle \alpha_{is} \rangle_o$ is obtained when the integration is carried out at equilibrium internuclear distance. For diatomic molecules the semi-minor axes of the polarizability ellipsoid may be related by $a \neq b = c$, where the axis a orients along the bond axis. If it is a σ-type bond, then all perpendicular polarization components will be equal, and the corresponding tensor components will be

$$\begin{aligned}
\alpha_{cc} &= \alpha_{bb} = \alpha_\perp\\
\alpha_{aa} &= \alpha_\parallel.
\end{aligned} \tag{10.48}$$

In the principal axes system, the tensor is diagonal, an argument that conforms with symmetry operations if the electric field vector of the radiation is along axis a, which is the bond axis. The charge distribution along the bond axis will be invariant with respect to reflection on mirror planes containing the bond axis. But reflections in ac, bc, and ab planes will change the sign of the induced electric dipole components μ_b, μ_a, and μ_c, respectively. This means all the off-diagonal tensor components vanish, yielding a diagonal tensor. One should then obtain the polarizability tensor at equilibrium configuration, a diagonal tensor, as

$$\tilde{\alpha} = \begin{pmatrix} \alpha_\perp & 0 & 0 \\ 0 & \alpha_\perp & 0 \\ 0 & 0 & \alpha_\parallel \end{pmatrix}, \tag{10.49}$$

which is the Rayleigh-rotational Raman tensor for a diatomic molecule. This corresponds to the static polarizability.

To continue with equation (10.47), we drop the quantum number K when it comes to the analysis of a diatomic molecule. The amplitude A_{is} in a pure rotational Raman transition is

$$A_{is} = \left\langle \psi_{JM} \left| \begin{matrix} \alpha_\perp (\cos\theta_{ci}\cos\theta_{cs} + \cos\theta_{bi}\cos\theta_{bs}) \\ + \alpha_\parallel \cos\theta_{ai}\cos\theta_{as} \end{matrix} \right| \psi_{J'M'} \right\rangle. \tag{10.50}$$

The operator part of the integral is written to conform with the direction of the field applied, which is the bond axis a. If the incident light is Z-polarized and one also chooses to detect the Z-polarized light, then the Raman intensity will be designated by $|A_{zz}|^2$. Recall that scattering intensities are given by the quadratic products of the polarizability tensor components relative to the space-fixed axes. To obtain an expression for A_{zz} we may write α_{zz} in terms of the molecule-fixed axes using the appropriate Euler angles (see Box 5.1)

$$\alpha_{ZZ} = \alpha_{aa}\cos^2\theta_{aZ} + \alpha_{bb}\cos^2\theta_{bZ} + \alpha_{cc}\cos^2\theta_{cZ}. \quad (10.51)$$

This expression can be reduced by using

$$\cos^2\theta_{aZ} + \cos^2\theta_{bZ} + \cos^2\theta_{cZ} = 1. \quad (10.52)$$

One obtains

$$\begin{aligned}\alpha_{ZZ} &= \alpha_{\parallel}\cos^2\theta + \alpha_{\perp}\left(1-\cos^2\theta\right)\\ &= A\left(\alpha_{\parallel}+2\alpha_{\perp}\right)+B\left(\alpha_{\parallel}-2\alpha_{\perp}\right),\end{aligned} \quad (10.53)$$

where A and B are constants. The amplitude for the rotational transition is now

$$\begin{aligned}A_{ZZ} &= \left\langle \psi_{JM}\left|\alpha_{\perp}\left(1-\cos^2\theta\right)+\alpha_{\parallel}\cos^2\theta\right|\psi_{J'M'}\right\rangle\\ &= \frac{1}{3}\left\langle \psi_{JM}|\psi_{J'M'}\left(\alpha_{\parallel}+2\alpha_{\perp}\right)+\frac{2}{3}\psi_{JM}\left|P_2\left(\cos\theta\right)\right|\psi_{J'M'}\right\rangle\\ &\quad \left(\alpha_{\parallel}-2\alpha_{\perp}\right),\end{aligned} \quad (10.54)$$

in which the operator

$$P_2\left(\cos\theta\right) = \frac{1}{2}\left(3\cos^2\theta-1\right)$$

is the second-order Legendre function. The first term in the integral will vanish unless $J = J'$ and $M = M'$, which is due to the orthogonality of rotational eigenstates belonging to the same electronic state. This term generates an undisplaced line, which is the Rayleigh line. The second term will vanish unless $J = J'$ or $J = J' \pm 2$. The ΔJ has to be 0 or ± 2 in order that a spherical harmonics matrix element exists. It is the $\cos^2\theta$ term of $P_2(\cos\theta)$ that restricts the change of J to values other than ± 2. Indeed, $\Delta J = 0, \pm 2$ is the selection rule for rotational Raman transitions, a result that we saw earlier also in the discussion of rotational Raman scattering in classical perspective. We thus have a line for $\Delta J = 0$ (Q-branch or Rayleigh line), anti-Stokes Raman lines corresponding to $\Delta J = +2$ (S-branch), and Stokes Raman lines corresponding to $\Delta J = -2$ (O-branch). This follows the convention of labeling rotational lines series by O,P,Q,R,S indicating respective change of the J quantum number by $-2,-1,0,+1,+2$.

The rotational Raman transitions occur very close to the Rayleigh frequency, obviously due to low energy of separation of the rotational levels compared to much larger separation of the vibrational energy levels. Regarding the frequency of separation of rotational Raman lines, one may recall that the rotational term value for a rigid diatomic molecule (equation (7.13)) in the ground vibrational state is

$$F_{\text{rot}}(J) = B_e J(J+1),$$

where B_e is the rotational constant corresponding to the principal moment of inertia I_b. Since the term value for the $J+2$ level is

$$F_{\text{rot}}(J+2) = B_e J(J+2)(J+3),$$

the transition wavenumber difference for a J to $J+2$ transition is

$$|\Delta\tilde{\nu}| = F_{\text{rot}}(J+2) - F_{\text{rot}}(J) = 4B_e\left(J+\frac{3}{2}\right), \quad (10.55)$$

where the tilde on ν denotes wavenumber. For $J = 0,1,2,\cdots$, $|\Delta\tilde{\nu}| = 6B_e, 10B_e, 14B_e, \cdots$, meaning the wavenumber difference between the Rayleigh line and the first, the second, the third, … rotational lines will be, respectively, $6B_e, 10B_e, 14B_e, \cdots$. The lines in the pure rotational Raman spectrum are therefore separated by $4B_e$ (Figure 10.10). The J-distribution of line intensity is obtained from the population ratio of the rotational levels P_J corresponding to a transition. The population distribution has already been covered in detail earlier (equation (7.28)). Using the same argument, one can draw the appearance of a pure rotational Raman spectrum for a rigid heteronuclear diatomic as shown in Figure 10.10.

For homonuclear diatomics the changes in intensity distribution depend on the nuclear spin angular momentum, nuclear spin statistical weight, and J-dependent symmetry of the rotational wavefunction with respect to a π-rotation of the molecule (see Box 10.2). The result is a characteristic alternation of line intensities of consecutive $J \rightarrow J + 2$ rotational transitions. A textbook example of this is the rotational Raman spectrum of H_2 which exists in both ortho and para forms. The rotational energy levels of ortho-hydrogen are all odd-J levels with the spin statistical factor of 3, but those of para-hydrogen are all even-J rotational levels having a spin statistical factor of 1. This means the J quantum number for each rotational energy level of ortho-hydrogen is odd and each is threefold degenerate, in contrast to the non-degenerate and even-J rotational levels of para-hydrogen. The rotational Raman intensities corresponding to the ortho form will be three times more intense than those corresponding to para-hydrogen, and because J is odd and even, respectively, the line intensities will alternate as 3:1. Similarly, as the lower panel of Figure 10.10 shows, one would observe 2:1 (even:odd) alternation in the intensity of rotational Raman lines of diatomic molecules with nuclear spin I=1 (Boson). However, the overall intensity distribution is still given by the population distribution function P_J. The $4B_e$ spacing is observed uniformly for changes in heteronuclear diatomics if centrifugal distortion is accounted for; that is, when the rotor is non-rigid. The idea is a modification of the rigid-rotor

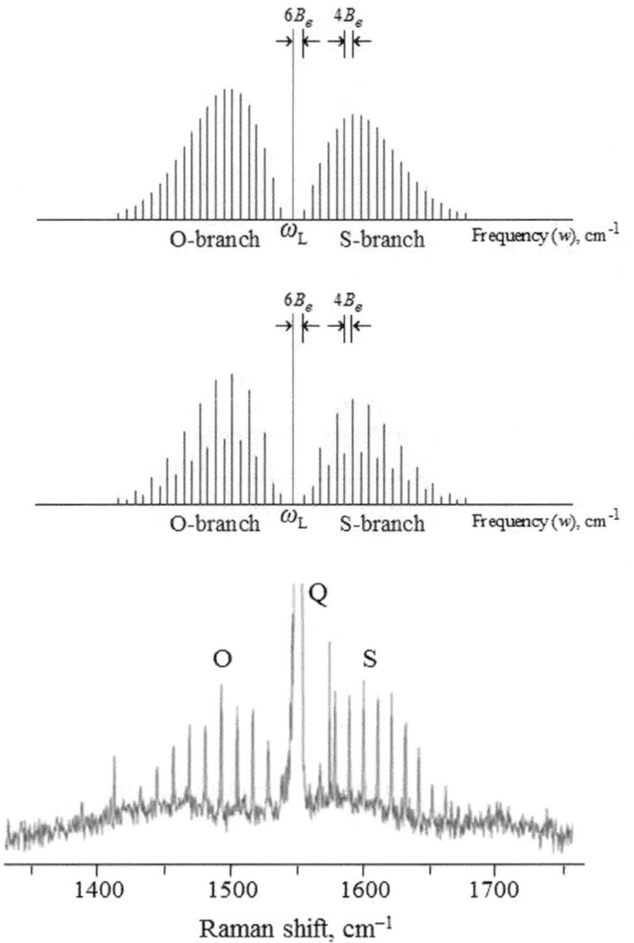

FIGURE 10.10 *Top*, rotational Raman spectrum of a typical heteronuclear diatomic molecule. The spacing between the Rayleigh line and the first rotational line $(J = 0 \to J = 2)$ is $6B_e$, but all subsequent lines are uniformly apart by $4B_e$. *Middle*, the rotational Raman spectrum of a homonuclear diatomic, typified by $^{14}N_2$, shows the same line spacing; however, the intensities of lines alternate as 1:2 (odd J to even J). *Bottom*, an experimental spectrum of N_2 showing the 1:2 intensity ratio. The alternating spacing pattern is convoluted with the population distribution. In both calculated and experimental spectra, the anti-Stokes lines have lower intensity compared to the Stokes lines. This discrepancy arises due to the quantum mechanical model, according to which the intensity depends on the population in the initial state that is excited. Since higher J-levels are progressively less occupied, the anti-Stokes lines are relatively less intense. The Stokes:anti-Stokes intensity ratio is $\frac{I_{\text{Stokes}}}{I_{\text{anti-Stokes}}} = \frac{(\nu_0 - \nu_k)^4}{(\nu_0 + \nu_k)^4} e^{\frac{hc\tilde{\nu}_k}{k_B T}}$ where $\tilde{\nu}_k$ is the transition wavenumber of the kth mode. (The experimental spectrum of N_2 is reproduced from Tuschel, 2014, with permission from the journal *Spectroscopy*.)

rotational term value when the nuclei are stretched out. The rotational term in this case is

$$F_{\text{rot}}(J) = B_e J(J+1) - DJ^2(J+1)^2 + \frac{E}{2} J^3(J+1)^3 - \cdots. \quad (10.56)$$

The higher order terms are insignificant really; even D is at least four-orders of magnitude smaller than B_e. Centrifugal stretching therefore modulates the term value only marginally. The incorporation of the term nonetheless modifies equation (10.55) to

$$|\Delta\tilde{\nu}| = (4B_e - 6D)\left(J + \frac{3}{2}\right) - 8D\left(J + \frac{3}{2}\right)^3. \quad (10.57)$$

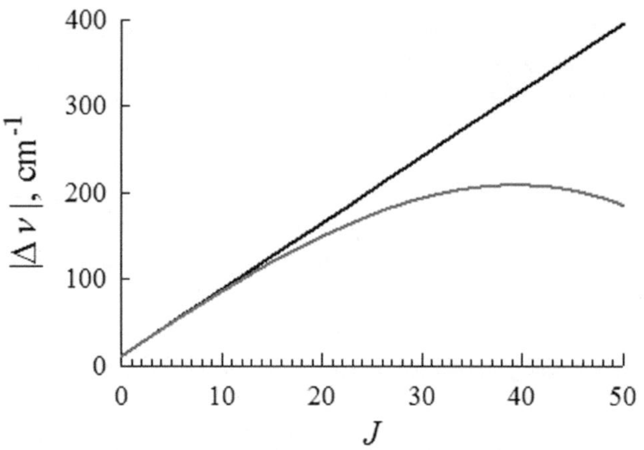

FIGURE 10.11 Calculated spacing between rotational lines in a Raman spectrum. The line in *black* shows the usual spacing between a J-line minus the $J = 0$ line if centrifugal distortions are not there. The line in *red* shows the same spacing when centrifugal distortions are taken into account.

The simulated spacing between rotational lines of a typical heteronuclear diatomic molecule as a function of J in the 0 to 50 range is shown in Figure 10.11. We notice that $\Delta\tilde{\nu}$ for $J \to J+2$ transition, for example, vanishes (saturates) for J values nearing 40. For still larger J, the wavelength separation of successive rotational lines appears to 'fold' back, which should not happen.

As a selection rule, the only factor one finds for the occurrence of pure rotational Raman transitions is the molecular polarizability. Since all molecules are more or less polarizable when allowed to interact with an electromagnetic field, they should show rotational Raman spectra. However, it is also stipulated that $\alpha_{\parallel} - \alpha_{\perp} \neq 0$ for the occurrence of rotational transitions. We mentioned earlier that spherically symmetric systems $(I_a = I_b = I_c)$ yield completely polarized Rayleigh-scattered light, but not Raman. The polarizability tensor for these molecules is spherically symmetric where $\alpha_{xx} = \alpha_{yy} = \alpha_{zz}$, so $\alpha_{\perp} - \alpha_{\parallel}$. Thus, spherical molecules like methane do not have pure rotational Raman spectrum. Asymmetric top molecules $(I_a \neq I_b \neq I_c)$, on the other hand, exhibit complex rotational Raman spectrum, the complexity increases with the extent of symmetry and transitions involving higher J values.

10.6.2 Vibration-Rotation Raman Spectra

We take up the general transition integral

$$\langle \alpha_{is}(Q) \rangle = \langle \psi_v \psi_{JKM} | \alpha_{is}(Q_o) | \psi_{v'} \psi_{J'K'M'} \rangle + \langle \psi_v \psi_{JKM} | \sigma_{is} Q_j | \psi_{v'} \psi_{J'K'M'} \rangle, \quad (10.58)$$

and proceed as before. The total vibrational wavefunction ψ_v in the harmonic oscillator approximation is the product of the oscillator states for each normal mode of vibration. So the first term involving the equilibrium polarizability tensor component will be nonzero only if $v' = v$ for all normal modes. The integral will vanish even if one of the normal modes shows a change in the vibrational quantum number. This is due to the orthogonality of harmonic oscillator wavefunctions. It follows that the only term needed to include to calculate the vibration-rotation integral is $\sigma_{is}Q_j$, where σ_{is} is the component of the polarizability derivative tensor $\sigma^{(k)} = \left(\partial\alpha / \partial Q_k\right)_0$,

$$\sigma_{is} = \sum_{a,b} \sigma_{ab} \cos\theta_{ai} \cos\theta_{bs}. \quad (10.59)$$

Simply getting rid of the first term involving $\alpha_{is}\left(Q_0\right)$ by the use of the orthogonality condition of the oscillator wavefunction does not imply that the Rayleigh scattering will not occur. It will, because the molecular polarizability also oscillates at the incident light frequency. Rayleigh scattering will always occur.

Consider a diatomic molecule. If the incident light is Z-polarized and we choose to observe the Z-polarized scattered light (*ZZ* configuration), then the Raman amplitude will be

$$A_{ZZ} = \left\langle v|Q|v'\right\rangle \left\langle JM|\sigma_{ZZ}|J'M'\right\rangle$$
$$= \left\langle v|Q|v_{\pm 1}\right\rangle \left\{ \frac{1}{3}\langle JM|J'M'\rangle\left(2\sigma_\perp + \sigma_\parallel\right) \right. \quad (10.60)$$
$$\left. + \frac{2}{3}\left\langle JM|P_2(\cos\theta)|J'M'\right\rangle\left(\sigma_\parallel + \sigma_\perp\right) \right\}.$$

The change in the vibrational quantum number is according to $\Delta v = \pm 1$, but the rotational quantum number may not change ($\Delta J = 0$) or will change by ± 2 ($\Delta J = \pm 2$). As indicated earlier, $\Delta J = 0$ corresponds to the Q-branch rotational lines whose intensities depend on the trace of the polarizability derivative tensor $\sigma_{aa} + \sigma_{bb} + \sigma_{cc} = \sigma_\parallel + 2\sigma_\perp$, and $\Delta J = -2$ and $+2$ yield O- and S-branches, respectively, depending on the magnitude of $\sigma_\parallel - \sigma_\perp$. Generally,

$$\sigma_\parallel + 2\sigma_\perp > \sigma_\parallel - \sigma_\perp,$$

and the ratio

$$\frac{\sigma_\parallel - \sigma_\perp}{\sigma_\parallel + 2\sigma_\perp}$$

is called Raman anisotropy which is a measure of the anisotropy of polarizability. Described below are specific cases.

Since the rotational wavefunction of symmetric tops (equation (7.42)) is

$$\psi_{\text{rot}} = \Theta_{JKM} e^{im\phi} e^{iK\chi},$$

the angle χ needs to be considered so that the quantum number K, which is the projection of J onto the molecular symmetry axis, enters the rotational integral in equation 10.60. If we still choose to use our configuration of 'excite with Z-polarization and detect Z-polarized light' (Figure 10.9), then the analysis becomes considerably simplified, because in the A_{ZZ} configuration there is no $\cos\chi$ dependence for any of the vibrational coordinates that have diagonal tensors. Under these situations, the vibration-rotation Raman amplitude for tops is the same as that for diatomic molecules. In general, however, vibrations along a coordinate Q_k can give rise to polarizability derivatives having nondiagonal components. If we consider the totally symmetric modes of vibration, we would find that the overall symmetry of tops is not changed during such motion, which means the principal axes of rotation do not change during vibration. For a symmetric top, these axes are the symmetry axis plus any two axes perpendicular to the symmetry axis. The polarizabilities can still be labeled as $\alpha_\parallel$ and $\alpha_\perp$, or $\sigma_\parallel^{(k)}$ and $\sigma_\perp^{(k)}$ if the derivative is taken with respect to the kth normal coordinate. Again, if the *ZZ* configuration is adopted, the results obtained are the same as that for a diatomic molecule. With *ZX* configuration, however, the extra K-selection rules will have to be applied.

Spherical tops, CH_4 for example, are isotropic molecules for which $\alpha_\parallel = \alpha_\perp$. They will show only the Q-branch rotational structure when the measurements are taken in the *ZZ* configuration. If *ZX* configuration is used for measurement, there will be no rotational structure at all, because

$$A_{ZX} = \left\langle JKM|\sigma_\parallel - \sigma_\perp(\cos\theta_{cZ}\cos\theta_{cX}|J'K'M'\right\rangle. \quad (10.61)$$

There is no anisotropy of the polarizability and the scattering is completely polarized.

With regard to the nontotally symmetric modes of vibration during which the molecular symmetry changes, σ_{lm} exists for some l and m. For example, consider three arbitrary normal coordinates Q_j, Q_k, and Q_l, and the variation of polrizability with respect to these coordinates in the neighborhood of the equilibrium nuclear positions, as shown in Figure 10.12. Since the polarizability in the vicinity of the equilibrium position

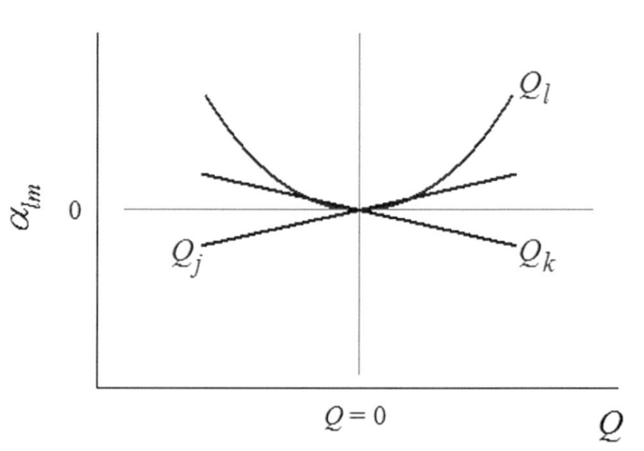

FIGURE 10.12 Schematic of the variation of the polarizability with normal coordinates. Since the polarizability derivative for the Q_l mode at $Q = 0$ vanishes, the mode is Raman inactive.

does not change during the motion involving the Q_l coordinate, we have

$$\left(\frac{\partial \alpha_{lm}}{\partial Q_l}\right)_0 = \alpha_{lm}^{(l)} = 0, \quad (10.62)$$

rendering the lth normal mode Raman inactive.

A vibrational motion Q_k must bring about a change in the principal axes of the polarizability ellipsoid if $\alpha_{lm}^{(k)}$ is to exist. Indeed, the vibration sets the principal axes into oscillation, the frequency of which is the frequency of the normal mode itself. If the principal axes are a, b, c at the equilibrium nuclear configuration, the vibrational motion will cause rotation of a, b to a', b', so that $\alpha_{lm}^{(k)} = \alpha_{ab}^{(k)}$ only.

For centrosymmetric molecules that have a center of inversion symmetry, a vibrational mode may be either Raman-active or IR-active, but not both. Also, a vibrational mode that destroys the center of symmetry is not Raman-active. A commonly cited example is that of CO_2 that has four modes of vibration – symmetric stretch (Q_1), antisymmetric stretch (Q_2), and a degenerate pair of bending motions (Q_{3a} and Q_{3b}) that have the same vibrational frequency (see Figure 9.12). From the pictorial depictions in Figure 10.13 we see that the mode Q_1 is Raman active, because the two polarizability (tensor) derivatives – one for each bond – are additive. This yields an overall molecular polarizability derivative with respect to this mode of motion. Modes Q_2 and Q_3 are Raman inactive for the same reason that their respective polarizability derivatives at the equilibrium position vanish. We also note that the individual bond dipole derivatives for the Q_1 mode are not zero; however, since the two dipole vectors are directionally opposite, they cancel each other out, so the dipole moment derivative at the equilibrium position does not exist. Hence the Q_1 mode is infrared inactive as discussed earlier (see equation (9.96)). This result in fact forms a general rule, a mutual exclusion selection rule if one would, according to which 'there is no stretching vibrational mode that is both infrared- and Raman-active'. The mutual exclusion is seen here for the Q_2 mode as well; the two bond dipole derivatives add up to yield a non-zero value rendering this mode infrared-active (see equation (9.96)), but the two polarizability derivatives cancel out so that the overall molecular polarizability derivative vanishes, thus resulting in Raman inactivity. The mode Q_3 is infrared-active with X and Y polarization because the electric dipole moment derivatives perpendicular to the molecular axis are not zero, but the polarizability derivatives are both zero which render Raman inactivity.

Another commonly cited example of symmetry destruction and Raman inactivity of centrosymmetric molecules is provided by the bending mode of benzene (Figure 10.14). The axes b and c are not rotated for both $+Q$ and $-Q$ motions. Although the polarizability α_{bc} along $+Q$ and $-Q$ configurations away from the nuclear positions are the same, there is a turning point at the equilibrium configuration $Q = 0$, implying zero derivative.

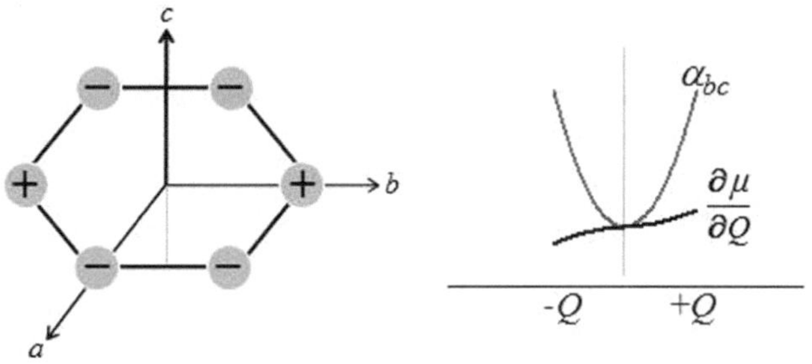

FIGURE 10.14 A bending mode of benzene and the corresponding polarizability and dipole derivative changes along the nuclear coordinates. Axis c projects out of the plane of the paper.

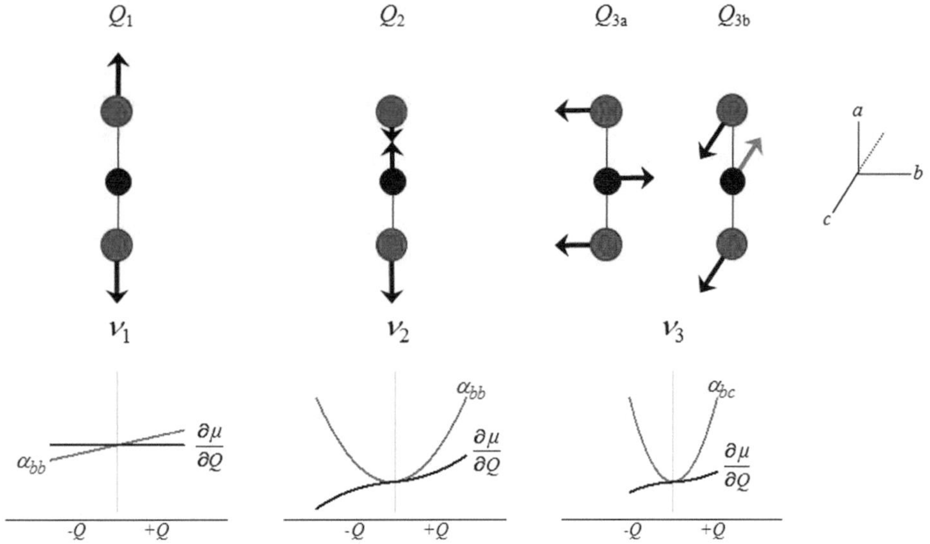

FIGURE 10.13 Changes in polarizability and dipole derivative at the equilibrium nuclear configuration for the three vibrational modes of CO_2. The axis c is perpendicular to the plane of the paper.

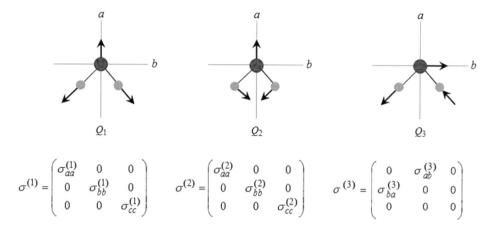

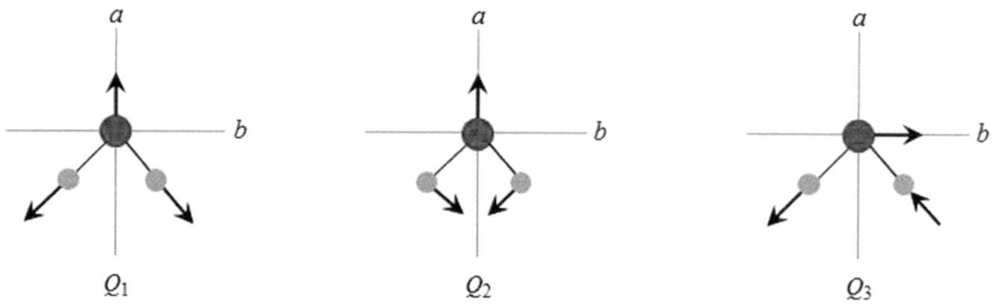

$$\sigma^{(1)} = \begin{pmatrix} \sigma^{(1)}_{aa} & 0 & 0 \\ 0 & \sigma^{(1)}_{bb} & 0 \\ 0 & 0 & \sigma^{(1)}_{cc} \end{pmatrix} \qquad \sigma^{(2)} = \begin{pmatrix} \sigma^{(2)}_{aa} & 0 & 0 \\ 0 & \sigma^{(2)}_{bb} & 0 \\ 0 & 0 & \sigma^{(2)}_{cc} \end{pmatrix} \qquad \sigma^{(3)} = \begin{pmatrix} 0 & \sigma^{(3)}_{ab} & 0 \\ \sigma^{(3)}_{ba} & 0 & 0 \\ 0 & 0 & 0 \end{pmatrix}$$

FIGURE 10.15 Diagonal derivative Raman tensors for water.

To end this section, we look at the Raman tensor of water, a standard example. The three vibrational modes and the existing derivatives of the polarizability tensor $\tilde{\alpha}$ are shown in Figure 10.15. For the two totally symmetric modes, Q_1 and Q_2, we have the diagonal derivative tensor $\tilde{\sigma}$, so all three modes are Raman active. The nontotally symmetric mode Q_3 is also Raman active because $\sigma^{(3)}_{ab} = \sigma^{(3)}_{ba}$ exists, which is due to the rotation of the principal axes a and b during this mode.

The Raman selection rules for water vapor can be extracted by approximating the water molecule to a symmetric prolate top. Let the respective vibrational state of the initial and final states of the molecule be

$$\begin{aligned} \psi_{v'} &= \psi^{(Q_1)}_{v'=0}\psi^{(Q_2)}_{v'=0}\psi^{(Q_3)}_{v'=0} = |000\rangle \\ \psi_{v} &= \psi_{v=0}\psi_{v=0}\psi_{v=1} = |001\rangle. \end{aligned} \tag{10.63}$$

In these states there can be a one-quantum transition for the mode Q_3. The Raman amplitude is

$$\begin{aligned} A^{(3)}_{zz} &= \langle 000|Q_3|001\rangle\langle JKM|\sigma^{(3)}_{ab}\cos\theta_{za}\cos\theta_{zb}|J'K'M'\rangle \\ &= \sqrt{\frac{\hbar}{2\omega_3}}\langle JKM|\sin^2\theta\cos\chi\sin\chi|J'K'M'\rangle\sigma^{(3)}_{ab}. \end{aligned} \tag{10.64}$$

The θ-integral gives $\Delta J = 0, \pm 1$, and for the χ-integral we write

$$\int_0^{2\pi} e^{iK\chi}e^{-iK'\chi}\cos\chi\sin\chi dx,$$

and then by using

$$\cos\chi = \frac{e^{i\chi} + e^{-i\chi}}{2}$$

$$\sin\chi = \frac{e^{i\chi} - e^{-i\chi}}{2}$$

one obtains the selection rule $\Delta K = \pm 2$.

This selection rule is specific to the example presented here. The Raman polyatomic selection rules depend on factors including the point group of the molecule and the symmetry and degeneracy of ground and excited states. The value of K is often found to change by ± 1.

10.7 RAMAN TENSOR AND VIBRATIONAL SYMMETRY

Raman experiment can provide the symmetry of vibrational motions, a task not accomplished by conventional infrared spectroscopy. Advantage is taken of the manipulated polarization of the electric field vectors of both incident and detected radiation. The molecules whose transition moment is along the direction of the polarization of the incident electric vector couple with the field, and then emit radiation whose polarization can be determined. It is important to note that the induction of the dipole by the incident field is almost always instantaneously followed by scattering emission. The two-photon nature of the Raman process refers to the instantaneous nature of the incident and scattered photons. The consequence of this instantaneity is that the molecule that is deformed because of the induction of dipole does not have enough time to reorient before the emission of the scattered photon occurs. This implies that the nature of the Raman tensor of a Raman-active vibration in the molecule-fixed axes system can be obtained with ease. For example, suppose the polarizability principal axes of the molecule are a,b,c whose orientation relative to the laboratory coordinates X,Y,Z are given by θ, ϕ, χ, then the Raman scattering amplitude for the kth mode of vibration is

$$A^{(k)}_{is} \propto \sum\nolimits_{a,b} \sigma^{(k)}_{ai}\cos\theta_{ai}\cos\theta_{bs} \propto \mathbf{E}_i\sigma^{(k)}\mathbf{E}_s. \tag{10.65}$$

The tensor components are expressed in the molecule-fixed axes system, and the scattering amplitude is a function of θ, ϕ, χ. The Raman tensor will be diagonal for the totally symmetric vibrations, but off-diagonal symmetric components exist for nontotally symmetric modes, as was seen above for the Q_3 mode of water. The scattering amplitude can be obtained by expressing $\mathbf{E}_i$ and $\mathbf{E}_s$ in the molecule-fixed axes system.

10.8 SECONDARY OR COUPLED BANDS IN RAMAN SPECTRA

Since Raman scattering is essentially a vibrational spectroscopy, overtone, combination, and Fermi and Coriolis-coupled transitions should also occur if anharmonicity of the potential is accounted for. These transitions in fact do occur even though they are much weaker than the Raman fundamental transitions. In IR spectra, however, as we have discussed already, the overtones and coupled bands are rather strong. The much weaker overtone and coupled bands in the Raman case produce a sparse spectrum relative to the appearance of the corresponding IR spectrum.

Mechanical anharmonicity results in the appearance of overtone $(2\nu_k, 3\nu_k, 4\nu_k, \cdots)$ and combination bands of the type $\left(\nu_k + \nu_l\right)$ provided the kth and l^{th} modes are Raman active. In other words, the fundamentals ν_k and ν_l must be allowed. When electrical anharmonicity of the potential is considered in addition, an interesting feature of the molecular polarizability derivative emerges. Electrical anharmonicity requires weighing the change in polarizability with respect to the second derivative or even higher derivative for that matter. This means additional Raman tensors that have components of the form

$$\sigma_{l,m}^{(j,j)} = \left(\frac{\partial^2 \alpha_{lm}}{\partial Q_j^{\,2}}\right)_0, \; \sigma_{l,m}^{(j,k)} = \left(\frac{\partial^2 \alpha_{lm}}{\partial Q_j \partial Q_k}\right)_0, \cdots \quad (10.66)$$

will involve. It may so happen that the first derivative polarizability tensor does not exist (Raman inactive), but the second derivative polarizability does (Raman active). Then a fundamental Raman is not allowed and will not occur, but overtone and combination transitions may arise. For the kth mode, for example, the fundamental ν_k will be missing, but overtones like $2\nu_k, 3\nu_k, \cdots$ can be Raman active. However, as mentioned earlier, an allowed fundamental, say ν_j, can combine with the ν_k transition to yield a combination band $\nu_j + \nu_k$. Other combination possibilities, $\nu_j + 2\nu_k$, $2\nu_j + \nu_k$, and so on, can also occur.

The vapor phase Raman spectrum of CO_2 provides an example of the occurrence of some of these secondary transitions (Figure 10.16). From earlier analyses we learned that only the symmetric –C–O stretching, which is represented by the Q_1 normal mode of vibration (Σ_g^+ symmetry) is Raman active, and Q_2 and Q_3 are not (Figure 10.13). The CO_2 spectrum accordingly is expected to yield only the ν_1 fundamental. But the infrared-allowed Q_2 mode has an overtone, $2\nu_2$ associated with Σ_g^+ symmetry. We then have a Raman-active fundamental ν_1 and an infrared-active overtone $2\nu_2$, both having Σ_g^+ symmetry. These two transitions interact to produce Fermi resonance leading to the appearance of ν_1 at 1284.8 cm^{-1} and $2\nu_2$ at 1387.5 cm^{-1}, the Fermi shift being 103 cm^{-1} (Figure 10.16). We notice that the Fermi-coupled $2\nu_2$ band is stronger than the ν_1 band, which arises due to the so-called 'vibrational borrowing' – a phenomenon of mixing of electronic states due to nuclear distortion. The ν_1 and $2\nu_2$ bands also show respective hot bands at 1264.4 and 1408.8 cm^{-1} that can be studied further by variable temperature measurements.

10.9 SOLUTION PHASE RAMAN SCATTERING

Both infrared and Raman spectral lines are observable in the solution (condensed) phase. The lines however broaden out to ~1–10 cm^{-1} or larger (FWHM) depending on the extent of collisions, intramolecular hydrogen bonding interactions, and solute-solvent interactions prevailing. Figure 10.17 shows line-broadening in the Raman spectrum of methanol in the liquid phase relative to that in the gas phase. Generally, the perturbation due to surrounding solvent molecules on the vibrational states of the solute molecules is taken to be small, which allows for a first-approximation use of the selection

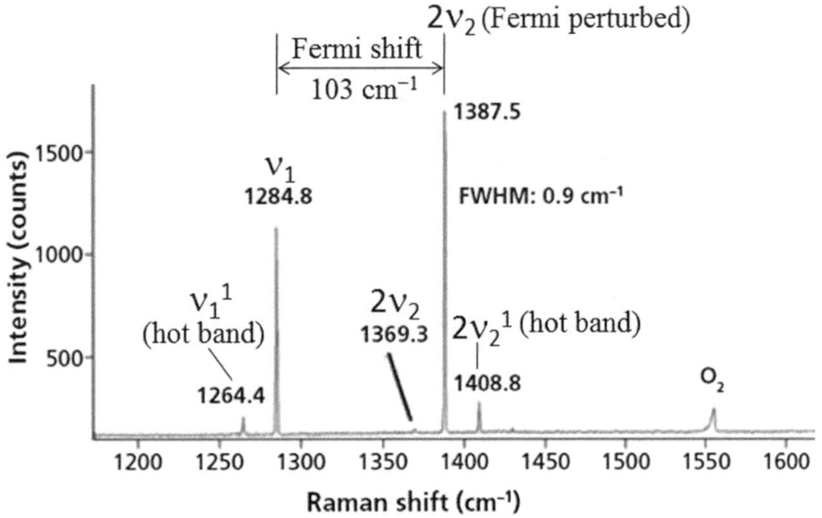

FIGURE 10.16 Raman spectrum of gas-phase CO_2 showing some secondary transitions. The band at the extreme right (~1555 cm^{-1}) is due to oxygen impurity. (Reproduced after some modification from Tuschel, 2014, with permission from the journal *Spectroscopy*.)

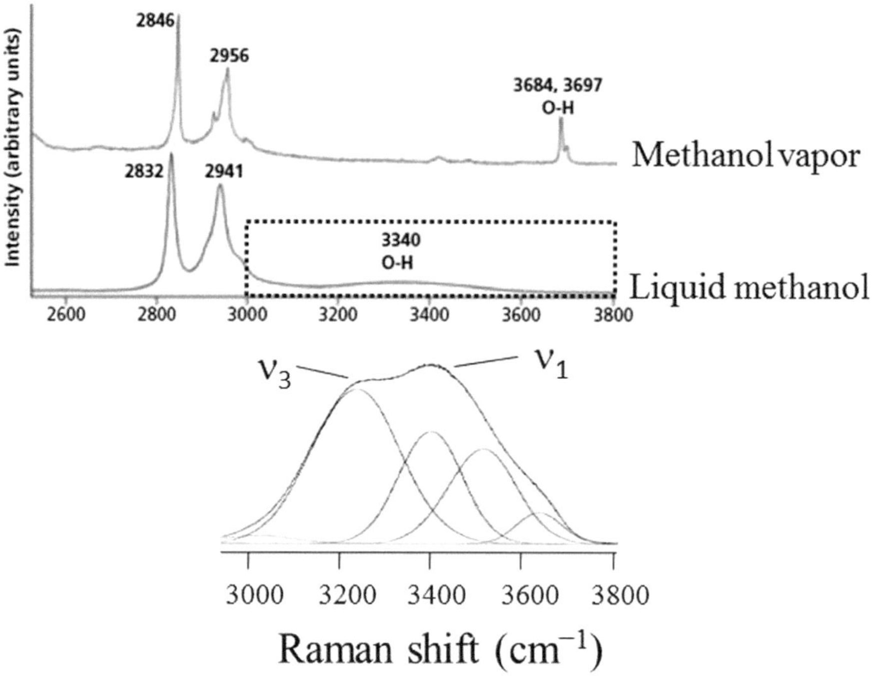

FIGURE 10.17 Raman spectra of CH_3OH in liquid phase and vapor form. The ν_1 band due to –OH stretching vibration is amplified by using the Raman vibrational spectrum of pure water. The bands under the envelope are obtained by deconvolution. (Reproduced from Tuschel, 2014, with permission from *Spectorscopy*. The lower spectrum of pure water is reproduced with ν_1 and ν_3 assignments from Carey and Korenowski, 1998, with permission from AIP.)

rules that are derived for freely moving isolated molecules. This approximation should be used with caution (see below) because the solution phase Raman lines are often shifted by tens of cm^{-1} due to the solvation effect. The Raman spectra of aqueous solutions invariably show a water band at ~1633 cm^{-1}, which can be as broad as 100 cm^{-1}. This corresponds to the bending mode Q_2 of water. The Raman spectrum of liquid water alone in the near-infrared region is a convoluted broad band due to –OH stretching vibrations, Q_1 and Q_3. Figure 10.17 also shows the broad Raman bands (ν_1 and ν_3) of pure water due to –OH stretching, which can be deconvoluted into five minor bands that may originate from different H-bonded states of water. As the liquid spectrum of methanol shows, these broad ν_1 and ν_3 bands of water appear routinely in Raman spectra of all liquid phase samples. The enormous bandwidth of water Raman scattering is due to the strength of multiple hydrogen bonding in water.

It may be mentioned that the quantized states of angular momentum in the solution phase are quenched, and hence there are no selection rules as such. In this situation, a procedure called 'averaging over all orientations' is used, which involves taking the Raman intensity $\left|A_{is}\right|^2$ for a particular molecule, and then averaging the intensity over all orientations to obtain $\left\langle \left|A_{is}\right|^2 \right\rangle$. The motional-averaging of a quantity is required under many situations, in the case of rotational magnetic moment in Stark and Zeeman effects, for example (see Chapter 11). The procedure of averaging spectra in solution is equivalent to calculating the matrix between different rotational states of freely moving isolated molecules existing in the low-pressure vapor phase. Putting another way, if the rotational motion of each molecule can be described by the same set of quantized levels of rotation, then we can calculate the intensity by squaring the appropriate matrix elements. Otherwise, and equivalently for the condensed phase, we first calculate the intensity, which is the square of the amplitude, for each individual molecule and then take the orientational average. Since different molecules are described by different orientations of the molecule-fixed axes with respect to the laboratory axes, the rotational state of a molecule is identified by the associated $\Omega(\theta, \phi, \chi)$. Then the total scattering intensity will be the sum of contributions from all molecules, the sum being over all possible values of θ, ϕ, χ

$$\left\langle \left|A_{is}\right|^2 \right\rangle = \frac{\int \left|A_{is}(\Omega)\right|^2 d\Omega}{\int \Omega d\Omega}. \tag{10.67}$$

10.10 RESONANCE RAMAN SCATTERING

Light scattering characteristically is a weak process, in the sense that most of the incident photons just pass through the sample without even interacting with the medium. The scattering yield – the ratio of the number of photons scattered to the number incident, is just about 1×10^{-3}, out of which the elastically scattered photons is about 10^3-fold larger than the inelastically scattered ones. Thus, the yield for the normal Raman process is ~10^{-6}, which amounts to only one of a million incident photons exchanging energy with the medium. Clearly, Raman signals are exceedingly weak under ordinary incident light intensity. An intense light source, a laser beam for example, could be a factor to increase scattering, but the signals can be enhanced by several orders of magnitude if the frequency of the exciting light is on resonance with a symmetry-allowed electronic transition of the molecule. This is called resonance Raman scattering, and we know that a laser source not only will provide more intensity but can also be tuned tightly to the resonance frequency of an electronic transition.

Since each electronic state contains vibrational energy levels within, the resonance Raman process refers to vibronic state functions that are represented by the products of electronic and vibrational states within the Born-Oppenheimer approximation. The three vibronic states relevant for the discussion of resonance Raman scattering are shown in Figure 10.18, in which the initial vibronic state $|i\rangle = |g\rangle|m\rangle$ is taken to the relevant excited vibronic state $|f\rangle = |g\rangle|n\rangle$. In the following we will obtain quantum mechanical expressions to understand resonance enhancement in Raman scattering.

To adapt to a quantum mechanical treatment, one may use a time-dependent induced dipole moment $\boldsymbol{\mu}_{if}(t)$ whose polarizability will evaluate the transition of the perturbed wavefunctions $|i\rangle$ to $|f\rangle$. This $\boldsymbol{\mu}_{if}(t)$ is induced by the excitation source and is distinct from the permanent dipole moment $\boldsymbol{\mu}(t)$ of the molecule, the interaction of which with the electric field of the light results in the conventional absorption and emission involving the unperturbed or zeroth-order wavefunctions. Both, however, add up to represent the transition moment from the state $|i'\rangle$ to the state $|f'\rangle$

$$\boldsymbol{\mu} = \boldsymbol{\mu}_{if\,(\text{unperturbed})} + \boldsymbol{\mu}_{if\,(\text{perturbed})}. \qquad (10.68)$$

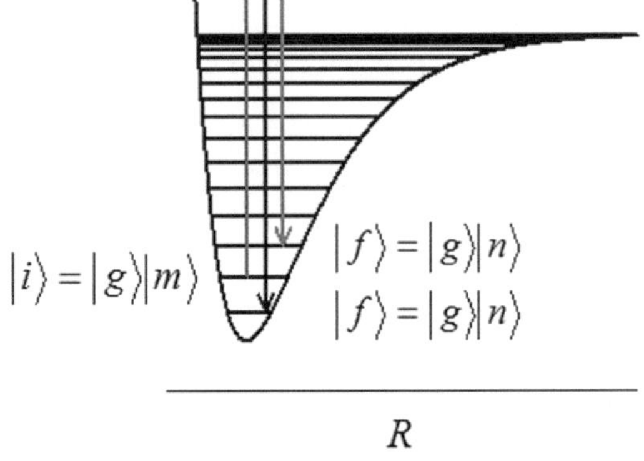

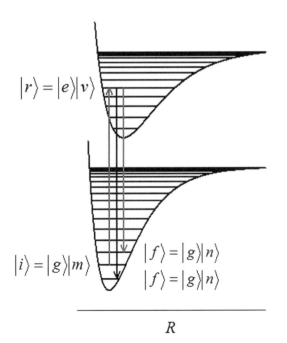

FIGURE 10.18 Vibronic transitions in resonance Raman scattering.

This time-dependent transition dipole $\boldsymbol{\mu}_{if}(t)$ which we assume linear in **E** is

$$\boldsymbol{\mu}_{if}(t) = \tilde{\boldsymbol{\alpha}}(t)\mathbf{E}(t).$$

The polarizability tensor written in the component form is

$$\mu_{\rho,if} = \alpha_{\rho\sigma} E_{\sigma}, \qquad (10.69)$$

with $\rho, \sigma = x, y, z$, the molecule-fixed axis system. By using

$$\mathbf{E}(t) = \mathbf{E}_{o}(t)\cos\omega t,$$

and employing the Hermitian property of the dipole, i.e., $\alpha_{if} = \alpha^{*}_{fi}$, we write

$$\mu_{if\,(\text{perturbed})} = \frac{1}{2}\left(\alpha_{\rho\sigma,if} e^{i\omega t} + \alpha^{*}_{\sigma\rho,fi} e^{-i\omega t}\right) E_{o}(t). \qquad (10.70)$$

Since μ_{if} is the transition electric dipole for the transition between states $|i\rangle$ and $|f\rangle$, we have the following transition integral involving time-dependent perturbed wavefunctions i' and f'

$$\mu_{if} = \langle f'|\boldsymbol{\mu}|i'\rangle, \qquad (10.71)$$

where $\boldsymbol{\mu}$ is the electric dipole moment operator. The perturbed wavefunctions can be expanded into respective wavefunctions of the zeroth, first, second, and higher orders

$$\begin{aligned} \psi_{i}' &= \psi_{i}^{(0)} + \psi_{i}^{(1)} + \psi_{i}^{(2)} + \cdots \\ \psi_{f}' &= \psi_{f}^{(0)} + \psi_{f}^{(1)} + \psi_{f}^{(2)} + \cdots. \end{aligned} \qquad (10.72)$$

The zeroth-order terms, $\psi_{i}^{(0)}$ and $\psi_{f}^{(0)}$, represent the unperturbed wavefunctions, and according to the formalism of the time-dependent perturbation theory we can write down each of the perturbed wavefunctions as a linear combination of the corresponding unperturbed wavefunction. Thus, the first-order time-dependent perturbed wavefunctions are

$$\begin{aligned} \psi_{i}^{(1)}(t) &= \sum_{r} c_{ir}^{(1)}(t)\,\psi_{r}^{(0)} e^{-i\omega_{r}t} \\ \psi_{f}^{(1)}(t) &= \sum_{r} c_{fr}^{(1)}(t)\,\psi_{r}^{(0)} e^{-i\omega_{r}t}. \end{aligned} \qquad (10.73)$$

The use of the index r in the linear combinations indicates that the sum runs over all vibronic states of the system. This in fact is the origin of calling the perturbation approach the 'sum over vibronic states' in the treatment of resonance Raman scattering. The 'sum over vibronic states' approach treats the effect of the electric field on both electronic and nuclear motions simultaneously. The calculations can be limited to the first-order wavefunctions only.

One now solves for the coefficients $c_{ir}(t)$ and $c_{fr}(t)$ in the equations above by following the procedure worked out explicitly in Chapter 3 (see section 3.3). The derivation using those steps gives the coefficients in a straightforward manner. We have

$$c_{ir}(t) = -\frac{i}{\hbar}\int_0^t V_{ir} e^{-i\omega_{ir}t} dt + \delta_{ir}$$
$$c_{rf}(t) = -\frac{i}{\hbar}\int_0^t V_{rf} e^{-i\omega_{rf}t} dt + \delta_{rf}, \tag{10.74}$$

in which V_{ir} and V_{rf} are integrals containing the $\boldsymbol{\mu}_{if}(t)\cdot\mathbf{E}(t)$ operator,

$$V_{ir} = -\langle r|\boldsymbol{\mu}_{if}\cdot\mathbf{E}_o(t)|i\rangle \cos\omega t$$
$$= \langle -r|\boldsymbol{\mu}_{if}(t)\cdot\mathbf{E}_o(t)|i\rangle\left(\frac{e^{i\omega t}+e^{-i\omega t}}{2}\right)$$

and similarly,

$$V_{rf} = -\langle f|\boldsymbol{\mu}_{if}(t)\cdot\mathbf{E}_o(t)|r\rangle\left(\frac{e^{i\omega t}+e^{-i\omega t}}{2}\right). \tag{10.75}$$

The solutions for the coefficients $c_{ir}(t)$ and $c_{fr}(t)$ are obtained identically as given in Box 3.2. Working them out explicitly yields

$$c_{ir}(t) = \frac{\langle r|\boldsymbol{\mu}_{if}(t)\cdot E_o(t)|i\rangle}{2\hbar}\left[\frac{e^{i(\omega+\omega_{ir})t}-1}{\omega+\omega_{ir}} - \frac{e^{i(\omega-\omega_{ir})t}-1}{\omega-\omega_{ir}}\right] + \delta_{ir}$$
$$c_{rf}(t) = \frac{\langle f|\boldsymbol{\mu}_{if}(t)\cdot E_o(t)|r\rangle}{2\hbar}\left[\frac{e^{i(\omega+\omega_{rf})t}-1}{\omega+\omega_{rf}} - \frac{e^{i(\omega-\omega_{rf})t}-1}{\omega-\omega_{rf}}\right] + \delta_{rf}. \tag{10.76}$$

Substitution of these coefficients into equation (10.73) gives the wavefunctions perturbed to first order

$$\psi_i' = |i^{(1)}\rangle = |i'\rangle = \left[|i\rangle + \sum_{r\neq i}\frac{\langle r|\boldsymbol{\mu}_{if}(t)\cdot E_o(t)|i\rangle}{2\hbar}\left\{\frac{e^{i(\omega+\omega_{ir})t}-1}{\omega+\omega_{ir}} - \frac{e^{i(\omega-\omega_{ir})t}-1}{\omega-\omega_{ir}}\right\}|r\rangle\right]e^{-i\omega_i t}$$
$$\psi_f' = |f^{(1)}\rangle = |f'\rangle = \left[|f\rangle + \sum_{r\neq f}\frac{\langle f|\boldsymbol{\mu}_{if}(t)\cdot E_o(t)|r\rangle}{2\hbar}\left\{\frac{e^{i(\omega+\omega_{rf})t}-1}{\omega+\omega_{rf}} - \frac{e^{i(\omega-\omega_{rf})t}-1}{\omega-\omega_{rf}}\right\}|r\rangle\right]e^{-i\omega_f t}. \tag{10.77}$$

Note that $|i\rangle = |i^{(0)}\rangle$ and $|f\rangle = |f^{(0)}\rangle$ are the unperturbed zeroth-order wavefunctions. We now use the rotating wave approximation (see Boxes 3.2 and 3.3) to remove the $e^{i\omega_{ir}t}$ and $e^{i\omega_{rf}t}$ terms, because it is only the frequency of the incident light ω, and not ω_{ir} or ω_{rf}, that induces the transition dipole moment $\boldsymbol{\mu}_{if}$. The result is

$$|i'\rangle = \left[|i\rangle + \frac{1}{2\hbar}\sum_{r\neq i}E_o(t)\mu_{ir}\left\{\frac{e^{i\omega t}}{\omega+\omega_{ir}} - \frac{e^{-i\omega t}}{\omega-\omega_{ir}}\right\}|r\rangle\right]e^{-i\omega_i t}$$
$$|f'\rangle = \left[|f\rangle + \frac{1}{2\hbar}\sum_{r\neq f}E_o(t)\mu_{rf}\left\{\frac{e^{i\omega t}}{\omega+\omega_{rf}} - \frac{e^{-i\omega t}}{\omega-\omega_{rf}}\right\}|r\rangle\right]e^{-i\omega_f t}. \tag{10.78}$$

With these workouts, the integral $\mu_{if} = \langle f'|\boldsymbol{\mu}|i'\rangle$ founded in equation (10.71) yields

$$\mu_{if\,(\text{unperturbed})} = \langle i|\boldsymbol{\mu}|f\rangle$$

$$\mu_{if\,(\text{perturbed})} = \frac{1}{2\hbar}\sum_r \mu_{ir}\mu_{rf}\left(\frac{e^{i\omega t}}{\omega+\omega_{rf}} - \frac{e^{-i\omega t}}{\omega-\omega_{rf}}\right)E_o(t)$$
$$+\frac{1}{2\hbar}\sum_r \mu_{rf}\mu_{ir}\left(\frac{e^{-i\omega t}}{\omega+\omega_{ir}} - \frac{e^{i\omega t}}{\omega-\omega_{ir}}\right)E_o(t) \tag{10.79}$$

This appears fine, but one is specifically interested in the induced dipole $\mu_{if\,(\text{perturbed})}$; the $\mu_{if\,(\text{unperturbed})}$ term may be left out. The $\mu_{if\,(\text{perturbed})}$ term compares with equation (10.70), showing

$$\alpha_{if} = \frac{1}{\hbar}\sum_r\left(\frac{\mu_{ir}\mu_{rf}}{\omega+\omega_{rf}} - \frac{\mu_{rf}\mu_{ir}}{\omega-\omega_{ir}}\right). \tag{10.80}$$

This equation can be written to indicate the directional polarizability components explicitly. If the Cartesian components of the dipole moment operators are represented with $\rho = x, y, z$ and $\sigma = x, y, z$, the polarizability components are obtained as

$$\alpha_{\rho\sigma,if} = \frac{1}{\hbar}\sum_r\left(\frac{\langle i|\mu_\sigma|r\rangle\langle r|\mu_\rho|f\rangle}{\omega+\omega_{rf}+\frac{i\Gamma}{2}} - \frac{\langle i|\mu_\rho|r\rangle\langle r|\mu_\sigma|f\rangle}{\omega-\omega_{rf}+\frac{i\Gamma}{2}}\right). \tag{10.81}$$

The appearance of the Γ-factor in the frequency denominator, call it the damping constant, means the loss of phase coherence which is similar to the time constant T_2 used generally in spectroscopy, giving rise to the electronic linewidth that we discussed in Chapter 3 (see equations (3.36) and (3.37))

$$\Gamma \equiv \frac{1}{T_2} \equiv \frac{1}{\Sigma_i \tau_i} + \frac{1}{\Sigma_j \tau_j},$$

where the index i represents the processes that contribute to the loss of phase coherence of the vibronic state, and j represents the processes that lead to spontaneous decay of the excited state. In fact, the Γ-factor could have been incorporated already in equation (10.76), because the excited-state populations $N|c_{ir}(t)|$ and $N|c_{rf}(t)|$, where N is the number of excited molecules, decay exponentially with a time constant $\Gamma / 2$ (see Chapter 3).

The expression for $\alpha_{\rho\sigma,if}$ above, call it the $\rho\sigma^{\text{th}}$ matrix element of the polarizability tensor, represents the Raman amplitude for $|f\rangle \leftarrow |i\rangle$ transition and is termed the Kramers-Heisenberg-Dirac (KHD) formula, where ω is the frequency of the exciting light and ω_{ir} is the resonance frequency. If the incident light frequency is tuned closer to the resonance frequency, $\omega - \omega_{ir} \sim 0$, then the second term, called the resonance term, dominates so that the first anti-resonant term can be neglected, and we have

$$\alpha_{\rho\sigma,if} \approx \frac{1}{\hbar}\sum_r \left(\frac{\langle i|\mu_\rho|r\rangle\langle r|\mu_\sigma|f\rangle}{\omega - \omega_{ir} - \frac{i\Gamma}{2}} \right), \tag{10.82}$$

the sum running over all vibronic states. The resonance Raman intensity is

$$I_{\text{Raman}} \propto \left|\alpha_{\rho\sigma,if}\right|^2. \tag{10.83}$$

Although the KHD formula accurately relates the matrix elements of the polarizability tensor to the resonance condition, the vibronic-state functions are not identified in terms of the respective electronic and vibrational states distinctly. Since Raman effect arises from the change of the molecular vibrational level during scattering, the vibrational states involved have to be identified. In the following, we look at the dependence of the electric transitions dipole integrals on the nuclear distance (Q) in the KHD formula with reference to normal modes of vibration Q_k. It will be helpful to recapitulate and formalize the symbols and labels required to identify the vibronic states. The amplitude of the transition moment is

$$\left\langle i|\mu_f|r \right\rangle = \mu_{f,ir}. \tag{10.84}$$

Similarly, $\mu_{f,rf}$, $\mu_{\sigma,ir}$, and $\mu_{\sigma,if}$ are amplitudes of the corresponding transition moments,

- $i = g,m$ represents the ground electronic state g and the vibrational state m,
- $r = e,v$ represents the electronic state e and the vibrational state v, and
- $f = g,n$ implies the ground electronic state g and the vibrational state n.

A vibronic state is a function of the set of electronic coordinates ζ and the set of nuclear coordinates Q. The coordinates locate the positions of the electrons and nuclei. A vibronic state could be

$$\psi_i = \theta_g(\xi,Q)\phi_j^{(g)}, \tag{10.85}$$

in which the $\theta_g(\xi,Q)$ function is the ground-state electronic wavefunction, and $\phi_j^{(g)}$ is the jth vibrational-state wavefunction in the ground electronic state.

With these designations, the KHD formula can be written as

$$\left(\alpha_{\rho\sigma}\right)_{gm,gn} = \frac{1}{\hbar}\sum_{ev}\left[\frac{\left(\mu_\sigma\right)_{gm,ev}\left(\mu_\rho\right)_{ev,gn}}{\omega + \omega_{ev,gn} + \frac{i\Gamma}{2}} + \frac{\left(\mu_\rho\right)_{gm,ev}\left(\mu_\sigma\right)_{ev,gn}}{\omega - \omega_{gm,ev} + \frac{i\Gamma}{2}} \right]. \tag{10.86}$$

Consider one of the dipole transition moments, $\left(\mu_\sigma\right)_{gm,ev}$ for example, which we will write for convenience in the usual integral format

$$\begin{aligned}\left(\mu_\sigma\right)_{gm,ev} &= \int\left(\theta_g\phi_m^{(g)}\right)^* \mu_\sigma\left(\theta_e\phi_v^{(e)}\right)d\xi dQ \\ &= \int\phi_m^{(g)*}\left[M_\sigma(\theta)\right]_{eg}\phi_v^{(e)}dQ,\end{aligned} \tag{10.87}$$

where

$$\left[M_\sigma(\theta)\right]_{eg} = \int\theta_g\mu_\sigma\theta_e d\xi$$

is the integration over the electronic coordinate at the nuclear configuration Q^*. But the integral over Q is not carried out when the dependence of the electronic transition moment on Q is not known. Instead, the nuclear motion is taken as a perturbation in the electronic Schrödinger equation, and the ground-state electronic wavefunctions at equilibrium nuclear configurations are used as unperturbed (zeroth-order) wavefunctions. We will not attempt to elaborate on this, reserving it for further research studies, but will give the available first-order perturbation result due to Albrecht (1961)

$$M_{eg} = M_{eg}^{(0)} + \sum_s \lambda_{es}(Q) M_{gs}^{(0)}, \tag{10.88}$$

where

$$\lambda_{es}(Q) = \frac{\sum_k \left(\int\theta_e^{(0)}\mathcal{H}'(Q)\theta_s^{(0)}\right)^{(k)} Q_k}{E_s^{(0)} - E_e^{(0)}}. \tag{10.89}$$

The index s refers to all excited states except the specific excited state 'e', and $M_{gs}^{(0)}$ incorporates all electronic transition moments involving the ground-state g and the excited-states s, all evaluated at the ground-state nuclear configuration. The summation in the equation above extends over all normal modes Q_k, and the integral represents the perturbation energy per unit displacement of the kth normal mode when the electronic states $\theta_s^{(0)}$ and $\theta_e^{(0)}$, both given under the ground-state equilibrium nuclear configuration, mix under nuclear vibrational perturbation (Albrecht, 1961).

The purpose of the discussion above is to indicate how the results of vibronic theory are incorporated into the KHD equation. The simplest form available to present the polarizability tensor is

$$\alpha_{\rho\sigma,if} = A + B \tag{10.90}$$

$$A = \frac{1}{\hbar}\sum_{v} \frac{\langle i|\mu_\rho|e\rangle\langle e|\mu_\sigma|i\rangle}{\omega_{ei} - \omega + \frac{i\Gamma}{2}} \langle v_i^{(i)}|v_k\rangle\langle v_k|v_f^{(i)}\rangle \tag{10.91}$$

$$B = -\sum_{s}\sum_{v_k} \left[\begin{array}{c} \langle i|\mu_\sigma|e\rangle\left\langle e|\frac{\partial \mathcal{H}_{el}}{\partial Q_k}|s\right\rangle\langle s|\mu_\rho|g\rangle \\ +\langle g|\mu_\rho|e\rangle\left\langle e|\frac{\partial \mathcal{H}_{el}}{\partial Q_k}|s\right\rangle\langle s|\mu_\rho|g\rangle \end{array} \right] \times$$

$$\frac{2\pi\langle v_i^{(i)}|Q_k|v_f^{(i)}\rangle}{\omega_{es}\left(\omega_{ei} - \omega + \frac{i\Gamma}{2}\right)}. \tag{10.92}$$

The denotations of various symbols and their sub- and superscripts need a review and are summarized below.

i and e	the ground and resonant excited states
s	excited states other than e
$v_i^{(i)}$ and $v_f^{(i)}$	the initial and the final vibrational wavefunctions of the ground electronic state
v_k	vibrational level of the Q_k normal mode in the excited electronic state e
$\frac{\partial \mathcal{H}_{el}}{\partial Q_k}$	Herzberg-Teller perturbation term which expresses the change in the electronic Hamiltonian due to nuclear vibrations in the Q_k normal mode of vibration (see Chapter 11)
$\langle v_i^{(i)}|v_k\rangle\langle v_k|v_f^{(i)}\rangle$	Franck-Condon overlap factor between the ground and the excited state.

In the form of the polarizability tensor given by equation (10.90) the term A contains the Franck-Condon overlap factor. If the electronic transition in resonance with the Raman excitation is symmetry-allowed, and if the excitation frequency ω approaches the electronic resonance frequency ω_{ei}, then the Raman intensity will increase dramatically. This is often called the A-factor enhancement. We see that the enhancement occurs only for totally symmetric vibrational modes which are Franck-Condon active and are coupled with the electronic transition. The occurrence of the enhancement requires that the Franck-Condon factor is not zero, meaning the minimum of the excited-state potential is shifted with respect to the minimum of the ground state surface. The Franck-Condon overlap is a topic for the next chapter, but for now we state that a vibrational transition $v_i^{(i)}$ to $v_f^{(i)}$ occurs via Franck-Condon overlaps with the vibrational state v_k of the excited electronic state. The subscript k refers to the kth normal mode. If the initial and the final excited-state electronic potentials have identical geometry and minima at the same equilibrium nuclear configuration, then the vibrational states in the initial and excited potentials would be identical, leading to vanishing Franck-Condon overlap. In such a case Raman scattering will not occur through the A-term. The intensity of Raman scattering is thus related to the extent of the shift in the equilibrium molecular geometry in the excited state with respect to the ground state.

The B-term in the polarizability expression shows the transition moment for not only the $i \rightarrow e$ transition, but also the $s \rightarrow g$ transition where s is another excited state. The involvement of two or more allowed electronic transitions results in the so-called vibronic borrowing of intensity from one to another electronic transition. The electronic states – all defined at the equilibrium nuclear geometry, are perturbed to the first-order by nuclear vibrations via the Herzberg-Teller matrix element $\left(\frac{\partial \mathcal{H}_{el}}{\partial Q}\right)$. The vibrations that couple different electronic excited states are enhanced if they are Raman-active modes within the resonant electronic absorption band.

Clearly, resonance Raman scattering is particularly useful to study complex systems like crystals, proteins, and mixtures of compounds rather than diatomics and small molecular systems. The selection rules specific to resonance-enhanced Raman scattering require that the electronic transitions are allowed. Consider, for example, the e and s excited electronic states. If the ground electronic state g is totally symmetric, then the character-based symmetry species Γ_e and Γ_s should correspond to different symmetry species of x,y,z translations. For example, if $\Gamma_e = \Gamma_x$ and $\Gamma_s = \Gamma_y$, then

$$\Gamma_e \otimes \Gamma_s = \Gamma_{xy}.$$

For the kth normal mode the species Γ_k must correspond to Γ_{xy}. In other words, a totally symmetric representation occurs in $\Gamma_e \otimes \Gamma_s \otimes \Gamma_k$.

It should be obvious that a resonance Raman spectrum does not always reflect transitions of all Raman active modes. Since the excitation is on resonance with an electronic absorption, only vibrational modes of the absorbing chromophore that are coupled to electronic transitions giving rise to the absorption spectrum will be observable in the resonance Raman spectrum. For example, resonance excitation in the Soret absorption band of the heme in myoglobin gives Raman lines associated with heme vibrational modes alone. Figure 10.19 shows picosecond-resolved temporal response of the in-plane vibrational modes (ν_2, ν_4, and ν_{10}) immediately after laser photolysis of the CO ligand from CO-bound myoglobin (MbCO). Since the frequency of the ν_2 band (A_{1g}) is known to be sensitive to the core size of the heme porphyrin ring, the time evolution of CO-dissociated spectrum indicates that the porphyrin core expands almost instantaneous to photodissociation. To obtain more details of the time evolution of heme vibrational modes, the resolution has to be extended to subpicosecond regime. Femtosecond-resolved resonance Raman spectra have, of course, been studied.

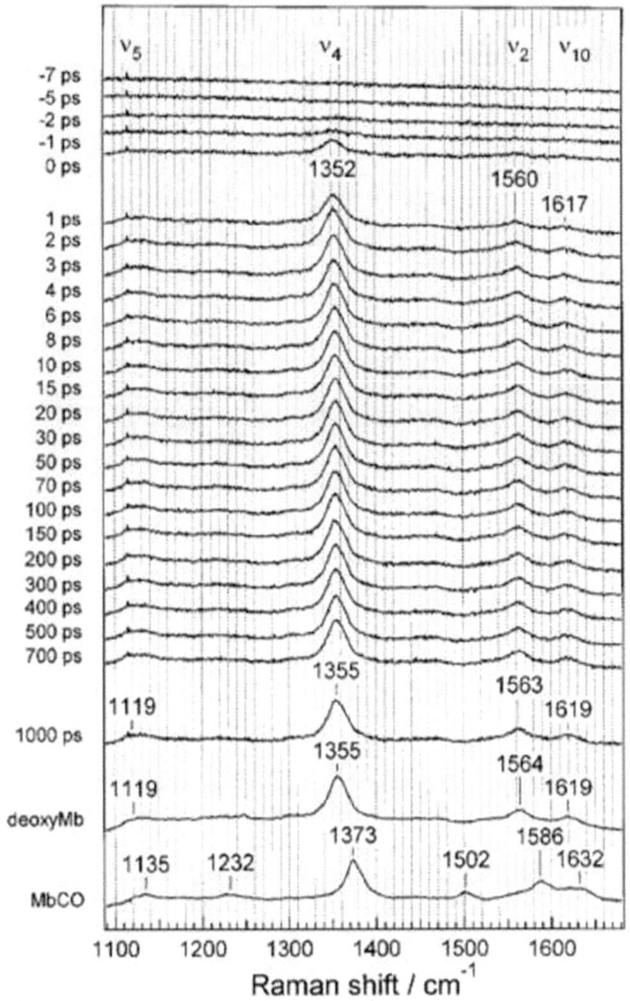

FIGURE 10.19 Time-resolved resonance Raman spectrum of the immediate photoproduct (Mb*) obtained by laser photodissociation, MbCO $\rightarrow$ Mb* + CO. The evolution of the in-plane heme vibrational modes are shown. (Reproduced from Mizutani and Kitagawa, 2001, with permission from the American Chemical Society.)

10.11 SUNDRIES AND OUTLOOK

A description of all the phenomenal advances in Raman spectroscopy during the past few decades is beyond the scope of this book. We only mention that closer on the heels of the experimental development of resonance Raman in 1970s was the discovery of surface-enhanced Raman spectroscopy (SERS) (Fleischmann et al., 1974), a technique in which the scattering molecules are finely deposited on a rough metallic surface. Resonance excitation of the localized surface plasmons produces near-field light on the metallic surface. This strongly enhanced light produces very intense Raman scattering from the deposited scatterers. Amazingly, SERS signals may be enhanced by a factor ~10^{13}-fold, which means Raman scattering can be studied at the single molecule level. Raman spectroscopy is also combined with microscopy. Such a spectro-microscopic approach provides complete vibrational spectroscopic information of each point of the sample scanned. This forms the basis of Raman imaging.

Box 10.1

The polarizability is a molecular property because the electron clouds are constantly perturbed by an applied electric field **E**. To a first approximation, the 'bulk polarization' density P may be taken to vary linearly with strength of the electric field **E**, so the polarizability is a derived quantity

$$P \propto \mathbf{E}.$$

This proportionality, however, need not be applicable at all field strengths. When **E** is very large the effect is not at the level of the electron cloud perturbation alone, but there will be atomic displacements as well. The polarizability property is thus studied at small **E** fields, where the induced polarization may be assumed linear with the field strength, and we write

$$\mathbf{P}_i = \chi_{ij}\mathbf{E}_j,$$

which implies that χ_{ij} is a measurable quantity that linearizes the ith component of the induced dipole to the jth component of the electric field. Note that $\mathbf{P}_i$ and $\mathbf{E}_j$ are two vectors, so χ_{ij} is a tensor (see Box 7.1).

How does χ_{ij} appear? To see this, we expand P in a Taylor series about the zero-value of the electric field, $\mathbf{E} = 0$. The expansion is

$$P(E) = (P)_{E=0} + (E-0)\left[\frac{\partial P(E)}{\partial E}\right]_{E=0} + \frac{(E-0)^2}{2!}\left[\frac{\partial^2 P(E)}{\partial E^2}\right]_{E=0} + \cdots$$

The first term in the expansion is 'zero' because $P = 0$ when $\mathbf{E} = 0$. The third and higher order terms are small, so we neglect them. The second term remaining is

$$E\left[\frac{\partial P(E)}{\partial E}\right] \equiv \chi_{ij}E$$

Now we have χ_{ij} as the electric susceptibility tensor. It is a bulk property. For an individual molecule we would write

$$\mu_i = \alpha_{is}\mathbf{E}_s$$

where μ_i is the ith component of the induced dipole vector, α_{is} is the molecular polarizability tensor, and $\mathbf{E}_s$ is the sth component of the electric field vector. In a Raman-type of light scattering experiment, the electric field, both exciting and measuring, is generally given in the lab-fixed Cartesian system (X, Y, Z), and the induced dipole is given in the molecule-fixed axes system $(x, y, z$, which are a, b, c, in the principal axes system).

Box 10.2

The rotation of a molecule, a diatomic molecule for the simplest case, involves an interchange of the nuclei in every π-rotation

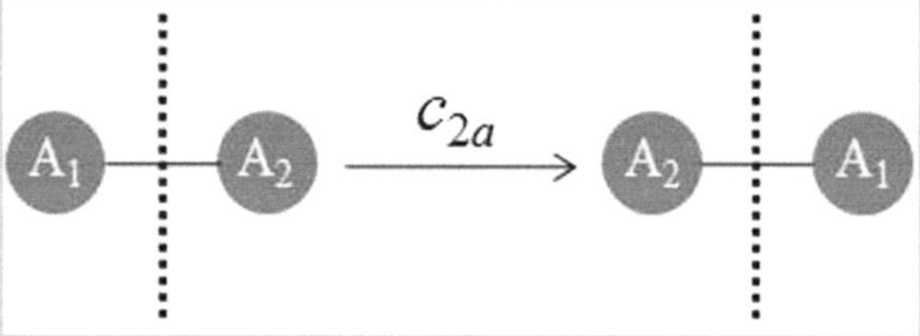

The basic question one asks is how the interchange of nuclei during rotation affects the total molecular wavefunction ψ? We can write for ψ as a product of electronic, vibrational, rotational, and nuclear spin functions

$$\psi_{\text{total}} = \psi_{\text{el}}\psi_{\text{vib}}\psi_{\text{rot}}\psi_{\text{ns}}$$

The ψ_{el} and ψ_{vib} functions are not expected to be affected, so we need to analyze how $\psi_{\text{rot}}\psi_{\text{ns}}$ will determine the symmetry of ψ_{total}. Let us consider the symmetry property of ψ_{ns} first. Denoting nuclear spin angular momentum (spin, in short) as **I**, each nucleus has $2I+1$components. If $I = 1/2$ as for a proton, the two nuclei A_1 and A_2 will produce $(2I+1)(2I+1) = 4$ product spin functions, which are

$$\alpha(1)\alpha(2)$$
$$\beta(1)\beta(2)$$
$$\alpha(1)\beta(2)$$
$$\beta(1)\alpha(2)$$

Application of nuclear spin interchange operator P gives

$$\alpha(1)\alpha(2) \rightarrow \alpha(2)\alpha(1)$$
$$\beta(1)\beta(2) \rightarrow \beta(2)\beta(1)$$
$$\alpha(1)\beta(2) \rightarrow \beta(2)\alpha(1)$$
$$\beta(1)\alpha(2) \rightarrow \alpha(2)\beta(1)$$

Clearly, the first two product functions do not interchange spin, but the latter two do – one reproducing the other. Whenever a product spin function interchanges the nuclear spins to produce another spin function, and vice versa, we take a linear combination of the two product functions to generate two combined product functions. For the present case we will have

$$\frac{1}{\sqrt{2}}\left[\beta(1)\alpha(2)+\beta(2)\alpha(1)\right]$$
$$\frac{1}{\sqrt{2}}\left[\alpha(1)\beta(2)-\beta(1)\alpha(2)\right]$$

The four spin functions of the two hydrogen nuclei may now be grouped according to whether a function is symmetric or antisymmetric with respect to nuclear interchange

Symmetric	Antisymmetric
$\alpha(1)\alpha(2)$	$\frac{1}{\sqrt{2}}\left[\alpha(1)\beta(2)-\beta(1)\alpha(2)\right]$
$\beta(1)\beta(2)$	
$\frac{1}{\sqrt{2}}\left[\beta(1)\alpha(2)+\beta(2)\alpha(1)\right]$	

This is exactly what we did earlier in the spin interchange of two electrons to generate triplet- and singlet-state spin functions (see equation (5.56)). Here we are considering the proton spin vis-à-vis electron spin there. As a rule, the number of symmetric product spin functions is $(2I+1)(I+1)$, and that of antisymmetric product spin functions is $(2I+1)I$. They are called nuclear spin-state degeneracy,g_N, which determines rotational level degeneracy for the obvius reason that a π-rotation interchanges the nuclei. For $^{14}N_2$ $(I = 1)$, for example, there will be six symmetric and three antisymmetric spin functions, and for $^{16}O_2$ $(I = 0)$ we would have just one symmetric and no antisymmetric function.

Now consider the symmetry property of ψ_{rot}, the rotational wavefunctions. The rotational levels are determined by the quantum number J, and ψ_{rot} are Legendre polynomials

$$\psi_{J=0} = 1$$
$$\psi_{J=1} = \cos\theta$$
$$\psi_{J=2} = 3\cos^2\theta - 1$$
$$\psi_{J=3} = \frac{1}{2}\left(5\cos^3\theta - 3\cos\theta\right)$$
$$\psi_{J=4} = \frac{1}{8}\left(35\cos^4\theta - 30\cos\theta\right)\cdots$$

If these functions are subjected to a c_2 rotation operation, those of odd J change sign and those of even J do not. In so far as the symmetry of the total molecular wavefunctions is concerned, we stipulated at the beginning that the total wavefunction is

$$\psi_{total} = \psi_{rot}\psi_{ns},$$

so ψ_{total} will be symmetric or antisymmetric depending on symmetric or antisymmetric property of ψ_{rot} and ψ_{ns}. We should also require that ψ_{total} be symmetric with respect to nuclear interchange if the two nuclei of the homodiatomic molecule are bosons ($I = 0,1,2,\cdots$). In the case of fermions ($I = 1/2, 3/2, 5/2,\cdots$) ψ_{total} is required to be antisymmetric. For example, the ^{14}N nucleus ($I = 1$) is a boson, for which ψ_{total} should be symmetric; accordingly, the ground state of the $^{14}N_2$ molecule is Σ_g^+, which is a symmetric function. To obtain a symmetric ψ_{total}, the ψ_{rot} states of $^{14}N_2$ with even J have to combine with symmetric spin functions ψ_{ns} that have a g_N value of $(2I+1)(I+1) = 6$. The ψ_{total} function of $^{14}N_2$ will be symmetric also when the odd-J rotational wavefunctions (antisymmetric) combines with the antisymmetric ψ_{ns} functions that have a g_N value of $(2I+1)\times I = 3$. Thus the intensity of the rotational lines in the $^{14}N_2$ spectrum will alternate in the 3:6 or 1:2 as J progresses from 1. This intensity pattern is shown in Figure 10.10 in the text.

To recapitulate, the total molecular wavefunction of homonuclear diatomics with fermionic nuclei is required to be antisymmetric with respect to the nuclear interchange during molecular rotation. For bosonic nuclei the requirement is symmetric. The appropriate combination of even or odd-J symmetry with spin functions, both with respect to nuclear interchange, is determined. The nuclear spin statistical weight g_N of the appropriate symmetry of spin functions then determines the intensity of the rotational lines corresponding to these J.

PROBLEMS

10.1 Raman spectrum of a solid is highly sensitive to temperature and pressure. For example, the figure below shows the temperature dependent shift of the 1332 cm^{-1} (T_{2g}) Raman line of diamond taken from one of the earliest measurements reported by Krishnan (1946).

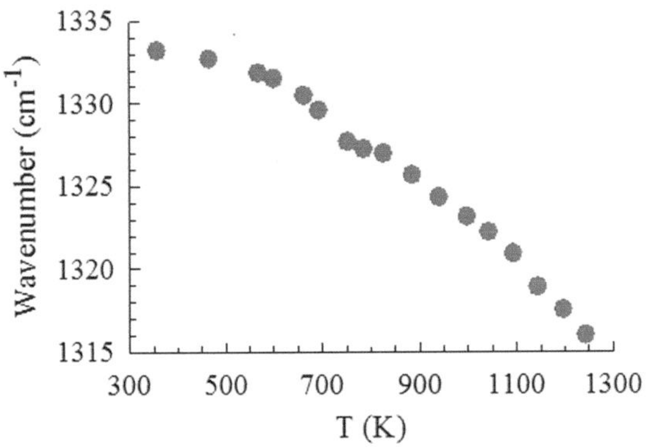

Why does the frequency, which is the C–C stretching in this case, shift with temperature? Find some applications of temperature-dependent Raman shifts.

10.2 The integrated intensity I_s of Stokes Raman lines of a solid is often found to decrease with temperature. Note that I_s may increase in certain instances. What could be the reason for the decrease of I_s with temperature? What inference can be drawn from such observations? (A good starting point to answer this question is to think about the derivative of polarizability.)

10.3 Although equation (10.21) in the text is given in an ordinary way, Stokes to anti-Stokes intensity ratios are often used to determine equilibrium vibrational temperature. What is the basis for this?

10.4 Similar to that for Raman lines, the Brillouin phonon lines are also sensitive to temperature and pressure. Predict the dependence of phonon frequency shift on temperature as a material undergoes expansion.

10.5 Consider the normal modes of ethylene we discussed in the preceding Chapter.
(a) Which of these modes are IR-active alone, Raman-active alone, and both IR- and Raman-active?
(b) Show the Raman scattering tensor for each mode of ethylene. Justify.

10.6 Inspect the representations of the polarizability ellipsoid for CO_2 shown in Figure 10.7. The representation that the ellipsoid turns spherical with symmetric stretching $\overleftarrow{O} = C = \overrightarrow{O}$ may seem not right to some. In fact, some textbooks (incorrectly) show the opposite representation, that is, a compressed ellipsoid for symmetric stretching. Explain why the ellipsoid should be elongated when the bonds compress symmetrically.

10.7 Prove that the electric dipole moment operator is Hermitian.

10.8 Suppose one wishes to study the ν_3 symmetric stretch band of water in the Raman spectrum. The geometry of the experiment may consist of Z-polarization of the incident light, and observation of the scattered light may be along X. Let one take two spectra, one with the analyzer set at Z and another set at Y. What will be the ratio of scattering strength of the two settings?

10.9 With reference to Figure 10.11 in the text, no explanation is given as to why the spacing between rotational lines in the Raman spectrum is not linear with J if centrifugal distortions are not neglected. Discuss why.

10.10 A centrosymmetric molecule may be either IR-active or Raman-active, but not both simultaneously. Provide an example (not of CO_2) and explain why this is so.

10.11 It is said that a correct Raman experiment can yield information about the symmetry of vibrations, which IR spectroscopy may fail to provide. Explain how.

10.12 It can be said that overtones and combination modes are not as frequently observed in the Raman spectrum as in vibrational spectra, and even when observed in the former, are weak. Explain why this is so. Note, this question is appropriate only in the context of non-resonance Raman spectroscopy. The argument may fail when resonance occurs in Raman excitation.

BIBLIOGRAPHY

Albrecht, A. C. (1961) *J. Chem. Phys.* 34, 1476.

Berne, B. J., and R. Pecora (2000) *Dynamic Light Scattering With Applications to Chemistry, Biology and Physics*, Dover Publications, Inc.

Carey, D. M. and G. M. Korenowski (1998) *J. Chem. Phys.* 108, 2669.

Damzen, M. J., V. I. Vlad, V. Babin, and A. Mocofanescu (2003) *Stimulated Brillouin Scattering: Fundamentals and Applications*, CRC Press, Taylor & Francis Group.

Fleischmann, M., P. J. Hendra, and A. McQuillan (1974) *Chem. Phys. Lett.* 26, 163.

Krishnan, K. S. (1946) *Proc. Indian Acad. Sci. A* 24, 45.

Long, D. A. (2002) *The Raman Effect: A Unified Treatment of the Theory of Raman Scattering by Molecules*, John Wiley & Sons Ltd.

Mizutani, Y. and T. Kitagawa (2001) *J. Phys. Chem. B* 105, 10992.

Tuschel, D. (2014) *Spectroscopy* 29, 9.

11 Electronic Spectra

Electronic transitions cause simultaneous alteration of vibrational and rotational energies and in many cases nuclear hyperfine interactions where the changes in the respective quantum numbers are not selective, although the electronic transitions themselves must be allowed by selection rules. The electronic spectrum in the case of moderate to large-size molecules could be quite complex due to multiple vibrational modes of different symmetries and equally complex rotor expressions. Electronic transitions in the case of small (diatomic) molecules however form a rich source of information about the equilibrium nuclear geometry and dissociation energies in different excited states, the vibrational quantum numbers in the ground and excited electronic states enabling assignment of a vibrational band, the rotational fine structure, and the coupling of electronic and rotational angular momenta. Provided here is a brief description of dipole transitions in the rovibronic spectrum. We analyze band progressions and coupling of angular momenta, bond dissociation, distinction between bound and unbound electronic potentials, Zeeman effect on rotational transitions, and rotational magnetooptic effects.

11.1 ENERGY TERM-VALUE FORMULAS FOR MOLECULAR STATES

By spectroscopic principles, the energy of a molecular state is partitioned into contributions from electronic, vibrational, and rotational states. The total energy term of a state is given by

$$T = T_{el} + G + F. \tag{11.1}$$

The electronic energy is partitioned into

$$T_{el} = T_0 + A\Lambda\Sigma, \tag{11.2}$$

the quantities Λ and Σ ($\Omega = \Lambda + \Sigma$) being the projections of electron (Λ) and spin (Σ) angular momenta onto the internuclear axis, and A is the spin-orbit coupling parameter. If the electron spin is neglected, then $T_{el} = T_0$, yielding the vibrational energy term

$$G = \omega_e\left(v+\frac{1}{2}\right) - x_e\omega_e\left(v+\frac{1}{2}\right)^2 + y_e\omega_e\left(v+\frac{1}{2}\right)^3 + z_e\omega_e\left(v+\frac{1}{2}\right)^4 + \cdots, \tag{11.3}$$

where $x_e\omega_e$, $y_e\omega_e$, and $z_e\omega_e$ are, respectively, the first, second, and third anharmonicity constants. The common quantity ω_e in these three constants is the harmonic vibrational frequency given in wavenumber (1 cm^{-1} = 29.98×10^9 Hz). Specifically, these are infinitesimal vibrations about the equilibrium configuration, which means $x_e\omega_e,\ y_e\omega_e,\ z_e\omega_e \ll \omega_e$. The harmonic vibration frequency or the fundamental frequency is denoted by

$$\omega_e = \frac{1}{2\pi c}\sqrt{\frac{k}{\mu}} = \frac{\beta}{c}\sqrt{\frac{D_e}{2\pi^2\mu}}, \tag{11.4}$$

in which k is the force constant, μ is the reduced mass of the molecule, and D_e is the dissociation energy. Clearly, the parameter β is related to k by

$$\beta = \sqrt{\frac{k}{2D_e}}. \tag{11.5}$$

In fact, β characterizes the potential, say a Morse potential

$$V(R) = D_e\left[1 - e^{-\beta(R-R_e)}\right]^2. \tag{11.6}$$

Keep in mind that the Morse potential is rather basic, not satisfactory to account for anharmonicity higher than the first order.

The rotational energy is

$$F = B_vJ(J+1) - D_vJ^2(J+1)^2 + H_vJ^3(J+1)^3 + \cdots, \tag{11.7}$$

where the first rotational constant B_v is given by

$$B_v = B_e - \alpha_e\left(v+\frac{1}{2}\right) + \gamma_e\left(v+\frac{1}{2}\right)^2 + \cdots. \tag{11.8}$$

The first term B_e is the rotational constant corresponding to the equilibrium internuclear separation R_e, and hence the centrifugal distortion is ignored. It is interesting that $B_v < B_e$ if anharmonicity is considered because R_e increases slightly for an anharmonic potential. Therefore, we should consider α_e, which, for a vibrational potential energy function containing up to quartic terms, is given by

$$\alpha_e = \frac{24B_e^3R_e^3g}{\omega_e^3} - \frac{6B_e^2}{\omega_e}, \tag{11.9}$$

where the cubic anharmonicity constant g is obtained from the electronic potential of a diatomic molecule

$$E(R) = E(R_e) + f(R-R_e)^2 - g(R-R_e)^3 + j(R-R_e)^4. \tag{11.10}$$

DOI: 10.1201/9781003293064-11

Normally, the coefficient g that determines the asymmetry of the potential is rather small, so the value of α_e should also be small relative to that of B_e. An approximate value of $x_e\omega_e$ is obtained from

$$x_e\omega_e \sim \frac{3\hbar^2}{8\pi^4\omega_e^2c^2}\left\{\frac{5g^2\hbar}{4\pi\omega_e^2\mu c}-j\right\}, \quad (11.11)$$

in which μ denotes the reduced mass, and j is quartic anharmonicity. The B_v expansion (equation (11.8)) is generally terminated at the α_e term, higher terms most often are redundant. The constants D_v and H_v in the rotational energy term F (equation (11.7)) are centrifugal stretching constants that depend on the vibrational state. The rotational constant D_v is

$$D_v = D_e + \beta_e\left(v+\frac{1}{2}\right) - \delta_e\left(v+\frac{1}{2}\right)^2 + \cdots. \quad (11.12)$$

The constant H_v is rarely needed, so we leave it out. Hence, the rotational energy term has the forms

$$F = B_vJ(J+1) \quad \text{without centrifugal distortion}$$
$$F = B_vJ(J+1) - D_vJ^2(J+1)^2 \quad \text{with centrifugal distortion.} \quad (11.13)$$

11.2 DIPOLE TRANSITIONS IN THE ELECTRONIC-VIBRATIONAL-ROTATIONAL SPECTRA

A rigid rovibronic state is generally represented by the product of electronic, vibrational, and rotational wavefunctions

$$\psi_e(x)\psi_{ev}(q)\psi_{JMK}(\Omega),$$

the coordinates being x, q, and $\Omega - 0, \phi, \chi$, respectively. This simplified representation assumes that the molecule is a rigid rotor for which $\psi_{JMK}(\Omega)$ has only angular dependence and that the moment of inertia is determined only at the equilibrium internuclear configuration. The simplification reduces the rotational energy term value to the one that neglects centrifugal distortion

$$F = B_vJ(J+1).$$

Similarly, the dependence of the vibrational motion on molecular rotation is assumed negligible, which amounts to ignoring Coriolis interactions. These assumptions of separation of motions allow for solving the vibrational and rotational motions exactly by considering the respective wavefunction for the harmonic oscillator and the rigid rotor. The solutions provide for an approximate but useful representation of the spectral transitions.

The electric transition dipole operator is $\boldsymbol{\mu}(x,q)\cdot\mathbf{E}$, which for a z-polarized light of unit intensity takes the form

$$\sum_\alpha \mu^{(\alpha)}(x,q)G_{\alpha Z}(\Omega).$$

The polarization direction of the electric field vector can be given in the laboratory-fixed Cartesian axes and the direction cosines for the orientation of the molecule-fixed frame $\alpha(=x,y,z)$ with respect to the Z-axis. A transition to some final (excited) state, say $\psi_{e'}(x)\psi_{e'v'}(q)\psi_{J'M'K'}(\Omega)$, is

$$\mu_{if} = \sum_\alpha \iiint \psi_e^*(x,q)\psi_{ev}^*(q)\psi_{JMK}^*(\Omega)\mu^{(\alpha)}(x,q)\times$$
$$G_{\alpha Z}(\Omega)\psi_{e'}(x,q)\psi_{e'v'}(q)\psi_{J'M'K'}(\Omega)dxdqd\Omega$$
$$= \sum_\alpha \iint \psi_{ev}^*(q)\mu_{ee'}^{(\alpha)}(q)\psi_{e'v'}(q)\psi_{JMK}^*(\Omega)$$
$$G_{\alpha Z}(\Omega)\psi_{J'M'K'}(\Omega)dqd\Omega \quad (11.14)$$

where the quantity $\mu_{ee'}^{(\alpha)}(q)$, called the electronic transition dipole, represents the integration over the electronic coordinates

$$\mu_{ee'}^{(\alpha)}(q) = \int \psi_e^*(x,q)\mu^{(\alpha)}(x,q)\psi_{e'}(x,q)dx. \quad (11.15)$$

Note that the assumption of unit intensity of light is the reason why $\mathbf{E}$ is not shown as a part of the operator. The transition moment integral shown in the second line of equation (11.14) is thus the product of two separate integrals, one over the nuclear coordinate and the other over the angular coordinates. This simplification rests on the assumption that $\psi_{JMK}(\Omega)$ is independent of q, corresponding to a rigid rotor state. Similarly, the vibrational motion is independent of the rotational motion. The advantage of these separations is that one can work with eigenfunctions of exactly solvable problems such as the rigid rotor and the harmonic oscillator. But the electronic factor in the μ_{if} integral should not be used as it is, because $\mu_{ee'}^{(\alpha)}(q)$ is the quantity valid at the equilibrium configuration only.

To solve for the vibrational integral small oscillations q of the nuclei about their equilibrium positions (q_o) must be allowed, which means we expand the dipole function about the equilibrium configuration of the nuclei in a Taylor series in the normal vibrational coordinates of any of the two electronic states involved in the transition

$$\mu_{ee'}^{(\alpha)}(q) = \left\{\mu_{ee'}^{(\alpha)}\right\}_o + \sum_i\left\{\frac{\partial\mu_{ee'}(q)}{\partial q_i}\right\}_o q_i +$$
$$\frac{1}{2!}\sum_{i,j}\left\{\frac{\partial^2\mu_{ee'}(q)}{\partial q_i\partial q_j}\right\}_o q_iq_j + \cdots. \quad (11.16)$$

For transitions in diatomic molecules, only the first constant term and the first derivative will be needed.

For description of the transition amplitudes in rotational, vibrational-rotational, and electronic-vibrational-rotational transitions, the following three cases may be considered.

1. $e = e'$ and $v = v'$ (pure rotational transition). Since the two vibrational quantum numbers v and v' are the same, the second term in the Taylor expansion

vanishes, and we obtain the transition amplitude for the rotational transition

$$\sum_{\alpha}\mu_{0}^{(\alpha)}\int\psi_{JKM}^{*}(\Omega)G_{\alpha Z}(\Omega)\psi_{J'K'M'}d\Omega.$$

It is plain that the occurrence of a pure rotational transition requires the existence of a permanent dipole moment under equilibrium nuclear configuration. For a diatomic molecule the dipole is along the internuclear axis, i.e., one of the molecular axes α, so that $G_{\alpha Z}=\cos\theta$.

2. $e=e'$ and $v\neq v'$ (vibrational-rotational transition). The vibrational spectral amplitude for a diatomic molecule should be

$$\int\psi_{ev}^{*}(q)\left[\left\{\mu_{ee'}^{(\alpha)}\right\}_{0}+\sum_{i}\left\{\frac{\partial\mu_{ee'}(q)}{\partial q_{i}}\right\}_{0}q_{i}\right]\psi_{ev'}(q)dq\int\psi_{JMK}^{*}(\Omega)G_{\alpha Z}(\Omega)\psi_{J'K'M'}d\Omega. \tag{11.17}$$

But because of the constant term $\left\{\mu_{ee'}^{(\alpha)}\right\}_{0}$ and the orthogonality of the harmonic oscillator functions $\int\psi_{ev}^{*}\psi_{ev'}dq=0$, the leading term of the vibrational transition moment vanishes, allowing consideration of the first derivative alone. The amplitude is effectively

$$\sum_{\alpha,i}\left\{\frac{\partial\mu_{ee'}^{(\alpha)}(q)}{\partial q_{i}}\right\}_{0}\int\psi_{ev}^{*}(q)q_{i}\psi_{ev'}(q)dq\int\psi_{JMK}^{*}(\Omega)G_{\alpha Z}(\Omega)\psi_{J'K'M'}d\Omega. \tag{11.18}$$

The dipole derivative with respect to the nuclear coordinate must not vanish for the occurrence of vibrational-rotational transitions. Additionally, the vibrational integral would not vanish only if $v'=v\pm1$ when harmonic oscillator wavefunctions are used. This is due to the orthogonality of the wavefunctions, which is also why the use of the first term in the Taylor expansion yields zero value for the integral, necessitating the use of the first derivative of the dipole.

3. $e\neq e'$ and $v\neq v'$(electronic-vibrational-rotational transition). The transition amplitude is

$$\sum_{\alpha}\left\{\mu_{ee'}^{(\alpha)}\right\}_{0}\int\psi_{ev}^{*}(q)\psi_{e'v'}(q)dq\int\psi_{JMK}^{*}(\Omega)G_{\alpha Z}(\Omega)\psi_{J'K'M'}d\Omega \tag{11.19}$$

In this case, the orthogonality of vibrational wavefunctions is not applicable, and the integral $\int\psi_{ev}^{*}(q)\psi_{ev'}(q)dq$ does not vanish because ψ_{ev} and $\psi_{e'v'}$ are eigenfunctions of separate vibrational energy operators. This vibrational integral is now called the Franck-Condon integral, the square of which is the Franck-Condon factor that will be discussed in the following. The coordinate α may correspond to any molecular axis if the dipole moment is along the axis. If α coincides with the bond axis of a diatomic molecule $G_{\alpha Z}=\cos\theta$ in the rotational integral, and when α is perpendicular to the internuclear axis $G_{\alpha Z}=\sin\theta\cos\phi$ or $\sin\theta\sin\phi$ (see Box 5.1).

In the form presented above, only the constant term $\mu_{ee'}^{(\alpha)}$ has been used in the electronic part of the integral, but the first derivative of the electric dipole in the Taylor expansion also produces the so-called 'vibronic progression' of bands depending on the magnitude of the derivative of the electronic transition moment with respect to the internal coordinates. The discussion of this idea starts with whether an electronic transition is allowed or not. Within the assumption of the separability of electronic, vibrational, and rotational wavefunctions, as we are doing here, an electronic transition is allowed if

$$\Gamma_{e}'\otimes\Gamma_{e}\supset\Gamma(\mu_{\alpha})\ \text{and}\ \Gamma_{v}'=\Gamma_{v'}, \tag{11.20}$$

in which we have used character-based symmetry species Γ, and where $\supset$ means $\Gamma_{e'e}$ includes $\Gamma(\mu_{\alpha})$. However, even if the transition is not electronically allowed, it can vibronically be. This can happen because of the electronic-vibrational selection rule

$$\Gamma_{e}'\otimes\Gamma_{e}\otimes\Gamma_{v'}\otimes\Gamma_{v}\supset\Gamma(r_{\alpha}), \tag{11.21}$$

where $\Gamma(r_{\alpha})$ represents the symmetry of the internal displacement coordinates $r_{\alpha}(\alpha=a,b,c)$. These transition displacement coordinates are discussed in Chapter 9 (for instance, see Figure 9.4). Such vibronic transitions, albeit electronically forbidden, can have a significant intensity if a transition-allowed electronic state couples (mixes) with another transition-forbidden excited electronic state (intensity borrowing) as already mentioned in the presentation of the resonance Raman excitation earlier. Hence, the electronic transition moment has a strong dependence on the nuclear coordinates. Note that these do not represent two different effects, but the same effect described in two different ways, and the effect depends on the magnitude of the second term in the Taylor expansion of the electronic transition moment (equation 11.16). The electronic-vibrational integral in this case is

$$\sum_{\alpha,i}\left(\frac{\partial\mu_{ee'}^{(\alpha)}(q)}{\partial q_{i}}\right)_{0}\int\psi_{ev}^{*}(q)q_{i}\psi_{e'v'}(q)dq\int\psi_{JMK}(\Omega)G_{\alpha Z}(\Omega)\psi_{J'K'M'}d\Omega, \tag{11.22}$$

and this term must be added to equation (11.19) to obtain a complete transition amplitude for an electronic-vibrational-rotational transition. This term is called Herzberg-Teller term in electronic spectra. The extent of occurrence of Herzberg-Teller vibronic transition depends on the magnitude of the derivative of the electric transition moment evaluated at the equilibrium nuclear configuration.

The UV-spectrum of benzene provides a classic example where Herzberg-Teller vibronic transitions are detected.

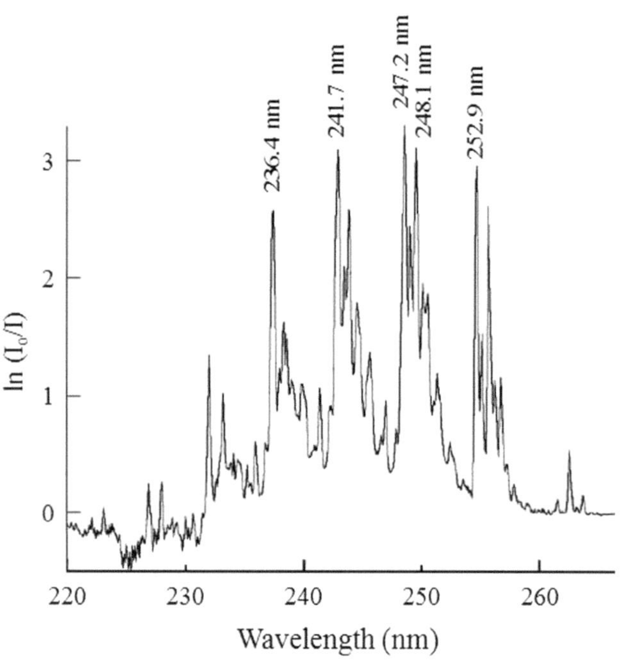

FIGURE 11.1 Ultraviolet spectrum of benzene vapor showing vibronic transitions. (Reproduced from Shastri, et al., 2002, with permission.)

A transition from the ground state (A_{1g}) to the first excited electronic state (B_{2u}) is normally not allowed, because the direct product table of D_{6h} gives

$$A_{1g} \otimes B_{2u} = B_{2u},$$

and the corresponding character table shows $\Gamma_{x,y} = E_{1u}$ and $\Gamma_z = A_{2u}$. Since $A_{1g} \otimes B_{2u} \notin E_{1u}A_{2u}$ (read $\notin$ to mean 'not a set of') the transition is not allowed. But the vibronic absorption transitions do occur and band progressions are also observed. The vacuum-UV absorption spectrum of gas-phase benzene is shown in Figure 11.1.

11.3 ELECTRONIC TRANSITION DIPOLE WITH NUCLEAR CONFIGURATIONS

The dependence of the electronic transition dipole on nuclear configuration $\mu(Q)$ need not be the same from one to another vibrational level. To see this, we take the Born-Oppenheimer average of the electric transition dipole at different nuclear configuration and then calculate the dipole moment for the vth vibrational level. The electronic wavefunction $\psi(x,Q)$ has dependence on both electronic (x) and nuclear (Q) coordinates. Let $\psi(x,Q)$ be the electronic wavefunction at a given Q, and let $\boldsymbol{\mu}^{(\alpha)}(x,q)$ be the dipole operator. Then

$$\langle\mu(Q)\rangle = \int \psi^*(x,Q)\boldsymbol{\mu}^{(\alpha)}(x,Q)\,\psi(x,Q)\,dx \qquad (11.23)$$

is the mean value of the dipole transition moment at a fixed value of Q obtained by averaging over all electronic coordinates. If we consider now the vibrational states $\psi_v(Q)$ then the probability that the nuclei will be found to move within dQ of Q is

$$|\psi_v(Q)|^2 = \psi_v^*(Q)\,\psi_v(Q)\,dQ, \qquad (11.24)$$

and the distribution of the dipole moment would be

$$\mu(Q)|\psi_v(Q)|^2\,dQ,$$

which can be integrated to obtain the expectation value of the dipole moment of the vth vibrational level

$$\langle\mu(Q)\rangle_v = \int|\psi_v(Q)|^2\,\mu(Q)\,dQ. \qquad (11.25)$$

For example, the $\mu(Q)$ function and the probability $|\psi_v(Q)|^2$ for harmonic oscillator wavefunctions with $v = 0$, 1, and 2 are sketched in Figure 11.2. The multiplication of the two functions provides the form of the integrand. It is clear that $\langle\mu(Q)\rangle_v$ is significant only where $|\psi_v(Q)|^2$ has a reasonably considerable value near the equilibrium nuclear configuration Q_e. Notice that values of $\mu(Q)$ near Q_e are favorable for $v = 0$ and $v = 2$, not for $v = 1$. In the latter case, the significant part of the measured dipole would occur only where $Q > Q_e$, Q_e being the equilibrium nuclear configuration.

11.4 FRANCK-CONDON FACTOR

The intensity of a $v' \leftarrow v''$ vibrational sub-band of an $e' \leftarrow e''$ electronic transition band depends on the quantum number corresponding to the vibrational level v' in the excited electronic state e'

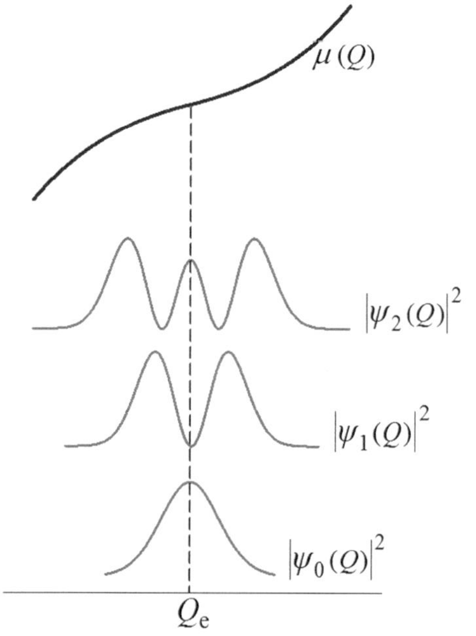

FIGURE 11.2 The $\mu(Q)$ function and the probability $|\psi_v(Q)|^2$ for Hermite functions corresponding to $v = 0$, 1, 2. Notice the unfavorable values of $\mu(Q)$ around Q_e for $v = 1$.

$$q_{v''v'} = \left| \int_{-\infty}^{+\infty} \psi^*_{e''v''}(q)\, \psi_{e'v'}(q)\, dq \right|^2 . \qquad (11.26)$$

This is called the Franck-Condon (FC) factor that amounts to the square of the vibrational overlap integral, which means the relative intensities of vibrational sub-bands contained within an electronic transition will depend on the relative magnitudes of the vibrational overlap integral. We see that only vibrational wavefunctions in the ground and excited electronic states are involved in the integral. The indices on e (e'' and e') used in $\psi_{e''v''}$ and $\psi_{e'v'}$ are used only to indicate the respective correspondence of v'' and v' with e'' and e'. We might as well write the FC factor as

$$q_{v''v'} = \left| \langle v'' | v' \rangle \right|^2 . \qquad (11.27)$$

The FC factor is interpreted with reference to the BO approximation that electrons are fuzzier and move much faster than the nuclei do, which can be seen as an electronic transition with stationary nuclei – often called a vertical transition. Consequently, the transition occurs only to that edge (or area) of the excited potential surface in which the initial-state vibrational wavefunction will have a finite density. This is a condition for obtaining a finite value of the FC overlap integral. For an analysis of overlap of vibrational wavefunction, the lower regions of the electronic surface that simulate the harmonic oscillator may be considered. The extent of the overlap of the two vibrational wavefunctions depends on two force constants – the force constant of the harmonic oscillator in the initial state and that in the excited state

$$\delta = \sqrt{\frac{\omega''}{\omega'}} \text{ with } \omega = \sqrt{\frac{k}{m}} . \qquad (11.28)$$

The shift in the equilibrium nuclear configuration from the initial (R''_e) to the final excited state (R'_e) is often given as an offset parameter

$$\Delta = R'_e - R''_e ,$$

where Δ is the offset. For illustration, Figure 11.3 considers four combinations of δ and Δ, and the overlaps of $v'' = 0$ with $v' = 0, 1, 2, \cdots$ in each case. Let us look at them one by one.

1. $\delta = 1 \Delta = 0$. The vibrational wavefunctions in the ground-state oscillator ($\psi_{1,v''}$) are orthogonal. Since the excited-state oscillator ($\psi_{2,v'}$) is identical to the ground-state oscillator, a wavefunction on either of these two oscillators will be mutually orthogonal. Therefore, a transition $1, v'' \rightarrow 2, v'$ can occur only from $v'' = 0$ to $v' = 0, v'' = 1$ to $v' = 1$, $v'' = 2$ to $v' = 2$, and so on. In other words, the transition must conform to the $\Delta v = 0$ selection rule or else the integral vanishes

$$\int_{-\infty}^{+\infty} \psi_{1,0}(Q)\, \psi_{2,0}(Q)\, dQ \neq 0,$$

but

$$\int_{-\infty}^{+\infty} \psi_{1,0}(Q)\, \psi_{2,1}(Q)\, dQ = 0. \qquad (11.29)$$

This is similar to what has been mentioned earlier in the context of vibrational transitions in non-vibronic cases that overtone transitions are forbidden in the limit of harmonicity. The response to the inquiry of the limit of harmonic approximation remains open.

2. $\delta = 1, \Delta < 0$. Although both oscillators have the same force constant, the equilibrium nuclear configurations are different. Unlike the case (1) above, a wavefunction on one oscillator is not orthogonal to any of the wavefunctions on the other oscillator because $\Delta \neq 0$. Thus, a transition from the $v'' = 0$ level is allowed to any v' level subject to non-vanishing density of the ground-state wavefunction in the excited state potential. However, the condition of $\Delta \neq 0$ causes the maximum of the two Gaussians (wavefunctions) corresponding to $v'' = 0$ and $v' = 0$ not to overlap. The maximum of the function for $v'' = 0$ is away from R'_e, the equilibrium internuclear distance in the excited state. As a result, the value of the overlap integral will be

$$\int_{-\infty}^{+\infty} \psi_{1,0}(Q)\, \psi_{2,0}(Q - \Delta)\, dQ,$$

which is smaller than that in the case (1), i.e., when $\delta = 1, \Delta = 0$. Since the value of the Franck-Condon integral provides the relative strength of a transition, the $v'' = 0 \rightarrow v' = 0$ transition is weaker in the case of $\Delta < 0$. Note that Δ can be positive also, as for totally symmetric vibrations.

3. $\delta < 1, \Delta = 0$. In general, the value of δ could be <1 or >1. But in this case of $\delta < 1$, the excited potential will be wider. The interpotential orthogonality of vibrational wavefunctions does not hold, so vertical transitions from $v'' = 0$ to $v' = 0, 1, 2, \cdots$ are all allowed. The relative intensities of the allowed transitions are determined by respective overlap integrals.

4. $\delta < 1, \Delta < 0$. In this case, transitions can occur from $v'' = 0$ to all excited vibrational states localized at the edge of the excited potential (referring to the slope of the potential surface), and the largest overlap of the ground state wavefunction will occur with the excited vibrational level whose edge is vertically right above the ground-state vibrational level. In other words, the vibrational levels in the excited electronic potential must correspond to the same nuclear configuration as that of the lower electronic potential, even though the equilibrium configurations in the two potentials are different ($\Delta \neq 0$). The overlaps of wavefunctions are shown in Figure 11.3. Generally, values of the FC factor for different ground and excited states are collectively presented in a grid, a cell in the grid is occupied by the value of the square of the FC integral for a given pair of vibrational wavefunctions. The grids in the profile are contained in a parabola called Condon parabola, whose geometry and the position of

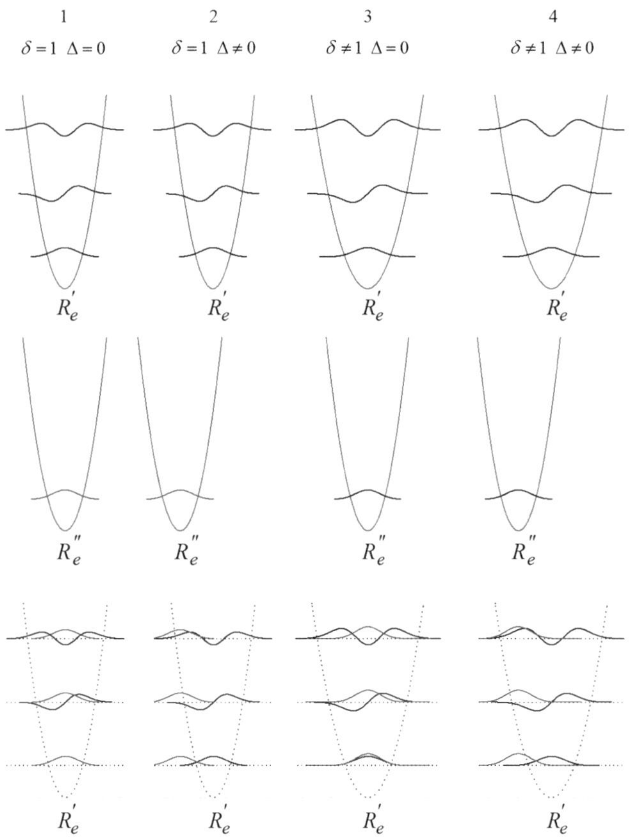

FIGURE 11.3 Combinations of δ and Δ, and FC overlaps of $v'' = 0$ and $v' = 0, 1, 2$. The bottom row depicts the overlaps.

the vertex depend on the value of δ and Δ. In a typical two-dimensional Cartesian presentation of the Condon parabola, the quantum numbers for ground- and excited-state wavefunctions are plotted along *xy*. In the case $\delta = 1 \Delta = 0$ above, the FC factors for different pairs of vibrational quantum numbers fall in a straight line of slope 1. This is the hallmark of the orthogonality of wavefunctions in the two potential surfaces. However, as δ grows smaller than 1 at fixed $\Delta = 0$, i.e., the situation in the case (3) above, the linearity of the FC factors spreads out tangentially along the parabola whose vertex is still positioned at the origin of the *xy* plane. If, on the other hand, Δ is allowed to increase from zero at fixed $\delta = 1$, then the parabola becomes wider and the vertex moves away symmetrically from the *xy* origin. The reader will be benefitted by drawing the Condon parabola for a few combinations of δ and Δ.

Slant (non-vertical) vibronic transitions, during which the transient nuclear configurations of the lower and the upper electronic surfaces are not the same, do not occur or occur extremely rarely. The far-UV spectrum of ethylene was earlier thought to provide an example of non-vertical transition. The bonding of the two sp^2 carbons in ethylene produces a π^* antibonding molecular orbital so that a $\pi \rightarrow \pi^*$ electronic

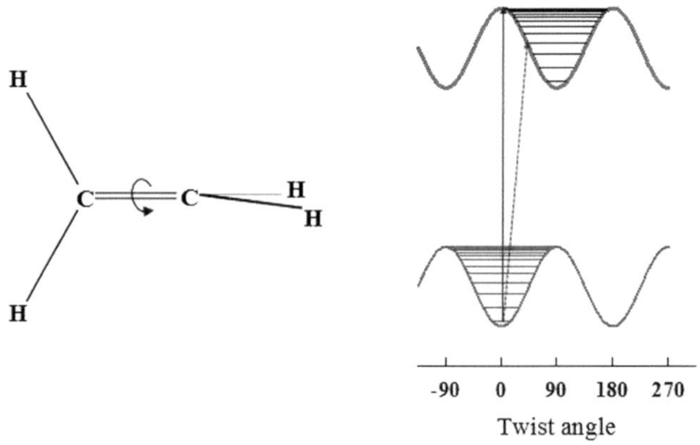

FIGURE 11.4 The possibility of non-vertical transition in ethylene.

transition could possibly occur in the UV region. Ethylene is a planar molecule in the lower electronic state but in the π^* excited state the two $-CH_2$ halves are twisted by 90° relative to each other. The twist arises entirely due to the antibonding nature of the π^* molecular orbital. Figure 11.4 shows the rotor minima and the intervening barriers to rotation in a lower and an excited electronic state. We understand that the torsion of a rotor is a normal mode that can be used for the analysis of the FC effect. The cusp in the excited-state potential ($\theta = 90°$) is vertically right above the ground-state planar minimum ($\theta = 0$). The wavefunctions at the cusp are those corresponding to a free-torsion ethylene, which means there is no bound vibrational wavefunction in the upper state that can overlap with a vibrational wavefunction from the lower surface during an electronic transition. Thus, for a long time the transition was thought to be nonvertical by which the vibrational band of maximum intensity does not correspond to an identical nuclear configuration for the lower and upper electronic surfaces. The difficulties of explaining the temporal nuclear configuration in the nonvertical FC effect prevail, although it is now thought that vertical FC transition indeed occurs for ethylene and the vibrational band structure can be assigned on the basis of C=C stretching (ν_2), CH_2-scissoring (ν_3), and torsional (ν_4) vibrations. That the vertical Franck-Condon transition indeed occurs in ethylene has been seen in the first principles simulation of the UV-absorption spectrum, which reasonably reproduces the experimental spectrum. The vibrational progression characterized by the series of doublet structures riding on a broad continuum and the long tail structure are observed in the experimental spectrum also (Figure 11.5).

11.5 PROGRESSION OF VIBRATIONAL ABSORPTION IN AN ELECTRONIC BAND

The vibrational band structure in an electronic absorption appears as a 'progression' of vibrational absorption – the progression consisting of many sub-bands whose appearance and resolution depend on the extent of density of the initial vibrational wavefunctions (v'') in the excited-state electronic potential. The overlap of the ground-state vibrational wavefunction with the excited-state vibrational level straight above yields the most intense band. The band intensity on either side of this intense band should fall because the FC factor decreases progressively as one moves away from the vertical transition. A sub-band in the vibrational progression may at times split into a doublet, which in the case of ethylene arises from superimposed Rydberg series of lines ($\pi \rightarrow 3s$) on the $\pi \rightarrow \pi^*$ bands (see Figure 11.5).

Juxtaposed on the vertical FC absorption is the 'vertical emission'. While the absorption occurs to a temporal nuclear configuration in the excited state, which is identical to the ground-state nuclear configuration, the reverse applies in the case of emission. The vertical fluorescence emission occurs by the same principle of the vertical absorption transition. In the case of fluorescence, the initial excited vibrational state $v' = n$ relaxes to $v' = m\ (m < n)$ by intramolecular vibrational relaxation in cases of gas-phase isolated molecules or by both intra- and intermolecular vibrational relaxation in the condensed phase. The edge of the excited-state potential and the vibrational state from which emission transition will occur depend on the offset Δ of ground- and excited-state potential surfaces. If Δ is very small, the initial excited vibrational state can relax to one of much lower vibrational states from which emission can occur. If the equilibrium nuclear configurations are very different, or say the excited-state equilibrium bond length of a diatomic molecule is significantly different from that in the ground state, then the emission transition is expected to occur from one of the upper vibrational states of the excited molecule. In the case of iodine vapor, for example, the transition from the excited triplet state ($B^3\Pi_u^+$) to the singlet ground state ($X^1\Sigma_g^+$) occurs by fluorescence emission from $v' = 32$ to $v'' = 0,1,2,\cdots,22$ in a series of band progression (Figure 11.6).

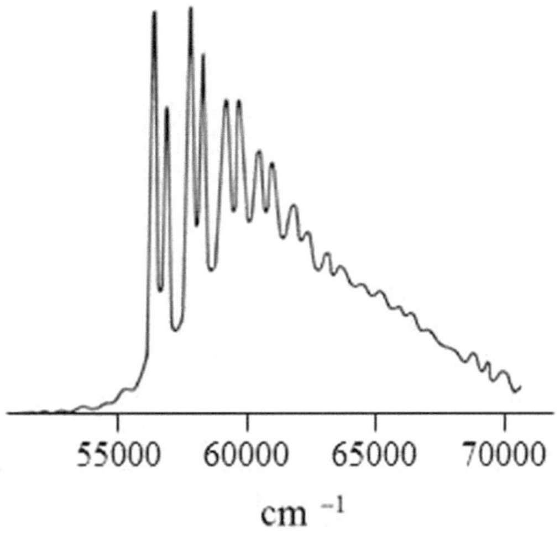

FIGURE 11.5 Experimental UV-absorption spectrum of ethylene showing a broad continuum on which rides the doublets due to vibrational progression. (Reproduced from Geiger and Wittmaack, 1965, Copyright © *Z. Naturforschg*.)

11.6 ANALYSIS OF VIBRATIONAL BANDS

The set of resolved vibrational bands is a rich source of spectroscopic information. The analytical exercise begins typically with the assignment of the vibrational bands that requires some effort, especially when the spectrum contains a large number of vibrational transitions. This is a modest way of submitting that large polyatomic vibrational analyses with current technologies are still far from adequate to understand the reality of polyatomic vibrations. Calculation of the FC factors, which can be evaluated analytically, and matching them against the vibrational band intensities often assist in the assignment of the bands according to the pair of v'' and v' involved. Customarily, a band is identified as $v' \leftarrow v''$ with the vibrational indices (quantum numbers) given explicitly.

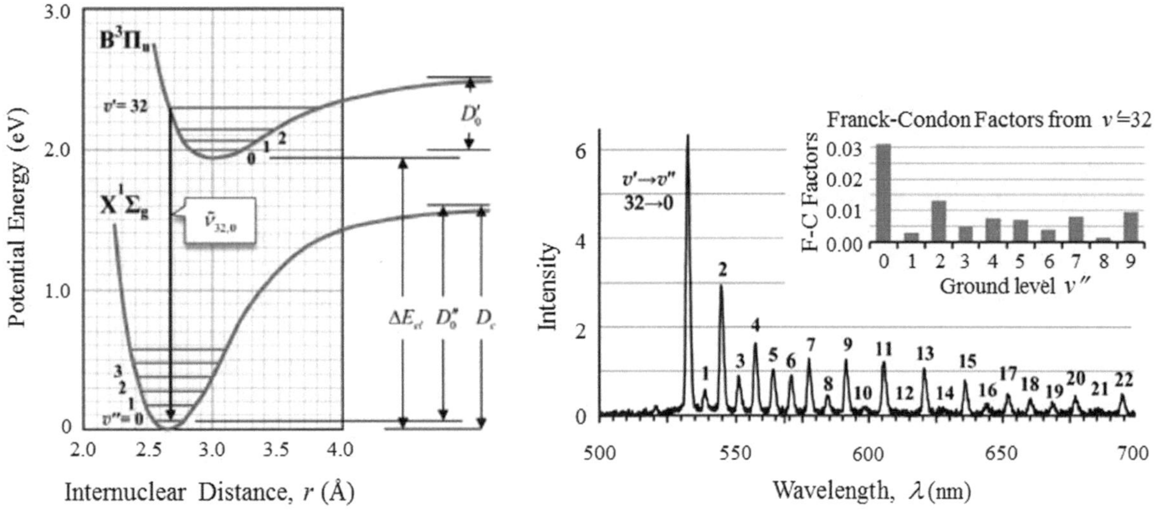

FIGURE 11.6 *Left*, potential energy curves for I_2 molecular states X and B such that $B \rightarrow X$ emission produces observable fluorescence. *Right*, an experimental fluorescence spectrum due to $v'' = 0,1,2,\cdots 22 \leftarrow v' = 32$ emission transitions. The FC factors calculated for ten transitions (from $32 \rightarrow 0$ to $32 \rightarrow 9$) are shown in the inset. (Adapted from arXiv:1507.02600v1 [Physics.ed-ph] 9 Jul 2015 with kind permission of authors S. B. Bayram and M. V. Freamat.)

Since the band structure is expected to reflect vibrational progression, a band identified correctly allows assignment of others in a sequentially progressive manner. For example, if the most intense band happens to arise from $v' = 12 \leftarrow v'' = 0$, written as $0 \rightarrow 12$, then $0 \rightarrow 11, 0 \rightarrow 10, \cdots$ bands will correspond to progressively lower energy. The assignment may not be straightforward when there are a large number of bands, the appearance of which may not be in order.

The correctness of vibrational band assignment can be checked by entering the assigned wavenumbers in a table and then entering the wavenumber difference of successive pairs of vibrational indices. In the schematic representation (Table 11.1) these wavenumber differences are shown by Δ' and Δ'', not to be confused with Λ that we have used earlier to denote the offset of the two electronic surfaces. If the columns Δ'' formed by the wavenumber differences for successive pairs of v'' indices are equal, and also the rows Δ' formed by the wavenumber differences between the successive pairs of v' indices are equal, then the correctness of band assignment is assured. A tabulated version of the v' vs v'' matrix is customarily called Deslandres table (Table 11.1).

The analysis is continued to the calculation of the energy term values. When the rotational bands in the vibronic spectrum are not sufficiently resolved the rotational term value can be dropped from equation (11.1) so that the total energy terms corresponding to the two vibronic states, neglecting spin, are

$$T' = T'_{el} + G'$$

$$T'' = T''_{el} + G'', \qquad (11.30)$$

and the transition wavenumber difference for the $v' \leftarrow v''$ transition is

$$|\Delta\tilde{v}| = T' - T'' = \omega_{el} + G_{v'} - G_{v''} \qquad (11.31)$$

TABLE 11.1
Representation of Deslandres table. The wavenumbers corresponding to the assigned $v' \leftarrow v''$ transitions are denoted by $a_i, b_i, c_i, d_i, \ldots (i = 0,1,2,\ldots)$, and the wavenumber differences are denoted by $a_{i+1} - a_i, b_{i+1} - b_i, c_{i+1} - c_i, \ldots$ and $b_i - a_i, c_i - b_i, d_i - c_i, \ldots$ Correctness of band assignment is confirmed if the columns $\Delta''_{01}, \Delta''_{12}, \Delta''_{23}, \ldots$ are equal and so are the rows $\Delta'_{01}, \Delta'_{12}, \Delta'_{23}, \ldots$

		v''							
		0	Δ''_{01}	**1**	Δ''_{12}	**2**	Δ''_{23}	**3**	$\cdots$
	0	a_0	$b_0 - a_0$	b_0	$c_0 - b_0$	c_0	$d_0 - c_0$	d_0	$\cdots$
	Δ''_{01}	$a_1 - a_0$		$b_1 - b_0$		$c_1 - c_0$		$d_1 - d_0$	$\cdots$
	1	a_1	$b_1 - a_1$	b_1	$c_1 - b_1$	c_1	$d_1 - c_1$	d_1	$\cdots$
v'	Δ''_{12}	$a_2 - a_1$		$b_2 - b_1$		$c_2 - c_1$		$d_2 - d_1$	$\cdots$
	2	a_2	$b_2 - a_2$	b_2	$c_2 - b_2$	c_2	$d_2 - c_2$	d_2	$\cdots$
	Δ''_{23}	$a_3 - a_2$		$b_3 - b_2$		$c_3 - c_2$		$d_3 - d_2$	$\cdots$
	3	a_3	$b_3 - a_3$	b_3	$c_3 - b_3$	c_3	$d_3 - c_3$	d_3	$\cdots$
	$\vdots$	$\vdots$	$\vdots$	$\vdots$	$\vdots$	$\vdots$	$\vdots$	$\vdots$	

in which

$$\omega_{el} = \tilde{v}_{el} = T'_{el} - T''_{el}.$$

For a simple analysis we may restrict the anharmonicity to the first order alone, in which case equation (11.2) gives the vibrational energy terms as

$$G' = \omega'_e\left(v' + \frac{1}{2}\right) - \chi'_e\omega'_e\left(v' + \frac{1}{2}\right)^2$$

$$G'' = \omega''_e\left(v'' + \frac{1}{2}\right) - \chi''_e\omega''_e\left(v'' + \frac{1}{2}\right)^2. \qquad (11.32)$$

These term values can be used to estimate graphically the anharmonicity of potential surfaces and dissociation energies by the Berge-Sponer method developed in 1926. Combining equations (11.31) and (11.32) we obtain

$$|\Delta\tilde{v}|_{v'} = G_{v'+1} - G_{v'} = \omega'_e - 2\chi'_e\omega'_e(v'+1)$$

$$|\Delta\tilde{v}|_{v''} = G_{v''+1} - G_{v''} = \omega''_e - 2\chi''_e\omega''_e(v''+1). \qquad (11.33)$$

Values of $|\Delta\tilde{v}|_{v'}$ and $|\Delta\tilde{v}|_{v''}$ are calculated from results of vibronic absorption and fluorescence, respectively. The straight-line plots of $|\Delta\tilde{v}|_{v''}$ vs $(v''+1)$ and $|\Delta\tilde{v}|_{v'}$ vs $(v'+1)$ yield the fundamental frequencies ω''_e and ω'_e, respectively. The area under the curve is the dissociation energy (heat of dissociation) with respect to $v' = 0$ and $v'' = 0$, and is obtained as

$$D_0 = hc\left(\frac{\omega_e^2}{4\chi_e\omega_e}\right). \qquad (11.34)$$

We also know that the dissociation energy is scaled from the bottom of the potential, and is

$$D_e = D_0 + \frac{\omega_e}{2} - \frac{\omega_e\chi_e}{4} + \cdots \approx D_0 + \frac{\omega_e}{2}. \qquad (11.35)$$

In the two formulas above, 'prime(s)' is not added to ω_e and χ_e to imply the generalization of the result to both potentials.

11.7 ANALYSIS ROTATIONAL BANDS

Obtaining resolution of rotational structure in vibronic spectra warrants some effort, but can be achieved by exciting rotationally cooled molecules produced by molecular expansion in supersonic jet (see Box 11.1). For illustration, we reproduce in Figure 11.7 a rotationally resolved excitation spectrum of pyrene cooled in a supersonic jet.

In a high-resolution electronic spectrum, each vibrational band is observed to have a large number of rotational bands that can be analyzed using the appropriate rotational energy terms. We assume that there is no contribution of electronic and nuclear angular momentum in both lower and upper (excited) states so that the only angular momentum is due to the nuclear rotation. Insertion of equation (11.12) into equation (11.7) yields

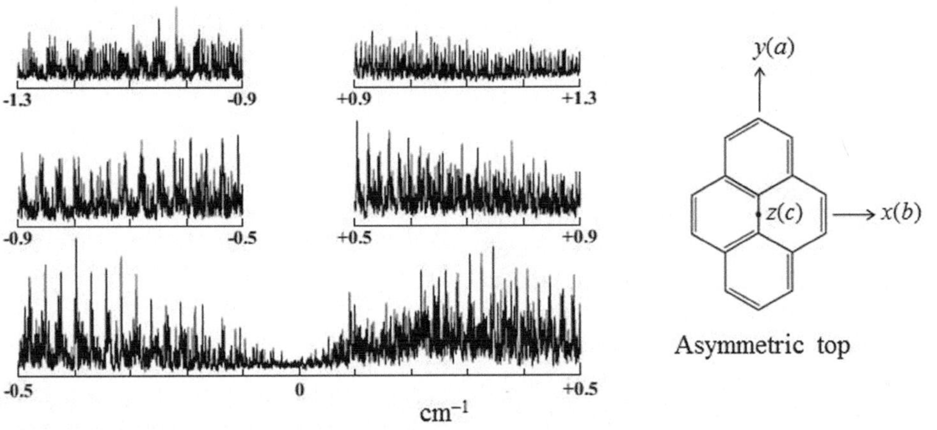

FIGURE 11.7 The rotational structure in a fluorescence excitation spectrum of supersonic jet-cooled pyrene with carrier argon gas. Pyrene is an asymmetric top, for which the rotational (principal) axes a, b, c are collinear with x, y, z molecular axes. (Reproduced from Baba et al., 2009, with AIP Publishing license.)

$$F(J)=\left[B_e-\alpha_e\left(v+\frac{1}{2}\right)+\gamma_e\left(v+\frac{1}{2}\right)^2\right]J(J+1)-$$

$$\left[D_e+\beta_e\left(v+\frac{1}{2}\right)-\delta_e\left(v+\frac{1}{2}\right)^2 J^2(J+1)^2\right]. \qquad (11.36)$$

Since the usual electric dipole selection rule ($\Delta J = 0, \pm 1$) applies, one expects the appearance of the P, Q, R branches in the rotational structure. Further, because we have already neglected the electronic and nuclear angular momenta so that the electronic states are of Σ-symmetry (i.e., $\Omega = 0$), the selection rule is restricted to $\Delta J = \pm 1$, which is the reason why we do not expect to observe the $\Delta J = 0$ transition. This is called the b-type selection rule which forbids the appearance of the Q-branch structure in the rotational band structure. But the expected frequency at which the Q-branch ($\Delta J = 0$) could have occurred otherwise is traditionally termed the 'band center'.

The frequencies (or rotational transition wavenumber, $\tilde{v}$) of J-specific lines in each of P, Q, and R branches is obtained from equation (11.1), similar to the way the vibrational wavenumber difference was obtained in equation (11.31). Here, we discount the centrifugal effect of rotation and write

$$\Delta J=-1 \quad |\Delta\tilde{v}|_P(J)=\omega_o-\left(B_e'+B_e''\right)J+\left(B_e'-B_e''\right)J^2$$

$$\Delta J=0 \quad |\Delta\tilde{v}|_Q(J)=\omega_o+\left(B_e'-B_e''\right)J+\left(B_e'-B_e''\right)J^2$$

$$\Delta J=+1 \quad |\Delta\tilde{v}|_R(J)=\omega_o+2B_e'+\left(3B_e'-B_e''\right)J+\left(B_e'-B_e''\right)J^2, \qquad (11.37)$$

in which

$$\omega_o=\left(T_{el}'-T_{el}''\right)+\left(G'-G''\right) \quad \omega_o=\tilde{v}_o. \qquad (11.38)$$

The frequency terms in these equations show a strong dependence of the rotational band positions on the difference of the rotational constant of the rotor in the lower and upper electronic states. We recall (see equation (7.12) and Figure 7.3) that

$$B_e=\frac{h}{8\pi^2 c\mu R_e^2}.$$

If the equilibrium configuration R_e is significantly different in the excited state, then the moment of inertia (μR_e^2) will change, called the inertial defect, and the magnitude of the $\left(B_e'-B_e''\right)$ term will determine the density of rotational band frequencies in P and R branches. When $B_e' > B_e''$ the P-branch lines are densely spaced forming a so-called 'band head', and the R-branch lines spread out to form the 'band-tail'. The reverse – the band-head for the R-branch and the band-tail for the P-branch should be observed when $B_e'' > B_e'$. The band-head effect that causes the rotational lines to close in is manifested strongly for rotational lines corresponding to low values of J, understandibly due to higher density of rotational levels as J runs lower. This is the reason why resolution of low-J rotational lines, which appear as a band-head, is often not achieved. This phenomenological dependence of $\Delta\tilde{v}(J)$ on J also provides a formula to determine the J-value at which a band-head will appear if the inertial defect is known, and vice versa. If we treat J as a continuous variable, the differentiation of $\Delta\tilde{v}(J)$ for the P-branch, for example, gives

$$\frac{d|\Delta\tilde{v}|_P(J)}{dJ}=-\left(B_e'+B_e''\right)+2\left(B_e'-B_e''\right)J. \qquad (11.39)$$

Since the resolution of rotational frequencies diminish at low-J

$$\frac{d|\Delta\tilde{v}|_P(J)}{dJ}=0,$$

which implies

$$J_P=\frac{B_e'+B_e''}{2\left(B_e'+B''\right)}. \qquad (11.40)$$

Similarly,

$$J_R = \frac{-\left(3B_e' - B_e''\right)}{2\left(B_e' - B_e''\right)} \tag{11.41}$$

So the appearance of the band-head depends on the molecular distortion, and hence the initial defect produced on excitation. This indeed is a learning point that the light-pulse parameters used for the initial excitation do matter in practice. The ground- and the first-excited electronic states, S_0 and S_1, repectively, of an approximately asymmetric top, pyrene for example, have very similar molecular structure and potential energy curves, and hence band-head features are not observed (Figure 11.7).

11.8 ELECTRON-NUCLEAR ROTATIONAL COUPLING AND SPLITTING OF ROTATIONAL ENERGY LEVELS

The electron spin and orbital angular momenta have so far been neglected in our discussion of the energy of a molecular state. Their incorporation into the analyses requires the inclusion of the $A\Lambda\Sigma$ term (equation (11.2)) which involves the extent of the mutual coupling of the electron spin and orbital angular momenta, their respective coupling with the molecular axis, and the coupling of their sum with the rotational angular momentum of the molecule. We will restrict the analysis to diatomic molecules and hence it is useful to briefly collate the information on the angular momenta, the molecular terms, and the electronic states. The rotational angular momentum vector **O** of the molecule is in a direction perpendicular to a plane containing the molecule (Figure 11.8); the total electron orbital and spin angular momenta are **L** and **S**, and their projections onto the molecular axis are **Λ** and **Σ**, respectively. As described earlier (see Figure 5.4), **Λ** is quantized in the unit of $\hbar$ so that $\mathbf{\Lambda} = \hbar\Lambda$, where $\Lambda = 0, 1, 2, \cdots$. Similarly,

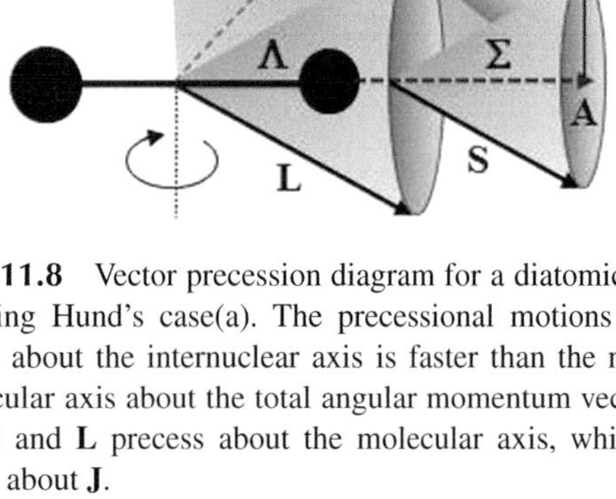

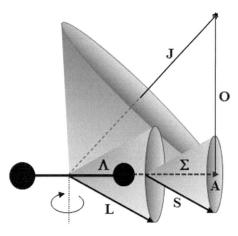

FIGURE 11.8 Vector precession diagram for a diatomic molecule representing Hund's case(a). The precessional motions of **L** and **S** vectors about the internuclear axis is faster than the nutation of the molecular axis about the total angular momentum vector **J**. The vectors **S** and **L** precess about the molecular axis, which in turn precesses about **J**.

Σ is quantized ($\mathbf{\Sigma} = \hbar\Sigma$) with $\Sigma = S, S-1, \cdots, -S+1, -S$. The **Λ** and **Σ** vectors sum up to define the total electron angular momentum along the internuclear axis

$$\mathbf{\Omega} = \mathbf{\Lambda} + \mathbf{\Sigma}, \tag{11.42}$$

so the quantized values of **Ω** are $\Lambda + S, (\Lambda + S) - 1, \cdots, |\Lambda - S|$. A diatomic molecular state is given by the molecular term symbol $2S+1_{\Lambda\Omega}$, and symbols $\Sigma, \Pi, \Delta, \cdots$ are used for Λ according to its allowed values of $0, \pm 1, \pm 2, \pm 3, \cdots$, respectively. The value of Λ is always an integer, but the value of Ω can be an integer or half an integer depending on whether Σ is an integer or a half-integer. When the total number of electrons is an odd number, then S is odd and so is Σ. One needs to be cautious to discriminate the two usages of the letter symbol Σ, whether a nuclear axis projection of the **S** vector (italic Σ) or a molecular state (normal-case Σ) as for $\Lambda = 0$ here.

11.8.1 Hund's Cases

The effect of various modes of coupling of internal angular momenta on the molecular rotational energies is analyzed by a set of angular momentum coupling schemes first described by Hund. Hund's original work is in German, but the cases have been extensively described by others later (Steinfeld, 1989: 161–166; Townes and Schawlow, 1975: 177–181; Herzberg, 1950: 219–226). The analysis is useful only for diatomic and small linear molecules, and is discussed under a few major cases according to the dominating interactions.

Hund's Case(a). This case is applicable when $\Lambda \neq 0$ and $S \neq 0$. The interactions are depicted in Figure 11.8, where **A** is a vector along the molecular axis whose end-over-end rotation gives rise to the molecular rotational angular momentum vector **O**. Note that some literature denote this vector by **R** or **N** adding on to our confusion. The vector **N** is used to represent the addition of vectors **Λ** and **O**; hence **O**~**N** if the magnitude of **Λ** is very small. It is also important to note that the vector **N** is shown in some literature as **K** (see, for example Herzberg, 1950: 221). The vector **O** is distinct, because it represents the molecular rotation, but we take the stand that the interactions between **L** and **A**, and **S** and **A** are the strongest of all other interactions. The axial angular momentum vector **Ω**(=**Λ**+**Σ**) and the rotating angular momentum vector **O** add vectorially to produce the total angular momentum vector **J**, about which the vectors **Ω** and **O** precess. Because **LA** and **SA** interactions are stronger than any other interactions, the individual precessions of **L** and **S** about **A** is faster than the precession of **A** about **J**. This also is the reason why **L** and **S** would not precess about **J**. Much slower molecular rotation compared with the precession of **L** and **S** about the molecular axis is a hallmark of Hund's case(a). The allowed values of the quantum number J are $\Omega + O = \Omega + 0, \Omega + 1, \Omega + 2, \cdots$, where, as shown explicitly, $O = 0, 1, 2, \cdots$. Neglecting the centrifugal distortion and oscillator anharmonicity, the rotational energy is given by using the argument that J and Ω are respectively similar to J and K quantum numbers of symmetric tops (see equation (7.50)). Thus

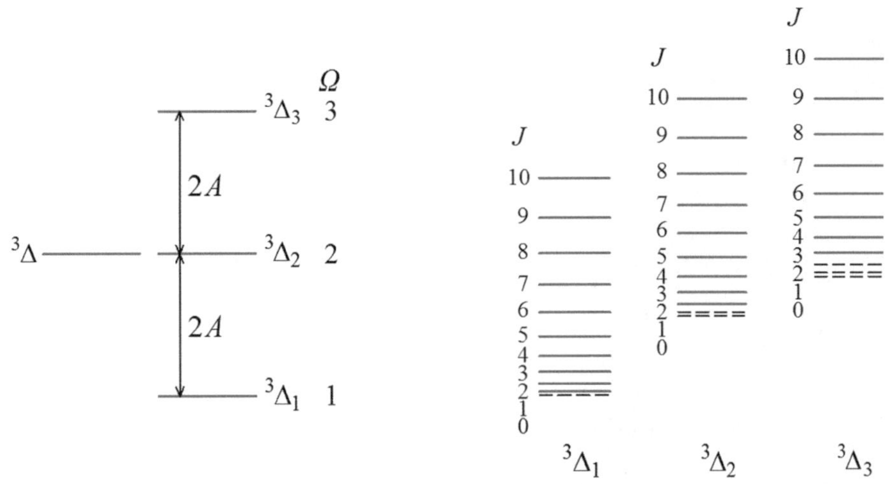

FIGURE 11.9 The substates of $^3\Delta$ (*left*), and the quantum number J and rotational energy levels for a $^3\Delta$ molecular state (*right*) obtained from Hund's case(a). The broken bars in black are the rotational levels which must not occur due to the restriction $J \geq \Omega$.

$$\begin{aligned} E_{\text{rot}} &= F(J) = B_e\left[J(J+1) - \Omega^2\right] + A\Omega^2 \\ &= B_e\left[J(J+1)\right] + \left(A - B_e\right)\Omega^2 \end{aligned} \tag{11.43}$$

The 'new' rotational constant A (not A_e), which is essentially the spin-orbit coupling constant and arising due to the inclusion of the electronic angular momentum, is rather too large to carry forward in the context of rotational spectroscopy. Hence the non-vibronic pure rotational premise will drop the $A\Omega^2$ term from equation (11.43) to yield

$$E_{\text{rot}} = F(J) = B_e\left[J(J+1)\right] + B_e\Omega^2. \tag{11.44}$$

But the study of vibronic transitions must retain A, so the rotational energy turns out to be

$$F(J) = B_e\left[J(J+1)\right] + \left(A - B_e\right)\Omega^2. \tag{11.45}$$

To see how $\left(A - B_e\right)\Omega^2$ might affect rotational structure, consider first the spin-orbit coupling constant A alone. A molecular state (ND, NH, OH^+, ScLi, TiC, VN, for example) splits according to

$$T_{el} = T_o + A\Lambda\Sigma$$

For a $^3\Delta$ state we have $\Lambda = 2, S = 1, \Sigma = 1, 0, -1$. Thus the $^3\Delta$ state has three substates, namely, $^3\Delta_3$, $^3\Delta_2$, and $^3\Delta_1$, as shown in Figure 11.9.

Then we consider the possible values of the quantum number J in equation (11.45). Since $J = \Omega + O = \Omega + 0,\ \Omega + 1,\ \Omega + 2, \cdots$, and $\Omega = 1, 2, 3$ for the $^3\Delta$ state, the values of J for the substates of $^3\Delta$ are

$$^3\Delta_1 \quad \Omega = 1 \quad J = 1, 2, 3, 4, \cdots$$

$$^3\Delta_2 \quad \Omega = 2 \quad J = 2, 3, 4, 5, \cdots$$

$$^3\Delta_3 \quad \Omega = 3 \quad J = 3, 4, 5, 6, \cdots.$$

The rotational energy levels in Figure 11.9 are positioned according to the splitting of $^3\Delta$ by A.

The derivation of the rotational branch structure is easy. Since the molecular states account for electronic angular momentum (non-$^1\Sigma$ states), the electric dipole selection rule $\Delta J = 0, \pm 1$ is applicable, which should give rise to P, Q, and R branches corresponding to $\Delta J = -1, 0$, and $+1$, respectively. But J derives from Ω, meaning each Ω gives a set of P, Q, R branches, such that the $^3\Delta$ state will produce three sets of P, Q, R branches.

The analysis provided here can be identically carried out for any molecular state. In the case of the half-integer value of Ω in the term symbol $2S+1\Lambda_\Omega$, the J-values obtained will also be half-integers. The same selection rule $\Delta J = 0, \pm 1$ applies, and each value of Ω gives rise to the three rotational branches P, Q, R.

Hund's Case(b). This case generally applies when $\Lambda = 0$ and $S \neq 0$. Having $\Lambda = 0$ must mean that one of the Σ-molecular states, the $^3\Sigma$ state of O_2 for example, should be treated under this case. However, some molecules for which $\Lambda \neq 0$ can also be considered with some discount. The idea is simple; it is the coupling of **S** to **Λ** that causes the precession of **S** about the internuclear axis, and if Λ tends to zero then spin-orbit coupling will grow smaller, and hence **S** will not precess about the axis eventually. This does not mean that **L** does not precess about the nuclear axis; **L** is still coupled to the molecular vector **Λ**, and hence precesses about the latter rather rapidly. The corresponding **Λ**-vector that may be small in magnitude adds up to the rotational angular momentum vector **O** to produce the angular vector **N** (Figure 11.10). It is to **N** that **S** couples strongly, as a result of which **N** and **S** add up to form the angular momentum **J**, but both **N** and **S** precess individually about **J**. Note that **O** and **S** can directly couple to form **J** when $\Lambda = 0$ (see Figure 11.10).

The main difference between Hund's cases(a) and (b) should be clear. In case (a) **Ω**(= **Λ** + **Σ**) and **O** combine to form **J**, but in case (b) **Ω** (=**Λ**) and **O** combine to form **N**, which

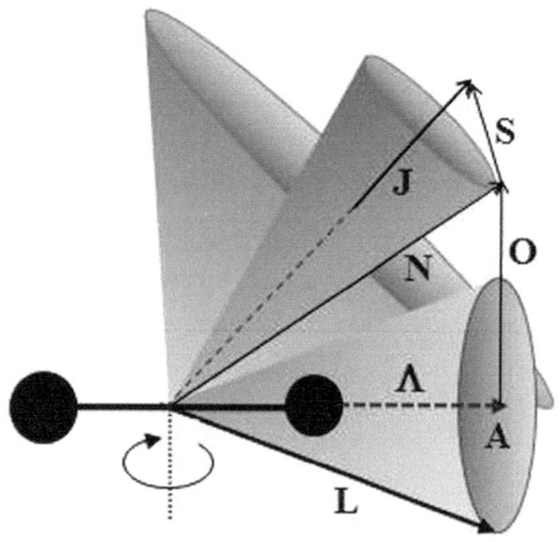

FIGURE 11.10 Vector precession diagram for Hund's case(b). The precessional motion of the **L** vector about the internuclear axis is faster than the nutation of the molecular axis about the total angular momentum vector **J**. But the precession of **N** and **S** about **J** is slower than the nutation of the molecular axis. When $\Lambda = 0$, **O** and **S** can directly couple to form **J**, about which the vectors rotate. Note that the vector **N** is also denoted by **K** in some literature.

then adds to **S** to give rise to **J**. This difference arises due to much weaker spin-orbit interaction in Hund's case(b). Now, whereas J is given by $\Omega + O$ in case (a), the allowed values of J in case (b) are $J = N + S = N+S, N+S-1, \cdots N-S$, the value of N being 0, 1, 2, ⋯. Each rotational level in the latter case therefore splits into $2S+1$ levels.

Regarding the rotational energy levels in case (b), the simplest energy terms are obtained when $S = 1/2$ as in a $^2\Sigma$ state. The coupling between the spin magnetic moment and the magnetic field produced by the rotation of the molecule produces a magnetic interaction energy term

$$E_M = \gamma \mathbf{S} \cdot \mathbf{N}, \tag{11.46}$$

where γ is spin-rotation coupling constant. The vector product $\mathbf{S} \cdot \mathbf{N}$ can be written (see for example, equation (4.42)) in terms of S, N, and J

$$\mathbf{S} \cdot \mathbf{N} = \frac{J(J+1) - S(S+1) - N(N+1)}{2}, \tag{11.47}$$

Since the allowed values of J are $N + \frac{1}{2}$ and $N - \frac{1}{2}$ when $S = 1/2$, one obtains

$$E = \begin{cases} \frac{\gamma}{2} N & \text{for } J = N + \frac{1}{2} \\ -\frac{\gamma}{2}(N+1) & \text{for } J = N - \frac{1}{2} \end{cases}. \tag{11.48}$$

So the rotational energy levels, without centrifugal distortion, are given by including the $\mathbf{S} \cdot \mathbf{N}$ magnetic interaction energy

$$F_1(N) = B_v N(N+1) + \frac{\gamma}{2} N$$

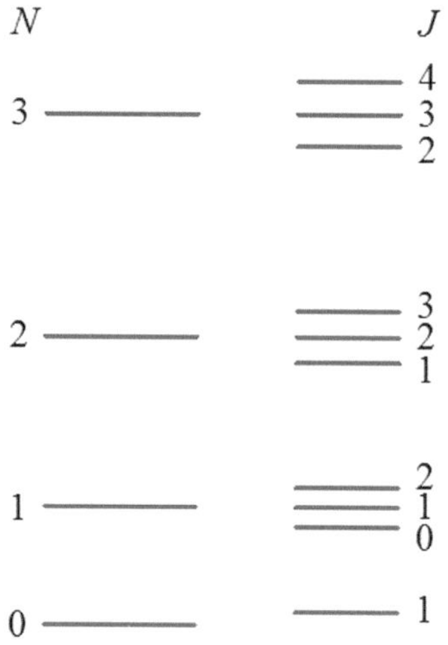

FIGURE 11.11 Splitting of ground-state rotational energy level into triplet fine structures for Hund's case(b).

$$F_2(N) = B_v N(N+1) - \frac{\gamma}{2}(N+1). \tag{11.49}$$

When $S > \frac{1}{2}$, magnetic interactions between electrons come into play. In the widely cited example of the $^3\Sigma$ state of O_2, the magnetic moments of the two parallel spins interact. For each rotational quantum number N there are three values of J

$$\begin{aligned} F_1(N) &\approx B_v N(N+1) + \gamma(N+1) - \frac{2\lambda(N+1)}{2N+3} && \text{for } J = N+1 \\ F_2(N) &\approx B_v N(N+1) && \text{for } J = N \\ F_3(N) &\approx B_v N(N+1) - \gamma N - \frac{2\lambda N}{2N-1} && \text{for } J = N-1 \end{aligned}, \tag{11.50}$$

where γ is spin-rotation coupling constant, and λ is the spin-spin coupling constant that reflects the extent of magnetic interaction between the two electron spins. More precise forms of rotational energy levels are also available, which are beyond the scope of our discussion here.

The splitting for the ground-state rotational level of molecules such as O_2, NH, and OH^+, to which Hund's case(b) applies, is shown in Figure 11.11. Notice the splitting of the rotational levels N into fine-structure triplets $J = N-1, N, N+1$. As depicted in the figure, the separations $J = (N+1) - J = N$ and $J = N - J = (N-1)$ need not be uniform for all N.

Hund's Case(c). This is similar to case (a) and applies to molecular states containing heavy nuclei such as I_2 where the spin-orbit interaction in the atomic limit is rather large. Consequently, the extent of coupling of **L** and **S** vectors is far in excess of the coupling of **L** and the molecular axis vector **A**. The vector addition of **L** and **S** forms the vector $\mathbf{J}_\alpha$, which then couples with the molecular axis to form **Ω**. Finally, the vector addition of **Ω** and **O** produces the total angular momentum **J** (Figure 11.12). Because the **LS** coupling is very large, the

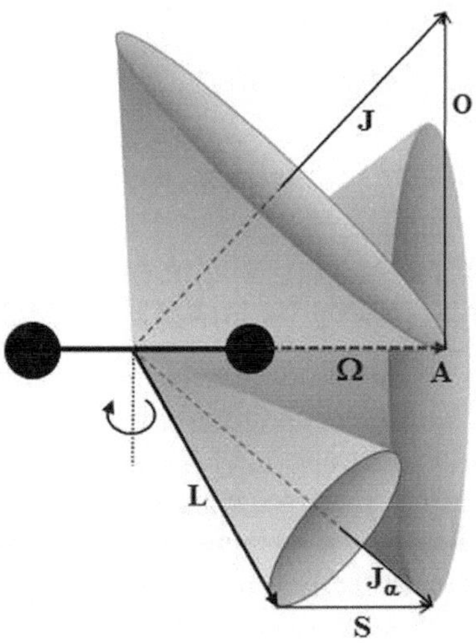

FIGURE 11.12 Vector diagram for Hund's case(c). The vector $\mathbf{J}_\alpha$ results from the addition of **L** and **S** vectors. The precession of $\mathbf{J}_\alpha$ about the internuclear axis is much slower than the precession of **L** and **S** vectors about $\mathbf{J}_\alpha$.

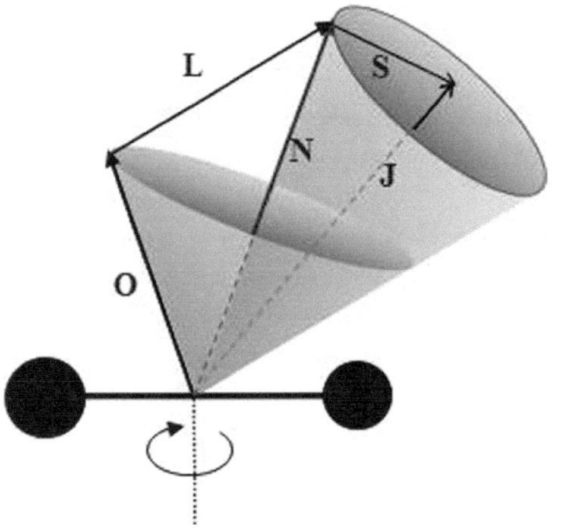

FIGURE 11.13 Hund's case(d). The vector **L** couples to the vector **O** to form vector **N**, which adds on to **S** to form **J**.

spin-orbit coupling is also too large, as a result of which the splitting of rotational energy levels in a single electronic state does not appear.

Hund's Case(d). This case applies to molecular states where an electron is excited to a higher shell such that the coupling of **L** and **O** is much larger compared to the coupling of **L** with the molecular axis vector **A**. The vector addition of **L** and **O** forms **N**, about which the forming vectors would rotate. If the spin vector **S** also exists, then **S** and **N** vectors add up to form the total angular momentum **J** (Figure 11.13).

A word of caution – the Hund coupling schemes should be considered instructive rather than rules to analyze the splitting of rotational energy levels. A particular Hund's case befitting for lower rotational states may not apply for higher rotational states of the same molecule. Still another case may apply for still higher rotational states. This need not be surprising because the molecular rotational frequencies increase at higher rotational states, due to which the coupling between electronic angular momenta and the molecular axis becomes weaker. Uncoupling of electron orbital angular momentum **L** can produce magnetic fields that would go on to interact with electron and nuclear magnetic moments. Such factors often render the analysis of rotational energy levels laborious. Consideration of both microwave rotational and vibronic spectra may be helpful in some occasions. There also are situations wherein none of the Hund's cases is applicable to a molecule; then an intermediate case should be sought.

11.8.2 Λ-Type Doubling

The effect of splitting of a doubly degenerate rotational level due to the decoupling of the electron spin angular momentum **S** from the molecular axis **A** is called Λ-type doubling, by which each J-level splits into two components. Yet, this sort of definition provides little insight into the cause of splitting. So we again review qualitatively the basic elements of electronic angular momenta and their interactions with the molecular axis vector so as to be clearer. The component of the orbital electron angular momentum **Λ** that is parallel to the axial vector is quantized. One may realize that **Λ** itself has two components, +**Λ** and −**Λ**, resulting from alternate rotation of the entire system about **J** (Figure 11.14). The value of $|\Lambda|^2$ is, however, quantized. We have also learned that $\mathbf{\Lambda} + \mathbf{\Sigma} = \mathbf{\Omega}$, which is the parallel or coaxial component of the total electronic angular momentum. The important point here is that $\mathbf{\Lambda} + \mathbf{\Sigma} = \mathbf{\Omega}$ holds only when spin-orbit coupling is non-vanishing (see Box 11.2), and it will not vanish as long as molecular rotational frequency is not leading. In a situation where molecular rotation is too rapid, in higher rotational states (larger rotational quantum number), for example, spin-orbit coupling is dominated by molecular rotation, as a result of which the electron spin begins to uncouple from the molecular axis. Consequently, **Ω** vanishes as the end-over-end rotation of the molecule turns faster.

The discussions above indicate that the two states of electronic angular momenta, (+Λ,+Σ) and (−Λ, −Σ), should be degenerate when molecular rotation is exceedingly slow, as it happens at the lower rotational levels J. As the frequency of end-over-end rotation picks up, the degeneracy of ±Λ and ±Σ are lifted, yielding separation into +Λ, +Σ and −Λ, −Σ. The separation energy of Λ-type doubling is, however, too small in comparison with rotational energies. Nevertheless, the degenerate ±(Λ,Σ) do split into +Λ, +Σ and −Λ, −Σ, and such splitting can be measured by pure rotational spectroscopy rather than rovibronic measurements.

Let us consider a singlet state ($S = 0$), so we have only $\pm\Lambda$. The doubly degenerate rotational level is represented by two linearly independent eigenfunctions, which is a consequence of the fact that both $+\Lambda$ and $-\Lambda$ orientations do not produce states that are eigenfunctions of the operator corresponding to a reflection operation in a mirror plane containing the two nuclei. This necessitates the use of a linear combination of the two states corresponding to the two **Λ**-orientations

$$\psi_1 = \Theta_{J,\Lambda,M}(\theta)\Phi_M(\phi) + \Theta_{J,-\Lambda,M}(\theta)\Phi_M(\phi)$$

$$\psi_2 = \Theta_{J,\Lambda,M}(\theta)\Phi_M(\phi) - \Theta_{J,-\Lambda,M}(\theta)\Phi_M(\phi), \quad (11.51)$$

where ψ_1 is symmetric to the reflection operation (Λ^+) and ψ_2 is not (Λ^-). The rotational level corresponding to Λ^+ or ψ_1 is called a 'positive' rotational level and the other one corresponding to Λ^- or ψ_2 is the 'negative' rotational level. They are distinguished on the basis of parity which is essentially the symmetry of reflection of the total eigenfunction.

The doubly degenerate J-levels for a $\Lambda = 1$ case, typically a symmetric top, and the splitting due to Λ-uncoupling is shown in Figure 11.14. For $\Lambda = 1$ and $\Sigma = 0$, the splitting energy is given by

$$E = q_\Lambda J(J+1), \quad (11.52)$$

which in Hund's case(b) is

$$E = q_\Lambda N(N+1). \quad (11.53)$$

The legend to Figure 11.14 narrates these results.

11.9 SELECTION RULES FOR ELECTRONIC TRANSITIONS IN DIATOMIC MOLECULES

Whether an electric dipole transition moment exists or vanishes is subject to a set of selection rules which can be obtained by calculating the dipole transition integral if the eigenfunctions involved are fully known (see equation (4.14), for example). We will not calculate the integrals here, but list the selection rules appropriate for the two electronic states involved in the transition. The selection rules apply generally to all electric dipole transitions irrespective of how the electronic angular momenta are mutually coupled in the two molecular states between whom a dipole transition is investigated. However, a few specific coupling-case-dependent selection rules are

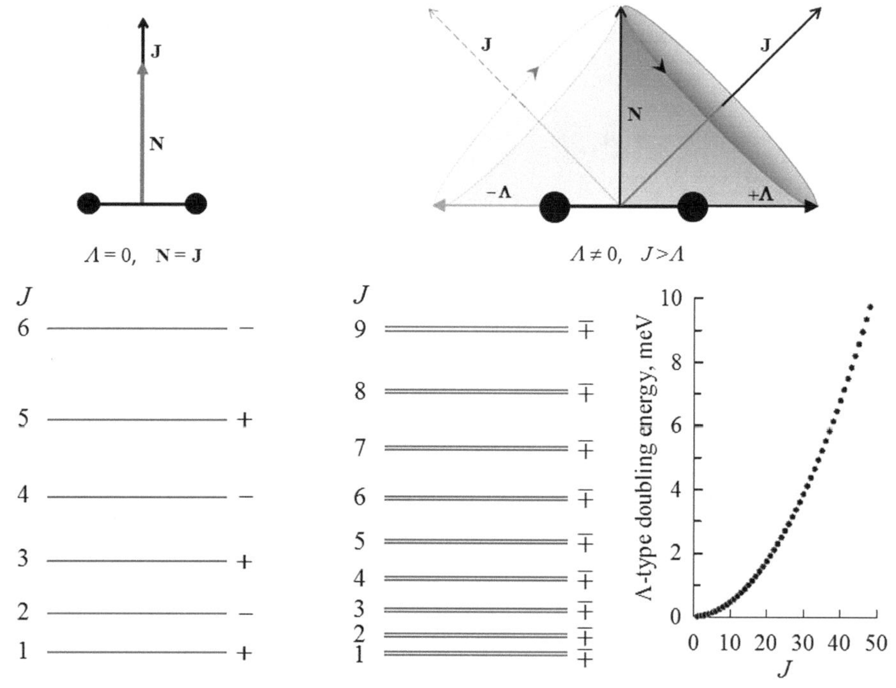

FIGURE 11.14 The degeneracy of energy levels and parity of rotational eigenfunctions corresponding to the eigenvalues. *Left*, vector diagram when $\Lambda = 0$, the $^1\Sigma^+$ state for example, for which the rotational levels corresponding to each J are degenerate. They are, however, distinguished according to whether the corresponding eigenfunctions change sign with respect to reflection in a plane containing the nuclei when the molecule rotates. *Right*, vector diagram for a symmetric top portraying the removal of degeneracy of Λ when the molecule rotates. Alternating anticlockwise (+Λ) and clockwise (−Λ) precession of the **L** vector sets in. This gives rise to splitting of the degenerate J levels, shown in the lower part for a system with $\Lambda = 1$ (Π state). The doubling energy in the graph refers to splitting of the $\pm J$ double degeneracy. The Λ-splitting energy for J-values from 1 to 50 was calculated taking q_Λ=1000 MHz (0.00414 meV) (Townes and Schawlow, 1975). The meaning of the constant q_Λ is conveyed by equations (11.52) and (11.53). It works in the following way (see equation (11.54)). Consider $\psi = \psi_{el}(1/r)\psi_v\psi_{rot}$. When $\Lambda = 0$ and continues to stay unchanged with respect to a reflection in a plane containing the nuclei, then ψ_{rot} alone determines the parity of the rotational levels. Accordingly, even-J (including 0) levels carry a +sign and odd-J levels a −sign. If ψ_{el} happened to change its sign with respect to a mirror plane reflection, then even J-levels will be labeled with a −sign and the odd-J levels will have a +sign. When $\Lambda \neq 0$, as in the case of a Π molecular state, one of the two linear combinations ψ^+ and ψ^- of $\Theta_{J,\Lambda,M}$ and $\Theta_{J,-\Lambda,M}$ does not change sign (ψ^+) upon reflection but the other one (ψ^-) does. Accordingly, each J-level will be doubly degenerate and is denoted by both + and −signs.

also required in addition to the general selection rules. These rules are not only instructive, but also form the basis of small-molecule spectroscopy. They are discussed in some detail in the following.

11.9.1 Symmetry-Based General Rules for Electronic Transitions

There are three general rules based on the symmetry of the total wave function

$$\psi = \psi_{\text{el}}\left(\frac{1}{r}\right)\psi_{\text{vib}}\psi_{\text{rot}} \quad (11.54)$$

with respect to reflection (i.e., interchange of nuclei), and inversion operations in the molecule. Let us briefly review these operations to see the outcome on the properties of the wavefunctions. The first thing to realize is that the symmetry operations can affect the properties of ψ_{el} and ψ_{rot}, not $(1/r)\psi_{\text{vib}}$, because the vibrational wavefunctions depend only on the internuclear separation, and hence will remain unchanged with respect to these operations for a homonuclear diatomic at least.

Reflection. This operation is analyzed in a molecule-fixed frame. A wavefunction is denoted with a + sign for ψ^+ or a – sign for ψ^-, subject to whether or not the function changes sign with respect to a reflection operation σ in a plane containing the nuclei, i.e., C_∞ axis. The total wavefunction ψ will not change, meaning ψ^+ or + remain unchanged, if both ψ_{el} and ψ_{rot} do not change. If either or both ψ_{el} and ψ_{rot} change, then the total wavefunction will change to ψ^- or –. The reflection also determines the parity of the rotational energy levels (Figure 11.15). Let us examine how it works by taking the cases of $\Lambda = 0$ (Σ-state) and $\Lambda = 1$ (Π-state).

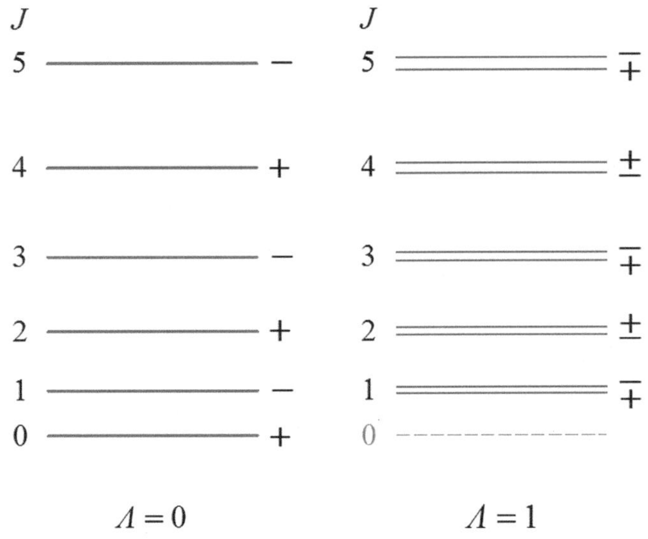

FIGURE 11.15 Rotational level symmetry. *Left*, a rigid rotor; *right*, a symmetric top. Since $\Lambda = 1$ is assumed for the latter, the rotational state $J = 0$ (i.e., $< \Lambda$) does not exist (shown by the faded broken line).

$\Lambda = 0.$

$$\sigma\left|\psi_{\text{el}}\right\rangle = +\left|\psi_{\text{el}}\right\rangle \qquad \Sigma^+\text{-state} \qquad \sigma\left|J\right\rangle = (-1)^J\left|J\right\rangle$$

$$\sigma\left|\psi_{\text{el}}\right\rangle = -\left|\psi_{\text{el}}\right\rangle \qquad \Sigma^-\text{-state} \qquad \sigma\left|J\right\rangle = (-1)^{J+1}\left|J\right\rangle \quad (11.55)$$

The first of these two operations indicate that when $\left|\psi_{\text{el}}\right\rangle$ does not change with respect to the reflection, then the parity of the rotational levels is determined by ψ_{rot} alone. For J= 0, 1, 2, ⋯ the rotational energy levels have odd and even parity (– and +) when J is odd and even, respectively, as indicated in Figure 11.15. The second operation indicates that when $\left|\psi_{\text{el}}\right\rangle$ changes sign with respect to reflection then the parity of the rotational levels is + for even J, and – for odd J.

$\Lambda = 1.$

In this case only a small number of reflection planes exist and the changes in + and – signs of $\left|\psi_{\text{el}}\right\rangle$ are indistinguishable so that each J value corresponds to a doubly degenerate energy level + and – as discussed earlier in the context of Λ-type doubling.

Inversion. A homonuclear state is labeled g or u in accordance with the relevant eigenvalue, +1 or –1, when the inversion operation is carried out in the spaced-fixed axes. However, ψ_{el} may or may not change sign with respect to a reflection operation carried out on both g and u states. For instance, the states Σ_g^+ and Σ_u^+ are the results if ψ_{el} remains unchanged in the reflection, but the two states will be Σ_g^- and Σ_u^- if ψ_{el} changes sign with respect to reflection.

Exchange of Nuclei. Whether or not a wavefunction changes sign with respect to an interchange of the nuclear positions is denoted by s and a, meaning symmetric and antisymmetric operations, respectively. As was the case with reflection, the effect of nuclear exchange operation depends on the properties of both electronic and rotational states. For a homonuclear diatomic molecule with quantum number $\Lambda = 0$ (Σ-state) the nuclear exchange accounts for both reflection (+ and –) and inversion symmetry (g and u). If the nuclear exchange operator is X_N, then we should check if ψ_{el} does not change sign with respect to reflection as shown in the following.

$$X_N\left|J\Sigma_g^+\right\rangle = (-1)^J\left|J\Sigma_g^+\right\rangle \qquad \text{symmetric}, s$$

$$X_N\left|J\Sigma_u^+\right\rangle = (-1)^{J+1}\left|J\Sigma_u^+\right\rangle \qquad \text{antisymmetric}, a$$

$$X_N\left|J\Sigma_u^-\right\rangle = (-1)^J\left|J\Sigma_u^-\right\rangle \qquad \text{symmetric}, s$$

$$X_N\left|J\Sigma_g^-\right\rangle = (-1)^{J+1}\left|J\Sigma_g^-\right\rangle \qquad \text{antisymmetric}, a$$

Because a rotational level of the Σ^+ state can be + (even J) or – (odd J) and that of the Σ^- state can be + (odd J) or – (even J), the symmetries s and a are not specific to + and – rotational levels. These symmetry properties of rotational levels for many molecular states have been elegantly described by Herzberg in the form of compact diagrams, a few of which are redrawn in Figures 11.16 and 11.17.

11.9.2 Selection Rules

1. While the electronic transitions involve the dipole operator, the character of the dipole moment μ has to be determined. It is easy to see that a reflection operation in the molecule-fixed axes system attaches a – character to μ, leading to the selection rule

$$+ \leftrightarrow - \quad \text{allowed}$$

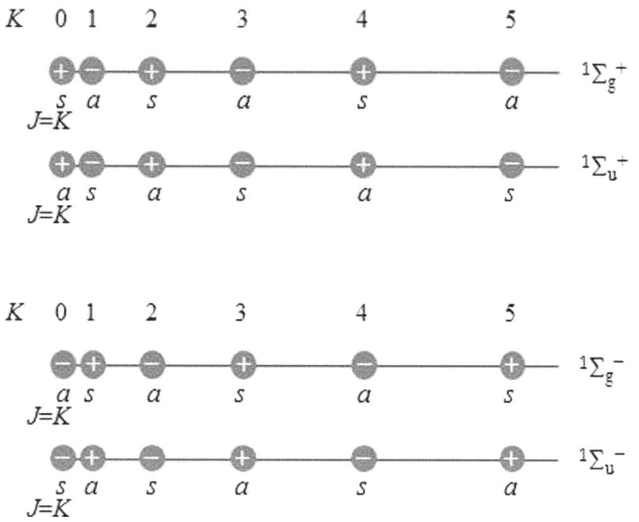

FIGURE 11.16 Herzberg diagram of symmetry properties of rotational levels for the molecular states Σ of a homonuclear diatomic molecule.

$$\begin{cases} + \leftrightarrow + \\ - \leftrightarrow - \end{cases} \quad \text{not allowed}$$

2. The effect of the nuclear interchange operation is easiest to see in the case of a homonuclear system for which μ does not change sign, leading to the selection rule that both terms involved in a transition are either symmetric or antisymmetric

$$\begin{cases} s \leftrightarrow s \\ a \leftrightarrow a \end{cases} \quad \text{allowed}$$

$$s \leftrightarrow a \quad \text{not allowed}$$

3. If the two nuclei of a diatomic molecule are equivalent with regard to the charge content, then the selection rule is

$$g \leftrightarrow u \quad \text{allowed}$$

$$\begin{cases} g \leftrightarrow g \\ u \leftrightarrow u \end{cases} \quad \text{not allowed}$$

4. Based on the quantum number J alone, the selection rule is $\Delta J = 0, \pm 1$.

11.9.3 Selection Rules Pertaining to Hund's Coupling Cases

These selection rules rely upon changes in the spin multiplicity, the quantum number Λ, the spin component Ω along the internuclear axis, and the quantum number K. One must exercise caution concerning the use of symbols here. As noted earlier, the vector **N** used to depict Hund's cases (see Figure 11.10, for example) may be designated by **K** in certain other references. The latter could be more appropriate

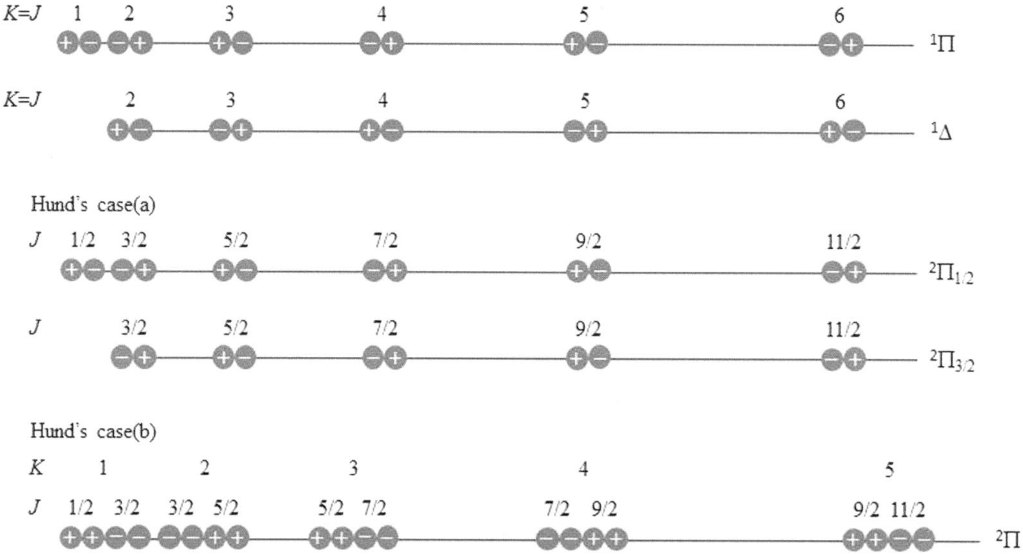

FIGURE 11.17 Herzberg diagram of symmetry properties of rotational eigenfunctions for Π and Δ molecular states of heteronuclear diatomics.

perhaps, because the quantum number K corresponds to $\mathbf{K}$. Since the quantum number K is important and the designation is widely used, we should also associate quantum number K to the vector $\mathbf{N}$. The quantum number K is the total quantum number minus the contribution from the electron spin. Given a value of Λ, we have

$$K = \Lambda,\ \Lambda + 1,\ \Lambda + 2,\ \cdots.$$

The selection rules are also particular regarding the nature of the coupling between electron spin and rotational angular momenta with respect to the two electronic states involved in the dipole moment transition. Should we include the spin contribution (quantum number S), the total quantum number J arises from K and S. In that case

$$J = K + S, K + S - 1, K + S - 2, \ldots.$$

Rather than describing them in text, it is more convenient to list these selection rules in a table (Table 11.2).

A particularly simple way of recognizing the rotational band structures P, Q, R to conform with the selection rules is to analyze the rotational level diagrams as prescribed by Hezberg (1950). The diagram for ${}^1\Sigma^- \leftrightarrow {}^1\Sigma^-$ and ${}^1\Sigma_g^- \leftrightarrow {}^1\Sigma_u^-$ transitions drawn in Figure 11.18 shows the relevant electronic states by two horizontal bars corresponding to the lower and upper electronic states, and each bar is labeled with the symmetry and the corresponding rotational quantum number J. The two bars are then connected with respect to each J in consistence with appropriate symmetry and the ΔJ selection rule. The oblique lines connecting the two electronic states reveal the transitions corresponding to the rotational band structures according to the ΔJ outcome. The diagram for these single-spin multiplicity transitions clearly shows the appearance of P and R band structures and the absence of the Q band structure. The diagrams for the other possible transitions ${}^1\Sigma_u^- \leftrightarrow {}^1\Sigma_g^-$, ${}^1\Sigma_u^+ \leftrightarrow {}^1\Sigma_g^+$, and ${}^1\Sigma_g^+ \leftrightarrow {}^1\Sigma_u^+$ can be similarly drawn.

The diagrams for ${}^1\Sigma_u^- \leftrightarrow {}^1\Pi$ and ${}^1\Sigma^+ \leftrightarrow {}^1\Pi$ transitions are depicted in Figure 11.19. In the upper state ${}^1\Pi$, the degeneracy of the rotational levels $\pm J$ resulting from the degeneracy of $\Lambda(=\pm 1)$ is shown (see also Figure 11.14). The rotational state $J = 0$ is absent because a rotational level with $J < \Lambda$ does not occur. Since $\Lambda \neq 0$ in the upper electronic state, we would have Q-branches appearing ($\Delta J = 0$). However, the absence of $J' = 0$ leads to the absence of the $P(1)$ and $Q(0)$ rotational bands corresponding to ΔJ values of -1 and $+1$, respectively. The same diagram applies to the ${}^1\Sigma_g^- \leftrightarrow {}^1\Pi_u$ transition except that the rotational levels of the two electronic states will bear the appropriate symmetry and antisymmetry labels, s and a. It is clear that the diagrams for ${}^1\Sigma^- \leftrightarrow {}^1\Pi$ and ${}^1\Sigma^+ \leftrightarrow {}^1\Pi$ are the same except for the reversal of + and – signs, for odd and even values of J, repectively. This reversal has been discussed earlier and is ascribed to whether ψ_{el} changes sign with respect to reflection; if it does (Σ^-), then the rotational levels

TABLE 11.2
Summary of Selection Rules for Hund's Coupling Cases

	$\Delta\Sigma$	$\Delta\Lambda$	$\Delta\Omega$	Δs	ΔJ	ΔK	allowed	not allowed
General rule:								
Cases (a) and (b)		0,±1		0‡	0,±1		$\Sigma^+ \leftrightarrow \Sigma^+, \Sigma^- \leftrightarrow \Sigma^-,$ $\Sigma^+ \leftrightarrow \Pi, \Sigma^- \leftrightarrow \Pi,$ $\Pi \leftrightarrow \Pi, \Pi \leftrightarrow \Delta,$ $\Delta \leftrightarrow \Delta, \cdots$	$\Sigma^+ \leftrightarrow \Sigma^-, \Sigma \leftrightarrow \Delta,$ $\Pi \leftrightarrow \Phi, \cdots$
Particular rules: when both ground and excited states belong to the same Hund's case								
Case (a) alone	0	0,±1	0,±1$	0	0,±1		$(a)2_{\Pi_{\frac{1}{2}}} \leftrightarrow (a)2_{\Pi_{\frac{1}{2}}},$ $(a)2_{\Pi_{\frac{3}{2}}} \leftrightarrow (a)2_{\Pi_{\frac{3}{2}}},$ $(a)2_{\Pi_{\frac{1}{2}}} \leftrightarrow (a)2_{\Delta_{\frac{1}{2}}},$ …	$(a)2_{\Pi_{\frac{1}{2}}} \leftrightarrow (a)2_{\Pi_{\frac{3}{2}}}$ $(a)2_{\Pi_{\frac{3}{2}}} \leftrightarrow (a)2_{\Delta_{\frac{1}{2}}}$ $(a)3_{\Pi_0} \leftrightarrow (a)3_{\Pi_1}$ …
	0	0,±1	0#	0	±1		$(a)2_{\Pi_{\frac{3}{2}}} \leftrightarrow (a)2_{\Pi_{\frac{3}{2}}}$ …	
Case (b) alone		0,±1	0		±1	±1	$(b)\Sigma^+ \leftrightarrow (b)\Sigma^+,$ $(b)\Sigma^- \leftrightarrow (b)\Sigma^-, \cdots$	
		0	0	0	0,±1	0,±1	$(b)\Pi \leftrightarrow (b)\Pi,$ $(b)\Delta \leftrightarrow (b)\Delta$	$(b)\Sigma \leftrightarrow (b)\Sigma$
Case (c) alone			0,±1		±1	0,±1		

Notes:
‡ The selection rule $\Delta S = 0$ tends to be loose when nuclear charge increases (heavy atom limit), leading to increased $\mathbf{S}$ and $\mathbf{\Lambda}$ spin-orbit coupling, implying a breakdown of Russel-Saunder's coupling.
$ This selection rule appears redundant when $\Delta\Lambda = 0, \pm 1$ and $\Delta\Sigma = 0$ already exist. However, $\Delta\Omega = 0, \pm 1$ holds even under strong spin-orbit coupling at an instant that the former selection rules tend to be ineffective.
# When only $\Delta\Omega = 0$ holds; now $\Delta J = 0$ transition is not allowed, and hence the Q-branch of the rotationl structure is absent.

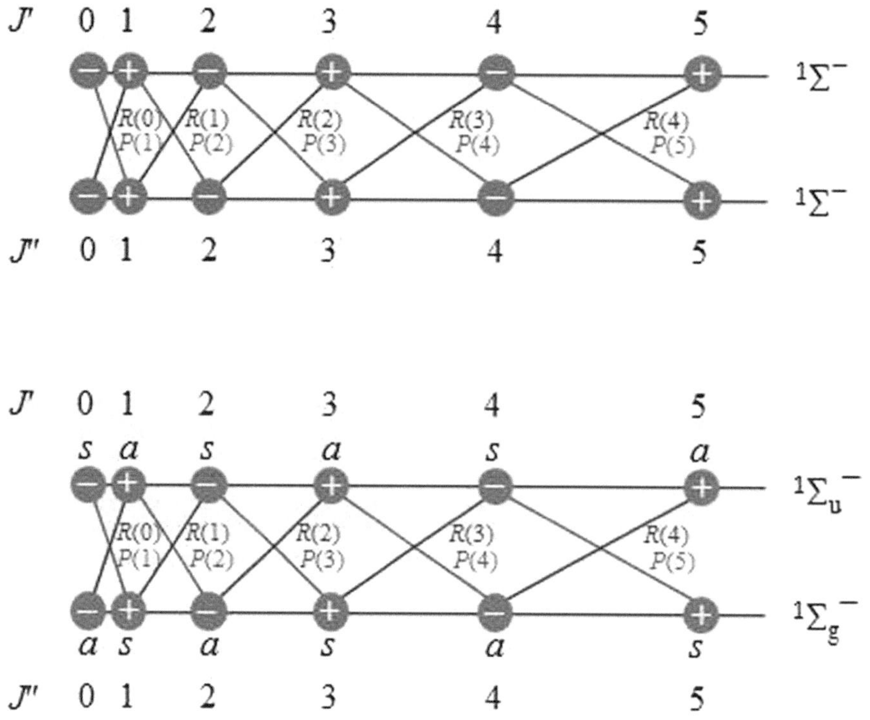

FIGURE 11.18 Herzberg diagram depicting the appearance of P- and R-band structures in the ${}^1\Sigma^- \leftrightarrow {}^1\Sigma^-$ and ${}^1\Sigma_g^- \leftrightarrow {}^1\Sigma_u^-$ transitions.

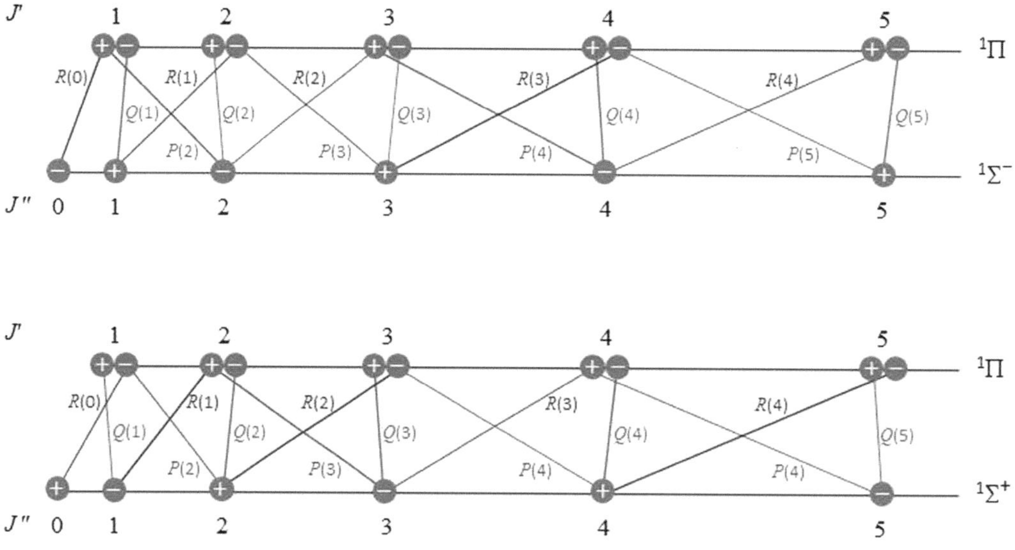

FIGURE 11.19 Herzberg diagram for ${}^1\Sigma^- \leftrightarrow {}^1\Pi$ and ${}^1\Sigma^+ \leftrightarrow {}^1\Pi$ transitions showing the P, Q, R lines.

are labeled with a – sign for even J, and if not, then they are denoted by a – sign for odd J.

Let us consider the highly resolved ${}^1\Pi \rightarrow {}^1\Sigma^+$ emission spectrum of aluminum hydride (AlH) that has been studied for at least 70 years now (Zeeman and Ritter, 1954; Grimaldi et al., 1966; Pelissier and Matrieu, 1977; Bauschlicher and Linghoff, 1988; Zhu et al., 1992; Ram and Bernath, 1996; Szajna and Zachwieja, 2009). Nine electronic states of AlH can be identified, of which six are singlet states, including the ${}^1\Sigma^+$ ground state, and three are triplet states (Figure 11.20). The ${}^1\Pi \rightarrow {}^1\Sigma^+$ emission spectrum in the visible region (~425 nm) occurs from the two bound vibrational levels $v' = 0,1$ in the ${}^1\Pi$ state to $v'' = 0,1,2,3, \cdots$ levels in the ground state. The rotational structure in the $0-0$ (i.e., $v' = 0 \rightarrow v'' = 0$) vibrational level transition is also shown in Figure 11.20. Note that $0-0$, $1-1$, $2-2$, $\cdots$ bands are called diagonal bands, and $0-1$, $1-0$, $1-2$, $\cdots$ are off-diagonal bands. The intensities of the off-diagonal bands are too low, often not even observed in spectra. Each band, whether on- or off-diagonal, will have P, Q, and R branches. We recall that the P, Q, and R branches in the emission spectrum correspond to the change of J as an increase by 1, no change, and a decrease

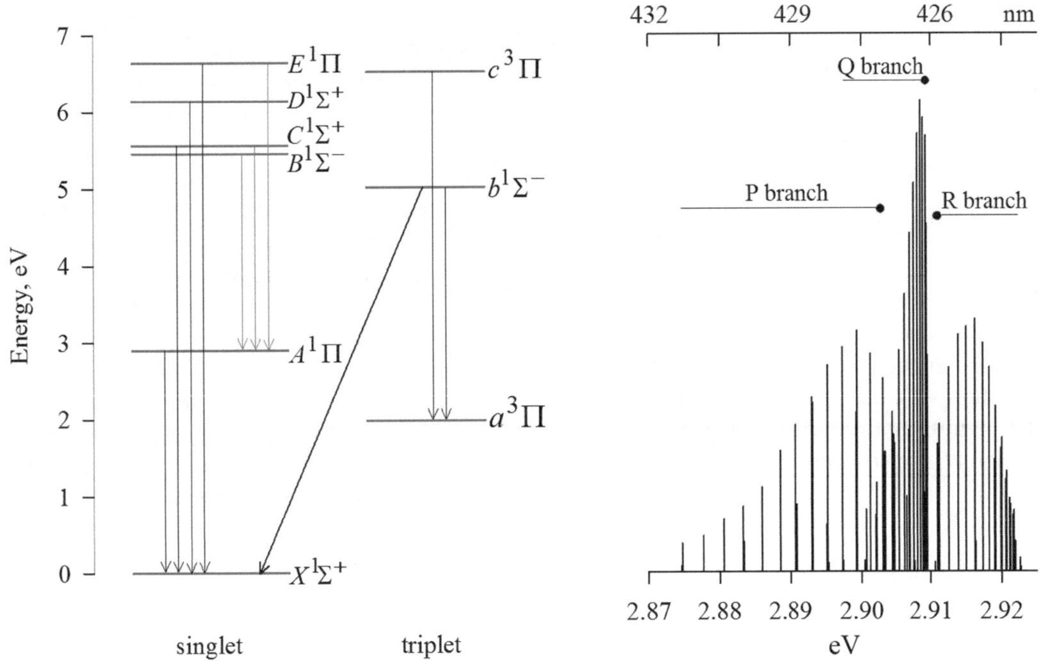

FIGURE 11.20 *Left*, energy level diagram for AlH with symmetry labels for various energy eigenstates and observable fluorescence transitions. The singlet (capital prefix before the molecular eigenstate) and triplet (small-case prefix before the molecular eigenstate) levels are labeled. *Right*, schematic of the spectrum showing all three combinations of transitions, singlet→singlet, triplet→singlet, and triplet→triplet, are permissible. The state $A^1\Pi$ is implicated in predissociation (see Chapter 12).

by 1, respectively. Conversely, the P, Q, and R branches in the absorption spectrum ($^1\Pi \leftarrow {}^1\Sigma^+$) correspond to a decrease of J by 1, no change of J, and an increase of J by 1, respectively. The AlH emission spectrum reveals the characteristic band head formation for all rotational branches. However, band progression is limited, not only for the $0 \rightarrow v''$ band but also for the $1 \rightarrow v''$ bands, meaning the appearance of fewer $J' \rightarrow J''$ emission lines. We can actually count the rotational lines in each branch of the $0-0$ transition (Figure 11.20); there are 17 in each. This limitation arises from a break-off in the rotational line intensities and is due to 'perturbation' of the excited state.

11.10 PERTURBATION MANIFESTS IN VIBRONIC SPECTRA

An abrupt cutoff in the intensity of rotational line progression is only one 'apparently abnormal' result; the J-dependent separation of rotational lines in a band may often deviate from the Deslandres-type of linearity, and rotational lines may split in an abnormal manner so as to appear too dense in certain frequency ranges. This means the fine structure in a vibronic spectrum most often is not uniform by the number, the splitting, and the line intensities. Even if not uniform, there still appears a resonance-like behavior; for example, the line splitting increases and decreases, both rapidly, as though there is a center frequency for this to happen. These irregularities, call non-ideal manifests, are due to perturbations of energy levels – a perturbant energy level may alter or modify the eigenstate and the energy eigenvalue of the level that it perturbs. We should be clear that such perturbations arise not from an external potential, but from the myriad of possible potential surfaces for the molecule. Some of these surfaces may have been worked out explicitly for certain molecules, and some others may be hidden otherwise. Thus, except for the well-studied molecules for which the perturber and the perturbant both are understood well, it may not often be clear as to which energy level acts as a perturber.

Although we have considered degenerate and non-degenerate perturbations in a somewhat arbitrary manner, we can have a general form of perturbation – something in between the two, because in the formidable energy level structures in the potential functions, even for a simple diatomic, the appropriateness of degeneracy or non-degeneracy is a difficult asking to determine. For example, two vibrational or rotational levels may be positioned so closely that a discretion between degeneracy and non-degeneracy of the levels is difficult to make. So, let us briefly look at the matrix elements for a general perturbation. To be simple we will take two unperturbed states $\psi_1^{(o)}$ and $\psi_2^{(o)}$ with the reaspective eigenvalues $E_1^{(o)}$ and $E_2^{(o)}$. Let us label the perturbed eigenfunctions and eigenvalues as ψ_1 and ψ_2, and E_1 and E_2. It is shown easily, after solving for the appropriate integrals, that

$$\begin{vmatrix} W_{11} - E & W_{12} \\ W_{21} & W_{22} - E \end{vmatrix} = 0, \qquad (11.56)$$

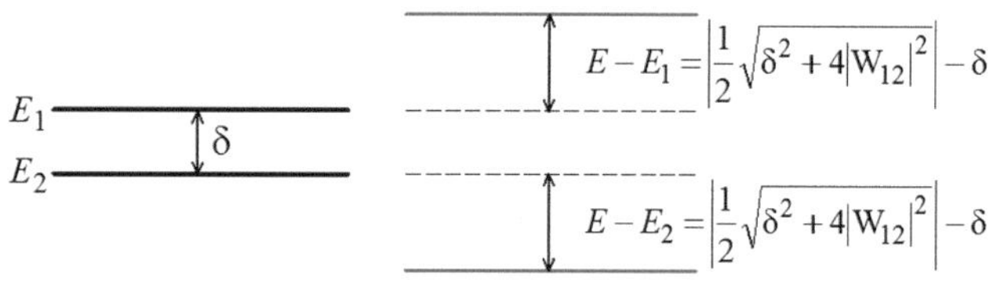

FIGURE 11.21 Perturbation of vibronic levels.

where W_{11} and W_{22} are the matrix elements of the diagonal, and E is the extent to which energy is shifted from the unperturbed case. In the zeroth-order energy difference between perturbed and unperturbed states, only the diagonal terms W_{11} and W_{22} are considered. The perturbation calculation, however, warrants inclusion of W_{12} and W_{21}, so equation (11.56) becomes

$$\begin{vmatrix} E_1 - E & W_{12} \\ W_{21} & E_2 - E \end{vmatrix} = 0, \tag{11.57}$$

in which E_1 and E_2 are now the unperturbed energies. The solution of the determinant is obtained as

$$E = \frac{1}{2}\left(E_1 + E_2\right) \pm \sqrt{\delta^2 + 4\left|W_{12}\right|^2}, \tag{11.58}$$

where $\delta = E_1 - E_2$. The interacting energy levels E_1 and E_2 corresponding to J_1 and J_2 may belong to two different electronic states as shown in Figure 11.21. Clearly the two unperturbed levels mutually repel, and the perturbed ones gain and lose energy of the same magnitude. Stronger coupling between the two initial levels (or larger magnitude of $|W_{12}|$) leads to greater perturbation separation. The perturbed eigenfunctions and energy levels then lead to perturbations or abnormalities in the spectrum.

11.10.1 Rotational Perturbation and Kronig's Selection Rules

The terms in the rotational perturbation potential are essentially the angular momentum terms, including $\mathbf{R}\cdot\mathbf{L}$ and $\mathbf{R}\cdot\mathbf{S}$ discussed earlier. These terms do not enter into the Hamiltonian in the calculation of energy levels of a molecule under the Born-Oppenheimer approximation, but they now act as perturbation potentials for rotational levels. Considering rotation and spin-orbit coupling, we get

$$\mathcal{H}'_{\text{rot}} = B\{\mathbf{J} - \mathbf{L} - \mathbf{S}\}^2 + A\mathbf{L}\cdot\mathbf{S}. \tag{11.59}$$

Accordingly, rotational perturbation introduces interactions of states of different angular momenta, $\mathbf{\Omega}$ and $\mathbf{\Lambda}$. The rotational part of the matrix element W_{ij} can be written as

$$\left\langle \Omega S \Lambda J v p \mathcal{L} \middle| \mathcal{H}'_{\text{rot}} \middle| \Omega' S' \Lambda' J' v' p' \mathcal{L}' \right\rangle, \tag{11.60}$$

in which v refers to the vibrational state as in the Franck-Condon factor $q_{vv'}$, p is parity, and $\mathcal{L}$ is symmetry. We will not discuss this aspect in detail, but simply state that this leads one to Kronig's selection rules of perturbation that are based on symmetry properties and quantum numbers of the two electronic states involved. Kronig's original work was published in German, but Herzberg (1950) has described the selection rules in some detail. The selection rules are listed as follows.

1. $\Delta J = 0$, meaning the rotational state characterized by the quantum number J must be the same in the two elelctronic states.
2. $\Delta S = 0$, which stipulates that the spin multiplicity of the two interacting states must be the same. This is not a rigorous selection rule because the spin multiplicity may change just as singlet-to-triplet transition may occur in emissive transitions.
3. $\Delta\Lambda = 0, \pm 1$, which means a Σ-state can interact with a Σ- or a Π-state, but not with a Δ-state. The spin-multiplicity rule narrows this down. The matrix element will not vanish if only ${}^1\Sigma^+$ interacts with ${}^1\Sigma^+$ or ${}^1\Delta$, and so on. This rule is not stringent in the sense that Λ may not be defined for certain electronic states, implying that the rule may hold strictly in Hund's cases(a) and (b). When Λ is not defined as for Hund's case(c), the selection rule turns out to be $\Delta\Omega = 0, \pm 1$. Perturbation in accord with the $\Delta\Lambda = 0$ selection rule can occur when both interacting states belong to the same electronic species, and is called homogeneous perturbation. Perturbations with $\Delta\Lambda = \pm 1$ can occur between two surfaces that do not belong to the same electronic state, which is referred to as heterogeneous perturbation and which is weaker for low rotational energies.
4. $+ \leftrightarrow +, - \leftrightarrow -, + \leftrightarrow -$, which implies that the reflection symmetry of the two states involved in the transition is the same. Only those levels that have the same parity can perturb each other.
5. $s \leftrightarrow s$, $a \leftrightarrow a$, $s \nleftrightarrow a$, which applies to a homonuclear diatomic such that the symmetry of nuclear interchange in the two interacting states must be the same.

These selection rules do not involve the vibrational quantum number, although the Franck-Condon principle is not violated. The manifest of rotational perturbation is both line-shift and line-splitting, even though the two may not occur concurrently. These perturbation effects arise when rotational levels of two electronic states tend to overlap. The perturbation is, of course, subject to the selection rules given above.

11.10.2 Frequency Shift and Λ-doubling in Rotational Perturbation

Suppose a system experiences the selection rule-allowed rotational perturbation. The manifest of rotational perturbation is both line-shift and line-splitting, even when the two do not occur concurrently. These perturbation effects arise when rotational levels of two electronic states tend to overlap.

The effect of line-shift is analyzed by preparing a Schmid-Gerö plot, in which the total energy

$$T_{el} + G + F = T_{el} + G + B_v J\left(J + 1\right) \tag{11.61}$$

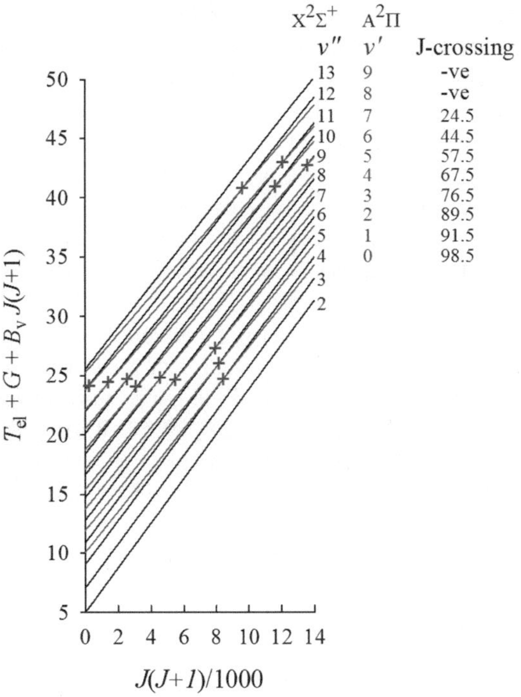

FIGURE 11.22 A representative part of the Schmid-Gerö plot of perturbation of the $X^2\Sigma^+$ and $A^2\Pi$ states of CN. The plot has been constructed using a part of data from Kotlar et al. (1980). The J-levels belonging to the v' and v'' levels and the values of J-crossings (indicated by + signs in red) are listed in the column labeled J-crossing.

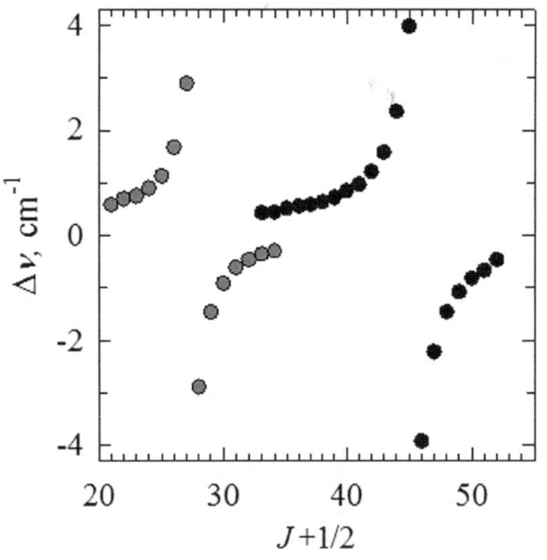

FIGURE 11.23 A plot of rotational frequency shift prepared by graphing $\Delta\nu = \nu_{\text{perturbed}} - \nu_{\text{unperturbed}}$ with $J+1/2$ for the $A^2\Pi_{1/2}$, $v'=7$ state of CN. The perturbation crossing occurs between $v'=7$ of $A^2\Pi_{1/2}$ and $v''=2$ of $X^2\Sigma^+$ states. Shifts for e- and f-parity levels are indicated in black and gray circles, respectively. Data taken from Kotlar et al. (1980).

for different vibrational levels of the perturbant and the perturbed electronic states are plotted with $J(J+1)$. Figure 11.22 shows a part of the analysis of the $A^2\Pi - X^2\Sigma$ 'red bands' of the CN radical based on literature data (Kotlar et al., 1980). The idea is to check for J-crossings. Mixing of two eigenfunctions occurs at the point of the crossing (equal energy) of two rotational levels of the same J-value – one rotational level belongs to v' vibrational level of $A^2\Pi_{1/2}$ and the other belongs to the v'' level of $X^2\Sigma$. The plot has been prepared by analyzing the spectral bands.

The regular unperturbed frequency for each J can be calculated from $\nu_{\text{unperturbed}} = B_v J(J+1)$, where B_v is the rotational constant. A plot of the frequency shift ($\nu_{\text{perturbed}} - \nu_{\text{unperturbed}} = \Delta\nu$) vs J then shows that the actual frequencies are shifted (peak-shift) as one passes through the J-crossing regions. Figure 11.23 shows this for perturbation in the $v'=7$ level of the state $A^2\Pi_{1/2}$, the perturbation satisfying the $\Delta\Lambda = \pm 1$ selection rule. Since $\Lambda = 1$ for the $A^2\Pi_{1/2}$ state, each J-level is doubly degenerate bearing + (even) and – (odd) signs, called respectively e and f parity. As mentioned earlier, parity refers to whether the total eigenfunction changes sign with reference to the reflection operation. The figure clearly shows that J-crossings occur at lower values of J for e-parity (+) and at higher values of J for f-parity (–). The frequency shifts as a result of perturbation of the $A^2\Pi_{1/2}$ state is within ~4 cm^{-1}, indicating small perturbation. More recent analysis of the perturbation in the $A^2\Pi(v=0)$ state of $^{12}C^{18}O$, a problem being investigated for a century now, records higher frequency shifts, to the extent of ~9 cm^{-1} (Trivikram et al., 2017).

The line-splitting in rotational perturbation is essentially a Λ-doubling effect. In the $^2\Pi - {}^2\Sigma$ perturbation for example, where the latter is perturbed, one has $\Delta\Lambda = +1$, which means a shift of only one Λ-component resulting in Λ-type doublets. The selection rule $\Delta\Lambda = +1$ and Λ-doubling appear mutually inclusive, because the derivation of both perturbation and Λ-doubling theories is fundamentally the same.

Extra lines in a rotational progression may also arise due to mixing of the characters of the perturbant-perturbed pair. If A' and B'' are two stacks of rotational levels between which rotational transitions occur in a non-perturbation scenario, but then A' is perturbed by another rotation series c such that c acquires some character of A' due to perturbation mixing, the $c \leftrightarrow B''$ transition can also occur in both absorption and emission. So we have $c \leftrightarrow B''$ transitions in addition to the $A' \leftrightarrow B''$ transitions because of c perturbing A'. Note that the mixing causes the loss of the intial identities of A' and c, the perturbed and perturber, respectively. It is this mixing of their characters that renders c establishing transitions with B' and thus giving rise to extra rotational lines.

11.10.3 Vibrational Perturbation

In the discussion of molecular eigenstates earlier we obtained the full form of the Schrödinger equation and mentioned that the off-diagonal nonadiabatic coupling terms (equation (5.11)) are generally ignored because of the smallness of their magnitudes. Interestingly, these are the terms that produce perturbations and they cannot be neglected for accurate spectroscopic calculations. Those ignored terms that contain the kinetic energy of nuclei that move in a potential set up by

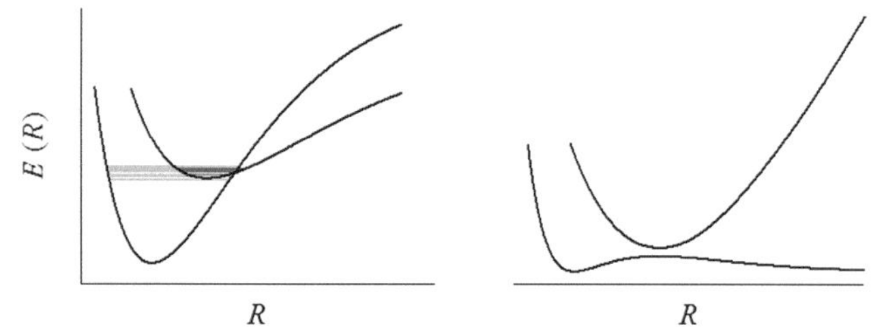

FIGURE 11.24 *Left*, vibrational levels closer to the region of intersection and at the same energy height can mix and perturb the levels across the two electronic states. *Right*, avoided crossing may produce a maximum in the lower potential surface.

electrons form the vibrational perturbation Hamiltonian. From equation (5.11) we have

$$\mathcal{H}'_{\text{vib}} = -\sum_{j=1}^{3N} \frac{\hbar^2}{2M_j}\left[\Delta_j^{\,2} + 2\Delta_j\Delta_j\right]. \qquad (11.62)$$

This coupling Hamiltonian has already been written as T_R earlier (equation 5.10).

As in the case of rotational perturbation (equation (11.60)) the matrix elements for vibrational perturbation are

$$\left\langle \Omega S \Lambda J v p \mathcal{L} \right| \mathcal{H}'_{\text{vib}} \left| \Omega' S' \Lambda' J' v' p' \mathcal{L}' \right\rangle, \qquad (11.63)$$

where the quantum numbers, the parity, and the symmetry functions are described in the same way as was done for the rotational perturbation matrix elements. There is no rigorous selection rule for the vibrational quantum number v'. However, the appearance of $\delta_{vv'}$, which is analogous to the Franck-Condon factor, suggests that there should be an 'apparent selection rule' for the perturbed-perturber vibrational levels. Since v and v' belong to two different electronic states, they can mutually perturb and mix strongly only if they are placed in the same energy height in the two potential energy surfaces which are crossing each other (Figure 11.24). The mutual perturbation of v and v' is subject to satisfying the Kronig's selection rules listed earlier, which stands obvious from the vibrational perturbation matrix elements in the preceding equation. As a consequence of perturbation, the energies of both vibrational levels shift, which will be manifested as equal-energy shift for all J-levels in both rotational stacks. So the shift of rotational lines is nearly uniform for all $J - J'$ pairs. Nonetheless, there is a very small dependence of the energy shift on the J-value and this dependence causes a little change in the rotational constant B_e. We may note that the vibrational perturbations are normally homogeneous ($\Delta\Lambda = 0$) because heterogeneous perturbations hardly involve lower rotational levels.

Vibrational perturbation occurs when two electronic states (surfaces) having the same Λ and the same symmetry, called the states of the same electronic species, tend to cross each other. There is a rule due to Hund initially, called 'noncrossing rule', which states that two potential energy surfaces of the same electronic species must not cross. One might argue that the degeneracy of the potential energy at the crossing point can be removed by certain terms in the Hamiltonian, so the surfaces will avoid and repel each other. This is true, especially for systems with cold rotational and vibrational motions, implying the importance of motional dynamics that lead to a violation of the noncrossing rule. Nevertheless, whether the surfaces do cross or avoid crossing is determined by the strength of interaction between the two surfaces – crossing occurs if the interaction is weak resulting in perturbation, and crossing will be avoided if the interaction is sufficiently strong. In the latter case, i.e., avoided crossing, two new potential surfaces appear, often with a distinct potential maximum in the conic region of the newly born lower surface (Figure 11.24). Textbook examples of avoided crossing include $A^1\Sigma^+$ and $B^1\Sigma^+$ surface-crossing for LiH, AlH, BH, and alkali halides. Regarding the strength of interaction of the surfaces and the nuclear dynamics in the region of the crossing, we have already discussed the quantitative and probabilistic aspects at length earlier (see Chapter 5). Such analyses have important applications in chemical kinetics as well.

11.10.4 Predissociation

This is a well-known effect of vibrational-rotational perturbation that has been studied for well over a century now. When an electronic transition takes place from a low-energy molecular state to a high-energy bound state such that the latter surface intersects or is very close to a nearby repulsive surface, the overlap of the vibrational eigenfunctions of the bound and repulsive surfaces mix the two states in the crossing region so as to produce a nonradiative transition of the molecular system from the bound to the repulsive state. The molecule dissociates, of course, and the process is called predissociation. Note that a distinction of predissociation from an orthodox direct dissociation can be experimentally achieved by excitation to a repulsive surface directly. The perturbation interaction is between the initial excited bound-state i and the final repulsive state f.

It is important to distinguish the properties of the two potential surfaces corresponding to i and f states with respect to nuclear motions. The vibrational and rotational energy levels below the asymptote of the bound-state potential are all discrete and well-defined by the respective eigenfunctions. The vibrational and rotational levels above the asymptote corresponding to the ascending region are, however, not defined discretely. This lack of discreteness of energy levels holds all along the course of the repulsive potential. It is assumed, however, that these regions of the potential surfaces have a continuous range of energy values. The assumption is valid because an excitation to the continuum region of an electronic surface yields a continuous spectral band as against discrete ones resulting from excitation to discrete vibrational levels. If an excitation is carried out to an upper surface that has no minimum (repulsive), only a continuous spectrum with no discrete bands will be obtained. In fact, discrete bands joining a continuous band, the latter toward the lower wavelength region of the spectrum, is rather commonly observed,

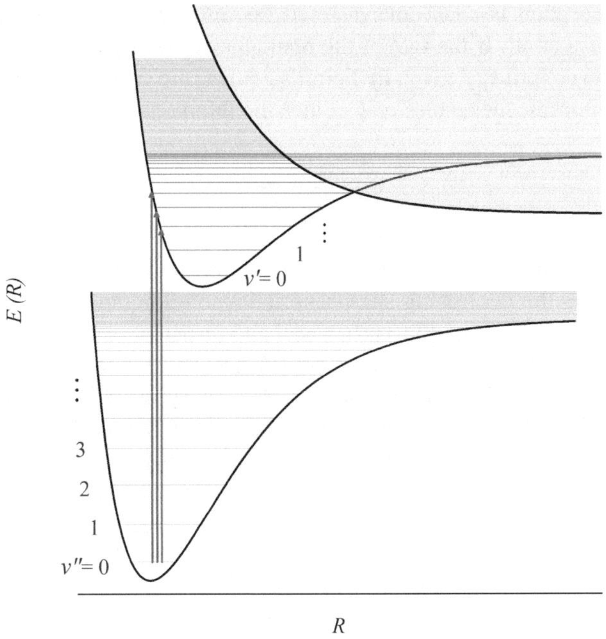

FIGURE 11.25 Discrete and continuous vibrational states. There is no discrete state in the repulsive surface. Franck-Condon transitions from the $v'' = 0$ state are indicated by green arrows.

and this serves as an important spectroscopic marker of molecular structure and decomposition in the excited state.

Consider the above description in the picture of the energy levels of the potential surfaces depicted in Figure 11.25, and notice the discreteness of the bound-state energy levels vis-à-vis the series of continuum states. The two highest-energy vibrational states of the bound surface overlap with the continuum states of the repulsive surface. How many discrete energy levels will overlap with the continuum is determined by the depth at which the bound and repulsive surfaces cross. The continuum of yet another molecular state may also overlap with the discrete states of the bound molecule. This overlap is a qualitative description of the perturbation by which the state $|i\rangle$ can make a nonradiative transition or undergo internal conversion to a state $|f\rangle$ in the continuum of states (Figure 11.26). Since the perturbation Hamiltonian is $\mathcal{H}' = \mathcal{H}'_{\text{vib}} + \mathcal{H}'_{\text{rot}}$, the rate of the nonradiative transition from $|i\rangle$ to $|f\rangle$ is given by

$$\omega_{i\to f} = k_{\text{NR}} = \frac{2\pi}{\hbar^2}\langle|f|\mathcal{H}'|i|\rangle^2 \rho\left(v_f\right), \tag{11.64}$$

in which $\rho\left(v_f\right)$ is the density of states $|f\rangle$ in the continuum of the repulsive states and γ is the nonradiative transition probability. As a result of the $i \to f$ nonradiative transition the molecule dissociates, and the rate constant k_{NR} is the rate of predissociation. The k_{NR} expression above is called Fermi-Golden rule (see Box 11.3).

The perturbation couples $|i\rangle$ and $|f\rangle$ states of the same spin **s**, same symmetry $\mathcal{L}$ (s and a), and same parity p (+ or even and – or odd), but Λ and Ω both differing by $0, \pm 1$

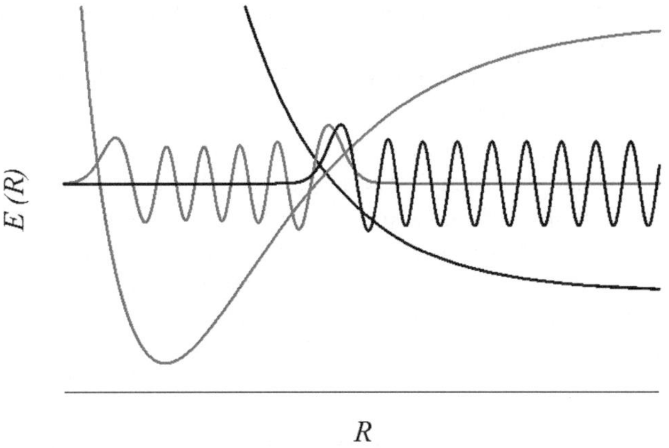

FIGURE 11.26 Overlap of vibrational wavefunctions of the bound molecular state ($v' = 10$) and a repulsive state. The Hermite function has been corrected for anharmonicity. The vibrational wavefunction of the latter state is called a continuum vibrational wavefunction which is a sine function, and whose frequency and amplitude become constant at some distance away from the edge of the repulsive potential.

$$\langle f|\mathcal{H}'|i\rangle = q_{vE}\left\langle \Omega \pm 1 s\, \Lambda \pm 1 l\, E\, p\mathcal{L}|\mathcal{H}'_{\text{rot}}|\Omega s\, \Lambda J\, v\, p\,\mathcal{L}\right\rangle. \tag{11.65}$$

These restrictions follow from Kronig's selection rules of perturbation listed earlier. The same set of selection rules should be applicable for predissociation because predissociation occurs due to perturbation only. The quantum number v of the discrete $|i\rangle$ state takes the form of translational energy of the dissociated atoms. Strictly, the nuclear motions in the $|f\rangle$-state continuum is a sine function of the displacement of a mass point about some position, which is much larger than the internuclear separation in a stable molecular state. This serves as the form of the eigenfunction of nuclear motions in the continuum state – a sine wave whose frequency is related to the energy of the state scaled with reference to the asymptote of the surface. The frequency increases as the energy of the state above the asymptote rises, because the relative kinetic energy of the dissociated parts of the molecule increases, so that

$$\lambda \propto \frac{1}{0.5 m v^2}. \tag{11.66}$$

The quantum number l for the continuum states arises due to the fact that the molecule dissociates into atoms in the set of continuum states itself. Comparison of the spherical harmonics part of the total eigenfunction of the atom (equations (4.9)–(4.11)) with the rigid rotor eigenfunction, which also is a spherical harmonics (equation (7.23)) would show that the quantum number m can have values $-l, -l+1, \cdots, l-1, l$ for the former and $-J, -J+1, \cdots, J-1, J$ for the latter. So the rotational quantum number J in the state $|i\rangle$ can be approximated with the orbital angular momentum of the motion of the dissociated atoms in the state $|f\rangle$. This comparison also enables assigning symmetry (s and a) and parity (+ and –) properties to the functions in the dissociated molecular state.

The rate of predissociation is calculated by the Fermi golden rule (equation (11.64)), according to which it is not only the $\langle f|\mathcal{H}^i|i\rangle$ matrix elements but also the density of the continuum states that determine the rate of the nonradiative transition from $|i\rangle$ to $|f\rangle$. To understand how the density of the continuum states might change along the course of the repulsive potential, let us express $\rho(v_f)$ in terms of the density of the continuum states, which is the number of states per unit energy $\rho(E)$, and invoke the formula

$$E = \frac{\hbar^2 (mv)^2}{2m}, \tag{11.67}$$

where mv is the momentum of a particle (the dissociating atoms in the case here). Then

$$dE = \frac{\hbar^2 mv\; dmv}{m}, \tag{11.68}$$

which means the higher the energy of a state in the continuum, the larger is the density of continuum states (also compare with equation (11.66)). It depends at which energy height a discrete state overlaps with a continuum – the higher the energy of a continuum state that overlaps with a discrete bound state, the larger is the value of k_{NR}. This leads one to consider the level of the repulsive potential surface at which the crossing of the bound and unbound surfaces takes place.

11.10.5 Diffused Molecular Spectra

A hallmark of the occurrence of predissociation is the appearance of diffused molecular spectra in both absorption and emission measurements. In particular, line-broadening and the loss of rotational structures in the absorption spectrum is a good indication of predissociation. The lines in the absorption or emission spectrum broaden because the probability of predissociation, $\langle f|\mathcal{H}'|i^2\rangle \sim k_{NR}$, is larger than the probability of radiative emission. If the mean lifetime of the overlapped state before predissociation is shorter than the time for rotation of very small molecules (~10^{-10} to 10^{-11} s), then the uncertainty broadening of the rotational bands will exceed the Doppler line-broadening due to thermal motions at ordinary temperature. A complementary consequence of a lack of rotational absorption is weak and diffused fluorescence or no fluorescence at all. If the rate of the nonradiative transition (internal conversion) k_{NR}, given by Fermi's golden rule exceeds the rate of fluorescence emission k_{FL}, then the fluorescence yield

$$\phi_F = \frac{k_{FL}}{k_{FL} + k_{NR}} \tag{11.69}$$

diminishes and may not even be detectable. Figure 11.27 shows the decreasing sharpness and resolution, and eventual disappearance of the rotational band structure successively in the vibrational bands $(16,0)$, $(17,0)$, $(18,0)$, and $(19,0)$ of the $B^3\Sigma_u^- - X^3\Sigma_g^-$ transition in molecular sulfur S_2 (Wheeler et al., 1998). In this notation of labeling vibrational bands, the two numbers within parenthesis represent vibrational levels in the v' (X-state) and v'' (B-state), respectively. The increasing diffuseness with higher wavenumber (from left to right in Figure 11.27), and eventual break-off of the rotational bands is due to increasing rate of predissociation k_{NR}, that is increasing overlap of the bound and continuum wavefunctions. The continuum-state wavefunctions for these overlaps are associated with the repulsive state $1^5\Sigma_u^-$. It may so happen that another set of bound-state wavefunctions overlaps with the continuum states of another repulsive surface, in which case a second predissociation process will occur; similarly, a third predissociation may crop up. Indeed, in the case of S_2 there are two predissociations for the $B^3\Sigma_u^-$ state depending on the amount of vibrational excitation within the $B^3\Sigma_u^- - X^3\Sigma_g^-$ transition – one occurs between (10,0) and (16,0) absorptions and another above the (16,0) transition (Wheeler et al., 1998). The continuum-state wavefunctions for these two predissociations are associated with the repulsive states $1^1\Pi_u$ and $1^5\Sigma_u^-$, respectively.

One may also notice that the vibrational bands corresponding to an excitation to the given vibrational levels are least affected by predissociation. This can be understood by comparing the time-scales for three events – the mean lifetime of the discrete-continuum overlapped state (τ_1), the time of molecular rotation ($\tau_2 \sim 10^{-10} - 10^{-11}$s), and the time of vibrational motion ($\tau_3 \sim 10^{-13}$s). Normally $\tau_3 > \tau_1 > \tau_2$, which means vibrational motion is not resolved in the lifetime of the overlapped state but rotational motion is. The order $\tau_3 > \tau_1 > \tau_2$ almost always prevails, even for higher rotational frequency at large values of J.

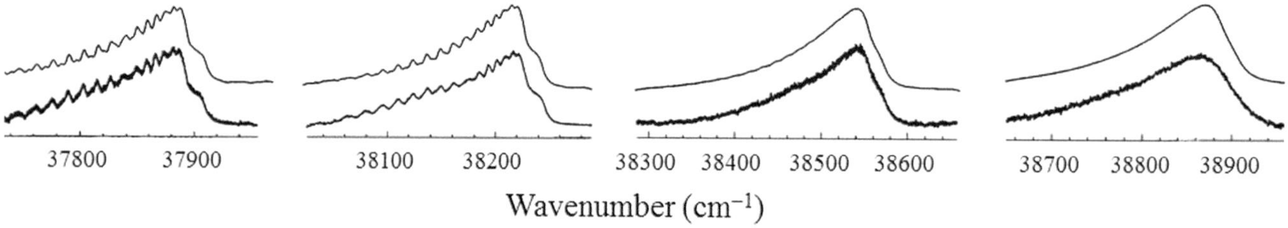

FIGURE 11.27 Appearance and successive disappearance of rotational structures in vibrational bands observed for molecular sulfur. From *left* to *right* are (16,0), (17,0), (18,0), and (19,0) vibrational bands with their respective rotational band structures. In each panel, the spectrum on top (thin line) is the simulated spectrum, and the thick-lined one below corresponds to what is observed. (Reproduced from Wheeler et al., 1998, with license from AIP.)

11.11 STARK EFFECT IN ROTATIONAL TRANSITIONS: OBSERVATION AND SELECTION RULES

The energy of interaction of a molecular electric dipole with an external electric field has the same form as discussed for the case of an atom (equation (4.86) and Box 4.7)

$$\mathcal{H}_{\text{Stark}} = -\langle \mathbf{E} \cdot \boldsymbol{\mu} \rangle, \quad (11.70)$$

but since the molecules are rotating, the interaction is averaged over the motion. The applicable order of energy correction depends on several factors, including the energy separation or energy spacing of the J levels in comparison to Stark perturbation energy, whether a molecule has an averaged permanent electric dipole component along the **J** vector, and whether the molecule is linear or a top. Here we will limit the discussion for the concepts alone, the details regarding the type of the top and the factors determining higher order corrections are left for further study and research.

The basic factor pertains to a match of the J-level energy spacing and the perturbation energy. It is only when the latter is smaller that the perturbation theory is applicable and the energy can be corrected for. This should mean that perturbation corrections can become convenient only when higher J-levels are approached, which also is one of the reasons why conventional microwave spectroscopy, where the resonance frequency is much smaller and rotational bands are close to each other, is often inconvenient to study electric field effects; instead, one relies on electric resonance experiments (see below) to detect perturbation.

Consider first a symmetric top rotating about its symmetry axis. The top can have a component of the rotational motion-averaged electric dipole moment lying along the direction of the rotational angular momentum vector **J**. One may then say that the molecule has a permanent component of averaged dipole moment **μ** fixed in space, not rotating. It is true for all symmetric tops that there is a direction-fixed permanent dipole moment even when an external electric field is absent. The existence of the motion-averaged directional component of **μ** is a consequence of the degeneracy of the K-energy levels. The quantum number K which quantizes the projection of **J** onto the symmetry axis exists even in the absence of electric field. This is unlike the quantum number M, which is the projection of **J** onto the z-axis defined only in the presence of an external field applied along z. If the total angular momentum vector corresponds to say $J = 1$, we would have $2J + 1 = 3$ possible states ($K = 0, \pm 1$) of the same energy. This is called K-degeneracy which allows for the existence of an electric dipole component along the direction of **J** (see Townes and Schawlow, 1975, for a qualitative discussion). Such an argument cannot, of course, be extended to a linear molecule for which K is not defined, nor does M exist unless an external field is turned on. Thus, a permanent dipole does not exist for a linear molecule, but the degeneracy of the $\pm K$ levels in symmetric tops allows for having a space-fixed electric dipole. In the case of an asymmetric rotor the component of averaged **μ** along the direction of **J** turns out to be zero, so such molecules do not possess a permanent dipole.

When a permanent component of **μ** along **J** is available, as for a symmetric top, the energy of interaction is simply

$$\mathcal{H}_{\text{Stark}} = \mu E \cos\theta, \quad (11.71)$$

where θ is the angle between the **J** vector and the electric field applied along the z-direction of the laboratory frame. We now have the quantum number M and the corresponding energy levels defined, so that

$$\mathcal{H}_{\text{Stark}} = -\mu E \frac{M}{J}.$$

At the same time, the component of **μ** along **J** can be related to the angular momentum **K** which is defined by projecting **J** onto the molecular symmetry axis. We then have

$$\mu = \mu \frac{K}{J}$$

$$\mathcal{H}_{\text{Stark}} = -\mu E \frac{MK}{J^2}. \quad (11.72)$$

Since $J = \hbar\sqrt{J(J+1)}$, where $\hbar$ is the unit if not specified by frequency, we write for the change in energy due to Stark interaction as

$$\Delta E \propto -\frac{MK}{J(J+1)}. \quad (11.73)$$

The proportionality is μE, implying that the perturbation is first order in the electric field. As a result, the M-degeneracy is completely lifted yielding $2J + 1$ energy levels for a given value of J. This is the first-order energy, but the calculation of higher-order energy is not difficult. Let us look at the second order. When $\mu = 0$ as for a linear molecule or an asymmetric rotor, the first-order energy change vanishes. But then the applied electric field induces a component of electric dipole moment along the direction of the **J**-vector. This newly produced component of **μ** in turn will interact with the same applied electric field $\mathbf{E}_z$ to yield an interaction energy which is quadratically proportional to μE

$$\Delta E \propto \frac{(\mu E)^2}{\frac{1}{2} I\omega^2} = \frac{(\mu E)^2}{hB\, J(J+1)}. \quad (11.74)$$

Now, instead of obtaining $2J + 1$ energy levels as in the case of the first-order perturbation we should have only $J + 1$ energy levels because $(+E)^2 = (-E)^2$. This is indeed an interesting result, because barring the case of $M = 0$ all $\pm M$ levels are equivalent and degenerate, which means for $J = 2$ say, we obtain $M = \pm 2, M = \pm 1$, and $M = 0$. Thus, there will not be five but three energy levels, of which ± 2 and ± 1 are degenerate. If $J = 3$, then ± 3, ± 2, and ± 1 are degenerate, so we have four energy levels 3, 2, 1, 0. This semi-qualitative discussion explains the appearance of $J+1$ energy levels for a given value of J when the Stark interaction energy is second-order. Higher-order effects, even but not odd powers of μE, can also be considered for the calculation of rotational constant and dipole moment.

The first-order perturbation result is

$$E_j^{(o)} = \int \psi_j^{(o)*} \mathcal{H}_{\text{Stark}} \psi_j^{(o)} d\tau, \tag{11.75}$$

in which $\psi_j^{(o)}$ are the wavefunctions for the unperturbed state of the molecule. The matrix elements are

$$\langle JKM | \mathcal{H}_{\text{Stark}} | JKM \rangle = -\mu E \langle JKM | \cos\theta | JKM \rangle. \tag{11.76}$$

Since the electric dipole moment components are fixed in the rotating molecule and the external electric field is along one of the space-fixed axes – customarily Z, the dipole moment requires expression in the space-fixed axes. This can be done by using the direction cosines (see Box 5.1). The right-hand side of equation (11.76) above can then be written as

$$\begin{aligned} &-\mu E \langle J | \cos\theta_{Fg} | J \rangle . \langle JK | \cos\theta_{Fc} | JK \rangle . \langle JM | \cos\theta_{Zg} | JM \rangle \\ &= -\frac{\mu E\, KM}{J(J+1)}, \end{aligned} \tag{11.77}$$

which, identical to the result in equation (11.72), is the first-order Stark energy. In the direction cosines above, $F = X, Y, Z$, and $g = a, b, c$, the principal axes of inertia. In some textbooks and literature $\cos\theta_{Fg}$ is written as ϕ_{Fg}. The dipole moment components of an asymmetric rotor along the Z-axis of the laboratory frame, for example, will be

$$\mu_Z = \mu_a \cos\theta_{Za} + \mu_b \cos\theta_{Zb} + \mu_c \cos\theta_{Zc}. \tag{11.78}$$

The matrix elements are diagonal as pertinent for a first-order correction.

The second-order correction is applicable for molecules including linear ones and the asymmetric rotors lacking a permanent electric dipole, but the polarization induced by the external field on the molecule produces an induced dipole. If the electric field is along the Z-direction, the operator $\mathcal{H}_{\text{Stark}}$ is written using the μ_Z formula above, so we have

$$\mathcal{H}_{\text{Stark}} = -E \sum_{g=a,b,c} \cos\theta_{Zg} \mu_g. \tag{11.79}$$

The second-order approximation is

$$E_i^{(2)} \sim \sum_{j \neq i} \frac{\langle \psi_i^{(0)} | \mathcal{H}_{\text{Stark}} | \psi_j^{(0)} \rangle \langle \psi_j^{(0)} | \mathcal{H}_{\text{Stark}} | \psi_i^{(0)} \rangle}{E_i^{(0)} - E_j^{(0)}}. \tag{11.80}$$

Note that this kind of perturbation correction is based on the assumption that there are no degenerate zero-order states. The energy change obtained from the second-order correction is the total energy change and hence all the zero-order states that interact through perturbation are considered.

The perturbation can be extended to higher-order if the first-order correction vanishes. In such cases, additional even-order energy terms can be added to the second-order term; the odd-order terms vanish due to the same reason that K-degeneracy is lacking (see Golden and Wilson, 1948, for the derivation that $E^{(3)} = 0$). Denoting the strength of the applied field with ε, the terms are written as

$$E(\varepsilon) = E^{(0)} + E^{(2)} + E^{(4)} + \cdots.$$

In the case of a linear rigid rotor for example, the second- and fourth-order perturbation energies calculated by Wijnberg (1974) are

$$E^{(0)}(J,M) = J(J+1)$$

$$E^{(2)}(J,M) = \frac{J(J+1) - 3M^2}{2J(J+1)(2J-1)(2J+3)} \lambda^2$$

$$E^{(4)}(J,M) = \left\{ \begin{aligned} &-\frac{(J-M+1)(J+M+1)(J-M+2)(J+M+2)}{8(2J+1)(2J+3)^3(2J+5)(J+1)^2} \\ &+\frac{(J-M)(J+M)(J-M-1)(J+M-1)}{8(2J-3)(2J-1)^3(2J+1)J^2} \\ &-E^{(2)}(J,M) \left[\begin{aligned} &\frac{(J-M)(J+M)}{4(2J-1)(2J+1)J^2} \\ &+\frac{(J-M+1)(J+M+1)}{4(2J+1)(2J+3)(J+1)^2} \end{aligned} \right] \end{aligned} \right\} \lambda^4, \tag{11.81}$$

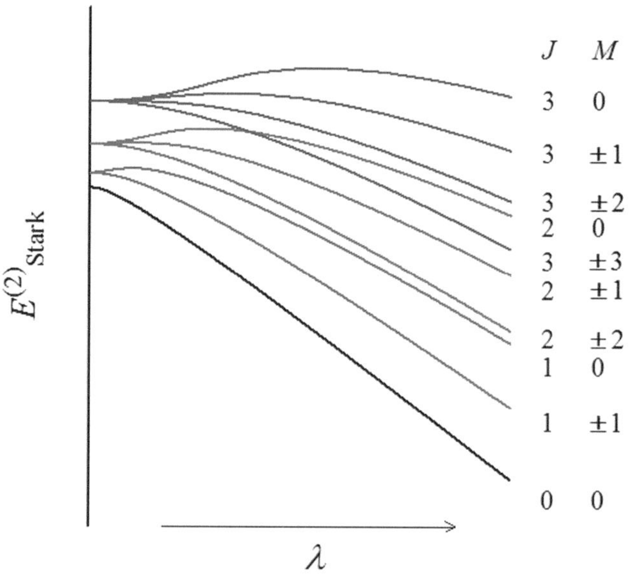

FIGURE 11.28 Second-order energy level pattern for several combinations of quantum numbers J and M as a function of the perturbation parameter $\lambda = \dfrac{\mu E}{hB}$ in the range 0–100.

where the perturbation parameter $\lambda = \frac{\mu E}{hB}$. The permitted values of M are $-J, -J+1, \cdots, J-1, J$, and we use the numerical data of Peter and Strandberg (1957) to illustrate the second-order energy level patterns for linear rotors in Figure 11.28.

The first- and second-order frequency shifts are most conveniently measured by Stark-modulated microwave spectroscopy, in which a square-wave electric field is applied by employing a conducting plate placed in the waveguide of the spectrometer containing the sample at low pressure. The microwave is obtained from a klystron source and is passed into the waveguide where both the external electric field and the microwave field will be present. If the microwave field is parallel to the applied electric field, the transitions involve $\Delta M = 0$, and if the latter is perpendicular to the former, transitions of the type $\Delta M = \pm 1$ occur. This is the selection rule for Stark-effect rotational transition. When the external field is turned off an observation of normal frequency is made, and when the field is turned on the Stark shift can be observed.

Stark effect in microwave spectroscopy is widely used to measure the ground-state dipole moment μ of molecules. The parallel configuration of the external and the microwave electric fields ($\Delta M = 0$) suffices for such measurements. Since the Stark interaction energy is proportional to μE when $K \neq 0$, but to $(\mu E)^2$ when $K = 0$, the slope of the magnitude of the frequency shift with varying electric field provides the dipole moment. One of the first measurements of dipole moment, that of NO, is found in the work of Burrus and Graybeal (1958). The procedure is the same whether the perturbation is to first- or second-order. The Stark effect in the rotational spectrum can be studied in both ground and electronically excited states by the molecular beam electronic resonance method where strong electric fields can be applied (Bichsel et al., 2007). The higher-order perturbation expansions, as in equation (11.81), are particularly useful in studies of the strong-field limit. Stark-modulated electronic spectroscopy can be used to resolve rotational structure in the excited state and determine excited-state dipole moment.

11.12 ZEEMAN EFFECT ON ROTATIONAL ENERGY LEVELS AND SELECTION RULES

The energy of interaction of the dipole of a rotating molecule in the presence of an external magnetic field is the dot product of the motion-averaged dipole moment and the field,

$$E_{\text{Zeeman}} = -\langle \boldsymbol{\mu} \cdot \mathbf{B} \rangle, \tag{11.82}$$

where $\mathbf{B}$ is the magnetic field. The total dipole moment has contributions from electron orbital angular momentum (μ_L), electron spin angular momentum (μ_S), molecular rotational angular momentum (μ_{rot}), and nuclear spin angular momentum (μ_N), so the complete Zeeman energy is

$$\mathcal{H}_{\text{Zeeman}} = -\langle \boldsymbol{\mu}_L \cdot \mathbf{B} + \boldsymbol{\mu}_S \cdot \mathbf{B} + \boldsymbol{\mu}_{\text{rot}} \cdot \mathbf{B} + \boldsymbol{\mu}_N \cdot \mathbf{B} \rangle. \tag{11.83}$$

The Hamiltonian can be simplified considerably just by a comparison of electronic and nuclear magnetic moments with respect to electron and nuclear masses that we have discussed earlier in Chapter 4. Briefly, the magnetic moment $\boldsymbol{\mu}$ is generally γ-times the angular momentum, where γ is the gyromagnetic ratio which for all of electron orbitals, electron spin, nuclear spin, and nuclear rotation motions are respectively

$$\boldsymbol{\mu}_e = \gamma_e \mathbf{L}, \boldsymbol{\mu}_s = \gamma_s \mathbf{S}, \boldsymbol{\mu}_{\text{N}} = \gamma_I \mathbf{I}, \text{ and } \boldsymbol{\mu}_{\text{rot}} = \gamma_{\text{rot}} \mathbf{J},$$

with

$$\gamma_e = g_l \frac{e\hbar}{2m_e c}, \gamma_s = g_s \frac{e\hbar}{2m_e c}, \gamma_I = g_N \frac{e\hbar}{2m_N c}, \text{ and} \quad \gamma_{\text{rot}} = g_{\text{rot}} \frac{e\hbar}{2m_N c}. \tag{11.84}$$

The quantities $\mu_e = e\hbar / 2m_e c$ and $\mu_N = e\hbar / 2m_N c$ are Bohr and nuclear magnetons, respectively. The respective g-factors for the electron and the proton are $g_s \approx 2.003$ and $g_N = 1$. The value of g_{rot} depends on the relative motions of electrons and nuclei; a large value of g_{rot} could arise from free rotation of the nuclei in a bath of 'sluggish electrons'. The appearance of m_e in the denominator of the formulas for $\boldsymbol{\mu}_e$ and $\boldsymbol{\mu}_s$, but m_N in that for $\boldsymbol{\mu}_N$ and $\boldsymbol{\mu}_{\text{rot}}$ suggests that the contribution of $\boldsymbol{\mu}_N$ and $\boldsymbol{\mu}_{\text{rot}}$ to the total magnetic moment will be ~1850 times less than the contribution of $\boldsymbol{\mu}_L$ and $\boldsymbol{\mu}_S$. If we neglect the contribution of the nuclear spin and the rotational motion, the effective interaction energy becomes

$$\mathcal{H}_{\text{Zeeman}} = -\langle \boldsymbol{\mu}_L \cdot \mathbf{B} + \boldsymbol{\mu}_S \cdot \mathbf{B} \rangle. \tag{11.85}$$

This shows that the effect of the magnetic field on molecular spectra arises predominantly from electronic angular momentum.

The Zeeman effect may be discussed for two cases – when the electronic angular momentum is vanishing and when the existing electronic angular momentum is strongly coupled to the internuclear axis, that is Hund's case(a). First, if there is no electronic angular momentum as in the $^1\Sigma$ state of a molecule, in which case $\lambda = 0$ and $\Sigma = 0$, then the interaction energy is

$$\mathcal{H}_{\text{Zeeman}} = -\langle \boldsymbol{\mu}_{\text{rot}} \cdot \mathbf{B} + \boldsymbol{\mu}_N \cdot \mathbf{B} \rangle.$$

Such an interaction is rather weak, and so is the splitting of spectral lines that resonate in the long meter-wave region. The analysis requires the consideration of the total rotational angular momentum J and the total nuclear spin angular momentum I_{T} of the molecule, the latter is more appropriately called resultant spin because it is obtained by vector addition of the angular momentum of individual nucleons. Note that in NMR literature the total nuclear spin is still written as I, even though $I = 1/2$ is the proton spin strictly. The resultant nuclear spin for the hydrogen molecule, for example, is $2I$ ($I_{\text{T}} = I + I$), $I = 1/2$ being the proton spin. The allowed values of the quantum number I_{T} in general are given by $2I, 2I-1, 2I-2, \cdots, 0$, so that the total nuclear spin of the hydrogen molecule can be in states 0 or 1. An external magnetic field applied along $+z$ splits the energy level for $I_T = 1$ into $M_T = 2I_T + 1 = 3$ levels, $-1, 0, +1$. Concurrenly, the magnetic field quantizes the total angular momentum J into M_J such that $M_J = -1, 0, +1$ for $J = 1$. This simultaneous quantization of I_T and J in terms of M_T and M_J happens when the applied magnetic field is strong enough to parallel the Paschen-Back or normal Zeeman effect that we studied in Chapter 4. Although the context of the nuclear spin here is I_T, and not I, the application does not change; it is I_T for the resultant nuclear spin, and I_1 and I_2 separately for the individual spins. What this means is illustrated by the Zeeman-energy expressions as follows. To a first approximation the Zeeman energy can be written as

$$E_{\text{Zeeman}} = -\mu_N g_{\text{rot}} M_J B - \mu_N g_{N(\text{T})} M_T B, \qquad (11.86)$$

where $g_{N(\text{T})}$ should be read as the g-factor of the nuclei taken together, meaning that the resultant nuclear spin angular momentum I_T is related to the magnetic moment by

$$\boldsymbol{\mu}_N = g_{N(\text{T})} \sqrt{I(I+1)} \mu_N. \qquad (11.87)$$

In the case where the two nuclei are spin-uncoupled by a strong magnetic field, the second term in equation (11.86) is split to account for the contribution of the individual nuclei

$$E_{\text{Zeeman}} = -\mu_N g_{rot} M_J B - \mu_N g_{N(1)} M_{I(1)} B - \mu_N g_{N(2)} M_{I(2)} B \qquad (11.88)$$

The Zeeman splitting into M_T and M_J levels of hydrogen in the state $J = 1$ in a moderately strong magnetic field ($< 1\text{T}$) is shown in Figure 11.29, where the transitions occur in the frequency of the radio wave region, and each transition occurs according to the following selection rules

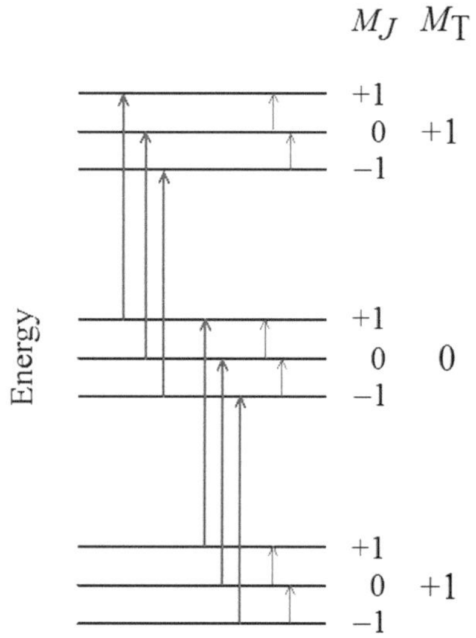

FIGURE 11.29 Zeeman transitions for H_2 in the $J = 1$ state.

TABLE 11.3
Electronic Magnetic Moment in a Few Molecular Eigenstates

States	μ_Ω (in Bohr magneton)
$^1\Sigma$	0
$^3\Sigma_1$	2.003
$^2\Pi_{1/2}$	0.001
$^2\Pi_{3/2}$	2.003
$^1\Delta_2$	2.000
$^3\Delta_3$	4.003

$$\Delta M_J = 0 \quad \text{and} \quad \Delta M_T = \pm 1$$

$$\Delta M_J = \pm 1 \quad \text{and} \quad \Delta M_T = 0.$$

Both conditions in each line have to be satisfied simultaneously. A first-order energy correction to these energy levels can be made by considering the interaction of nuclear and rotational angular momenta and the nuclear spin-spin interaction.

In Hund's case(a), $\boldsymbol{\Lambda}$ and $\boldsymbol{\Sigma}$ are strongly coupled to the nuclear axis, so one has

$$\boldsymbol{\mu}_{\Omega=\Lambda+\Sigma} = \frac{e\hbar}{2m_e c}\left(g_l \Lambda + g_s \Sigma\right) \cong \mu_e\left(\Lambda + 2.003\Sigma\right), \qquad (11.89)$$

where $\mu_e = e\hbar / 2m_e c$ is the Bohr magneton. If Λ and Σ values are known one can estimate the total electron magnetic moment along the molecular axis, but care should be exercised to choose only the positive value of Λ (see Box 11.2). The μ_Ω values for a few randomly chosen molecular eigenstates are listed in Table 11.3.

To obtain the time-averaged μ_Ω along the direction of $\mathbf{B}$ we look at the vector diagram for Hund's case(a) where the molecular axis, aligned with $\boldsymbol{\Omega}$ (=$\boldsymbol{\Lambda}$+$\boldsymbol{\Sigma}$), precesses about $\mathbf{J}$. The component of $\boldsymbol{\mu}$ in the direction of $\mathbf{J}$, call it $\boldsymbol{\mu}_J$, is given with reference to the angle θ between $\boldsymbol{\mu}$ and $\mathbf{J}$

$$\langle \mu_J \rangle = \mu_\Omega \cos\theta \qquad (11.90)$$

Substitution for μ_Ω from equation (11.89) yields the time-averaged magnitude of $\boldsymbol{\mu}_J$

$$\langle \mu_J \rangle = \frac{\mu_e\left(\Lambda + 2.003\Sigma\right)\Omega}{\sqrt{J(J+1)}}. \qquad (11.91)$$

In the presence of the field $\mathbf{B}$, the vector $\boldsymbol{\mu}_J$ precesses about it so that the time-average of the magnetic moment in the direction of $\mathbf{B}$ is

$$\langle \mu \rangle = \langle \mu_J \rangle \cos\theta, \qquad (11.92)$$

in which θ is now the angle between $\boldsymbol{\mu}_J$ and $\mathbf{B}$

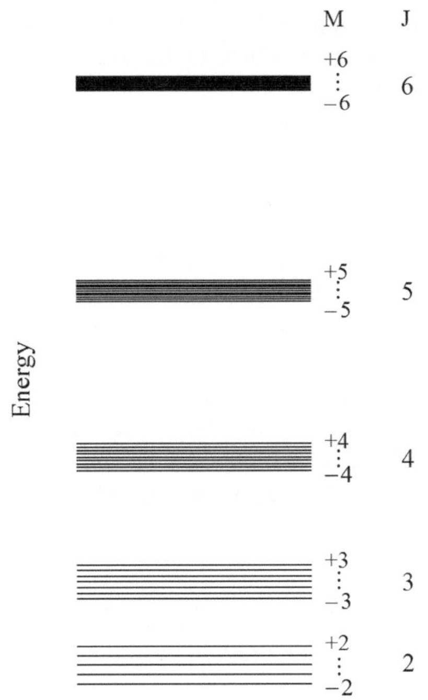

FIGURE 11.30 Zeeman splitting of rotational energy levels in a molecular state.

$$\langle \mu \rangle = \langle \mu_J \rangle \frac{M}{\sqrt{J(J+1)}}$$

$$= \frac{\mu_e (\Lambda + 2.003\Sigma)\Omega M}{J(J+1)}. \tag{11.93}$$

Thus the Hamiltonian is

$$\mathcal{H}_{\text{Zeeman}} = -\langle \boldsymbol{\mu} \rangle \cdot \mathbf{B} = \frac{\mu_e (\Lambda + 2.003\Sigma)\Omega MB}{J(J+1)}, \tag{11.94}$$

and the total splitting of energy is

$$\Delta E = 2\frac{\mu_e (\Lambda + 2.003\Sigma)\Omega MB}{J(J+1)}, \tag{11.95}$$

implying that the splitting is inversely related to the angular momentum J (Figure 11.30). Each J-level splits into $2J+1$ equally spaced levels as per the possible values of M $(-J, -J+1, \cdots, J+1, J)$, and transitions between the levels can occur according to the selection rule $\Delta M = 0, \pm 1$. The selectivity $\Delta M = 0$ applies when $\Delta J = \pm 1$, and $\Delta M = \pm 1$ applies if $\Delta J = 0$.

These selection rules are identical to those applicable to the Stark effect transitions discussed in the preceding section. The rule $\Delta M = 0$ is proper if the applied magnetic field **B** is parallel to the electric field of the microwave radiation which excites only rotational transitions. If **B** is perpendicular to the microwave electric field, the selection rule $\Delta M = \pm 1$ is appropriate. An absorption line component in a given branch of rotational structure is denoted by a σ-component line if $\Delta M = \pm 1$ or a π-component line if $\Delta M = 0$. The schematic in Figure 11.31 provides an example of the appearance of σ- and π-components in a few sets of P-branch lines.

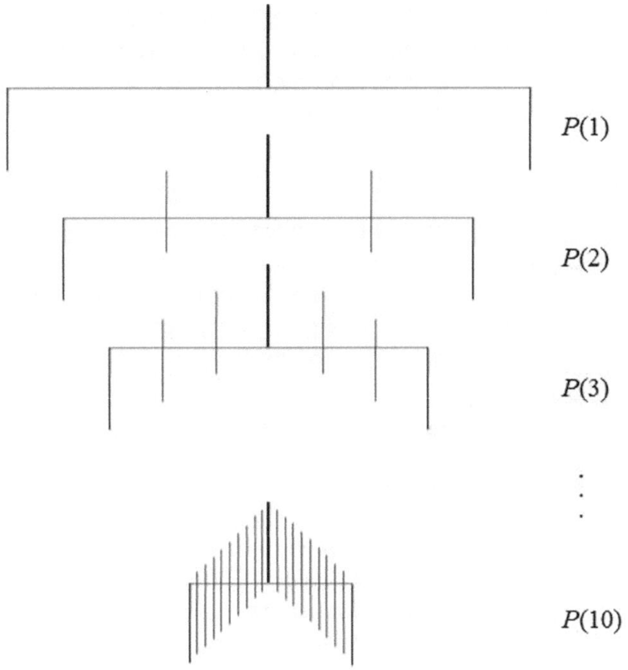

FIGURE 11.31 Schematic of Zeeman-splitting of P-branch rotational lines. The lines below the horizontal frequency axis correspond to $\Delta m = \pm 1$ (σ-components obtained by polarization of the microwave electric field perpendicular to the magnetic field) and those above correspond to $\Delta m = 0$ (π-components obtained by polarizing the microwave electric field parallel to the magnetic field).

The theoretical intensities of both parallel ($\Delta m = 0$) and perpendicular ($\Delta m = \pm 1$) lines of P, Q, and R branches in the case of the ${}^1\Sigma \rightarrow {}^1\Pi$ transition of CO were given by Crawford (1934) in one of the earliest discussions of the Zeeman effect in diatomic spectra. The study of Zeeman effect of the rotational states by radiofrequency absorption in molecular beam has also been described decades back (Hendrie and Kusch, 1957). Such experiments provide information about the g-factor of rotational magnetic moment of individual rotational states. The Zeeman effect of rotational structure can be studied in the ultraviolet region as well. These studies provide a wealth of information about the excited-state electronic structure and angular momentum coupling effects.

11.13 MAGNETOOPTIC ROTATIONAL EFFECT

The phenomenon of the rotation of linearly polarized light passing through a sample subjected to a magnetic field with the field direction parallel to the propagation of the light beam is called magnetooptic rotation. A typical experimental setup (Figure 11.32) consists of an initial polarizer (P1) which polarizes the incident light beam. This plane-polarized light passes through a solenoid-system magnetic field with the field direction perpendicular to the direction of the electric field polarization (σ-configuration). The magnetic field hosts the sample (S) of molecules that rotate the plane of polarization, and the extent of rotation is measured by using a second polarizer (P2) which acts as an analyzer. The degree of rotation of the plane of the electric field polarization is

$$\theta = VBL, \tag{11.96}$$

where B is the strength of the applied magnetic field, L is the optical path length through the sample, and V is the Verdet constant, which is the magnetooptic rotatory characteristic. Denoting the intensities of light entering and exiting the sample with $I(0)$ and $I(L)$, respectively,

$$I(L) = I(0)\cos^2(\theta + \alpha), \tag{11.97}$$

in which α is the difference of the relative angles of orientation of the two polarizers. The importance of the rotation angle θ lies in the fact that it represents the rotation of the plane of polarization of the incident light near the resonance frequencies of absorption lines of the sample, implying that θ is a representation of dispersion – defined as the frequency dependence of the refractive index of the sample.

To see how the Zeeman-split absorption lines (frequencies) will be dispersive, one can consider the electric field of the incident light which is already linearly polarized along, say, x-direction. This plane-polarized light is the resultant of two electric vectors rotating in opposite directions and in phase (Figure 11.32). The circularly polarized components combine to give the linearly polarized electric vector

$$\mathbf{E}_x = \frac{1}{2}\left(\mathbf{E}_+ + \mathbf{E}_-\right) = \frac{E_0}{2}\left\{e^{i(\omega t - kz)} + e^{-i(\omega t - kz)}\right\}, \tag{11.98}$$

where E_0 is the amplitude of the resultant wave, and the wavenumber $k = 2\pi / \lambda$, which is related to the refractive index of the medium $n(\omega)$ by

$$k = \frac{\omega n(\omega)}{c}.$$

The frequency dependence of n means this is not a static dielectric constant, but depends on the electric field-induced oscillation (polarization) of the electron cloud from its equilibrium position. The resonance frequency corresponding to this oscillation is $\omega_0 = \sqrt{q/m}$, where q and m denote the restoring force of displacement of the electrons and the electron mass, respectively. It can also be shown (see Born and Wolf, 1970: 87) that the mean polarizability α ($P = \alpha E$) that we saw in the preceding chapter is given by

$$\alpha = \frac{e^2}{m\left(\omega_0^{\,2} - \omega^2\right)} = \frac{3}{4\pi N}\frac{n^2 - 1}{n^2 + 2} \tag{11.99}$$

in which e is the electron charge, ω is the circular frequency of the electric field of light, N is the number of electrons per unit volume, and n is the refractive index of the medium through which the light is passing. The relation above, also known as Lorentz-Lorenz formula, shows the variation of refractive index on the frequency of the incident light which is the basis of the dichroic rotation.

The prerequisite for the observation of magnetooptic rotation is that the molecules under investigation show the Zeeman effect; for then only a Zeeman transition would show dichroic behavior, implying that the $\Delta M = \pm 1$ transitions will register

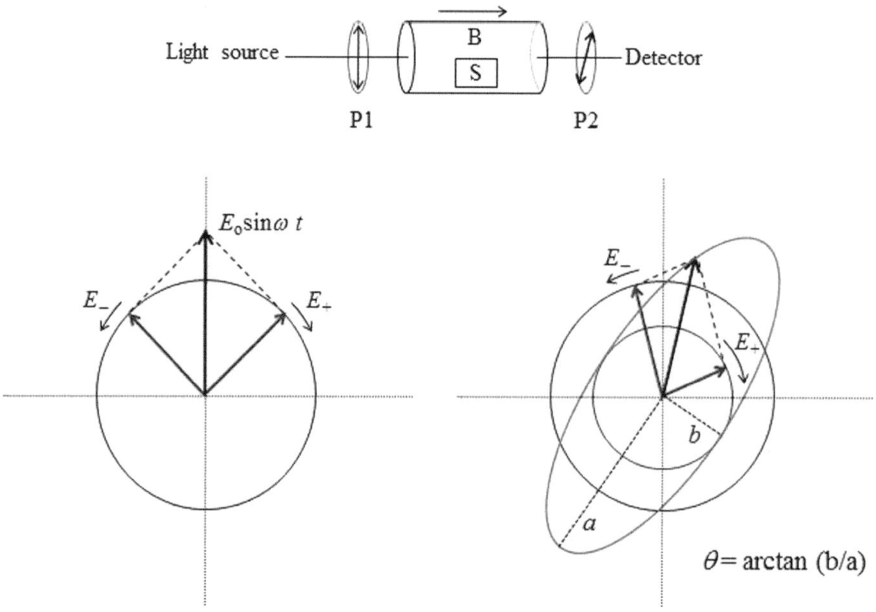

FIGURE 11.32 Magnetooptic rotation. *Top*, the setup for the detection of the effect elicited by a sample S. *Lower left*, the resolution of the polarized electric field vector into the right ($\boldsymbol{E}_+$) and left ($\boldsymbol{E}$-) circularly polarized components, both rotating in phase. *Lower right*, the transmitted polarized light is elliptical. The angle θ is the rotation of the plane of polarization.

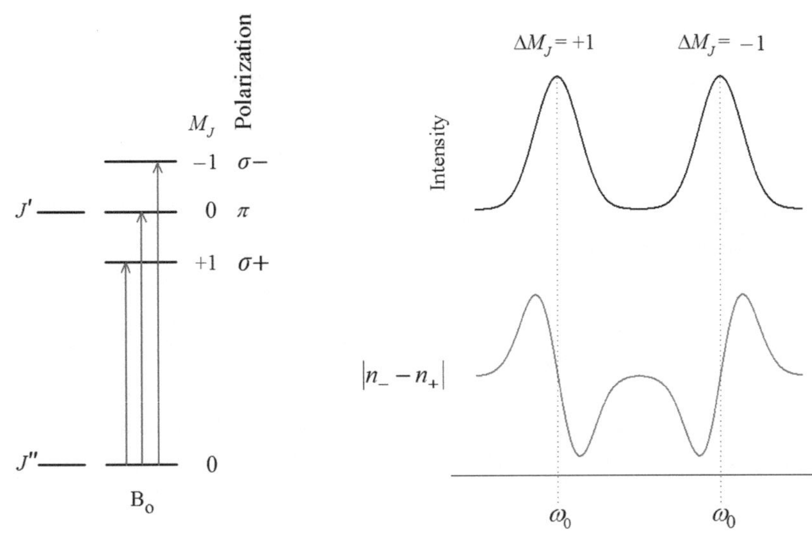

FIGURE 11.33 *Left*, normal Zeeman effect for a $J'' \rightarrow J'$ rotational transition. *Right*, schematic of the magnetooptic dichroic signal for a Zeeman-active rotational transition $J'' = 0 \rightarrow J' = 1$. See text for further discussion.

different refractive indices for the $\mathbf{E}_+$ and $\mathbf{E}_-$ components of the electric field vector. The dichroic effect is detected by the angle of rotation of the plane of polarization of the exciting light. Denoting the refractive indices corresponding to the electric field components $\mathbf{E}_+$ and $\mathbf{E}_-$ by n_+ and n_-, the angle θ introduced in equation (11.96) can be written as

$$\theta = \frac{\pi L}{\lambda_0}\left(n_- - n_+\right), \qquad (11.100)$$

where $\lambda_0 = c/\upsilon$. The angle θ can be written explicitly in terms of the minor and major axes of the ellipse produced by different degrees of rotation of the $\mathbf{E}_+$ and $\mathbf{E}_-$ components (Figure 11.32)

$$\theta = \tan^{-1}\frac{b}{a}. \qquad (11.101)$$

To illustrate magnetooptic lines, let us consider a $\Delta J = \pm 1$ rotational transition such that the upper state is Zeeman-split. As shown in Figure 11.33, microwave absorption will normally produce three lines according to the selection rule $\Delta M = 0, \pm 1$. But as discussed in the preceding section, the $\Delta M = 0$ rotational absorption is a π-component line (both magnetic field and microwave field are parallel) and $\Delta M = \pm 1$ absorption are σ-component lines (the magnetic field is perpendicular to the electric field). Since the **B** field is required to be perpendicular to the plane-polarized electric field for the observation of magnetooptic rotation, the effect will be shown by the $\Delta M = \pm 1$ or σ-component transitions, σ_+ for $\Delta M = +1$ and σ_- for $\Delta M = -1$. It may also be recalled from earlier discussions in Chapter 4 that the $\Delta M = \pm 1$ transition in the normal Zeeman effect corresponds to the two oppositely circular components of the electric field. Accordingly, the Zeeman transitions corresponding to $\Delta M = +1$ and $\Delta M = -1$ will be associated with the refractive index components n_+ and n_- of the electric field. As equation (11.99) shows, the refractive index undergoes dispersion near the resonance frequency ω_0. Both Zeeman lines will show dispersion behavior of the respective refractive index, dispersion of n_- near ω_0 of the $\Delta M = -1$ transition and dispersion of n_+ near ω_0 of the $\Delta M = -1$ transition (Figure 11.33). To note is that the zero-point of dispersion, i.e., the point at which the refractive index intersects the frequency axis, does not exactly match the resonance frequency ω_0, but is shifted a little to the higher frequency side. In principle, the real part of the refractive index should be both positively and negatively infinite at ω_0, but in practice one observes dispersion of only finite amplitudes which is due to the phenomenon of energy-damping by various intramolecular motions as well as through energy-dissipative molecular collisions. Since Zeeman energy-splitting (see equation (11.95)) is inversely proportional to $J(J+1)$, the higher-J rotational levels are not quite amenable to magnetooptic study, but are useful to investigate the band head regions (lowest J-values) of rotational structure.

The description of magnetooptic effect produced here is rather rudimentary and classical in nature. A detailed quantum mechanical theory of the effect in small molecules has been provided by Groenewedge (1962), and Hutchinson and Hameka (1965), which the interested readers are encouraged to look up.

Box 11.1 Supersonic beam

Atoms and/or molecules initially held in a reservoir at high pressure are released into a vacuum chamber. The reservoir is normally held at room temperature and the pressure could be set to ~1 bar (atmospheric pressure) or somewhat more, so the molecules have random thermal motion but little mass flow. On releasing the molecules into the vacuum chamber through a nozzle or a pulsed

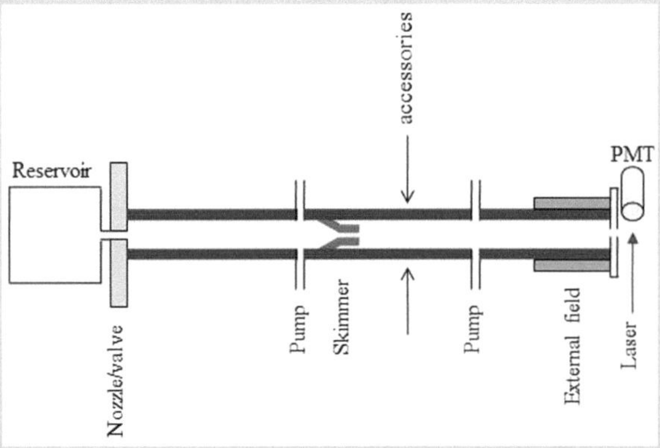

FIGURE 11.34 A tentative design of the supersonic jet assembly.

valve they undergo isentropic expansion and quickly form a mass-flow supersonic beam traveling down along the axis of the vacuum chamber (Figure 11.34).

The speed of the molecular jet can be equal to or greater than the speed of sound, hence it is called a supersonic beam. The beam density or number density – defined as the number of atoms or molecules per m^3, the beam flux, and the drop in the translational temperature at some point downstream – depends critically on factors including the mechanical design of the nozzle or the pulsed valve, collimation of the beam, inclusion of an engineered skimmer that would transmit the central part of the beam along the skimmer axis, and sufficient diffusion pumping to sustain low pressure, typically 10^{-7} bar down the vacuum chamber. As the gas continues to expand in the chamber, the gas pressure reduces to a point where intermolecular collisions hardly occur. This situation corresponds to the terminal speed of the jet and minimum temperature achievable. Accessories like a magnetic or an electric field (see Figure 11.34) down the beam path can be included for studies of Zeeman and Stark effects.

The mechanical design and the geometry of the orifice connecting the gas reservoir and the vacuum chamber, and the skimmer downstream are related to the temperature, pressure, and the molecular number density by the isentropic equation of state of an ideal gas, helium for example, through

$$\frac{T}{T_0}=\left(\frac{P}{P_0}\right)^{\frac{\gamma-1}{\gamma}}=\left(\frac{\rho}{\rho_0}\right)^{\gamma-1}=\frac{1}{1+0.5(\gamma-1)M^2}, \qquad (11.B1.1)$$

where the denominators T_0 and P_0, and ρ_0 are the initial reservoir temperature and pressure, and number density of molecules, respectively, and $T, P,$ and ρ are corresponding values in due time course in the chamber. This equation is valid only when the gas expands in the vacuum chamber reversibly and adiabatically. The quantity γ (see below) is the C_p / C_v heat capacity ratio. We can work out that the Mach number M will vary downstream depending on the length of the vacuum cylinder from the nozzle that the beam has traveled at pressure P in the vacuum chamber. The result is

$$M \propto (x/d)^{\gamma-1}, \qquad (11.B1.2)$$

where x is the length the beam has traveled from the point of the nozzle or the valve, and d is its diameter. Clearly, the constant of the proportionality will depend on the characteristic of the gas, the C_p / C_v ratio. The value of M will, however, increase irrespective of the value of C_p / C_v, although to different extents for different gases because the pressure, the temperature, and the molecular number density generally drop downstream. At the terminal speed of the beam, where virtually no intermolecular collision occurs, the eventual Mach number is

$$M_{\mathrm{T}} = 2.05\epsilon^{-(1-\gamma)/\gamma}\left(\frac{\lambda_0}{D}\right)^{(1-\gamma)/\gamma}, \qquad (11.B1.3)$$

where $\lambda_0 = \langle \lambda_0 \rangle$ is the mean free-flight distance – the distance in free-flight before another collision occurs, of the molecules (Smalley et al., 1977). The skimmer, when used and which may come somewhere upstream, removes shock waves at the expense of the skimmer blockade that can degrade the beam quality in terms of speed, number density, and temperature.

The theory of supersonic expansion is based on the principles of thermodynamics and ideal gas expansion (Wall, 2016). In the vacuum chamber the gas undergoes adiabatic expansion, which means the gas neither absorbs nor gives off heat $(\Delta Q = 0)$; it is rather the kinetic energy of the gas molecules which is used up for the work of expansion. An important consequence of this adiabatic expansion is a decrease in the temperature of the gas beam. Indeed, the temperature of the gas molecules can drop from ~293 to ~20 K or even lower within ~10 mm of travel down a well-designed supersonic beam apparatus. The speed of the mass flow of the gas can be derived easily. The work of gas expansion is pressure-volume work. If P_0V_0 is the work done in the reservoir and P_1V_1 is that in the vacuum chamber, then

$$\Delta w = P_1V_1 - P_0V_0. \qquad (11.B1.4)$$

Let the internal energy of the gas molecules in the reservoir where no mass flow occurs be U_0, and in the flowing chamber where mass flow occurs be U_1. Note that $U_0 \neq U_1$ in general because of the pressure-volume difference. So the change in the internal energy in the two regions with respect to the PV work solely is

$$\Delta U = U_1 - U_0. \qquad (11.B1.5)$$

In the expanding chamber though, the internal energy U_1 is also modulated by the kinetic energy corresponding to the total mass of the flowing gas m_T having the velocity v. Thus the effective change in the internal energy is actually

$$\Delta U = \left(U_1 + \frac{1}{2}m_T\right) - U_0. \qquad (11.B1.6)$$

Because the supersonic expansion is adiabatic one can use the first law of thermodynamics ($\Delta w = \Delta U$) and combine equations (11.B1.4) and (11.B1.6) to obtain the speed of the gas flow. The result is

$$v = \sqrt{\frac{2}{m}\left\{\left(U_0 + P_0V_0\right) - \left(U_1 + P_1V_1\right)\right\}}, \qquad (11.B1.7)$$

where m is the mass of the gas molecule. One may now invoke the following definitions involving pressure and volume

$$H = U + PV$$

$$C_v = \left(\frac{\partial U}{\partial T}\right)_v$$

$$C_p = \left(\frac{\partial H}{\partial T}\right)_p$$

to facilitate rewriting the gas-flow speed as

$$v = \sqrt{2C_pT_0 - T_1}, \qquad (11.B1.8)$$

where $C_p = \frac{k_B\gamma}{m(\gamma-1)}$, and $\gamma = \frac{C_p}{C_v}$ as was mentioned earlier with reference to equation (11.B1.1). Note that $C_p \neq C_v$ because the internal ebergy is not equal to the enthalpy ($U \neq H$) for gases, and C_v is the quantity that measures the heat energy that must be withdrawn from the gas in order to cool it to a certain value. Further, γ continues to increase from representing monatomic to higher-atomic gases, implying larger flow speed for small molecules.

The dynamics of a gas molecule in the supersonic jet refers to translational, rotational, and vibrational motions. The average kinetic energy of molecules which depend on their C_v values is stored into kinetic energy levels corresponding to translational, rotational, and vibrational motions (Figure 11.35).

The initial event in supersonic expansion of gases is a fall in the translational temperature which is reflected as a narrowing distribution of molecular velocity as the gas flows downstream; the width of the velocity distribution is a measure of the translational temperature.

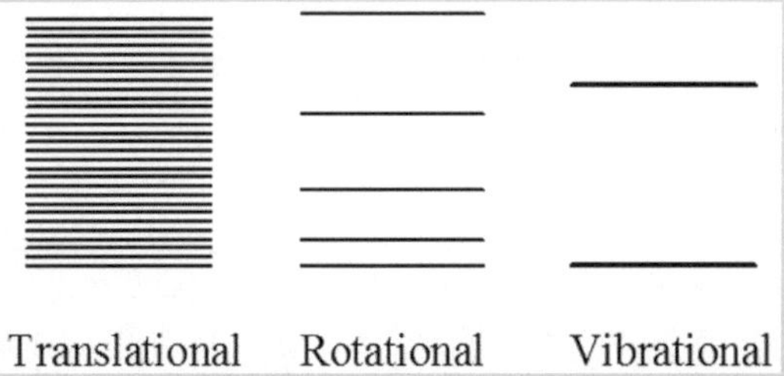

FIGURE 11.35 Translational, rotational, and vibrational energy levels.

In a performing supersonic apparatus this temperature may drop to a few Kelvins when the Mach number M touches ~10 or higher. The beam state is now described by how rapidly rotational and vibrational temperatures equilibrate with the translational temperature. As the molecular collisions decrease the rotational degrees of freedom equilibrate with translation. The rate of this equilibration should be fast as one can make out from the energy level spacing in Figure 11.35. The translational-vibrational equilibration is a far slower process and may not reach completion even in the ultralow density region of the beam where no molecular collision is expected. Thus molecules at different dynamic states are obtained as the beam travels downstream, and in the collision-free region finally a single molecule is isolated which is translationally entirely cold, rotationally ultracold, and vibrationally cold. A molecule will stay in this free-flight state until it hits the wall of the chamber. In this way spectroscopic experiments can be carried out in differently cold states of the molecule.

One more important property of supersonic beams is the slow rate of phase equilibration, which means the molecules stay in the gas phase for a long time even at ultracold temperature in which a solid state should have been there.

Box 11.2 Quantum numbers Λ and Σ can be positive or negative

The total of the quantum numbers Λ and Σ should be thought to arise from spin-orbit coupling in which the z-component of the orbital angular momentum intercats with the z-component of the spin angular momentum, thus generating the total electronic angular momentum along the internuclear axis. Before the spin-orbit coupling the z-projections of Λ, which are $+\Lambda$ and $-\Lambda$, are degenerate and so is the degeneracy of $+\Sigma$ and $-\Sigma$. Spin-orbit coupling cancels the respective degeneracy so as to produce the following four states

$$\left.\begin{matrix} +\Lambda, +\Sigma \\ -\Lambda, -\Sigma \end{matrix}\right\} \text{Group 1}$$

$$\left.\begin{matrix}+\Lambda,-\Sigma\\-\Lambda,+\Sigma\end{matrix}\right\}\text{Group 2}$$

Groups 1 and 2 are degenerate, and for each of the two groups the total angular momentum $\Omega = \Sigma + \Lambda$ is evaluated by taking only $+\Lambda$, but not $-\Lambda$. But the fact still remains that $+\Lambda$ and $-\Lambda$ are degenerate, and so are $+\Sigma$ and $-\Sigma$. These degeneracies hold when the molecular rotation tends to the freezing limit or too slow, but the pair $\pm(\Lambda,\Sigma)$ splits into $+(\Lambda,\Sigma)$ and $-(\Lambda,\Sigma)$ when the rotational velocity increases. This is called Λ-doubling.

Box 11.3

Let us assume a transition from an initial state i to a final state f. To set up the form of the matrix element we consider equation (3.12)

$$V_{if} = \int \psi_f^*(x,t)V(t)\psi_i(x,t)d\tau$$
$$= \langle f|V(t)|i\rangle. \quad (11.B3.1)$$

Since the perturbation Hamiltonian in the present case of predissociation is

$$\mathcal{H}' = \mathcal{H}'_{\text{vib}} + \mathcal{H}'_{\text{rot}}, \quad (11.B3.2)$$

we write the matrix elements as $\langle f|\mathcal{H}'|i\rangle$.

Now the function $\psi_i(x,t)$ can be expanded as a complete orthonormal set of unperturbed time-dependent functions

$$\psi_i(x,t) = \sum_j a_{ij}(t)\phi_j(x,t)e^{i\omega_j t}, \quad (11.B3.3)$$

where a_{ij} are mixing coefficients. We can use the same procedure described in Chapter 3 to solve the time-dependent Schrödinger equation

$$i\hbar\frac{\partial}{\partial t}\psi_i(x,t) = \left[\mathcal{H}^0(x) + \mathcal{H}'(x,t)\right]\psi_i(x,t). \quad (11.B3.4)$$

Since the $\phi_j(x,t)$ functions are orthonormal, we have

$$\frac{\partial a_{if}}{\partial t} = \frac{2\pi}{ih}\sum_j a_{ij}(t)\langle f|\mathcal{H}'|i\rangle. \quad (11.B3.5)$$

At time $t=0$, the system is in the pure state $\psi_i(x,t)$, and the coefficient $a_{if} = a_{ij} = 1$. We assume that a_{ii} is reasonably large at all times $t > 0$, so $a_{ij} = a_{ii} \sim 1$

$$\frac{\partial a_{if}}{\partial t} = \frac{2\pi}{ih}\langle f|\mathcal{H}'|i\rangle$$

$$a_{if}(t) - a_{if}(0) = \frac{2\pi}{ih}\int_0^t \mathcal{H}'_{fi}(t)e^{\left[\frac{2\pi i}{h}(E_f - E_i)t\right]}dt. \quad (11.B3.6)$$

At time $t=0$, $a_{if}(0) = 0$, so the form of a_{if} at some point of time t would be

$$\partial a_{if} = \frac{2\pi}{ih}\int_0^t \mathcal{H}'_{fi}(t)e^{\left[\frac{2\pi i}{h}(E_f - E_i)t\right]}dt. \quad (11.B3.7)$$

Note that $|a_{if}|^2$ is the probability of moving the system from the stationary state $\psi_i(r)$ to the stationary state $\psi_f(r)$. We now use the procedure described in Box 3.2 to write

$$|a_{if}|^2 = \frac{4}{h^2}|\mathcal{H}'_{fi}|^2\int_{-\infty}^{+\infty}\frac{\sin^2\pi(\nu_{fi} - \nu)t}{(\nu_{fi} - \nu)^2}d\nu, \quad (11.B3.8)$$

where ν is the frequency of the perturbation field. Notice that we chose the limits of integration from $-\infty$ to $+\infty$, because we assume a broad density of the final stationary states. The $\pm$ limits of the integration also casts it in the form of a standard integral.

Letting $\pi(\nu f_i - \nu)t = \theta$

$$|a_{if}|^2 = \frac{4}{h^2}|\mathcal{H}'_{fi}|^2\pi t\int_{-\infty}^{+\infty}\frac{\sin^2\theta}{\theta^2}d\theta = \frac{4\pi^2}{h^2}|\mathcal{H}'_{fi}|^2 t. \quad (11.B3.9)$$

Since $|a_{if}|^2$ is the transition probability and the transition takes place to a broad density of final states f, the total probability of transition is the sum of all $|a_{if}|^2$.

At this point it will be useful to inspect Figure 3.3, which shows the probability amplitude for a discrete stationary state. When transition takes place to many final states, then we have a density distribution $\rho(\nu_f)$ consisting of many of these amplitudes spreading over $\pi(\nu f_i - \nu)t$. So the total transition probability is

$$p(t) = \sum_f |a_{if}|^2. \quad (11.B3.10)$$

If the density of the final states, say density of states per unit energy or unit frequency, is $\rho(\nu_f)$, then

$$p(t) = \frac{4\pi^2}{h^2}|\mathcal{H}'_{fi}|^2 t\rho(\nu_f). \quad (11.B3.11)$$

The rate of upward transition, or relaxation downward, is then given by

$$\omega_{i\to f} = \frac{2\pi}{\hbar^2}|\langle f|\mathcal{H}'|i\rangle|^2\rho(\nu_f) \quad (11.B3.12)$$

This expression is called Fermi's golden rule. It expresses a first-order perturbation besides giving the nonradiative rate of relaxation, and finds wide use in the chemistry of small molecules – in the description of a host of molecular states, and rate calculations in electron transfer processes and chemical reactions.

TABLE PROBLEM 11.1

		T_0 (cm^{-1})	ω_e (cm^{-1})	$\omega_e x_e$(cm^{-1})	B_e (cm^{-1})
NO	$^2\Sigma^{+*}$	43965.7	2371.3	14.48	1.9952
	$^2\Pi_{\frac{1}{2}}$	0	1904.03	13.97	1.7046
MgH^+	$^1\Sigma^{+*}$	35905	1132.7	6.8	4.330
	$^1\Sigma^+$	0	1695.3	30.2	6.411
RbH	$^1\Sigma^{+*}$	18906.4	244.6	−4.1	1.231
	$^1\Sigma^+$	0	936.77	14.15	3.020

States labeled with * are the first-excited states.

PROBLEMS

11.1 Produced in the Table above are spectroscopic data for NO, MgH^+, and RbH from Herzberg's compilation.

(a) Determine the equilibrium bond length in both ground and electronically excited states of each diatomic molecule.

(b) Estimate the zero-point energy in the excited state of each molecule.

(c) Consider the transitions $v' = 0 \leftarrow v'' = 0$ in the electronic spectrum, such that transitions take place from J=0, 1, and 2 in the $v'' = 0$ region. Calculate the energies of the transitions at sufficiently low temperature, say a little above liquid helium temperature, and show the peaks in a line diagram.

11.2 Our description of dipole transitions in the electronic-vibrational-rotational spectra has grossly overlooked the motional reality of molecules. The assumptions of the rigidity of a rotor and the neglect of Coriolis interactions have been made only to make the problem exactly solvable at the expense of a great deal of information. List the information lost and indicate if some of them can still be salvaged within the assumptions made.

11.3 Reconsider the table of spectroscopic data in problem 11.1 above. Since we are focusing on the $v' = 0 \leftarrow v'' = 0$ transition, a harmonic approximation is fair enough.

(a) Show that the Franck-Condon factor for the $v' = 0 \leftarrow v'' = 0$ transition can be expressed by

$$F_{oo} = \left[\frac{2(\omega''_e / \omega'_e)}{1+(\omega''_e / \omega'_e)^2}\right] e^{-D/\left[1+(\omega''_e/\omega'_e)^2\right]}$$

where $D = \Delta r''$, $\Delta = q'_e - q''_e$, and $r = \omega''_e / \hbar$.

(b) Calculate F_{oo} for the ground-to-excited states of NO, MgH^+, and RbH.

11.4 A closer look at the R and P branches of rotational structures of a diatomic molecule, say CO (see Figure 8.6), would reveal that the spacing between rotational lines progressively increases with successive increase in line frequency, that is, with successively higher value of J. Walk from lower to higher energy (equivalently, from higher to lower wavenumber) to observe the increasing spacing. This is observed in both R and P branches.

(a) Why does this happen?

(b) Suppose we could measure the width of individual rotational lines to a very high level of accuracy. Do we expect to see a progressive increase in FWHM with progressively higher values of J? Justify.

(c) Suppose we measured rotational structures by both FC absorption and FC emission. How would the rotational line positions (structure) change?

(d) What function(s) describe the R and P envelopes containing the rotational lines?

(e) Could our description of dipole transitions in electronic spectra, where the molecule is assumed to be rigid, account for the rotational structures in vibrational bands?

11.5 Classifying molecules according to Hund's cases is not clear-cut. Nevertheless, lighter molecules generally tend to get included into case (b), while heavier molecules tend to approach case (a). What is the basis for these suggestions? One may use the argument of rotational constant as one of the bases.

11.6 This question pertains to Fermi's golden rule. Suppose a laser pulse placed a molecule at time $t \approx 0$ in a discrete vibrational level in a bound excited electronic surface, call it state $|i\rangle$ such that $|i\rangle$ overlaps with the continuum of a repulsive surface $|f\rangle$. Because of the overlap we expect the system to be in state $|f\rangle$ at time $t = t$ with the probability

$$P_{|f\rangle}(t) = \frac{|\mathcal{H}_{fi}|^2}{\hbar^2}\left(\frac{\sin^2 x}{x^2}\right)t^2$$

where $|\mathcal{H}_{fi}|$ is the matrix element and $x = (\omega_{fi} - \omega)t/2$, ω_{fi} being $(\omega_f - \omega_i)$.

Show that the rate of transition of the molecule from $|i\rangle$ to $|f\rangle$ can be written as

$$k_{fi} \sim \frac{2\pi}{\hbar}|\mathcal{H}_{fi}|^2 \frac{dN(E)}{dE}$$

in which $\frac{dN(E)}{dE}$ denotes the density of $|f\rangle$ states within a very small region of energy E. This region could be the point where the bound and repulsive excited surfaces cross each other. If needed, use the standard integral

$$\int_{-\infty}^{+\infty} \frac{\sin^2 x}{x^2} dx = \pi.$$

11.7 Consider placing an asymmetric top which may have a dipole moment of 1×10^{-3} D in an electric field whose strength varies from 1000 to 10000 V cm^{-1} at an interval of 1000 V cm^{-1}. Graph the fourth order perturbation energy for J=15 and 20 as a function of the applied field.

11.8 Draw an approximate stick diagram showing the changes in the P(4), Q(4), and R(4) branches of $^2\Pi_{\frac{1}{2}} \rightarrow 2\Sigma^{+^*}$ transition of NO in the presence of ~500 G magnetic field. This field is applied along $+z$ direction, and the plane of the microwave electric field is polarized both parallel and perpendicular to the magnetic field.

BIBLIOGRAPHY

Baba, M., M. Saitoh, Y. Kowaka, K. Taguma, K. Yoshida, Y. Semba, S. Kasahara, T. Yamanaka, Y. Ohshima, Y.-C. Hsu, and S. H. Lin (2009) *J. Chem. Phys.* 131, 224318.

Bauschlicher, C. W. and S. R. Linghoff (1988) *J. Chem. Phys.* 89, 2116.

Bichsel, B. J., M. A. Morrison, N. Shafer-Ray, and E. R. I. Abraham (2007) *Phys. Rev. A* 75, 023410.

Birge, R. T. and H. Sponer (1926) *Phys. Rev.* 28, 259.

Born, M. and E. Wolf (1970) *Principles of Optics*, Pergamon Press.

Burrus, C. A. and J. D. Graybeal (1958) *Phys. Rev.* 109, 1553.

Crawford, F. H. (1934) *Rev. Mod. Phys.* 6, 90.

Geiger, J. and K. Wittmaack (1965) *Z. Naturforschg.* 20a, 628.

Golden, S. and E. B. Wilson (1948) *J. Chem. Phys.* 16, 669.

Grimaldi, F., A. Lecourt, H. Jefbvre-Brion, and C. M. Moser (1966) *J. Mol. Spectrosc.* 20, 341.

Groenewedge, M. P. (1962) *Mol. Phys.* 5, 541.

Hendrie, J. M. and P. Kusch (1957) *Phys. Rev.* 107, 716.

Herzberg, G. (1950) *Molecular Spectra and Molecular Structure. I. Spectra of Diatomic Molecules.* D. Van Nostrand Company, Inc.

Hutchinson, D. A. and H. F. Hameka (1965) *J. Mol. Spectrosc.* 18, 141.

Kotlar, A. J., R. W. Field, J. I. Steinfeld, and J. A. Coxon (1980) *J. Mol. Spectrosc.* 80, 86.

Pelissier, P. M. and J. P. Matrieu (1977) *J. Chem. Phys.* 67, 5963.

Peter, M. and M. W. P. Strandberg (1957) *J. Chem. Phys.* 26, 1657.

Ram, R. S. and P. F. Bernath (1996) *Appl. Opt.* 35, 2879.

Shastri, A., K. Sunanda, P. Saraswathi, and N. C. Das (2002) *Technical Report BARC/2002/E/009.*

Smalley, R. E., L. Field, and D. H. Levy (1977) *Acc. Chem. Res.* 10, 139.

Steinfeld, J. I. (1989) *Molecules and Radiation*, MIT Press.

Szajna, W. and M. Zachwieja (2009) *Eur. Phys. J. D.* 55, 549.

Townes, C. H. and A. L. Schawlow (1975) *Microwave Spectroscopy*, Dover Publications, Inc.

Trivikram, T. M., R. Hakalla, A. N. Heays, M. L. Niu, S. Scheidegger, E. J. Salumbides, E. J. Salumbides, N. D. Oliviera, R. W. Field, and W. Ubachs (2017) *Mol. Phys.* 115, 3178.

Wall, T. E. (2016) *J.. Phys. B.* 49, 243001.

Wheeler, M. D., S. M. Newman, A. J. Orr-Ewing (1998) *J. Chem. Phys.* 108, 6594.

Wijnberg, L. (1974) *J. Chem. Phys.* 60, 4632.

Zeeman, P. B. and G. J. Ritter (1954) *Can. J. Phys.* 32, 555.

Zhu, Y., R. Shehadeh, and E. Grant (1992) *J. Chem. Phys.* 97, 883.

12 Vibrational and Rotational Coherence Spectroscopy

There is a myriad of molecular states (energy surfaces) calculable by combining atomic states – greater the multiplicity of the atom states, larger is the number of molecular eigenstates. Ordinarily, we are more familiar with the ground state, and perhaps only a few of the low-lying excited states of diatomic molecules that are needed to explain the basic principles of spectroscopic transitions considered so far. The characterization of the excited states – both bound and repulsive, many of which interact with each other and even with the ground state by surface crossing to produce points of degenerate energy, is necessary to understand the nature of excited states of molecules, the principles of molecular dissociation, the dynamics of atom recombination, the quantal phenomena like the coherence of an excited state wave packet, and the properties of a free particle not constrained to a potential. The following is a brief description of coherence effect in the spectroscopy of bound and continuum states. The physical meaning of coherence is discussed by the density matrix formalism with an illustration of density operator description of an optical experiment. The density matrix is carried over to the next chapter to analyze theory and experiments in NMR.

12.1 ULTRASHORT TIME OF SPECTROSCOPY

We start with the discussion by noting the strategy of placing the ground electronic state on one of the upper surfaces almost instantaneously (Franck-Condon excitation) which will enable probing the evolution of the excited state at times short enough to match nuclear vibrations at least. A particular advantage of working at this timescale is the avoidance of molecular collision all through the measurement so that Doppler-broadening of spectral lines is completely eliminated. This strategy requires ultrashort light (laser) pulses, and hence falls in the realm of pico- to femtosecond spectroscopy of small molecules formed of two or a few atoms where the motion of the wavefunction can be described by a simple time-distance space; the time is the real time and the distance may refer to internuclear separation or bond isomerization coordinates. It is warranted at this stage to compare the meaning of ultrafast regime in the case of a small molecule vis-à-vis a large molecule – as large as a protein, because the minds of many spectroscopist chemists, especially the so-called theoreticians, are oriented only to pico- and femtosecond regimes when ultrafast timescales are in picture. Here comes the difference of the motion of a fly and an elephant. We see that ultraselection of one of thousands of bonds in a protein molecule is beyond the scope of any discussion. Furthermore, a protein molecule has hardly rotated in the sub-nanosecond timescale; we know from the theory of condensed-phase diffusion that it takes tens of nanoseconds or longer for a segmental diffusion in proteins. These apparent arguments must place the 'ultrafast' definition applicable to a protein-like large molecule to the micro- to nanosecond regime. Nevertheless, ultrashort laser pulses are necessary for studies of small molecules because the dynamics to probe fall in the sub-picosecond regime. Ultrashort light pulses, as short as less than 10-femtosecond duration are deliverable by the existing technology, but not much of molecular dynamics observation is expected because the nuclei of a diatomic molecule have hardly moved in a few femtoseconds. The current attosecond spectroscopy probes electron ejection and thus falls in the subatomic regime of dynamics. Even so, laser pulses of tens of femtosecond duration are necessary to probe the excited-state dynamics at molecular and atomic levels.

One should also distinguish the consequences of excitation of the ground molecular state by a continuous wave (CW) light versus an ultrashort light pulse. Both would yield vertical Franck-Condon excitation, but there will be differences with regard to the properties of the excited state. If more than one excited vibrational levels are produced, which invariably is the case, the excited-state wave packet is a superposition of the individual excited-state wavefunctions. However, these individual wavefunctions will have variable phase relationships if a CW light was used for excitation. Normally, CW laser light is monochromatic (narrow spectral bandwidth), coherent, and intense, and hence excitation of fewer vibrational levels of the ground state is expected. In such a case, the spatial distributions of the sum of the component amplitudes will not change with time. This corresponds to the case of spatial delocalization of the excited state where the amplitude of the wave packet does not change as it travels with time. On the other hand, an ultrashort laser pulse, produced by compressing an already short pulse, acquires a broad spectral width and hence contains many frequencies in addition to the carrier frequency. Suppose we desire to carry out a singled-out excitation $v' = 5 \leftarrow v'' = 0$. This is achievable only when the laser bandwidth is narrow enough so as not to overlap with the frequencies associated with $v' = 6 \leftarrow v'' = 0$ and $v' = 4 \leftarrow v'' = 0$ transitions. In fact, short laser pulses are never monochromatic. Thus Franck-Condon excitation by ultrashort laser pulses invariably excites several vibrational states coherently in proportion with the pulse bandwidth, creating one or more individually coherent wave packets. The pulse width is proportional to the levels excited – the shorter the pulse width, the higher is the number of ground-state levels excited.

DOI: 10.1201/9781003293064-12

12.2 WAVE PACKET

There are two forms of wave packet here, the light pulse itself is one that goes on to create the second one by exciting the ground vibrational states. What is the difference between the two? The electromagnetic field, the laser pulse in the context, originates from the excitation state of a charge. This definition of the electromagnetic field is operational in the realm of classical physics (Chapter 1). On the other hand, vibrational states or vibrational wavefunctions are associated with nuclear motions that will instantaneously carry along the electron positions. While the electromagnetic wave involves an excited charge, the vibrational wavefunction is due to a normal mode involving the relative motions of two nuclei. Summarily, the light wave has no attribute of a material particle but the vibrational wave does. This distinction is important because the foundation of the former is classical and hence does not satisfy the de Broglie hypothesis, but the latter does. Thus, the light pulse is a classical wave packet and the vibrational states form a quantum mechanical wave packet. The fundamental difference between the two forms is also easy to see. Consider the light-wave packet at the bottom of Figure 1.7 which shows $|\Psi(z)|$ at $t = 0$. To conform with the convention, let us change the axis from z to x so that $|\Psi(x,0)|$ decreases on either side of $x = 0$ due to destructive interference of the component plane waves. It is also useful to analyze $|\Psi(x,0)|$ in the k-space (momentum space) for which we write the wave packet in the general form

$$\Psi(x,0) = \frac{1}{\sqrt{2\pi}} \int g(k) e^{ikx} dk, \tag{12.1}$$

in which the function $g(k)$ is the Fourier transform of $\Psi(x,0)$. If x_o and k_o are the respective values at time $t = 0$ and

$$g(k) = |g(k)| e^{i\alpha(k)}, \tag{12.2}$$

then the expansion of $\alpha(k)$ about k_o in a Taylor series yields

$$\Psi(x,0) \propto \int_{-\infty}^{+\infty} |g(k)| e^{i(x-x_0)(k-k_0)} dk = \int_{-\infty}^{+\infty} |g(k)| e^{i\Delta x \Delta k} dk, \tag{12.3}$$

according to which the variation of $|g(k)| e^{i\Delta x \Delta k}$ with k will measure the variation of $\Psi(x,0)$ with k. The analysis, akin to that performed in Chapter 1 for the wave packet of light gives

$$\Delta k \Delta x \geq 1, \tag{12.4}$$

where Δx is the width of the wave packet. Using the de Broglie relation, the inequality can be cast in the form

$$\Delta p \Delta x \geq \hbar, \tag{12.5}$$

in which Δp and Δx are respective uncertainties in the momentum and the position of the particle. We see that the quantum wave packet is described by the position-momentum uncertainty principle.

12.3 COHERENCE

Both classes of wave packets – the classical and the quantal – are coherent. In the class of the light wave, the concept of coherence refers to the same phase or a fixed phase between the electric fields of the component waves at different spatial locations (spatial coherence) and time (temporal coherence). In general, a light wave always has some coherence, but the laser light by the virtue of being stimulated emission of radiation is invariably coherent.

The double-slit experiment introduced in the senior high-school year is a good starting point to understand coherence. The two diffracting light waves periodically interfere constructively and destructively, producing the respective bright and dark fringes on the observation screen. Recall that the experimental arrangement consists of a light source, two diffracting slits S_1 and S_2, and an observation screen. If the slit S_1 is open and the slit S_2 is blocked, or the reverse, the two intensity patterns due to the electric field of the light on the screen are identically Gaussian having all the frequencies contained in the source light. The maxima of the two Gaussians are shifted according to the geometry adopted for the observation. The region of interest is the overlap of the two Gaussians that represents the superposition of the light waves due to the sources S_1 and S_2. In this region, the two waves interfere constructively and destructively producing the respective bright and dark fringes. For further discussion it will be useful to consider the analytical representation of interference. Let the two waves from the sources S_1 and S_2 produce the respective electric fields E_1 and E_2 on the screen. Since the intensity of light is proportional to the field squared, the corresponding intensities I_1 and I_2 are

$$I_{1(2)} \propto |E_{1(2)}|^2, \tag{12.6}$$

and the resultant intensity on the screen is

$$I \propto (E_1 + E_2)^2 = |E_1|^2 + |E_2|^2 + 2E_1E_2. \tag{12.7}$$

If the path difference of the two waves from the slit positions to a point on the screen is Δl, then a phase difference δ modulates the third term to alter the intensity

$$I \propto |E_1|^2 + |E_2|^2 + 2E_1E_2 \cos\delta, \tag{12.8}$$

where $\delta = 2\pi\Delta l / \lambda$ with λ being the wavelength of light. It is the cross term $2E_1E_2 \cos\delta$ that bears on the interference phenomenon – constructive interference when $\Delta l = m\lambda$ ($\lambda = 0,1,2,\cdots$), and destructive interference when $\Delta l = (m+1/2)\lambda$ ($\lambda = 0,1,2,\cdots$). The intereference of the amplitudes is one of the signs of coherence. We discuss below the manifestation of the coherence of quantum wave packets.

12.3.1 Linear Superposition and Interference

It is emphasized that it is the cross term(s) in the superposition of two or more waves which gives rise to the interference effect. Let us look at the physical meaning of interference by considering simply the linear superposition of two orthogonal normalized eigenstates $|\psi_1\rangle$ and $|\psi_2\rangle$

$$|\psi\rangle = \lambda_1|\psi_1\rangle + \lambda_2|\psi_2\rangle. \quad (12.9)$$

Since $|\psi_1\rangle$ and $|\psi_2\rangle$ are normalized, $|\psi\rangle$ is also normalized. The coefficients λ_1 and λ_2 are respective probability coefficients for finding $|\psi\rangle$ in the two eigenstates $|\psi_1\rangle$ and $|\psi_2\rangle$. Let these two eigenstates represent an observable property A, meaning the property is measurable in any of the two eigestates. If the system is in the state $|\psi_1\rangle$ and we make measurements of the property A in the eigenbase $\{|u_n\rangle\}$, the mean value found is

$$\langle A\rangle = \int \psi_1^* A\psi_1 d\tau. \quad (12.10)$$

But each individual measurement yields an eigenvalue a_n corresponding to an eigenvector $|\phi_n\rangle$, with $n = 1,2,3,\cdots$. The probability of finding a_n in a measurement is given by Postulate 4 (see equation (2.50)), according to which

$$p_1(a_n) = |\langle u_n|\psi_1\rangle|^2. \quad (12.11)$$

Similarly, if the system is in the state $|\psi_2\rangle$ and a measurement of the observable A is made in the same eigenbase $\{|u_n\rangle\}$ the probability of finding a_n is

$$p_2(a_n) = |\langle u_n|\psi_2\rangle|^2.$$

Because the superposed state $|\psi\rangle$ is normalized (equation (12.10)) the sum of the probabilities of finding the state in $|\psi_1\rangle$ and $|\psi_2\rangle$ will be given by the sum of the square of the probability coefficients

$$\left(|\lambda_1|^2 + |\lambda_2|^2\right) = 1. \quad (12.12)$$

Note that λ_1 and λ_2 are probability amplitudes (coefficients) and $|\lambda_1|^2$ and $|\lambda_2|^2$ are respective probabilities.

If we consider the superposed state $|\psi\rangle$ in the same eigenbase $\{|u_n\rangle\}$ for making a measurement instead of using the $|\psi_1\rangle$ and $|\psi_2\rangle$ states individually, the probability that the eigenvalue a_n corresponding to the eigenvector $|u_n\rangle$ will be measured is given by Postulate 4. Accordingly,

$$\begin{aligned} p(a_n) &= |\langle u_n|\psi\rangle|^2 = |\lambda_1\langle u_n|\psi_1\rangle + \lambda_2\langle u_n|\psi_2\rangle|^2 \\ &= \lambda_1^2|\langle u_n|\psi_1\rangle|^2 + \lambda_2^2|\langle u_n|\psi_2\rangle|^2 \\ &\quad +2\left\{\lambda_1\lambda_2^*\langle u_n|\psi_1\rangle\langle u_n|\psi_2\rangle^*\right\}, \end{aligned} \quad (12.13)$$

in which λ_1 and λ_2 are real numbers (probability amplitudes) that give a complex constant $\lambda = \lambda_1 + i\lambda_2$. We see again that the third term above, which was dubbed the 'cross term' earlier contains the interference effect. The phase difference between λ_1 and λ_2 are contained in the cross term in the guise of $\lambda_1\lambda_2^*$.

Let us translate this concept of linear superposition and interference to a real-life and real-time spectroscopic scenario. The experiment typically involves exciting a molecule with an ultrashort laser pulse and then using another ultrashort pulse in tandem to probe the wave packet dynamics as a function of time (Figure 12.1). The requirement of measurement with ultrashort light pulses is obvious because the nuclear dynamic processes occur in the sub-picosecond regime. It may be realized that measurements of molecular structure, which can be carried out under steady-state conditions using long pulse-time or continuous irradiation, not only by ultrashort light, are relatively more accurate as compared to the results of ultra-short measurements. This difference is understood from the time-energy uncertainty relation

$$\Delta E\Delta t \geq \frac{\hbar}{2} \quad (12.14)$$

that we have discussed at length earlier. The measurement of relative dynamics of nuclear wave packet motion at ultrashort time-scale provides less energy accuracy, or equivalently, more spread in the energy spectrum. If the molecule is excited by a very short pulse of light, the excitation is highly localized in both time and space, but the calculation of the spread of energy ΔE requires consideration of several wavefunctions of flanking frequencies whose coherent superposition creates a

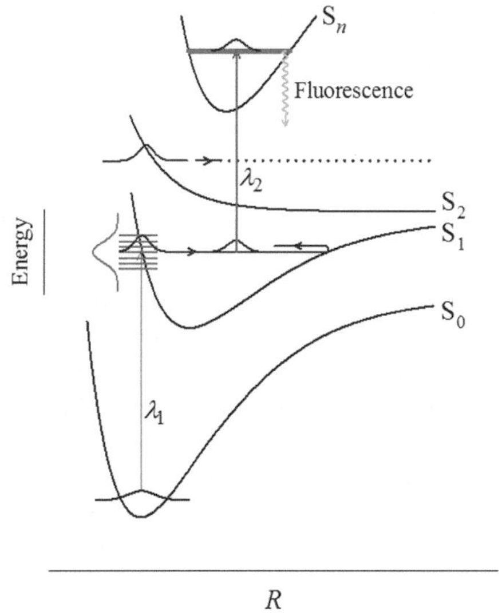

FIGURE 12.1 The summary of a typical pump-probe experiment involving electronic surfaces and vibrational states. While the pump pulse λ_1 is excitatory, the probe pulse prepares the excited state by taking it into another state S_n for observation by fluorescence or absorption. The wave packet formed on the surface S_1 and the shuttling motion of this wave packet between the edges of the same surface is also shown. Note that the shuttling motion occurs only within its lifetime in the excited surface. While the probe pulse prepares for detection the frequency characteristic of the wave packet is not lost, and thus the detected signal faithfully represents the characteristic of the wave packet.

position-dependent wave packet. The number of vibrational states required to form the wave packet will depend on the spectral band width of the light pulse and the slope of the edge of the excited potential surface.

12.3.2 Vibrational Coherence

We could carry out an analysis in a simple manner using only two vibrational levels in the excited state; the concept and the mathematical derivations can be extended to many levels in the excited state of a molecule. The excited states v_i' and v_j' may correspond to some arbitrary vibrational levels, say $v' = i, j$ or any other quantum number. To see the quantum interference in the wave packet formed of v_i' and v_j' in the excited vibronic state (S_1) a transition from any of the vibrational levels v'' of the ground vibronic state (S_o) is required. Let a laser pulse carry out the excitation from v'' to $v' = i$ and $v' = j$ (Figure 12.1). The excitation produces a coherent superposition of the vibrational states, and the superposed wavefunction (see equation (12.9)) is

$$|\psi\rangle = \lambda_1 |v_i'\rangle + \lambda_2 |v_j'\rangle$$

$$\cong a_{12}(v_i')|v_i'\rangle e^{-i\omega_1 t} + a_{12}(v_j')|v_j'\rangle e^{-i\omega_2 t}, \qquad (12.15)$$

where ω_1 and ω_2 are the frequencies corresponding to the vibrational states $|v_i'\rangle$ and $|v_j'\rangle$. The front factor coefficient a_{12} is a coupling factor proportional to the electronic transition dipole, and hence is given by the product of the electric field of the excitation pulse and the electric dipole moment,

$$a_{12} \propto E_{12}\mu_{12}. \qquad (12.16)$$

But the electric dipole moments for the v_i' and v_j' states should be different, so $a_{12}(v_i') \neq a_{12}(v_j')$ which is overt from the functional dependence of the coefficient.

In principle, the excitation pulse also causes the two electronic states S_o (ground) and S_1 (excited) to oscillate coherently at a given internuclear configuration R. The oscillation can cover several periods when the pulse is intense, because increasing the pulse intensity leads to longer pulse duration. The coherent oscillation between the two vibronic states is not desirable since it can interfere with the ensuing dynamics of the nuclear wave packet in the excited potential. If interference does take place, then the wave packet created by the pulsed laser excitation will be deformed. Pulse width of the excitation light, typically tens of femtoseconds (FWHM) easily achievable with the present technology, is shorter than the timescale of nuclear motion, meaning the interference effect can be prevented by using such short pulses. Nevertheless, coherences between electronic states as well as wave packet deformation can also be studied.

The coherence of the wave packet $|\psi\rangle$ in equation (12.15) should in principle be reflected by the interference of the frequencies of the component vibrational states. In practice, $|\psi\rangle$ is interrogated in real time by probing an observable, fluorescence for example, by exciting the entire wave packet to the same fluorescent state v contained in a higher electronic surface S_n. The fluorescence emission from the state v is normally recorded in a time-integrated (continuous) mode. The transition integral for this probe-excitation is

$$\langle v|\mathbf{E}_{23} \cdot \boldsymbol{\mu}_{23}|\psi\rangle,$$

where $\mathbf{E}_{23}$ is the electric field of the excitation laser pulse called probe pulse, and $\boldsymbol{\mu}_{23}$ is the transition dipole moment. We will use $a_{23} = E_{23}\mu_{23}$, which essentially is the part of the Hamiltonian that causes the transition of the wave packet to the final state from which fluorescence emission is to be measured. Since the energy of fluorescence emission is roughly the same as the energy of excitation, the fluorescence signal that would be measured is approximately the square of the excitation energy expressed by the transition integral for probe excitation

$$I(t) \sim \left|\langle v|\mathbf{E}_{23}\mu_{23}|\psi\rangle\right|^2. \qquad (12.17)$$

Substituting for $|\psi\rangle$ from equation (12.15) yields for the signal intensity

$$I(t) \sim a_{12}(v_i')^2 a_{23}(v_i')^2 + a_{12}(v_j')^2 a_{23}(v_j')^2 +$$

$$2a_{12}(v_i')a_{23}(v_i')a_{12}(v_j')a_{23}(v_j')\cos(2\pi\Delta\omega t). \qquad (12.18)$$

The first two terms are phase-independent, and the third cosinusoidal cross term represents the coherence beats that are detectable in the time-domain fluorescence signal. The cross term shows the connection of interference to coherence. The frequency of the coherence beat in this simple case of the wave packet formed of just two vibrational levels is directly proportional to the frequency difference of the two vibrational levels. To note, the time-axis of the fluorescence signal will not be due to the real-time fluorescence but is generated by inserting variable delays between pump and probe pulses, meaning each fluorescence data point corresponds to the experimental pump-probe delay set *a priori*. The coherence beats detected in this manner by perpendicularly polarized emission are drawn in Figure 12.2, which also contains faster oscillations due to the motion of the wave packet between the inner and outer turning points of the bound electronic (excited) potential.

Regarding these faster oscillations, the motion of the nuclear wave packet is a corollary of the nuclear vibrational mode itself. If the excited-state wave packet is located somewhere in the harmonic (lower) part of the excited potential then the position of the wave packet will oscillate between the classical turning points as would a classical particle (Figure 12.3). The wave packet remains localized, implying that the component frequencies are all in phase and the oscillation should sustain in the limit of the lifetime of the excited state. This oscillation of the wave packet corresponds to the oscillation of a coherent state observed in the theory of harmonic oscillator. If, however, the wave packet is created in a

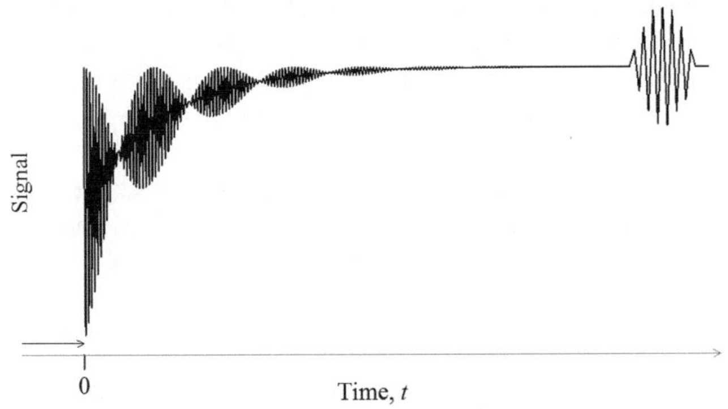

FIGURE 12.2 A simulated transient showing the rotation-vibration coupling. The rapid oscillations earlier in time are due to vibrational motion of the wave packet between the inner and outer points in the potential to which the excitation was carried out. At longer time the rotating dipoles realign. The period in the rotational regime corresponds to pure rotational coherence.

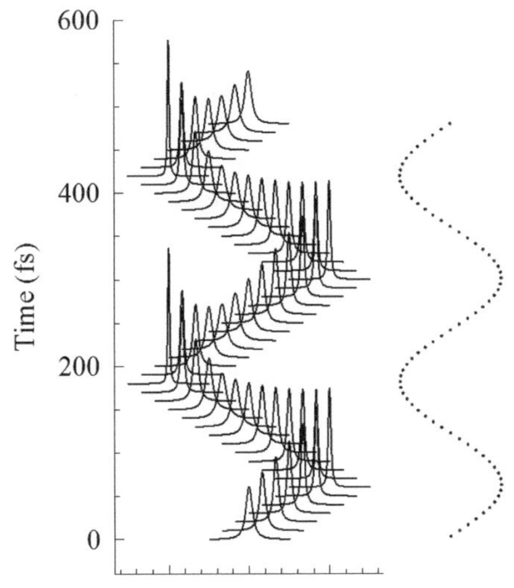

FIGURE 12.3 Simulated oscillation of the wave packet in a harmonic potential where the coherence sustains within the lifetime of the excited state, and the position of the wave packet shuttles between the end points of the potential. Shown alongside is the corresponding harmonic wave.

zone of anharmonic (upper) part of the excited potential, the anharmonicity effect will set in to delocalize the initial spatial localization of the wave packet. As a result, the wave packet dephases to produce the respective oscillations of the component vibrational states. This corresponds to a breakdown of the wave packet. However, due to the quantized nature of nuclear vibrations, the wave packet rephases partially or completely at some later time. If the initial wave packet was created by superposition of several vibrational states, these states may rephase in a way to yield two or more wave packets which may rephase completely at longer time. We see that such delocalization and relocalization do not happen in a classical sense, so such effects must be quantal in nature.

12.3.3 Rotational Coherence

As before, we carry out an excitation from the ground-state rotational level to excited-state rotational levels. If the excitation occurs to only two rotational levels, for the simplicity of description again, the linear superposition in equation (12.15) still describes the rotational coherence. One can be more specific regarding the form of the superposed rotational states by specifying the J, K, and M quantum numbers. To be simple, we will consider a $|J_0''K_0''M_0''\rangle$ ground-state rotational level of a prolate symmetric top (see Figure 7.11) whose electric transition dipole moment $\boldsymbol{\mu}$ directs along the rotational symmetry axis. Let the excitation take place from $|J_0''K_0''M_0''\rangle$ contained in the ground vibronic state $|S_0''v_0''\rangle$ to the rotational levels $|J_{0+1}{}'K_0'M_0'\rangle$, $|J_0'K_0'M_0'\rangle$ and $|J_{0-1}{}'K_0'M_0'\rangle$ contained in the excited vibronic state $|S_1'v_1'\rangle$. These excitations are in accord with the rotational selection rules $\Delta J = 0, \pm 1$, $\Delta K = 0$ and $\Delta M = 0$. In the case of rotational wavefunctions, it is convenient to write the phase factor as $e^{-iE_{\text{rot}}t/h}$ instead of $e^{-i\omega t}$ because the rotational energy relates the J, K, M quantum numbers and the rotational constants A_e, B_e, C_e. As discussed earlier, the rotational energy of a prolate symmetric top is

$$E_{\text{rot}} = B_e J(J+1) + (A_e - B_e)K^2,$$

which can be inserted into the phase factor to express the rotationally coherent superposed state or the rotational wave packet as

$$\begin{aligned}|\Psi(t)\rangle = \Big[a_1 |J_{0-1}{}'K_0{}'M_0{}'\rangle e^{-i2\pi J_0'(J_0'-1)B_e t} + \\ a_2 |J_0{}'K_0{}'M_0{}'\rangle e^{-i2\pi J_0'(J_0'+1)B_e t} + \\ a_3 |J_{0+1}{}'K_0{}'M_0{}'\rangle e^{-i2\pi(J_0'+1)(J_0'+2)B_e t} \Big] e^{-i2\pi(\nu_{ev}+(A_e-B_e)K^2\}t} |S_1{}'v_1{}'\rangle\end{aligned} \quad (12.19)$$

The inclusion of ν_{ev}, which is the frequency corresponding to the energy of the excited vibronic state $|S_1'v_1'\rangle$, in the multiplicative exponential accounts for the energy offset between the lower vibronic level in the ground state $|S_0''v_0''\rangle$ and the excited upper vibronic state $|S_1'v_1'\rangle$. As before, this wave packet at any instant of its evolution can be willfully (excitationally) placed in a vibronic state, say a higher $|S_1 v_1\rangle$ or a lower $|S_0''v_0''\rangle$, both of which are fluorescent. The fluorescence signal intensity at some delay time between the pump and the probe pulse is

$$I(t) \sum\nolimits_{J_f K_f M_f} \left| \langle \Psi(t) | \mathbf{E} \cdot \boldsymbol{\mu} | \mathrm{S}_1 v_1; J_f K_f M_f \rangle \right|^2, \quad (12.20)$$

where $J_f K_f M_f$ denote the fluorescent rotational levels corresponding to J, K, and M quantum numbers. The form of the signal intensity follows the same general pattern as in equation (12.18), and is given by

$$I(t) \sim \left| \langle \Psi(t) | \mathbf{E} \cdot \boldsymbol{\mu} | \phi_f \rangle \right|^2, \quad (12.21)$$

in which the state ϕ_f represents any state that relaxes by emitting fluorescence or some other means (observable). If the excited rotational states are limited to two levels only, then equations (12.19) and (12.20) can be combined to obtain a general expression for the time dependence of the evolution of rotational coherence

$$I(t) \sim \left[a + b\cos\left(\omega_{12}t\right)\right], \tag{12.22}$$

where a and b are constants and the cosine term represents the quantum beat. Notice the recovery of the cosine term that contains the interference effect. The constants a and b (for a two-level excited rotation) will depend on a_1, a_2, $\omega_{12} \sim \left|E_1 - E_1\right| / \hbar$, and the respective integrals for the excited states with the final fluorescence states linked by the fluorescence emission dipole operator. The dipole operator $\mathbf{E} \cdot \boldsymbol{\mu}$ is written in equation (12.21) in the general form. Since the motion is rotational, one expects the evolution of rotational coherence to depend on the polarization state of the dipole operator. Specifically, the polarization of the detecting light with respect to that of the excitation and the direction of the emission dipole moment with respect to the absorption dipole modulate the phase of the rotational coherence.

The question that follows is where is this fluorescence signal along the time domain? The fluorescence from rotational states, if isolated in a pure rotational form, would occur at times much longer than the time for fluorescence decay due to vibrational transitions. The reason is simple, the rotational fluorescence should occur in nano- or sub-nanosecond regime because rotational motions are about three-orders of magnitude slower than the vibrational motions (~10^{-13} s). Consequently, the rotational quantum beats appear much later than the decay of vibrational coherence beats as we saw in Figure 12.2. The slow rotational motion, in the 10^6–10^9 Hz range, indeed offers the advantage of measuring Doppler free optical signals.

A distinct feature of rotational coherence is a regular period of constructive interference of frequencies of all rotational levels excited. The origin of this periodic return of coherence beats is the existence of a population distribution of thermally averaged rotational levels. Such a distribution almost always exists irrespective of the temperature at which the excitation is carried out. For example, an ensemble of molecules is not homogeneous for this purpose if the molecules do not exist in the same rovibronic ground state. Consequently, the excitation involves a very large number of rotational states with the result that several excited rotational wave packets are created. If j is the number of wave packets formed then the time dependence of the signal measured by using the probe pulse will be

$$I(t) \sim A + \Sigma_j \alpha_j \cos\omega_j t, \tag{12.23}$$

where ω_j is the frequency of the quantum beat corresponding to the jth superposed level (wave packet), and A and α are constants. Specifically, α depends on the rotational quantum number, the temperature, and the polarization of the electric fields of both excitation and detection lights.

A large number of $\cos\omega_j t$ waves are expected to produce destructive interference so as not to modulate the resultant amplitude with time. This does not happen, however, one rather observes periodic recurrence of quantum beats. It turns out that the quantum beat frequency ω_j for all the rotational wave packets are frequency-correlated, which means

$$\omega_j = 2\pi m_i \omega_0, \tag{12.24}$$

suggesting that the quantum beat frequency of the jth wave packet is an integral ($m_i = 1, 2, 3, \cdots$) multiple of the fundamental frequency ω_0. Thus, we get

$$\cos\left(2\pi m_i \omega_0 t\right) = 1 \tag{12.25}$$

for the times $t = n / \omega_0$ with $n = 0, 1, 2, \cdots$. Whenever $\cos\left(2\pi m_i \omega_0 t\right) = 1$ occurs there is a constructive interference at that time t, and hence a quantum beat. Regarding the sign of α_j in equation (12.23) the polarization of pump and probe lights can be chosen appropriately to obtain the same sign for almost all α_j. The resultant beats are often also called 'thermally-averaged rotational beats'. In practice, a wiggly ringing pattern, which arises from centrifugal distortion of the rotating molecule, follows the rotational beats.

12.3.4 Coherence Decay

The total waveform envelope decays at longer probe times, which happens because of the decay of the coherence by both excited-state lifetime and the spread of the wave packet with time. Let the vibrational wavefunction prepared by the pump pulse is a one-dimensional Gaussian wave packet of the form

$$\Psi(x,t) = \frac{\sqrt{a}}{(2\pi)^{3/4}} \int_{-\infty}^{\infty} e^{\frac{-a^2}{4}(k-k_0)^2} e^{i\{kx-\omega(k)t\}} dk, \tag{12.26}$$

where the dispersion $\omega(k) = \hbar k^2 / 2m$, $a = 2\Delta x = 1/\Delta k$, m is the mass of the particle, and Δx and Δk are respective widths of the wave packet in x and k spaces. Note that this function has the same general form as given earlier (equation (12.1) and (12.3)). The general solution of the integral in equation (12.26) can be found in Cohen-Tannoudji et al. (1977), and is

$$\int_{-\infty}^{\infty} e^{-\alpha^2(\alpha+\beta)^2} d\xi = \frac{\sqrt{\pi}}{\alpha}, \tag{12.27}$$

in which the principal value of the argument (Arg α) lies in the $-\pi/4, \pi/4$ interval

$$-\frac{\pi}{4} < \text{Arg}\ \alpha < \frac{\pi}{4}.$$

The probability density of the wave packet can also be seen as a Gaussian function in the guise of a bit of complex expression (Cohen-Tannoudji et al., 1977)

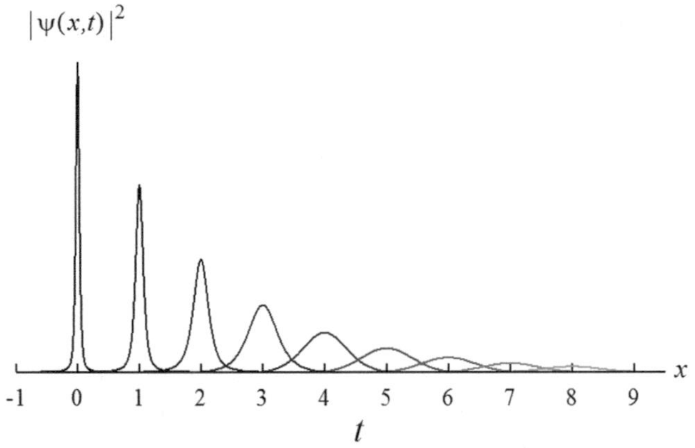

FIGURE 12.4 The spread of a traveling wave packet at distances 0 to 8 units from its origin as it travels at a constant velocity v_0. The change in the width in the time interval Δt is $v_0\ \Delta t$.

$$\left|\Psi(x,t)\right|^2 = \sqrt{\frac{2}{\pi a^2}}\frac{1}{\sqrt{1+\frac{4\hbar^2t^2}{m^2a^4}}}e^{\left[-\frac{2a^2\left(x-\frac{\hbar k_0}{m}t\right)^2}{a^4+\frac{4\hbar^2t^2}{m^2}}\right]}, \quad (12.28)$$

with the width of the wave packet

$$\Delta x(t) = \frac{a}{2}\sqrt{1+\frac{4\hbar^2t^2}{m^2a^4}}. \quad (12.29)$$

The center of the Gaussian is $x = v_0 t$, where the velocity $v_0 = \hbar k_0 / m$.

For a small time interval Δt, the width of the wave packet traveling at the velocity v_0 will change by $\Delta x = v_0\ \Delta t$. The simulated width of a wave packet traveling along the x-direction from 0 to 8 arbitrary units is shown in Figure 12.4. The wave packet continues to spread at $t \to \infty$. It is this spread of the wave packet which is one of the factors contributing to the decay of the waveform. One may note however that the spread of the wave packet is much slower than the time of its oscillation in the bound potential.

In addition to the spreading, the excited-state lifetime which is expected to decay exponentially also contributes to the damping of the wave packet. Analytically, we incorporate the decay of the lifetime by multiplying the superposed wavefunction with a phenomenological damping constant Γ / t which is carried over to the expression for signal intensity. For example, the signal intensity given in equation (12.23) may be written as

$$I(t) \sim \{A + \Sigma_j \alpha_j \cos(\omega_j t)\} e^{-\Gamma/t}. \quad (12.30)$$

12.4 WAVE PACKET OSCILLATION

Generally, the wave packet spends relatively more time in the classical turning point regions of the (an)harmonic potential and travels rapidly through the intermediate region before

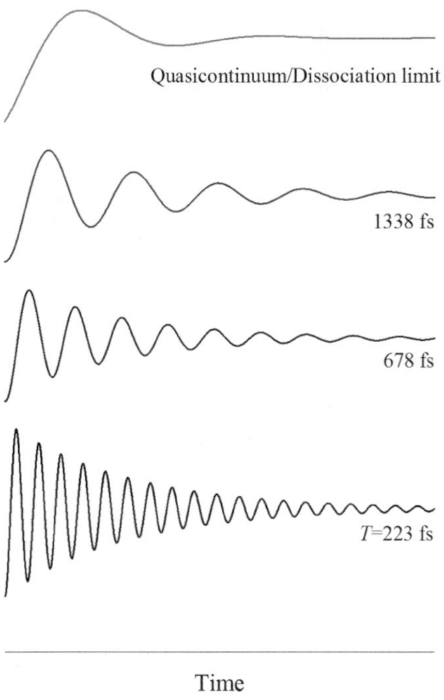

FIGURE 12.5 Computed wave packet oscillation time with increasing vibrational level around which the wave packet is prepared.

reflecting back from the turning point. This is called wave packet oscillation whose period $T = 1 / \nu$ can be calculated if spectroscopic parameters are available from prior steady-state measurements. The idea is to determine the v' level to which the vibronic transition $v' \leftarrow v''$ is excited by the pump pulse. The large bandwidth of the short pump pulse will cause excitation to several v' levels around that v' level which is presumably known from spectroscopic parameters. This allows determining vibrational spacings in the region of the excitation potential. For instance, if a pump beam happens to be the second harmonic of a YAG laser (532 nm) that excites the transition $v' \sim 4 \leftarrow v'' = 0$, meaning excitation around $v' = 4$, then the vibrational spacing would be simply

$$\Delta h\nu \approx \left[\left(T_e' - T_e''\right) + \left(G_v' - G_v''\right)\right] - \left[\left(T_e' - T_e''\right) + \left(G_v' - G_{v\pm1}''\right)\right]. \quad (12.31)$$

Extension of the formula yields $G''_{v\pm2}$ from $G''_{v\pm1}$. Again, if the vibrational spacing in the excited state amounts to say 150 cm^{-1}, then $T = \frac{1}{\nu} \cong 223.4\,\text{fs}$, which is the period of the wave packet oscillation. This period can be compared directly with the peak-to-peak or trough-to-trough time in the real-time transient measured (Figure 12.5).

It is clear to see what the pattern of oscillation would be when higher levels of v' are roped into the wave packet – the time period of wave packet oscillation will be increasingly longer because the vibrational spacing becomes narrower. Figure 12.5 depicts the oscillations for three v' levels around each of which oscillation occurs. These three levels can be prepared by changing the pump wavelength – higher energy will create the wave packet around a higher vibrational level, and the oscillation period will be longer. This picture tells what to expect for quasi-continuum and continuum levels in a repulsive or unbound potential – there is little sign of oscillation (Figure 12.5), suggesting that excitation has occurred to the dissociating limit. In the calculation of these transients, we have ignored the possible phase distortions, albeit a phenomenological exponential damping in each case has been incorporated.

12.5 FREQUENCY SPECTRUM OF TIME-DOMAIN COHERENCE

We now come to an important concept regarding the frequency spectrum or the power spectrum generated by Fourier transformation (see Box 12.1) of the time-dependence of coherence evolution. This is indeed distinct from a steady-state frequency-domain spectral measurement carried out in general. Suppose we perform a rovibrational or rovibronic measurement on a molecule under steady-state conditions by shining light of appropriate frequencies and then recording the final signal by the use of a photomultiplier tube. Plainly, we obtain a frequency-domain (or wavelength-domain) spectrum. Such a spectrum corresponds to the effective energy of the radiation utilized to produce the transition. This is not the case with the energy of transition(s) in a coherence spectrum. The characteristics of the pump radiation, except for its carrier frequency and the spectral band width, is of no consequence. What really matters is how many vibrational and rotational levels are involved in the excited state, although the initial ground-state level, say $v'' = 0$, may be predefined. Note that both duration and frequency of the pump pulse can be tuned to a very high accuracy with present-day technology. The frequency bandwidth, which is also called the spectral bandwidth, is inversely proportional to the pulse duration; more vibrational levels in the excited electronic state will be linearly superposed to produce a wave packet as the excitation pulse time becomes shorter. As soon as the wave packet is created, the properties of the excitation light are of less relevance and further dynamics are based on the dynamics of the wave packet alone. Clearly, the basis vibrational wavefunctions of the excited state which produce the wave packet are going to be different according to the pump frequency and time. The central point here is that transitions within the vibrational levels in a specified wave packet will occur strictly according to the vibrational selection rule. If, for example, six vibrational levels of the excited vibronic state are superposed to a wave packet, then the dynamics of the wave packet will display vibrational transitions of kinds fundamental ($v' \rightarrow v' \pm 1$), first overtone ($v' \rightarrow v' \pm 2$), second overtone ($v' \rightarrow v' \pm 3$), and so on. To note is that these vibrational transitions occur within the levels that form the wave packet. The Fourier transform of a coherence-time function will show the transitions within the vibrational levels that superpose to define the wave packet. One might like to call the intra-coherence transitions as single-, double-, or multiple-quantum transitions, all happening within the energy states of the wave packet.

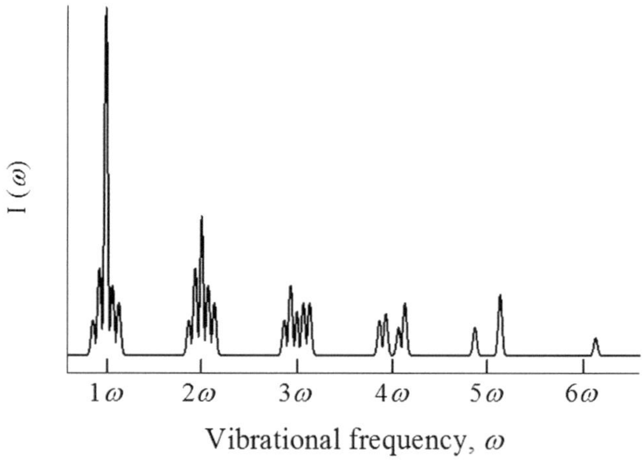

FIGURE 12.6 Vibrational intensity spectrum obtained by Fourier transformation of the oscillatory vibrational part of the transient shown in Figure 12.2. The origin of vibrational multiplets lies in the characteristic of the excitation laser pulse. The weakening and/or disappearance of the vibrational lines at higher frequency arise partly from rotation-vibration coupling, as though rotational motion dominates vibrational.

This much of explanation should be convincing that the Fourier transform of a time-coherence function is not the same as the steady-state frequency-domain function. This result is identically applicable to the Fourier-transformed frequency spectrum of rotational coherence. We conclude that the transitions in coherence spectroscopy are strictly within the coherence basis functions.

A frequency-domain spectrum obtained by Fourier transformation of a synthetic vibrational coherence dataset, the time domain data of Figure 12.2 in fact, is depicted in Figure 12.6. Bands appear at 1ω, 2ω, 3ω, $\cdots$ corresponding respectively to the fundamental, first, second, ... overtone transitions. The overtone clusters are weaker in intensity due to rotation-vibration coupling. The excitation must be around a vibrational level in the anharmonic region because overtone transitions occur only when anharmonicity sets into the potential. Clearly, the spectral bandwidth of the pump beam is large enough to cause excitation into several vibrational levels, producing overtones. If the coherence involved only two excited vibrational levels, the fundamental band will appear alone.

12.6 ASSIGNMENT OF VIBRATIONAL BANDS

Vibrational coherence spectroscopy offers no particular advantage over the traditional continuous wave method in the assignment of the excited-state vibrational quantum numbers. Yet, ideal conditions of the slope and the equilibrium internuclear distance in the excited potential relative to that of the ground-state potential can create favorable Franck-Condon factors so as to yield a long progression in v'. The pump laser pulse can be chosen to have an optimized bandwidth and a carrier frequency for excitation from $v'' = 0$. The excitation

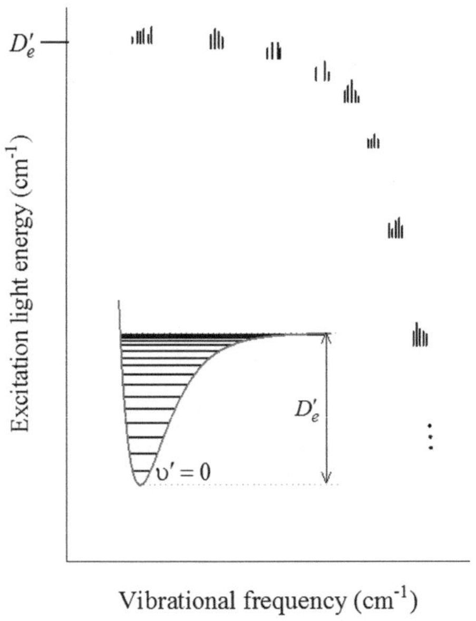

FIGURE 12.7 Qualitative impression of vibrational progression in transient spectra. Each line in the diagram represents intensity (amplitude). The inset shows the dissociation limit, D_e'.

from a single vibrational level in the ground vibronic state ensures only a single v' progression and no overlapping progression that can arise if excitation occurs from more than one vibrational level.

A schematic vibrational progression, each cluster of lines containing the fundamental and overtones, is approximated in Figure 12.7. The larger anharmonicity toward the dissociation limit (D_e') is reflected in the appearance of more vibrational clusters. In practice, one can follow the progression to trace out the cluster having the highest vibrational frequency which can be safely labeled as $v' = 0$. A word of caution here, this exercise needs some care and one can take help of the fact that vibrational levels follow a power series in $v+1/2$. Then starting from $v' = 0$ each cluster of lines with successively lower frequency is assigned to quantum numbers $v' = 1, 2, 3, \cdots$.

12.7 PURE ROTATIONAL COHERENCE

Although the time-evolution of a rotational coherence in ~10^{-9} s regime is observed in a rovibronic spectrum (later than the vibrational oscillations), it is useful to measure pure rotational coherence in the nano- to picoseconds regime, especially for simple systems with lower symmetry – the lowest being a symmetric top. Before we discuss the information content of the coherence spectrum it should be mentioned that the measurement involves a simple pump-probe methodology using molecules isolated in a jet-cooled beam (Box 11.1). By using a polarized light pulse, the cooled isolated molecules are excited from a rovibrational level of the ground electronic state, and the time dependence of fluorescence emission from the excited rovibronic state to some rovibronic ground state is analyzed by employing a polarization analyzer set in a way that the polarization of the emitted fluorescence is parallel to the polarization of the excitation light. This configuration could be parallel or perpendicular to the direction of the rotor electric dipole, the reason for which will be clear in a moment. The only condition is that the polarization of the emission light must be the same as that of the excitation light. Figure 12.8 indicates the experimental approach involving excitation to a rovibrational manifold, and the signal detection by absorption or fluorescence emission. The transient fluorescence signal due to the transition $\left\{\left|\psi_{v_1}\psi_{J_1K_1M_1}\right\rangle\right\} \to \left\{\left|\psi_{v_2}\psi_{J_2K_2M_2}\right\rangle\right\}$ is also shown. There is the signature of quantum beats (spikes) as the fluorescence decays. The point here is the occurrence and the origin of these quantum beat signals which eventually die off as the fluorescence decays.

The theory that can relay the classical motion of a rotor to the quantum coherence was developed by Felker, Baskin, and Zewail (Felker and Zewail, 1987; Baskin and Zewail, 1989), the summary of which is presented here. A rotating molecule, a symmetric top to be simple, precesses about the total angular momentum vector **J** (Figure 12.8). The two angular frequencies of rotation are ω_2, which denotes the frequency of rotation of the molecule about its own symmetry axis, and ω_1, which describes the precession about the **J** vector. Note that some may prefer using 'nutation' instead of 'precession' but the meaning of nutation and a compounded motion of rotation about its own axis and simultaneous precession about another vector, **J** in this case, are the same. Importantly, the two frequencies ω_1 and ω_2 are related to the rotational constants as

$$A_e = \frac{\hbar}{4\pi I_a} \text{ and } B_e = \frac{\hbar}{4\pi I_b}, \tag{12.32}$$

where I_a and I_b are moment of inertia about the symmetry axis and perpendicular to the symmetry axis, respectively (see Herzberg, 1945). These relations in angular frequency are

$$\omega_1 = |\mathbf{J}|\frac{1}{I_b} = J\hbar\frac{4\pi B_e}{\hbar} = 4\pi B_e J$$

$$\begin{aligned}\omega_2 &= |\mathbf{J}|\cos\theta\left(\frac{1}{I_a}-\frac{1}{I_b}\right) = K\hbar\left(\frac{4\pi A_e}{\hbar}-\frac{4\pi B_e}{\hbar}\right) \\ &= 4\pi\left(A_e - B_e\right)K,\end{aligned} \tag{12.33}$$

in which J is the magnitude of rotational angular momentum in the unit of $\hbar$. The vector **K** is the projection of **J** onto the symmetry axis and θ is the angle between these two vectors. Since θ is a constant the magnitude of $K\hbar$ is also a constant.

The two motions – rotation about the symmetry axis of the top and simultaneous precession about the vector **J** (Figure 12.8), produce the quantum beats in the transients of rotational coherence, and the beats are detected via the light radiated by the oscillating dipole in the excited state of the top. We once again see the operation of the concept that an oscillating dipole emits secondary radiation. If the dipole is oriented along the symmetry axis a then the radiation emitted by the dipole is modulated only by the precession motion at

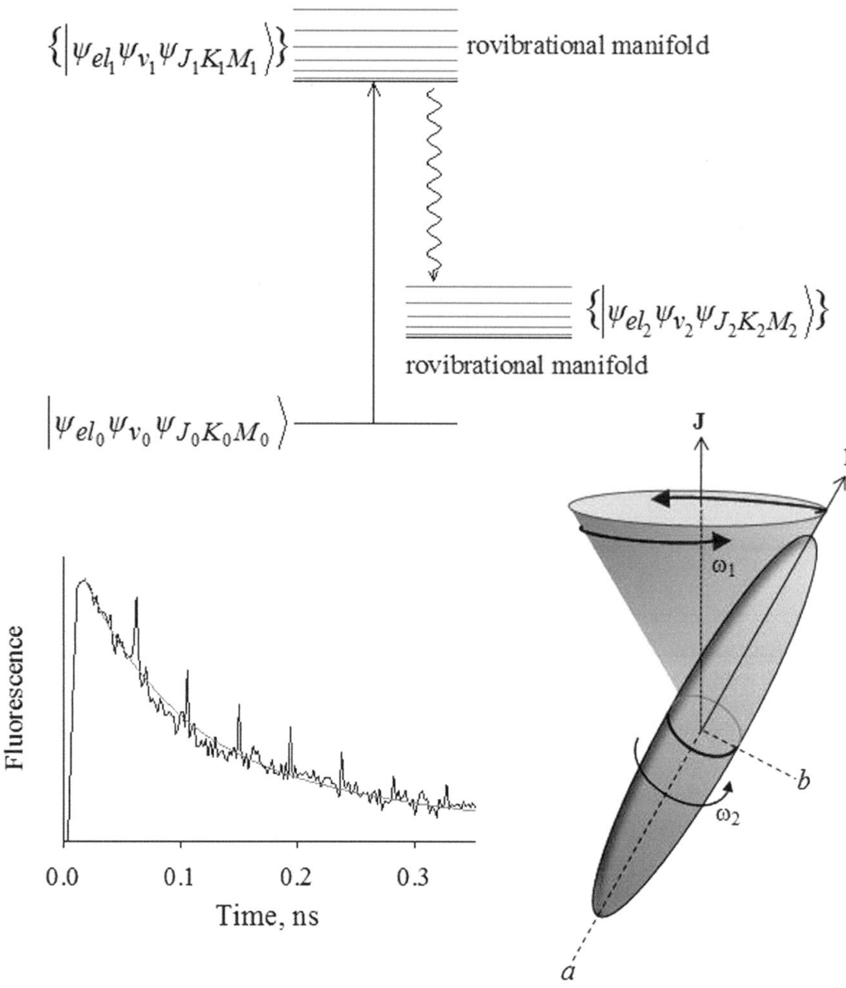

FIGURE 12.8 *Top*, generation and detection of rotational coherence in a typical pump-probe experiment. The labeled figure is self-explanatory. *Lower left*, fluorescence decay intervened by coherence beat signals that appear as spikes. The decay here appears multiexponential due to the synthetic nature of the data set; in reality, however, small molecules may show just one exponential at such a short timescale. *Lower right*, the precessional motion of a symmetric top includes its rotation about the symmetry axis at frequency ω_2 and the simultaneous rotation about the total rotational angular momentum vector **J** at the frequency ω_1. These two rotational motions produce rotational quantum beats.

both $\omega_1 = 4\pi B_e J$ and $2\omega_1 = 8\pi B_e J$. But if the dipole is along the *b*-axis, which would be perpendicular to the symmetry axis, then it also undergoes rotation in addition to the precession. In this case, only ω_2 needs to be considered and the dipole-radiated light is modulated at the frequency

$$2\omega_2 = 8\pi\left(A_e - B_e\right)K, \tag{12.34}$$

because a π-rotation of the dipole about the symmetry axis will yield its original position, even though only half a period of precession is completed. The model clarifies the rationale of using 'parallel' and 'perpendicular' configurations of the polarizer settings in the experiment. Both excitation and emission (or absorption) polarizers are parallel to each other irrespective of whether the transition is parallel, in which the dipole is parallel to the symmetry axis, or perpendicular, in which the dipole is perpendicular to the symmetry axis.

In the parallel transition, the molecular dipoles in the excited state rephase at the fundamental precession frequency

$$\omega_1 = 4\pi B_e J = 4\pi B_e', \tag{12.35}$$

where the quantum number J is an integer. This rephasing should happen for all the excited rotor molecules because the individual precession frequency is expected to be the fundamental frequency ω_1 or some integral multiple of it, $n\omega_1$, $n = 1, 2, 3, \cdots$. Because the quantum number K is an integer the rephasing of the dipoles in the perpendicular transition will occur at

$$\begin{aligned} 2\omega_2 &= 8\pi(A - B)K = 8\pi\left(A_e' - B_e'\right) \\ &= 8\pi A_e' - 8\pi B_e' \\ &= 8\pi A_e' - 2\omega_1, \end{aligned} \tag{12.36}$$

which means the total rephasing can occur only when ω_1 and ω_2 are comparable, the rephasing will be partial otherwise. We have mentioned above that a π-rotation of the perpendicularly oriented dipole will generate the original direction of the dipole only at half the precession period.

Each consecutive rephasing of the dipole gives rise to a beat which individually shows up in the fluorescence transient as a spike (Figure 12.8), and the spike recurs with great precision of time. This recurrence time period is the fundamental quantity one wishes to extract from pure rotational coherence spectra. Once the recurrence time is extracted, we can immediately derive the values of rotational constants A_e' and B_e'. However, detailed characterization of the direction of the transition dipole, the angle θ, and the molecular structure in the excited state needs studies of polarization-specific beat frequency and amplitudes.

Although the principle of rotational coherence analysis is straightforward if the molecule is a symmetric top, difficulties arise when the effect of asymmetry sets in. The assumption that asymmetric rotors can be approximated to a symmetric top if two of the three moment of inertia components can be equated may not always work. If not, a combination of experiments and simulation could be useful. Nonetheless, rotational coherence experiments measure only the excited-state structure and turns out to be a nice technique to probe the excited structures of symmetric tops.

12.8 DENSITY OPERATOR, COHERENCE, AND COHERENCE TRANSFER

In the discussion above the wave packet coherence and its dynamics have been presented in a rather rudimentary way. The dynamics of the state of a quantum mechanical system, or equivalently the equation of motion of the state, are more conveniently described by working with density operators, often in abstract space. The density operator approach is rather inescapable to obtain solution of quantum coherence. In the description of the state of a system it is useful to distinguish whether the system is 'pure or homogeneous' for which only one state vector $|\psi\rangle$ can describe the system, or an 'inhomogeneous statistical mixture' in which case the system is describable by not just one state but a mixture of them $|\psi_1\rangle$, $|\psi_2\rangle$, $|\psi_3\rangle$, $\cdots$, $|\psi_n\rangle$. An average operator called the density operator can represent both homogeneous and inhomogeneous states, but the density operator formalism turns out to be obligatory to analyze the latter. It is this formalism which can cohere a set of eigenstates in the eigenbase of the Hamiltonian to a coherent state, a wave packet for example, and is at the heart of all coherence spectroscopy, including NMR. Detailed treatment of the density operator and superoperator approach can be found in several papers and textbooks, including those of Fano (1957), Böhm (1979), and Ernst et al. (1988).

12.8.1 HOMOGENEOUS AND STATISTICAL MIXTURE OF STATES OF A SYSTEM

Some of the quantum mechanical ideas of the density operator, also known as the density matrix, can be understood by considering the classic example of a beam of photons which can be polarized and isolated in a pure homogeneous form as against an unpolarized beam where the polarization status of the photons may be considered inhomogeneous. These two situations can be created and analyzed experimentally. Let the photons from some light source traveling along the x-direction be polarized by an angle θ to the z-axis by placing a crystal or a polaroid film on the path of the beam. All photons entering through the polarizer will exit with the same polarization angle θ, so the system is homogeneous. If we did not use the polarizer the k photons could have any of θ_1, θ_2, θ_3, $\cdots$ angles of polarization, each with respect to the z-axis. At a given instance, a photon can have any one of these polarization angles, not all polarization angles occurring simultaneously. In this sense, the inhomogeneous photon mixture can be considered as a mixture of pure states that differ from each other by the angle θ. A measurement of the polarization will associate a probability for the occurrence of one angle, another probability for another angle, and so on. In other words, a heterogeneous mixture entails many pure states, and at a given instance the system is found in one of the pure states, not in all of them, and there is a probability associated with each of these pure states. The system is then said to contain statistical mixture of states and is hence inhomogeneous.

Note that it is the electric field vector **E** the direction of which determines the polarization. The θ-dependence of **E** in classical optics is given by

$$\mathbf{E}(\theta) = \mathbf{E}_o\left(\cos\theta\hat{z} + \sin\theta\hat{y}\right), \tag{12.37}$$

in which $\hat{z}$ and $\hat{y}$ are unit vectors. In quantum mechanics, the θ-polarized state of the photon is given as a linear superposition of the two orthogonal z- and y-polarization states

$$\mathbf{E} = \frac{1}{\sqrt{2}}(\mathbf{E}_z + \mathbf{E}_y).$$

The analysis of the photons of both polarized (homogeneous) and unpolarized (inhomogeneous) systems involves allowing the beams to pass through an analyzer polaroid whose axis is set along the z-axis, and then measuring the transmitted intensity by using a photon detector. The analyzer transmits a fraction of the beam ($= \cos^2\theta$) rejecting the $\sin^2\theta$ component. The intensity measured is commensurate with the field squared

$$I \sim \left|\mathbf{E}(\theta)\right|^2 = \mathbf{E}_o^{\,2}. \tag{12.38}$$

If the analyzer is oriented along the z-axis so as to detect z-polarization of the photons, the intensity detected will be

$$\left|\mathbf{E}(\theta)\cdot\hat{z}\right|^2 = \mathbf{E}_o^{\,2}\cos^2\theta. \tag{12.39}$$

Now consider the polarized photons (homogeneous system). Since all of them have the same polarization angle θ with respect to the z-axis, we can use the same state vector $|\theta\rangle$

for each photon. If the operator for the measurement of z-polarization of the photons is denoted by P_z, we get

$$P_z|z\rangle = 1|z\rangle$$

$$P_z|y\rangle = 0|y\rangle, \tag{12.40}$$

where the eigenkets $|z\rangle$ and $|y\rangle$ form an orthogonal basis set. The state vector for each photon can now be written in the form of a linear superposition

$$|\theta\rangle = a_z|z\rangle + a_y|y\rangle. \tag{12.41}$$

Then we calculate the expectation value of P_z to obtain

$$\langle P_z\rangle = \langle\theta|P_z|\theta\rangle = |a_z|^2. \tag{12.42}$$

The mean value $|a_z|^2 = \cos^2\theta$ implies that the normalized state vector ($\langle\theta|\theta\rangle = 1$) in the basis set $\{|z\rangle,|y\rangle\}$ is

$$|\theta\rangle = \cos\theta|z\rangle - \sin\theta|y\rangle. \tag{12.43}$$

Then we take up the analysis of the unpolarized photons represented by state vectors $|\theta_1\rangle, |\theta_2\rangle, |\theta_3\rangle, \cdots, |\theta_k\rangle$ with the corresponding probabilities $p_1, p_2, p_3, \cdots, p_k$ such that $\Sigma_k p_k = 1$. These set of vectors need not be orthogonal; however, each can be chosen normalized separately $(\langle\theta_1|\theta_1\rangle = 1, \langle\theta_2|\theta_2\rangle = 1, \cdots)$ in the basis $\{|z\rangle, |y\rangle\}$. This ensures that the procedures of measurement, analysis, and the calculation of the expectation value presented above for the general state vector $|\theta\rangle$ of the polarized photons can be identically carried out for each individual state vector $|\theta_k\rangle$ of the unpolarized photon beam. However, each individual state of the unpolarized system is a pure state associated with a probability, requiring weighting of each individual result by the corresponding probability factor and then summing over all the k results for the k number of states. If the probability of finding a_z in a measurement of z-polarization is $P(a_z)$ then the probability function with reference to each individual state vector $|\theta_k\rangle$ will have the form

$$P_k(a_z) = \langle\theta_k|P_z|\theta_k\rangle, \tag{12.44}$$

where $k = 1, 2, 3, \cdots$. By weighting $P_k(a_z)$ with the probability corresponding to the kth state vector p_k and then summing over k, one obtains

$$P(a_z) = \Sigma_k p_k P_k(a_z) = \langle P_z\rangle. \tag{12.45}$$

This is the result for finding the z-polarization of a statistical mixture of photon states contained in the unpolarized beam. Notice that $P(a_z) = \langle P_z\rangle$, because the analyzer gives only one result which is the z-polarization. In most cases, where the system is described by a statistical mixture of states, a measurement may give one of n number of possible results. In such cases, the general form of the expectation value of a property P is

$$\langle P\rangle = \Sigma_n a_n P(a_n), \tag{12.46}$$

and this is the expectation value for the property that will be obtained for the statistical mixture.

Here is the summary of comparison of pure and mixed states. The state function of a pure state is a linear superposition of the component states

$$|\psi\rangle = \Sigma_i \sqrt{\lambda_i}\,|i\rangle, \tag{12.47}$$

where λ_i represents normalization of the individual component state $|i\rangle$, but all components exist simultaneously. The occurrence of the component states does not involve probability. A mixed state is represented by a mixture of state functions $|\psi_j\rangle\, j = 1,2,3,\cdots$ such that the system can be present in the $|\psi_1\rangle$ pure state with a probability p_1, $|\psi_2\rangle$ pure state with a probability p_2, and so on, not that $|\psi_j\rangle$ occur simultaneously. Thus, a probability distribution p_i over the $|\psi_j\rangle$ states characterizes the mixed state.

12.8.2 Density Operator

The method of obtaining a result on a statistical mixture of states by a weighted sum of probabilities has no caveat as such. However, a more useful and advantageous approach deploys the density operator to represent the mixed states, and of course a pure state as well. Given a ket $|\psi\rangle$, the density operator ρ is constructed mathematically by placing the corresponding bra next to it, implying the product of the ket and the bra of the vector

$$\rho = |\psi\rangle\langle\psi|. \tag{12.48}$$

Since $\langle\psi|\psi\rangle = 1$, one can make out that

$$\rho^2 = |\psi\rangle\langle\psi|\,\psi\rangle\langle\psi| = |\psi\rangle 1\langle\psi| = |\psi\rangle\langle\psi| = \rho, \tag{12.49}$$

which means two successive actions of the density operator on an arbitrary ket vector is equivalent to acting only once. So ρ can be considered a geometric projection operator applicable to the case of a pure state. The projection of some arbitrary vector $|i\rangle$ is given by

$$P|i\rangle = |\psi\rangle\langle\psi|i\rangle = c|\psi\rangle, \tag{12.50}$$

showing that the action of the projector on the vector $|i\rangle$ gives $|\psi\rangle$ proportional to the scalar product $\langle\psi|i\rangle = c$. For example, the total angular momentum somewhere in space can be projected by using the projection operator onto an axis, say the molecular symmetry axis, implying that projectors are required for resolution of a spectral observable. We will see later in this section the use of projectors in detection of spectroscopic signals.

The projector can also be used to project a vector onto a subspace. For a set of normalized orthogonal vectors

$\{|\phi_i\rangle, i = 1, 2, \cdots, n\}$ that span a subspace $\mathcal{E}_n$ of space $\mathcal{E}$, a projector can be defined to have the form

$$P = \sum_{i=1}^{n} c_i |\phi_i\rangle\langle\phi_i|. \quad (12.51)$$

If $|\psi\rangle$ is another vector such that $|\psi\rangle \in \mathcal{E}$, then P will also act on $|\psi\rangle$ to produce a linear superposition of the projection of $|\psi\rangle$ onto each $|\phi_i\rangle$

$$P|\psi\rangle = \sum_{i=1}^{n} c_i |\phi_i\rangle\langle\phi_i|\psi\rangle = \sum_{i=1}^{n} c_i k_i |\phi_i\rangle, \quad (12.52)$$

which is equivalent to the projection of $|\psi\rangle$ onto $\mathcal{E}_n$.

An important property of the projection operator is the resolution identity, according to which the projection operators over all states in an orthogonal basis sum up to give the identity operator I

$$\sum_{i=1}^{n} |\phi_i\rangle\langle\phi_i| = I. \quad (12.53)$$

Now we look at the emergence of ρ from the more familiar characterization of a system by its state function. The system can be in the pure state and represented by a single state function. Let $|\psi(t)\rangle$ be the normalized state function which can be expanded in an orthonormal basis of the state space $\{|i\rangle, i = 1, 2, 3, \cdots, n\}$, where n is the dimension of the vector space

$$|\psi(t)\rangle = \sum_{i=1}^{n} c_i(t)|i\rangle$$

$$\sum_{i=1}^{n} |c_i(t)|^2 = 1, \quad (12.54)$$

ensuring that $|\psi(t)\rangle$ is normalized

$$\langle\psi(t)|\psi(t)\rangle = \sum_{i=1}^{n} |c_i(t)|^2 = 1. \quad (12.55)$$

The expectation value for some observable A can be written down easily, and we have

$$\langle A\rangle(t) = \langle\psi(t)|A|\psi(t)\rangle = \Sigma_i\Sigma_j c_i^*(t)c_j(t)A_{ij}, \quad (12.56)$$

where A_{ij} are the matrix elements of the observable A. Ordinarily, the expectation value of a physical observable associated with a Hermitian operator such as ρ is $\langle A\rangle = \langle\psi|A|\psi\rangle$ (see equation (12.42)). But the appearance of the coefficients $c_i(t)c_j^*(t)$ in equation (12.56) is due to the expansion of $|\psi(t)\rangle$ in the basis of the state space, and a comparison of equations (12.54) and (12.56) shows at once that

$$c_i^*(t)c_j(t) = \langle j|\psi(t)\rangle\langle\psi(t)|i\rangle = \langle j|\rho(t)|i\rangle = \rho_{ji}(t), \quad (12.57)$$

suggesting that the coefficients represent the matrix elements $\rho_{ji}(t)$ of the operator $|\psi(t)\rangle\langle\psi(t)|$. Indeed, the density operator acts in the state space, Hilbert space. So, one can see that in the state space basis $\{|i\rangle, i=1, 2, 3, \cdots, n\}$, the density operator $\rho(t)$ is always associated with the density matrix of elements $\rho_{ji}(t)$. It is for this reason that the density operator formalism is referred to as density matrix, meaning the observable which is a physical property is contained in the matrix.

Since $|\psi(t)\rangle$ is normalized (equation (12.54)), the sum over the diagonal elements of the density matrix $\rho_{ii}(t)$, which is the trace of the matrix, must also be equal to 1

$$\Sigma_i\rho_{ii}(t) = tr\{\rho(t)\} = 1. \quad (12.58)$$

For a pure state where $\rho^2 = \rho$, the trace of the density matrix satisfies

$$tr\{\rho(t)\} = tr\{\rho^2(t)\} = 1. \quad (12.59)$$

We should also look at the expectation value of an observable (equation (12.56)) in terms of the trace of the density matrix because all observable results are invariably expressed by the trace. By substituting for the coefficients $c_i(t)c_j^*(t)$ in the density matrix (see equation (12.57)) the expectation value can be obtained in terms of the trace

$$\begin{aligned}\langle A\rangle(t) &= \langle\psi(t)|A|\psi(t)\rangle = \Sigma_i\Sigma_j c_i^*(t)c_j(t)A_{ij}\\ &= \Sigma_i\Sigma_j\langle j|\rho(t)|i\rangle\langle i|A|j\rangle\\ &= \Sigma_j\langle j|\rho(t)A|j\rangle\\ &= tr\{\rho(t)A\} \quad (12.60)\end{aligned}$$

Density Operator in Pure and Mixed States. For a pure state that is described by the general state function $|\psi(t)\rangle$, the definition of the density operator is

$$\rho(t) = |\psi(t)\rangle\langle\psi(t)| = \Sigma_i\Sigma_j c_i(t)c_j^*(t)|i\rangle\langle j|. \quad (12.61)$$

In the mixed state, $\rho(t)$ can be understood as an ensemble average operator. Since the system is represented by different state functions $|\psi_k(t)\rangle$, each expandable in the orthonormal basis $\{|i\rangle, i=1, 2, 3, \cdots, n\}$, and each associated with the probability p_k of its representation of the system, an average density operator (Figure 12.9) is written as

$$\rho = \Sigma_k p_k\rho_k,$$

$$\begin{aligned}\rho(t) &= \Sigma_k p_k|\psi_k(t)\rangle\langle\psi_k(t)| = \Sigma_k p_k\Sigma_i\Sigma_j c_i^{(k)}(t)c_j^{(k)*}(t)|i\rangle\langle j|\\ &= \sum_i\sum_j \overline{c_i(t)c_j^*(t)}|i\rangle\langle j|. \quad (12.62)\end{aligned}$$

The second step is made to show the weighted sum of probabilities of the density matrix, where the overbar indicates that it is the ensemble average density matrix whose elements correspond to the values of the averaged eigenstate coefficients.

A list of the summary of the properties of the density operator follows.

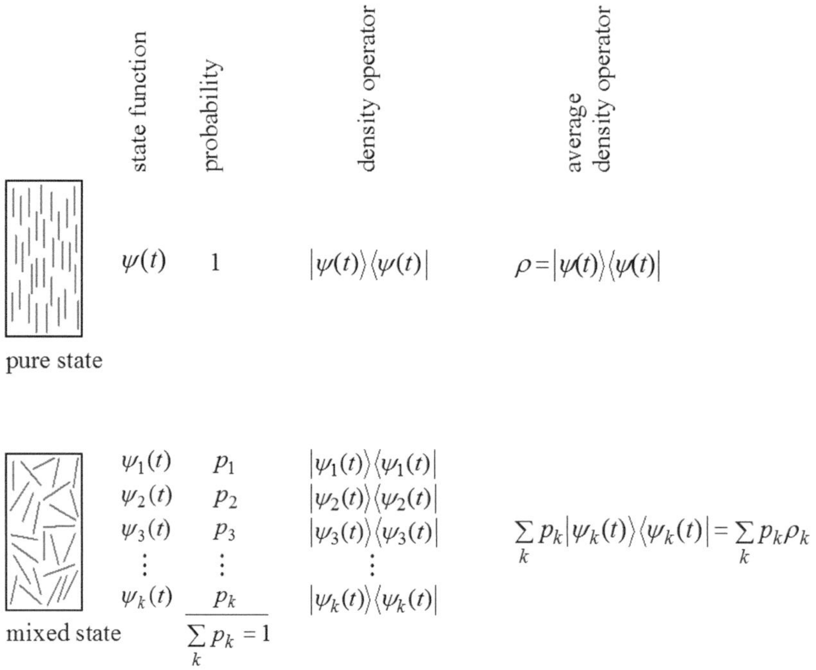

FIGURE 12.9 A pictorial depiction of the derivation of the average density operator for pure and mixed states.

1. The trace of the density operator is normalized, $tr\{\rho(t)\} = 1$. But $tr\{\rho^2(t)\} < 1$ for the mixed state, implying that the diagonal elements for the mixed-state matrix lie between 0 and 1.
2. The density operator is Hermitian $\langle i|\rho(t)|j\rangle = \langle j|\rho(t)|i\rangle^*$, and its time evolution is given by the time-dependent Schrödinger equation

$$i\hbar\frac{\mathrm{d}}{\mathrm{d}t}\rho(t) = \frac{\mathrm{d}}{\mathrm{d}t}\rho(t)|\psi(t)\rangle\langle\psi(t)|$$

$$\begin{aligned}\frac{\mathrm{d}}{\mathrm{d}t}\rho(t) &= \frac{1}{i\hbar}\left(\frac{\mathrm{d}}{\mathrm{d}t}|\psi(t)\rangle\right)\langle\psi(t)| + |\psi(t)\rangle\left(\frac{\mathrm{d}}{\mathrm{d}t}\langle\psi(t)|\right)\\ &= \frac{1}{i\hbar}\mathcal{H}(t)|\psi(t)\rangle\langle\psi(t)| + \frac{1}{(-i\hbar)}|\psi(t)\rangle\langle\psi(t)|\mathcal{H}(t)\\ &= \frac{1}{i\hbar}\left[\mathcal{H}(t),\rho(t)\right]\end{aligned} \tag{12.63}$$

Note that ρ $(= \Sigma_k p_k\rho_k)$ is the average of the individual density operators, so the same form of the time evolution is applicable to each ρ_k

$$i\hbar\frac{\mathrm{d}}{\mathrm{d}t}\rho_k(t) = \frac{1}{i\hbar}\left[\mathcal{H}(t),\rho_k(t)\right]. \tag{12.64}$$

3. The expectation value of the observable A (see equation (12.60)) is

$$\langle A\rangle(t) = tr\{\rho(t)A\},$$

and the time-evolution of the observable operator A is given by the same form given above

$$\frac{\mathrm{d}}{\mathrm{d}t}A(t) = \frac{1}{i\hbar}\left[\mathcal{H}(t),\rho_k(t)\right]. \tag{12.65}$$

4. The trace of A is indifferent to unitary transformation

$$tr\{A\} = tr\{U^{-1}AU\}, \tag{12.66}$$

where $U^{-1} = U^\dagger$.

Physical Meaning of the Density Operator. In the equation of motion for the density operator (equation (12.64))

$$\frac{\mathrm{d}}{\mathrm{d}t}\rho(t) = \frac{1}{i\hbar}\left[\mathcal{H}(t),\rho(t)\right],$$

the Hamiltonian in the commutator $\left[\mathcal{H}(t),\rho(t)\right]$ indeed represents a superoperator because it acts on another operator, and the commutator superoperator relationship is written as

$$[\mathcal{H},\rho] \equiv \hat{\hat{\mathcal{H}}}\rho = \mathcal{H}\rho - \rho\mathcal{H}, \tag{12.67}$$

where $\hat{\hat{\mathcal{H}}}$ is a superoperator because it acts on the density operator ρ, and the two operators commute. The corresponding matrix of a superoperator is of $n^2 \times n^2$ dimension. Since both $\mathcal{H}$ and ρ commute and are Hermitian, we say that an eigenbase $\{j\}$ of $\mathcal{H}$ is also the eigenbase of ρ, provided that both operators have only nondegenerate eigenvalues. In this eigenbase, a diagonal element of the density matrix is

$$\rho_{rr} = \langle r|\rho(t)|r\rangle = |c_r(t)|^2 = p_r, \tag{12.68}$$

where p_r is the probability that the system occupies the eigenstate $|r\rangle$. It can then be stated that the diagonal element $\rho_{rr} = p_r$

is the population of the eigenstate $|r\rangle$, and hence ρ_{rr} can be zero or a positive number. Whatever is the value, it does not change with time, $\rho_{rr}(t) =$ constant. For a system of mixed states, we have to express the population probability as an average probability

$$\rho_{rr} = \sum_k p_k \left|c_r^{(k)}(t)\right|^2 = \sum_k p_k \left(\rho_k\right)_{rr} = \overline{\left|c_r(t)\right|^2}. \quad (12.69)$$

An off-diagonal element of the density matrix in the same eigenbase $\{j\}$ is

$$\rho_{rs} = \langle r|\rho(t)|s\rangle = c_r(t)c_s^*(t), \quad (12.70)$$

which represents a coherent superposition $c_r(t)|r\rangle + c_s(t)|s\rangle$ of the eigenstates $|r\rangle$ and $|s\rangle$ in the state function $|\psi(t)\rangle$. For a system of mixed states, $c_r(t)c_s^*(t)$ averaged over all of $|\psi_k(t)\rangle$ functions is

$$\rho_{rs} = \sum_k p_k c_r^{(k)}(t)c_s^{(k)*}(t) = \overline{c_r(t)c_s^*(t)}. \quad (12.71)$$

Coherent superposition is often simply referred to as coherence, and it will be useful to mention the following characteristic properties of coherence.

1. A coherent state is a linear superposition of some eigenstates, and is not a probability-based mixing of these states. The linear superposition or coherence produces interference effects that arise from both magnitudes and relative phases (phase difference) of the coefficients of the superposed states.
2. The coherence in the density matrix appears as a complex number. The off-diagonal element ρ_{rs} represents the complex amplitude of the coherence corresponding to the density operator $|r\rangle\langle s|$, so that $\overline{c_r(t)c_s^*(t)}$ represents the sum of complex amplitudes of the coherence ρ_{rs} for all $|\psi_k(t)\rangle$ functions. It is possible that $\overline{c_r(t)c_s^*(t)} = 0$, which would mean that the interference effect in the coherence of eigenstates $|r\rangle$ and $|s\rangle$ averages out to zero, and hence there is no coherence. But this is not what we would like to have, because the off-diagonals need to exist for the observation of coherence.
3. A coherence oscillates with time at the Bohr frequency defined by the energy eigenvalues of the Hamiltonian. If the eigenstates $|r\rangle$ and $|s\rangle$ belong to the eigenbase $\{j\}$ of $\mathcal{H}$, which we will assume to be time-independent for the present, then

$$\mathcal{H}|r\rangle = E_r|r\rangle \text{ and } \mathcal{H}|s\rangle = E_s|s\rangle. \quad (12.72)$$

To look at the time evolution of the coherence we consider the equation of motion for the density operator (equation (12.63)) and simplify it to obtain the expression for $\rho_{rs}(t)$ as follows

$$i\hbar\frac{\mathrm{d}}{\mathrm{d}t}\rho_{rs}(t) = \left[\mathcal{H}, \rho_{rs}(t)\right] \quad (12.73)$$

$$= \mathcal{H}\rho_{rs}(t) - \rho_{rs}(t)\mathcal{H}$$

$$= \mathcal{H}\langle r|r\rangle\langle s|s\rangle - \langle r|r\rangle\langle s|s\rangle\mathcal{H}$$

$$= \langle r|\mathcal{H}|r\rangle\langle s|s\rangle - \langle r|r\rangle\langle s|\mathcal{H}|s\rangle$$

$$= E_r\langle r|r\rangle\langle s|s\rangle - E_s\langle r|r\rangle\langle s|s\rangle$$

$$= \rho_{rs}\left(E_r - E_s\right)$$

$$\rho_{rs}(t) = e^{-\frac{i}{\hbar}(E_r - E_s)t}\rho_{rs}(0). \quad (12.74)$$

This shows the oscillatory nature of both real and imaginary parts of the coherence.

4. A coherence can involve transitions between the superposed eigenstates in accordance with the selection rules. These four properties of eigenstate coherence provide the key to the area of coherence spectroscopy.

Reduction of the Density Operator. The density operator $\rho(t)$ of a system as such is expected to describe the quantum system globally including all possible sub-systems. The subsystems of an electron, for example, could be its spin and space variables, or separate degrees of freedom of a system may be identified with sub-systems. Most often, however, all the degrees of freedom or sub-systems are not needed for a measurement; for example, if the interest lies only in the rotation of a particle, the translational degrees of freedom need not be considered so that an appropriately reduced density operator with regard to the desired expectation value may be formulated. This possibility of reduction of a system comes as a boon because it spares the difficulty of working with the full density operator. The idea is to construct a reduced density operator $\sigma(t)$ for the variable of interest while dumping the others as lattice variables.

One can find $\sigma(t)$ with particular reference to the nuclear spin variables which are of interest in NMR spectroscopy (Ernst, 1957). We assume that the base functions $|\alpha\rangle$ of the global system are represented by product functions of the two sub-systems – nuclear spin functions $|s\rangle$, which depend only on the nuclear spin coordinates, and lattice functions $|f\rangle$, which depend on the lattice variables alone

$$|\alpha\rangle = |f\rangle|s\rangle = |fs\rangle. \quad (12.75)$$

Let the operator for the expectation value be Q. Since this operator acts only on the spin variables defined in a subspace of the whole space of the system and the observable is measured using the whole space, the expectation value is denoted by $\langle\tilde{Q}\rangle(t)$ instead of the usual designation $\langle Q\rangle(t)$. Let us evaluate $\langle\tilde{Q}\rangle(t)$ according to equation (12.60)

$$\langle\tilde{Q}\rangle(t)=\sum_{\alpha}\sum_{\alpha'}\langle\alpha|\rho(t)|\alpha'\rangle\langle\alpha'|Q|\alpha\rangle$$
$$=\sum_{s,s'}\sum_{f,f'}\langle sf|\rho(t)|f's'\rangle\langle s'f'|Q|sf\rangle.$$

The operator Q acts on the spin functions alone. If the lattice functions are orthonormal, then the matrix $\langle f'|Q|f\rangle=\delta_{ff'}$ is diagonal

$$\langle\tilde{Q}\rangle(t)=\sum_{s,s'}\sum_{f,f'}\langle sf|\rho(t)|f's'\rangle\langle s'|Q|s\rangle\delta_{ff'}. \quad (12.76)$$

We will now obtain the partial trace of the lattice variables, an operation that requires carrying out the summation over only one of the two pairs of variables s,s' and f,f'; one can choose any pair making the other redundant (see Cohen-Tannoudji et al., 1977). Furthermore, the distinction between s and s' or f and f' is not made. The presence of the diagonals $\delta_{ff'}$ in the form above (equation (12.76)) allows us to write

$$\langle\tilde{Q}\rangle(t)=\sum_{s,s'}\left[\sum_{f}\langle s'f|\rho(t)|fs\rangle\right]\langle s'|Q|s\rangle$$
$$=\sum_{s,s'}\langle s|\sigma(t)|s'\rangle\langle s'|Q|s\rangle$$
$$=\sum_{s}\langle s|\sigma(t)Q|s\rangle$$
$$=tr\{\sigma(t)Q\}, \quad (12.77)$$

where $\sigma(t)$ is the reduced density operator representing the spin functions. The second line of the above may also be expressed as

$$\Sigma_f\langle s'f|\rho(t)|fs\rangle-tr_f\{\rho(t)\}-\sigma(t), \quad (12.78)$$

in which tr_f denotes a partial trace (see Box 12.2) over the lattice variables.

It will be helpful to keep a definition in mind – in a system S consisting of two sub-systems a and b, the reduced density matrix for a, σ_a, is formed by taking the partial trace on b and vice versa

$$\sigma_a=tr_b\{S_{ab}\}$$
$$\sigma_b=tr_a\{S_{ab}\}. \quad (12.79)$$

Corresponding to the reduced density operator $\sigma(t)$ there also exists a reduced Hamiltonian $\mathcal{H}^s$, which, for the present case of nuclear spin, acts on the spin variables alone and is obtained by taking partial trace over the lattice variables

$$\mathcal{H}^s=\langle f|\mathcal{H}|f\rangle=tr_f\{\mathcal{H})\}. \quad (12.80)$$

It is this reduced $\mathcal{H}^s$ that will be the commutator $\left[\mathcal{H}^s,\sigma(t)\right]$ in the equation of motion for $\sigma(t)$.

12.8.3 Time Evolution of the Density Operator

The equation of motion for the reduced density operator $\sigma(t)$

$$i\hbar\frac{\mathrm{d}}{\mathrm{d}t}\sigma(t)=\left[\mathcal{H}(t),\sigma(t)\right], \quad (12.81)$$

generically written as

$$\frac{\mathrm{d}}{\mathrm{d}t}\sigma=\mathcal{L}(\sigma) \quad (12.82)$$

is called Liouville-von Neumann equation or master equation. Specifically, $\mathcal{L}(\sigma)$ is called the Liouvillian, perhaps due to the similarity of this equation with the Liouville equation of classical mechanics. The Hamiltonian in the equation is a superoperator designated by $\hat{\hat{\mathcal{H}}}$. In general, any superoperator $\hat{\hat{S}}$ can be represented in a basis set of operators $\{B_s; s=1,2,3,\cdots,n^2\}$ by

$$\hat{\hat{S}}=\Sigma_{jk}s_{jk}\hat{\hat{B}}_j^{\mathrm{L}}\hat{\hat{B}}_k^{\mathrm{R}}, \quad (12.83)$$

in which $\hat{\hat{B}}^{\mathrm{L}}$ and $\hat{\hat{B}}^{\mathrm{R}}$ are left- and right-translational superoperators (see below). The representation with another operator A is shown by

$$\hat{\hat{S}}A=\Sigma_{jk}s_{jk}B_jAB_k. \quad (12.84)$$

With regard to the superoperator $\hat{\hat{\mathcal{H}}}(t)$, the density operator equation of motion can be treated in two ways. First, if we take the commutator superoperator $\hat{\hat{\mathcal{H}}}$ as such, we have by definition

$$\hat{\hat{\mathcal{H}}}=\hat{\hat{\mathcal{H}}}^L-\hat{\hat{\mathcal{H}}}^R$$
$$\hat{\hat{\mathcal{H}}}^L\sigma=\mathcal{H}\sigma$$
$$\hat{\hat{\mathcal{H}}}^R\sigma=\sigma\mathcal{H}$$

$$[\mathcal{H},\sigma]\equiv\hat{\hat{\mathcal{H}}}\sigma=\mathcal{H}\sigma-\sigma\mathcal{H}=\hat{\hat{\mathcal{H}}}^L\sigma-\hat{\hat{\mathcal{H}}}^R\sigma, \quad (12.85)$$

in which $\hat{\hat{\mathcal{H}}}^L$ and $\hat{\hat{\mathcal{H}}}^R$ are left- and right-translation superoperators, the difference of which describes $\hat{\hat{\mathcal{H}}}$. The equation of motion can now be written as

$$i\hbar\frac{\mathrm{d}}{\mathrm{d}t}\sigma(t)=\hat{\hat{\mathcal{H}}}^L\sigma(t)-\hat{\hat{\mathcal{H}}}^R\sigma(t)=\mathcal{H}^{(')}\sigma(t)-\sigma(t)\mathcal{H}^{('')}. \quad (12.86)$$

Clearly, the Hamiltonian operators to the left and the right of $\sigma(t)$ are distinct. If we arbitrarily define two energy states, say for a vibronic transition, the Hamiltonians will be

$$\mathcal{H}^{(')}=T_{el}+V^{(')}\,\mathcal{H}^{('')}=T_{el}+V^{('')}$$

where the indices 'double-prime' and 'single prime' refer to ground and excited states, respectively, and the corresponding

Hamiltonians will describe the eigenvalue properties. The solution of equation (12.86) is

$$\sigma(t) = e^{-\frac{i}{\hbar}\mathcal{H}^{(')}(t-t_0)}\sigma(t_0)e^{\frac{i}{\hbar}\mathcal{H}^{('')}(t-t_0)}, \tag{12.87}$$

where $\sigma(t_0)$ is the initial density operator. This formally describes a quantum Liouville equation which shows mixed time-development of the density operator because the evolution of the operator from the right and the left are distinguished by the involvement of $\mathcal{H}^{('')}$ and $\mathcal{H}^{(')}$, respectively. Note that these Hamiltonians operate on the ground and excited state wavefunctions only according to their designations. The quantum Liouville equation is comparable to the time-dependent Schrödinger equation in the coordinate space, and can thus be used to solve time-dependent evolution of states, coherence for example. The time-development of the state then provides the time evolution of the density operator.

The second way to look at the Liouville equation relies on the supposition that a density operator will evolve from an earlier one to all later times t and the solution should have the form

$$\sigma(t) = \hat{\hat{R}}\sigma(0), \tag{12.88}$$

where $\hat{\hat{R}}$ is a unitary superoperator (see Cohen-Tannoudji et al., 1977), and $\sigma(0)$ is the density operator at $t = 0$. An important property of the master equation by virtue of its form is the 'divisibility', which means a solution desired from an initial time to some final time can be split into solutions involving intermediate times. A solution from $t = t$ to $t = \tau_n$ can be obtained by first solving from $t = t$ to $t = \tau_1$, then from $t = \tau_1$ to $t = \tau_2$, and then from $t = \tau_2$ to $t = \tau_3$, and so on. If the Hamiltonian in the master equation is considered time-independent in the time interval τ, then we have

$$\begin{aligned}\hat{\hat{R}}(\tau)\sigma(t) &= e^{-i\hat{\hat{\mathcal{H}}}\tau}\sigma(t)\\ &= e^{-\frac{i}{\hbar}\mathcal{H}\tau}\sigma(t)e^{\frac{i}{\hbar}\mathcal{H}\tau}\\ &= \sigma(t+\tau).\end{aligned} \tag{12.89}$$

If the density operator is to be transformed from $t = t$ to $t = \tau_n$, then by the divisibility property of the master equation the unitary transformation can be carried out in a time-ordered successive manner. The following is an example,

$$\begin{aligned}\hat{\hat{R}}_n\hat{\hat{R}}_{n-1}\cdots\hat{\hat{R}}_2\hat{\hat{R}}_1\sigma(t) &= e^{-i\hat{\hat{\mathcal{H}}}_n\tau_n}e^{-i\hat{\hat{\mathcal{H}}}_{n-1}\tau_{n-1}}\cdots e^{-i\hat{\hat{\mathcal{H}}}_2\tau_2}e^{-i\hat{\hat{\mathcal{H}}}_1\tau_1}\sigma(t)\\ &= e^{-\frac{i}{\hbar}\mathcal{H}_n\tau_n}e^{-\frac{i}{\hbar}\mathcal{H}_{n-1}\tau_{n-1}}\cdots e^{-\frac{i}{\hbar}\mathcal{H}_2\tau_2}e^{-\frac{i}{\hbar}\mathcal{H}_1\tau_1}\\ &\quad \sigma(t)e^{\frac{i}{\hbar}\mathcal{H}_1\tau_1}e^{\frac{i}{\hbar}\mathcal{H}_2\tau_2}\cdots e^{\frac{i}{\hbar}\mathcal{H}_{n-1}\tau_{n-1}}e^{\frac{i}{\hbar}\mathcal{H}_n\tau_n}\end{aligned}$$

which can be shown in schematic notation (Ernst et al., 1988) by

$$\begin{aligned}&\sigma(t)\xrightarrow{\mathcal{H}_1\tau_1}\sigma(t+\tau_1)\xrightarrow{\mathcal{H}_2\tau_2}\sigma(t+\tau_1+\tau_2)\cdots\cdots\\ &\xrightarrow{\mathcal{H}_n\tau_n}\sigma(t+\tau_1+\tau_2+\cdots+\tau_n).\end{aligned} \tag{12.90}$$

The divisibility of the density operator transformation thus implies that the effective coherence transfer during a process can be thought of as arising from a linear cascade of coherence transfers at intermediate times. The operators $e^{(i/\hbar)\mathcal{H}_k\tau_k}$ are referred to as propagators because the time evolution of the density operator is equivalent to its propagation according to the divisibility property.

12.8.4 Matrix Representation of the Unitary Transformation Superoperator

The matrix representation of the evolution of the density operator is particularly revealing with regard to coherence transfer and this is central to coherence spectroscopy. For the unitary transformation superoperator

$$\hat{\hat{R}}\sigma = R\sigma R^{-1}, \tag{12.91}$$

the matrix representation is obtained as the direct product of the matrix $\tilde{\mathbf{R}}$ and its complex conjugate $\tilde{\mathbf{R}}^*$

$$\tilde{\tilde{\mathbf{R}}} = (\tilde{\mathbf{R}})\otimes(\tilde{\mathbf{R}}^*). \tag{12.92}$$

Let us take the $\sigma(t)\xrightarrow{\mathcal{H}\tau}\sigma(t+\tau)$ transformation which represents equation (12.89) for the evolution of $\sigma(t)$ in the time interval τ

$$\sigma(t+\tau) = \hat{\hat{R}}(\tau)\sigma(t). \tag{12.93}$$

The matrix form of this evolution is

$$\tilde{\boldsymbol{\sigma}}(t+\tau) = \tilde{\tilde{\mathbf{R}}}(\tau)\tilde{\boldsymbol{\sigma}}(t) \rightarrow \sigma(t+\tau) = R(t)\sigma(t)R(t)^{-1}. \tag{12.94}$$

The $\hat{\hat{\mathrm{R}}}(\tau)$ superoperator matrix is of $n^2 \times n^2$ dimension. To be multipliable, we can take the $\tilde{\boldsymbol{\sigma}}(t)$ matrix as a $1\times n^2$ column vector and obtain $\tilde{\boldsymbol{\sigma}}(t+\tau)$ as a $1\times n^2$ column vector. If the matrix elements of the initial density operator are $\sigma_{rs}(t)$ and those of the transformed one are $\sigma_{tu}(t+\tau)$, then the matrix elements of the superoperator must be

$$R_{rs\,tu}(\tau) = R_{rt}(t)R_{us}^{-1}(t), \tag{12.95}$$

implying that the transformation of the initial matrix elements follows as

$$\sigma_{tu}(t+\tau) = \sum_{rs} R_{rs\,tu}(\tau)\sigma_{rs}(t). \tag{12.96}$$

At time t the transformation of the density matrix elements σ_{rs} to σ_{tu} in the time interval τ results in a coherence transfer under the influence of the superoperator

$$\hat{\hat{R}}(\tau)=e^{-i\hat{\hat{\mathcal{H}}}\tau}. \tag{12.97}$$

12.8.5 Matrix Representation of the Commutator Superoperator

To obtain the matrix representation of the commutator superoperator $\hat{\hat{S}}$, the relationship $\hat{\hat{S}}\sigma=S\sigma-\sigma S$ can be expressed using the unity operator E as

$$\hat{\hat{S}}\sigma=S\sigma E-E\sigma S$$

$$\tilde{\tilde{\mathbf{S}}}=\tilde{\mathbf{S}}\otimes\tilde{\mathbf{E}}-\tilde{\mathbf{E}}\otimes\tilde{\mathbf{S}}. \tag{12.98}$$

A simple example of a commutator superoperator matrix is that for the electron spin superoperator for which $\tilde{\mathrm{S}}$ is the Pauli matrix.

$$\hat{\hat{\mathbf{S}}}_x=\frac{\hbar}{2}\begin{pmatrix}0&1\\1&0\end{pmatrix}\otimes\begin{pmatrix}1&0\\0&1\end{pmatrix}-\begin{pmatrix}1&0\\0&1\end{pmatrix}\otimes\frac{\hbar}{2}\begin{pmatrix}0&1\\1&0\end{pmatrix}$$

$$=\frac{\hbar}{2}\begin{pmatrix}0&-1&1&0\\-1&0&0&1\\1&0&0&-1\\0&1&-1&0\end{pmatrix} \tag{12.99}$$

This result is identical to that for the superoperator matrix $\tilde{\tilde{\mathbf{I}}}_x$ for a spin-half nucleus found in Ernst, Bodenhausen and Wokaun (1988).

12.8.6 Partial Density Matrix

From a reductionist's viewpoint, all of the eigenstates of a state function need not be set up or prepared experimentally if the interest lies only in a limited set of outcomes of measurement of an observable. This reduction in fact holds for most of the cases; for example, if the interest lies in the measurement of a set of vibronic transitions, all vibrational levels in the ground electronic state need not be considered. One can then form a partial density matrix by projecting the initial density operator onto the manifold of eigenstates. The projection of $\sigma(t)$ is given by the projection operator.

Consider a pure state and suppose a measurement of an observable A on the state function $|\psi\rangle$ needs to be performed. Let the eigenbasis of the operator A be $\{|i\rangle; i=1,2,3,\cdots,n\}$ with the corresponding eigenvalue basis spectrum $\{f_i; i=1,2,3,\cdots,n\}$. If we want the discrete eigenvalue f_k or k as the outcome of the measurement, the corresponding eigenfunction $|k\rangle$ must be set up or experimentally prepared. The converse has the same meaning – if we obtained k as the outcome of a measurement, the state after the measurement will be $|k\rangle$. The probability of obtaining the eigenvalue k is

$$p_k=\langle\psi|k\rangle\langle k|\psi\rangle=\langle\psi|E_k|\psi\rangle=\mathrm{Tr}\{P_\psi E_k\}, \tag{12.100}$$

in which $P_\psi=|\psi\rangle\langle\psi|$ is the projector (density operator) for the input $|\psi\rangle$ function, that is,

$$P_\psi|k\rangle=|\psi\rangle\langle\psi|k=c|\psi\rangle, \tag{12.101}$$

where $c=\langle\psi|k\rangle$ can be a scalar quantity, but a complex number in general. The density operator for the state after the measurement of k is $P_k=|k\rangle\langle k|=\sigma\rho_k$, and

$$\sigma_k=P_k=|k\rangle\langle k|=\frac{E_kP_\psi E_k}{\mathrm{Tr}\{P_\psi E_k\}}. \tag{12.102}$$

To carry out the same measurement of A with k outcomes in a mixed state, the initial input density operator (see equation 12.62) is

$$\sigma_{in}=\Sigma_j p_j|\psi_j\rangle\langle\psi_j|=\Sigma_j p_j\sigma_j.$$

In the mixed-state case, the expression for the density operator, call it σ_f, after the measurement of eigenvalue k should be defined by considering all of the $|k\rangle$ eigenstates over the ensemble. This requires invoking conditional probability of each $|k\rangle$ in σ_f, and the expression for the reduced density operator works out to

$$\sigma_f=\frac{E_k\sigma_{in}E_k}{\mathrm{Tr}\{\sigma_{in}E_k\}}. \tag{12.103}$$

12.8.7 Density Operator Expression Using Irreducible Tensor Operator

A tensor operator is a tensor of operators. For example, a second-rank tensor operator is a tensor of nine operators. The Cartesian tensor components are mixed during a rotation transformation by a unitary operator, but the components that mix belong only to the asymmetric parts of the tensor, and these parts represent the irreducible parts of the tensor being rotated. To note is that irreducible tensors are most generally treated as spherical tensors in the angular momentum basis. By definition, an irreducible tensor operator T^k of rank k consists of a set of $2k+1$ operators T_q^k with $q=-k,-k+1,\cdots,k-1,k$; so T_q^k represents a basis of $2k+1$ components. Each T_q^k operator undergoes rotation transformation according to

$$\hat{\hat{R}}(\alpha,\beta,\gamma)T_q^k=R(\alpha,\beta,\gamma)T_q^kR^{-1}(\alpha,\beta,\gamma)=\sum_{q'}T_{q'}^kD_{q'q}^k(R), \tag{12.104}$$

where R is the rotation operator and α,β,γ are Euler angles. A unitary transformation essentially corresponds to a rotation of axes. Note the change of notation here, the Euler angles are denoted by α,β,γ to conform to the notation in the literature of tensor operator rotation. The matrix elements of R are the elements of the Wigner matrix $D_{q'q}^k(R)$ of order k, and they have to appear with respect to some basis. In 3D-rotation, $D_{q'q}^k(R)$ is the standard angular momentum basis in an irreducible operator subspace, implying that the components of the irreducible tensor operator T^k correspond to basis operators in the basis set of angular momentum. With respect to this basis, the indices k and q in the component operator T_q^k behave like the angular momentum quantum numbers l and m, the latter being a projection of the former. It is important to grasp that the irreducible tensor operators are not vectors, but integers representing the values of angular momentum quantum number. The reason for this is that they are operators corresponding to physical observables that must remain invariant to a rotation transformation.

The irreducible tensor operators have several useful properties, of which three are listed below.

1. The rotation of tensor operators is divisible in steps of the Euler angle rotation. In the case of a spin operator S, for example, a rotation about the z-axis by an angle γ followed by a rotation about x-axis by an angle β and then another rotation about the z-axis by an angle α can be written as

$$R(\alpha,\beta,\gamma) = e^{-i\alpha S_z} e^{-i\beta S_x} e^{-i\gamma S_z}$$

$$T_q^k \xrightarrow{\gamma S_z} \xrightarrow{\beta S_z} \xrightarrow{\alpha S_z} \sum_{q'} T_{q'}^k D_{q'q}^k(R) \qquad (12.105)$$

The form of time-evolution of irreducible tensor operators is similar to the divisible propagation of the density operator mentioned earlier (equation (12.90)). Suppose there are two eigenstates, and the basis states of the T_q^k components are the eigenstates of the molecular Hamiltonian $\mathcal{H}$, then

$$e^{\frac{-i\mathcal{H}t}{\hbar}} T_q^k e^{\frac{i\mathcal{H}t}{\hbar}} = \left(T_q^k e^{i\omega t}\right) e^{-\Gamma t}, \qquad (12.106)$$

where $\hbar\omega$ is the energy difference between the two eigenstates. The term $e^{-\Gamma t}$ is phenomenological in which Γ is a diagonal relaxation matrix whose elements describe the decay of the excited states.

2. Irreducible tensor operators under trace operation are orthogonal. If $T_{q_1}^{k_1}$ and $T_{q_2}^{k_2}$ are the operators then

$$Tr\left\{T_{q_1}^{k_1}\left(T_{q_2}^{k_2}\right)^{\dagger}\right\} = \delta_{k_1k_2}\,\delta_{q_1q_2}\,\delta_{12}. \qquad (12.107)$$

3. The product of two irreducible tensor operators can be expanded as a linear combination of the irreducible tensor operators with expansion coefficients that are Clebsch-Gordan coefficients. If $T_{q_1}^{k_1}$ and $T_{q_2}^{k_2}$ are the two individual tensor operators, then their product is

$$T_{q_1}^{k_1} T_{q_2}^{k_2} = \sum_{kq} \langle k_1k_2q_1q_2|kq\rangle T_q^k$$

$$T_q^k = \sum_{q_1q_2} T_{q_1}^{k_1} T_{q_2}^{k_2} \langle k_1k_2q_1q_2|kq\rangle, \qquad (12.108)$$

in which

$$\langle k_1k_2q_1q_2|kq\rangle = \langle kq|k_1k_2q_1q_2\rangle \qquad (12.109)$$

are Clebsch-Gordan coefficients. The coefficients are chosen to be real, whose values can be calculated by using the properties of spherical harmonics or by combining projection and ladder operators, but tables of these coefficients for various cases are found in literature (see, for example, Rose, 1957). An analytical expression of Clebsch-Gordan coefficients was given by Wigner, who introduced the 3-j symbols that convey more symmetric properties and are more useful to describe density operators. The coefficients are also discussed in the book by Margenau and Murphy (1964). Thus the integral in the right-hand side of equation (12.108) is written as

$$\langle k_1k_2q_1q_2|kq\rangle = (-1)^{k_2-k_1-q}\sqrt{2k+1}\begin{pmatrix} k_1 & k_2 & k \\ q_1 & q-q_1 & -q \end{pmatrix}, \qquad (12.110)$$

where $q = q_1 + q_2$ and $|k_1 - k_2| \le k \le k_1 + k_2$.

12.9 DENSITY MATRIX TREATMENT OF AN OPTICAL EXPERIMENT

In this section we briefly review the prototype of the density operator description of a typical optical experiment, the details of such description of NMR experiments will be given in detail in Chapter 13. The discussion here is elementary and should be taken as a guideline only. The formalism is similar to the density operator treatment of a quantum beat experiment originally given by Silverman, Haroche, and Gross (1978) using an atomic system, and Felker and Zewail (1987) for moderately large molecules.

Consider a standard pump-probe experiment in which an energy level of the ground electronic state $|g\rangle$ is excited by polarized light to the manifold of an upper electronic state $|e\rangle$, and the decay of the excited state to some final state $|f\rangle$ is observed by polarized fluorescence light (Figure 12.10). Going by the customary labels, the polarization of the excitation light is indicated by the polarization vector $\hat{e}_1$ and that of the fluorescence light by the polarization vector $\hat{e}_2$. Associated with the $|g\rangle$, $|e\rangle$, and $|f\rangle$ states are the respective projection operators that project onto these states. These are listed in Table 12.1.

Since the observable is a polarized fluorescence light, the intensity I of the fluorescence detected is given according to the property of the density operator (equation (12.60)). We write

$$I(\hat{e}_1,\hat{e}_2) = K\ Tr\left\{\sigma\theta^{\dagger}(\hat{e}_1,\hat{e}_2)\right\}, \qquad (12.111)$$

where $I(\hat{e}_1,\hat{e}_2)$ is the fluorescence intensity that has dependence on the polarization of both the excitation and the emitted

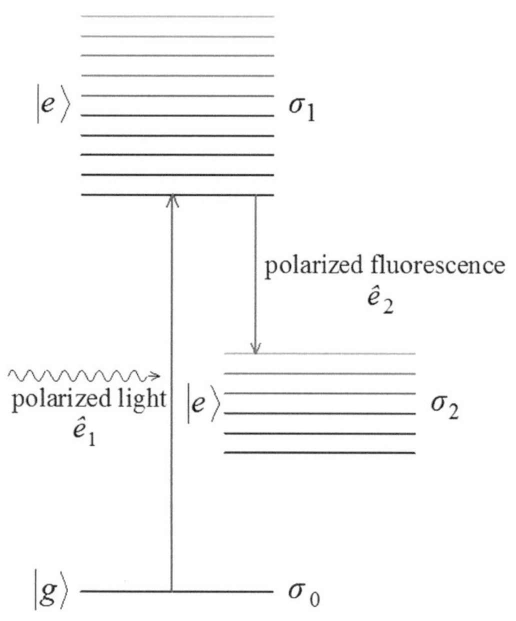

FIGURE 12.10 The excitation of the ground state to an excited state by polarized light, and the relaxation of the excited to another non-fluorescent ground state such that the emitted fluorescence is polarized.

TABLE 12.1
Partial density matrix and projection operators corresponding to the ground, excited, and final states in a standard pump-probe experiment.

Eigenstate	Partial density matrix	Projection operator
$\lvert g\rangle$	σ_g	P_g
$\lvert e\rangle$	σ_e	P_e
$\lvert f\rangle$	σ_f	P_f

light, σ is the density operator, and θ is the observable or probe operator. The constant K is introduced to account for the geometry of detection that may incorporate factors like the distance between the sample and the detector because the emitted electric field is a function of the fluorescence-emitting element and the detector. Except for the geometry of detection of light scattered by a medium, K is normally scaled to 1, and hence need not be carried forward. Because the observable is specific to the eigenstates $\lvert e\rangle$ and $\lvert f\rangle$, the former decaying to the latter by emitting fluorescence, the observable operator θ can be written as

$$\theta = P_e\left(\hat{e}_2\cdot\mathbf{D}\right)P_f\left(\hat{e}_2\cdot\mathbf{D}\right)^* P_e, \tag{12.112}$$

where $\mathbf{D}$ is the electric dipole operator. Inclusion of the dipole in this form yields a general form of the operator which allows for any orientation of the absorption and emission dipoles and any geometry of the excitation and emission polarizers. Specific dipole orientations and polarization geometries can be considered if transition characteristics are to be determined according to the selection rules specific to dipole orientations (Felker and Zewail, 1987).

The form of the observable operator θ suggests that the density matrix σ in equation (12.111) should be changed with the partial density matrix σ_e associated with the state $\lvert e\rangle$. But σ_e will clearly depend on the ground-state partial density matrix $\lvert\sigma_g\rangle$ because it is the excitation of $\lvert g\rangle$ with $\hat{e}_1$ polarization that prepares the state $\lvert e\rangle$. By using the projector P_g onto the ground-state density operator one obtains σ_e^+, which is called the instantaneous density operator after the excitation light pulse

$$\sigma_e^+\left(t\to 0,\hat{e}_1\right) = P_e\left(\hat{e}_1\cdot\mathbf{D}\right)P_g\sigma_g P_g\left(\hat{e}_2\cdot\mathbf{D}\right)^* P_e. \tag{12.113}$$

We may think of σ_e^+ as time-independent σ_e, or equivalently σ_e at time $t = 0$, which is still not under the influence of the molecular Hamiltonian. At a finite interval later the time-evolution of σ_e sets in, and the Hamiltonian acts upon the eigenstates of the $\lvert e\rangle$ manifold which belong to the eigenbase of the Hamiltonian. The time-evolution of $\sigma_e(t)$ projected onto the eigenbase manifold is

$$\sigma_e(t) = \left(e^{-\frac{i}{\hbar}\mathcal{H}t}\sigma_e^+\left(t\to 0,\hat{e}_1\right)e^{\frac{i}{\hbar}\mathcal{H}t}\right)e^{-\Gamma_e t} \tag{12.114}$$

where $e^{-\Gamma_e t}$, which is not a part of the calculation, describes the relaxation of the excited state $\lvert e\rangle$. Substitution for θ and σ from equations (12.112) and (12.114) produces the signal in the final form

$$I\left(t,\hat{e}_1,\hat{e}_2\right) = Tr\left\{\begin{matrix} e^{-\frac{i}{\hbar}\mathcal{H}t}\sigma_e^+\left(t\to 0,\hat{e}_1\right) \\ e^{\frac{i}{\hbar}\mathcal{H}t}P_e\left(\hat{e}_2\cdot\mathbf{D}\right)P_f\left(\hat{e}_2\cdot\mathbf{D}\right)^* P_e \end{matrix}\right\}e^{-\Gamma_e t}. \tag{12.115}$$

This description is a rather conventional outline of the density operator formalism of an optical signal, but makes clear that the operator description of the excited state is indispensable. The effect of the geometry of excitation and coherence transfer both are embodied in the manifold of the excited state, as a result of which the detected signal is modulated with time. A convenient approach is to expand all the operators in a base of irreducible tensor operators which are integer values of angular momentum quantum number and whose rotational transformations are simple enough to yield the final signal as a sum of a few terms. Let $\lvert J''M''\rangle^{(e)}$ and $\lvert J'M'\rangle^{(e)}$ be two states in the excited manifold which superpose to produce a coherence. The irreducible tensor operator representing σ_e^+ will be the product of the component irreducible tensor operators. The base consists of the tensor operators for $\lvert J''M''\rangle^{(e)}$ and $\lvert J'M'\rangle^{(e)}$. Using equations (12.108) and (12.110) we have

$$\sigma_e^+ = \sum_{J'^{(e)}J''^{(e)}}\sum_{k^{(e)}q^{(e)}}\alpha_{k^{(e)}q^{(e)}}^{J'^{(e)}J''^{(e)}}\times{}^{J'^{(e)}J''^{(e)}}T_{q^{(e)}}^{k^{(e)}}$$

where

$$^{J'(e)J''(e)}T^{k^{(e)}}_{q^{(e)}} = \sum_{M'M''}(-1)^{J^{(e)}-M^{(e)}}\sqrt{2k^{(e)}+1} \times \begin{pmatrix} J'^{(e)} & k'^{(e)} & J''^{(e)} \\ -M'^{(e)} & q^{(e)} & M''^{(e)} \end{pmatrix} |J'M'\rangle\langle J''M''|$$

and the coefficients α are

$$\alpha^{J'^{(e)}J''^{(e)}}_{k^{(e)}q^{(e)}} = Tr\left\{\sigma_e^+ ({}^{J'^{(e)}J''^{(e)}}T^{k^{(e)}}_{q^{(e)}})^\dagger\right\} \qquad (12.116)$$

The index 'e' within the small bracket is included to indicate that the expansion is specific to the excited state $|e\rangle$. The other operators σ_e, θ, P_e, and P_f are also expanded in a similar manner. For example, for the single rovibrational level in the ground state $|g\rangle$ from which excitation occurs one can write for the zero-time density operator as

$$\sigma_0 = \frac{1}{\sqrt{2k^{(e)}+1}} T_0^0, \qquad (12.117)$$

with T_0^0 being a zero-order irreducible tensor operator – a scalar.

The operator θ assumes the form

$$\theta = \Sigma \beta^{J'J''}_{kq} \times {}^{J'J''}T^k_q, \qquad (12.118)$$

in which the notations and indices used appear untidy, but a closer look makes them easy to use. These expansions can be simplified to obtain the signal in the final form as

$$I\left(t, \hat{e}_1, \hat{e}_2\right) \propto \text{coefficients } e^{-i\omega t}\left(e^{-\Gamma t}\right). \qquad (12.119)$$

The sum of the coefficients describe all optical transitions, the frequency ω corresponds to the energy difference between states $|J'M'\rangle$ and $|J''M''\rangle$, and the added parameter Γ is the decay constant.

Box 12.1 Fourier analysis of spectra

A molecular spectrum is obtained by analyzing some property of the electromagnetic field rather than the molecular impact itself. The interaction with the primary (source) electromagnetic field causes the molecule to radiate a secondary field or induce an alternating current in a receiver coil placed near the sample. The secondary effect is measured and analyzed by the spectrometer to produce a spectrum. We will call this secondary field $E(t)$ which is related to the frequency spectrum $E(\omega)$ by Fourier transformation.

A fundamental property of linear response theory that shows that the impulse response $E(t)$ and the frequency response $E(\omega)$ functions form a Fourier pair is that the $E(t)$ function is a linear superposition of basis functions

$$E(t) = \Sigma_i x_i(t) \quad \text{for} \quad i = 1, 2, 3, \cdots$$

and there exists a system operator Φ such that

$$\Phi\{\Sigma_i x_i(t)\} = \Phi\{x_1(t)\} + \Phi\{x_2(t)\} + \Phi\{x_3(t)\} + \cdots \qquad (12.B1.1)$$

To understand how Φ works let us consider the following arrangement (Figure 12.11) of a minimal spectrometer

In general, the amplitude detector will produce an amplitude spectrum of the secondary field coming out of the monochromator. The monochromator is the analogue of the operator Φ of equation (12.B1.1). This makes sense because the total electromagnetic field $E(t)$ emitted by the sample is an impulse response and is a superposition of the basis harmonic waves. The detector provides the response function, which is essentially the spectral amplitude as a function of frequency, $a(\omega)$. We see that the detector is an amplitude detector that responds according to the content of the frequency components of the electromagnetic field. If $E(t)_{\text{out}}$ is a monochromatic field of frequency ω_i then a sharp line of magnitude a_i will be produced at $\omega = \omega_i$. This line mathematically represents delta function, $a_i\delta(\omega - \omega_i)$, and the integral of the line from $-\infty$ to $+\infty$ of frequencies is proportional to the magnitude of a_i

$$a_i \propto \int_{-\infty}^{\infty} a_i \delta(\omega - \omega_i) d\omega$$

These considerations tell us that the frequency amplitude of the emitted (secondary) electromagnetic field $E(\omega)_{\text{out}}$ can be calculated by

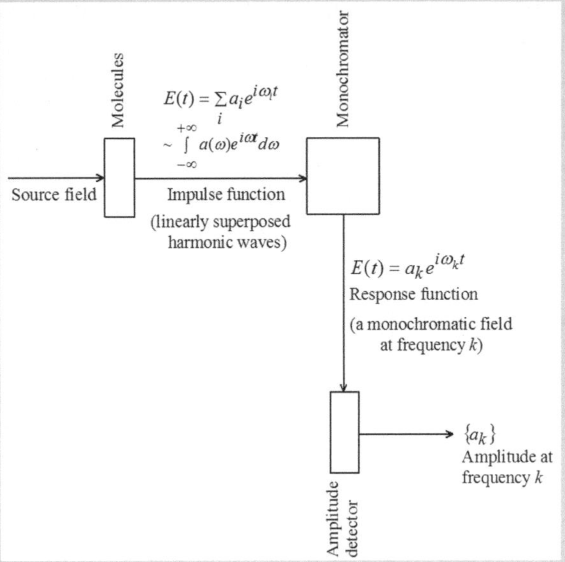

FIGURE 12.11 The backbone of a spectrometer.

$$E(\omega)_{\text{out}} = \frac{1}{\sqrt{2\pi}}\int_{-\infty}^{\infty} E(t)_{\text{in}}\, e^{-i\omega t}dt = \sqrt{2\pi}\sum_i a_i \delta(\omega-\omega_i)$$

The use of the front factor $1/\sqrt{2\pi}$ is a matter of convention having the purpose of maintaining a symmetry because the delta function is often represented analytically by a function that includes the $1/2\pi$ factor

$$\delta(\omega-\omega_i) = \frac{1}{2\pi}\int_{-\infty}^{\infty} e^{i(\omega-\omega_i)t}dt \qquad (12.B1.2)$$

To see this development further we recall the impulse function $E(t)$

$$E(t) = \Sigma_i a_i e^{i\omega_i t} \qquad (12.B1.3)$$

and convert it into a delta function. Note that the representation of the input function $E(t)$ in the form of a delta function has a special significance in Fourier spectroscopy. The delta function form of $E(t) = \sum_i a_i e^{i\omega_i t}$ is simply

$$E(t)e^{-i\omega t} = \Sigma_i a_i e^{i(\omega_i-\omega)t} \qquad (12.B1.4)$$

which can be written (see below) in the integral form

$$E(t) = \int_{-\infty}^{\infty} a_i(\omega_i) e^{i\omega_i t} d\omega_i \qquad (12.B1.5)$$

where a_i is set as equivalent to $a(\omega)d\omega$, which is the amplitude and the Σ sign is replaced by the integral. By integrating the delta function of $E(t)$ we obtain

$$\int_{-\infty}^{\infty} E(t)e^{-i\omega t}dt = \Sigma_i a_i \int_{-\infty}^{\infty} e^{i(\omega_i-\omega)t}dt \to 2\pi\Sigma_i a_i \delta(\omega_i-\omega)$$

where we have used the analytical form of the delta function (equation (12.B1.2)). This is the recipe to transform $E(t) = \sum_i a_i e^{i\omega_i t}$ to what we would measure finally – the amplitude.

If the field $E(t)$ is a monochromatic field $E(t) = ae^{-i\omega_0 t}$ where ω_0 is the single frequency, the frequency spectrum will be simply

$$E(\omega) = 2\pi a\delta(\omega-\omega_0)$$

which would appear as a sharp and narrow absorption-mode line at $\omega = \omega_0$. The intensity of this spectral component will be

$$2\pi|a|^2 = E(\omega)E^*(\omega) = |E(\omega)|^2 \qquad (12.B1.6)$$

that shows that the coefficient a defines the spectrum. The formula also shows that the signal-processing detector is a 'square-law detector' because the intensity of a spectral component is related to the square of the field $E(\omega)$.

In general, the spectrum of a field $E(t)$, which is a linear superposition of a set of basis monochromatic waves is defined by a set of discrete coefficients a_i, one a corresponding to the frequency of an individual monochromatic component. As the number of monochromatic waves in the field increases, or equivalently the size of the basis set becomes larger, the field becomes continuous. For a continuous field $E(t)$ the coefficients a_i take the amplitude form $a(\omega)d\omega$ which is the amplitude of a monochromatic wave whose frequency lies between ω and $\omega + d\omega$. This explains how we got the equation (12.B1.5) from (12.B1.4)

$$\Sigma_i a_i e^{i\omega t} = \int_{-\infty}^{\infty} a(\omega)e^{i\omega t}d\omega$$

Explicitly, a field $E(t)$ is defined by the integral

$$E(t) = \int_{-\infty}^{\infty} \frac{1}{\sqrt{2\pi}} E(\omega)e^{i\omega t}d\omega \qquad (12.B1.7)$$

and corresponding amplitude spectrum is

$$E(\omega) = \frac{1}{\sqrt{2\pi}}\int_{-\infty}^{\infty} E(t)e^{-i\omega t}dt \qquad (12.B1.8)$$

The spectral amplitude $E(\omega)$ is a Fourier tranform (FT) of the field $E(t)$ and the latter is a FT of the former. The existence of this inverse transformartion can be verified by substituting for the $E(t)$ integral in the $E(\omega)$ integral

$$E(\omega) = \frac{1}{2\pi}\int_{-\infty}^{\infty} e^{-i\omega t}dt\left\{\int_{-\infty}^{\infty} E(\omega')e^{i\omega' t}d\omega'\right\}$$

$$= \frac{1}{2\pi}\int_{-\infty}^{\infty} E(\omega')d\omega'\int_{-\infty}^{\infty} e^{i(\omega'-\omega)t}dt$$

$$= \int_{-\infty}^{\infty} E(\omega')d\omega'\delta(\omega'-\omega) = E(\omega)$$

In the last line above we have used the analytical representation of the delta function (equation (12.B1.2)). Shown in Figure 12.12 is the graphical portrayal of the $E(t)$ and $E(\omega)$ functions

In a typical spectral measurement, the $E(t)$ field is measured as a periodic time-varying function that looks like

$$E(t) = \cos(\omega_0 t) = \frac{1}{2}A(t)e^{i\omega_0 t} + \frac{1}{2}A(t)e^{-i\omega_0 t}$$

The $A(t)$ function is called envelope function which is an exponential most often. By comparing with equation (12.B1.7) the amplitude spectrum can be written as

$$E(\omega) = \frac{1}{\sqrt{2\pi}}\int_{-\infty}^{\infty} E(t)e^{i\omega t}dt = \frac{1}{\sqrt{2\pi}}\int_{-\infty}^{\infty} A(t)\cos(\omega_0 t)e^{i\omega t}dt$$

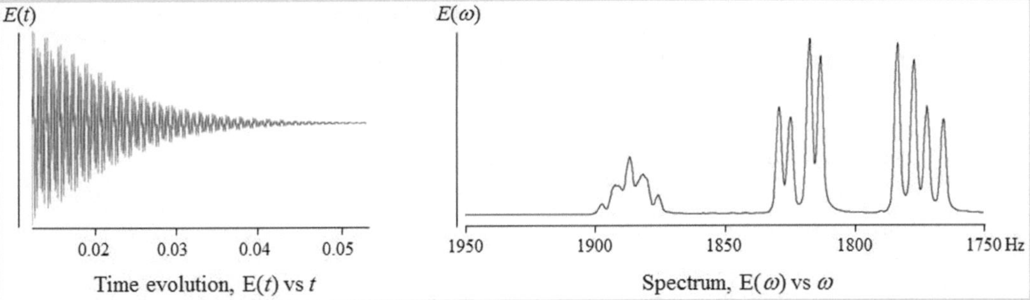

FIGURE 12.12 The $E(t)$ spectrum given by the $E(\omega)$ spectrum.

$$= \frac{1}{\sqrt{2\pi}} \left\{ \frac{1}{2} \int_{-\infty}^{\infty} A(t)\, e^{i(\omega_0+\omega)t} dt + \frac{1}{2} \int_{-\infty}^{\infty} A(t) e^{-i(\omega_0-\omega)t} dt \right\}$$

$$= \frac{1}{2} A(\omega+\omega_0) - \frac{1}{2} A(\omega-\omega_0). \qquad (12.B1.9)$$

Notice that $A(\omega)$ is the Fourier transform of the envelope function $A(t)$. As mentioned earlier (equation (12.B1.6)) the square of the amplitude spectrum gives the energy spectrum or the intensity distribution of frequency

$$I(\omega) = \frac{E(\omega)E^*(\omega)c}{8\pi} = \frac{c}{8\pi}\left|E(\omega)\right|^2$$

Note that $E^*(\omega) = E(-\omega)$, and it is the photomultiplier tube (PMT) of a spectrometer that measures $I(\omega)$.

TIME ANALOG OF SPECTRA

In Figure 12.13 below we take a HCl-like rotor which is set to a rotational motion starting at $t = 0$. This rotation is an impulse response that gives the time evolution of the dipole. The dipole changes direction periodically and would appear as an oscillating dipole to an observer who is not riding with the rotor frame. The oscillating dipole radiates a field – say $\mu \cos\omega_j t$, a cosine wave at frequency ω_j, but there will be phase shifts $\omega_j + \phi$ of different magnitudes of ϕ for different molecules. This is the way to think about rotational spectrum in time domain. If we consider many J levels in the sample we will also have many frequencies, not just ω_j, and the whole bunch of waves would look like a damped cosine wave

$$E(t) = A(t)\,\cos\omega_0 t \qquad (12.B1.10)$$

where $A(t)$ is an envelope function which can be an exponential or a Gaussian. For a rectangular envelope $A(t) = 1$. We understand that a basic calculation, the motion of a wave in vacuum for example, will not provide us with an envelope function. But it can nevertheless be included as a phenomenological damping function because of the necessity to incorporate a lifetime to a state.

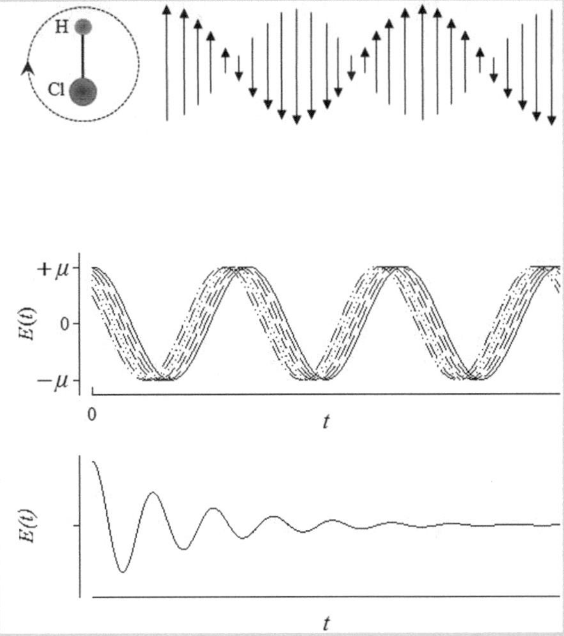

FIGURE 12.13 Oscillation of a rotator dipole (μ) with time, generating impulse response (*top*). The dipole of the jth molecule radiates a field $\mu\cos(\omega_j t)$, but a phase shift exists amongst different molecules. The phase shift θ is increased by 10° in the simulation shown (*middle*). The occurrence of the phase shift for different molecules results in a damped cosine wave (*bottom*).

The envelope function $A(t)$ in equation (12.B1.10) above can be taken as an exponential that is applicable to most cases in spectroscopy. We write

$$A(t) = e^{-\Gamma t} \text{ for } t \geq 0 \qquad (12.B1.11)$$

and state that the system responds to an impulse by generating a field

$$E(t) = e^{-\Gamma t}\,\cos\omega_0 t \qquad (12.B1.12)$$

where Γ is the damping rate constant. This $E(t)$ field is also the field generated by a harmonic oscillator (see below) and a two-level system in general.

The task now is to generate a Fourier transform of $E(t)$ as given in the equation (12.B1.12) above. It can be done easily by first calculating the Fourier transform of $A(t) = e^{-\Gamma t}$

$$A(\omega) = \frac{1}{\sqrt{2\pi}} \int_0^\infty e^{(-\Gamma + i\omega)t} dt = \frac{1/\sqrt{2\pi}}{i\omega - \Gamma}$$

and then obtaining

$$E(\omega) = \frac{1}{2}\left\{A(\omega + \omega_0) + A(\omega - \omega_0)\right\}$$

so that

$$E(\omega) = \frac{1}{\sqrt{8\pi}}\left\{\frac{1}{i(\omega + \omega_0) - \Gamma} + \frac{1}{i(\omega - \omega_0) - \Gamma}\right\}. \quad (12.B1.13)$$

The front factor $1/\sqrt{8\pi}$ comes from $1/2\sqrt{2\pi}$. Having obtained the expression for $E(\omega)$, we can just write down the equation for the energy spectrum

$$I(\omega) = E(\omega)E^*(\omega) = \frac{1}{8\pi}\frac{1}{(\omega - \omega_0)^2 + \Gamma^2}, \quad (12.B1.14)$$

which is a Lorentzian spectral line characterized by FWHM $= 2\Gamma$.

In another example we may like to take up the motion of an oscillator. Under the impulse the oscillator generates a field

$$E_v(t) = A_v(t)\cos\omega_0 t \quad (12.B1.15)$$

with the oscillator envelope function $A_v(t)$. Regarding the envelope function we may like to consider the following two cases (Figure 12.14).

1. EXPONENTIAL TIME ENVELOPE

For this case the impulse function

$$E_v(t) = e^{-\Gamma t}\cos\omega_0 t$$

is identical to the one discussed with reference to equation (12.B1.12). We should note that depending on the frequency of oscillation ω, the harmonic oscillator can be overdamped $(\omega^2 < \Gamma)$ or underdamped $(\omega^2 > \Gamma)$. The situation here is underdamped, and the Fourier transform is identical to the one obtained above for the rotational motion. The transform is

$$E_\omega(t) = \frac{1}{\sqrt{8\pi}}\left\{\left[\frac{1}{i(\omega + \omega_0) - \Gamma} + \frac{1}{i(\omega - \omega_0) - \Gamma}\right]\right\}$$

2. GAUSSIAN TIME ENVELOPE

The impulse function is

$$E_v(t) = e^{-\left(t^2/a^2\right)}\cos\omega_0 t \quad (12.B1.16)$$

First we calculate the Fourier transform of $A(t) = e^{-\left(t^2/a^2\right)}$.

$$\frac{1}{\sqrt{2\pi}}\int_0^\infty e^{-(t^2/a^2)i\omega t} dt = \frac{1}{\sqrt{2\pi}} e^{\frac{-a\omega^2}{4}} \int_{t=-\infty}^{t=\infty} e^{-1/a^2\left(t - i\omega a^2/2\right)} dt$$

$$= \frac{a}{\sqrt{2\pi}} e^{\frac{-a\omega^2}{4}} \int_{t=-\infty}^{t=\infty} e^{-u^2} du$$

$$A(\omega) = \frac{a}{\sqrt{2\pi}} e^{\frac{-a\omega^2}{4}} \quad (12.B1.17)$$

Notice that the FT of a Gaussian is a Gaussian. The spectral amplitude is

$$E(\omega) = \frac{a}{\sqrt{8\pi}}\left[e^{\frac{-a^2(\omega + \omega_0)^2}{4}} + e^{\frac{-a^2(\omega - \omega_0)^2}{4}}\right] \quad (12.B1.18)$$

so that the energy spectrum is

$$I(\omega) = E(\omega)E^*(\omega) \sim \frac{|a|^2}{8} e^{-a^2(\omega - \omega_0)^2/2}. \quad (12.B1.19)$$

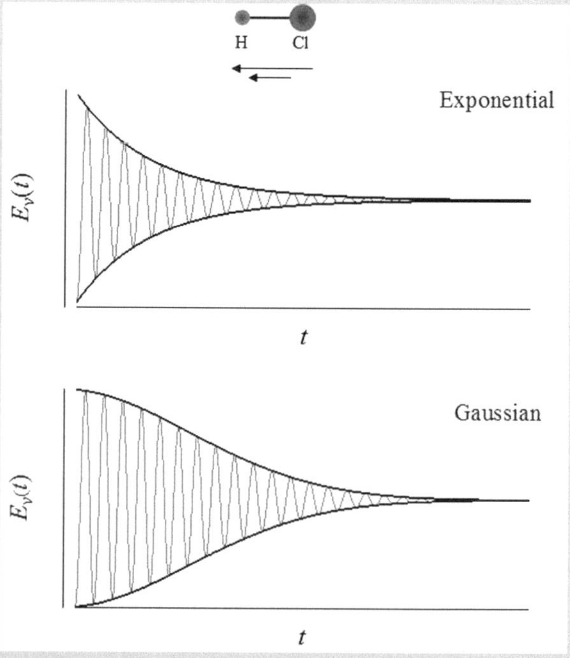

FIGURE 12.14 Two envelope functions for the damping of a harmonic oscillator motion.

Box 12.2

Suppose we have two sets $A=\{a,b,c,d\}$ and $B=\{i,j,k,l\}$, and a map f from A to B is to define.

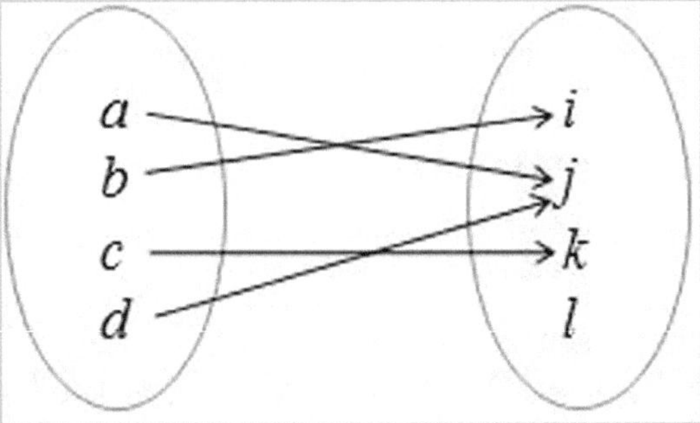

$$f(a)=j, f(b)=i, f(c)=k, \text{and } f(d)=j.$$

The image of f is $\{i,j,k\}$, and is written as

$$f(A)=\{i,\ j,\ k\}.$$

Now, let us define a density matrix ρ_{AB} in the global space $\varepsilon_A\otimes\varepsilon_B$, where ε_A and ε_B are subspaces corresponding to the reduced density matrices ρ_{A} and ρ_{B}. Then the mapping of ρ_{AB} onto ρ_{A} is called the trace of B, denoted by tr_B, and the mapping of ρ_{AB} onto ρ_{B} is tr_A. Let $\{|a_r\rangle\}$ and $\{|b_r\rangle\}$ be the respective eigenbases in subspaces ε_A and ε_B. Then, by definition, the density matrix ρ_{AB} in the subspace ε can be decomposed by

$$\rho_{\text{AB}}=\sum\nolimits_{rstu}c_{rstu}|a_r\rangle\langle a_s|\otimes|b_t\rangle\langle b_u|. \quad (12.\text{B}2.1)$$

If, for example, one wants to know the partial trace of A, then one has

$$tr_A\{\rho_{\text{AB}}\}=\sum\nolimits_{rstu}c_{rstu}|b_t\rangle\langle b_u|\langle a_s|a_r\rangle, \quad (12.\text{B}2.2)$$

where

$$\langle a_s|a_r\rangle=tr\langle a_r|a_s{}'\rangle.$$

As described in the text (see equation (12.78)), $tr_A\{\rho_{\text{AB}}\}$ is the reduced density operator (ρ_{B}=σ_{B}) in the space ε_B.

For an example of a partial trace, we may consider the singlet-state spin function of the hydrogen molecule (see equation (5.59)), which we rewrite here as

$$\frac{1}{\sqrt{2}}\left[|\alpha\rangle|\beta\rangle-|\beta\rangle|\alpha\rangle\right]=\phi_{\text{s}}. \quad (12.\text{B}2.3)$$

The density matrix can be written as

$$\rho_{\text{AB}}=|\phi_{\text{s}}\rangle\langle\phi_{\text{s}}|=\frac{1}{2}\big[|\alpha\beta\rangle\langle\alpha\beta|-|\alpha\beta\rangle\langle\beta\alpha|-$$
$$|\beta\alpha\rangle\langle\alpha\beta|+|\beta\alpha\rangle\langle\beta\alpha|\big]$$

$$tr_A\{\rho_{\text{AB}}\}=\frac{1}{2}\big[|\beta\rangle\langle\beta|\langle\alpha|\alpha\rangle-|\beta\rangle\langle\alpha|\langle\beta|\alpha\rangle-$$
$$|\alpha\rangle\langle\beta|\langle\alpha|\beta\rangle+|\alpha\alpha\rangle\langle\beta\beta|\big]. \quad (12.\text{B}2.4)$$

Using orthogonality of the spin functions, we write

$$tr_A\{\rho_{\text{AB}}\}=\frac{1}{2}\big[|\beta\rangle\langle\beta|+|\alpha\rangle\langle\alpha|\big]=\sigma_{\text{B}}, \quad (12.\text{B}2.5)$$

where $\sigma_{\text{B}}=\rho_{\text{B}}$ is the reduced density operator.

PROBLEMS

12.1 To what extent a vibrating bond having an equilibrium length of 130 pm will change its length in ~5 fs? At what temporal scale ultrashort laser pulses are useful in studies of chemical dynamics? What are the advantages of having pulses shortened to the attosecond regime?

12.2 Consider two one-dimensional plane waves, one is moving in the $+x$ and the other in the $-x$ direction. The two waves can be represented by

$$\psi_1(x,t)\propto\cos(kx-\omega t+\alpha)$$

$$\psi_2(x,t)\propto\cos(kx-\omega t-\alpha)$$

where $\alpha=\beta=0$ at $t=0$.
They may also be represented by

$$\psi_1(x,t)=A_1e^{i(kx-\omega t)}$$

$$\psi_2(x,t)=A_2e^{i(kx-\omega t)}.$$

Which of these two representations is appropriate for the superposition of $\psi_1(x,t)$ and $\psi_2(x,t)$ so as to generate a quantum wave packet? Justify.

12.3 Take several one-dimensional plane waves of different wavenumbers within a momentum distribution (Gaussian) given by $\phi(k_x)$with a mean momentum $\langle k_x\rangle$.

(a) Show that a localized wave packet formed at $t=0$ can be represented by

$$\Psi(x,0)=\frac{1}{\sqrt{2\pi}}e^{i\langle k_x\rangle x}\int_{-\infty}^{+\infty}\phi(k_x)e^{i(k_x-\langle k_x\rangle)x}d\left(k_x-\langle k_x\rangle\right).$$

(b) Draw a graph of $|\Psi|$with the distance x.

(c) Also draw the momentum distribution function $\phi(k_x)$.

(d) How are the widths of the wave packet function and the momentum distribution function related? A mathematical derivation, which is a long one here, is not needed.

(e) What is the significance of the widths of $|\Psi|$ and $\phi(k_x)$ functions?

(f) Show that the wave packet will dephase if the term $e^{i(k_x-k_x)x}$ in the integral above oscillates. Is this the precise mechanism by which a wave packet might dephase?

12.4 Consider the following single laser-pulse experiment (not a concrete example) in which an ideal diatomic molecule is excited to three rovibronic levels $v'=0\,J'=60$, $v'=1\,J'=60$, and $v'=5\,J'=60$, one at a time. The molecule is presumably bound in the excited state. Suppose dispersed fluorescence from each of the excited rovibronic manifold to the rovibronic manifold of the intermediate electronic surface (S_1) is detected, as shown below.

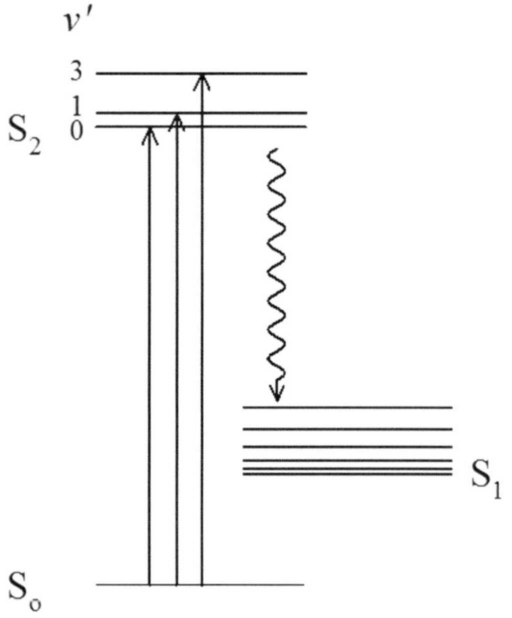

(a) Draw an approximate line diagram of the fluorescence spectrum for emission from each excited rovibronic state.

(b) Provide the fluorescence expectation value (signal) in density matrix representation.

12.5 Prove that the density operator is Hermitian.

BIBLIOGRAPHY

Baskin, J. S. and A. H. Zewail (1989) *J. Phys. Chem.* 93, 5701.

Böhm, A. (1979) *Quantum Mechanics*, Springer.

Cohen-Tannoudji, C., B. Diu, and Franck Laloë (1977) *Quantum Mechanics* (Vol I), John Wiley & Sons.

Ernst, R. R., G. Bodenhausen, and A. Wokaun (1988) *Principles of Nuclear Magnetic Resonance in One and Two Dimensions*, Oxford University Press.

Fano, U. (1957) *Rev. Mod. Phys.* 29, 74.

Felker, P. M. and A. H. Zewail (1987) *J. Chem. Phys.* 86, 2460.

Herzberg, G. (1945) *Molecular Spectra and Molecular Structure* (Vol II), Van Nostrand Reinhold Company, Inc.

Margenau, H. and G. M. Murphy (1964) *The Mathematics of Physics and Chemistry* (Vol II), D. Van Nostrand Co.

Rose, M. E. (1957) *Elementary Theory of Angular Momentum*, John Wiley & Sons.

Silverman, M. P., S. Haroche, and M. Gross (1978) *Phys. Rev. A* 18, 1517.

13 Nuclear Magnetic Resonance Spectroscopy

The idea of both electron and nuclear spins, s and I being the respective spin angular momentum, was introduced contemporarily by Pauli to explain the electron-nuclear hyperfine structure in atomic spectra (Chapter 4). For both electron and proton, the z-projection of the spin angular momentum vector is $\pm\hbar/2$. Note that the angular momentum vector takes the unit of frequency if $\hbar$ is removed, and the use of $\hbar$ should be construed according to the context. In an ensemble under ordinary conditions, these two quantized energy levels are degenerate in the sense that the random orientations of different spin angular momentum vectors in the population produce a canceling effect. The two energy levels, however, split in the presence of a magnetic field applied along the space z-axis due to the Zeeman effect and resonate with microwave frequency in the case of the electron spin (ESR), and with radiofrequency in the case of the nuclear spin (NMR). The advances in NMR have reached an extent of maturity in terms of both theoretical and experimental scholarships, facilitating the search of phenomena at greater depths of molecular science. Such a large scope of NMR can be discussed at different levels of theoretical treatment. The initial sections in this chapter provide the basic concepts briefly in a semiclassical manner with the objective of walking with beginners to a level adequate for basic understanding of NMR. The later sections describe NMR experiments in the perspective of density operator, the understanding of which is indispensable to not only explore the richness of the subject, but also implement new spectroscopic methods.

13.1 NUCLEAR SPIN OF DIFFERENT ELEMENTS

The nuclear spin is not always $\hbar/2$ (or $1/2$ in the frequency unit); it is strictly the spin of a proton nucleus (^1H) or a neutron. The nuclear spin varies from one to another element, and even within the isotope of the same element, but some elements may have the same spin. For example, values of I for ^1H, ^{11}B, ^{13}C, and ^{17}O are, respectively, $1/2, 3/2, 1/2$ and $5/2$, and those for ^1H, ^2H, and ^3H are $1/2$, 1, and $1/2$. In nuclei other than ^1H, the total angular momentum of each nucleon (proton or neutron) couples in a way to produce the effective ground-state nuclear spin. Thus, the spins of ^{14}N and ^{15}N are 1 and $1/2$, which suggests that the nucleon number A (or the atomic mass number) and the nuclear charge Z_{eff} (the total charge of protons minus the total charge of the inner shielding electrons) could be the factors determining the nuclear spin. Indeed so, an examination of the values of A and Z_{eff} of different elements and isotopes, and comparing them with corresponding nuclear spins tabulated in literature identifies the following

1. Nuclei with odd A have half-integral spin.
2. Nuclei with even A and even Z_{eff} have zero spin.
3. Nuclei with even A and odd Z_{eff} have integral spin.

The reader should verify these characteristics by consulting literature values.

13.2 EXCITED-STATE NUCLEAR SPIN

The nuclear spin mentioned refers to the ground-state spin implicitly. There are excited states of nuclei; for example, the ground and excited nuclear spins of both ^{57}Fe and ^{119}Sn are characterized by $I_{ground} = 1/2$ and $I_{excited} = 3/2$, respectively, and the energy difference between these states is 14.4 keV (~3×10^{18} Hz) for ^{57}Fe and 23.9 keV (~5×10^{18} Hz) in the case of ^{119}Sn. Importantly, these are γ-ray frequencies, implying that 'nuclear resonance' studied by Mossbauer spectroscopy are very high-energy transitions. 'Nuclear magnetic resonance', on the other hand, employs nuclear Zeeman effect that splits only the ground-state spin energy in the presence of an external magnetic field. Such splitting amounts to ~2×10^{-6} eV (~5×10^{8} Hz), which can resonate only with radiofrequency. It is seen clearly now that only the ground-state nuclear spin of quantized energy is relevant to NMR.

13.3 NUCLEAR SPIN ANGULAR MOMENTUM AND MAGNETIC MOMENT

The magnitude of the angular momentum vector is quantized by the spin quantum number, simply call spin I,

$$I^2 = \hbar^2\left[I(I+1)\right], \tag{13.1}$$

and the projection of the **I** vector onto the z-axis gives $I_z = m\hbar$ where $m(=I, I-1, I-2, \cdots, -I)$ is the magnetic quantum number in conformity with the fact that I_z is realized only in the presence of the external magnetic field B_0 (Figure 13.1). For a spin-half nucleus $(I = 1/2)$ there are clearly two z-projections, $I_z = \hbar/2$ (β-state) and $I_z = -\hbar/2$ (α-state). The α- and β-states refer to the eigenstates (see Box 4.1) given by

$$I_z|\alpha\rangle = \frac{\hbar}{2}|\alpha\rangle \quad \text{and} \quad I_z|\beta\rangle = -\frac{\hbar}{2}|\beta\rangle, \tag{13.2}$$

where the vectors

$$\alpha = \begin{pmatrix}1\\0\end{pmatrix}, \beta = \begin{pmatrix}0\\1\end{pmatrix} \tag{13.3}$$

DOI: 10.1201/9781003293064-13

are represented in a 2-D spin space. In the NMR parlance one says that the α-state is lower in energy, but the kind of representation shown in Figure 13.1 is apparently confusing to the beginners because we are used to drawing a ground-state level lower in the vertical scale of energy in spectroscopy. This need not be alarming, however; since B_o is applied along $+z$ direction, the $+z$ projection of the spin vector is lower in energy. Naturally, something that aligns itself with a force is lower in energy.

The magnetic moment $\boldsymbol{\mu}$ associated with I is given using the nuclear gyromagnetic ratio γ (Chapter 4), which is a constant for a given nucleus

$$\boldsymbol{\mu} = \gamma \mathbf{I} = \gamma\hbar\sqrt{I(I+1)}$$

$$\gamma = g_N \frac{e\hbar}{2m_N c} \quad (13.4)$$

The $\boldsymbol{\mu}$ and $\mathbf{I}$ vectors are collinear, but they can be parallel or antiparallel depending on the sign of the value of γ that can be positive or negative conditioned to whether the sign of the nuclear g-factor is positive or negative. When positive, $\boldsymbol{\mu}$ and $\mathbf{I}$ vectors are parallel, and when negative they are antiparallel.

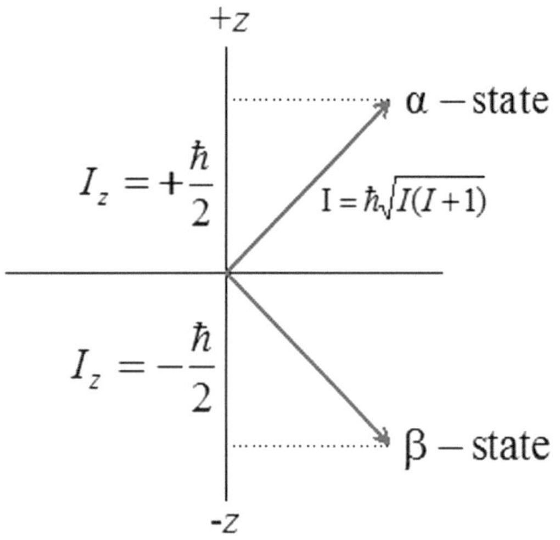

FIGURE 13.1 Quantization of angular momentum. Definition of I_z, and α- and β-states according to the direction of the static magnetic field.

13.4 ZEEMAN SPLITTING OF NUCLEAR ENERGY LEVELS

When the nuclei are placed in an external magnetic field applied along $+z$ direction, the Zeeman Hamiltonian is

$$\mathcal{H}_{Zeeman} = -\boldsymbol{\mu} \cdot \mathbf{B}_o$$

which is the energy E of the magnetic moment. Since μ_z is defined by the magnetic quantum number m, the use of equation (13.4) yields

$$\mu_z = -\gamma\hbar m$$

$$E = -\gamma\hbar m B_o. \quad (13.5)$$

The energy according to the m (=5/2, 3/2, 1/2, −1/2, −3/2, −5/2) states of an $I = 5/2$ nucleus with B_o are shown in Figure 13.2. The splitting energy for each value of m amounts to $|\gamma\hbar B_o|$ from which the frequency of the transition can be determined. The two-headed arrows in the figure show the allowed transitions that can occur only between the neighboring energy levels. This conforms with the selection rule that NMR transitions are restricted to $\Delta m = \pm 1$, so equation (13.5) written in terms of frequency is

$$\omega_o = |\gamma| B_o. \quad (13.6)$$

The frequency ω_o is called Larmor precession frequency.

13.5 LARMOR PRECESSION OF ANGULAR MOMENTUM

A very useful way of looking at the NMR phenomenon is to consider the classical motion of the spin vector $\mathbf{I}$ in the presence of a magnetic field. Analogous to the manner in

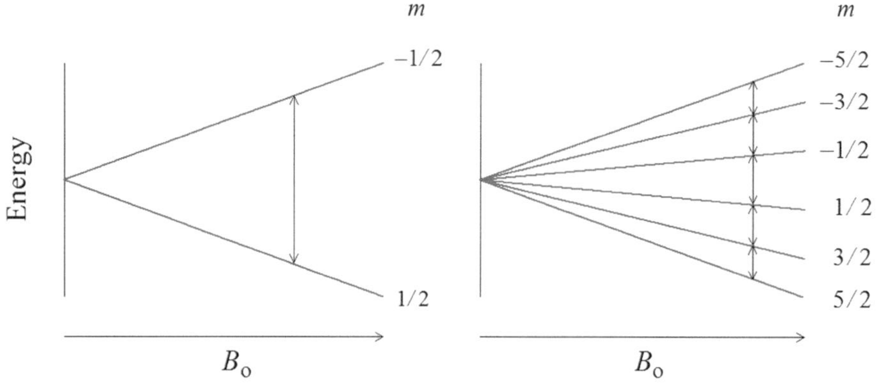

FIGURE 13.2 Splitting of spin energy levels proportional to the static field strength. *Left*, splitting for a spin-half nucleus (^{1}H, for example); *right*, splitting in the case of a spin-5/2 nucleus ([11]Be, for example).

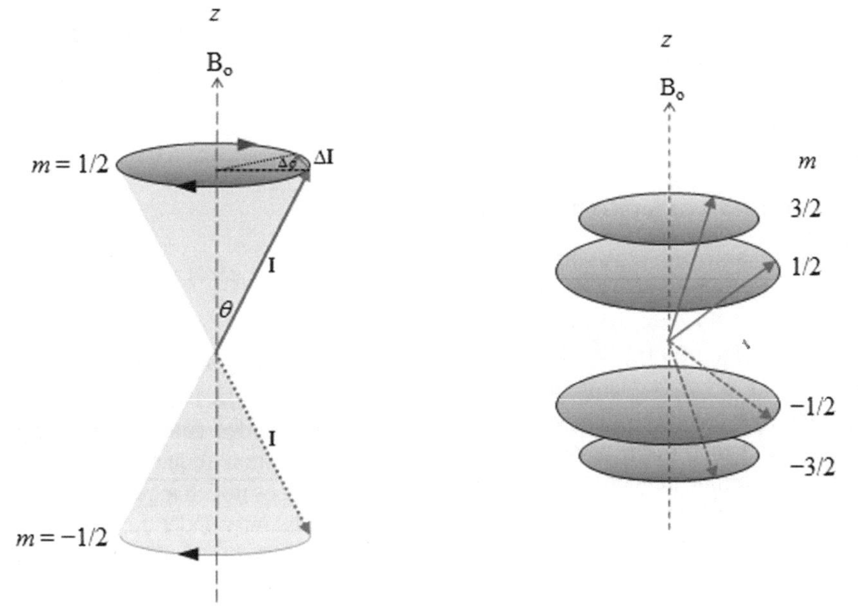

FIGURE 13.3 Larmor precession of magnetic moments for nuclei of $I = 1/2$ (*left*), and $I = 3/2$ (*right*) showing quantization of dipolar energy according to the allowed values of $m\,(=2I+1)$.

which an electric field exerts a torque on an electric dipole moment that we discussed in Chapter 4, the torque $\boldsymbol{\tau}$ produced on the vector $\boldsymbol{\mu}$ due to the static field $\mathbf{B}_o$ tends to align the two vectors, because the energy of $\boldsymbol{\mu}$ is lowest when it is aligned with $\mathbf{B}_o$. They cannot be collinear, however, because the torque will then vanish. The torque here is

$$\boldsymbol{\tau} = \boldsymbol{\mu} \times \mathbf{B}_o = \mu B_o \sin\theta. \tag{13.7}$$

But the magnetic moment arises from the spin angular momentum, so the torque due to $\mathbf{B}_o$ is also exerted on $\mathbf{I}$. The torque on $\mathbf{I}$ can be treated according to Newton's second law of motion

$$\boldsymbol{\tau} = \frac{d\mathbf{I}}{dt}. \tag{13.8}$$

While $\mathbf{I}$ is a constant provided that there is no magnetic field, the effect of the torque due to $\mathbf{B}_o$ is to change $\mathbf{I}$ in the direction perpendicular to the plane defined by $\mathbf{B}_o$ and $\mathbf{I}$. If $\Delta\mathbf{I}$ is a small change of $\mathbf{I}$ in the time interval Δt subtending an angle $\Delta\phi$ (Figure 13.3), then

$$\boldsymbol{\tau} = \frac{\Delta\mathbf{I}}{\Delta t} = \frac{I\sin\theta\Delta\phi}{\Delta t} = \mu B_o \sin\theta = |\gamma B_o| I \sin\theta = \omega_o I \sin\theta, \tag{13.9}$$

showing that the torque sets the $\boldsymbol{\mu}$ vector into Larmor precession at frequency ω_o. The energy of the nuclear dipole is independent of the direction of its precession – clockwise and anticlockwise for positive and negative values of γ, respectively. However, the energy is quantized by the magnetic quantum number m as $E = -\gamma\hbar m B_o$, and the half-angle of the cone of precession (see Chapter 4) is

$$\theta = \cos^{-1}\frac{m}{\sqrt{I(I+1)}}. \tag{13.10}$$

13.6 TRANSITION TORQUE MECHANICS

The m-dependence of the angle θ is clearly a quantized condition. If θ changes, so would the magnetic dipolar energy. The alteration in θ will occur when a second, much weaker, magnetic field $\mathbf{B}_1$ having a frequency ω_{rf} equal to the Larmor frequency is brought in the form of radiofrequency radiation. Following the same principle of classical mechanics, the $\mathbf{B}_1$ field exerts a torque on $\boldsymbol{\mu}$. If $\mathbf{B}_1$ is applied along a transverse axis, x or y of the laboratory frame, the torque on $\boldsymbol{\mu}$ will alternately increase and decrease the angle θ because $\boldsymbol{\mu}$ is already precessing due to the first torque exerted by $\mathbf{B}_o$. As a result, $\boldsymbol{\mu}$ wobbles a little, resulting in a compounded motion called nutation. But the nutation motion can be avoided if $\mathbf{B}_1$ is allowed to rotate at Larmor frequency, which is achieved by containing $\mathbf{B}_1$ in a frame rotating at ω_o (see below). The relative orientation of $\boldsymbol{\mu}$ and $\mathbf{B}_1$ vectors is now constant. It is more useful if the frequency of the rotating frame is phase-shifted to $\omega_o + \phi$ by some angle ϕ so that $\mathbf{B}_1$ is always oriented perpendicular to the plane containing $\boldsymbol{\mu}$ and $\mathbf{B}_o$. In this condition, the torque due to $\mathbf{B}_1$ will cause $\boldsymbol{\mu}$ to tilt away from $\mathbf{B}_o$, altering the angle θ, thus changing its dipolar energy. This is essentially an NMR transition, although the extent to which the transition is tuned to $\Delta E\left(=|\gamma|\hbar B_o\right)$ will depend on how long $\mathbf{B}_1$ is applied. Off-resonance condition will appear if the B_1 field is away from the optimum. In practice, the applied radio wave of frequency ω_{rf} has the $\mathbf{B}_1$ field oscillating linearly along x or $y$axis of the laboratory frame. The linearly oscillating $\mathbf{B}_1$ can be resolved into two counter-rotating components, and the component whose direction of rotation matches the direction of precession of the spin is retained, the other one is not considered further.

13.7 SPIN POPULATION AND NMR TRANSITION

13.7.1 Static Field Dependence of Signal Intensity

The description up to this point has referred to a single nucleus which cannot have all the quantized energy states simultaneously. A single nucleus with $I = 1/2$ can be in the energy state corresponding to the m_I value of +1/2 or −1/2, but not both. Similarly, a nucleus with $I = 3/2$ can be in only one of the four states relating to the four possible values of m, and a transition of the single nucleus would occur when irradiated with radio wave matching the Larmor frequency. One should recognize that all possible transitions of a nucleus with $I > 1/2$ are isoenergetic because of the restriction imposed by $\Delta m = \pm 1$. In the case of $I = 3/2$, for example, the allowed transitions are $3/2 \leftrightarrow 1/2$, $1/2 \leftrightarrow -1/2$, and $-1/2 \leftrightarrow -3/2$; all three transitions will have the same transition frequency as follows from the relation

$$\omega_o = |\gamma| B_o \Delta m = |\gamma| B_o. \tag{13.11}$$

The appearance of a single transition for a nucleus allows population distribution in an ensemble of N-nuclei only between the low- and high-energy levels, α- and β-states as for ^1H. The spin populations in the two levels turn out to be almost the same; there is only a minor excess of spins in the α-state. The reason for the near-equal population in the two levels is the radiofrequency resonance energy which is much lower than molecular rotational energy or thermal energy. For a sample containing N-nuclei, the ratio of the number of nuclei in the α-state (N_α) to that in the β-state (N_β) is given by the Boltzmann factor

$$\frac{N_\alpha}{N_\beta} = e^{\Delta E / k_B T}, \tag{13.12}$$

which is ~1.000007 for a ^1H sample placed in a B_o field of 1 Tesla (=10^4 Gauss) at T=300 K. The ratio N_α / N_β is a positive exponential, implying that higher the energy gap ΔE between the α- and β-states, more will be the number of spins in the α-state. Since $\Delta E = |\gamma| \hbar B_o$, the N_α population rises with both γ and B_o, leading to higher signal intensity. This shows the advantage of higher static field.

13.7.2 Nuclear Receptivity

The nuclear spin transition, classically viewed as the effect of the torque exerted by $\mathbf{B}_1$, is essentially the absorption of radio wave energy by the nuclei. It is useful to introduce a quantity R called the rate of absorption of radio wave energy by a nucleus, which, we can guess already, will depend on the nuclear property γ, the temperature T, and the respective strength of $\mathbf{B}_o$ and $\mathbf{B}_1$,

$$R \propto \frac{N\gamma^4 B_0^2 B_1^2 g(\omega)}{T}. \tag{13.13}$$

The factor $g(\omega)$, analogous to the absorption coefficient $\alpha(\omega)$ discussed earlier (equations (3.34) and (3.35)) is a Lorentzian lineshape factor

$$g(\omega) \propto \frac{1/T_2}{(1/T_2)^2 + (\omega - \Omega)^2}, \tag{13.14}$$

in which T_2 is the decay time of spin coherence in the transverse plane and Ω is a measure of the extent to which a nucleus is shielded from B_o. The expression for the rate of energy absorption (equation (13.13)) assumes that the RF absorption does not change the ratio N_α / N_β to a large extent, which is valid only under the condition of lower strength of B_1. Furthermore, R is not a directly measurable quantity; instead, the signal amplitude S is obtained by measuring the rate of change of the transverse magnetization $d\mathbf{M}/dt$ by using a receiver coil placed in the transverse plane. This allows rewriting the expression as

$$\frac{d\mathbf{M}}{dt} \sim \frac{R}{B_1} \sim S \propto \frac{N\gamma^4 B_0^2 B_1 g(\omega)}{T}, \tag{13.15}$$

showing that the intensity of the NMR signal S is proportional to the nuclear gyromagnetic ratio γ, the static field strength B_o, the RF field strength B_1, and the number of nuclei N in the sample. It is also known empirically that

$$B_1 \sim \frac{1}{\sqrt{\gamma^2 T_2^2}}. \tag{13.16}$$

Insertion of this into the expression for signal intensity (equation (13.15)) yields

$$S \propto N|\gamma^3|. \tag{13.17}$$

To check this proportionality, one can measure signal intensities of different samples (isotopes) of the same volume and atomic concentration. In the strict sense, it is not N but the natural isotopic abundance C which should be considered, because it is the abundance of the NMR-sensitive nuclei in the total number of nuclei that will contribute to the signal

$$S \propto C|\gamma^3|. \tag{13.18}$$

The number $C|\gamma^3|$ is called nuclear receptivity. It is also obvious that lower nuclear receptivity will make the nucleus less suitable for NMR because the signal intensity may even be buried in the noise. Receptivity of a nucleus is given relative to the receptivity of ^1H and ^{13}C, denoted by D^P and D^C, respectively. For example, values of D^P for ^1H, ^{13}C, ^{15}N, ^{19}F, and ^{31}P nuclei are 1, 1.76×10^{-4}, 3.85×10^{-6}, 0.834, and 0.0665, respectively. The very small value of D^P for ^{13}C, particularly ^{15}N, arises due to extremely low natural abundance of the two isotopes, 1.108 and 0.37%, respectively. This is the reason for the requirement of isotopic enrichment of large molecules like peptides and proteins whose aqueous sample concentrations are tens of micromoles, at the most a few millimoles, per liter.

13.7.3 Macroscopic Magnetization

In a sample of N nuclei, all spin vectors of the N_α population are in the direction of $\mathbf{B}_0$ precessing about it at the same Larmor frequency, but they are not in phase. The spins of the N_β population, although opposite to the direction of $\mathbf{B}_0$, also precess at the same Larmor frequency and are out of phase individually (Figure 13.4). A qualitative description of a NMR transition and relaxation processes is achieved by considering the macroscopic magnetization $\mathbf{M} = \mathbf{M}_z$ formed by summing up all $\boldsymbol{\mu}$ vectors in N_α and N_β populations (Figure 13.4). In the absence of $\mathbf{B}_1$ there will be no $\mathbf{M}_x$ and $\mathbf{M}_y$ components of the macroscopic magnetization because the out-of-phase precession of the dipole vectors cancel out the individual transverse components. If now the radio wave of frequency – call it the carrier frequency ($\omega_{rf} = \omega_0$), is turned on such that its $\mathbf{B}_1$ field is along x or y-axis and rotates at Larmor precession frequency (see below), then the torque $\tau = \mathbf{B}_1 \times \mathbf{M}$ will tip $\mathbf{M}$ away from the z-axis. The angle by which $\mathbf{M}$ tilts depends on the time the torque is effective; if the radiofrequency (RF) field is on for a very short time the tilt angle will be very small, but in general the time is optimized to obtain a tilt by 90°. Suppose $\mathbf{B}_1$ is along x and the tilt angle is β with respect to the z-axis, then

$$\mathbf{M} \xrightarrow{\beta_x} \mathbf{M}_z \cos\beta + \mathbf{M}_y \sin\beta, \qquad (13.19)$$

indicating the appearance of transverse magnetization. If the duration of the RF field is adjusted to obtain a tilt angle of 90°, then the entire magnetization appears as M_y at time $t = 0$. The tipping of $\mathbf{M}_z$ completely to $\mathbf{M}_y$ (or $\mathbf{M}_x$ if $\mathbf{B}_1$ was along y) is equivalent to a tuned on-resonance NMR transition in the sense that the signal intensity is maximum. But any non-zero value of β also produces a change in the energy of the nuclear dipole causing a transition, although the transition is detuned in this case and hence the signal intensity will be less.

13.8 BLOCH EQUATIONS AND RELAXATION TIMES

The macroscopic magnetization $\mathbf{M} = \left(\sum_N \boldsymbol{\mu}_N\right)$ is said to be at equilibrium in the presence of the static field $\mathbf{B}_0$ alone. Since the field is along z direction the $\mathbf{M}$ vector also directs along z, so at equilibrium $\mathbf{M} = \mathbf{M}_z$ but $\mathbf{M}_x = \mathbf{M}_y = 0$. When the radiofrequency field $\mathbf{B}_1$ is brought in, the total magnetic field is $\mathbf{B} = \mathbf{B}_0 + \mathbf{B}_1$, and the effective torque τ experienced by the equilibrium magnetization can be written in the form of the equation of motion

$$\frac{d\mathbf{M}(t)}{dt} = \tau = \mathbf{M}(t) \times \gamma \mathbf{B}(t) = \gamma \mathbf{M}(t) \times \mathbf{B}(t). \qquad (13.20)$$

Ideally, suppose the duration of $\mathbf{B}_1$ is optimized to have the magnetization tilted by 90° to the transverse plane, then $\mathbf{M}_z = 0$ and $\mathbf{M}_x = \mathbf{M}_y \neq 0$. If the $\mathbf{B}_1$ field was removed immediately after generating the transverse components, the spin magnetization will relax to the equilibrium value, which entails two distinct processes: restoration of the initial value of the longitudinal magnetization M_z, and the decay of the transverse magnetizations M_x and M_y both to zero. The former is called longitudinal relaxation (T_1) and the latter transverse relaxation (T_2). Bloch's phenomenological assumption was that these relaxation processes are given by the first order differential equations

$$\frac{dM_z}{dt} = -\frac{M_z - M}{T_1}$$

$$\frac{dM_x}{dt} = -\frac{M_x}{T_2}$$

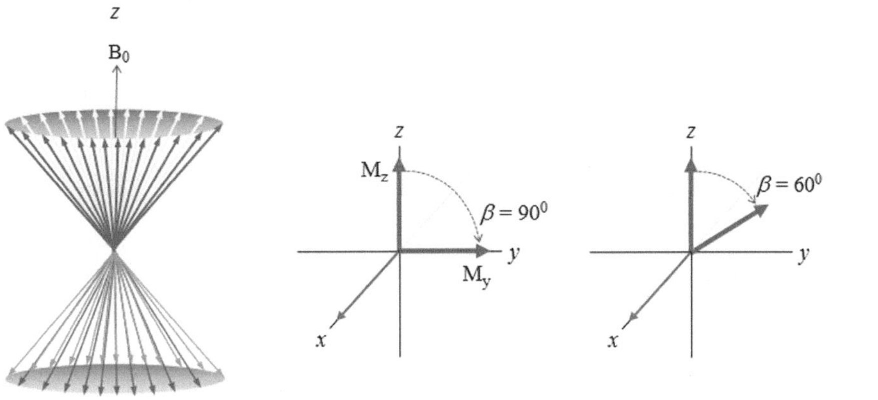

FIGURE 13.4 Definition of the macroscopic magnetization which is rotated by a RF pulse. The slight excess of spins along the $+z$ axis (*left*) yields a net macroscopic magnetization vector along $+z$. The effects of 90° and 60° pulses (the $\mathbf{B}_1$ field is along x) are depicted in the *center* and the *right* panel, respectively.

$$\frac{dM_y}{dt} = -\frac{M_y}{T_2}, \tag{13.21}$$

where T_1 and T_2 are longitudinal- and transverse relaxation times, respectively. These two relaxations are indeed distinct. The longitudinal relaxation T_1 involves energy flow from the spin system to various degrees of freedom of the surrounding lattice, and hence is also called spin-lattice relaxation. The transverse relaxation T_2 is the time of spin dephasing or the time of loss of the phase coherence. The transverse magnetization M_x or M_y at time $t = 0$ may be considered as a coherent superposition of the individual μ vectors. With time the phase coherence is lost, which means disappearance of the transverse magnetization. The spin dephasing occurs due to spin interactions within different nuclei, and hence is also called spin-spin relaxation.

At this stage, the form of the $\mathbf{B}_1$ field should be introduced. This is conventionally an x-polarized field such that the field amplitude oscillates linearly along x. We know from earlier discussions that the field vector of a linearly polarized light can be decomposed into two circularly polarized components, one rotating clockwise and the other anticlockwise, both rotating at the same frequency ω_0, which is the Larmor frequency or carrier frequency. Here, the amplitude vector $\mathbf{B}_1^{\text{o}}$ is along the x-direction, which implies that the two counter-rotating circular components are in the xy plane

$$\mathbf{B}_1(t) = \mathbf{B}_1^{\text{o}} \cos\omega_{\text{rf}} t\, \hat{e}_x - \mathbf{B}_1^{\text{o}} \sin\omega_{\text{rf}}\, \hat{e}_y. \tag{13.22}$$

The circularly polarized component that rotates in the same direction as that of Larmor precession is retained for further consideration and analysis. Note that Larmor precession can be clockwise or anticlockwise depending on the sign of γ. Accordingly, the clockwise or anticlockwise circular component of $\mathbf{B}_1^{\text{o}}$ is retained. The total field is then

$$\mathbf{B}_1 = \mathbf{B}_0 + \mathbf{B}_1 = \mathbf{B}_1^{\text{o}} \cos\omega_{\text{rf}} t\, \hat{e}_x - \mathbf{B}_1^{\text{o}} \sin\omega_{\text{rf}} t\, \hat{e}_y + \mathbf{B}_0\, \hat{e}_z. \tag{13.23}$$

Combining the four equations above ((13.20)–(13.23)) one gets the Bloch equations of motion for M_x, M_y, and M_z components of magnetization

$$\frac{dM_x}{dt} = \gamma\left(B_0 M_y + B_1 \sin\omega_{\text{rf}} t\, M_z\right) - \frac{M_x}{T_2}$$

$$\frac{dM_y}{dt} = -\gamma\left(B_0 M_x - B_1 \cos\omega_{\text{rf}} t\, M_z\right) - \frac{M_y}{T_2}$$

$$\frac{dM_z}{dt} = -\gamma\left(B_1 \sin\omega_{\text{rf}} t\, M_x + B_1 \cos\omega_{\text{rf}} t\, M_y\right) - \frac{(M_z - M)}{T_1}. \tag{13.24}$$

13.9 THE ROTATING FRAME

Up to now we have been working with a static or laboratory frame xyz, and the Bloch equations are written in this frame where the time dependence of the transverse components contain both magnetic fields, $\mathbf{B}_0$ and $\mathbf{B}_1$. One can sense already that the motion of the magnetization is going to be complicated. Also, consider the fact that $\mathbf{B}_0 \gg \mathbf{B}_1$, which would seem to mar the importance of $\mathbf{B}_1$; however, it is the latter which is important, not $\mathbf{B}_0$, because a transition is induced by $\mathbf{B}_1$ alone. These arguments might sound strange because the spin energy levels do not exist if there is no $\mathbf{B}_0$, but the point is to find an analytical strategy by which we can clearly see the effect of $\mathbf{B}_1$ alone and the relaxation of excited spins. This is central to understanding the motion of the magnetization under the influence of RF.

The strategy is the transformation of the laboratory-fixed frame xyz to a frame $x'y'z$ that rotates about the z-axis at the frequency ω_{rf} of the $\mathbf{B}_1$ field. Mathematically, a time-dependent rotation matrix $\mathbf{R}^{\text{r}}(t)$ operates to set the fixed frame to rotation (Figure 13.5). Since the z-axis remains the same, the rotation matrix contains the elements of a two-dimensional rotation (see Box 7.1) with the diagonal $zz = 1$,

$$\begin{pmatrix} x' & y' & z \end{pmatrix} = \begin{pmatrix} x & y & z \end{pmatrix} \begin{pmatrix} \cos\theta & \sin\theta & 0 \\ -\sin\theta & \cos\theta & 0 \\ 0 & 0 & 1 \end{pmatrix}, \tag{13.25}$$

in which $\theta = \omega_{\text{rf}} t$.

In the frame rotating at the RF frequency (ω_{rf}), the circular component of the amplitude vector $\mathbf{B}_1^{\text{o}}$ of the RF field will obviously be static. Now consider the Larmor frequency ω_0 of the vector $\mathbf{M}$ in the rotating frame where the frequency ω_0 has to be seen with reference to the frequency ω_{rf} of rotation. If $\omega_0 = \omega_{\text{rf}}$, then the effective precession frequency will be $\Omega = \omega_0 - \omega_{\text{rf}} = 0$, rendering $\mathbf{M}$ stationary, but if $\omega_0 \neq \omega_{\text{rf}}$, the $\mathbf{M}$ vector will precess at the offset frequency Ω, which can be positive or negative. The energy levels of an isolated spin system in the rotating frame with respect to the three offset conditions are depicted in Figure 13.6. Note that the magnitude of Ω is very small even if it is not zero, because ω_0 cannot be very different from ω_{rf}, which is why we call Ω the offset frequency. The importance of Ω is seen by writing the Larmor formula in the rotating frame, $\Omega = \gamma\omega_0$, according to which $B_0 = 0$ if $\Omega = 0$, which would appear to mean that the B_0 of the static frame xyz is reduced to zero in the rotating frame $x'y'z$.

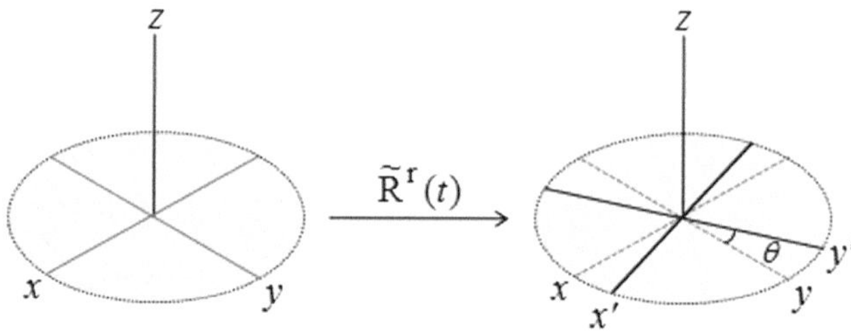

FIGURE 13.5 Transformation of the laboratory frame (xyz) to a rotating frame ($x'y'z$) by the time-dependent rotation matrix. The time dependence of the angle $\theta(t)$ by which the frame has rotated with reference to the static frame is given by $\omega_{\text{rf}} t$, where ω_{rf} is the RF frequency.

Even if the situation $\Omega \neq 0$ occurs, the strength of B_o is substantially reduced in the rotating frame; hence, the B_o field in the rotating frame is called the reduced field, ΔB_o.

The zero offset frequency ($\Omega = \omega_o - \omega_{rf} = 0$), also called on-resonance condition, is the best condition one wishes to achieve because having $\Delta B_o = 0$ in the rotating frame leaves us with the B_1 field alone for interaction with the magnetization **M**. This is the main result of the transformation to the rotating frame in which the torque force experienced by the spin magnetization is due to B_1 alone. The rotation of the **M** vector under the influence of a RF field applied in the form of a short pulse of a few microseconds or as a continuous wave is shown in Figure 13.7. Note that the direction of rotation of a spin vector in NMR might appear confusing at times, even though the rotation direction does not influence results and conclusions. In the vector model of spin rotation, the left-hand rule is used, according to which the thumb points to the direction of the $\mathbf{B}_1$ field and the curling fingers give the direction of rotation of the magnetization vector. In the transverse plane of the rotating frame, the left-hand thumb points to the direction of the $\mathbf{B}_o$ field and the transverse magnetization rotates in the direction of the curling fingers. By this convention, a magnetization will rotate in the following manner

$$+\mathbf{M}_z \xrightarrow{(90^o)_x} +\mathbf{M}_y \rightarrow +\mathbf{M}_x \rightarrow -\mathbf{M}_y \rightarrow -\mathbf{M}_x. \quad (13.26)$$

At a higher level of NMR description, it is the spin angular momentum operator, to be called density operator later, which rotates in the operator subspace under the influence of another operator. The operator subspace is isomorphous to the Cartesian axes system, where the operator rotation is conventionally described by the right-hand rule by which the thumb coincides with the direction of the rotation vector and the rotating operator rotates in the direction of the curling fingers (Figure 13.7).

If $\Omega \neq 0$, which often is the case, there will be Larmor precession in the rotating frame ($\Omega = \gamma\Delta B_o$) implying that $\Delta B_o \neq 0$. Under this condition we have two magnetic fields, $\mathbf{B}_1$ and $\Delta\mathbf{B}_o$, which together define the effective magnetic field B_{eff} in the rotating frame

$$B_{eff} = \sqrt{B_1^2 + \Delta B_o^{\,2}}. \quad (13.27)$$

Normally B_{eff} is expressed in terms of frequency ω_{eff} (Figure 13.8). Using $\omega_{eff} = \gamma B_{eff}$, $\omega_{rf} = \gamma B_1$, and $\Omega = \gamma\Delta B_o$ in the above expression, we get

$$\omega_{eff} = \sqrt{\Omega^2 + \omega_{rf}^{\,2}}. \quad (13.28)$$

The significance of the off-resonance effect is that the spin vector now rotates about this tilted effective field ω_{eff} (Figure 13.8), and not ω_{rf}, as was the case for the on-resonance condition.

13.10 BLOCH EQUATIONS IN THE ROTATING FRAME

The magnetization relaxation times T_1 and T_2 can be incorporated into the equation of motion of **M** (equation 13.20) to obtain the Bloch's equations in a compact form. In the laboratory frame we have

$$\frac{d\mathbf{M}(t)}{dt} = \gamma\mathbf{M}(t)\times\mathbf{B}(t) - \tilde{\mathbf{R}}\{\mathbf{M}(t) - \mathbf{M}\} \quad (13.29)$$

where $\mathbf{M} = \mathbf{M}_o$ is the equilibrium magnetization, and $\widetilde{\mathbf{R}}$ is the relaxation matrix given by

$$\widetilde{\mathbf{R}} = \begin{pmatrix} 1/T_2 & 0 & 0 \\ 0 & 1/T_2 & 0 \\ 0 & 0 & 1/T_1 \end{pmatrix}. \quad (13.30)$$

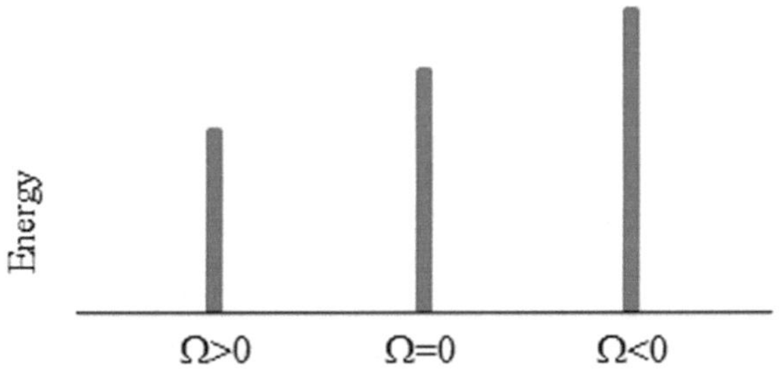

FIGURE 13.6 The relative energy shift of a non-interacting spin with reference to the carrier frequency ω_{rf} and the Larmor precession frequency ω_o in the rotating frame; $\omega_{rf} = \omega_o$ *(middle)*, $\omega_{rf} < \omega_o$ *(left)*, $\omega_{rf} > \omega_o$ *(right)*.

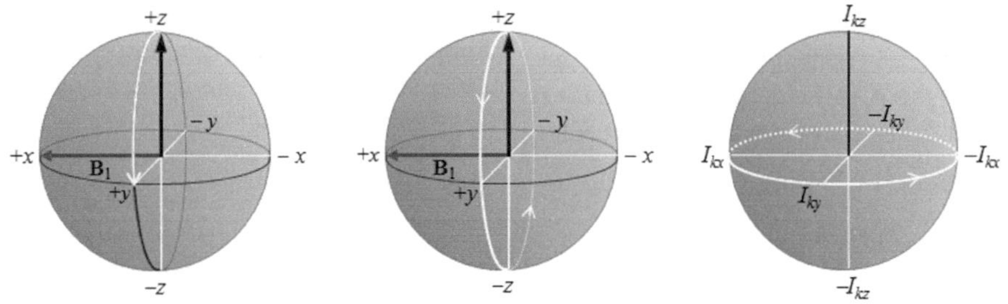

FIGURE 13.7 Conventional directions of rotation of classical magnetization vector under the influence of a 90° pulsed B_1 field *(left)* and continuous B_1 field *(middle)*. In the density operator formalism of spin evolution, the operator rotation is described by the right-hand rule by which the thumb points to the Cartesian operator about which the rotation occurs *(right)*.

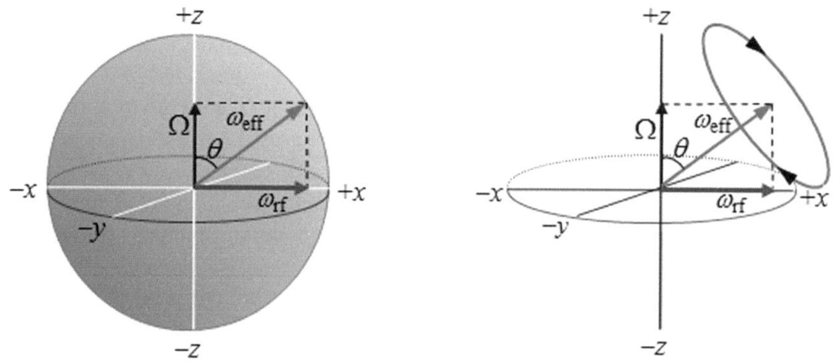

FIGURE 13.8 The tilted effective magnetic field ω_{eff} in the rotating frame. If an exact on-resonance condition is achieved, then $\Delta B_o = \Omega = 0$ and $\omega_{\text{eff}} = \omega_{\text{rf}}$, hence no tilted field. The rotation of the spin magnetization about the tilted ω_{eff} is shown at the *right*.

The total magnetic field in the rotating frame is

$$\mathbf{B}(t) = \mathbf{B}_o + \mathbf{B}_1(t). \tag{13.31}$$

In the rotating frame, the applied RF field is conventionally given with a phase ϕ with respect to the $+x$ axis; for example, $\phi = \pi/2$ if the field is applied along $+y$. The general form is

$$\mathbf{B}_1(t) = \mathbf{B}_1^o \cos(\omega_{\text{rf}} t + \phi)\hat{e}_x - \mathbf{B}_1^o \sin(\omega_{\text{rf}} t + \phi)\hat{e}_y. \tag{13.32}$$

The procedure for obtaining the rotating frame from the laboratory frame has been described above (equation (13.25)). We now also transform the time dependent magnetization vector $\mathbf{M}(t)$ from the laboratory frame to the rotating frame using the same rotation matrix $\widetilde{\mathbf{R}}^{\text{r}}(t)$

$$\mathbf{M}^{\text{r}}(t) = \mathbf{M}(t)\widetilde{\mathbf{R}}^{\text{r}}(t)$$

$$\begin{pmatrix} M_x^{\text{r}} \\ M_y^{\text{r}} \\ M_z^{\text{r}} \end{pmatrix} = \begin{pmatrix} M_x \\ M_y \\ M_z \end{pmatrix} \begin{pmatrix} \cos\theta & \sin\theta & 0 \\ -\sin\theta & \cos\theta & 0 \\ 0 & 0 & 1 \end{pmatrix}, \tag{13.33}$$

in which the superscript ‘r’ refers to the rotating frame. The Bloch equation in the rotating frame is then

$$\frac{d\mathbf{M}^{\text{r}}(t)}{dt} = \gamma \mathbf{M}^{\text{r}}(t) \times \mathbf{B}^{\text{r}} - \widetilde{\mathbf{R}}\{\mathbf{M}^{\text{r}}(t) - \mathbf{M}\}. \tag{13.34}$$

Notice that the time dependence of $\mathbf{B}^{\text{r}}$ is not shown because $\mathbf{B}_o$ is static, and the frame is already rotating at the frequency of $\mathbf{B}_1$. In the rotating frame, the transverse components of $\mathbf{B}^{\text{r}}$ are obtained from the phase of the $\mathbf{B}_1(t)$ field, but the longitudinal component is the reduced field ΔB discussed in the preceding section. These field components are

$$B_x^{\text{r}} = B_1 \cos\phi$$

$$B_y^{\text{r}} = B_1 \sin\phi$$

$$B_z^{\text{r}} = -\frac{\Omega}{\gamma}. \tag{13.35}$$

With the terms for the respective components of $\mathbf{M}^{\text{r}}(t)$ that are $\mathbf{B}^{\text{r}}$, relaxation times, and the equilibrium magnetization $\mathbf{M}$, the Bloch equations in the rotating frame appear as

$$\frac{dM_x^{\text{r}}(t)}{dt} = \gamma\left[-M_y^{\text{r}}\frac{\Omega}{\gamma} - M_z^{\text{r}} B_1 \sin\phi\right] - \frac{M_x^{\text{r}}}{T_2}$$

$$\frac{dM_y^{\text{r}}(t)}{dt} = \gamma\left[-M_x^{\text{r}}\frac{\Omega}{\gamma} + M_z^{\text{r}} B_1 \cos\phi\right] - \frac{M_y^{\text{r}}}{T_2}$$

$$\frac{dM_z^{\text{r}}(t)}{dt} = -\gamma\left[M_x^{\text{r}} B_1 \sin\phi - M_y^{\text{r}} B_1 \cos\phi\right] - \frac{\left(M_z^{\text{r}} - M\right)}{T_1}. \tag{13.36}$$

13.11 RF PULSE AND SIGNAL GENERATION

NMR transitions can be induced by scanning the sample with a continuous radio wave, which is what was done earlier, but the use of RF pulses and Fourier transformation of the time-domain signal, a luminous idea of Ernst and Anderson (Ernst and Anderson, 1966; Ernst, 1966; Aue et al., 1976), are invariably practiced now. The pulsed RF field has a large bandwidth of radiofrequency which can excite resonances of all nuclei in a molecule. If the Larmor frequency ν_0 is 500 MHz, a pulse of 8 μs duration for example, will have 4,000 different frequencies in a range of 125 kHz. As discussed in the preceding chapter, the frequency bandwidth is commensurate with the pulse time, so shorter pulse times will have wider frequency components. The pulse duration τ_p to rotate the magnetization about the direction of an applied field by an angle β is given by (Ernst et al., 1988)

$$\beta = -\gamma B_1 \tau_p. \tag{13.37}$$

If the RF field in the rotating frame is applied along +y direction, written as $(\pi/2)_y$, the field is

$$\mathbf{B}_1(t) = 2\mathbf{B}_1^o \cos(\omega_{\text{rf}} t + \phi)\hat{e}_y. \tag{13.38}$$

The factor 2 is included to implicate the left and right circularly polarized components of the amplitude vector. The RF is removed immediately after the specified pulse time; the pulse cut-off time can be counted as zero-time t_0, because the pulse duration is just a few microseconds. The effect of a $(\pi/2)_y$ pulse at time t_0 will be

$$M \xrightarrow{(\pi/2)_y} M_z \cos\beta + M_x \sin\beta, \tag{13.39}$$

with which the three magnetization components can be written as

$$\begin{aligned} M_x(0) &= M_x \sin\beta \\ M_y(0) &= 0 \\ M_z(0) &= M_z \cos\beta. \end{aligned} \tag{13.40}$$

In the simplest experiment (Figure 13.9), the magnetization is rotated through 90° by applying a $(\pi/2)_y$ pulse of a specified pulse time to create the transverse magnetization. For a perfect pulse with rotation angle $\beta = \pi/2$, the magnetization components of equation (13.40) immediately after the pulse are

$$\begin{aligned} M_x(0) &= M_x \\ M_y(0) &= 0 \\ M_z(0) &= 0. \end{aligned} \tag{13.41}$$

At times $t > t_0$, the transverse component evolves by free precession of $M_x(0)$ about B_0 in the rotating frame (equation (13.27)). Starting from $M_x(0)$, the magnetization should precess at the Larmor frequency $\omega_0 = -\gamma B_0$, but since the xy plane is also rotating at the carrier frequency (ω_{rf}) the relative motion of the two gives the net precession frequency of the magnetization $\omega_0 - \omega_{rf} = \Omega$, which is called chemical shift. In principle, magnetizations of different nuclei will precess at different Ω because each nucleus may be differently shielded from the effect of B_0. It is also clear that $M_x(0)$ in the rotating frame will be motionless and not evolve if $\Omega = 0$, which means the precession frequency of the magnetization reflects the chemical shift. This precession is free of the $\mathbf{B}_1$ field which is turned off already at t_0 and hence is called free precession (precession in the absence of the $\mathbf{B}_1$ field).

The time evolution of $M_x(0)$ by free precession to $M_{xy}(t)$ can be decomposed into $M_x(t)$ and $M_y(t)$

$$\begin{aligned} M_x(0) &\xrightarrow{\Omega t} M_x \sin\beta\left[\cos(\Omega t) + \sin(\Omega t)\right] \\ M_x(t) &= M_x \sin\beta \cos(\Omega t) e^{-t/T_2} \\ M_y(t) &= M_x \sin\beta \sin(\Omega t) e^{-t/T_2}. \end{aligned} \tag{13.42}$$

The inclusion of the exponential term containing the time T_2, although phenomenological, indicates relaxation of the transverse magnetization simultaneous with chemical shift evolution. A magnetization precessing freely at the frequency Ω induces an alternating current (AC) of the same frequency in a receiver coil situated along the x-axis of the rotating frame. This is the free induction signal that decays exponentially to zero due to the transverse relaxation T_2. The signal acquired is therefore called the free induction decay (FID). Since the transverse magnetization in the complex form is

$$M_{xy}(t) = M_x(t) + iM_y(t), \tag{13.43}$$

a complex FID is recorded by detecting the induced voltage due to both M_x and M_y in the rotating frame, called quadrature detection. The quadrature procedure employs simultaneous phase shifters (±45°) to record both magnetizations to produce x and y channel signals, which are then mixed to obtain the complex FID. Each individual magnetization with respective frequency Ω produces a FID of that frequency. If there are k ($k = 1, 2, 3, \cdots$) chemically different spins – meaning the individual nuclei are differently shielded from B_0, there should in principle be k different precession frequencies, $\Omega_k = \omega_{0,k} - \omega_{rf}$, that would give rise to a composite FID (Figure 13.9).

A complex Fourier transformation of the time-domain FID signal $S(t)$ yields the complex spectrum in the frequency domain

$$\begin{aligned} S(\omega) &= \int_0^\infty S(t) e^{-i\omega t} dt \\ S(\omega) &= v(\omega) + iu(\omega), \end{aligned} \tag{13.44}$$

where $v(\omega)$ corresponds to the magnetization component 90° out-of-phase with respect to the direction of the RF field, and $u(\omega)$ is the in-phase component. These components work out to

$$\begin{aligned} v(\omega) &= M_z \frac{1/T_2}{(1/T_2)^2 + (\omega - \Omega)^2} \\ u(\omega) &= M_z \frac{(\omega - \Omega)}{(1/T_2)^2 + (\omega - \Omega)^2}. \end{aligned} \tag{13.45}$$

The $v(\omega)$ term corresponds to a Lorentzian signal, and is called absorption-mode signal while the $u(\omega)$ term produces a dispersive signal, called dispersion-mode lineshape (Figure 13.10). This distinction of lineshape was also possible to achieve by setting the hardware appropriately in the first-generation continuous-wave NMR. Although the FWHM of the Lorentzian should be $2/T_2$ (Figure 13.10), it is mostly not. The inhomogeneity of the magnetic field through the sample volume often broadens the line, called inhomogeneity broadening, resulting in underestimation of the actual T_2. The field can be shimmed to a large extent of homogeneity, but not absolutely. The linewidth therefore gives only an apparent relaxation time T_2^*.

In the description above the receiver was taken to share x- and y-axes of the rotating frame by a phase shifter hardware.

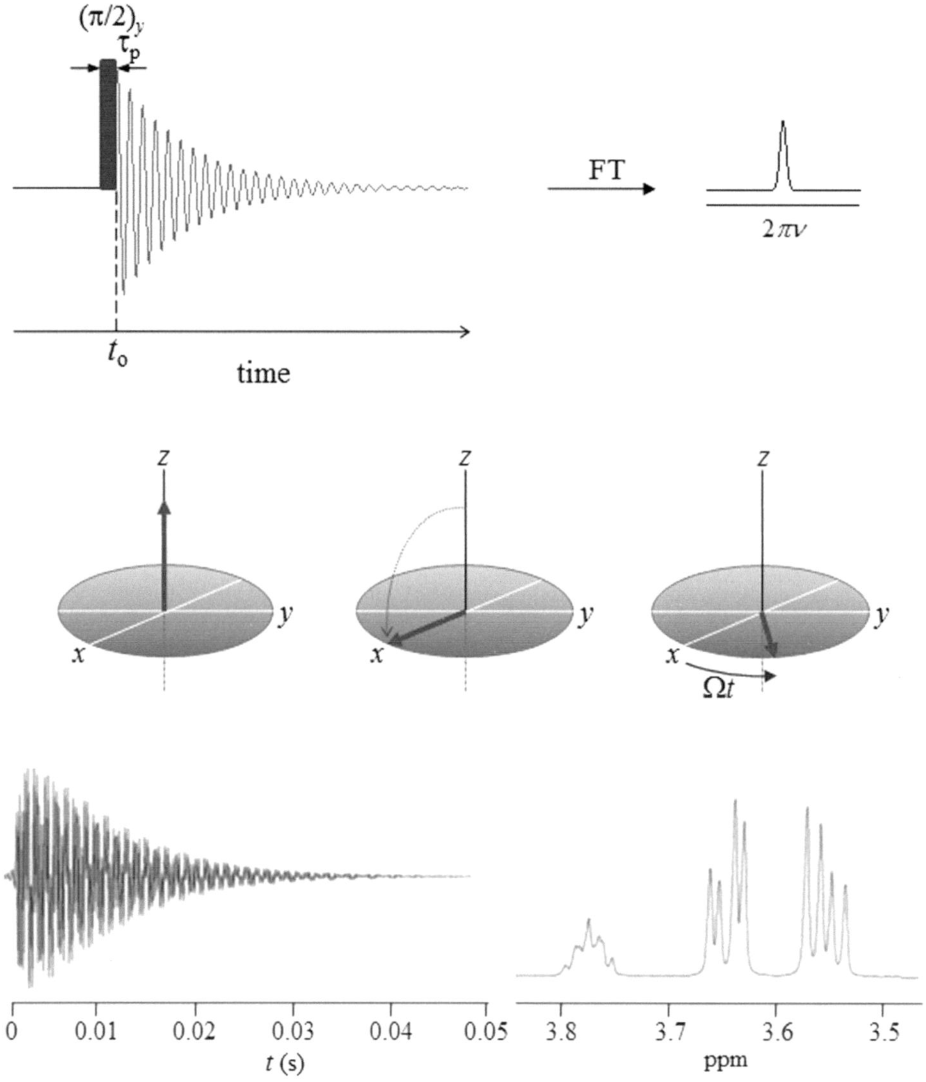

FIGURE 13.9 *Top*, a simulated 5 kHz AC signal due to the evolution of the transverse component of the magnetization ($T_2 = 0.4$ s) following a 90° RF pulse. Note the instantaneous rotation of the magnetization vector into the *xy* plane due to the 90° pulse, so the time domain signal depicted in the upper panel starts at $t = 0$. *Middle*, the evolution of the magnetization with time. *Bottom*, the FID and the corresponding spectrum (500 MHz) of 5% glycerol.

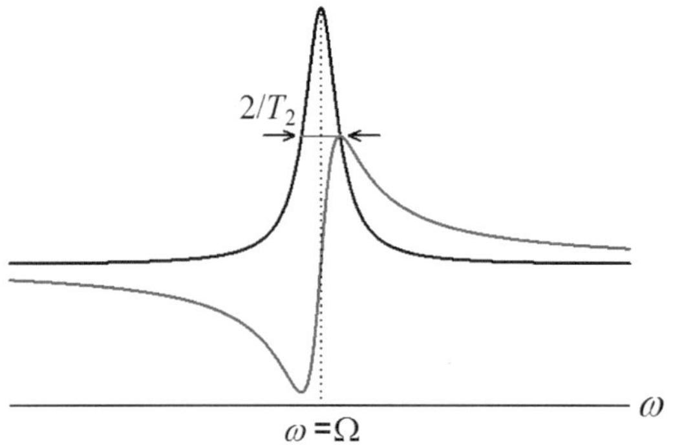

FIGURE 13.10 Absorption and dispersion mode signals, in black and cyan lines, respectively.

But the receiver can be aligned along any of the two transverse axes ($\pm x$ or $\pm y$), a feature readily available in the later generations of FT-NMR spectrometers. The absorption and dispersion mode lineshapes, corresponding to *x*-alignment of the receiver but different phases of the transverse magnetization at the beginning of the FID signal, are shown in Figure 13.11. Note that in this example we have aligned the receiver along $+x$ for all measurements. Pure absorption mode lineshape will be obtained if the magnetization is along $+x$ (positive absorption) or *x* (negative absorption). Dispersion mode lineshape are obtained if the magnetization is along $+y$ (positive dispersion) or *y* (negative dispersion). If the receiver is aligned along $+y$ axis, the lineshapes obtained are absorptive for *y* magnetization and dispersive for *x* magnetization. These comments are depicted in Figure 13.11, where the receiver is always *x*-aligned, and the magnetization vectors in the *xy* plane are represented by bold arrows.

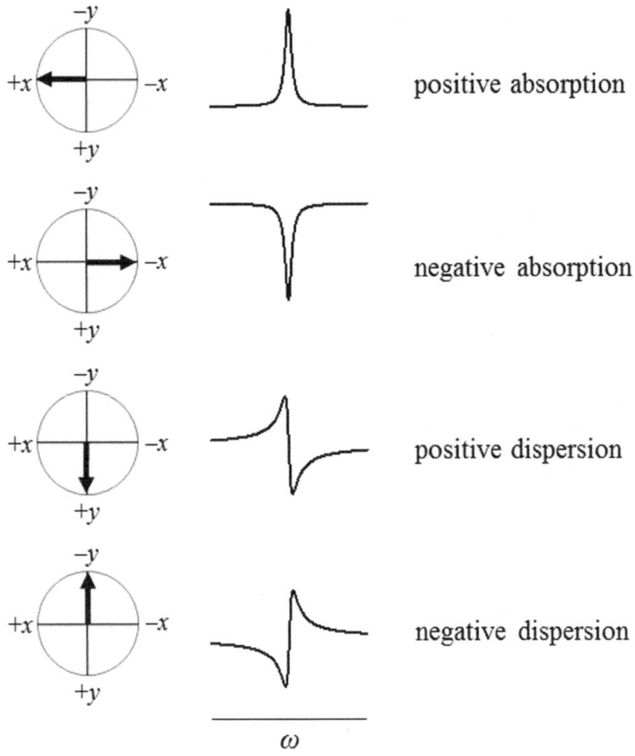

FIGURE 13.11 Lineshapes in the Fourier-transformed spectrum when the receiver phase is kept constant at $+x$ to detect $\pm x$ and $\pm y$ magnetizations.

13.12 ORIGIN OF CHEMICAL SHIFT: LOCAL SHIELDING

Chemical shift arises from magnetic shielding of nuclei in the presence of the same external static field $\mathbf{B}_0$, which is employed to prepare the energy eigenstates, although $\mathbf{B}_0$-independent molecular effects also contribute. Before considering the molecular effects, it is useful to analyze the direct magnetic induction at the nuclear site due to the presence of the static field. The $\mathbf{B}_0$-induced motion of electrons around the nucleus generates a tiny secondary magnetic field, which changes the actual strength of B_0 at the nuclear site to an effective field B_{eff}, given by

$$B_{\text{eff}} = B_0 - B_0\sigma = B_0(1-\sigma). \tag{13.46}$$

The formula implicates a quantity called shielding constant σ for the change of the transition frequency from $\omega_0 = |\gamma| B_0$ to $\Omega = |\gamma| B_0 (1-\sigma)$. The frequency Ω is called chemical shift frequency. The meaning of the quantity σ is also clear, it is the extent to which the magnetic field originating from the electron motion augments or retards the primary field B_0.

The magnetic induction at the nucleus is best understood by taking a free atom in the presence of an external magnetic field. Consider classically what will happen if negatively charged electrons q^- already revolving in space-fixed orbits with certain velocity $\mathbf{v}$ are introduced into the static magnetic field $\mathbf{B}_0$. By elementary concepts of magnetism, the electron experiences a force

$$\mathbf{F} = q\mathbf{v} \times \mathbf{B}_0, \tag{13.47}$$

which is perpendicular to the linear velocity vector. Since the force acts constantly while the electron moves with a constant velocity $\mathbf{v}$, the result will be a circular motion of the electron in the uniform magnetic field (Figure 13.12). The direction of the electron circulation follows the right-hand rule in which $\mathbf{B}_0$ points along the thumb of the right hand and the curling fingers correspond to the electron circulation direction. One might like to call this electron motion 'circular diamagnetic motion' since it produces a diamagnetic effect at the nuclear site (see below). Using the mass of the electron m_e we can calculate the radius of the circular path of motion by Newton's second law

$$F = m_e a$$

$$qvB_0 = m_e \frac{v^2}{r}$$

$$r = \frac{mv}{qB_0}. \tag{13.48}$$

It is convenient to consider the motion of all electrons together as the motion of a charge cloud whose angular velocity

$$\boldsymbol{\omega} = \frac{e}{2m_e}\mathbf{B}_0 \tag{13.49}$$

coincides with the direction of $\mathbf{B}_0$. We can also show that the linear velocity of the charge cloud is

$$\mathbf{v} = \boldsymbol{\omega} \times \mathrm{r} = \omega r \sin\theta. \tag{13.50}$$

The motion of the electrons imply the existence of a current density $\mathbf{j}$ which is proportional to charge density ρ_e and the

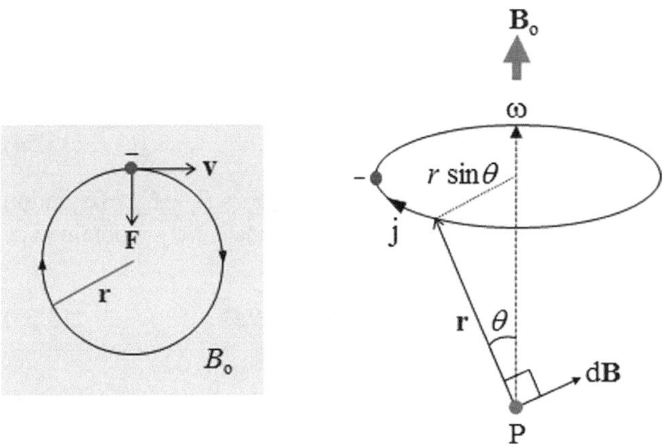

FIGURE 13.12 *Left*, circular motion of a negative charge, the electron, in a uniform magnetic field. The shaded area of the figure is a cross section of the magnetic field B_0 whose direction is perpendicular to the plane of the paper. *Right*, the angular velocity vector coincides with the field axis, and the secondary field dB produced at the nuclear site (point P) is perpendicular to the distance $\mathbf{r}$ from P to the circular path of the electron charge.

velocity **v**. Using equations (13.49) and (13.50), the current density is obtained as

$$\mathbf{j} = -e(\boldsymbol{\omega}\times\mathbf{r})\rho_e = \frac{e^2}{2m_e}(\mathbf{B}_0\times\mathbf{r})\rho_e. \tag{13.51}$$

The magnetic induction at the nuclear site (point P in Figure 13.12) due to a linear current element dl of current i is given by Biot-Savart law (see Resnick and Halliday, 1966) as

$$dB = \frac{\mu_0 i}{4\pi}\frac{dl\sin\theta}{r^2}, \tag{13.52}$$

in which μ_0 is the permeability of free space. Now consider a volume element dV of the current density j instead of the linear element dl of current i, and write the Biot-Savart formula in the vector form

$$d\mathbf{B} = -\frac{\mu_0}{4\pi r^3}(\mathbf{j}\times\mathbf{r})dV.$$

Substitution for **j** from equation (13.51) yields

$$d\mathbf{B} = -\frac{\mu_0 e^2}{8\pi m_e r^3}(\mathbf{B}_0\times\mathbf{r})\times\mathbf{r}\rho_e dV,$$

and the strength of magnetic induction at the nucleus (point P) is

$$dB = -\frac{\mu_0 e^2}{8\pi m_e r}B_0\rho_e\sin\theta dV. \tag{13.53}$$

We are interested in the z-component of the induced field **B** because the static field $\mathbf{B}_0$ is oriented along z so that the two fields can be combined. The z-component of the induced field is $-dB\sin\theta$ for the volume element dV. The total z-component for the entire 'spherical' volume of the current density is obtained by integrating dB over the total space

$$B_z = -\frac{\mu_0 e^2}{8\pi m_e}B_0\int\frac{\rho_e}{r}\sin\theta dV = -B_0\left(\frac{\mu_0 e^2}{8\pi m_e}\int\frac{\rho_e}{r}\sin\theta dV\right). \tag{13.54}$$

Since the effective field $B_{\text{eff}} = B_0 - B_z = B_0 - B_0\sigma$ (equation (13.46)), the diamagnetic shielding constant σ_d is obtained as

$$\sigma_d = \frac{\mu_0 e^2}{8\pi m_e}\int\frac{\rho_e}{r}\sin\theta dV. \tag{13.55}$$

The integrated form of this in spherical polar coordinate gives the Lamb term

$$\sigma_d = \frac{\mu_0 e^2}{3m_e}\int_0^\infty r\rho_e dr. \tag{13.56}$$

The shielding is diamagnetic because B_z is negative and therefore the shielding is positive, implying that the magnetic induction acts to diminish B_0.

The derivation shown here applies to atoms where electrons are not delocalized, which suggests that shielding in atoms is exclusively diamagnetic and leads to low-frequency NMR transition. In fact, it turns out that diamagnetic shielding in free atoms increases with atomic number. For molecules, where atoms are bonded and the electrons are delocalized, the electrical potential is no longer spherical. The presence of many nuclei act to attract the electrons toward them which hinders the diamagnetic motion (circulation) of electrons around a given nucleus. Because the atoms are bonded, the hindrance to electron motion around the nucleus prevails for all atoms in the molecule. It would appear as though multiple nuclei oppose the diamagnetic effect σ_d, giving rise to a second-order paramagnetism in molecules. In other words, unlike the case of exclusive diamagnetic shielding as for free atoms, nuclei in molecules also experience an opposite shielding – called paramagnetic shielding σ_p in addition to the diamagnetic shielding σ_d.

A classical description of paramagnetic shielding is difficult to provide. In the initial quantum calculations of Ramsey (1950), it was assumed that all nuclei but the one whose shielding is to be calculated have no magnetic moment, and the magnetic moment **μ** of the nucleus under consideration is parallel to $\mathbf{B}_0$. An electronic Hamiltonian is constructed which contains the vector potential from **μ** and $\mathbf{B}_0$, and the electrostatic potential energy function for all electrons in the molecule. The electronic wavefunctions are formed by a linear combination of the eigenstates that would have been there if there was no $\mathbf{B}_0$. The energy of the whole molecule is now calculated using the second-order perturbation theory, but only those energy terms E'_λ which are linear with respect to μ and B_0 in the Zeeman energy, $E_{\text{Zeeman}} = |\mu B_0|$, are retained. These energy terms are proportional to the overall shielding constant σ_λ for the pertinent nucleus

$$E'_\lambda = \sigma_\lambda|\mu B_0|$$
$$\sigma_\lambda = \sigma_d + \sigma_p. \tag{13.57}$$

The shielding constant thus calculated contains contributions from both diamagnetic and paramagnetic shielding, and the σ_d term here is the same σ_d that is obtained for the free atom in Lamb's formula (equation (13.56)). It should, however, be stressed that the two shielding terms are not independent, and a clear separation of σ_d and σ_p and their quantification individually are not achieved; a change in the value of σ_d also alters the value of σ_p and vice versa.

13.13 LONG-RANGE SHIELDING

The shielding σ_d and σ_p considered thus far are also called local molecular shielding, σ_d^{local} and σ_p^{local}; by local one means not long-range molecular shielding. In molecules, several factors that include ring current, electric dipoles, paramagnetic centers, bond magnetic anisotropy, hydrogen bonds, and solvent molecules in the proximity of an atom can shift the NMR transition frequency. These are called long-range shielding. The overall shielding of a nucleus is the sum of σ_d^{local}, σ_p^{local}, and any kind of long-range shielding the nucleus may be experiencing. Thus, the total shielding in general is

$$\sigma = \sigma_d^{\text{local}} + \sigma_p^{\text{local}} + \sigma_r + \sigma_e + \sigma_m + \sigma_H + \sigma_{hfs} + \sigma_s + \cdots \tag{13.58}$$

where the long-range shielding terms are

σ_r ring-current
σ_e electric field
σ_m neighboring bond magnetic anisotropy
σ_H hydrogen bonding
σ_{hfs} hyperfine sheilding
σ_s solvent effect

13.13.1 Ring Current Effect, σ_r

The diamagnetic circulation of delocalized electrons of unsaturated rings (Figure 13.13) in molecules generates a substantial current because the nuclei may not prevent the motion of the electrons. If $\mathbf{B}_0$ is perpendicular to the plane of the ring, the electron motion will be along the curling fingers of the right hand, hence the current flows in the opposite direction. The magnetic induction due to the current produces a large magnetic moment in the direction opposite to that of $\mathbf{B}_0$, and the secondary magnetic field $\mathbf{B}$ so generated is parallel to $\mathbf{B}_0$ outside the ring and antiparallel inside the ring (see the magnetic flux lines in Figure 13.13). This implies that $\mathbf{B}_{\text{eff}}$ experienced by nuclei outside the planar ring is larger than $\mathbf{B}_0$, and the relation $B_{\text{eff}} = B_0(1-\sigma)$ would indicate a negative value of σ, suggesting high-frequency shift or deshielding of these nuclei. Recall that shielding means relatively low magnetic field experienced by a nucleus leading to a low-frequency shift of the NMR transition, and deshielding means a higher field and hence a high-frequency shift. Indeed, the protons of imidazole or benzene show a high-frequency shift. If a nucleus is positioned above or below the plane of a ring or inside the ring then shielding results, and hence a low-frequency shift. In the case of porphyrin, the ring current produces high-frequency shift for the protons outside the ring and low-frequency shift for those inside.

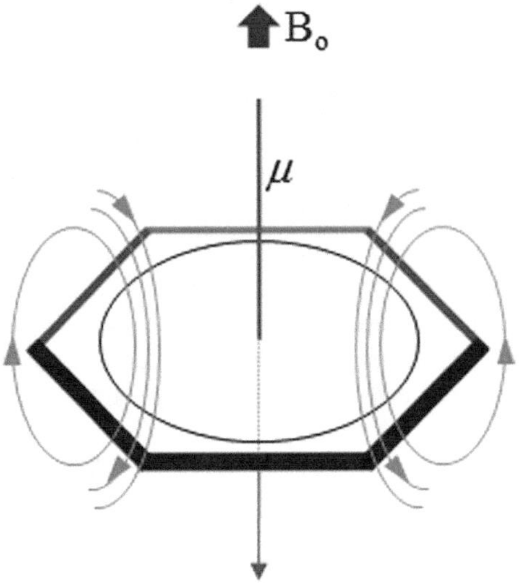

FIGURE 13.13 Ring current magnetization in a benzene that augments the total magnetic field for protons outside the ring, but retards the field inside the ring.

The ring current effect is highly sensitive to the vertical distance of a nucleus from the plane of the ring. It is also sensitive to the distance of a nucleus from the symmetry axis $d(\approx \mu)$ that passes vertically through the center of the ring. In both cases, the shielding increases rapidly with the distance, meaning the transition frequency shifts to lower values. The size of the ring current effect is directly proportional to the area of the ring and the number of π-electrons. For example, the area of the benzene ring (~6.15 Å²) is smaller than that of the porphyrin ring (~7.1 Å²), and this is reflected in significantly larger shielding for the latter. Larger shielding is indeed observed for not only naked porphyrin, but also heme porphyrin-containing protein systems such as myoglobin and cytochrome *c*. In a large ring that can be represented as a system of several rings fused, the total ring current shift can often be equated to the sum of the effects of the individual rings. In systems naphthalene, indole, and porphyrin, for example, the effect of each ring is added up to obtain the ring current shift.

Another interesting characteristic of ring current shift is its temperature independence, which is readily tested with a ring-containing molecule that does not undergo structural and conformational changes within a selected range of temperature. The ring current effect appears isotropic because the observed shift is an average of all orientations of the ring with respect to the direction of $\mathbf{B}_0$. However, no effect will be observed if the ring plane is constrained to be parallel to $\mathbf{B}_0$, implying that ring current shift of NMR transitions is actually anisotropic.

13.13.2 Electric Field Effect, σ_e

An internal electric field of small magnitude arises in molecules due to the presence of strongly polar groups such as C=O, C–F, and N=O, which act as source dipoles $\delta^+ \rightarrow \delta^-$. If such a dipole is taken as a point charge q, then the familiar Coulomb's law provides the electric field

$$\mathbf{E} = \frac{1}{4\pi\varepsilon_0 \varepsilon} \frac{q}{r^2} \hat{\boldsymbol{i}}, \tag{13.59}$$

in which ε_0 and ε are respectively the free-space permittivity and the dielectric constant of the medium, r is the distance from q to the nucleus whose shielding is to be determined, and $\hat{\mathrm{i}}$ is a unit vector from the observation nucleus to q. The magnitude of this Coulomb field, presumably uniform, enters into the shielding expression as a power series in E

$$\sigma_e = AE + BE^2 + \cdots \tag{13.60}$$

There are higher order terms that are not important. The coefficient A is the nuclear shielding polarizability tensor or the linear electric field coefficient, and B is the nuclear shielding hyperpolarizability tensor or the quadratic electric field coefficient. Values of both A and B differ from one dipole bond source to another.

In addition to the terms AE and BE^2, a third term called van der Waals shielding could arise if the bond containing the nucleus under consideration and another nearby bond mutually induce time-dependent electric dipoles, and the field so generated is included as an averaged quadratic field $\langle E^2 \rangle_{\text{vdw}}$

$$\sigma_e = \left(AE + BE^2\right) + \langle E^2 \rangle_{\text{vdw}}. \tag{13.61}$$

Note that the van der Waals shielding field is not a part of the power series expansion shown in equation (13.60); this is shown by placing the first two terms within braces. The value of $\langle E^2 \rangle_{\text{vdw}}$ has a r^{-6} dependence (characteristic of dispersion interaction), r being the distance between the time-varying partner dipole to the nucleus being shielded. The $\langle E^2 \rangle_{\text{vdw}}$ value also depends on the polarizability of the interacting bond.

Suppose there is no van der Waals shielding field, and the quadratic electric field term is also negligible – a situation that can definitely prevail as the size of the molecule becomes smaller. In fact, the earliest calculation of σ_e by Marshall and Pople (1958) used the linear electric field alone, the quadratic term was introduced later by Buckingham (1960). If the isotropy of chemical shift in solution NMR is also invoked, the electric field shielding to first order is simplified to

$$\sigma_e = A_x E_x + A_y E_y + A_z E_z, \tag{13.62}$$

in which A_x, A_y, and A_z are the polarizabilities along x,y,z axes of a molecular frame. Now consider the polarization of a bond to which the shielding nucleus is attached. It is the N–H bond in Figure 13.14, and one may be interested in the shielding of H due to q generated by a strong polar bond. The bond polarization affecting the proton occurs along the N–H bond only; hence, we can use the Coulomb formula and the shielding polarizability $A_{\|}$, which is A_x in equation (13.62) above, to write

$$\sigma_e = \frac{1}{4\pi\varepsilon_0 \varepsilon} A_{\|} \frac{q\cos\theta}{r^2}, \tag{13.63}$$

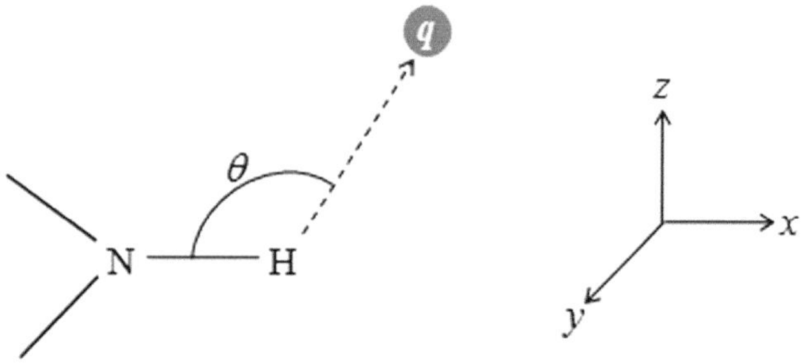

FIGURE 13.14 The geometry for shielding of an amide proton by the linear electric field due to a point charge dipole q.

where θ is the angle between the vector connecting the proton to the point dipole q and the N–H bond vector.

13.13.3 Bond Magnetic Anisotropy, σ_m

The localized electrons in chemical bonds act as a source of secondary magnetic field. The origin of the bond magnetic field is the same magnetic induction due to circulation of the electrons in the presence of the static field $\mathbf{B}_0$. The magnitude of the induced magnetic moment is proportional to the strength of the field B_0 and the proportionality is the bond magnetic susceptibility tensor χ_{ij}

$$\mu_i = \chi_{ij} B_{0,j}, \tag{13.64}$$

in which $i, j = x, y, z$. Since the static field is always along z ($j = z$), the χ_{iz} component of the tensor describes the induced magnetic moment in the ith direction. The secondary magnetic field B generated by the magnetic induction is given by

$$B = \mu_0 \left(\mu + B_0\right), \tag{13.65}$$

where μ_0 is the electric permittivity of free space. Importantly, the involvement of the proportionality tensor for the dependence of the induced magnetization on the external field (equation (13.64)) creates a directional restriction on shielding – only the anisotropy of bond susceptibility can cause shielding of a nearby nucleus, even though the molecules in the sample are tumbling.

Consider a magnetically induced bond as a point magnetic dipole $\boldsymbol{\mu}$ along with a nearby nucleus X whose shielding we are interested to evaluate, and let $\mathbf{r}$ be the position vector between $\boldsymbol{\mu}$ and X (Figure 13.15). With respect to the direction of $\mathbf{B}_0$, only two orientations of $\mathbf{r}$ will result in shielding at the nucleus X

$\mathbf{r} \parallel \mathbf{B}_0$ nucleus X is shielded (low-frequency shift)
$\mathbf{r} \perp \mathbf{B}_0$ nucleus X is deshielded (high-frequency shift).

The point magnetic dipole may be taken to be the center of the bond so that a shielding cone of half-angle 54.44° arises on either side of it along the bond axis (Figure 13.15). If the bonds are cylindrically symmetric, which most of them are, the two transverse principal susceptibilities are equal. This suggests that all nuclei situated within a cross section of the conic are deshielded to the same extent. Similarly, nuclei on either side of the bond axis at equal proximity from the bond termini are shielded to the same extent. If the angle between $\mathbf{r}$ and the bond axis is θ, the shielding is given by

$$\sigma_m = \frac{1}{3r^3} \frac{\left(\chi_{\|} - \chi_{\perp}\right)\left(1 - 3\cos^2\theta\right)}{4\pi}. \tag{13.66}$$

13.13.4 Shielding by Hydrogen Bonding, σ_H

A proton when hydrogen bonded intramolecularly or intermolecularly shows a large change in chemical shielding.

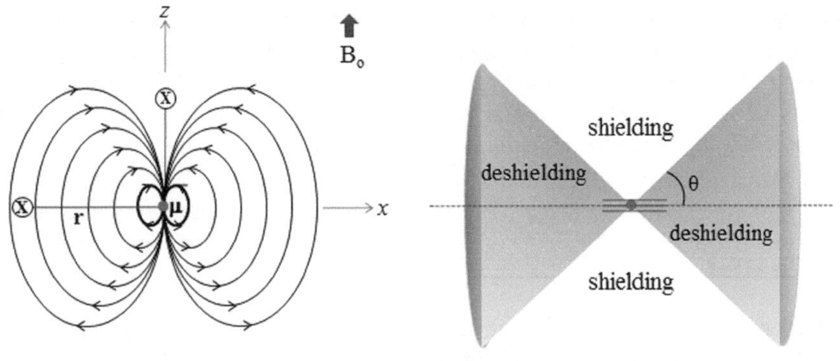

FIGURE 13.15 *Left*, magnetic field generated by a magnetic point dipole moment (brown circle at the center) which arises from the magnetic induction of the bond-associated electrons by B_o. The location of the nucleus X with respect to the point dipole is denoted by the vector **r**. The magnetic field in the transverse *xy* plane is along $+z$ direction, hence the nucleus X located in the plane experiences deshielding ($r \parallel B_o$). When $r \perp \mathbf{B}_o$, the nucleus X experiences a shielding effect. *Right*, the cone of shielding anisotropy due to a cylindrically symmetric triple bond.

The H-bonding all by itself cannot bring about such a sizable change in the NMR transition frequency; rather a multitude of factors, including magnetic anisotropies of the surrounding groups, ring currents of the nearby aromatic systems, local dihedral angles, local solvent composition, and the geometry of the H-bond itself, may contribute to the exquisite sensitivity of the bonded proton. To see the importance of these non-local contributions, let us consider qualitatively the electron density around the proton in the H-bond. Typically, a hydrogen bond X–H····Y is ~3 Å long, the length measured is the distance from the center of X to the center of Y through H. But the van der Waals radii of the donor and acceptor atoms (~1.55 Å) add up to cover the entire bond length, and this is true for all types of hydrogen bonds where the donor-acceptor atoms include N, C, and O. Hence the electron density around the proton is expected to be very large, producing substantial diamagnetic shielding (σ_d^{local}) and therefore very low-frequency transition, but this does not happen. A proton when hydrogen bonded may show a transition whose frequency is tens to hundreds of Hz higher relative to that of the same proton when not hydrogen-bonded. This must mean the operation of paramagnetic shielding (σ_p^{local}) as well as some of the surrounding effects like magnetic anisotropy and others listed above.

The details of the factors that contribute to lowering of the electron density in the hydrogen orbital are less known, but the geometry around the hydrogen appears to have a definite influence on the shielding of the H-bonded nucleus. In principle, the description of the hydrogen bond geometry should account for the disposition of atoms and bonds in addition to X–H····Y. For example, the angles θ and ϕ in the N–H···O=C type of intramolecular hydrogen bond (Figure 13.16) affect the electron density around the hydrogen and hence the shielding. In the simplified one-dimensional H-bond geometry, one considers the distances r_{OH} and R_{NO} and the angle $\theta = \angle$N-H···O (Figure 13.16). The geometry in this form can be used for theoretical studies of hydrogen-bonding and to describe hydrogen bond energy. The bent geometry has a net dipole and shorter r_{OH} and R_{NO} distances relative to those in the

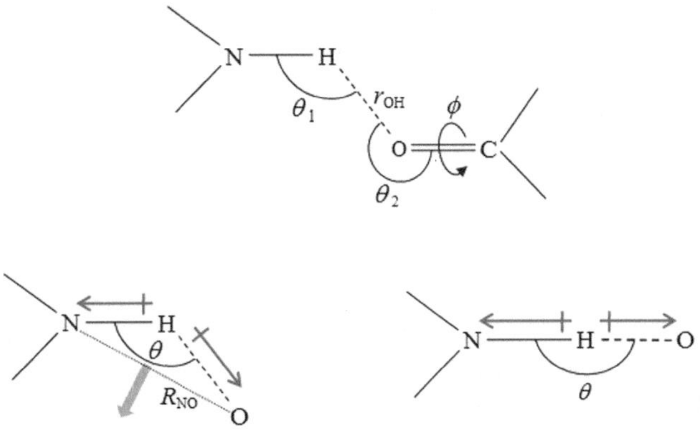

FIGURE 13.16 *Top*, angles that affect the electron density of the hydrogen orbital. *Bottom*, the reduced one-dimensional H-bond geometry showing two different bond directionalities. The thick arrow in blue (*left*) is the direction of the net bond dipole moment. In a symmetric linear geometry of the H bond (*right*) the net dipole moment tends to zero.

linear geometry, and the angle θ increases with deshielding (high-frequency shift) of the H-bonded proton. Because the opposite dipoles in the linear geometry tend to cancel out the net dipole in the bent geometry, the dipole decreases to a residual value which indicates that a decrease in the polarity of the N-H···O segment and hence a decrease in the overall directionality of the H-bond leads to deshielding of the proton.

13.13.5 Hyperfine Shielding, σ_{hfs}

This type of shielding occurs when an unpaired electron interacts with the resonating nucleus, so the shielding is dubbed hyperfine interaction shielding or hyperfine shielding. At this point the reader may revise the mechanism and the physical structure of electron-nuclear hyperfine interactions detailed for the hydrogen atom in Chapter 4. An unpaired electron somewhere in a molecule can get delocalized to a resonating nucleus through the intervening chemical bond or through space and establish the same hyperfine interaction that we already described. The key parameter of interest is the hyperfine coupling constant $\mathcal{A}$, which was called there simply a constant because the energy of hyperfine interaction is proportional to the spin density of the unpaired electron at the nucleus. We now see already that the hyperfine shielding of the nucleus $\sigma_{hfs} \propto \mathcal{A}$. Note that the hyperfine Hamiltonian is given by

$$\mathcal{H} = \mathcal{A}\mathbf{S}\cdot\mathbf{I},$$

where **S** and **I** are electron and nuclear magnetic moments. The connection of $\mathcal{A}$ might appear to complicate an NMR transition because the transition is expected to split commensurate with the value of $\mathcal{A}$. But the relaxation time of the electron (~10^{-13} s) is many orders of magnitude shorter than the relaxation time of the nucleus (~0.5 s). So, when RF excitation of the electron-coupled nucleus is carried out the electron relaxes almost instantaneously, causing a self-decoupling from the nucleus and thus rendering the NMR transition not to split. Yet, inefficient decoupling of the electron from the nuclear spin is often the case, resulting in broadening of hyperfine NMR lines, also called paramagnetic broadening.

When the shielding of a nucleus occurs by hyperconjugation or by delocalization of the unpaired electron through a bond, the shift in the NMR resonance frequency is called Fermi contact (FC) shift, which is proportional to the magnitude of the spin density of the unpaired electron (ρ) at the resonating nucleus. The change in ρ at the nucleus can occur by a direct overlap of the molecular orbital containing the unpaired electron with that containing the atomic orbital s of the resonating nucleus, leading to an increase of ρ. The overlap in some cases may be partial in the sense that the s orbital of the resonating atom is not involved – for example, in a $d_\pi - p_\pi$ overlap where the unpaired electron is in the d-orbital of a metal atom. In such a case, which is quite common for iron-ligand systems like iron porphyrin and heme-containing proteins, we invoke the spin polarization process. One of the paired electrons in the s orbital whose spin is parallel to the spin of the unpaired electron will tend to be in spatial proximity of the unpaired electron, causing the spin density of the other s electron to be larger at the nucleus. The difference of these two gives the unpaired spin density ρ at the nucleus which is now negative. The magnitude of isotropic contact shielding experienced by a nucleus, relative to its shielding had the molecule been diamagnetic, is given by the Bloembergen formula

$$\sigma_{\mathrm{c}} = -\left(\frac{\mathcal{A}}{\hbar}\right)\frac{g\mu_e S(S+1)}{3k_{\mathrm{B}}T\gamma} \tag{13.67}$$

in which $\mathcal{A}/\hbar$ is hyperfine coupling constant in rad s^{-1}, g is Lande g-factor which is assumed to be isotropic, μ_e is Bohr magneton, S is the total electron spin, k_{B} is Bolzmann constant, T is the absolute temperature, and γ is the gyromagnetic ratio of the nucleus. The factor $\mathcal{A}/\hbar$ is a number related directly to the unpaired spin density ρ at the observing nucleus, implying that contact shift is sensitive to the distribution of the unpaired electron(s) among different nuclei. Since the value of $\mathcal{A}$ changes among the resonating nuclei, the extent of shielding also varies; two protons may have resonance frequencies that are different by tens of thousands of Hz. Furthermore, the sign of $\mathcal{A}$, which can be positive or negative from one to another nucleus, determines the relative direction of shielding.

A different physical picture emerges when the hyperfine interaction occurs through space, called pseudo-contact interaction. Then it is a magnetic dipole-dipole interaction between the nuclear magnetic moment and the unpaired electron magnetic moment. An accurate evaluation of the magnitude of pseudo-contact shift (σ_{pc}) strictly requires the determination of spatial distribution of the electron spin which may be a difficult task to perform. What is used instead is the point-dipole approximation by which the position of the electron spin magnetic moment is restricted to the center of the paramagnetic atom. Since the unpaired electron spin density ρ at the nucleus is now irrelevant, the hyperfine coupling constant $\mathcal{A}$ is not included in the description. The involvement of the electron spin dipole requires the consideration of the dependence of electron spin relaxation time on the correlation time of molecular tumbling (τ_c) and the electron spin g-tensor. Note from above that the contact shielding expression assumes an isotropic g-factor. We will avoid the derivation of the expression but simply produce the general result used in literature (Wüthrich, 1976)

$$\sigma_{\mathrm{pc}} = \frac{2\mu_e S(S+1)}{3k_{\mathrm{B}}T} F\left(g, \frac{\Omega, \theta}{r^3}\right), \tag{13.68}$$

in which F is a function of the g-tensor anisotropy and polar coordinates Ω, θ, r of the nucleus with respect to the principal axes of the g-tensor whose origin lies at the electron spin point dipole. Pseudocontact interaction, being dipolar in origin, can shield nuclei that are many-bonds away from the electron spin dipole but closer to it spatially. This is of special significance in the case of paramagnetic polymers or protein chains where the folding of the chain can bring sequentially remote atoms to the proximity of the paramagnetic center.

13.13.6 Shielding from Solvent Effect, σ_s

The effect of solvent on solute nuclear shielding occurs by the same mechanisms discussed above, although which shielding mechanism(s) is appropriate depends on the polarity of the solvent and the solute, and the nature of the solute-solvent interaction, if they interact. A direct interaction can establish a H-bond where the solvent molecule can act as the donor and the solute the acceptor, and vice versa. One can then consider the hydrogen bonding effect σ_H in the magnetic shielding of the solute nucleus. The direct interaction may also entail a stacked arrangement in which the nitrogen of a solute amide can establish at least one contact with a solvent aromatic ring. Such contact distances are less than ~3.8 Å as seen in X-ray structures. Assuming their isotropy in the solution, both hydrogen bonding and ring current effects may be analyzed to determine the shielding of the solute nucleus.

Shielding at the solute nucleus can be large even if no direct solute-solvent interactions exist. When the solvent consists of aromatic rings, the operation of the ring current effect will obviously alter shielding of solute nucleus. In the case of alkyl group-containing polar solutes in a polar solvent, the former induces the so-called reaction field in which the solvent dipoles surrounding the solute molecule are oriented so as to cause electron drifts in the solute, thus causing a change in the shielding. Change in solute shielding can also occur if the solvent type has anisotropy of magnetic susceptibility.

The effect of water, which is almost always used as the solvent for biological molecules, has not been understood completely. Many potential H-bonding donors and acceptors are hydrogen-bonded to water, if not already intramolecularly, which is consistent with the argument that the lack of hydrogen bonding of peptide polar groups NH and CO is energetically expensive. A particularly inspiring example of the role of water in shielding is found in the NMR transitions of nuclei in α-helices and β-sheets of proteins. In general, the shielding of these nuclei is affected intramolecularly by local electrostatic energy (E_{loc}) arising from the dipolar interactions of the adjacent N–H and C=O dipoles (Figure 13.17), which becomes readily apparent from the linearity of a plot of resonance frequencies with the corresponding E_{loc} (Figure 13.18). But the slope of the plot changes if the nuclei whose local N–H and C=O dipoles are exposed to water. When the two dipoles of a peptide group are antiparallel as in the β-sheets structure (Figure 13.17) the two electric fields cancel each other out, leaving a residual dipole that can interact with a water dipole only marginally. In this case, the solvent water does not affect the shielding of the peptide group atoms. On the other hand, the peptide dipoles in a right-handed α-helix are parallel so that the magnitude of the resultant dipole is much larger, giving rise to a reinforced dipole. This field will be screened if a water dipole is brought closer, thus E_{loc} will change and hence the shielding of the nuclei. The extent of screening of the solute electric field depends on the extent of exposure of the solute dipoles to water.

13.13.7 Chemical Shift Scale

Up to now the shift of the NMR transition has been mentioned in terms of frequency. The frequency of a transition would, however, vary with the strength of the static field, because Larmor precession is proportional to the field strength, $\omega_0 = |\gamma| B_o$. Accordingly, the transition of a given nucleus measured at different field strength in different parts of the world will be reported differently. This is avoided by converting the linear frequencies to δ values scaled as ppm

$$\delta(\text{ppm}) = \frac{\nu_{\text{sample}} - \nu_{\text{reference}}}{\nu_{\text{reference}}} \times 10^6, \tag{13.69}$$

where $\nu_{\text{reference}}$ is the transition frequency of a reference molecule. Commonly used reference molecules are sodium trimethylsilyl propionate (TSP) or sodium trimethylsilyl propanesulfonate (DSS) for ^{1}H and ^{13}C, phosphoric acid or triphenyl phosphate for ^{31}P, and ammonium chloride for ^{15}N. All nuclei in a reference molecule are magnetically equivalent so that the molecule produces a single NMR line. The frequency $\nu_{\text{reference}}$ differs little from the Larmor frequency because nuclear shielding is so small that the formula $B_{\text{eff}} = B_o - B_o\sigma$ in equation (13.46) reduces to $B_{\text{eff}} \approx B_o$. This allows replacing the $\nu_{\text{reference}}$ in the denominator of the equation above with the Larmor frequency ν_o, which is the frequency of the spectrometer for the nucleus under study

$$\delta(\text{ppm}) = \frac{\nu_{\text{sample}} - \nu_{\text{reference}}}{\nu_o} \times 10^6 = \left(\sigma_{\text{reference}} - \sigma_{\text{sample}}\right) \times 10^6. \tag{13.70}$$

13.14 SPIN-SPIN COUPLING

The scalar interaction of closely neighboring nuclei that are separated by chemical bonds, also called through bond interaction, J, splits the resonance lines of the coupled nuclei. This is called spin-spin splitting or J-coupling and is commonly seen for those nuclei for which $I = 1/2$, although the splitting in quadrupolar nuclei can also be observed if the quadrupolar spectral broadening effect is removed. Consider only two chemically different but bonded spin-half nuclei ($I = 1/2$), which are distinguished by the respective chemical shifts that are far apart. In $CCl_3CH^{(A)}ClCH^{(X)}Cl_2$ for example, protons A and X that are separated by three bonds resonate at 6.67 and 4.95 ppm, respectively. For such a large difference of chemical shifts, also called first-order spectra, the two protons whose J-coupling is examined are said to represent an AX

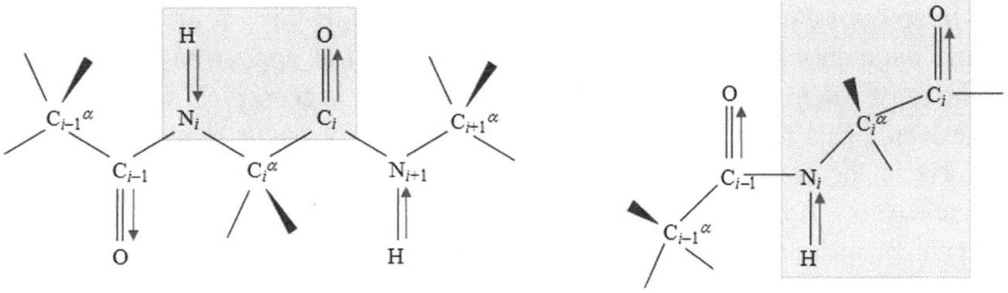

FIGURE 13.17 *Left*, the dipoles of the ith amino acid residue in a β-sheet structure counter and tend to cancel out, hence a nearby water dipole does not interact. *Right*, the dipoles of the ith residue in a right-handed α-helix reinforce each other to produce a dipole larger in magnitude whose electric field is screened if a water molecule is present nearby.

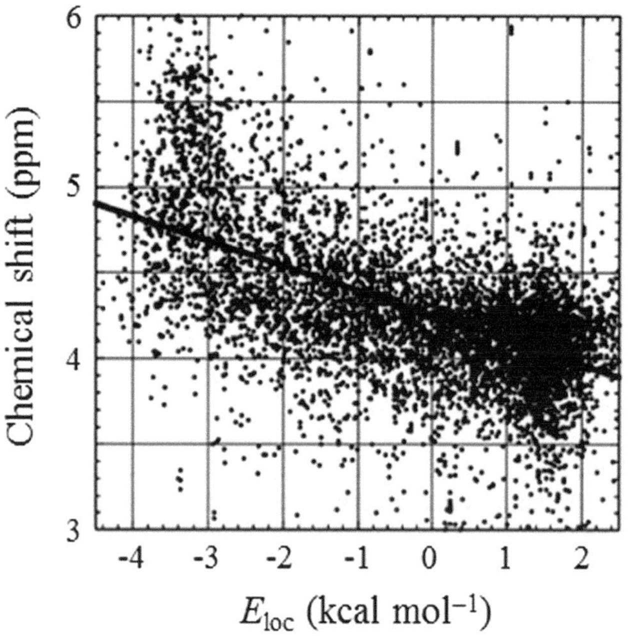

FIGURE 13.18 Correlation of chemical shift with E_{loc}. The chemical shift plotted is that of ^{1}H$^\alpha$ of individual alanine residues across a large set of proteins. (Reproduced with permission from Avbelj et al., 2004, © National Academy of Sciences, USA.)

system. The scalar interaction between them will split each other's NMR absorption identically into two, called spin-spin splitting. A little qualitative thought would enable one to relate the splitting to the two possible z-projections ($m_I = \pm 1/2$) of the angular momentum vector **I** in the presence of the static field $\mathbf{B}_o$. Analogous to the possible values of $m_I = 2I + 1$, the individual resonances would split into $2I + 1$ lines of equal intensity. The separation of the split lines in Hz, which are identical for the two spin-half nuclei A and X in consideration, is called the spin-spin coupling constant or scalar coupling constant J_{AX}.

The involvement of m_I implicates polarization states α and β of the nuclear spin in J-coupling. Consider the nucleus A whose spin polarization also polarizes the σ-bonded localized inner electrons, but the state of electron polarization due to the α nuclear state will be effectively opposite to that due to the β nuclear state, and vice versa. Both states of the electron polarization are simultaneously present because the population of α and β states of the nucleus are nearly equal. These two polarization states of the bonded electrons affect the nuclear shielding at X in an opposite manner – one state of the electron polarization augments and the other state retards the shielding σ to the same extent relative to the value of σ when no J-coupling existed. This means the resonance of X which occurred at x Hz in the absence of J-coupling will now occur at two frequencies $x - \Delta$ and $x + \Delta$ Hz, where $2\Delta = J_{AX}$. The same mechanism operates for the effect of nuclear spin states of X on the resonance absorption of A, splitting the resonance of A identically with $2\Delta = J_{AX}$.

The mechanism outlined in a rather crude way here indicates the coupling of nuclear spins mediated by localized electrons; the coupling is scalar because it is mediated through electrons. What would the coupled nuclear spin states look like? We can construct the coupled basis states for the AX system using the respective spin functions $|\alpha_A\rangle, |\beta\rangle_A, |a\rangle_X$, and $|\beta_X\rangle$. There are four energy states corresponding to the four product functions

$$|\alpha_A\alpha_X\rangle, |\alpha_A\beta_X\rangle, |\beta_A\alpha_X\rangle \text{and} |\beta_A\beta_X\rangle$$

within which to carry out the transitions (Figure 13.19). But all transitions are not allowed. The selection rule for NMR transition of a single nucleus is $\Delta m_I = \pm 1$, but for coupled spin states it is $\Delta m_T = \pm 1$, where $m_T = \Sigma m_I$. Note that the double-quantum transition is prohibited. This allows for four single-quantum transitions, namely $|\alpha_A\alpha_X\rangle \to |\beta_A\alpha_X\rangle, |\alpha_A\alpha_X\rangle \to |\alpha_A\beta_X\rangle, |\alpha_A\beta_X\rangle \to |\beta_A\beta_X\rangle$, $|\beta_A\alpha_X\rangle \to |\beta_A\beta_X\rangle$, and providing two transitions for each of nuclei A and X (Figure 13.19). In the absence of electron-nuclear hyperfine interaction, the two transitions for each nucleus will be equivalent, meaning the absence of J-coupling. The J-coupling energy appears only when the hyperfine interaction (see Chapter 4) comes into effect, the energy of which for the AX system is

$$E_J = J_{AX}\, m_A m_X, \tag{13.71}$$

where m refers to α (= 1/2) and β (= 1/2) only. Accordingly, hyperfine interaction will change the energy of the four coupled nuclear states as

$$E_J = \begin{cases} \frac{1}{4} J_{AX} & \alpha\alpha, \beta\beta \\ -\frac{1}{4} J_{AX} & \alpha\beta, \beta\alpha \end{cases}. \tag{13.72}$$

Because J is given in Hz, the frequency of transition for both A and X shift by $\pm\frac{1}{2}$ J$_{AX}$ relative to their individual chemical shift frequencies (Figure 13.19).

The J-coupling for a three non-equivalent spins AMX splits the resonance of each nucleus into four. The resonance due to A is split by interaction with X into two, and each of these two is split into two by the interaction with M. Similarly, X is split into two due to coupling with A and each of those two is split into two due to the coupling with M. Likewise, M also splits into four. The pictorial depiction of these splitting may appear to suggest in a naive way that the process is 'successive', which it is not, because spin coupling does not occur in succession. There are three pairs of related spins AM, AX, and MX, and the extent of pairwise coupling interaction plays an important role. The coupling energy depends on whether the two protons are attached to the same carbon (geminal coupling) or two adjacent carbons (vicinal coupling). The angle θ generated by carbon hybridization in the former and the dihedral angle ϕ in the latter are also factors that determine J-coupling. These are shown

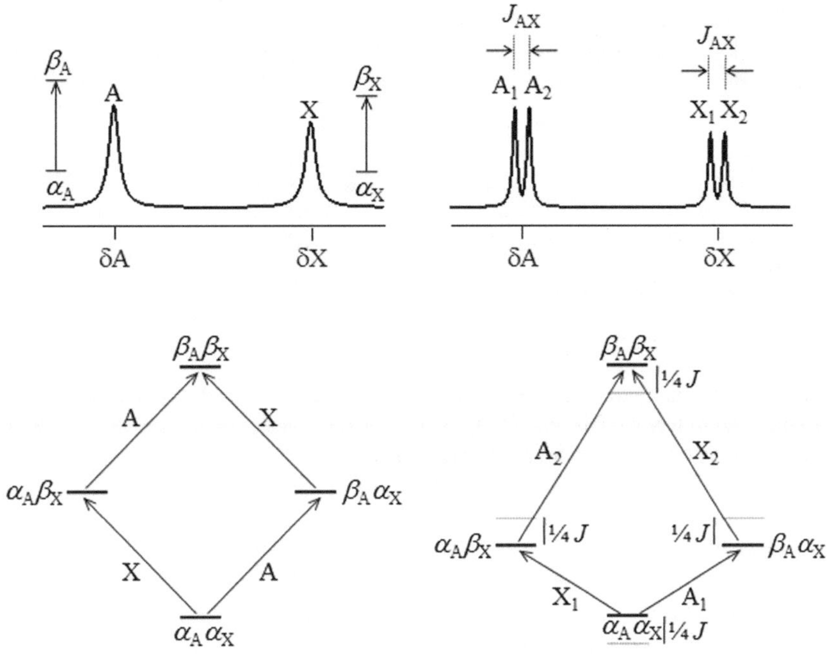

FIGURE 13.19 Effect of scalar coupling in a coupled AX system. *Left* (*top and bottom*), the individual spectral lines of the coupled A and X nuclei if there were no hyperfine coupling (*top*), in which case the intensity of the A line originates from both $\alpha_A\alpha_X \to \beta_A\alpha_X$ and $\alpha_A\beta_X \to \beta_A\beta_X$ transitions, and the intensity of the X line arises from both $\alpha_A\alpha_X \to \alpha_A\beta_X$ and $\beta_A\alpha_X \to \beta_A\beta_X$ transitions (*bottom*). *Right* (*top and bottom*), hyperfine coupling causes both A and X lines to split by J_{AX} (*top*), which occurs because of the shift in the energy of each coupled nuclear level by $\frac{1}{4}J$ (*bottom*). Note that these are single-quantum transitions.

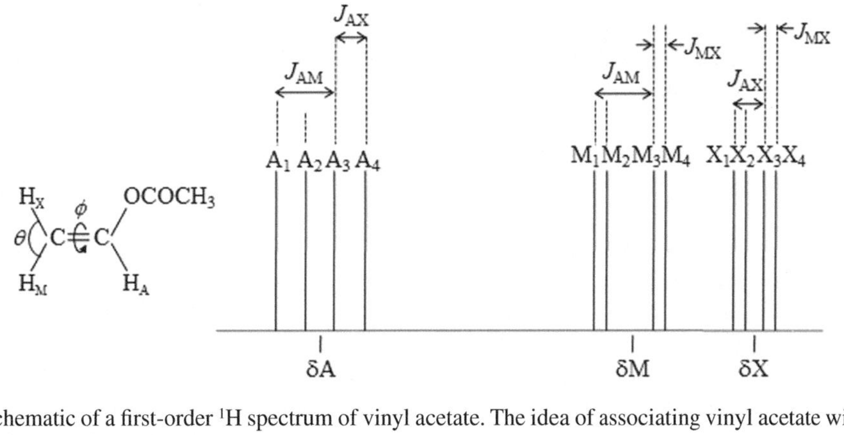

FIGURE 13.20 The schematic of a first-order ^{1}H spectrum of vinyl acetate. The idea of associating vinyl acetate with the AMX spin system is based on somewhat larger distance of the methyl protons, so they are presumably not coupled to the vinyl protons. One can then treat the vinyl protons labeled A, M, and X as a standalone three-spin system.

in Figure 13.20 for the AMX system vinyl acetate, for which ${}^3J_{AM} = 17.4$, ${}^3J_{AX} = 10.5$, and ${}^2J_{MX} \sim 3$Hz. Note that these scalar coupling values are just numbers relevant to an isotropic liquid, but J actually is orientation dependent, and hence is a tensor.

The spin product functions for the AMX system shows 12 possible transitions according to the selection rule $\Delta m_T = \pm 1$ (Figure 13.21). But they should also be understood with respect to pairwise interactions

$$E_J = J_{AM}\, m_A m_M + J_{AX}\, m_A m_X + J_{MX}\, m_M m_X, \quad (13.73)$$

indicating that the factor $\pm(1/4)\,J$ scales the splitting here too. Since $J_{AM} \neq J_{AX} \neq J_{MX}$ and only two appropriate couplings are considered to analyze the splitting, the amount of splitting for each resonance is distinct. Further, in the analysis of the splitting which pair of nuclei should be considered first is irrelevant. It is just a matter of convenience to show the overall splitting by successive splitting as indicated by the line diagrams for the splitting of the A resonance due to AM and AX couplings (Figure 13.21), there is no succession in reality.

Both chemical shift and J-splitting arise from magnetic shielding of nuclei, but the latter may be distinguished as

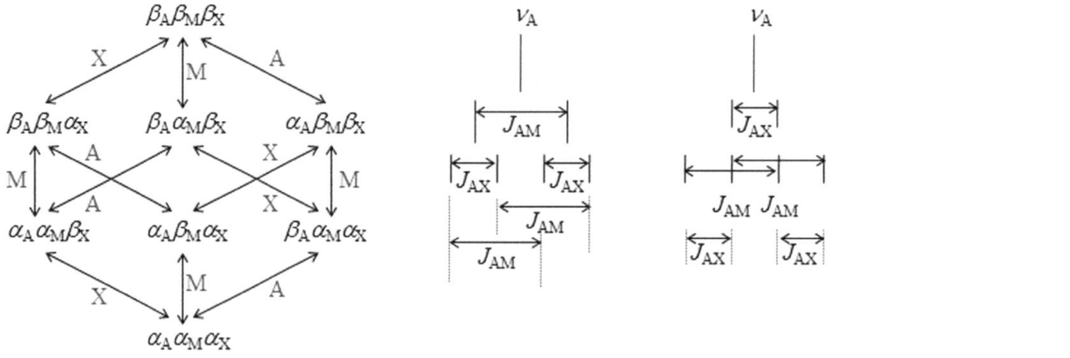

FIGURE 13.21 *Left*, the coupled spin functions of three nuclei showing the possible transitions according to the selection rule $\Delta m_T = \pm 1$. *Middle* and *right*, the line diagrams show that the four-line splitting of the A-resonance does not depend on whether splitting due to M or X is examined first, because it is a simultaneous event rather than sequential.

additional shielding due to mutually altered effective B_0 seen by the bonded nuclei. The mutual effect of the bonded nuclei is mediated mostly by σ-bonded electrons. Then we have to consider the same three terms in the electron-nuclear hyperfine interactions that we discussed while studying hydrogen atom spectra (Chapter 4). They are: one, the interaction of the magnetic moment due to electron orbital motion with the nuclear magnetic moment; two, the interaction of the electron spin magnetic moment with the nuclear magnetic moment known as Fermi contact interaction; and three, the dipole-dipole interaction between the electron spin magnetic moment and the nuclear magnetic moment. Note that the contribution of the Fermi contact term is the largest. Also, both coupled nuclei will experience these interactions with the same set of bonding electrons simultaneously. A quick review of the hydrogen atom hyperfine interactions studied already will also tell us that all these three interactions are proportional to the gyromagnetic ratio of the nucleus. They have to be, because all of these interactions directly involve the nuclear magnetic moment. This consideration gives an important result – spin-spin coupling is proportional to the gyromagnetic ratio. The change in shielding at one nucleus produced by another nucleus is proportional to the gyromagnetic ratio of the inducer nucleus and the field sensed by the former is proportional to its gyromagnetic ratio. The shift from the resonance frequency of a ^{1}H resonance due to J-coupling of a pair of bonded protons ^{1}H–^{1}H is $\Delta\nu \propto \pm|\gamma_H|^2$. For two directly bonded heteronuclei, ^{13}C–^{1}H for example, the shift from the respective resonance frequencies of both ^{1}H and ^{13}C will be $\Delta\nu \propto \pm\gamma_H \gamma_{^{13}C}$. Similarly, the shift in both ^{15}N and ^{1}H resonance frequencies in the ^{15}N–^{1}H group will be $\Delta\nu \propto \pm\gamma_H \gamma_{^{15}N}$.

Recognition of the splitting pattern involving magnetically equivalent nuclei is rather simple and widely discussed. Magnetic equivalence of a set of nuclei refers to their identical electronic environment which can occur by virtue of the molecular structure itself or by internal motion. For example, the pair of protons at positions 4 and 6 in the molecule 5-fluoro-1,2,3-tribromobenzene are magnetically equivalent by symmetry, the three protons of a methyl group in ethyl bromide become equivalent due to their collective rotation about the C–C bond in picosecond timescale, and the two protons of the methylene carbon of a compound $H_2C{=}R$ are equivalent owing to position and environment (Figure 13.22). The equivalent nuclei act as a unit. Since each nucleus has two spin functions, α and β, a unit of $-n$ equivalent nuclei will have 2^n spin product functions which can be grouped according to the value of $m_T = \Sigma m_I$. The eight product functions for the three protons of the methyl group form four groups corresponding to the m_T values of $-3/2, -1/2, 1/2$, and $3/2$, and the number of product functions belonging to these groups are 1, 3, 3, and 1, respectively (Figure 13.21). The split pattern observed for a scalar-coupled nucleus due to these equivalent nuclei follows the rule of $n+1$ frequencies, meaning $n+1$ lines with relative intensities of 1:3:3:1. The number of groups of product functions corresponds to the number of frequencies, and the binomial coefficients represent the relative intensities. Accordingly, in the spectrum of CH_3–CH_2–Br the resonance of the methylene protons split into four lines, showing the intensity pattern 1:3:3:1. The resonance for the methyl protons splits into 1:2:1.

Although spin-spin splitting is independent of B_0, it is sensitive to gyromagnetic ratios of the coupled nuclei, and the number, geometry, and conformation of the bonds connecting them. Coupling constants are indicated by $^1J_{ab}$, $^2J_{ab}$, $^3J_{ab}$, ..., where the superscript denotes the number of bonds that intervene the coupled nuclei subscripted as a and b. When the coupled nuclei are separated by a single bond, as for ^{13}C–^{1}H and ^{15}N–^{1}H,

$$^1J_{ab} \propto \gamma_a \gamma_b \Pi_{s_a s_b}, \tag{13.74}$$

in which $\Pi_{s_a s_b}$ is a factor that determines the mutual polarizability of the s orbitals of a and b atoms, suggesting the effect of hybridization on the s orbital character of the σ bond between a and b. The variation of $^1J_{^{13}C^1H}$ with carbon hybridization is shown by the numbers listed in Table 13.1. These coupling constants will vary again if an electronegative substituent is bonded to the ^{13}C atoms. Similarly, the geminal

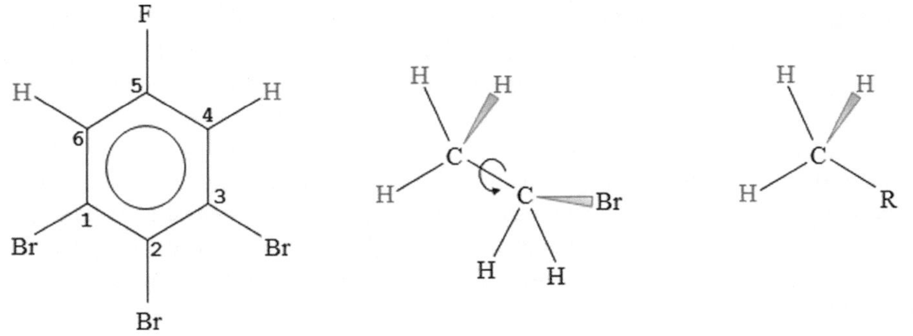

FIGURE 13.22 Structure and symmetry-dictated magnetic equivalence of protons shown in red.

TABLE 13.1
Commonly Observed Coupling Constants

Coupling type	Hybridization	J(Hz)
$^1J_{^{13}C^1H}$	sp	248.7
	sp^2	156.4
	sp^3	124.9
$^1J_{^{13}C^{13}C}$	sp^3	35
$^1J_{^{15}N^1H}$	sp^3	92
$^2J_{HH'}$	sp^2	0 to 2
	sp^3	−9 to −15
$^3J_{HH'}$	sp^2	17
	sp^3	2 to 14

coupling constant $^2J_{HH'}$ for two protons attached to a carbon depends on the angle θ between C–H and C–H′ bonds and the s character of the σ bond between carbon and hydrogens. Generally, $^2J_{HH'}$ increases as the angle θ decreases, and the absolute value of $^2J_{HH'}$ decreases from sp^3 to sp through sp^2 hybridization of the carbon.

The vicinal coupling $^3J_{HH'}$ in the H–C–C–H′ group depends on the dihedral angle of the C–C rotation, the C–C bond length, and any electronegative atom that may be bonded to a carbon. The dependence on the dihedral angle ϕ is given by the well-known Karplus equation

$$^3J_{HH'} = \mathrm{A}\cos^2\phi + \mathrm{B}\cos\phi + \mathrm{C}, \qquad (13.75)$$

with A, B, and C being empirical constants. The equation is also written by some as $^3J_{HH'} = \mathrm{A} + \mathrm{B}\cos\phi + \mathrm{C}\cos 2\phi$ by using the identity $\cos^2\phi = 1/2(1+\cos 2\phi)$. Although this equation is applicable only to the coupling of two vicinal protons, Karplus-type equations have been found for different types of couplings and between heteronuclei (see, for example, Thibaudeau et al., 1998). The Karplus relation is indispensable and used enormously in the determination of dihedral angles from J-correlated experiments. In fact, the myriad of NMR-determined molecular structures available today have relied on the Karplus equation.

When the number of bonds separating the coupled protons is more than three the coupling is said to be of the 'long-range' type. Indeed, $^4J_{HH'}$ and $^5J_{HH'}$ are measurable by suitably adjusting a parameter called 'mixing time' in a total J-type experiment to allow for the transfer of coherence; such couplings are, however, small. We also note that J-coupling will be observed only when the spin-spin splitting is sufficiently smaller than the chemical shift difference, a limit referred to as 'weak-coupling'. To the first order, this condition for spins a and b in angular frequency is

$$2\pi|J_{ab}| \ll |\Omega_a - \Omega_b|, \qquad (13.76)$$

called the 'weak coupling limit'. Most of the calculations and spectral analysis in NMR are carried out within this limit.

13.15 BASIC THEORY OF THE ORIGIN OF NUCLEAR SPIN RELAXATION

The relaxation times T_1 and T_2 that are developed phenomenologically through Bloch equations fundamentally refer to the times for the restoration of equilibrium Boltzmann population following a RF pulse. While T_1 is the time for recovery of the equilibrium longitudinal magnetization M_z, T_2 is the time for the disappearance of transverse magnetization M_x and M_y that occurs by equatorial spin dephasing, causing the loss of spin coherence. The two relaxation processes are simultaneous but clearly distinct – the longitudinal relaxation is a thermal transition from the excited level to the ground-state level, and therefore involves a change in the energy of the total spin dipolar system. In contrast, transverse relaxation is based on the excited-state lifetime in which interactions within the entire spin-dipole–spin-dipole system do not change the energy of the system. Since T_2 is related to the excited-state lifetime, it is T_2 that determines NMR linewidth.

Both T_1 and T_2 are processes involving nuclear spin magnetic moment, and can be induced only by magnetic fields that originate from within the molecule. Such an internal magnetic field is a time-varying or fluctuating magnetic field $B^o(t)$, which can be Fourier-transformed to obtain the frequency spectrum of the field $B^o(\omega)$. Now suppose there is a transverse magnetization M_{xy} which was created by a 90° pulse. There can be a frequency component of $B^o(\omega)$ in the xy

plane that would match the Larmor precession frequency $\omega = -\gamma B_0$ of M_{xy}. Then the matching component of $B^o(\omega)$ can exert a torque on M_{xy} to rotate it to M_z, the vector mechanics analog of a thermal transition from the excited to the ground state. This is T_1, which can be of the order of a second, that is five orders of magnitude longer than a typical 90°-pulsed excitation time for a proton. The reason T_1 can be so long is that the fluctuating molecular magnetic field $B^o(t)$ is very weak. To see how small $B^o(t)$ could be, let us do a short calculation. The angle β by which a magnetization can be rotated by the torque action of the B_1 field in time τ_p is given by

$$\beta = -\gamma B_1 \tau_p. \tag{13.77}$$

For rotating a proton magnetization (γ = 26.752×10^7 rad T^{-1} s^{-1}) by an angle $\beta = \pi/2$ in $\tau_p \sim 9\mu s$, the required strength of B_1 is ~652×10^{-6} T. If the same transverse magnetization created by the $\pi/2$ pulse earlier is to be rotated back to the longitudinal axis in ~1 s, which is roughly the value of T_1, the strength of the magnetic field needed would be ~652×10^{-11} T. This is the typical strength of an average molecular field $B^o(t)$. Note that the two magnetic fields of different field amplitudes can have the same frequency.

The effect of $B^o(t)$ on T_2 may not appear obvious, but can be understood in the following manner. Suppose the homogeneity of the external magnetic field B_1 is so good that the transverse magnetization of a proton spin precesses about B_o at the same Larmor frequency across the sample volume. This means that the spins have the same phase and the coherence sustains. Since the individual spin dipoles can interact mutually, the dipole-dipole interaction will create a local field at the site of a nucleus. The strength of this local field is not uniform across the sample. The same nucleus in one element of the sample volume may experience a slightly higher local field than in another element, which means the effective field experienced by a nucleus in different regions of the sample is $B_1 + \Delta B$. In other words, the fluctuating local field produces inhomogeneity in the magnetic field such that the same spin in different regions of the sample precesses at different Larmor frequency, and thus gives rise to the loss of phase coherence. The time needed for a complete loss of phase coherence is T_2. We see that the same fluctuating internal field $B^o(t)$ leads to both T_1 and T_2 relaxations.

What is the basis for the fluctuating internal molecular field $B^o(t)$? This is a local magnetic field seen by a nuclear spin due to its through-space interaction with other nuclear dipoles close by. The nucleus is a tiny magnet and hence produces a magnetic field in the space nearby. To see the strength of this field at some point nearby, consider a nuclear magnetic moment **μ** fixed at some angle with respect to the z-axis (Figure 13.23). The point p that can be a point nuclear dipole is connected to the midpoint of the first dipole **μ** such that the angle between the dipole vector and the position vector **r** connecting the point p is θ. Then the parallel and perpendicular components of the magnetic field due to **μ** at point p are

$$B^o_{\parallel} = \frac{\mu_o}{4\pi}\frac{2\mu}{r^3}\cos\theta$$

$$B^o_{\perp} = \frac{\mu_0}{4\pi}\frac{\mu}{r^3}\sin\theta, \tag{13.78}$$

in which μ_o is the permeability constant or permeability of free space that measures the degree of resistance to forming a magnetic field in vacuum. The factor 4π is included in the denominator to obtain the field in SI unit; if one chooses cgs unit, then 4π should be omitted. The factor 2 in the numerator of the $B^o_{\parallel}$ component appears because of the presence of two perpendicular components. The total magnetic field at p is

$$B^o = \sqrt{\left(B^o_{\parallel}\right)^2 + \left(B^o_{\perp}\right)^2} = \frac{\mu_0}{4\pi}\frac{1}{r^3}\sqrt{1+3\cos^2\theta}.$$

If the point dipole is surrounded by many nuclear spin dipoles, collectively called the lattice or bath as shown in Figure 13.23 for a cage of four dipoles, the contributions from all cage dipoles are summed up to obtain a mean-field approximated value

$$B_o = \sum_i B^0_i = \frac{1}{4\pi}\sum_i \frac{\mu_0}{r_i^3}\left(\sqrt{1+3\cos^2\theta_i}\right)_i. \tag{13.79}$$

In general, each spin dipole experiences a local field due to the field generated by neighboring dipoles. The interaction field produced is expected to vary from one to another nucleus.

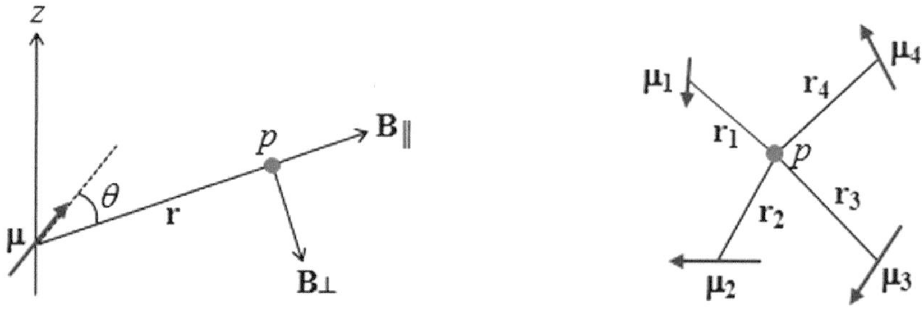

FIGURE 13.23 The local magnetic field at a point p produced by a single magnetic dipole (*left*), and by four dipoles caging the point dipole p (*right*).

As mentioned already, these local magnetic fields characteristically fluctuate with time $B^o(t)$, which should be obvious from our knowledge of diffusive motions of molecules. If the lattice motion is reduced it may take longer – many minutes, for the spins to relax as happens for solid samples. These motions can be characterized by the well-understood theory of time-correlation function that expresses the 'loss of memory' of the position of a particle (molecule) undergoing random motion. Briefly, consider a function

$$G\left[p(t),q(t)\right]=G(t), \tag{13.80}$$

p and q being coordinates and momenta. The function describes the position variable of the molecule at time t. The ensemble average of the product of the values of $G(t)$ at times $t=t_1$ and $t=t_2$, $\left\langle G(t_1)G(t_2)\right\rangle$, is defined as the autocorrelation function of $G(t)$. If we argue that the ensemble distribution function for the molecule at equilibrium does not change from one time to another, then we obtain the key identity

$$\left\langle G(t_1)G(t_2)\right\rangle=\left\langle G(0)G(t_2-t_1)\right\rangle. \tag{13.81}$$

If $t_1=t_2$, meaning no time displacement is allowed, the right-hand side of the above identity gives

$$\left\langle G(t_1)G(t_2)\right\rangle=\left\langle G^2\right\rangle, \tag{13.82}$$

and we say that $G(0)$ and $G(t_2-t_1)$ are correlated. Now if $\langle G^2\rangle$ is allowed to decay for a long time, a new value of $\langle G^2\rangle$ will be obtained, which means the initial correlated function at time $t=0$ has exponentially evolved to a quantity which is no longer correlated (Figure 13.24). What has happened is that the initial value of the autocorrelation function $\langle G^2\rangle$, in which $G(0)$ and $G(t_2-t_1)$ were fully correlated, has changed to a new value of the function $\langle G^2\rangle$, where $G(0)$ and $G(t_2-t_1)$ are no longer correlated. This exponential change of an autocorrelation function with a decay time constant τ_c expresses the 'memory loss' of the position $G(t)$. The time constant τ_c is called the correlation time of the molecule.

The autocorrelation $\langle G^2\rangle$ is also shown as $G(t)$ for convenience. Since $G(t)$ is an exponential, $\exp(-t/\tau_c)$, its Fourier transformation yields a Lorentzian function called the spectral density

$$J(\omega)=\int_{-\infty}^{\infty}G(t)e^{-i\omega t}dt=\frac{2\tau_c}{1+\omega^2\tau_c^{\,2}}, \tag{13.83}$$

which may look awkward due to the frequency coverage of many orders of magnitude, but can be plotted as a function of $\log\omega$ (Figure 13.25). In the initial flat part of the graph covering lower frequencies, the condition $\omega^2\tau_c^2\ll 1$ holds, which is called 'extreme narrowing'. This part is followed by a sharp decrease in $J(\omega)$ registering a midpoint that corresponds to τ_c, where $\omega^2\tau_c^2=1$. At higher frequencies $\omega^2\tau_c^2\gg 1$, and $J(\omega)$ collapses to zero. These three regions are especially relevant to study spin relaxation and will be developed below.

At this stage we should mention that the theory of spin relaxation can be considered at different levels of difficulty, starting from Bloch's phenomenological equations to higher levels of quantum mechanics. The most useful and comprehensive approach is perhaps a semiclassical one that inducts the mean-field approximated local magnetic field introduced above (see equation (13.79)) as a perturbation term to the spin interaction Hamiltonian, called the BPP model due to Bloembergen, Purcell, and Pound (1948), which is quite useful to discuss the relaxation of soft matter in solution.

The BPP model supposes that the interaction Hamiltonian can be written as a product of the spin and the motional $G(t)$ part

$$\mathcal{H}(t)=-\hbar\gamma I_{x,y}B^o(t)G(t), \tag{13.84}$$

where the operator part $I_{x,y}B^o(t)$ acts as a second-order perturbation Hamiltonian. Note again that the relaxation theory

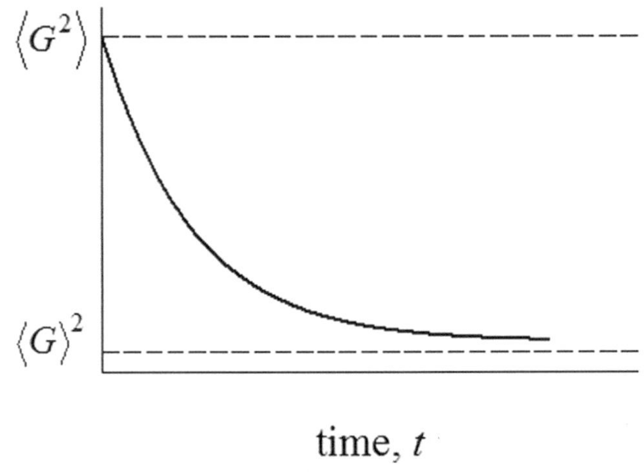

FIGURE 13.24 Decay of the autocorrelation function of $G(t)$ from an initial amplitude $\langle G^2\rangle$ to $\langle G\rangle^2$ with time.

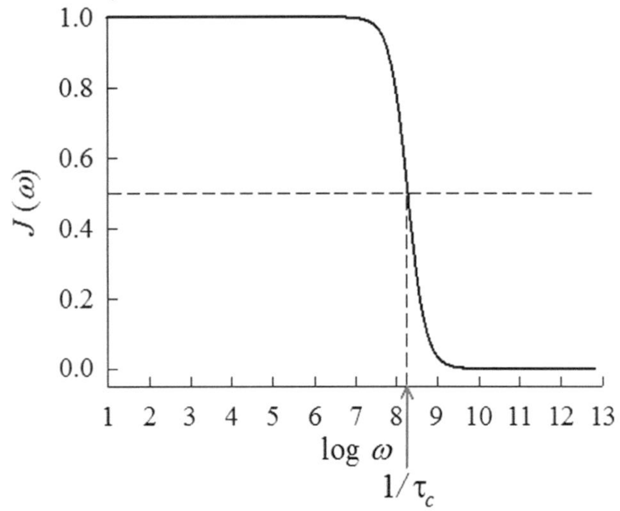

FIGURE 13.25 Spectral density.

appears in a different guise, but to achieve the purpose we may like to carry forward the Hamiltonian the way we put here. The application of the perturbation theory returns a transition probability W_1 for the spin-lattice relaxation

$$W_1 = \frac{1}{2}\gamma^2 \left(B^0\right)^2 J\left(\omega_0\right)$$

$$W_1 = \frac{1}{2}T_1^{-1}. \qquad (13.85)$$

This result accounts for both transverse components of the $B^{\circ}(t)$ field, that are $B_x^{\circ}(t)$ and $B_y^{\circ}(t)$, leading to the appearance of the front factor 1 / 2. The expression for the relaxation rate of *x*- or *y*-component individually remains the same except that the front factor becomes 1 / 4 for the individual components. Note also that the spectral density value $J(\omega_0)$ at the frequency ω_0 corresponds to the frequency of the transition, say $\omega_{\alpha\beta}$, under consideration. Combining equations (13.83) and (13.85), the spin lattice relaxation rate constant is written as

$$T_1^{-1} = \gamma^2 \left(B^{\circ}\right)^2 \frac{2\tau_c}{1+\omega_0^2\tau_c^{\,2}}. \qquad (13.86)$$

Spin-spin relaxation time T_2 cannot be derived by the perturbation approach because T_2 does not involve exchange of energy with the surrounding, but requires a density operator treatment. Here, we describe it only as the exponential decay of coherence between a pair of spin states $|\alpha\rangle$ and $|\alpha'\rangle$

$$\sigma_{\alpha\alpha'}(t) = \sigma_{\alpha\alpha'}(0)e^{-i\omega_{\alpha\alpha'}t - t/T_2}. \qquad (13.87)$$

Different pairs of spin states will have different T_2. For analysis, T_2 can be considered to contain two contributions – a secular (adiabatic) and a nonsecular (nonadiabatic). The former arises due to finite lifetime of the excited state producing the linewidth. The latter contribution arises from the same fluctuating molecular field $B^{\circ}(t)$ that gives rise to T_1, and therefore depends on $J(\omega_0)$. It does not entail energy exchange, and hence is only $J(0)$-dependent. The expression for T_2 is of the form

$$T_2^{-1} = \left(2T_1\right)^{-1} + \frac{1}{2}\gamma^2\left(B^{\circ}\right)^2 J(0). \qquad (13.88)$$

To calculate the variation of T_1 and T_2 with correlation time τ_c according to equations (13.86) and (13.88), let us consider a ^{1}H transition occurring at some frequency ω_0, say 500 MHz. The value of γ for ^{1}H is 26.752×10^7 rad T^{-1} s^{-1}, and let the value of the local random field be 1 mT. The logarithmic plot of T_1^{-1} vs τ_c appears as an inverted chevron whose inflection occurs when $\tau_c = \omega_0^{-1}$ (Figure 13.26). The inflection point corresponds to the midpoint of the spectral density map where the frequency is given by $\tau_c = \omega_0^{-1}$ (see Figure 13.25). The left limb represents extreme narrowing condition $\omega^2\tau_c^2 \ll 1$, that is the limit of fast motion, and the right limb corresponds to $\omega^2\tau_c^2 \gg 1$ when the motion is relatively slow. The variation of T_2^{-1} merges with T_1^{-1} in the fast-motion limit, but increases monotonously into the slow motion limit due to reduced spectral density $J(0)$. We should like to mention here that $J(0)$ in general also includes contributions from chemical exchange and other pseudo first-order processes. The T_2^{-1}, however, tends to level off at sufficiently large values of τ_c, that is, when rigid lattice condition appears.

Let us reconsider the condition $\tau_c = \omega_0^{-1}$ in which T_1 is minimum. The formula for T_1^{-1} (equation (13.86)) then reduces to

$$T_1^{-1} = \frac{\gamma^2 B_0^2}{\omega_0} \text{ or } T_1 = \frac{\omega_0}{\gamma^2 B_0^2}, \qquad (13.89)$$

which suggests that the minimum of T_1 increases with RF magnetic field. This is shown in Figure 13.26 at several resonance frequencies.

The direct proportionality of T_1 on the RF field strength is used to investigate the dynamics of molecular systems. The technique, called dispersion relaxometry studied more conveniently by field-cycling NMR, involves measurement of T_1^{-1}

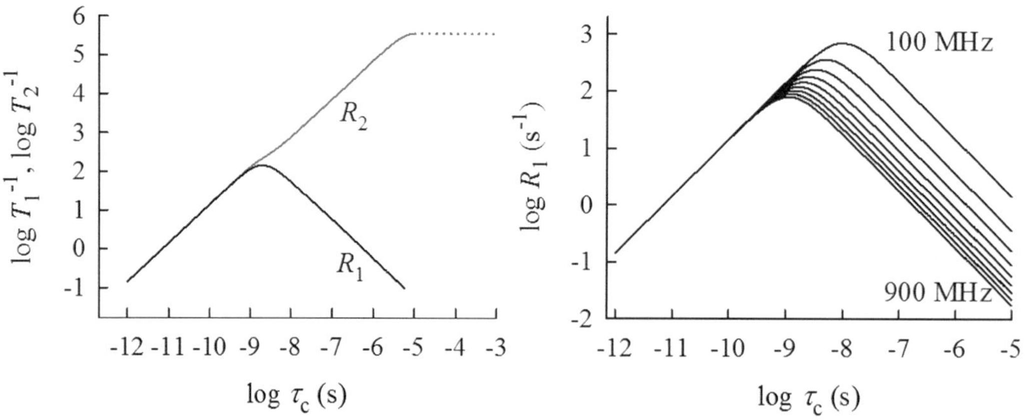

FIGURE 13.26 Double logarithmic plots for the dependence of $T_1^{-1}(=R_1)$ and $T_2^{-1}(=R_2)$ on τ_c. *Left*, inverted chevron for log T_1^{-1} vs log τ_c in which the left arm corresponds to extreme narrowing. In the slow motion limit T_2^{-1} breaks away from T_1^{-1}, and continues upward to level off at large values of τ_c. *Right*, the change of T_1 according to resonance frequency simulated from 100 to 900 MHz.

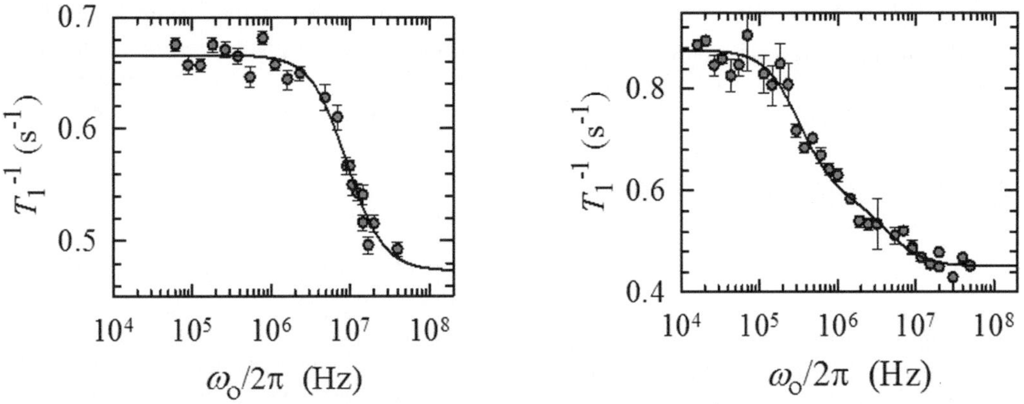

FIGURE 13.27 Longitudinal relaxation dispersion of ^{1}H of water in aqueous solutions of the proteins lysozyme (*left*) and bovine serum albumin (*right*). The solid lines describe one or more Lorentzian functions of the form

$$T_1^{-1} - T_b^{-1} = \alpha + \sum_n \beta_n \tau_{cn} \left[\frac{0.2}{1+(\omega\tau_{cn})^2} + \frac{0.8}{1+(2\omega\tau_{cn})^2} \right]$$

in which T_b^{-1} is the contribution of bulk water molecules to the observed relaxation, and can be determined by a single-frequency relaxation measurement of water alone. The parameter α represents contribution of loosely bound water molecules to the protein surface. The parameter β includes the dipolar coupling constant (ω_D), the fraction of protons in integral water molecules, and the labile protons in the total water molecules (f_I), and the generalized order parameter (S_I). The second term on the right-hand side of the equation above represents contributions from the integral water molecules buried inside the protein and the labile protons exchanging with the bulk water protons at a rate slow compared to the global correlation time of the protein, τ_c, but fast compared to the inherent relaxation rate of water. The number of Lorentzians applied is one for lysozyme and two for bovine serum albumin. The use of more than one Lorentzian to describe relaxation dispersion is based on multiple structural domains in some molecules.

of a spin system as a function of the frequency of the RF field strength. Obviously, the static field also needs to be changed each time the RF field strength is varied. In practice, the static field strength is changed in a manner that allows varying the RF field from tens of kHz to the MHz range. Analyses of such dispersions (Figure 13.27) provide information about molecular reorientational dynamics, correlation time distribution, and population heterogeneity.

13.16 MECHANISM OF SPIN RELAXATION

This refers to the physical processes by which the fluctuating molecular magnetic field is generated, leading to spin relaxation. There are several means to create local magnetic field at a nucleus and they work cumulatively. Some of these mechanisms are listed below.

13.16.1 Shielding Anisotropy

Our earlier mentions have indicated that the net static field B_0 felt at a nucleus is characterized by the nuclear shielding constant σ, whose value is determined by the orientation of the nuclear site with respect to the direction of B_0. Consider, for example, a ^{1}H nucleus σ-bonded to a carbon atom. The two extremes of the orientation of C–H with reference to $\mathbf{B}_0$ are parallel and perpendicular, and the corresponding shielding constants at the proton site are $\sigma_{\parallel}$ and $\sigma_{\perp}$. One then says that shielding is anisotropic. In an asymmetric surrounding, the three tensor components of the constant are σ_{xx}, σ_{yy}, and σ_{zz}, in which the principal axis and the molecule-fixed axis systems are parallel and collinear. In an axially symmetric system, $\sigma_{zz} = \sigma_{\parallel}$ and $\sigma_{xx} = \sigma_{yy} = \sigma_{\perp}$, so the trace of the tensor is

$$\sigma_{tr} = \sigma_{\parallel} + 2\sigma_{\perp}. \qquad (13.90)$$

In NMR literature, the shielding constants are written with subscript indices 1,2,3 corresponding to *x,y,z*, such that

$$\sigma = \sigma_{11}\cos^2\theta_1 + \sigma_{22}\cos^2\theta_2 + \sigma_{33}\cos^2\theta_3, \qquad (13.91)$$

where θ_1, θ_2, and θ_3 are the angles between the respective shielding components and the static field vector $\mathbf{B}_0$. While the anisotropy of σ certainly prevails in rigid crystalline systems, isotropic tumbling of molecules in solution averages out the shielding to yield an isotropic shielding constant, which is what we discussed earlier under chemical shift. The message here is that isotropic molecular tumbling modulates the actual $\mathbf{B}_0$ field at the nucleus so as to alter Larmor precession, thus facilitating its relaxation. Under rigid anisotropic conditions, the nuclear relaxation is very slow as in solids. Since this mode of spin relaxation is static field-dependent, the relaxation time constant T_1 due to the shielding anisotropy is

$$T_1^{-1} \propto \gamma^2 B_0^2 \left(\sigma_{\parallel} - \sigma_{\perp}\right)^2 \tau_c,$$

where τ_c is the same tumbling correlation time that appears in the spectral density function for an axially symmetric system.

13.16.2 Spin-Rotation Interaction

It has been said earlier in the context of coupling of angular momenta in rotational spectroscopy that the magnetic field produced by a rotating molecule can couple with the magnetic moment of a spin-half nucleus. This magnetic interaction energy is given by the vector product of the two magnetic moments

$$E = \gamma_{\text{rot}} \mathbf{s} \cdot \mathbf{N}, \tag{13.92}$$

where **N** is the molecular rotational angular momentum arising from its end-to-end rotation, such that it is perpendicular to the molecular axis. The quantity γ_{rot} is the spin-rotation coupling constant, not the gyromagnetic ratio. If the rotating molecule suffers a collision during its rotational motion, the spin-coupling will be interrupted, leading to relaxation of the nuclear spin. The time interval between two successive collisions defines the spin-rotation correlation time, τ_{sr}, which is distinct from the correlation time of molecular tumbling. The expression for T_1^{-1} contributed by spin-rotation interaction is

$$T_1^{-1} = \frac{1}{\hbar^2} 2Nk_B T \gamma_{\text{rot}}^{\ 2} \tau_{\text{sr}}, \tag{13.93}$$

where N is rotational angular momentum and γ_{rot} is the spin-rotation coupling constant.

13.16.3 Scalar Interaction

The premise here is the same as in J-coupling, but the respective relaxations of the coupled nuclei influence each other. For example, an amide proton in a ^{14}N–^{1}H group will relax differently from that in ^{15}N–^{1}H; the nitrogen nucleus in the former is quadrupolar, which relaxes faster than the dipolar ^{15}N. Consider the ^{14}N–^{1}H group alone, and we realize that the J-splitting of ^{1}H due to ^{14}N will be a null when T_1 of the latter is very small, meaning it relaxes fast. The ^{1}H resonance will however continue to split, from a broad line to sharp J-lines – eventually three, as the T_1 of the ^{14}N nucleus turns larger. This rate of ^{14}N relaxation is also called 'exchange time' by some. But the idea refers to anisotropy again, because scalar interaction is anisotropic in nature. In a rigid lattice the ^{14}N relaxation is expected to be slower which will produce the three signature line-splitting for the ^{1}H nucleus in the limit of sufficiently slower relaxation of the ^{14}N nucleus. We can therefore conclude that if isotropic molecular motion interferes with the relaxation of the ^{14}N nucleus, then the relaxation of the ^{1}H nucleus will also be affected. The expression for the spin-lattice relaxation time due to scalar interaction alone works out to

$$T_1^{-1} \propto J_{\text{AX}}^{\ 2} I_{\text{X}} \left(I_{\text{X}} + 1\right) \frac{\tau_{\text{sc}}}{1 + \left(\omega_{\text{X}} - \omega_{\text{A}}\right)^2 \tau_{\text{sc}}^{\ 2}}, \tag{13.94}$$

in which τ_{sc} is the scalar-coupling correlation time, I_{X} is the angular momentum of spin X, ω_{X} and ω_{A} are respective Larmor frequencies of spins X and A, and J_{AX} is the coupling constant. The use of the indices A and X is arbitrary, X could be the ^{14}N or ^{1}H nucleus for the example taken above. The main idea is how the relaxation of one nucleus induces the relaxation of the coupled nucleus.

13.16.4 Paramagnetic Effect

The effect of an unpaired electron interacting with a resonating nucleus through bond or through space is easy to understand qualitatively given that the magnetic moment of the electron is overwhelmingly large compared to that of the nucleus. The relaxation of the unpaired electron itself is sufficient to affect the nucleus whose relaxation is under consideration. This is the reason why a paramagnetic species is often included in a sample, so the relaxation of the nuclei nearby is shortened.

13.16.5 Dipole-Dipole Interaction

The dipole-dipole interaction involves through-space coupling of two intra- or intermolecular magnetic moments. The interactions are interrupted or modulated by molecular tumbling, causing relaxation of the dipole-coupled nuclei. The intermolecular dipolar interactions are also affected by molecular translational motion because collisions can disrupt the interactions leading to nuclear relaxation. The strength of dipole-dipole interaction has the familiar r^{-3} dependence, where r is the distance between the two dipoles, but the rate processes involving dipolar interactions are proportional to the square of the interaction strength. As such, the expressions for spin-lattice relaxation are given by

$$T_1^{-1} \propto \left(\frac{\mu_0}{4\pi}\right)^2 \frac{\gamma^4 \hbar^2 \tau_c}{r^6} \text{ intramolecular homonuclear, AA or XX}$$

$$T_1^{-1} \propto \left(\frac{\mu_0}{4\pi}\right)^2 \frac{\gamma_{\text{A}}^2 \gamma_{\text{X}}^2 \hbar^2 \tau_c}{r^6} \text{ intramolecular heteronuclear, AX}$$

$$T_1^{-1} \propto \left(\frac{\mu_0}{4\pi}\right)^2 \frac{\gamma_{\text{A}}^2 \gamma_{\text{X}}^2 \hbar^2 \tau_c}{\left(D_{\text{A}} + D_{\text{X}}\right) d_{\text{AX}}} \text{ intermolecular heteronuclear, AX.} \tag{13.95}$$

The expression for the intermolecular heteronuclear case is for the relaxation by a spin-half X nucleus. Interestingly, this expression has the features of a microscopic diffusion-controlled rate equation encountered in chemical kinetics, in which D_{A} and D_{X} are respective diffusion coefficients of the molecules containing the A and X nuclei, and d_{AX} is a critical distance between the two molecules such that a collision between them will occur when this distance is attained.

13.17 DIPOLAR INTERACTION AND CROSS-RELAXATION

Dipolar interaction is a through-space spin-spin interaction allowing the two spins to flip mutually, called cross-relaxation.

Note that auto- or self-relaxation refers to the two processes of spin-lattice and spin-spin relaxation. Physically, cross-relaxation gives rise to magnetization transfer which is manifested in a change in NMR line intensities of the two spins – called nuclear Overhauser effect (NOE).

The classical energy of interaction of two nuclear spins I_k and I_l can be analyzed by considering them as magnetic point dipoles, $\boldsymbol{\mu}_k$ and $\boldsymbol{\mu}_l$, provided the distance of their separation r is larger than the length of the individual dipoles (Figure 13.28). Since the magnetic field generated by the $\boldsymbol{\mu}_k$ dipole at the site of the $\boldsymbol{\mu}_l$ dipole is (in SI unit)

$$\mathbf{B}_{kl} = -\frac{\mu_0}{4\pi}\nabla\frac{\boldsymbol{\mu}_k\cdot\mathbf{r}}{r^3}, \tag{13.96}$$

vector differentiation gives the interaction energy as

$$E = -\frac{\mu_0}{4\pi}\left[\frac{\boldsymbol{\mu}_k\cdot\boldsymbol{\mu}_l}{r^3} - 3\frac{(\boldsymbol{\mu}_k\cdot\mathbf{r})(\boldsymbol{\mu}_l\cdot\mathbf{r})}{r^5}\right]. \tag{13.97}$$

It is more convenient to discuss dipolar interactions in terms of spin operators. The classical expression for the spin-spin interaction itself serves for the dipolar Hamiltonian, in which the point dipoles can be expressed as

$$\boldsymbol{\mu} = \gamma\hbar\mathbf{I}. \tag{13.98}$$

In the presence of the magnetic field $\mathbf{B}_0$ along z, the Hamiltonian for dipolar interaction between I_k and I_l reads

$$\mathcal{H}_{\mathrm{D}} = \mathbf{I}_k\widetilde{\mathbf{D}}_{kl}\mathbf{I}_l$$

$$= \gamma_k\gamma_l\hbar^2\frac{\mu_0}{4\pi}\left[\frac{\mathbf{I}_k\cdot\mathbf{I}_l}{r^3} - 3\frac{(\mathbf{I}_k\cdot\mathbf{r})(\mathbf{I}_l\cdot\mathbf{r})}{r^5}\right]$$

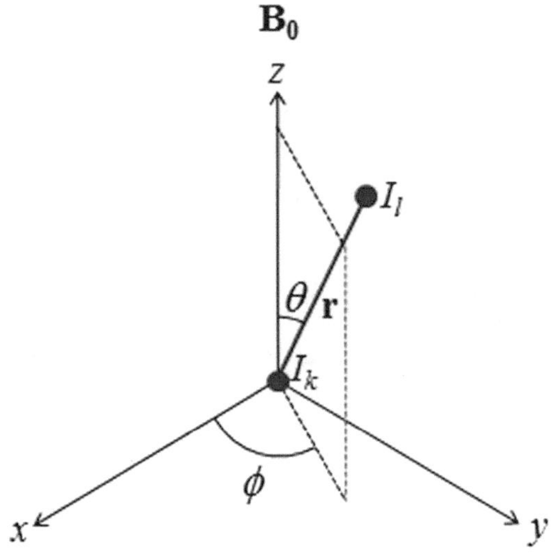

FIGURE 13.28 The definition of the distance between two nuclear spins I_k and I_l each taken as a point dipole.

$$= b_{kl}\left[\mathbf{I}_k\cdot\mathbf{I}_l - 3\frac{(\mathbf{I}_k\cdot\mathbf{r})(\mathbf{I}_l\cdot\mathbf{r})}{r^2}\right], \tag{13.99}$$

in which $\widetilde{\mathbf{D}}_{kl}$ is the dipolar coupling tensor, and b_{kl} is the dipolar coupling constant

$$b_{kl} = \frac{\mu_0\gamma_k\gamma_l\hbar}{4\pi r^3}, \tag{13.100}$$

expressed in SI unit of frequency (Hz). Since the two spins have been placed in an external magnetic field, the z-projections, i.e., the quantum number m, of each spin is defined. This allows for using the commutation relations of ladder operators of angular momentum

$$\begin{aligned}
\left[I^+, I_z\right] &= -I^+\\
\left[I^-, I_z\right] &= I^-\\
\left[I^+, I^-\right] &= 2I_z.
\end{aligned} \tag{13.101}$$

By taking individual scalar products of spin-spin and spin-unit vectors for I_k and I_l, and by using the ladder operators, an expression of six terms is obtained, each of which contains the spin operator and an orientation factor described by the angles θ and ϕ, which are second order spherical harmonics, ie., Y_l^m with $m = 0, \pm1, \pm2$ for $l = 2$. This appears a complicated senetence, but the meanings are clear. The six terms mentioned are also called *alphabetic terms* labeled A to F discussed in many textbooks. These terms are produced below in order to appreciate how the geometric factors determine dipolar interaction under conditions of isotropic molecular motions. Following equation (13.99)

$$\mathcal{H}_{\mathrm{D}} = b_{kl}\left[\mathrm{A+B+C+D+E+F}\right]$$

$$\begin{aligned}
&\mathrm{A} = -\left(I_{kz}I_{lz}\right)\left(3\cos^2\theta - 1\right) && Y_2^0\\
&\mathrm{B} = \frac{1}{4}\left(I_k^+I_l^- + I_k^-I_l^+\right)\left(3\cos^2\theta - 1\right) && Y_2^0\\
&\mathrm{C} = -\frac{3}{2}\left(I_{kz}I_l^+ + I_k^+I_{lz}\right)\sin\theta\cos\theta e^{-i\phi} && Y_2^{-1}\\
&\mathrm{D} = -\frac{3}{2}\left(I_{kz}I_l^- + I_k^-I_{lz}\right)\sin\theta\cos\theta e^{i\phi} && Y_2^{+1}\\
&\mathrm{E} = -\frac{3}{4}I_k^+I_l^+\left(\sin^2\theta e^{-2i\phi}\right) && Y_2^{-2}\\
&\mathrm{F} = -\frac{3}{4}I_k^- - \left(\sin^2\theta e^{2i\phi}\right). && Y_2^{+2}
\end{aligned} \tag{13.102}$$

The column at the right (spherical harmonics) reads the m-value ($m = 0, \pm1, \pm2$ for $l = 2$) corresponding to the angular part of each term. Each of the terms C to F contains the angle ϕ with respect to the x-axis. In a liquid sample, the terms vanish because isotropic motion (tumbling) averages out ϕ, so that $\exp(\pm ni\phi = 0$. The terms surviving are then A and B alone. Even these two will average out to zero if B_0 is weak, but survive for a homonuclear case if a strong B_0 field is used. The shortened dipolar Hamiltonian for homonuclear case in the limit of strong B_0 appears as

$$\mathcal{H}_{\mathrm{D}} = b_{kl}\left[\mathrm{A+B}\right]. \tag{13.103}$$

Retaining both A and B terms is subject to homo- or heteronuclear contexts. To decide upon which one to retain in which context, one can use the ladder operator commutation relation $\left[I^+, I^-\right] = 2I_z$ given in equation (13.101) to rewrite equation (13.103) as

$$\mathcal{H}_{\mathrm{D}} = \frac{b_{kl}}{2}\left[3I_{kz}I_{lz} - \mathbf{I}_k\mathbf{I}_l\right]\left(1-3\cos^2\theta\right). \tag{13.104}$$

In the case of heteronuclear dipolar coupling, only term A remains

$$\mathcal{H}_{\mathrm{D}} = b_{kl}I_{kz}I_{lz}\left(1-3\cos^2\theta\right), \tag{13.105}$$

where k and l are now heteronuclear spins.

13.18 EFFECT OF DIPOLAR INTERACTION ON NUCLEAR RELAXATION

The spin operator part of each term in the dipolar Hamiltonian affects a transition of the dipole-coupled spins k and l. There are four spin-product functions for the kl pair $\alpha_k\alpha_l$, $\beta_k\alpha_l$, $\alpha_k\beta_l$, and $\beta_k\beta_l$, which can be arranged qualitatively in an energy level diagram (Figure 13.29). There are six transitions each involving raising and lowering of spins, and are described by using the terms B to F of the dipolar Hamiltonian as

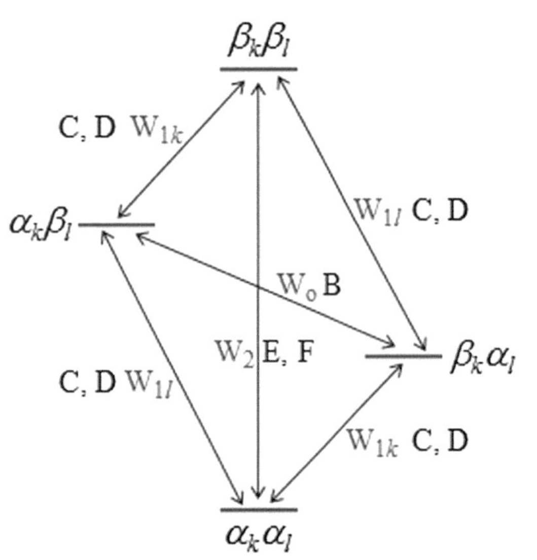

FIGURE 13.29 Association of terms B to F in the dipolar Hamiltonian with zero, single, and double quantum transitions in the case of a kl spin pair. Relaxation can occur by any of the transition pathways, but only single-quantum absorption is allowed.

depicted in the figure. The term A contains longitudinal two-spin order ($I_{kz}I_{lz}$), which is unobservable. It is important to note that each of the six 'alphabet' transitions are allowed in principle; however, the selection rule, according to which the total magnetic quantum number m_{T} can change only by ±1, prohibits zero-quantum ($\alpha_k\beta_l \rightarrow \beta_k\alpha_l$) and double-quantum ($\alpha_k\alpha_l \rightarrow \beta_k\beta_l$) absorptions. Even then the zero- and double-quantum pathways are available for spin relaxation whatever way the $\beta_k\alpha_l$ (or $\alpha_k\beta_l$) and $\beta_k\beta_l$ states are populated, and we know that these states are populated by the allowed single-quantum absorptions. Since we are concerned here with dipolar relaxation, we say that all pathways – zero-, single-, and double-quantum – are allowed relaxation pathways, even though only the single-quantum pathway can operate for the absorption transition.

One should also consider the fact that the relaxation transition is a slow time-dependent phenomenon, in contrast to the absorption transition which is nearly instantaneous, happening as soon as the RF pulse falls on the system. In the discussion of the two-level system in Chapter 3 we introduced the concept of probability of a transition in a general way, but the transition probability of relaxation is central in the present context because relaxation is time-dependent. Conventionally, the probabilities of zero-, single-, and double-quantum relaxation pathways are denoted by W_0, W_1, and W_2, respectively. The approach to finding the expressions for these probabilities is the same as described in Chapter 3. Let the total operator contain a time-independent Hamiltonian of kinetic and potential energy, and the time-varying field be represented by a dipolar Hamiltonian based on a fluctuating field of molecular origin

$$\mathcal{H} = \mathcal{H}_0 + \mathcal{H}_{\mathrm{D}}(t). \tag{13.106}$$

Let the eigenfunctions for ground and excited states be $\left|\Psi_1\right\rangle$ and $\left|\Psi_2\right\rangle$ with the respective coefficients $c_1(t)$ and $c_2(t)$. We suppose that the coupled spin system is initially in the excited state $\left|\Psi_2\right\rangle$, so that at time $t = 0$, $c_1(0) = 0$ and $c_2(0) = 1$. The probability of finding the system in the ground state $\left|\Psi_1\right\rangle$ turns out to be

$$\left|c_1(t)\right|^2 = \left(2\pi b_{kl}\right)^2 \sum_m \left\langle \left|\left\langle \psi_2\right| A^m \left|\psi_1\right\rangle\right|^2 \right\rangle \int_0^t \left(\int_0^{t-t'} \left\langle Y_2^m(t) Y_2^m(t+\tau)\right\rangle e^{(-i\omega_{21}\tau)} d\tau \right) dt'. \tag{13.107}$$

The term b_{kl} is the interaction constant, and τ represents the decay constant. The operator A^m is actually a spin operator containing angular momentum ladder operators. For example, if $m = 0, \pm 1, \pm 2$, then

$$A^0 = \sqrt{\frac{4}{5}}\left[I_{kz}I_{lz} - \frac{1}{4}\left(I_k^+I_l^- + I_k^-I_l^+\right)\right]$$

$$A^{\pm 1} = -\sqrt{\frac{3}{10}}\left[I_k^{\pm}I_{lz} - I_{kz}I_l^{\pm}\right]$$

$$A^{\pm 2} = -\sqrt{\frac{3}{10}}. \tag{13.108}$$

The angular parts are specified by the symbols of spherical harmonics

$$Y_2^0 = \sqrt{\frac{5}{4}}(1 - 3\cos^2\theta$$

$$Y_2^{\pm 1} = \sqrt{\frac{15}{2}}\cos\theta\sin\theta e^{\pm i\phi}$$

$$Y_2^{\pm 2} = -\sqrt{\frac{15}{8}}\sin^2\theta e^{\pm 2i\phi}. \tag{13.109}$$

The way the spherical harmonics enter into the integral implies that the angular part of the dipolar operator represents the autocorrelation time of molecular tumbling. At time long enough, the second integral in equation (13.107) can be integrated over τ from $-\infty$ to $+\infty$ (see equation (13.83)). The probability for $c_1(t)$ works out to

$$|c_1(t)|^2 = (2\pi b_{kl})^2 \sum_m \left\langle \left\langle \psi_2 |A^m| \psi_1 \right\rangle^2 \right\rangle \int_{-\infty}^{\infty} G(\tau) e^{-i\omega_{21}\tau} d\tau. \tag{13.110}$$

The time τ in the orientation part of the above is the decay time (constant), and $G(\tau)$ is the autocorrelation function whose Fourier transformation yields the spectral density $J(\omega_{21})$. The equation can be rewritten as

$$\frac{1}{t}|c_1(t)|^2 = W = (2\pi b_{kl})^2 \sum_m \left\langle \left\langle \psi_2 |A^m| \psi_1 \right\rangle^2 \right\rangle J(\omega_{21}). \tag{13.111}$$

The left-hand side now represents the probability per unit time – also called the transition rate. The integrals in the spin operator part are

$$\left\langle \psi_2 |A^0| \psi_1 \right\rangle^2 = \frac{4}{5} \times \frac{1}{16}$$

$$\left\langle \psi_2 |A^{\pm 1}| \psi_1 \right\rangle^2 = \frac{3}{10} \times \frac{1}{4}$$

$$\left\langle \psi_2 |A^{\pm 2}| \psi_1 \right\rangle^2 = \frac{3}{10}. \tag{13.112}$$

The exponential factor in the orientation integral in equation (13.110) changes according to the type of the relaxation transition. For heteronuclear dipolar coupling of spins k and l, the frequency for the single-quantum relaxation is ω_k, but the frequencies for zero- and double-quantum transitions are $\omega_l - \omega_k$ and $\omega_l + \omega_k$, respectively. Therefore, Fourier transformation of the respective autocorrelation functions yields the spectral densities $J(\omega_k)$, $J(\omega_l - \omega_k)$, and $J(\omega_l + \omega_k)$. In writing these spectral densities we have assumed the relaxation of the kth spin alone. To monitor the relaxation of the spin, one can use the corresponding spectral densities $J(\omega_l)$, $J(\omega_k - \omega_l)$, and $J(\omega_k + \omega_l)$. Putting all these together the following three transition rates are obtained

$$W_0 = \frac{1}{20}(2\pi b_{kl})^2 J(\omega_l - \omega_k)$$

$$W_1 = \frac{3}{40}(2\pi b_{kl})^2 J(\omega_k)$$

$$W_2 = \frac{3}{10}(2\pi b_{kl})^2 J(\omega_l + \omega_k). \tag{13.113}$$

It would turn out in the description below that the spin-lattice relaxation of each dipole-coupled spin can be written as the sum of the three respective transition probabilities

$$T_{1k}^{-1} = W_0 + 2W_{1k} + W_2$$

$$T_{1l}^{-1} = W_0 + 2W_{1l} + W_2. \tag{13.114}$$

Substitution for W_0, W_1, and W_2 from equation (13.113) yields the dipolar spin-lattice relaxation rate constants for spins k and l induced by the dipole-dipole interaction between them (Abragam, 1961)

$$T_{1k,\text{dd}}^{-1} = \frac{1}{20}(2\pi b_{kl})^2 \left[J(\omega_l - \omega_k) + 3J(\omega_k) + 6J(\omega_l + \omega_k) \right] + c^2 J(\omega_k)$$

$$T_{1l,\text{dd}}^{-1} = \frac{1}{20}(2\pi b_{kl})^2 \left[J(\omega_k - \omega_l) + 3J(\omega_l) + 6J(\omega_l + \omega_k) \right] + d^2 J(\omega_l) \tag{13.115}$$

The angular frequencies ω_k and ω_l in these expressions must correspond to spin-respective Larmor frequencies. The form of spectral densities in the equations is determined by the number of exponentials by which the correlation function decays. In the simplest case, a single exponential may be assumed to obtain the spectral density as

$$J(\omega_0) = \frac{2\tau_c}{1 + \omega_0^2 \tau_c^2}, \tag{13.116}$$

The dipolar spin-lattice relaxation rate constants (equation 13.115) can now be written as

$$T_{1k,\text{dd}}^{-1} = \frac{1}{10}\tau_c (2\pi b_{kl})^2 \left[\frac{1}{1 + (\omega_{0l} - \omega_{0k})^2 \tau_c^2} + \frac{3}{1 + \omega_{0k}^2 \tau_c^2} + \frac{6}{1 + (\omega_{0l} + \omega_{0k})^2 \tau_c^2} \right]$$

$$T_{1l,\mathrm{dd}}^{-1} = \frac{1}{10}\tau_c \left(2\pi b_{kl}\right)^2 \left[\frac{1}{1+\left(\omega_{0k}-\omega_{0l}\right)^2 \tau_c^2} + \frac{3}{1+\omega_{0l}^2 \tau_c^2} + \frac{6}{1+\left(\omega_{0k}+\omega_{0l}\right)^2 \tau_c^2}\right] \quad (13.117)$$

Let us go two steps back to equation (13.115). The extra terms $c^2 J(\omega_k)$ and $d^2 J(\omega_l)$ are not fuzzy; they, in fact, account for the chemical shift anisotropy, the most common being an axially symmetric chemical shift tensor, in which case the constant c would be

$$c = \omega_k \frac{\left(\sigma_{\parallel} - \sigma_{\perp}\right)}{\sqrt{3}}. \quad (13.118)$$

A similar expression applies for the constant d. The value of c varies with the types of nuclei coupled. For example, in $^1H-^{15}N$ heteronuclear coupling in a protein backbone we have $\sigma_{\parallel} - \sigma_{\perp} = -160$ ppm, where $\sigma_{\parallel} = \sigma_{zz}$ and $\sigma_{\perp} = \sigma_{xx} = \sigma_{yy}$.

Analogous to the expressions for dipolar spin-lattice relaxation rate constants, we can obtain expressions for dipolar spin-spin relaxation rate constants for dipole-coupled spins k and l. For purely intramolecular conditions in the limit of isotropic random motion of the molecule, the formulas for $T_{2,\mathrm{dd}}^{-1}$ incorporating zero- (flip-flop), single-, and double-quantum relaxations are given as

$$T_{2,\mathrm{dd}}^{-1(0)} = \frac{1}{40}\left(2\pi b_{kl}\right)^2 \left[2J\left(\omega_{0k}-\omega_{0l}\right) + 3J\left(\omega_{0k}\right) + 3J\left(\omega_{0l}\right)\right]$$

$$T_{2,\mathrm{dd}}^{-1(1)} = \frac{1}{40}\left(2\pi b_{kl}\right)^2 \left[J\left(\omega_{0k}-\omega_{0l}\right) + 4J(0) + 3J\left(\omega_{0k}\right) + 3J\left(\omega_{0l}\right) + 6J\left(\omega_{0k}+\omega_{0l}\right)\right]$$

$$T_{2,\mathrm{dd}}^{-1(2)} = \frac{1}{40}\left(2\pi b_{kl}\right)^2 \left[3J\left(\omega_{0k}\right) + 3J\left(\omega_{0l}\right) + 12J\left(\omega_{0k}+\omega_{0l}\right)\right], \quad (13.119)$$

in which the spectral density may again correspond to a single-exponential correlation function (equation 13.116). These expressions for $T_{2,\mathrm{dd}}^{-1}$, however, will not hold strictly if external random magnetic fields operate to contribute to transverse relaxation (Macura et al., 1981). When such conditions prevail, the spin-spin relaxation rates $1/T_2^{(0)}$, $1/T_2^{(1)}$, and $1/T_2^{(2)}$ will fall into the $2:3:6$ ratios in the limit of fast motion $\left(\omega_0\tau_c \ll 1\right)$, but $0:1:4$ in the slow motion limit $\left(\omega_0\tau_c \gg 1\right)$.

Dipolar relaxation and chemical shift anisotropy are major sources of relaxation and steady-state NOE enhancement of ^{15}N nucleus in the $^1H-^{15}N$ format, implying that the spectral densities also define the transverse relaxation time (T_2) and steady-state NOE. One may also note that the expressions for both T_1^{-1} and T_2^{-1} in homo- and heteronuclear cases may differ by the numerical values of coefficients. But in the limit of extreme narrowing condition, $T_{1,\mathrm{dd}}^{-1}$(homonuclear) $= T_{1,\mathrm{dd}}^{-1}$ (heteronuclear).

13.19 SPIN CROSS-RELAXATION: SOLOMON EQUATIONS

An analysis of the rate of change of populations of the dipole-coupled spin states $\alpha_k\alpha_l$, $\beta_k\alpha_l$, $\alpha_k\beta_l$, and $\beta_k\beta_l$ would show the occurrence of cross-relaxation between the coupled spins kl giving rise to nuclear Overhauser effect (NOE). With reference to Figure 13.29, let us denote the population of the states at some instant by P and at Boltzmann equilibrium by P^o. The differential equations describing the population change of each of the four states can be written down by a close look at the relaxation pathways. For the $\alpha_k\alpha_l$ state, we can write

$$\frac{dP_{\alpha_k\alpha_l}}{dt} = -\left(W_{1k} + W_{1l} + W_2\right)P_{\alpha_k\alpha_l} + W_0 P_{\alpha_k\beta_l} + W_0 P_{\beta_k\alpha_l} + W_2 P_{\beta_k\beta_l}. \quad (13.120)$$

If the deviation of the population at some instant from the Boltzmann equilibrium is

$$\Delta P_{\alpha_k\alpha_l} = P_{\alpha_k\alpha_l} - P^0_{\alpha_k\alpha_l},$$

the differential equation for the deviated population will be

$$\frac{d\Delta P_{\alpha_k\alpha_l}}{dt} = -\left(W_{1k} + W_{1l} + W_2\right)\Delta P_{\alpha_k\alpha_l} + W_0\Delta P_{\alpha_k\beta_l} + W_0\Delta P_{\beta_k\alpha_l} + W_2\Delta P_{\beta_k\beta_l}. \quad (13.121)$$

Similarly, the differential equations for the deviated populations of the other three spin states are

$$\frac{d\Delta P_{\beta_k\alpha_l}}{dt} = -\left(W_0 + W_{1k} + W_{1l}\right)\Delta P_{\beta_k\alpha_l} + W_0\Delta P_{\alpha_k\beta_l} + W_{1k}\Delta P_{\alpha_k\alpha_l} + W_{1l}\Delta P_{\beta_k\beta_l}$$

$$\frac{d\Delta P_{\alpha_k\beta_l}}{dt} = -\left(W_0 + W_{1k} + W_{1l}\right)\Delta P_{\alpha_k\beta_l} + W_0\Delta P_{\beta_k\alpha_l} + W_{1k}\Delta P_{\beta_k\beta_l} + W_{1l}\Delta P_{\alpha_k\alpha_l}$$

$$\frac{d\Delta P_{\beta_k\beta_l}}{dt} = -\left(W_{1k} + W_{1l} + W_2\right)\Delta P_{\beta_k\beta_l} + W_{1k}\Delta P_{\alpha_k\beta_l} + W_{1l}\Delta P_{\beta_k\alpha_l} + W_2\Delta P_{\alpha_k\alpha_l}. \quad (13.122)$$

It is also easy to see that the time dependence of macroscopic magnetic moment of each spin I_k and I_l can be determined by

$$I_{kz}(t) = \left(P_{\alpha_k\alpha_l} + P_{\alpha_k\beta_l}\right) - \left(P_{\beta_k\alpha_l} - P_{\beta_k\beta_l}\right)$$

$$I_{lz}(t) = \left(P_{\alpha_k\alpha_l} + P_{\beta_k\alpha_l}\right) - \left(P_{\alpha_k\beta_l} - P_{\beta_k\beta_l}\right). \quad (13.123)$$

If I_k^o and I_l^o are the equilibrium magnitudes of the magnetic moments I_k and I_l, then

$$\Delta I_{kz} = I_{kz} - I^{o}_{kz}$$

$$\Delta I_{lz} = I_{lz} - I^{o}_{lz}. \qquad (13.124)$$

Combining the equations above, one can write

$$\frac{d\Delta I_{kz}(t)}{dt} = -\left(W_0 + 2W_{1k} + W_2\right)\Delta I_{kz}(t) - \left(W_2 - W_0\right)\Delta I_{lz}(t)$$

$$\frac{d\Delta I_{lz}(t)}{dt} = -\left(W_0 + 2W_{1l} + W_2\right)\Delta I_{lz}(t) - \left(W_2 - W_0\right)\Delta I_{kz}(t). \qquad (13.125)$$

If we use the substitutions

$$W_0 + 2W_{1k} + W_2 = \rho_k$$

$$W_0 + 2W_{1l} + W_2 = \rho_l$$

$$W_2 - W_0 = \sigma_{kl} \qquad (13.126)$$

to write equation (13.125) in compact forms, we get Solomon equations

$$\frac{d\Delta I_{kz}(t)}{dt} = -\rho_k \Delta I_{kz}(t) - \sigma_{kl}\Delta I_{lz}(t)$$

$$\frac{d\Delta I_{lz}(t)}{dt} = -\rho_k \Delta I_{lz}(t) - \sigma_{kl}\Delta I_{kz}(t). \qquad (13.127)$$

Ionel Solomon's work (1955) came as the discovery of spin cross-relaxation of dipolar-coupled nuclei.

It is convenient to write Solomon equations in the matrix form

$$\begin{pmatrix} \Delta \dot{I}_{kz} \\ \Delta \dot{I}_{lz} \end{pmatrix} = -\begin{pmatrix} \rho_k & \sigma_{kl} \\ \sigma_{lk} & \rho_l \end{pmatrix}\begin{pmatrix} \Delta I_{kz} \\ \Delta I_{lz} \end{pmatrix}. \qquad (13.128)$$

One also easily identifies the diagonals of the relaxation matrix with autorelaxations (T_1^{-1}) of the dipole-coupled spins. It is emphasized that the autorelaxation rate constant, which we have been writing as ρ_k or ρ_l, is indeed the direct dipolar relaxation of spin k by the dipolar coupled spin l, and vice versa; so, ρ_k and ρ_l may also be written as ρ_{kl} or ρ_{lk}, respectively. Nevertheless, they are generally shown implicitly as ρ_k or ρ_l. Hence, the two parts of equation (13.115) in the preceding section are effectively

$$T_{1k}^{-1} = \rho_k = W_0 + 2W_{1k} + W_2$$

$$T_{1l}^{-1} = \rho_l = W_0 + 2W_{1l} + W_2. \qquad (13.129)$$

The off-diagonals of the rate matrix in equation (13.128) are the cross-relaxation rates which give rise to NOE.

The dipolar autorelaxation of a spin need not necessarily arise from just one coupled spin. It also happens that the autocorrelation of a spin is contributed by more than one dipolar-coupled spins. Suppose $k,l,m,n,\ldots$ is such a group of multispin dipolar system, then the relaxation of spin l, for instance, will be

$$\rho_l = \rho_{lk} + \rho_{lm} + \rho_{ln} + \cdots + \rho_l^*, \qquad (13.130)$$

in which ρ_l^* represents additional relaxation mechanism – chemical shift anisotropy, for example. Clearly, one need not construe that NOE depends only on the distance between the two coupled spins; the nuclei in the effective proximity will also contribute to the NOE. So, Solomon equations can be written down for more than two coupled nuclei. Writing all of them individually is laborious, but is convenient in the matrix form. If there are n interacting spins ($n = 1,2,3,\cdots$), the general form of the equation follows from equation (13.127),

$$\frac{d\Delta I_{nz}(t)}{dt} = -\sum_{n\neq r}\rho_{nr}\Delta I_{nz}(t) - \sum_{n\neq r}\sigma_{nr}\Delta I_{rz}(t), \qquad (13.131)$$

which in the matrix form reads

$$\frac{d\Delta \mathrm{M}_z(t)}{dt} = -\mathbf{R}\Delta\mathbf{M}_z(t),$$

where $\Delta\mathbf{M}_z(t)$ is a $(n\times 1)$ column vector. The matrix $\mathbf{R}$ represents a $(n\times n)$ matrix containing the ρ_{nn} diagonals and σ_{nr} off-diagonals

$$\begin{pmatrix} \Delta\dot{\mathbf{M}}_{1z} \\ \vdots \\ \Delta\dot{\mathbf{M}}_{nz} \end{pmatrix} = -\begin{pmatrix} \rho_{11} & \cdots & \sigma_{1n} \\ \vdots & \ddots & \vdots \\ \sigma_{n1} & \cdots & \rho_{nn} \end{pmatrix}\begin{pmatrix} \Delta\mathbf{M}_{1z} \\ \vdots \\ \Delta\mathbf{M}_{nz} \end{pmatrix}. \qquad (13.132)$$

One should be clear by now that the differential equations, the eigenvalues, and the eigenvectors in Solomon equations bear on the familiar linear algebraic equations of chemical kinetics for reversible systems. One then recognizes that cross-relaxation influences all rate constants in the dipolar relaxation matrix. To illustrate, let us take the two-spin system kl for which the relaxation matrix is given in equation (13.128). Written in terms of transitions probabilities, the four elements yield (Ernst et al., 1988)

$$W_{1k} = \frac{3}{4}q_{kl}J_{kl}\left(\omega_{ok}\right)$$

$$W_{1l} = \frac{3}{4}q_{kl}J_{kl}\left(\omega_{ol}\right)$$

$$W_0 = \frac{1}{2}q_{kl}J_{kl}\left(\omega_{ok} - \omega_{ol}\right)$$

$$W_2 = 3q_{kl}J_{kl}\left(\omega_{ok} + \omega_{ol}\right) \qquad (13.133)$$

where ω_o is Larmor frequency in the laboratory frame. The constant q_{kl} is given by

$$q_{kl} = \left(\frac{\mu_0}{4\pi}\right)^2 \frac{1}{10}\gamma_k^{\,2}\gamma_l^{\,2}\hbar^2 r_{kl}^{\,-6},$$

and the spectral density functions are Lorentzian

$$J_{kl}\left(\omega_{\text{o}k}\right) = \frac{2\tau_c^{kl}}{1+\omega_{0k}^{\,2}\left(\tau_c^{kl}\right)^2}. \quad (13.134)$$

Two more quantities – cross-relaxation rate constant R_{c} and leakage rate constant R_{L} – influence the peak intensities in spectra. They are better understood in the context of diagonal and cross peaks in two-dimensional NOE spectra. We define them here simply for a two-spin homonuclear system as

$$R_{\text{c}} = 2\left|W_2 - W_0\right|$$

$$R_{\text{L}} = R_1^{\text{ext}} + 2W_1 + W_0 + W_2 - 2\left|W_2 - W_0\right|,$$

where R_1^{ext} signifies contribution by any other interacting spin to the relaxation of both k and l to the same extent. The relevance of R_{L}, which in the absence of R_1^{ext} diminishes to zero, that is $\omega_0\tau_c \gg 1$ (slow motion limit), surfaces only in the limit of fast motion.

13.20 NUCLEAR OVERHAUSER EFFECT (NOE)

The cross-relaxation effect prevents independent relaxation of dipolar-coupled spins. This is one way of stating the NOE effect, seen at once from Solomon equations. To illustrate, consider the first of the two in equation (13.127) for dipolar-coupled spins k and l

$$\frac{d\Delta I_{kz}(t)}{dt} = -\rho_k \Delta I_{kz}(t) - \sigma_{kl}\Delta I_{lz}(t).$$

Assume that the spins are under steady-state condition. We drop the time dependence and use equation (13.124) to write

$$-\rho_k\left(I_{kz} - I_k^{\text{o}}\right) - \sigma_{kl}\left(I_{lz} - I_l^{\text{o}}\right) = 0. \quad (13.135)$$

If the spin l is selectively irradiated by a weak RF pulse at frequency $\omega_{\text{o},l}$ so as to equalize its populations in α and β states, then $I_{lz} = 0$. Note that I_{lz} is the z-component of the macroscopic magnetization of spin l, and I_l^{o} is its equilibrium magnitude. The use of $I_{lz} = 0$ in equation (13.135) yields

$$I_{kz} = I_k^{\text{o}} + \frac{\sigma_{kl}}{\rho_k} I_l^{\text{o}}$$

$$\frac{I_{kz}}{I_k^{\text{o}}} = 1 + \frac{\sigma_{kl}}{\rho_k}\frac{I_l^{\text{o}}}{I_k^{\text{o}}}. \quad (13.136)$$

Since the initial equilibrium magnetization is proportional to the gyromagnetic ratio, we obtain

$$\frac{I_{kz}}{I_k^{\text{o}}} = 1 + \frac{\gamma_l \sigma_{kl}^{\text{NOE}}}{\gamma_k \rho_k} = 1 + \eta_{kl}. \quad (13.137)$$

where η_{kl} is the NOE enhancement factor which determines the additional intensity due to the dipolar interaction between k and l. This is called steady-state NOE; steady state, because the temporal buildup of NOE is not included in this development.

13.20.1 Positive and Negative NOE

In practice, one can irradiate the spin l continuously with a weak RF pulse so as not to perturb spin k, and then look at the spectral intensity of the latter by exciting the entire range of frequencies using a non-selective 90º pulse. We will come to this shortly, but let us briefly look at the meaning of equation (13.137). One, the intensity of the k line will be 'in addition to the normal' depending on what the value of the factor η_{kl} is. Accordingly, NOE can be positive or negative. Positive NOE would mean a larger difference of α_k and β_k populations at the expense of a reduced difference of α_l and β_l populations. Negative NOE would result from a smaller difference of α_k and β_k populations, which can occur if the population difference of the spin l is made smaller (Figure 13.30). Two, if the system is homonuclear ($\gamma_l = \gamma_k$), then the NOE enhancement will be determined by the relative magnitudes of cross-relaxation and autorelaxation of the nucleus under observation. If the cross-relaxation is weak, which means larger spatial separation of spins k and l, the NOE enhancement will be weak too. Thus, the efficacy of NOE transfer depends on the spatial distance between the two nuclei and the correlation time of the molecule,

$$\eta \propto \frac{\tau_{\text{c}}}{r^6}, \quad (13.138)$$

where we have included only a proportional sign because other nuclei in proximity also affect η.

Experimentally, steady-state NOE enhancement η_{kl} is measured accurately by comparing two 1D spectra – one is a normal reference spectrum recorded without perturbing the system, and the other one is recorded by first saturating a spin, say l, with a l-selective weak pulse for a time τ_m long enough for complete saturation and NOE propagation (Figure 13.31). One looks at the line intensity of spin k in the two spectra; the intensity is proportional to I_k^{o} in the former and I_{kz} in the latter, where I_{kz} is the z-component of spin k and magnetization of I_k^{o} is its equilibrium value. Using equation (13.137) we write

$$\eta_{kl} = 1 - \frac{I_{kz}}{I_k^{\text{o}}} = \eta_k\{l\}. \quad (13.139)$$

The difference of such two spectra will have signals only from that nucleus whose intensity changed by positive or negative enhancement and the nucleus that was saturated. The same exercise can be carried out by saturating spin k. Often, η_{kl} is

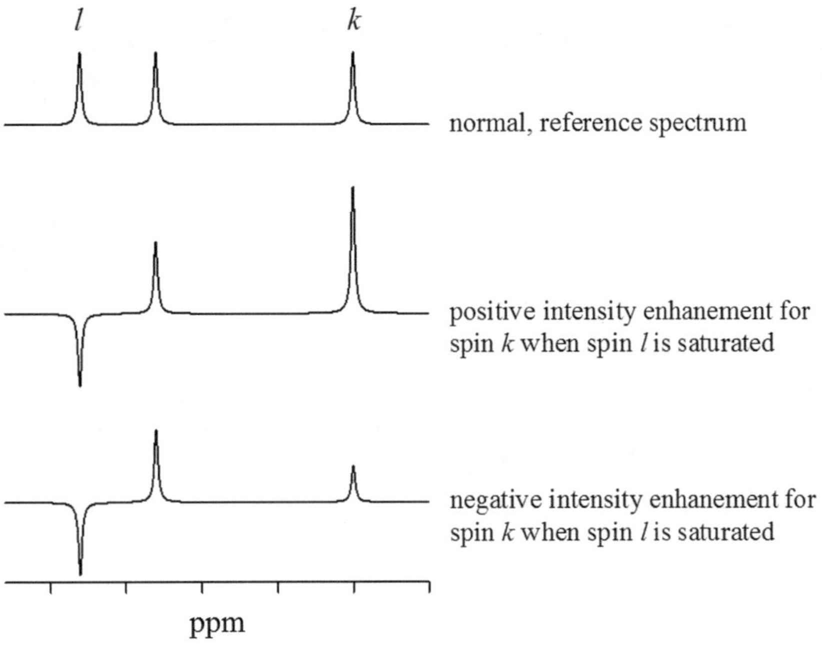

FIGURE 13.30 Steady-state NOE, positive and negative enhancement.

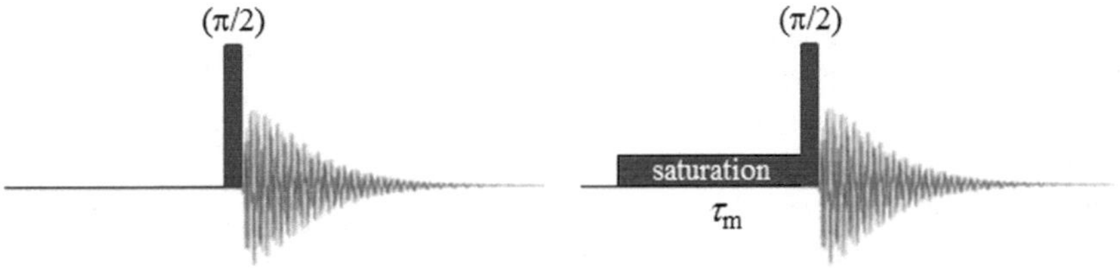

FIGURE 13.31 The pulse sequence for steady-state NOE measurement. *Left*, the usual reference measurement. *Right*, presaturation of a spin for time τ_m.

written in a way to indicate which spin is saturated and whose NOE enhancement is being observed. If the resonance line of spin k is observed by saturating l, then the enhancement factor is written as $\eta_k\{l\}$.

13.20.2 Direct and Indirect NOE Transfer

The total value of $\eta_k\{l\}$ can be segregated into direct and indirect components if an intermediary spin comes into the play. In general,

$$\eta_k\{l\} = \eta_{\max}\frac{r_{kl}^{-6}}{r_{kl}^{-6}+\sum_n r_{kn}^{-6}} - \eta_{\max}\sum_n\left[\frac{f_n\{l\}r_{kn}^{-6}}{r_{kl}^{-6}+\sum_n r_{kn}^{-6}}\right], \tag{13.140}$$

where the first term represents the direct NOE enhancement of spin k which is due to the effect of spins in addition to l, all of which form the multispin dipolar system. This term is called the extreme narrowing term. The second term arises due to a third intermediary spin which is in close proximity of both k and l, and which mediates the transfer of NOE. This indirect term is also called the spin-diffusion term. Figure 13.32 illustrates the direct and indirect contributions to the total NOE enhancement $\eta_k\{l\}$ for a dipolar system of three spins k,l,m as the $k-l$ distance is varied from 0.5 Å (extreme narrowing) to the limit of 5 Å (spin diffusion). For large molecules, the NOE enhancements are negative and $T_2 \ll T_1$, which results in broad NOE lines. In fact, when τ_c gets longer with increasing size of molecules the NOE propagates to other nuclear sites, which is why it is called spin diffusion.

To look at the τ_c dependence of NOE for a pair of homonuclear dipolar coupled spins, let us expand the NOE enhancement factor (equation (13.136)) in terms of transition probabilities (equation (13.126)) and then spectral densities (equation (13.133)). The enhancement factor is

$$\eta_{kl} = \frac{\gamma_l}{\gamma_k}\frac{\sigma_{kl}}{\rho_{kl}} = \frac{\gamma_l}{\gamma_k}\left[\frac{W_2 - W_0}{W_0 + 2W_{1k} + W_2}\right]. \tag{13.141}$$

For a homonuclear system we get

$$\eta = \frac{5+\omega_0^2\tau_c^2 - 4\omega_0^4\tau_c^4}{10+23\omega_0^2+23\omega_0^2\tau_c^2+4\omega_0^4\tau_c^4}. \tag{13.142}$$

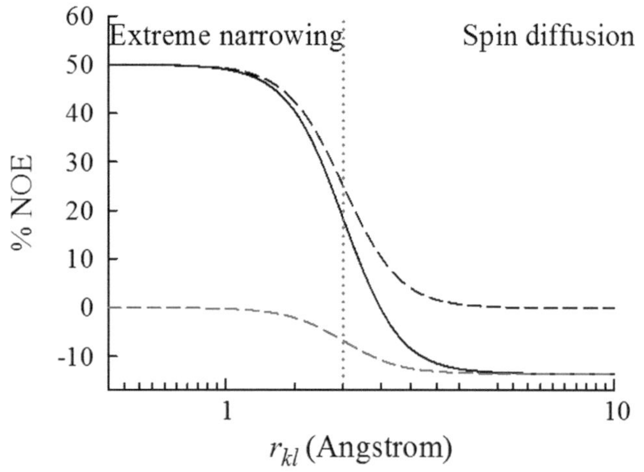

FIGURE 13.32 The total NOE (black solid line) is the resultant of the direct NOE (red dash line) and indirect NOE (blue dash line). The dotted line demarcates the NOE limits.

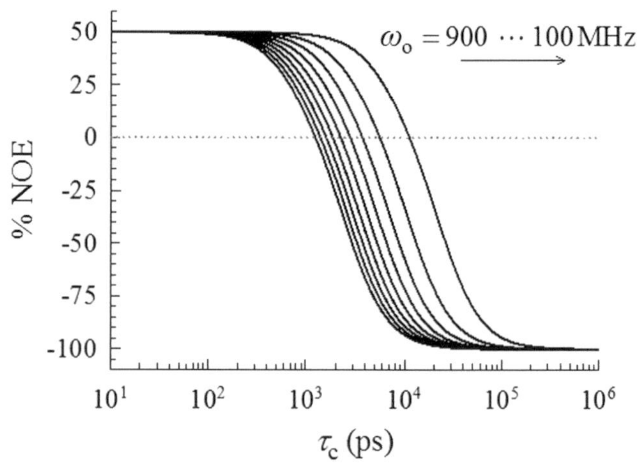

FIGURE 13.33 Computed curves to show the Larmor frequency dependence of the variation of NOE with tumbling time.

The variation of η with respect to τ_c at different Larmor frequencies is shown in Figure 13.33. We notice that the change in the value of NOE is Larmor frequency-dependent, which is obvious from the equations above. This should mean that the measurement of NOE at lower static field strength B_o is advantageous, especially for small molecules of shorter τ_c, because the smallness of τ_c is compensated by lower field strength. Put another way, NOE will turn negative for small molecules of the kind of organic molecules, whose mass could be roughly < 200 Da, say, as the strength of the B_o field is increased. We do not want the NOE to be negative. This argument is applicable for large molecules (longer τ_c) as well; however, resolution and sensitivity of resonances call for higher field strength. Hence, a trade-off between NOE sign and spectral resolution comes into the play. In any case, NOE = 0 when $\omega_0\tau_c$ approaches 1.12. Some literature and textbooks appear to state that NOE = 0 when $\omega_0\tau_c = 1$, which is not quite right.

13.20.3 Rotating Frame Overhauser Effect

The result that NOE = 0 for $\omega_0\tau_c \approx 1.12$ sounds disastrous. What if we had only one field strength and the τ_c of the molecule works out to $\omega_0\tau_c \approx 1.12$? This often is the case for medium-size molecules. Fortunately, there is another version of NOE that comes to the rescue, which is called rotating frame Overhauser effect, ROE. Note that NOE is a laboratory frame experiment in contrast to ROE, which refers to a rotating frame that we are already familiar with – the frame rotates at the same frequency that the RF field does. So, the magnetization appears static. Whereas cross-relaxation of longitudinal magnetizations of the dipolar-coupled spins gives rise to NOE, it is the cross-relaxation of the transverse magnetization of the dipolar coupled spins that yields ROE. In practice, the initial equilibrium magnetization is transformed to transverse magnetization by a 90_x RF pulse (B_1). Immediately after applying the pulse, the B_1 field is phase-shifted by 90° and left on for some hundreds of milliseconds of mixing time (τ_m). Under this condition the spins do not precess and are said to be 'locked', which amounts to freeing the spins from the main polarizing field B_o and placing them under the much weaker spin-locking B_1 field. Now, $\gamma B_1 = \omega_{o(\text{spin lock})} \ll \gamma B_o = \omega_o$. They will, however, relax longitudinally, and also cross-relax when dipolar-coupled. Since the spin-locked Larmor frequency is very small, one expects the extreme narrowing condition ($\omega_{o(\text{spin lock})}\tau_c \ll 1$) to hold. Consequently, the NOE enhancement will be large and positive. The spectrum is then recorded by a 90° detection pulse. If we assume the correlation time and the spectral density of a spherical top, the cross-relaxation time may be written as

$$\sigma_{kl}^{\mathrm{ROE}} = W_1 - W_0 = \frac{\hbar^2\mu_0^{\,2}\gamma^4\tau_c}{10\pi^2 r_{kl}^{\,6}}\left\{2 + \frac{3}{1+\omega_0^2\tau_c^{\,2}}\right\}. \quad (13.143)$$

A comparison of NOE and ROE at different length of τ_c is the following

Extreme narrowing limit

$$\omega_0\tau_c \ll 1 \qquad \sigma_{kl}^{\mathrm{NOE}} = \sigma_{kl}^{\mathrm{ROE}} = \frac{\hbar^2\mu_0^{\,2}\gamma^4\tau_c}{2\pi^2 r_{kl}^{\,6}} \qquad \eta_{kl} = 0.5$$

Spin diffusion limit

$$\omega_0\tau_c \gg 1 \qquad \sigma_{kl}^{\mathrm{ROE}} = \frac{\hbar^2\mu_0^{\,2}\gamma^4\tau_c}{5\pi^2 r_{kl}^{\,6}} \qquad \eta_{kl} = -1$$

Intermediate region

$$\sigma_{kl}^{\mathrm{NOE}} = \frac{-\sigma_{kl}^{\mathrm{ROE}}}{2}$$

Null-point

$$\omega_0\tau_c \approx 1.12 \qquad \sigma_{kl}^{\mathrm{NOE}} = 0,\ \sigma_{kl}^{\mathrm{ROE}} > 0$$

(irrespective of τ_c values)

13.20.4 Transient NOE

Steady-state NOE cannot directly quantify the distance between the kl dipolar-coupled spins even though $\eta \propto \tau_c / r_{kl}^6$. The measurement of η_{kl} then provides the qualitative information that the spins k and l are dipolar-coupled and in close spatial proximity. The quantification of the distance r_{kl} requires direct measurement of σ_{kl}^{NOE} or σ_{kl}^{ROE} by transient NOE experiments in 1D or 2D.

Let us define transient NOE by taking just two spins, k and l. Recall the Solomon equations of motion for the deviated z-magnetization from the equilibrium value (equations (13.127) and (13.128))

$$\begin{pmatrix} \Delta \dot{I}_{kz} \\ \Delta \dot{I}_{lz} \end{pmatrix} = -\begin{pmatrix} \rho_k & \sigma_{kl} \\ \sigma_{lk} & \rho_l \end{pmatrix} \begin{pmatrix} \Delta I_{kz} \\ \Delta I_{lz} \end{pmatrix}.$$

We wish to evaluate the time evolution of the matrix elements. Unlike the simple coupled differential equations of chemical kinetics involving the equilibrium of two states, these equations are different because of the simultaneous operation of both autorelaxation and cross-relaxation. Consider the 1D transient NOE pulse sequence (Figure 13.34) in which the second 90° pulse inverts the longitudinal magnetization of the spin l to create the following initial ($t = 0$) condition

$$\Delta I_{kz}(0) = I_{kz}(0) - I_{kz}^{o}(0) = 0$$
$$\Delta I_{lz}(0) = I_{lz}(0) - I_{lz}^{o}(0) = -2I_{lz}^{o}. \quad (13.144)$$

Let the autorelaxation rates be equal, $\rho_k = \rho_l = \rho$, which holds in the extreme narrowing limit ($\omega_0 \tau_c \ll 1$) in particular. The solutions are obtained as (Macura et al., 1981)

$$\Delta I_{kz}(t) = I_{kz}(t) - I_{k}^{o}(t) = I_{lz}^{o}\left[e^{-(\rho+\sigma_{kl})t} - e^{-(\rho-\sigma_{kl})t}\right]$$
$$\Delta I_{lz}(t) = I_{lz}(t) - I_{l}^{o}(t) = -I_{lz}^{o}\left[e^{-(\rho+\sigma_{kl})t} + e^{-(\rho-\sigma_{kl})t}\right]$$
$$(13.145)$$

that state that the change of the longitudinal magnetization of both spins with time is biexponential (Figure 13.35), and each exponential has contributions from both ρ and σ. The figure shows the change of $\left[I_z(t) - I_z^{o}(t)\right] / I_z^{o}$ with time, calculated by using $\sigma_{kl} = -0.75\ s^{-1}$ and $\rho_k = \rho_l = \rho = 1.25\ s^{-1}$. The NOE intensity on the spin k rises exponentially and then decreases by the second exponential. The increase of the intensity with time, call it mixing time τ_m, is called the NOE build-up rate. The initial increase of the intensity of the k spin is proportional to the cross-relaxation rate constant σ_{kl}, which is the rationale for initial rate approximation, i.e., $\sigma_{kl}^{NOE} \sim$ rate of NOE build-up.

It is also possible to define a transient NOE-enhancement factor $\eta^{(tr)}(\tau_m)$. With reference to a standard 1D spectrum

$$\eta^{(tr)}(\tau_m) = \frac{\Delta I_{kz}(\tau_m) - \Delta I_{kz}^{(ref)}(\tau_m)}{I_{lz}^{o}}, \quad (13.146)$$

which, in the initial rate approximation, can be written as

$$\eta^{(tr)}(\tau_m) \sim 2\kappa\sigma_{kl}^{NOE}\tau_m, \quad (13.147)$$

where the scaling factor $\kappa(0 \le \kappa \le 1)$ is included to account for the signal loss.

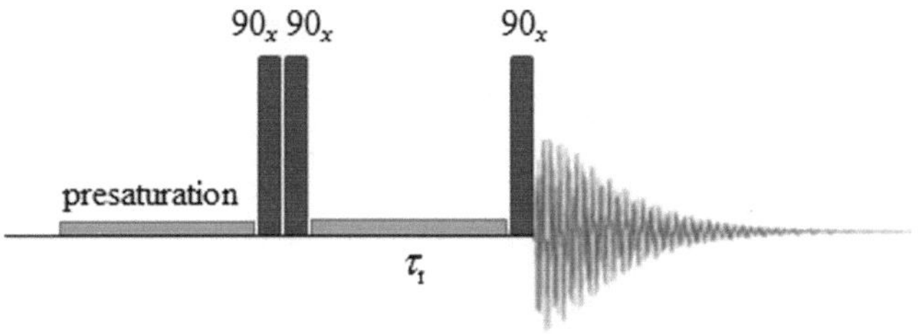

FIGURE 13.34 The pulse sequence for a 1D transient NOE experiment. The initial presaturation consists of a long low-power pulse which allows for selectively presaturating the solvent. The magnetization is inverted at the end of the second 90° pulse, which is followed by the mixing delay τ_m. The gray box covering the mixing delay represents another low-power long pulse used to suppress the solvent. The final 90° pulse creates detectable magnetization. The use of two 90° pulses instead of having a single 180° inverting pulse serves for phase cycling (see below). The experiment can be repeated with variable τ_m for determination of NOE build-up (σ_{kl}). In the 2D version of this experiment, the same pulse sequence may be run with incremental delay between the first and the second pulse.

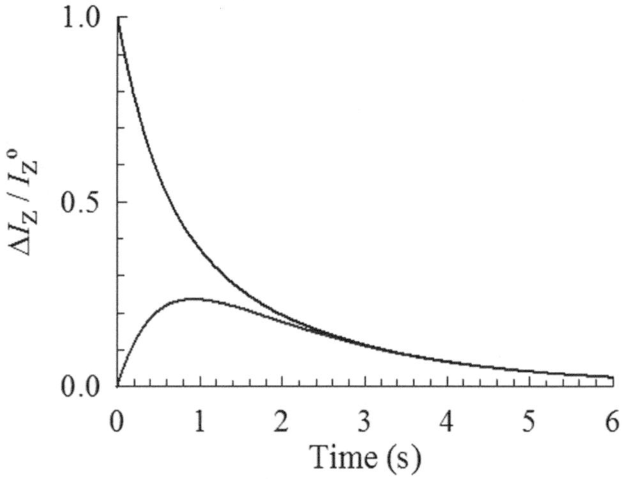

FIGURE 13.35 The solutions of Solomon equations providing a biexponential decay of longitudinal magnetization due to autorelaxation (black line), and a rising exponential followed by a decay due to cross-relaxation (red line). The slope of the rising exponential at short times is a good approximation of the rate of NOE build-up.

13.21 CHEMICAL EXCHANGE

There are three interactions that affect the details of a NMR resonance line: one, *J*-coupling interaction (through bond), which splits a resonance line as we have already discussed; two, dipole-dipole interaction (through space), covered in the preceding section; and three, chemical exchange (through reaction network), which is the subject of the present discussion. To project chemical exchange, let us consider a flipping phenyl or tyrosyl ring that can be a part of some molecular structure, or an amide proton NH that can exchange with solvent protons or deuterons (Figure 13.36). The flipping of the ring is the prototype of a first-order chemical exchange process, because protons 2 and 3 in the initial orientation of the ring, say 0° in the *xy* plane, find themselves at positions 6 and 5, respectively, if the ring is flipped by 180°. A successive flip by 180° will return the protons to where they were initially. The flipping event can then be associated with a dynamic equilibrium existing intramolecularly. The magnetic environment of protons 2 and 3 in state A need not be necessarily the same as in state B. We say that the nucleus is magnetically distinct in its occupancy of states A and B. The event of –NH proton exchange is a one-sided intermolecular exchange, because the solvent protons or deuterons are in large excess of that of the solute. We can, in fact, tune it to be a reversible exchange process by modulating the solvent condition, by varying the pH and percentage of the solvent constituents. For illustration though, we let the exchange reaction be unidirectional in the case of –NH proton exchange. Such chemical exchange events affect the frequency, the width, the shape, and often the splitting of the relevant NMR resonance.

For an introductory note on the effect of exchange on lineshape, consider an isolated nuclear spin in exchange between states A and B $(A \rightleftharpoons B)$. The rate constants k_{AB} and k_{BA} provide the residence times of the spin in state A (k_{AB}^{-1}) and state B (k_{BA}^{-1}). Let the chemical shift frequency of the spin in states A and B be ν_A and ν_B, respectively. Then, whether the NMR lines will appear distinctly at ν_A and ν_B will depend on how $\Delta\nu = \nu_A - \nu_B$ compares with k_{AB} and k_{BA}. It is convenient and easier to depict pictorially the dependence of the resonance positions on the magnitude of the rate constants (Figure 13.37), although calculations can be done to determine the lineshapes accurately as described later. The lifetimes of the nucleus in states A and B are determined by the time constants (k_{AB}^{-1} and k_{BA}^{-1}), the lifetime turns smaller as the rate constants become larger. In the context of exchange NMR, the lifetime of a spin, often in the range of microseconds to seconds, is generally referred to as NMR timescale, and the difference of frequencies of the two resonances under slow-exchanging condition is called the chemical shift timescale ($\Delta\nu$). It is easy to see that under slow-exchanging condition the NMR timescale is larger with respect to the chemical shift timescale, which causes the two resonances to appear distinctly at ν_A and ν_B. As the two timescales approach each other, the resonance lines move closer increasingly. In the intermediate exchange regime, the two timescales are comparable and the resonance lines eventually coalesce. In the fast exchange limit the coalesced line sharpens, indicating that the lifetime of the nucleus in the two states is much less compared to the chemical shift timescale. One may note that the changes in the two frequencies ν_A and ν_B need not be proportional. As shown in Figure 13.37, if $k_{AB} \ll k_{BA}$ then the lifetime of the spin in state A will be longer than that in state B. This causes larger shift of ν_B relative to that of ν_A. Similarly, the shift of ν_A will be larger than that of ν_B if $k_{AB} \gg k_{BA}$.

For a one-sided first-order reaction $A \rightarrow B$, the simplest situation arises under slow-exchange limit, in the case of HD exchange reaction at ~pH 4 for example. The resonance intensity corresponding to state A that appears at frequency ν_A will exponentially decrease to a level commensurate with the concentration of deuterium in the solvent. But these are very simple examples involving simple systems. The linewidth, absorptivity and dispersivity, distortion in the lineshape, and perturbed *J*-splitting are some of the manifestations to be expected depending on the exchange rate constant. Some of

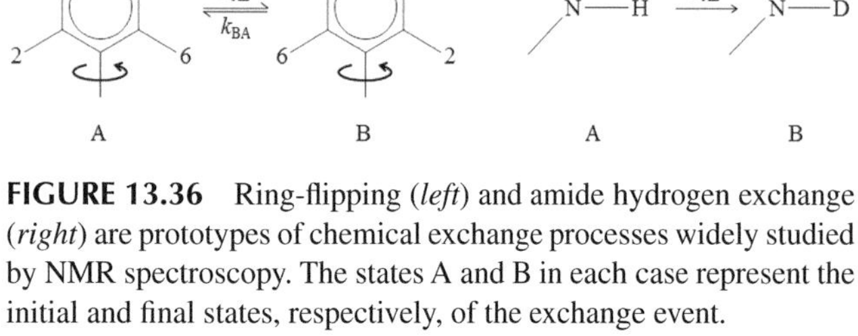

FIGURE 13.36 Ring-flipping (*left*) and amide hydrogen exchange (*right*) are prototypes of chemical exchange processes widely studied by NMR spectroscopy. The states A and B in each case represent the initial and final states, respectively, of the exchange event.

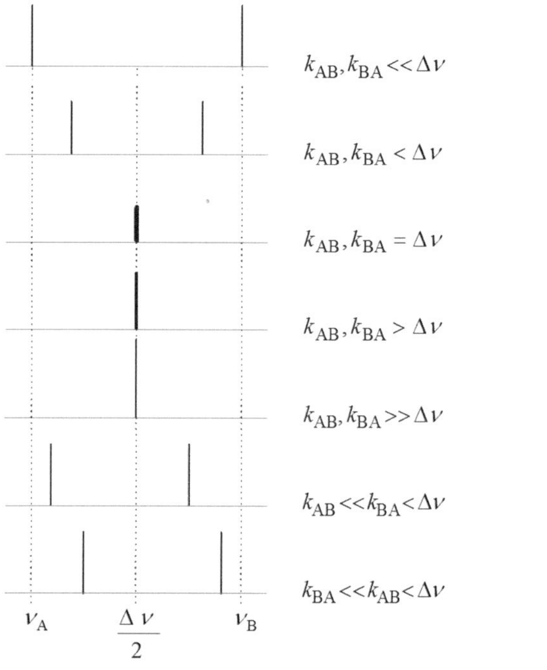

FIGURE 13.37 The dependence of A and B resonance lines under exchanging condition $(A \rightleftharpoons B)$ on the magnitude of the forward and reverse rate constants.

these aspects will be discussed as and when they surface; the experimenter, however, has to interpret the data appropriately.

13.21.1 Effect of Chemical Exchange on Line Shape

Let us now look at the mathematical structure of the magnetizations based on spin relaxation and the kinetic matrix. When it comes to spin relaxation the Bloch equations are considered; however, the apparatus of magnetization exchange contained in the kinetic matrix needs to be incorporated into the Bloch equations, because the magnetization is exchanging. This addition of the kinetic matrix elements to the conventional Bloch equations results in the modified Bloch equations, the solution of which gives the NMR lineshape.

Consider the exchange of a magnetization between just two states A and B

$$\mathrm{A} \underset{k_{\mathrm{BA}}}{\overset{k_{\mathrm{AB}}}{\rightleftharpoons}} \mathrm{B}$$

The first-order coupled differential equations and the kinetic matrix K can be obtained from the elementary principles of chemical kinetics, and we write

$$\frac{d[\mathrm{A}]}{dt} = -k_{\mathrm{AB}}[\mathrm{A}] + k_{\mathrm{BA}}[\mathrm{B}]$$

$$\frac{d[\mathrm{B}]}{dt} = k_{\mathrm{AB}}[\mathrm{A}] - k_{\mathrm{BA}}[\mathrm{B}]$$

$$\mathrm{K} = \begin{pmatrix} -k_{\mathrm{AB}} & k_{\mathrm{BA}} \\ k_{\mathrm{AB}} & -k_{\mathrm{BA}} \end{pmatrix}, \quad (13.148)$$

in which the square brackets on A and B represent respective concentrations.

The matrix for exchange among three states

$$\mathrm{A} \underset{k_{\mathrm{BA}}}{\overset{k_{\mathrm{AB}}}{\rightleftharpoons}} \mathrm{B} \underset{k_{\mathrm{CB}}}{\overset{k_{\mathrm{BC}}}{\rightleftharpoons}} \mathrm{C}$$

is obtained as

$$\mathrm{K} = \begin{pmatrix} -k_{\mathrm{AB}} & k_{\mathrm{BA}} & 0 \\ k_{\mathrm{AB}} & -k_{\mathrm{BA}} - k_{\mathrm{BA}} & k_{\mathrm{CB}} \\ 0 & k_{\mathrm{BC}} & -k_{\mathrm{CB}} \end{pmatrix} \quad (13.149)$$

The kinetic matrix can be set up easily for j number of states appearing in linear or cyclic schemes. The recipe to set up the matrix is as follows, the diagonals represent the negative of the microscopic rate constants which specify the spin 'going out' from the relevant states, and the off-diagonals are positive of those rate constants that specify the spin(s) 'accumulating' in.

It is also easy to show that the eigenvector solution of the differential equations takes the general form

$$[\mathbf{A}](t) = e^{\kappa t}[\mathbf{A}](0), \quad (13.150)$$

where $\mathbf{A}$ is the vector whose elements are the states $\mathrm{A}_j, j = A, B, C, \ldots, J$ or 1, 2, 3, …, J and $\mathbf{K}$ is the $J \times J$ matrix.

To see the modified Bloch equations, let us take the simple two-state magnetization exchange $\mathrm{A} \rightleftharpoons \mathrm{B}$. In the conventional Bloch equations, the magnetizations $\mathbf{M}_{\mathrm{A}}$ and $\mathbf{M}_{\mathrm{B}}$ are written as

$$\frac{d}{dt}\mathbf{M}_{\mathrm{A}}(t) = \gamma(1-\sigma_{\mathrm{A}})\mathbf{M}_{\mathrm{A}}(t) \times \mathbf{B}(t) - \mathbf{R}_{\mathrm{A}}\{\mathbf{M}_{\mathrm{A}}(t) - \mathbf{M}_{\mathrm{A}}^{\mathrm{o}}\}$$

$$\frac{d}{dt}\mathbf{M}_{\mathrm{B}}(t) = \gamma(1-\sigma_{\mathrm{B}})\mathbf{M}_{\mathrm{B}}(t) \times \mathbf{B}(t) - \mathbf{R}_{\mathrm{B}}\{\mathbf{M}_{\mathrm{B}}(t) - \mathbf{M}_{\mathrm{B}}^{\mathrm{o}}\}, \quad (13.151)$$

where, as mentioned earlier, $\mathbf{B}(t)$ consists of both static and RF magnetic fields, σ is the shielding constant, and $\mathbf{R}$ the relaxation matrix,

$$\mathbf{R} = \widetilde{\mathbf{R}} = \begin{pmatrix} 1/T_2 & 0 & 0 \\ 0 & 1/T_2 & 0 \\ 0 & 0 & 1/T_1 \end{pmatrix}.$$

The modified Bloch equations are obtained by adding the product of the off-diagonal rate constant specifying the 'accumulation' of the magnetization for which the Bloch equation is written, and the magnetization of the state that 'goes out'. This might sound a bit confusing. To enlighten, look at the modified equations explicitly for states A and B

$$\begin{aligned}\frac{d}{dt}\mathbf{M}_{\mathrm{A}}(t) &= \gamma(1-\sigma_{\mathrm{A}})\mathbf{M}_{\mathrm{A}}(t) \times \mathbf{B}(t) \\ &\quad - \mathbf{R}_{\mathrm{A}}\{\mathbf{M}_{\mathrm{A}}(t) - \mathbf{M}_{\mathrm{A}}^{\mathrm{o}}\} + k_{\mathrm{AB}}\mathbf{M}_{\mathrm{B}}(t) \\ \frac{d}{dt}\mathbf{M}_{\mathrm{B}}(t) &= \gamma(1-\sigma_{\mathrm{A}})\mathbf{M}_{\mathrm{B}}(t) \times \mathbf{B}(t) \\ &\quad - \mathbf{R}_{\mathrm{B}}\{\mathbf{M}_{\mathrm{B}}(t) - \mathbf{M}_{\mathrm{B}}^{\mathrm{o}}\} + k_{\mathrm{BA}}\mathbf{M}_{\mathrm{A}}(t).\end{aligned} \quad (13.152)$$

In the case of a three-state exchange of the spin ($\mathrm{A} \rightleftharpoons \mathrm{B} \rightleftharpoons \mathrm{C}$), the terms to add to the expressions for $\dot{\mathrm{M}}_{\mathrm{A}}(t)$, $\dot{\mathrm{M}}_{\mathrm{B}}(t)$, and $\dot{\mathrm{M}}_{\mathrm{C}}(t)$ can be found out from the rate matrix (equation (13.149)). They are $k_{\mathrm{BA}}\mathbf{M}_{\mathrm{B}}(t)$, $k_{\mathrm{AB}}\mathbf{M}_{\mathrm{A}}(t) + k_{\mathrm{CB}}\mathbf{M}_{\mathrm{C}}(t)$, and $k_{\mathrm{BC}}\mathbf{M}_{\mathrm{B}}(t)$, respectively. Thus, the modified Bloch equation can be given a general form (Ernst et al., 1988)

$$\begin{aligned}\frac{d}{dt}\mathbf{M}_j(t) &= \gamma(1-\sigma_j)\mathbf{M}_j(t) \times \mathbf{B}(t) \\ &\quad - \mathbf{R}_j\{\mathbf{M}_j(t) - \mathbf{M}_j^{\mathrm{o}}\} + \sum_r k_{jr}\mathbf{M}_r(t),\end{aligned} \quad (13.153)$$

where $j = \mathrm{A, B, C, \ldots, J}$, and $\mathbf{M}_j^{\mathrm{o}}$ is the equilibrium z-magnetization of the jth state given by

$$M_j^{\circ} = M^{\circ}\frac{[\mathrm{J}](t)}{\sum_k [\mathrm{K}]},$$

in which M° is the overall equilibrium magnetization, [J] is the spin concentration in the jth state, and the denominator represents the sum of the concentrations of all species.

After a 90° RF pulse, the time evolution of the longitudinal and transverse components of the spin magnetization in the rotating frame can be written down easily. For the two-state exchange case (A $\rightleftharpoons$ B) these equations for the spin in site A are

$$\frac{d}{dt}M_{\mathrm{A}}^{+} = \left(i\Omega_{\mathrm{A}} - \frac{1}{T_{2\mathrm{A}}} - k_{\mathrm{AB}}\right)M_{\mathrm{A}}^{+} + k_{\mathrm{BA}}M_{\mathrm{A}}^{+}$$

$$\frac{d}{dt}M_{\mathrm{A}z} = \left(-\frac{1}{T_{1\mathrm{A}}} - k_{\mathrm{AB}}\right)\left\{M_{\mathrm{A}z} - M_{\mathrm{A}}^{\circ}(t)\right\} + k_{\mathrm{BA}}M_{\mathrm{B}z}, \quad (13.154)$$

where the subscript '+' refers to the transverse component

$$M_{\mathrm{A}}^{+} = M_{\mathrm{A}x} + iM_{\mathrm{A}y},$$

and $\Omega_{\mathrm{A}} = (1-\sigma_A)B_{\mathrm{o}}$ is the chemical shift frequency.

The time evolution of the transverse and longitudinal magnetizations of the spin in state B, M_{B}^{+} and $M_{\mathrm{B}z}$, can be written in an analogous manner. Equations for spin exchange within three states A $\rightleftharpoons$ B $\rightleftharpoons$ C can also be written similarly. A general matrix form for expressing the time dependence of the magnetizations is

$$\frac{d}{dt}\mathbf{M}^{+}(t) = \mathbf{L}^{+}\mathbf{M}^{+}(t)$$

$$\frac{d}{dt}\mathbf{M}_z(t) = \mathbf{L}\left\{\mathbf{M}_z(t) - \mathbf{M}^{\circ}(t)\right\} + \mathbf{KM}_z^{\circ}(t). \quad (13.155)$$

The notations adopted here are those of Ernst and coworkers (1988). The transverse, the longitudinal, and equilibrium magnetizations of each state j are placed in the $\mathbf{M}^{+}$, $\mathbf{M}_z$, and $\mathbf{M}^{\circ}$ vectors, respectively; so they are $1\times j$ matrices. The matrices $\mathbf{L}^{+}$ and $\mathbf{L}$ can be written as

$$\mathbf{L}^{+} = i\boldsymbol{\Omega} - \boldsymbol{\Lambda} + \mathbf{K}$$

$$\mathbf{L} = -\mathbf{R} + \mathbf{K},$$

where $\boldsymbol{\Omega}$ is a diagonal chemical shift matrix, $\boldsymbol{\Lambda}$ is a diagonal transverse relaxation matrix, $\mathbf{R}$ is a longitudinal relaxation matrix, and $\mathbf{K}$ is the kinetic rate matrix – all of these are $j\times j$ matrices.

An important part of these expressions with regard to the dynamics of the system should be understood clearly. When the system is in dynamic equilibrium, meaning $k_{\mathrm{AB}} = k_{\mathrm{BA}}$ for a A $\rightleftharpoons$ B exchange of the spin for example, the spin concentrations in A and B do not change, implying that the equilibrium longitudinal magnetizations M_{A}° and M_{B}° do not exchange. Therefore, $\mathbf{KM}_z^{\circ}(t) = \mathbf{KM}_z^{\circ} = 0$, which allows for dropping out the $\mathbf{KM}_z^{\circ}(t)$ term from equation (13.155). Nulling $\mathbf{KM}_z^{\circ}$ is a consequence of microscopic reversibility of the equilibrium that renders $\mathbf{KM}_z^{\circ}(t) = 0$; it does not mean that $\mathbf{K} = 0$, the rate matrix always persists. This analysis tells us that under dynamic exchange equilibrium, cross-relaxation of longitudinal magnetizations of the two states does not take place, hence $\sigma_{\mathrm{AB}} = \sigma_{\mathrm{BA}} = 0$. Then no NOE should build up, suggesting that pure exchange spectra will have no NOE peak. This aspect will be considered in some detail later.

Because $\mathbf{KM}_z^{\circ}(t) = 0$ under dynamic equilibrium, the Bloch equation for the longitudinal magnetization (equation (13.155)) reduces to

$$\frac{d}{dt}\mathbf{M}_z(t) = \mathrm{L}\left\{\mathbf{M}_z(t) - \mathbf{M}^{\circ}(t)\right\}, \quad (13.156)$$

where $\mathbf{L} = -\mathbf{R} + \mathbf{K}$.

These developments are described in some more detail in the monograph by Ernst, Bodenhausen, and Wokaun (1988). The matrix form of the equation above applied to the two-state equilibrium exchange A $\rightleftharpoons$ B can be written as

$$\begin{pmatrix}\dot{\mathbf{M}}_{\mathrm{A}z}\\ \dot{\mathbf{M}}_{\mathrm{B}z}\end{pmatrix} = \left[-\begin{pmatrix}\rho_{\mathrm{A}} & 0\\ 0 & \rho_{\mathrm{B}}\end{pmatrix} + \begin{pmatrix}-k_{\mathrm{AB}} & k_{\mathrm{BA}}\\ k_{\mathrm{AB}} & -k_{\mathrm{BA}}\end{pmatrix}\right]\left[\begin{pmatrix}\mathbf{M}_{\mathrm{AZ}}(t)\\ \mathbf{M}_{\mathrm{BZ}}(t)\end{pmatrix} - \begin{pmatrix}\mathbf{M}_{\mathrm{A}}^{\circ}(t)\\ \mathbf{M}_{\mathrm{B}}^{\circ}(t)\end{pmatrix}\right]$$

$$= \begin{pmatrix}k_{\mathrm{AB}} - \rho_{\mathrm{A}} & -k_{\mathrm{BA}}\\ -k_{\mathrm{AB}} & k_{\mathrm{BA}} - \rho_{\mathrm{B}}\end{pmatrix}\begin{pmatrix}\mathbf{M}_{\mathrm{AZ}}\\ \mathbf{M}_{\mathrm{BZ}}\end{pmatrix} \quad (13.157)$$

Interestingly, this equation is similar to the Solomon equation written in the matrix form (equation (13.128)), implying the similarity of the phenomena of dynamic exchange and NOE. In fact, both are detected by using the same RF pulse sequence (Figure 13.34). The NOE spectrum contains both NOE and exchange peaks. The former can be suppressed to isolate pure exchange peaks by filling up the mixing time, τ_{m}, with a pulse train which renders $\sigma_{\mathrm{AB}} = \sigma_{\mathrm{BA}} = 0$. We will look at the pulse sequence for pure exchange spectroscopy shortly.

Let us write the transverse magnetization of the A $\rightleftharpoons$ B exchange (equation (13.154)) in the matrix form

$$\begin{pmatrix}\dot{\mathbf{M}}_{\mathrm{A}}^{+}\\ \dot{\mathbf{M}}_{\mathrm{B}}^{+}\end{pmatrix} = \left\{i\begin{pmatrix}\Omega_{\mathrm{A}} & 0\\ 0 & \Omega_{\mathrm{B}}\end{pmatrix} - \begin{pmatrix}T_{2\mathrm{A}}^{-1} & 0\\ 0 & T_{2\mathrm{B}}^{-1}\end{pmatrix} + \begin{pmatrix}-k_{\mathrm{AB}} & k_{\mathrm{BA}}\\ k_{\mathrm{AB}} & -k_{\mathrm{BA}}\end{pmatrix}\right\}\begin{pmatrix}\mathbf{M}_{\mathrm{A}}^{+}\\ \mathbf{M}_{\mathrm{B}}^{+}\end{pmatrix}$$

$$= -\begin{pmatrix}-i\Omega_{\mathrm{A}} + T_{2\mathrm{A}}^{-1} + k_{\mathrm{AB}} & -k_{\mathrm{BA}}\\ -k_{\mathrm{AB}} & -i\Omega_{\mathrm{B}} + T_{2\mathrm{B}}^{-1} + k_{\mathrm{BA}}\end{pmatrix}\begin{pmatrix}\mathbf{M}_{\mathrm{A}}^{+}\\ \mathbf{M}_{\mathrm{B}}^{+}\end{pmatrix}. \quad (13.158)$$

The rate matrix or the $\mathbf{L}$ matrix can be solved by setting the determinant equal to zero. The eigenvalues are obtained as

$$\lambda_{\pm} = \frac{1}{2}\left(-i\Omega_A - i\Omega_B + T_{2A}^{-1} + T_{2B}^{-1} + k_{AB} + k_{BA}\right)$$

$$\pm\left[\left(-i\Omega_A + i\Omega_B + T_{2A}^{-1} - T_{2B}^{-1} + k_{AB} - k_{BA}\right)^2 + 4k_{AB}k_{BA}\right]^{0.5}. \tag{13.159}$$

The time dependence of $\mathbf{M}_A^+$ and $\mathbf{M}_B^+$ will appear as

$$\begin{pmatrix}\mathbf{M}_A^+\\ \mathbf{M}_B^+\end{pmatrix} = \begin{pmatrix} a_{AA}(t) & a_{AB}(t)\\ a_{BA}(t) & a_{BB}(t)\end{pmatrix}\begin{pmatrix}\mathbf{M}_A^+(0)\\ \mathbf{M}_B^+(0)\end{pmatrix}, \tag{13.160}$$

with the matrix elements

$$a_{AA}(t) = \frac{1}{2}\left[\left\{1+\alpha e^{(-\lambda_+ t)}\right\} + \left\{1-\alpha e^{(-\lambda_- t)}\right\}\right]$$

$$a_{BB}(t) = \frac{1}{2}\left[\left\{1+\alpha e^{(-\lambda_- t)}\right\} + \left\{1-\alpha e^{(-\lambda_+ t)}\right\}\right]$$

$$a_{AB}(t) = \frac{k_{BA}}{(\lambda_+ - \lambda_-)}\left[e^{(-\lambda_- t)} - e^{(-\lambda_+ t)}\right]$$

$$a_{BA}(t) = \frac{k_{AB}}{(\lambda_+ + \lambda_-)}\left[e^{(-\lambda_- t)} - e^{(-\lambda_+ t)}\right],$$

in which

$$\alpha = \frac{-i\Omega_A + i\Omega_B + T_{2A}^{-1} - T_{2B}^{-1} + k_{AB} - k_{BA}}{\lambda_+ - \lambda_-} \tag{13.161}$$

Fourier transformation of these time functions yields the lineshapes with varying exchange rates k_{AB} and k_{BA}. As mentioned above, the lifetimes of the spin in the two exchanging states, A and B, are determined by magnitudes of k_{AB} and k_{BA}. Accordingly, the exchange can be in slow ($\Omega \gg k$), intermediate ($\Omega \sim k$), or fast ($\Omega \ll k$) limits (Figure 13.38).

An analytical expression for lineshape alone is found in the work of McConnell (1958) who originally devised the modified Bloch equations. Letting $k_{AB} = k_{BA}$ and $M_{AZ} = M_{BZ} = 0.5M_Z$, the lineshape expression in linear frequency unit (ν) is obtained as

$$g(\nu) = \frac{\frac{2}{k_{AB}}(\nu_A - \nu_B)^2}{\left[\nu - 0.5(\nu_A + \nu_B)\right]^2 + \pi^2\left(\frac{1}{k_{AB}}\right)^2(\nu - \nu_A)^2(\nu - \nu_B)^2}. \tag{13.162}$$

It is convenient to define the x-axis of a plot in terms of ν such that

$$x = \frac{\nu - 0.5(\nu_A + \nu_B)}{0.5(\nu_A - \nu_B)}, \tag{13.163}$$

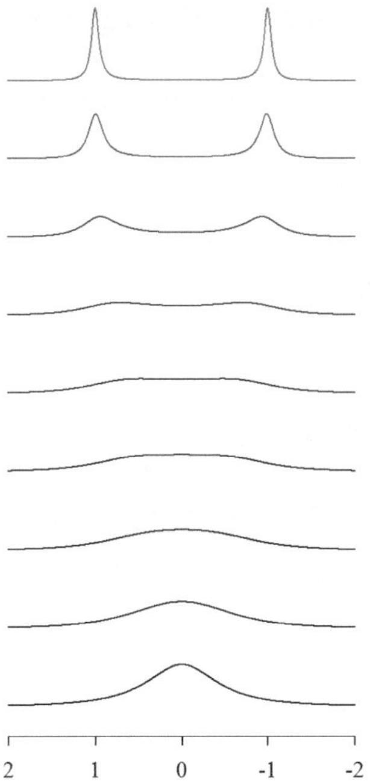

FIGURE 13.38 Calculated lineshape for a two-site magnetization exchange (A $\rightleftharpoons$ B) with β=10, 5, 2, 0.8, 0.7, 0.6, 0.5, 0.4 (equation (13.164)). The spectra from top to bottom correspond to conditions of increasingly faster exchange.

and then define the parameter

$$\beta = \frac{\pi(\nu_A - \nu_B)}{2k_{AB}}. \tag{13.164}$$

The idea is that the absorption lineshape expression thus obtained will take the form

$$g(x) = \frac{\frac{2}{k_{AB}}}{x^2 + \beta^2(x^2 - 1)^2}. \tag{13.165}$$

Spectra calculated for different values of β are shown in Figure 13.38. The value of β decreases when the rate constant k_{AB} increases, so the exchange equilibrium shifts from slow to fast through the intermediate limit (spectra from top to bottom) as the value of β decreases. To examine the lineshape above, one must rely on Fourier transformation of the time dependence of M_A^+ and M_B^+ to obtain the amplitudes correctly.

13.21.2 One-Sided Chemical Reaction

Unlike a system A $\rightleftharpoons$ B that already exists in dynamic equilibrium and can be probed by NOE-suppressed exchange spectroscopy, a one-sided reaction A $\rightarrow$ B has to be initiated by a transient mixing method and probed concurrently. The A $\rightarrow$ B type of reactions are generally initiated by stopped-flow mixing hardware suitably modified and interfaced to the NMR spectrometer.

Since only one rate constant k_{AB} describes the kinetics of the $A \rightarrow B$ reaction, the transverse magnetizations of the modified Bloch equation (13.154) can be written as

$$\frac{d}{dt}M_A^+ = \left(i\Omega_A - \frac{1}{T_{2A}} - k_{AB}\right)M_A^+$$

$$\frac{d}{dt}M_B^+ = \left(i\Omega_B - \frac{1}{T_{2B}}\right)M_B^+ + k_{AB}M_A^+. \qquad (13.166)$$

These magnetizations are created by the 90° pulse of a 1D experiment, and they evolve when the FID is acquired. Note that $M_A^+(t)$ means

$$M_{Ax}(t) + iM_{Ay}(t) = M_A^o \sin\beta\, e^{i\Omega t - \frac{t}{T_{2A}}}, \qquad (13.167)$$

where β is the pulse rotation angle. Therefore, both M_A^+ and M_B^+ components are present in tandem during acquisition, the former leading the latter. As a result, resonance lines of both A and B are present in the spectrum. The reactant lines appear broad due to the short lifetime of the reactant, and the product lines are dispersive because of the relatively slower formation of the product while the acquisition is progressing. These features of the spectrum are indeed observed in the Fourier analysis of the evolution of M_A^+ and M_B^+, which is the Fourier transform of the acquired signal itself (Ernst et al., 1988).

Now consider the slowness of the reaction. If the conversion of A to B is very slow the smallness of k_{AB} would make the modification of the Bloch equations redundant, and equation (13.166) above reduce to the decoupled forms

$$\frac{d}{dt}M_A^+ = \left(i\Omega_A - \frac{1}{T_{2A}}\right)M_A^+$$

$$\frac{d}{dt}M_B^+ = \left(i\Omega_B - \frac{1}{T_{2B}}\right)M_B^+. \qquad (13.168)$$

Under this condition, if the $A \rightarrow B$ reaction can be initiated transiently by stopped-flow conversion of A to B such that the isolated nucleus exists only in state B, then the acquisition signal will be only due to $\dot{M}_B^+$. Repeated acquisition of the signal, without reinitiating the reaction, then yields the kinetics of formation of B. The experiment consists of a simple time-arrayed acquisition of the B signal which can be done in a one- or multidimensional mode. Figure 13.39 exemplifies time-arrayed 1D appearance of the final state in the unfolding $(N \rightarrow U)$ and refolding $(U \rightarrow N)$ of a protein called barstar. The unfolding and folding reactions are poised in appropriate solvent conditions to lower down the reaction rates so as to render the measurements amenable. These experiments under ideal conditions can be performed in a variety of modes.

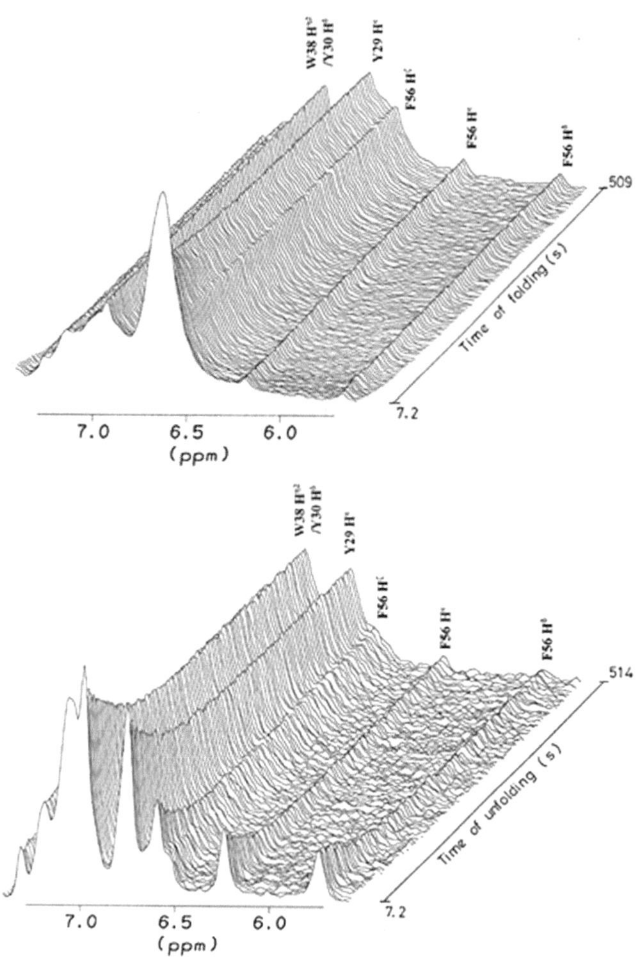

FIGURE 13.39 Real-time NMR measurement of slow folding and unfolding rates of the protein barstar. Each spectrum in the array is an average of 2 spectra. (Reproduced from Bhuyan and Udgaonkar, 1999, © American Chemical Society.)

In the sections above we have discussed the bare essentials of nuclear relaxation. It must be said that the facet of spin relaxation is overwhelming, encompassing a vast area of NMR spectroscopy, the discussion of all of which is beyond the scope of these short descriptions. Interested readers can consult dedicated monographs, such as Noggle and Schirmer (1971).

13.22 HAHN ECHO AND DOUBLE RESONANCE

It is appropriate at this stage to introduce a two-pulse or double resonance experiment that has a wide range of applications, and at the same time would serve as a prelude to two-dimensional NMR. Suppose we have to measure spin-spin relaxation time T_2, which we know as the spin dephasing time in the equatorial (xy) plane. The measurement involves a two-pulse experiment, although it may not be obvious at the first sight. In principle, the measurement could be done easily by using a 90° pulse to create the transverse magnetization in the rotating frame. If a $-x$ pulse is used, then the transverse magnetization vector is initially along the $+y$ axis, but will fan out into individual component vectors (dephase) with time. The dispersion of the magnetization should appear as an exponential free-induction decay whose decay time is characterized by T_2. This is actually apparent T_2, shown as T_2^*, and not the real T_2 – a realization that occurred first to Hahn (1950) in the seminal years of NMR development. The reason why a single 90° pulse followed by signal acquisition provides T_2^* is due to spatial inhomogeneity of the strength of the static field B_0 across the sample volume.

Usually, the sample is expected to be homogeneous – the concentration is thermodynamically an intensive property, so it is the spatial inhomogeneity of B_0, which is of concern. No matter how fervently one shims the field, the inhomogeneity persists to varying extent. When the sample volume is seen to consist of a certain number of volume elements, each of these elements experience a different strength of the static field. If the field inhomogeneity is assumed to be linear, and say b gauss per mm along the y-axis, then two volume elements apart by 1 mm along the y-axis will experience fields that differ by b gauss. The same argument holds for all volume elements in the sample. We assume that diffusion of the molecules from one volume element to another is slow with respect to the spin precession frequency. This means the precession frequency of the same spin magnetization vector will be different in two volume elements, as though the molecules in the sample are motionally frozen. In essence, the precession frequency of a spin magnetization in the transverse plane is inhomogeneous because the B_0 field is inhomogeneous. However, the total transverse magnetization of the spin sensed by the receiver is the resultant of its precession behavior in all the volume elements. If one or a few volume elements could be isolated to observe the spin precession, the fanning time (dephasing time) of the magnetization would be found to be much longer than what the resultant fanning time gives. A 90° pulse thus creates inhomogeneous precession frequency of a magnetization across the sample volume. This means the T_2 measured transiently after just a 90° pulse is not the real T_2, but T_2^*.

Hahn was able to show that the unseen signal could be made seen in the form of an echo signal if a second 90° pulse ($= \omega_1 t$, where t is the pulse time) is applied after a time τ of the initial pulse such that the response of the second pulse is maximum at time 2τ. One says that a Hahn echo is produced at time 2τ, by which some of the slow-precessing magnetizations are brought into the phase coherence. A third pulse applied at 3τ produces an echo at 4τ where the signal is maximum. Although Hahn used 90° pulses, Carr and Purcell later employed $180°_x$ pulses at times τ, 3τ, 5τ, $\cdots$, each pulse reversing the phase angle, to refocus the precessing spins at times 2τ, 4τ, 6τ, $\cdots$ (Figure 13.40). Meiboom and Gill then suggested replacing the $180°_x$ pulse by a $180°_y$ in order to remove pulse imperfections. Thus, the widely used Carr-Purcell-Meiboom-Gill (CPMG) method in all branches of NMR, including radiology and molecular dynamics, consists of a series of $180°_y$ refocusing pulses during which the amplitude of each successive echo decreases exponentially.

The reason why the echo height decreases should be understood clearly. Spin dephasing originates from two sources – one, the B_0 inhomogeneity that gives rise to inhomogeneity in Larmor precession of a spin magnetization located in different parts of the sample volume, and two, the random fluctuating field of molecular origin, since molecules are in constant motion often interacting with each other and with other surrounding molecules. The former is said to be an instrumental effect, and the latter is an intrinsic molecular effect. The CPMG pulses ameliorate the B_0 inhomogeneity problem by repeated refocusing of the dephased spins. These pulses are, however, innocuous to the intrinsic spin dephasing (coherence loss) due to the molecular random field effect. Therefore, the refocusing effect of the CPMG pulses continue to decrease with successive $180°_y$ pulses and eventually dies off exponentially.

The CPMG pulse sequence can be applied by different approaches. Measurement of T_2 by 1D spectra may use the sequence shown in Figure 13.41 which will produce a train of echoes. If an FID is acquired from each echo maximum, one scan of the experiment will result in n echoes and hence n acquisitions. The peak intensity decays with time as

$$I(\tau) = A + Be^{-\frac{\tau}{T_2}}, \tag{13.169}$$

where A is the baseline (steady-state) intensity as $\tau \to \infty$.

The CPMG sequence can also be used as a T_2 filter, which attenuates broad signals from relatively large molecules that tumble slow. The filter (Figure 13.42) can be used in both 1D and multidimensional versions of an experiment.

13.23 ECHO MODULATION AND J-SPECTROSCOPY

An important concept that forms the basis of J-spectroscopy – not to be confused with J-correlation spectroscopy, which will be discussed in detail later – is that chemical shift is modulated by spin echo, but J-coupling is not. To understand how this happens, let us consider a pair of homonuclear spins, which we will denote here by I and S, although they may be indicated by k and l without losing the generality – one has to simply keep track of the labels used for the spins. A nonselective 90° pulse excites both nuclei to create in-phase transverse magnetizations of I and S (Figure 13.43). The chemical shifts of I and S evolve with time as offsets from the RF transmitter frequency, also called carrier frequency. The chemical shift frequencies for spins I and S are

$$\Omega_I = (1 - \sigma_I) B_0$$

$$\Omega_S = (1 - \sigma_S) B_0. \tag{13.170}$$

If $\Omega_I > \Omega_S$, then the magnetization M_I at time τ precesses faster than the magnetization M_S. If now a $180°_y$ pulse is applied, the magnetizations will be rotated about the y-axis so as to place M_I behind M_S. However, because $\Omega_I > \Omega_S$, the

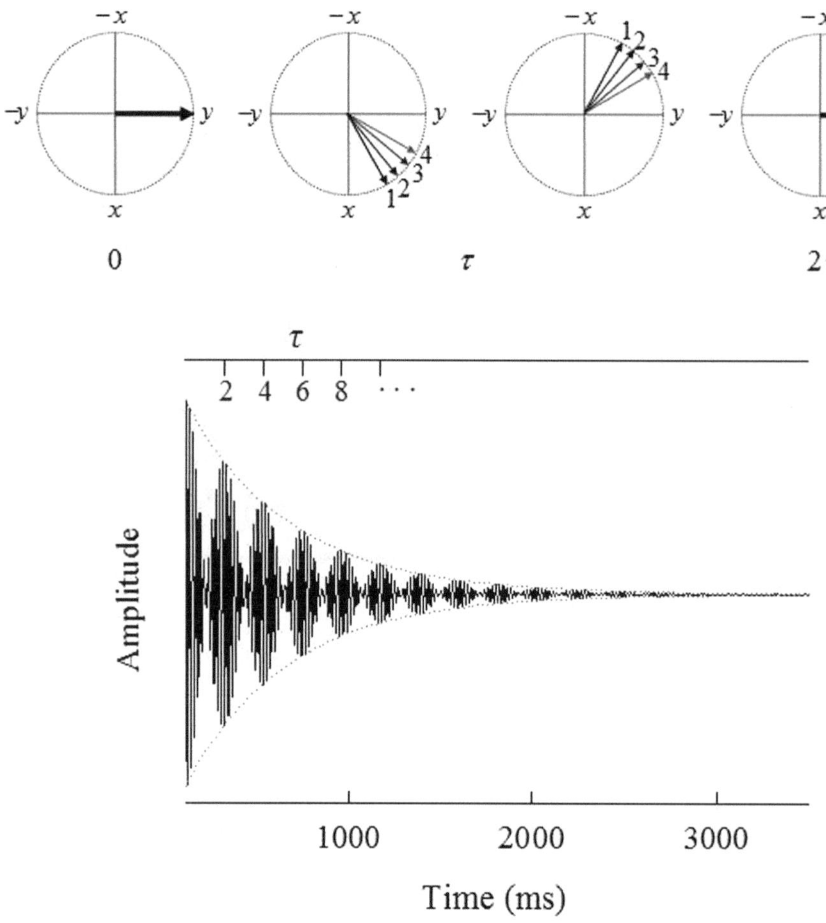

FIGURE 13.40 Pictorial presentation of the spin-echo of just four spin magnetizations. *Top*, the initial 90° pulse creates an in-phase magnetization (*far left*), which fans out due to the loss of phase coherence, producing different Larmor frequencies for the spins (*second from left*). Application of a 180°_{y} pulse at time τ reverses the order of precession (*third panel from left*) – the faster ones become slower and vice versa, read by the spin arrow labels 1, 2, 3, 4. At time 2τ, the spins are refocused (*far right*), because the forced reversal of the spins would not sustain for long. *Bottom*, the appearance of the exponentially decreasing acquisition signal (FID) with echo amplitudes at times $2\tau, 4\tau, 6\tau, \cdots$.

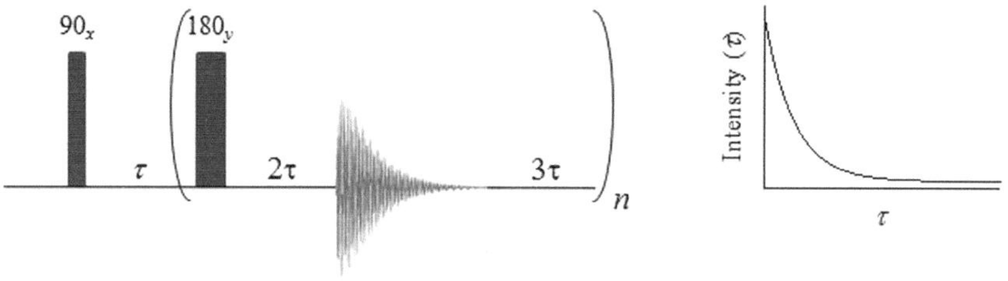

FIGURE 13.41 Measurement of T_2 using CPMG echo pulses (*left*) in 1D spectra. The exponential decay of the intensity is shown to the *right*.

two magnetizations will produce an echo at time 2τ. Since the chemical shifts are refocused, we say that chemical shift does not modulate the echo, although repeated refocusing will lead to the decay of the echo height gradually.

Now suppose the I and S homonuclei are J-coupled such that the coupled magnetizations of I are $M_I^{S_\alpha}$ and $M_I^{S_\beta}$, and those of S are $M_S^{I_\alpha}$ and $M_S^{I_\beta}$. Consider the precession of one of the pairs of these magnetizations in the transverse plane. A 90° initial pulse creates in-phase magnetizations $M_I^{S_\alpha}$ and $M_I^{S_\beta}$. Let the coupling constant be positive, so $M_I^{S_\beta}$ will start to precess at the frequency $\nu_I + \frac{1}{2}J_{IS}$ and $M_I^{S_\alpha}$ will precess at $\nu_I - \frac{1}{2}J_{IS}$. After time τ, $M_I^{S_\beta}$ will be ahead of $M_I^{S_\alpha}$ (Figure 13.43). If now a 180°_{y} refocusing pulse is applied, these vectors are rotated by 180°. However, $M_I^{S_\beta}$ still appears ahead of $M_I^{S_\alpha}$. The reason for this is that the $180°_y$ pulse inverts the populations of both homonuclei; what was $M_I^{S_\beta}$ before the pulse is $M_I^{S_\alpha}$ after. This interchange of levels is applicable for an $M_S^{I_\alpha}$ and $M_S^{I_\beta}$ pair as well. Because the faster-precessing magnetization is already ahead of the slower one, refocusing does not occur. Rather, the phase angle ϕ between the two vectors oscillates with time.

The variation of ϕ with time τ can be quantified if J_{IS} is known. Since the difference in frequency of the two J-coupled components is $2\pi J_{IS}$, the time-dependence of the phase angle can be written as $2\pi J_{IS}\tau$, which at the echo time of 2τ will be

$$\phi = 4\pi J_{IS}\tau. \tag{13.171}$$

The pulse scheme of Figure 13.41 can be used to measure a train of echoes, each at an interval of 2τ, such that an individual FID is acquired from each echo maximum. The absorption mode signal following Fourier transformation of each FID is

$$M_I(2\tau) = M_I^0 \cos(2\pi J_{IS}\tau)e^{-\frac{\tau}{T_2}}, \tag{13.172}$$

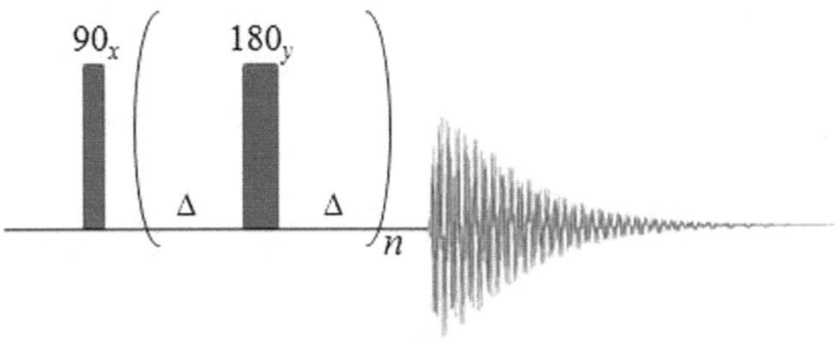

FIGURE 13.42 The CPMG echo sequence as a T_2 filter. Typically, $\Delta \sim 200\ \mu s$.

showing cosine modulation of the echo signal (Figure 13.44). The figure shows one of the two lines of the I spin, resulting from the J-coupling with the S spin. The two I lines have phases $\pm\phi$ that will vary with time τ as shown. Since chemical shift does not modulate echo but J-coupling does, the echo-modulated spectrum generated by the CPMG sequence is also called the J-resolved spectrum.

13.24 HETERONUCLEAR *J*-SPECTROSCOPY

In the homonuclear IS spin pair, the refocusing pulse inverts the population of both I and S, so a single $180°_y$ pulse of a reasonable bandwidth serves to produce an echo modulation. It is not the same for a pair of J-coupled heteronuclei IS though. Note that in the context of heteronuclei, I is the sensitive source nucleus, almost always ^1H, and S is the less sensitive nucleus – ^{13}C or ^{15}N. For example, if the nuclei are ^1H and ^{13}C the refocusing pulses appropriate for the nuclei differ by a frequency factor of ~4, as can be checked with the respective gyromagnetic ratios. The RF oscillator cannot generate both frequencies simultaneously or in quick succession. This is where the idea of separate channels for heteronuclei and double resonance comes in. Consider the pulse sequence in Figure 13.45 which excites the carbon nucleus by a 90° RF pulse of frequency appropriate for ^{13}C resonance. The objective here is to observe ^{13}C echo modulation. If the population of ^1H is not inverted then $^{13}C^{^1H_\alpha}$ and $^{13}C^{^1H_\beta}$ vectors will not modulate the echo, they will be refocused at 2τ instead. The refocusing can be prevented by inverting the ^1H populations, which can be done by applying a $180°_y$ pulse at the ^1H frequency (Figure 13.45). The faster precessing $^{13}C^{^1H_\beta}$ will continue to be the faster vector with respect to the slower $^{13}C^{^1H_\alpha}$. The echo modulation of the spin coupling of ^{13}C and ^1H

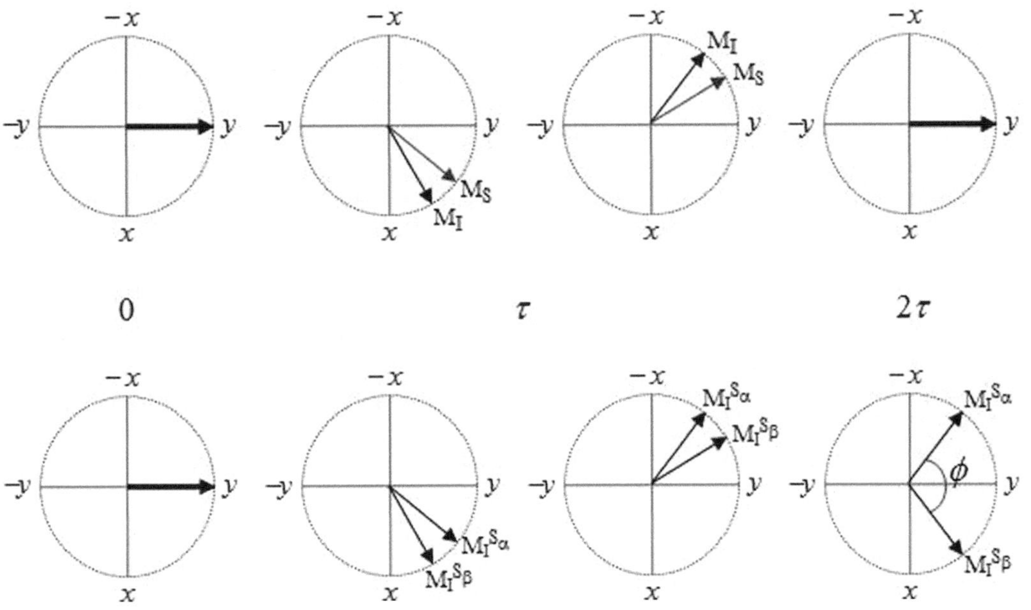

FIGURE 13.43 Vector precession diagrams depicting the refocusing of chemical shift (*top row*), and echo modulation due to J-coupling (*bottom row*). The occurrence of the phase angle ϕ and its development with time characterizes the echo modulation.

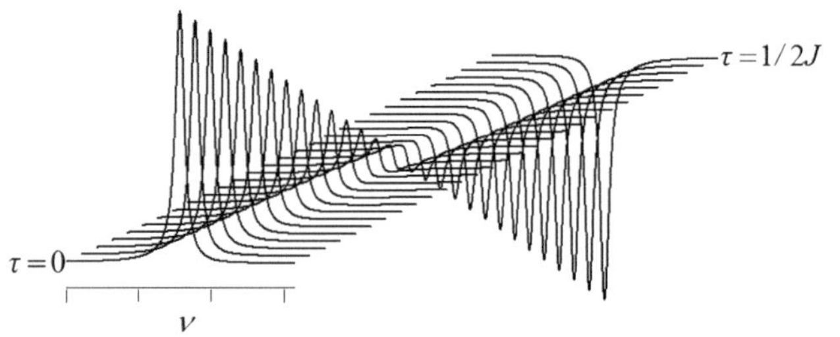

FIGURE 13.44 Echo-modulated spectrum shown for only one of the two J-split lines of the I spin due to its coupling with the S spin. The variation of the phase angle between the two J-coupled spins is shown with $\tau = J^{-1}$. For clarity, the modulation of only one of the two J-coupled lines is shown.

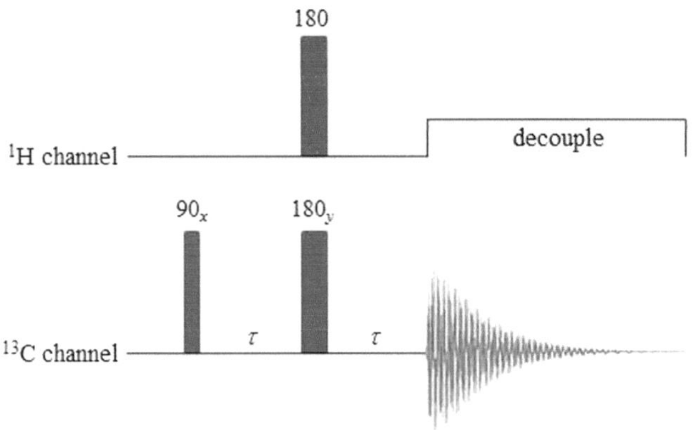

FIGURE 13.45 Pulse scheme to observe ^{13}C echo modulation for a pair of J-coupled CH spins. The decoupler is a ^{1}H RF pulse that irradiates the proton while ^{13}C acquisition is allowed.

can then be observed by acquiring an FID in the ^{13}C channel keeping ^{1}H decoupled (Figure 13.45).

13.25 POLARIZATION TRANSFER (INEPT AND REFOCUSED INEPT)

A terminology called 'spin tickling' appears in NMR jargon, which generally means irradiating a resonance selectively. If two spins I and S are coupled and one of them is selectively perturbed (irradiated), then the transition intensity of the coupled non-irradiated spin is also affected. This has to happen because the population of a product spin function is altered by selectively inverting one of the coupled spins. This concept is used to enhance the signal intensity of a spin of lower gyromagnetic ratio, ^{15}N for example, when it is coupled to a spin of higher gyromagnetic ratio, ^{1}H in most cases. Basically, the spin population of the I nucleus is inverted, so the low-energy state is populated more at the expense of the S nucleus. Based on their relative γ and hence Larmor frequency, let us call ^{1}H the sensitive nucleus I and ^{15}N the insensitive nucleus S. The lower the γ value, the lesser is the Boltzmann population difference

$$\frac{N_{\text{upper}}}{N_{\text{lower}}} = e^{-\frac{E}{k_B T}}.$$

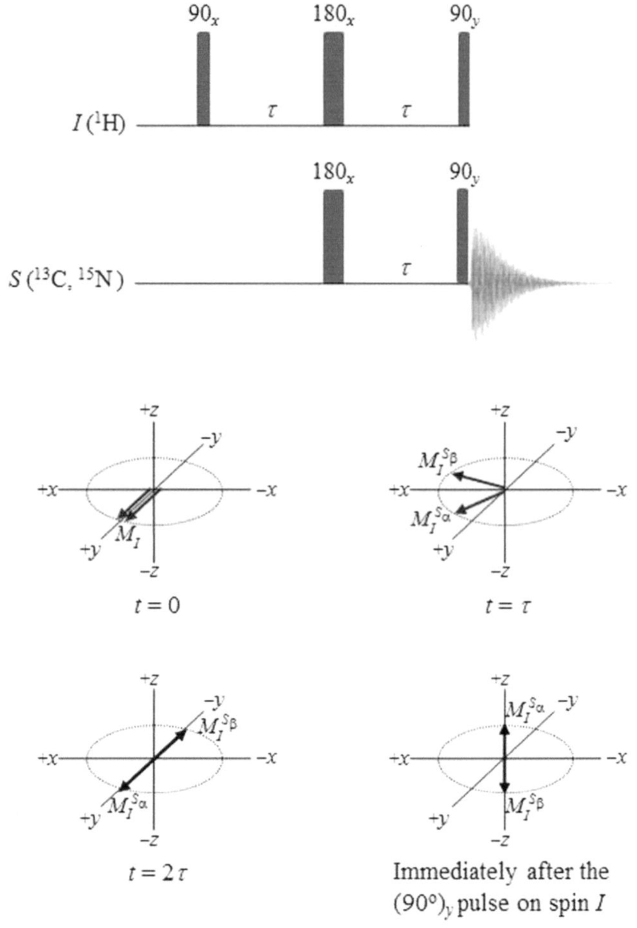

FIGURE 13.46 Pulse sequence for INEPT, and the evolution of the sensitive-spin magnetization vectors at $t = 0$, τ, 2τ, and just before signal acquisition.

Now consider the pulse sequence depicted in Figure 13.46, which is similar to the one used in heteronuclear spectroscopy. The delays shown by τ are not set arbitrarily, but to achieve the phase angle ϕ between the spin-coupled component magnetizations ($M_I^{S_\beta}$ and $M_I^{S_\alpha}$) equal to 90° at τ and 180° at 2τ, where

$$\tau = \frac{1}{4J_{\text{IS}}}.$$

The rationale for setting $\tau = 1/4J_{\text{IS}}$ is that the magnetization at the end of the second τ is

$$\gamma_I\left[I_y \cos\left(2\pi J_{IS}\tau\right) - 2I_x S_z \sin\left(2\pi J_{IS}\tau\right)\right]$$

where I_x, I_y, and S_z are component magnetizations of the respective spins I and S. Provided $\tau = 1/4J_{\text{IS}}$,

$$\cos\left(2\pi J_{IS}\tau\right) = 0$$

$$\sin\left(2\pi J_{IS}\tau\right) = 1, \qquad (13.173)$$

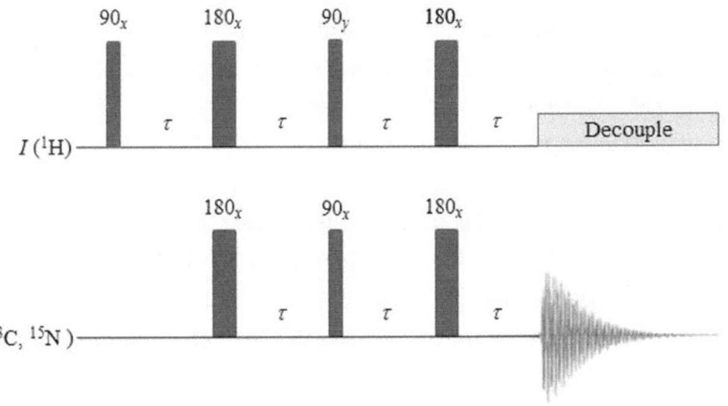

FIGURE 13.47 Pulse scheme for refocused INEPT.

so that the effective magnetization immediately before the $(90°)_y$ pulse is $-\gamma_I 2I_x S_z$.

The evolution of the magnetization vector shown in Figure 13.46 makes clear that the magnetization is transferred from the more sensitive (I) to the less sensitive (S) spin, leading to higher sensitivity of the latter by the factor γ_I / γ_S. The simultaneous application of the $(90°)_y$ pulse on I and S transforms the magnetizations as

$$-\gamma_S 2I_x S_z \rightarrow -\gamma_S 2I_z S_y.$$

We note that the signal is acquired in the S-channel alone, and this signal is proportional to γ_I, not γ_S, accounting for the sensitivity enhancement of the S spin by γ_I / γ_S. This is a popular sensitivity enhancement method called INEPT (Insensitive Nuclei Enhanced by Polarization Transfer).

The signal-to-noise ratio of an INEPT experiment can be improved by extending the INEPT pulse scheme to include the basic scheme for heteronuclear J-spectroscopy discussed in the previous section (Figure 13.45). The objective of this extension is to refocus the magnetization already transferred to the S spin via the INEPT sequence, and hence the scheme is called refocused INEPT (Figure 13.47). Besides regular 1D heteronuclear experiments involving ^{13}C, the INEPT and refocused INEPT sequences form parts of 2D heteronuclear pulse schemes to discuss later.

13.26 TWO-DIMENSIONAL *J*-RESOLVED SPECTROSCOPY

For a moment consider the two interactions – chemical shift and J-coupling of two spins, k and l. The chemical shifts ω_{ok} and ω_{ol}, where ω_o is the Larmor frequency, are determined by the extent of shielding of the respective spins. Regarding scalar coupling, let the splitting of the resonance of the respective spins (J_{kl} and J_{lk}) by the other coupled spin be identical ($J_{kl} = J_{lk}$) as it happens with the AX two-spin system. Generally, $2\pi|J_{kl}| \ll |\omega_{ok} - \omega_{ol}|$, a prevalent condition called weak coupling limit, in which J-spectroscopy is efficient for separation of scalar coupling from composite multiplets containing overlapped scalar coupling and chemical shifts.

Let us take up a homonuclear J-resolved experiment involving protons, the J-couplings among which we are

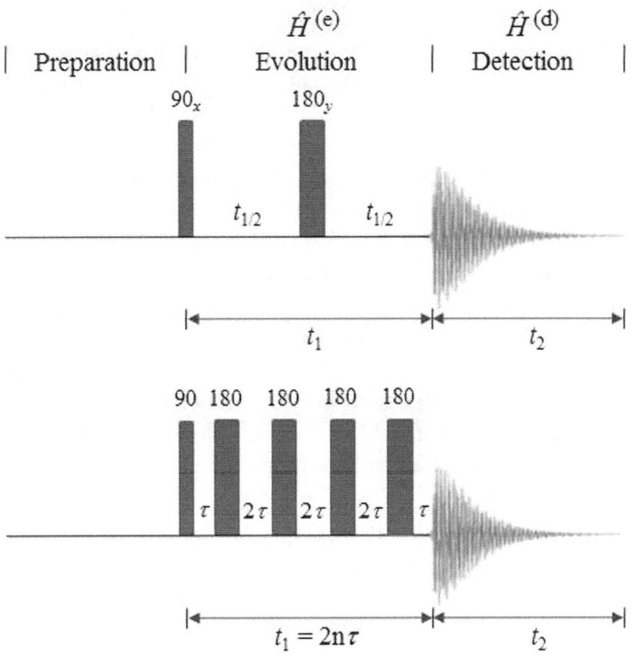

FIGURE 13.48 Pulse schemes for 2D J-resolved spectroscopy. There are two time domains, t_1 and t_2 – called the evolution and detection periods, the latter is a constant but the former is incremented so as to generate an evolution time domain. *Top*, a one-echo sequence where t_1 is incremented centered about the $180°_y$ pulse. *Bottom*, a scheme where the t_1 period contains a series of Carr-Purcell type of spin echoes produced by n 180° refocusing pulses. This type of spin-echo sequence suppresses translational diffusion.

interested to separate. All protons need not be coupled, but the J-split resonances along with all the chemical shift resonances appear together in the 1D proton spectrum. Since the idea is to resolve scalar coupling which modulates spin echo, the experiment includes the spin-echo sequence. The basic pulse scheme for the 2D J-resolved spectroscopy is shown in Figure 13.48. It may consist of the simplest of the 1D echo sequence shown (see Figures 13.41, 13.42, and 13.45), and one may choose to record the FID during time t_2 resulting from the decay of one echo alone. The idea however is to acquire several FIDs, each having the same signal detection time t_2 but different evolution time t_1 spreading through the spin-echo sequence.

How are t_1 and t_2 different in terms of the content of interactions? The distinction is made on the basis of the interaction Hamiltonians acting on the spins. Because spin-echo is modulated by scalar coupling, not by chemical shift, the Hamiltonian $\mathcal{H}^{(e)}$ acting through the evolution time t_1 contains the scalar coupling interaction alone. In the t_2 domain, the Hamiltonian $\mathcal{H}^{(d)}$ contains both interactions – scalar coupling and chemical shift. These interactions are

$$\mathcal{H}^{(e)} = \sum_{k<l} 2\pi J_{kl} I_{kz} I_{lz}$$

$$\mathcal{H}^{(d)} = \sum_k \Omega_k I_{kz} + \sum 2\pi J_{kl} I_{kz} I_{lz}. \qquad (13.174)$$

Notice the simplicity of the Hamiltonians. In $\mathcal{H}^{(e)}$, J_{kl} is a scalar quantity which is the coupling constant in hertz. The operators I_{kz} and I_{lz} are Cartesian spin rotation operators for spins k and l, respectively (see below). The rotation about the z-axis is indicated by the subscript 'z' of the spin indices k and l. Note that these are transverse magnetizations we are working with. Consider now the $2\pi J_{kl}$ term, whose value modulates the phase angle ϕ between k and l magnetizations as the evolution time t_1 is incremented. The 180º pulse in the middle of t_1 refocuses chemical shift and is hence eliminated. Thus, the frequency domain ω_1 corresponding to t_1 will have no chemical shift interaction, but scalar coupling alone. The detection Hamiltonian $\mathcal{H}^{(d)}$ contains both chemical shift and scalar coupling interactions, so the multiplets in the frequency domain ω_2 will contain both of these.

13.26.1 Absence of Coherence Transfer in 2D J-spectroscopy

Although we have started discussing 2D spectroscopy, the J-resolution achieved by generating two frequency dimensions does not strictly define the principles of two-dimensional spectroscopy. Imagine that we are developing a two-dimensional gel electropherogram in which the first dimension separates the components – say proteins or some organic compounds subjected to thin-layer chromatography, based on molecular expansion – and the second dimension separates each component already resolved in the first run as a function of pH or charge content. The first dimension provides information only about the degree of expansion of the components, but the second dimension maps migration due to both expansion and charge content. One can pit the molecular dimension applicable to electrophoresis with the scalar coupling of NMR (first dimension), and the molecular dimension and charge count together with scalar coupling and chemical shift of NMR (second dimension). It surely turns out that we have not done quantum mechanics to resolve the problem This is one of the answers as to why J-resoved 2D spectroscopy is not strictly a two-dimensional spectroscopy. Two dimensions can be defined by choosing two variables, as in Figure 13.49 for example, but a real two-dimensional NMR spectrum must strictly evolve by transferring a concept developed in the first dimension to the second. The concept is the coherence, so the 'transfer of coherence' generates the second dimension in a genuine 2D spectrum.

To illuminate on the lack of coherence transfer here, let us examine what happens in the J-resolved spectroscopy. The two Hamiltonians $\mathcal{H}^{(e)}$ and $\mathcal{H}^{(d)}$ commute in the weak coupling limit, $2\pi|J_{kl}| \ll |\omega_{0k} - \omega_{0l}|$, and the coherence that was developed during the evolution period (t_1) is identically preserved up to the detection period (t_2), although the precession frequency of the coherence in the evolution and detection periods are not the same. The preservation of the coherence means there was 'no coherence transfer' which is the hallmark of prototypical J-resolved spectroscopy in the weak coupling limit, where the transfer of transverse magnetization between different transitions do not occur. There is no need of coherence transfer in J-resolution, since the idea is only to separate the coupling interactions from chemical shifts. In the absence of coherence transfer, the number of peaks in the J-resolved 2D spectrum is the same as in the 1D spectrum. Thus, J-resolved 2D spectroscopy is not a correlation spectroscopy like COSY (to discuss shortly) in which different transitions, say two chemical shifts, are correlated by coherence transfer. It is reiterated that a 2D spectrum in the true sense must result from 'coherence transfer'; however, two-dimensional display of any spectrum can be presented based on the two variables involved.

Returning to 2D J-resolved spectroscopy, Fourier transformation of the time-domain signal $S(t_1,t_2)$ involves two successive 1D Fourier transformations. Normally, the t_2 signals are transformed first

$$s(t_1,t_2) \xrightarrow{\mathrm{FT}^{(2)}} \mathrm{F}^{(2)}\{s(t_1,t_2)\} \xrightarrow{\mathrm{FT}^{(1)}} \mathrm{F}^{(1)}\mathrm{F}^{(2)}\{s(t_1,t_2)\} = S(\omega_1,\omega_2), \tag{13.175}$$

to obtain a time-frequency data matrix that would yield a frequency-frequency matrix. Data plotting needs some care so as to convey the true meaning of the outcome of the experiment. We often come across oblique-axis stack-plots presented using different kinds of datasets. For example, a series of 1D spectra acquired as a function of time t of an event can be stack-plotted in an oblique pattern so as to reduce the overlap of resonance peaks in the display. The variable t then forms a dimension, even if it is not a Fourier transformed frequency dimension. This is helpful to visualize what happens in real time. However, the ω_1 and ω_1 dimensions in a 2D experiment truly reflects coherence transfer induced by an appropriate pulse – a 'mixing pulse', for instance.

The frequency data of a 2D J-resolved spectrum plotted as contours after Fourier transformation appear as shown in Figure 13.49. The appearance of the contours along the two frequency axes is a faithful representation of what is obtained. Often in practice, the diagonal is twisted anticlockwise by 45º to present the spectrum as a 'sheared' one. This is just an alignment of the frequency data matrix to obtain the multiplets parallel to the ω_1 axis, amounting to discount the spread of chemical shifts.

13.26.2 2D J-spectroscopy in Strong Coupling Limit

A concept we wish to introduce here is the distinction between the weak- and strong-coupling limits, both with respect to the chemical shift. Let us consider two spins k and l to examine these limits and the consequences thereof.

$$\begin{aligned} &\text{weak-coupling}\, 2\pi|J_{kl}| \ll |\omega_{ok} - \omega_{ol}| \\ &\text{strong-coupling}\, 2\pi|J_{kl}| \gg |\omega_{ok} - \omega_{ol}|. \end{aligned} \tag{13.176}$$

To see the difference between the two, we will need to weigh the product functions of the two coupled nuclei with regard to the acting Hamiltonian. In the weak-coupling limit, the

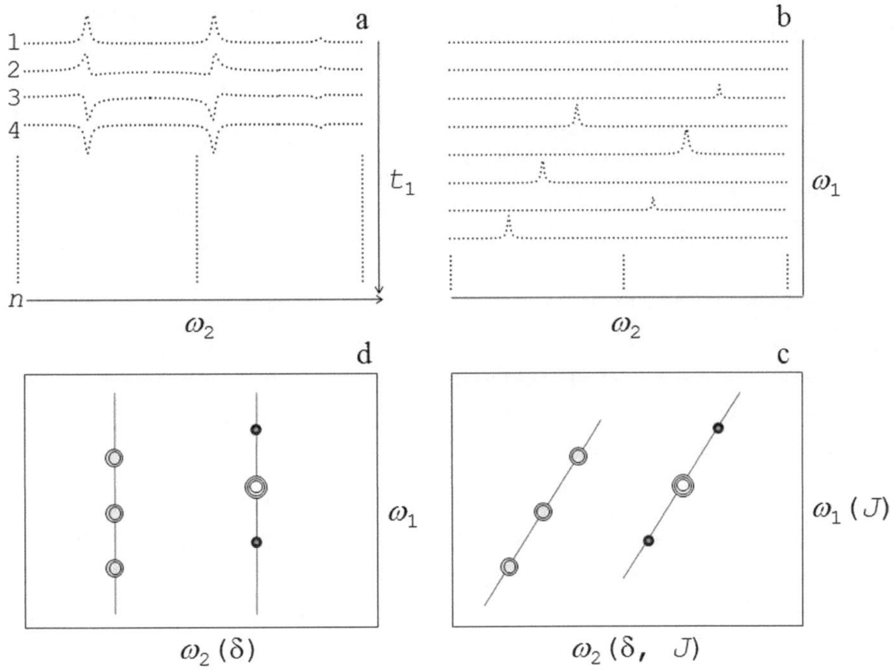

FIGURE 13.49 2D J-spectroscopy. (a) Time-frequency data matrix obtained after Fourier transformation of t_2. (b) Frequency-frequency data matrix obtained after the second Fourier transformation. (c) Contour plot of the 2D spectrum. (d) The contour spectrum after twisting the diagonal by 45°.

product functions are eigenfunctions of an effective evolution Hamiltonian

$$\mathcal{H}^{(e)} = \sum 2\pi J_{kl} I_{kz} I_{lz}, \quad (13.177)$$

which is not the case if the coupling is strong. In the strong-coupling case, the eigenfunctions are linear superposition of the product functions. If we have only two coupled spins, k and l, then the state vector will be

$$|\Psi\rangle = c_1|\alpha_k\alpha_l\rangle + c_2|\alpha_k\beta_l\rangle + c_3|\beta_k\alpha_l\rangle + c_4|\beta_k\beta_l\rangle. \quad (13.178)$$

A linearly superposed function is not an eigenfunction of the bare Hamiltonian that does not contain a persistent time-varying potential. The operator role of the naked Hamiltonian on $|\psi\rangle$ is therefore dismissed. However, a π pulse – call it a 'mixing pulse', represented by the operator

$$\mathcal{R} = e^{-i\beta I_x}, \quad (13.179)$$

where β is the pulse rotation angle, induces a coherence transfer or transfer of transverse magnetization between different transitions. For example,

$$|\alpha_k\beta_l\rangle \xrightarrow{\mathcal{R}_{\pi_y}} -|\beta_k\alpha_l\rangle$$

$$|\beta_k\alpha_l\rangle \xrightarrow{\mathcal{R}_{\pi_y}} -|\alpha_k\beta_l\rangle. \quad (13.180)$$

Going back to the question of what would happen to a J-resolved spectrum under a strong-coupling limit, one says that the chemical shift cannot be refocused by the refocusing pulses. Rather, a 180° RF pulse would induce coherence transfer within various connected transitions. Consequently, new signals in addition to those present under a weak-coupling limit would appear. The vector precession model (Figure 13.43) on which rested the explanation of the elimination of chemical shift, cannot fully describe the coherence transfer phenomenon, so one turns to the density matrix formalism.

13.27 DENSITY MATRIX METHOD IN NMR

What is generally known is that the motion of a magnetization vector under the influence of a RF pulse and during the precession period can be described by Bloch's vector model. The classical and semiclassical vector models are not only physically intuitive, but can also describe a multitude of experiments, the most vivid of which are the spin-echo phenomenon, J-resolution spectroscopy in the weak coupling limit, and polarization transfer from sensitive (higher γ) to insensitive (lower γ) nuclei. However, a complete and accurate description of coupled spins warrants quantum mechanical treatment of spin operators. This becomes amply clear when the phenomenon of spin coherence surfaces. Coherence transfer, which dictates the outcome of coupled-spin experiments, including 2D J-correlation spectroscopy (COSY) and multidimensional spectroscopy, is purely a quantum mechanical phenomenon that has no classical analogue and hence is not intuitive. Limited quantitative descriptions can still be provided with non-classical vector representations of density operators, but such descriptions are extremely laborious when the number of coupled spins or the spin operator products increase. Density matrix treatment of spin evolution is therefore necessary to detect the state of the spin system. Note the meaning of 'spin state' – it is a state function, the vector interpretation of which has severe limitations.

13.27.1 Outline of the Density Matrix Apparatus in NMR

The 1D NMR analogue of the density matrix result of an optical experiment outlined in Chapter 12 is shown in Figure 13.50. We are considering only the spin variables of interest, ignoring the lattice variables, and hence are using only the reduced density operator σ described there already. It is easy to follow after the reader familiarizes with the concepts discussed. The 90°_x pulse at the end of the preparation period produces a coherent non-equilibrium state. The preparation superoperator for an isolated spin k is

$$\hat{\hat{P}} = e^{\{-i\phi I_{kx}\}}, \tag{13.181}$$

where ϕ is the rotation angle, which is $\pi/2$ here. It is a superoperator because it acts on the density operator $\sigma(0)$. In schematic depiction of motion, it is customary to show the superoperator as

$$\hat{\hat{P}} \to \phi I_{kx}. \tag{13.182}$$

The creation of phase coherence is nearly instantaneous with the RF pulse, implying that the coherence evolution corresponds to the transformation of the density operator $\sigma(0)$ with time. The evolution occurs under the action of the evolution

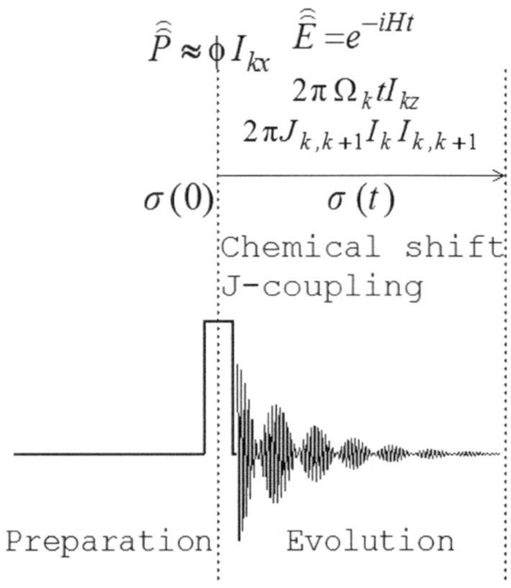

FIGURE 13.50 Outline of the density matrix description of a 1D NMR experiment with two scalar-coupled spins k and l under weak coupling limit.

superoperator $\hat{\hat{E}}(t)$ on the density operator. In NMR, the evolution of spin coherence always refers to the evolution of chemical shift (Ω) and scalar coupling (J) interactions whose transformations are carried out in a cascade, irrespective of the order in which they are carried out. The order of transformations of Ω and J does not matter because the chemical shift and scalar coupling terms in the Hamiltonian commute. The evolution superoperator is given by the spin Hamiltonian containing the chemical shift and scalar coupling terms

$$\hat{\hat{E}}(t) = e^{\{-i\mathcal{H}_0 t\}}$$
$$\mathcal{H}_0 = \sum_k \omega_{ok} I_{kz} + \sum_{k<l} 2\pi J_{kl} I_{kz} I_{lz}. \tag{13.183}$$

Notice that we are using only the I_z components of density operators here. Using I_z for the chemical shift evolution should be obvious (see below) because the chemical shift corresponds to rotation of the density operator about I_{kz}. The use of I_z in the scalar coupling term is due to the assumption of weak coupling between spins k and l.

Coming to the signal detection, which is concomitant with evolution of chemical shift and scalar coupling, we can write down the mean value of the observable $\langle D\rangle$ as

$$\langle D\rangle(t) = KTr\left\{\sigma(t) D\hat{\hat{E}}\hat{\hat{P}}\right\}, \tag{13.184}$$

where we have included a constant K determined by spin concentration and gyromagnetic ratio, and we call D the observable operator or probe operator or detector operator. Notice the use of time dependence here. The density operator $\sigma(t)$ gives the time evolution of the state of the system from $\sigma(0)$, but both $\hat{\hat{E}}$ and $\hat{\hat{P}}$ are shown as time independent. Expression of the observable with $\sigma(t)$ and time-independent D is often called Schrödinger representation. Another way to denote the observable, called Heisenberg representation, is to make $\hat{\hat{E}}$ time dependent and σ time independent. The transformation of time-dependent σ to its time independence is achieved by applying a unitary transformation to $\sigma(t)$. If $U(t)$ is the unitary operator, then equation (13.184) above can be written as

$$\begin{aligned}\langle D\rangle(t) &= KTr\left\{U(t)\sigma(0)U^{-1}(t)\hat{\hat{E}}D\hat{\hat{P}}\right\}\\ &= KTr\left\{\sigma(0)U^{-1}(t)\hat{\hat{E}}U(t)D\hat{\hat{P}}\right\}\\ &= KTr\left\{\sigma(0)\hat{\hat{E}}(t)D\hat{\hat{P}}\right\}.\end{aligned} \tag{13.185}$$

To describe the density matrix apparatus of 2D NMR one can take up only two homonuclear spins, k and $(k+1)$ or l (Figure 13.51). The interactions start with the same preparation pulse as they do in 1D NMR; the preparation superoperator excites all spin coherences instantaneous with the fall of the 90° pulse. In the first half of the evolution period t_1, the coherence evolves under the influence of $\hat{\hat{E}}(t_1)$, which in effect is the precession of the spin operators. As mentioned earlier, the transformation of the density matrix

$$\sigma(0) \rightarrow \sigma(t_1)$$

involves the evolution of chemical shift and J-coupling, irrespective of the order of the operation of the two,

$$\sigma(0) \xrightarrow{2\pi\Omega_k t_1 I_{kz}} \xrightarrow{2\pi J_{k,k+1} t_1 I_{kz} I_{k+1,z}} \sigma(t). \quad (13.186)$$

The chemical shift evolution here has considered only the kth spin operator, not the spin $k+1$ (or spin l). This simplification rests on the assumption of a homonuclear system in which the effect of chemical shift evolution will be symmetrical for the two nuclei. The evolution (or precession) flags the spins with respect to the individual precession frequency, called frequency labeling.

At the heart of 2D NMR is the mixing pulse given by the superoperator $\hat{\hat{R}}$ in Figure 13.51. The mixing pulse also rotates the spin operators as does the preparation pulse. However, the result of this mixing pulse-induced rotation is momentous in the sense that coherences are transferred to different transitions. As a result, a coherence that precessed at frequency ω during the evolution period t_1 will precess at $\omega \pm \Delta$ in the period t_2. For example, the coherence of spin k is transferred to the coherence of spin $k+1$ (or spin l) by the mixing pulse

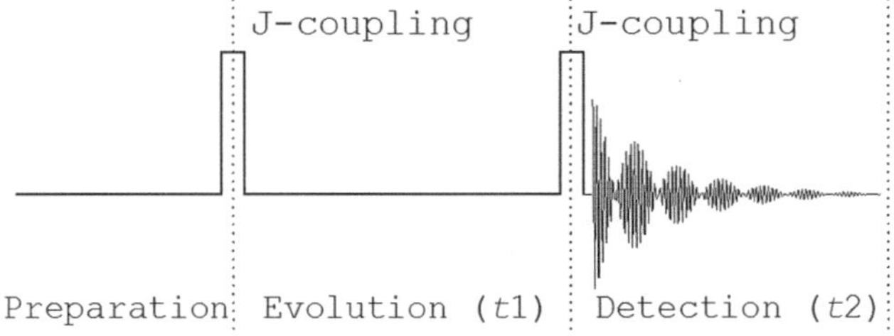

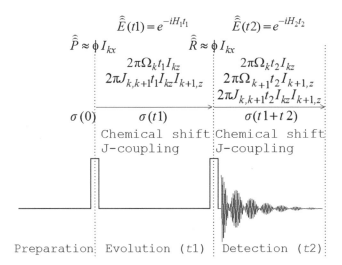

FIGURE 13.51 The summary of the density matrix description of a general 2D NMR experiment with two scalar-coupled spins k and $k+1$ (or l) under a weak coupling limit. We pay attention to the transformations of both chemical shift and scalar coupling in both t_1 and t_2 dimensions. The signal acquisition (t_2 time period) is allowed while evolution of both interactions in the mixed coherences continues.

$$2I_{ky}I_{k+1,z} \xrightarrow{(\pi/2)I_{kx}} -2I_{kz}I_{k+1,y}. \quad (13.187)$$

Please note that spin indices used may vary from one topic to another, for no reason; here k and $k+1$ are being used, the latter may also be read as l or any other alphabet or symbol as the reader wishes. We will come to these operators in some detail soon; they are being shown here only to set the outline of the density operator approach.

The transferred coherences precess in the period t_2 under the influence of the superoperator $\hat{\hat{R}}(t_2)$. Some coherences belonging to certain other transitions may not be transferred, but they also evolve in the t_2 period. The evolution of the density operator

$$\sigma(t_1) \rightarrow \sigma(t_1 + t_2),$$

is described again by chemical shift and scalar coupling evolution.

The expectation (mean) value of the observable in 2D NMR is written using the detection operator D

$$\langle D \rangle (t_1, t_2) = KTr\left\{\sigma(t_1, t_2) D\hat{\hat{E}}(t_1)\hat{\hat{E}}(t_2)\hat{\hat{P}}\hat{\hat{R}}\right\}. \quad (13.188)$$

The transfer of the time variables t_1 and t_2 from the density operator to the respective evolution operators yield the Heisenberg representation of the observable

$$\langle D \rangle (t_1, t_2) = KTr\left\{\sigma(0)\hat{\hat{E}}(t_1)\hat{\hat{E}}(t_2)\hat{\hat{P}}\hat{\hat{R}}\right\}. \quad (13.189)$$

13.27.2 Expression of Nuclear Spin Density Operators

The density operator of quantum systems can be expanded in a basis set of operators, the choice of which vary within the systems studied and depend on whether the Hamiltonian can be expressed in terms of the chosen operator.

Irreducible Tensor Operator. In the last chapter, the density operator of an electronic eigenstate was expanded in a base of irreducible tensor operators T_q^k, which are normally chosen for systems where two or more states superpose to produce a coherence. The nuclear spin density operator can also be expanded in the basis of tensor operators when the spin system consists of multiple spins. The expression for the tensor operators is identical to that already mentioned (see equation 12.108) and reproduced below

$$T_q^k = \sum_{q_1 q_2} T_{q1}^{k_1} T_{q2}^{k_2} \langle k_1 k_2 q_1 q_2 | kq \rangle.$$

According to this expression the irreducible tensor operator for two nuclear spins 1 and 2 will be the linear combination of products of tensor operators for the two individual spins

$$T_l^{m(1,2)} = \sum_{1,2} T_{l_1}^{m_1} T_{l_2}^{m_2} \langle l_1 l_2 m_1 m_2 | lm \rangle, \quad (13.190)$$

in which the symbols l and m correspond to the respective symbols k and q used in the equation initially. The coefficients $\langle l_1 l_2 m_1 m_2 | lm \rangle$, where $l = l_1 + l_2$ and $|m_1 - m_2| \le m \le m_1 + m_2$, are Clebsch-Gordan coefficients (see equation 12.110). We see that irreducible tensor operator expansion can be carried out when the density operator consists of many spin states 1, 2, 3, $\cdots$.

It is worth spending a little time to familiarize oneself with the meaning and the significance of l and m that have replaced the respective traditional symbols k and q, even though this was discussed earlier (see section 12.8.7). One may realize that the indices k and q in the component operator T_q^k imitate the angular momentum quantum numbers l and m. Summarily, l is the angular momentum quantum number and m is the projection of l onto the z-axis such that $m = -l, -l+1, \cdots, 0, \cdots l+1, l$. Thus m must also be a quantum number. The significance of the direct use of l and m here is the individual change of l and m during precession and rotation of spin operators. One needs to grasp this idea that l changes under free precession (evolution) of spins, leaving m as it is, but m can change under the influence of a RF pulse during which spin operators rotate but do not precess, or equivalently, the spin system does not evolve. The change of m refers to the order of coherence, P, which can be single-quantum coherence ($\Delta m = \pm 1$) or multiple quantum coherence ($\Delta m = \pm n$). This shows the advanatage of using irreducible tensor operator to expand the density operator.

However, an apparent difficulty arises regarding the structure of the spin Hamiltonian $\mathcal{H}_0$ (equation (13.183)) during the evolution period. Plainly, $\mathcal{H}_0$ cannot be described in terms of the irreducible tensor operators, because the latter are not vectors; they are rather integers which represent the magnitudes of l (see Chapter 12). This explains the inconvenience of the tensor expansion of the reduced density operator σ. Nevertheless, one can still retain the T_l^m basis for expansion of σ during the action of the pulse, but choose another basis set of operators to set up the evolution Hamiltonian in the transverse precession period. This means the expansion of σ has to be altered between the RF pulse and the evolution period, which means expansion in the T_l^m basis for the former and a suitable set of base operators for the latter.

Single-transition Angular Momentum Ladder Operators. These operators are suitable for expansion of σ during the evolution period if the preceding and the following pulse operators are expanded with tensor operators T_q^k. They are called single-transition ladder operators because each operator in the basis acts to raise or lower the magnetic quantum number m of only one transition, even though several coupled spins may be present. Consider, for example, two spins r and s (= 1, 2), whose product functions (eigenstates) are

$$|1\rangle = |\alpha\alpha\rangle \quad |2\rangle = |\alpha\beta\rangle \quad |3\rangle = |\beta\alpha\rangle \quad |4\rangle = |\beta\beta\rangle.$$

The operator definition including the conventional symbols is

$$I^{+(rs)} = I^{+(12)} = |1\rangle\langle 2| = I_x^{(12)} + iI_y^{(12)} = |\alpha\alpha\rangle\langle\alpha\beta| = I_1^\alpha I_2^+$$

$$I^{-(rs)} = I^{-(12)} = |2\rangle\langle 1| = I_x^{(12)} - iI_y^{(12)} = |\alpha\beta\rangle\langle\alpha\alpha| = I_1^\alpha I_2^-, \tag{13.191}$$

which indicates that

$$I^{+(12)}|2\rangle = |1\rangle$$
$$I^{-(12)}|1\rangle = |2\rangle. \tag{13.192}$$

The four product functions of the two spins yield 12 transitions, each of which is identified with a distinct ladder operator. The total of all step-up and step-down transitions in Figure 13.29 indeed count to 12. These 12 operators are formulated to $I^{\pm(rs)}$, where $r = 1, 2, 3$ and $s = 2, 3, 4$.

Explicitly,

$$I^{\pm(12)} = I_1^\alpha I_2^\pm$$
$$I^{\pm(34)} = I_1^\beta I_2^\pm$$
$$I^{\pm(13)} = I_1^\pm I_2^\alpha$$
$$I^{\pm(24)} = I_1^\pm I_2^\beta$$
$$I^{\pm(14)} = I_1^\pm I_2^\pm$$
$$I^{\pm(23)} = I_1^\pm I_2^\pm. \tag{13.193}$$

The set of single-transition operators also contains polarization operators $\mathrm{I}^{(rr)}$ such that

$$I^{(rr)} = |r\rangle\langle r|. \tag{13.194}$$

Accordingly, we will have four more operators

$$I^{(11)} = I_1^\alpha I_2^\alpha$$
$$I^{(22)} = I_1^\alpha I_2^\beta$$
$$I^{(33)} = I_1^\beta I_2^\alpha$$
$$I^{(44)} = I_1^\beta I_2^\beta, \tag{13.195}$$

tallying to $4^2 = 16$ operators, where the superscript '2' stood for the two weakly coupled spins.

Each transition is associated with a coherence order

$$p_{rs} = \Delta m_{rs} = m_r - m_s, \tag{13.196}$$

where m_r and m_s denote the respective magnetic quantum numbers (z-projection of l). The quantity $p_{rs} = \Delta m_{rs}$ is not a coherence transfer which occurs only under the influence of a pulse. It rather refers to the sustenance of the coherence, meaning a transfer already created by a mixing RF pulse. The coherence transfer sustains even after the evolution period, before another RF pulse falls. This suggests that the evolution of the density operator set by a Hamiltonian occurs only under a given coherence order created with respect to the preceding pulse. It is clear that the evolution Hamiltonian is the superoperator acting on the eigenbase of the single-transition ladder operator

$$e^{-i\hat{\hat{\mathcal{H}}}t}\sigma(t) = e^{-i\hat{\hat{\mathcal{H}}}t} I^{\pm(rs)}$$

$$i\hbar \frac{d}{dt} I^{\pm(rs)}(t) = \left[\hat{\hat{\mathcal{H}}}, I^{\pm(rs)}(t)\right]$$

$$= I^{\pm(rs)}\left(E_r - E_s\right)$$

$$I^{\pm(rs)}(t) = I^{\pm(rs)}(0) e^{-\frac{i}{\hbar}\Delta E.t}. \quad (13.197)$$

Cartesian Spin Operators. The most convenient to use and hence the popular basis operators for expansion of the density operator is the set of Cartesian angular momentum operators described in detail by Sorensen et al. (1983). In reality, multidimensional spectroscopy is deeply rooted in the use of the Cartesian spin operators. For just one spin k, the component angular momentum operators I_{kx}, I_{ky}, and I_{kz}, and their commutation relations are familiar

$$\left[I_{k\alpha}, I_{k\beta}\right] = I_{k\alpha} I_{k\beta} - I_{k\beta} I_{k\alpha} = i\hbar I_{k\gamma}, \quad (13.198)$$

where $\alpha, \beta, \gamma = x, y, z$, an extensively used pattern to label operator components (Ernst et al., 1988). Now the base operators used to expand $\sigma(t)$ of a spin-half system is defined by

$$B_s = 2^{(q-1)} \prod_{k=1}^{N} \left(I_{k\alpha}\right)^{a_{sk}}, \quad (13.199)$$

in which N is the total number of coupled spins and $\alpha = x, y, z$ (Sorensen et al., 1983). Notice that each operator in the basis set is a product of the single-spin angular momentum operators. As to how many operators will appear in the product operator is indicated by the symbol q in the above formula. The exponent a_{sk} is 1 for q nuclei and zero for the remaining $N - q$ nuclei.

The number of operators in a basis set is 4^N, where N is the number of spins present in the coupled spin system. The set of base operators for a lone nucleus is just $\left\{\frac{1}{2}, E, I_x, I_y, I_z\right\}$. The operators for a system of two weakly coupled spin-half nuclei 1 and 2 are shown in Figure 13.52 along with the corresponding non-classical vector and matrix representations. Note that each component operator can be represented by using the appropriate Pauli matrix $\frac{\hbar}{2}\sigma_\alpha$ with $\alpha = x, y, z$. One may choose to drop out $\hbar$ from the Pauli matrix if the results are expressed in the frequency unit instead of energy. The matrices are produced below

$$\sigma_x = \begin{pmatrix} 0 & 1 \\ 1 & 0 \end{pmatrix}, \sigma_y = \begin{pmatrix} 0 & -\mathrm{i} \\ \mathrm{i} & 0 \end{pmatrix}, \sigma_z = \begin{pmatrix} 1 & 0 \\ 0 & -1 \end{pmatrix}, E = \mathbf{1} = \begin{pmatrix} 1 & 0 \\ 0 & -1 \end{pmatrix}. \quad (13.200)$$

The matrix representation of the I_{1z} or I_{2z} operator, for example, is obtained from the direct product of σ_z and the unitary operator

$$I_{1z} = I_{2z} = 2 \times \frac{1}{2}\sigma_z \otimes \frac{1}{2}E = \frac{1}{2}\begin{pmatrix} 1 & 0 & 0 & 0 \\ 0 & -1 & 0 & 0 \\ 0 & 0 & 1 & 0 \\ 0 & 0 & 0 & -1 \end{pmatrix}. \quad (13.201)$$

Matrices for all base operators can be calculated similarly (Figure 13.52).

The operators in the operator basis set $\{B_S\}$ are called 'product operators' simply because the formula to construct these operators (equation (13.199) contains the product sign 'Π'. A large majority of multidimensional NMR experiments can be very clearly understood by checking as to how the product operators evolve starting from time $t = 0$ to longer times under the influence of the intermittent superoperators corresponding to RF pulses and evolution periods. Suppose we could exactly calculate the final outcome of the operator by working through the intermediate Hamiltonians, which will be discussed shortly, what kind of frequency-domain spectra we expect to see? Before going to the details of product operator transformations under the influence of the Hamiltonian superoperators, it is useful to classify the product operators for a weakly coupled system of two-spins 1 and 2 as follows (Ernst et al., 1988).

Longitudinal operator I_{1z}, I_{2z}

These correspond to z-magnetizations of the respective spins.

Transverse operator $I_{1x}, I_{2x}, I_{1y}, I_{2y}$

They are in-phase transverse magnetizations or x- and y-coherences, of which I_{1x} and I_{2x} produce absorptive in-phase, and I_{1y} and I_{2y} produce dispersive in-phase spectral lines (Figure 13.53). Since spins 1 and 2 are weakly coupled, each of the four lines split into two, $\Omega_1 \pm 2\pi J_{12}$ and $\Omega_2 \pm 2\pi J_{12}$.

Longitudinal operator of two-spin order $I_{1z}I_{2z}$

This is a longitudinal J-ordered product operator corresponding to zero net magnetization, and hence is not observable.

Antiphase operators $2I_{1x}I_{2z}, 2I_{1y}I_{2z}, 2I_{1z}I_{2x}, 2I_{1z}I_{2y}$

We see that one component operator of each of these antiphase product operators is a longitudinal z-magnetization. Since coherence occurs only for transverse magnetization, we can read an antiphase operator as a coherence that is antiphase with respect to the longitudinal magnetization of the component spin. For example, $2I_{1x}I_{2z}$ is an x-coherence of spin 1 whose phase is anti with respect to the z-magnetization of

$$\frac{1}{2}E \quad \frac{1}{2}\begin{pmatrix} 1 & 0 & 0 & 0 \\ 0 & 1 & 0 & 0 \\ 0 & 0 & 1 & 0 \\ 0 & 0 & 0 & 1 \end{pmatrix}$$

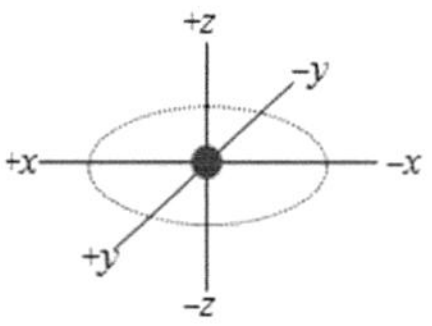

$$I_{1x} \quad \frac{1}{2}\begin{pmatrix} 0 & 0 & 1 & 0 \\ 0 & 0 & 0 & 1 \\ 1 & 0 & 0 & 0 \\ 0 & 1 & 0 & 0 \end{pmatrix}$$

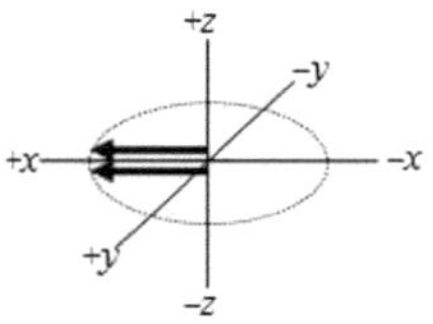

$$I_{1y} \quad \frac{1}{2}\begin{pmatrix} 0 & 0 & -i & 0 \\ 0 & 0 & 0 & -i \\ i & 0 & 0 & 0 \\ 0 & i & 0 & 0 \end{pmatrix}$$

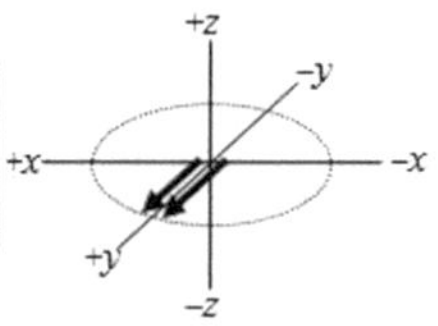

$$I_{1z} \quad \frac{1}{2}\begin{pmatrix} 1 & 0 & 0 & 0 \\ 0 & -1 & 0 & 0 \\ 0 & 0 & 1 & 0 \\ 0 & 0 & 0 & -1 \end{pmatrix}$$

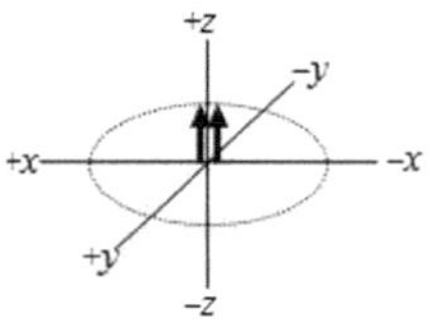

$$I_{2x} \quad \frac{1}{2}\begin{pmatrix} 0 & 0 & 1 & 0 \\ 0 & 0 & 0 & 1 \\ 1 & 0 & 0 & 0 \\ 0 & 1 & 0 & 0 \end{pmatrix}$$

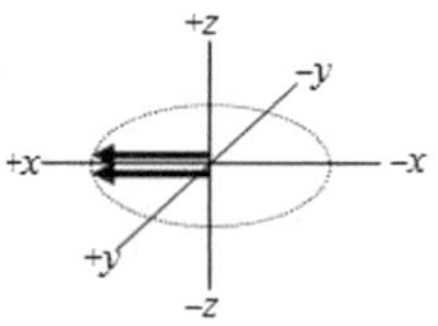

$$I_{2y} \quad \frac{1}{2}\begin{pmatrix} 0 & 0 & -i & 0 \\ 0 & 0 & 0 & -i \\ i & 0 & 0 & 0 \\ 0 & i & 0 & 0 \end{pmatrix}$$

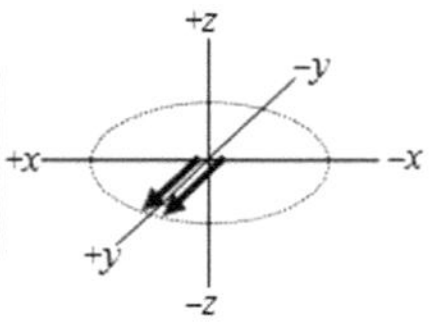

$$I_{2z} \quad \frac{1}{2}\begin{pmatrix} 1 & 0 & 0 & 0 \\ 0 & -1 & 0 & 0 \\ 0 & 0 & 1 & 0 \\ 0 & 0 & 0 & -1 \end{pmatrix}$$

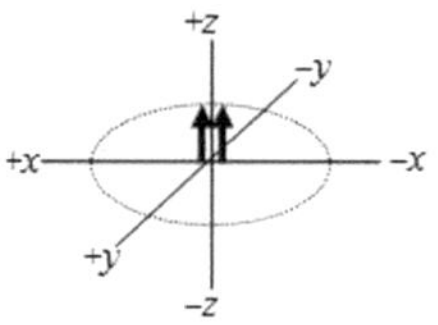

$$2I_{1x}I_{2x} \quad \frac{1}{2}\begin{pmatrix} 0 & 0 & 0 & 1 \\ 0 & 0 & 1 & 0 \\ 0 & 1 & 0 & 0 \\ 1 & 0 & 0 & 0 \end{pmatrix}$$

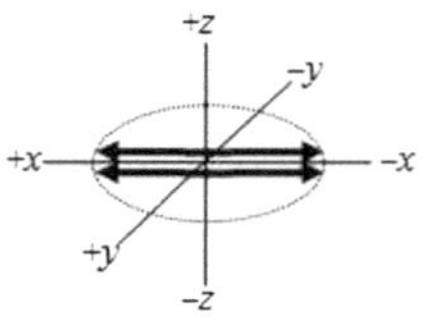

$$2I_{1x}I_{2y} \quad \frac{1}{2}\begin{pmatrix} 0 & 0 & 0 & -i \\ 0 & 0 & i & 0 \\ 0 & -i & 0 & 0 \\ i & 0 & 0 & 0 \end{pmatrix}$$

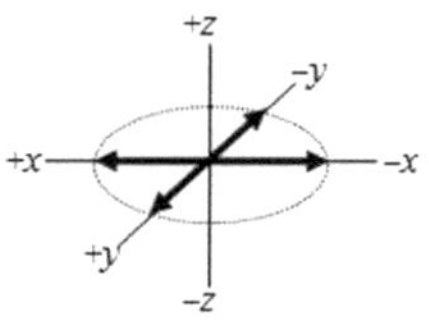

$$2I_{1x}I_{2z} \quad \frac{1}{2}\begin{pmatrix} 0 & 0 & 1 & 0 \\ 0 & 0 & 0 & -1 \\ 1 & 0 & 0 & 0 \\ 0 & -1 & 0 & 0 \end{pmatrix}$$

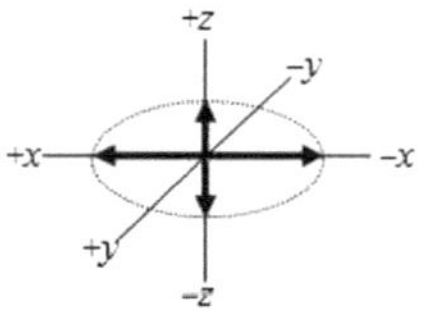

$$2I_{1y}I_{2y} \quad \frac{1}{2}\begin{pmatrix} 0 & 0 & 0 & -1 \\ 0 & 0 & 1 & 0 \\ 0 & 1 & 0 & 0 \\ -1 & 0 & 0 & 0 \end{pmatrix}$$

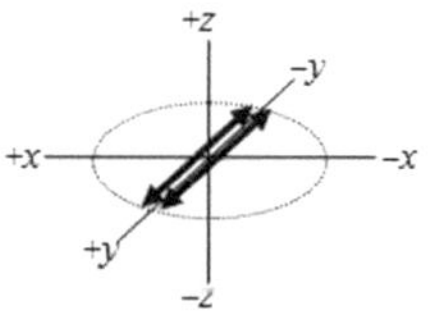

$$2I_{1y}I_{2x} \quad \frac{1}{2}\begin{pmatrix} 0 & 0 & 0 & -i \\ 0 & 0 & -i & 0 \\ 0 & i & 0 & 0 \\ i & 0 & 0 & 0 \end{pmatrix}$$

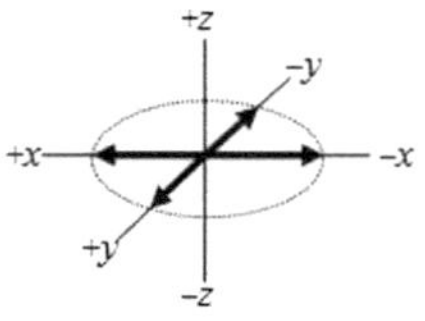

$$2I_{1y}I_{2z} \quad \frac{1}{2}\begin{pmatrix} 0 & 0 & -i & 0 \\ 0 & 0 & 0 & i \\ i & 0 & 0 & 0 \\ 0 & -i & 0 & 0 \end{pmatrix}$$

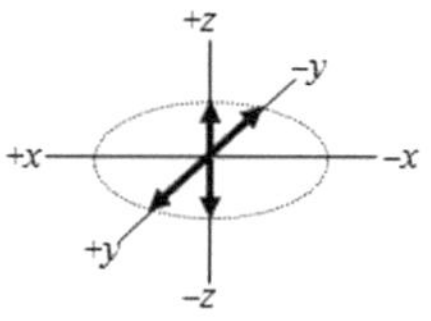

$$2I_{1z}I_{2x} \quad \frac{1}{2}\begin{pmatrix} 0 & 1 & 0 & 0 \\ 1 & 0 & 0 & 0 \\ 0 & 0 & 0 & -1 \\ 0 & 0 & -1 & 0 \end{pmatrix}$$

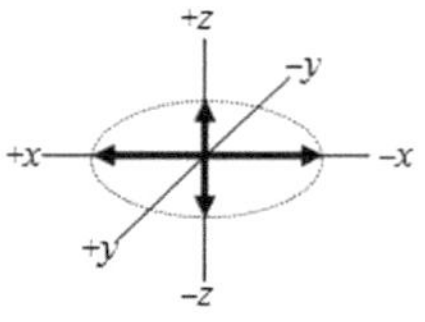

$$2I_{1z}I_{2y} \quad \frac{1}{2}\begin{pmatrix} 0 & -i & 0 & 0 \\ i & 0 & 0 & 0 \\ 0 & 0 & 0 & i \\ 0 & 0 & -i & 0 \end{pmatrix}$$

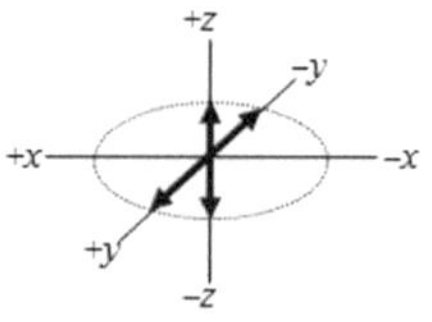

$$2I_{1z}I_{2z} \quad \frac{1}{2}\begin{pmatrix} 1 & 0 & 0 & 0 \\ 0 & -1 & 0 & 0 \\ 0 & 0 & -1 & 0 \\ 0 & 0 & 0 & 1 \end{pmatrix}$$

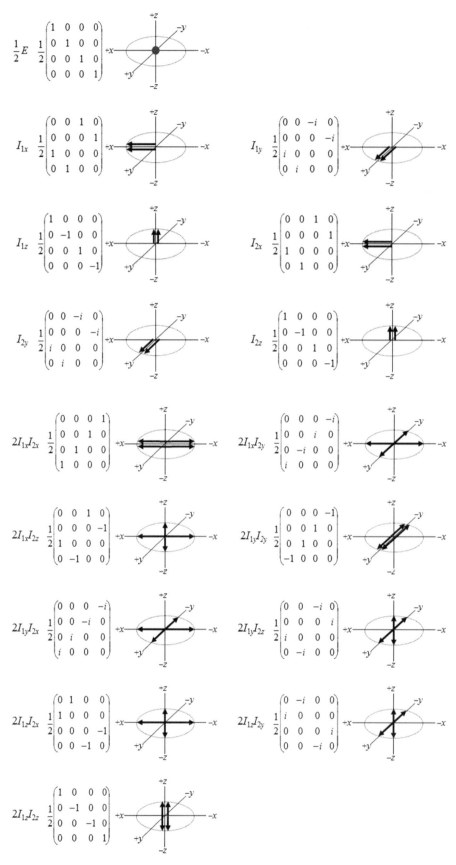

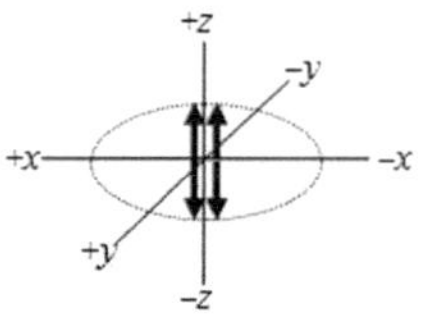

FIGURE 13.52 Matrix and vector representations of Cartesian product operators for two weakly coupled spins.

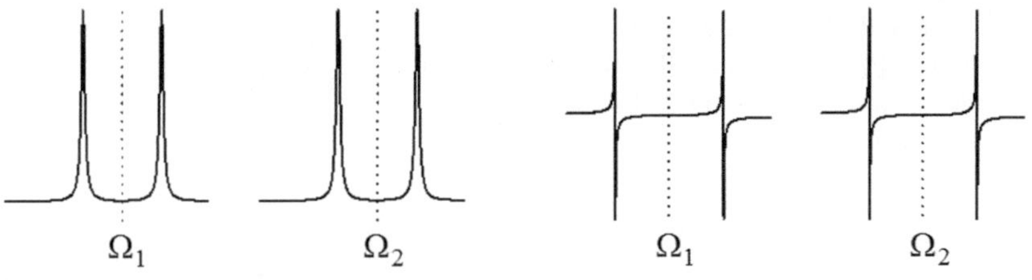

FIGURE 13.53 Absorptive and dispersive in-phase lines resulting from the four transverse magnetizations of two weakly coupled spins. The coupling splits each line into $\pm 2\pi J_{12}$.

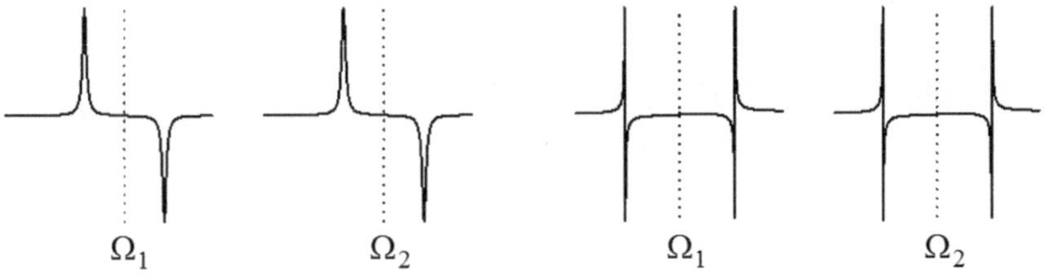

FIGURE 13.54 Absorptive and dispersive antiphase lines resulting from the four transverse magnetizations of two weakly coupled spins.

spin 2. Antiphase states are detectable, and the spectral lines appear as an absorptive antiphase doublet for $2I_{1x}I_{2z}$ and $2I_{1z}I_{2x}$, and a dispersive antiphase doublet for $2I_{1y}I_{2z}$ and $2I_{1z}I_{2y}$ (Figure 13.54).

Two-spin coherence operators $2I_{1x}I_{2x}, 2I_{1y}I_{2x}, 2I_{1x}I_{2y}, 2I_{1y}I_{2y}$

They do not contain longitudinal magnetization. The linear superposition of these operators yield zero- and double-quantum coherences, collectively called multiple quantum coherences.

$\frac{1}{2}\left(2I_{1x}I_{2x}+2I_{1y}I_{2y}\right)$ *x*-component of zero-quantum coherence

$\frac{1}{2}\left(2I_{1y}I_{2x}-2I_{1x}I_{2y}\right)$ *y*-component of zero-quantum coherence

$\frac{1}{2}\left(2I_{1x}I_{2x}-2I_{1y}I_{2y}\right)$ *x*-component of double-quantum coherence

$\frac{1}{2}\left(2I_{1y}I_{2x}+2I_{1x}I_{2y}\right)$ *y*-component of double-quantum coherence

The matrix representation of these coherences can be written down easily. For example, by using the corresponding matrices shown in Figure 13.52 we obtain for the *x*-component of the double-quantum coherence

$$\frac{1}{2}\left[\frac{1}{2}\begin{pmatrix}0&0&0&1\\0&0&1&0\\0&1&0&0\\1&0&0&0\end{pmatrix}-\frac{1}{2}\begin{pmatrix}0&0&0&-1\\0&0&1&0\\0&1&0&0\\-1&0&0&0\end{pmatrix}\right]=\frac{1}{2}\begin{pmatrix}0&0&0&1\\0&0&0&0\\0&1&0&0\\1&0&0&0\end{pmatrix}. \quad (13.202)$$

13.27.3 Transformations of Product Operators

We now use RF pulses and allow the evolution of interactions for expansion of the density matrix method outlined above in order to derive the observable product operators.

Effect of RF Pulses. The form of the RF Hamiltonian in the preparation superoperator is

$$\hat{\hat{P}}(t)=e^{-i\hat{\hat{\mathcal{H}}}_{rf}t}$$

$$\mathcal{H}_{rf}=-\sum_{k=1}^{N}\gamma_k \mathbf{B}_1\mathbf{I}_k, \quad (13.203)$$

where $\mathbf{B}_1$ is the applied RF field with an amplitude and frequency. In practice, the RF field can be chosen along one of the transverse axes ($\pm x$ or $\pm y$), and applied in the form of a pulse of duration τ_p to achieve a desired rotation of the spin magnetization vector (see equation (13.37)). For a homonuclear spin system, we can write

$$\mathcal{H}_{rf}=-\sum_{k=1}^{N}\phi I_{k\alpha}, \quad (13.204)$$

where $\phi=-\gamma B_{1,\alpha}\tau_p$ is the angle by which the magnetization operator rotates under the influence of the RF pulse. The rotation of spin operators with respect to the direction of the B_1 vector is given by the right-hand clockwise convention, by which the curling fingers point to the direction of the rotation. Importantly, the product operators define a space isomorphous to the Cartesian axes system. It is possible, in principle, to apply the RF pulse in any of the Cartesian directions and along both +ve and −ve directions. Generally, only transverse axes are used in the application of RF pulse, and if symmetry is desirable in a pulse train, then the '+' and '−' directions are modulated accordingly (see below). A quick succession of four $(90)_x$ pulses, for example, are applied symmetrically as $+x$ $-x$ $-x$ $+x$. Also note that a z-pulse which will rotate a spin operator in the transverse plane is equivalent to chemical shift evolution. This equivalence arises from the similarity of the RF Hamiltonian and the chemical shift or the Zeeman Hamiltonian under weak chemical shielding, $\sigma \ll 1$

$$\mathcal{H}_{\text{Zeeman}} = -\sum_{k=1}^{N}\left\{\gamma_k\left(1-\sigma_k^{zz}\right)\mathbf{B}_o\right\}I_{kz} = \sum_{k=1}^{N}\omega_{0k}I_{kz}$$

$$\mathcal{H}_{rf}(t) = \sum_{k=1}^{N}\omega_{0k}\phi_k I_{kz}. \tag{13.205}$$

Suppose we apply a RF pulse of phase ϕ along the x-direction of the Cartesian frame, then the transformation of the I_{kz} operator is

$$I_{kz}\xrightarrow{\phi I_{kx}} I_{kz}\cos\phi - I_{ky}\sin\phi. \tag{13.206}$$

The first of the two terms on the right-hand side resulting from this operation is an untransformed term as the coefficient I_{kz} declares. The untransformed term (magnetization) is of no use and hence can be discarded right at this point. The transformed $I_{ky}\sin\phi$ term is then subjected to further transformations. Following are some of the simple rotations of product operators of a single spin 1 under RF pulses

$$I_{1x}\xrightarrow{(90)_y} I_{1x}\cos(90) - I_{1z}\sin(90)$$

$$I_{1x}\xrightarrow{(90)_x} I_{1x}$$

$$I_{1y}\xrightarrow{(90)_y} I_{1y}$$

$$I_{1y}\xrightarrow{(90)_x} I_{1y}\cos(90) + I_{1z}\sin(90)$$

$$I_{1z}\xrightarrow{(90)_y} I_{1z}\cos(90) + I_{1x}\sin(90)$$

$$I_{1z}\xrightarrow{(90)_x} I_{1z}\cos(90) - I_{1y}\sin(90). \tag{13.207}$$

Transformations Under Evolution Superoperator. The time-independent spin Hamiltonian $\mathcal{H}_0$ contains the Zeeman- and scalar-coupling Hamiltonians

$$\mathcal{H}_0 = \mathcal{H}_{\text{Zeeman}} + \mathcal{H}_J. \tag{13.208}$$

In the weak coupling limit, one writes

$$\mathcal{H}_0 = \sum_k \omega_{ok}I_{kz} + \sum_{k<l} 2\pi J_{kl}I_{kz}I_{lz}. \tag{13.209}$$

We identify three useful properties of $\mathcal{H}_0$. One, the assumption of weak coupling allows retaining only the longitudinal magnetization in each of the component operators in $\mathcal{H}_J$. The operator $\mathcal{H}_{\text{Zeeman}}$ by definition already has the longitudinal magnetization. Consequently, the evolution of both chemical shift and scalar coupling are described by rotation of

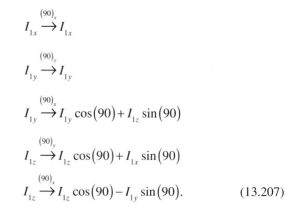

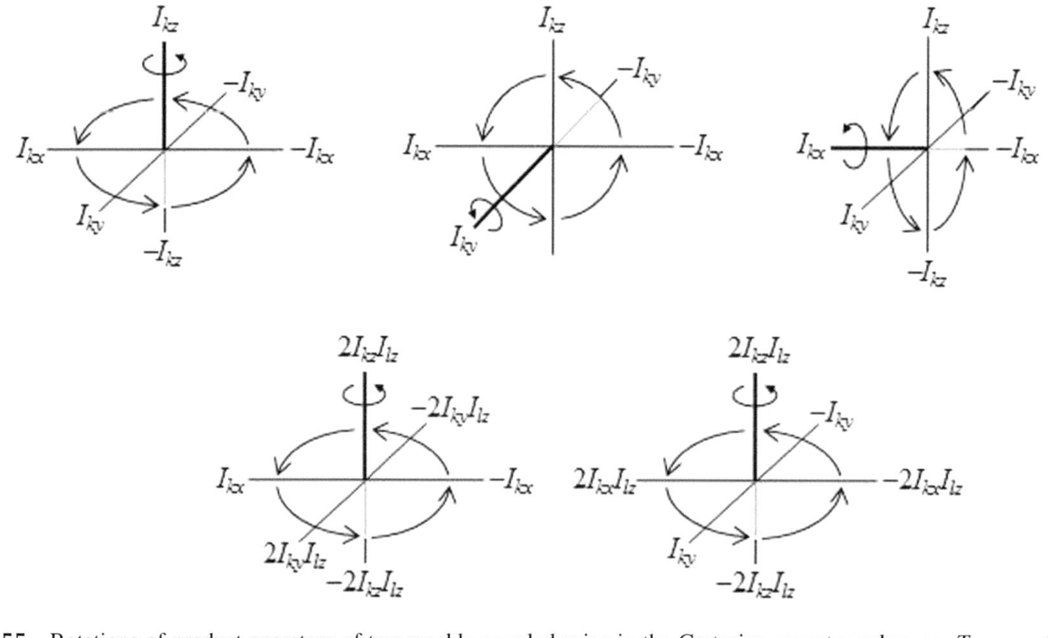

FIGURE 13.55 Rotations of product operators of two weakly coupled spins in the Cartesian operator subspace. *Top row*, the clockwise rotation about I_{kz} represents chemical shift evolution of spin k; the chemical shift of spin l will be given by rotation about I_{lz} in the $I_{l\pm\alpha}$ $(\alpha = x, y, z)$ operator subspace. The rotation about I_{kz} is also equivalent to a RF pulse along $+z$, called z-pulse, even though z-pulses are normally not used. However, rotations about I_{ky} and I_{kx} are the consequences of RF pulses along +x and +y, respectively. *Bottom row*, operator rotations about $2I_{kz}I_{lz}$ correspond to scalar coupling transformations for the two coupled spins. One should distinguish the contents of these two bottom-row figures; notice the operator transformations $2I_{ky}I_{lz} \to -I_{kx}$ in the left figure, but $2I_{kx}I_{lz} \to I_{ky}$ in the right.

the product operators in the Cartesian operator subspace as Figure 13.55 shows. Two, by writing the Hamiltonian more explicitly for a system of two weakly coupled spin-half nuclei 1 and 2, for example,

$$\mathcal{H}_0 = \omega_{o,1} I_{1z} + \omega_{o,2} I_{2z} + 2\pi J_{12} I_{1z} I_{2z}, \qquad (13.210)$$

we see that the chemical shift evolution of the two spins can be treated independently. Three, because $\mathcal{H}_{\text{Zeeman}}$ and $\mathcal{H}_J$ commute, the evolution of chemical shift and scalar coupling can be considered in any order – first chemical shift and then scalar coupling or the reverse. These three properties are useful in analytical calculations of operator transformations.

The chemical shift evolution of product operators of two spins 1 and 2 under the influence of $\mathcal{H}_{\text{Zeeman}}$ is shown below

$$I_{1x} \xrightarrow{\Omega_1 I_{1z} \tau} I_{1x} \cos\Omega_1\tau + I_{1y} \sin\Omega_1\tau$$

$$I_{1y} \xrightarrow{\Omega_1 I_{1z} \tau} I_{1y} \cos\Omega_1\tau - I_{1x} \sin\Omega_1\tau$$

$$I_{1z} \xrightarrow{\Omega_1 I_{1z} \tau} I_{1z}$$

$$\vdots \qquad \vdots$$

$$2I_{1x}I_{2z} \xrightarrow{\Omega_1 I_{1z} \tau} 2I_{1x}I_{2z} \cos\Omega_1\tau + I_{1y}I_{2z} \sin\Omega_1\tau$$

$$2I_{1y}I_{2z} \xrightarrow{\Omega_1 I_{1z} \tau} 2I_{1y}I_{2z} \cos\Omega_1\tau - I_{1x}I_{2z} \sin\Omega_1\tau. \qquad (13.211)$$

The scalar coupling evolutions are written as

$$I_{1x} \xrightarrow{2\pi J_{12} I_{1z} I_{2z} \tau} I_{1x} \cos(\pi J_{12}\tau) + 2I_{1y}I_{2z} \sin(\pi J_{12}\tau)$$

$$I_{1y} \xrightarrow{2\pi J_{12} I_{1z} I_{2z} \tau} I_{1y} \cos(\pi J_{12}\tau) - 2I_{1x}I_{2z} \sin(\pi J_{12}\tau)$$

$$2I_{1x}I_{2z} \xrightarrow{2\pi J_{12} I_{1z} I_{2z} \tau} 2I_{1x}I_{2z} \cos(\pi J_{12}\tau) + I_{1y} \sin(\pi J_{12}\tau)$$

$$2I_{1y}I_{2z} \xrightarrow{2\pi J_{12} I_{1z} I_{2z} \tau} 2I_{1y}I_{2z} \cos(\pi J_{12}\tau) - I_{1x} \sin(\pi J_{12}\tau). \qquad (13.212)$$

13.28 HOMONUCLEAR CORRELATION SPECTROSCOPY (COSY)

The transformations of spin product operators discussed above provide all necessary tools to work out the evolution or transformation of the density operator, which we take as the Cartesian product operators for a system of weakly coupled spins. The pulse sequence for a COSY experiment appears quite simple (Figure 13.56) – the first $(90)_x$ pulse prepares the system for coherence evolution through the evolution time t_1, at the end of which a $(90)_x$ pulse, called a mixing pulse, acts to transfer coherence between two product operators. If coherence transfer does not take place, cross peaks would not appear. The density operator at different stages of the sequence in Figure 13.56 are labeled σ_0, σ_1, σ_2, and σ_3. Since COSY is the gateway to the idea of multidimensional NMR, the operator

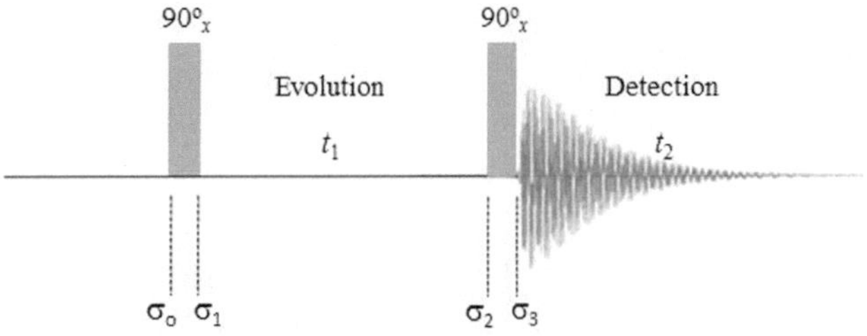

FIGURE 13.56 The COSY pulse sequence with the density operator labels σ_0, σ_1, σ_2, and σ_3 at different stages during the experiment.

transformations for a two weakly coupled spin system are explicitly worked out below as some other textbooks may also do. The exercise may seem somewhat unpleasant to carry out to the end, but worth it. The reader may closely compare the state of the density operator with time.

$$\sigma_o = I_{1z} + I_{2z} \qquad (13.213)$$

$$I_{1z} \xrightarrow{(90)_x} I_{1z} \cos\left(\frac{\pi}{2}\right) - I_{1y} \sin\left(\frac{\pi}{2}\right)$$

$$I_{2z} \xrightarrow{(90)_x} I_{2z} \cos\left(\frac{\pi}{2}\right) - I_{2y} \sin\left(\frac{\pi}{2}\right)$$

$$\sigma_1 = I_{1z} \cos\left(\frac{\pi}{2}\right) + I_{2z} \cos\left(\frac{\pi}{2}\right) - \left[I_{1y} \sin\left(\frac{\pi}{2}\right) + I_{2y} \sin\left(\frac{\pi}{2}\right)\right] \qquad (13.214)$$

The cosine functions represent untransformed operators (or magnetizations) which need not be considered further for precession under the influence of the spin Hamiltonian. The density operator will now be transformed ($\sigma_1 \rightarrow \sigma_2$) under the evolution superoperator during time t_1 as follows

$$-I_{1y} \sin\left(\frac{\pi}{2}\right) \xrightarrow{\Omega_1 t_1 I_{1z}} -I_{1y} \cos(\Omega_1 t_1) + I_{1x} \sin(\Omega_1 t_1)$$

$$-I_{2y} \sin\left(\frac{\pi}{2}\right) \xrightarrow{\Omega_2 t_1 I_{2z}} -I_{2y} \cos(\Omega_2 t_1) + I_{2x} \sin(\Omega_2 t_1)$$

$$-I_{1y} \cos(\Omega_1 t_1) + I_{1x} \sin(\Omega_1 t_1) \xrightarrow{2\pi J_{12} t_1 I_{1z} I_{2z}}$$
$$\left[-I_{1y} \cos(\Omega_1 t_1) + I_{1x} \sin(\Omega_1 t_1)\right] \cos(\pi J_{12} t_1) +$$
$$\left[2I_{1x}I_{2z} \cos(\Omega_1 t_1) + 2I_{1y}I_{2z} \sin(\Omega_1 t_1)\right] \sin(\pi J_{12} t_1)$$

$$-I_{2y} \cos(\Omega_2 t_1) + I_{2x} \sin(\Omega_2 t_1) \xrightarrow{2\pi J_{12} t_1 I_{1z} I_{2z}}$$
$$\left[-I_{2y} \cos(\Omega_2 t_1) + I_{2x} \sin(\Omega_2 t_1)\right] \cos(\pi J_{12} t_1) +$$
$$\left[2I_{1z}I_{2x} \cos(\Omega_2 t_1) + 2I_{1z}I_{2y} \sin(\Omega_2 t_1)\right] \sin(\pi J_{12} t_1)$$

Rearrangement yields

$$\sigma_2 = -\left[I_{1y}\cos(\Omega_1 t_1) + I_{2y}\cos(\Omega_2 t_1)\right]\cos(\pi J_{12}t_1)$$
$$+\left[I_{1x}\sin(\Omega_1 t_1) + I_{2x}\sin(\Omega_2 t_1)\right]\cos(\pi J_{12}t_1)$$
$$+\left[2I_{1x}I_{2z}\cos(\Omega_1 t_1) + 2I_{1z}I_{2x}\cos(\Omega_2 t_1)\right]\sin(\pi J_{12}t_1)$$
$$+\left[2I_{1y}I_{2z}\sin(\Omega_1 t_1) + 2I_{1z}I_{2y}\sin(\Omega_2 t_1)\right]\sin(\pi J_{12}t_1) \quad (13.215)$$

Next, $\sigma_2 \rightarrow \sigma_3$ transformation under the influence of the 90_x mixing pulse causes coherence transfer. For detection, both untransformed and transformed functions are preserved. The functions obtained instantaneously with the arrival of the mixing pulse are shown below

$$\sigma_3 = -\left[I_{1y}\cos(\Omega_1 t_1) + I_{2y}\cos(\Omega_2 t_1)\right]\cos(\pi J_{12}t_1)\cos\left(\frac{\pi}{2}\right)$$
$$+\left[I_{1x}\sin(\Omega_1 t_1) + I_{2x}\sin(\Omega_2 t_1)\right]\cos(\pi J_{12}t_1)$$
$$-\left[I_{1z}\cos(\Omega_1 t_1) + I_{2z}\cos(\Omega_2 t_1)\right]\cos(\pi J_{12}t_1)\sin\left(\frac{\pi}{2}\right)$$
$$+\left[2I_{1x}I_{2z}\cos(\Omega_1 t_1) + 2I_{1z}I_{2x}\cos(\Omega_2 t_1)\right]\sin(\pi J_{12}t_1)\cos\left(\frac{\pi}{2}\right)$$
$$-\left[2I_{1x}I_{2y}\cos(\Omega_1 t_1) + 2I_{1y}I_{2x}\cos(\Omega_2 t_1)\right]\sin(\pi J_{12}t_1)\sin\left(\frac{\pi}{2}\right)$$
$$+\left[2I_{1y}I_{2z}\sin(\Omega_1 t_1) + 2I_{1z}I_{2y}\sin(\Omega_2 t_1)\right]\sin(\pi J_{12}t_1)\cos^2\left(\frac{\pi}{2}\right)$$
$$-\left[2I_{1y}I_{2y}\sin(\Omega_1 t_1) + 2I_{1y}I_{2y}\sin(\Omega_2 t_1)\right]\sin(\pi J_{12}t_1)\sin\left(\frac{\pi}{2}\right)\cos\left(\frac{\pi}{2}\right)$$
$$+\left[2I_{1z}I_{2z}\sin(\Omega_1 t_1) + 2I_{1z}I_{2z}\sin(\Omega_2 t_1)\right]\sin(\pi J_{12}t_1)\sin\left(\frac{\pi}{2}\right)\cos\left(\frac{\pi}{2}\right)$$
$$-\left[2I_{1z}I_{2y}\sin(\Omega_1 t_1) + 2I_{1y}I_{2z}\sin(\Omega_2 t_1)\right]\sin(\pi J_{12}t_1)\sin^2\left(\frac{\pi}{2}\right). \quad (13.216)$$

The expansions of the operators to such details are not necessary, especially when software is available or can be written to derive the entire transformation from σ_0 to σ_3. The purpose here is instructive only.

The density operator σ_3 represents the state of the spin system at the threshold of data acquisition ($t_2 = 0$), so it is useful to distinguish the 18 terms contained in σ_3 (equation (13.216)) with respect to their observability. We see that the characteristics of the $\sigma_2 \rightarrow \sigma_3$ transformation relies on the phase of the operator terms in σ_2 with respect to the phase of the mixing pulse. There are four in-phase transverse operators in σ_2, which are I_{1y}, I_{2y}, I_{1x}, and I_{2x}. Since the mixing pulse is 90_x, where x is the phase, the I_{1y} and I_{2y} operators have transformed to the unobservable z-magnetizations, $-I_{1z}$ and $-I_{2z}$ of σ_3 (line 3 of equation (13.216)). The untransformed I_{1y} and I_{2y} magnetizations (line 1) will be observable as diagonals in the 2D spectrum. The I_{1x} and I_{2x} terms of σ_2 are not transformed by the 90_x pulse (line 2), but they are observable; their invariance to the mixing pulse leads to their appearance as diagonals. Accordingly, the transverse operators that are not affected by the mixing superoperator reproduce the 1D spectrum along the diagonal, and the lines are absorptive for I_{1x} and I_{2x} operators.

For the product operators $2I_{1x}I_{2z}$ and $2I_{1z}I_{2x}$ of σ_2, the component operators in each product operator are mutually antiphase, and one of the component operators in each is parallel to the phase of the mixing pulse, the I_{1x} component in the former and I_{2x} in the latter. These two terms are transformed by the 90_x pulse only partially, in the sense that only the z-magnetization of the respective component operators are rotated to yield $-2I_{1x}I_{2y}$ and $-2I_{1y}I_{2x}$ (line 5 in equation (13.216)). Each component operator in both of these has transverse magnetizations corresponding to zero- and double-quantum coherences, respectively, and hence are not observable ($\Delta m \neq \pm 1$). The untransformed part of $2I_{1x}I_{2z}$ and $2I_{1z}I_{2x}$ (line 4) is observable, and contributes to the diagonal.

Each of the antiphase terms $2I_{1y}I_{2z}$ and $2I_{1z}I_{2y}$ appearing in σ_2 (equation (13.215)) has a transverse component operator term (y-phase) which is perpendicular to the x-phase of the mixing pulse. A 'perpendicular term' of this kind transforms to a longitudinal operator of two-spin order, $2I_{1z}I_{2z}$ (line 8 of σ_3, equation (13.216)) that corresponds to 'zero net magnetization' (line 8), which means it is not observable. Such magnetizations are also partly transformed to unobservable 'zero'- and 'double'-quantum coherences, $-2I_{1y}I_{2y}$ and $-2I_{2y}I_{1y}$ (line 7). The untransformed part of the terms $2I_{1y}I_{2z}$ and $2I_{1z}I_{2y}$ of σ_2 appearing in σ_3 (line 6) is observable and appears as diagonal.

The terms $2I_{1y}I_{2z}$ and $2I_{1z}I_{2y}$ of σ_2 (equation (13.215)) also transform by coherence transfer to $-2I_{1z}I_{2y}$ and $-2I_{1y}I_{2z}$ (line 9 of σ_3) in which the coherences of spins 1 and 2 are swapped under the influence of the mixing pulse. They are the observable cross-peak multiplets. Notice that the component operators in both $2I_{1y}I_{2z}$ and $2I_{1z}I_{2y}$, as they appear in σ_2, are antiphase with respect to each other, and the phase

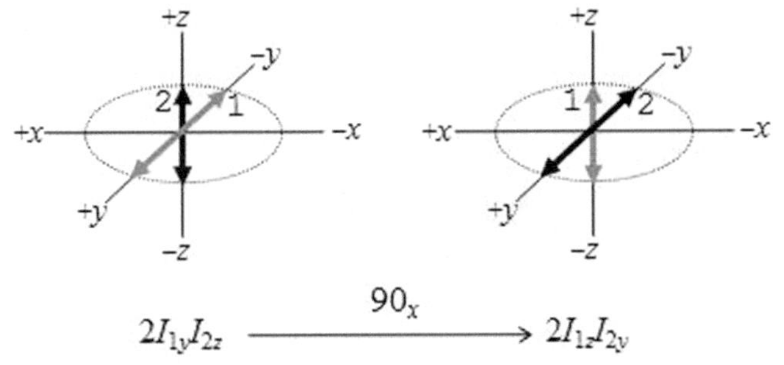

FIGURE 13.57 Apparent equivalence of coherence transfer depicted by non-classical vectors.

TABLE 13.2
Observability of Magnetizations and Characteristic Appearance of Spectral Lines in COSY

Terms in σ_3	Detection and appearance
$-I_{1y}, -I_{2y}$	Observable, diagonal multiplets
I_{1x}, I_{2x}	Observable, diagonal multiplets
$-I_{1z}, -I_{2z}$	Unobservable (longitudinal magnetization)
$2I_{1x}I_{2z}, 2I_{1z}I_{2x}$	Observable, diagonal multiplets
$-2I_{1x}I_{2y}, -2I_{1y}I_{2x}$	Unobservable (multiple quantum coherence)
$2I_{1y}I_{2z}, 2I_{1z}I_{2y}$	Observable, diagonal multiplets
$-2I_{1y}I_{2y}, -2I_{1y}I_{2y}$	Unobservable (multiple quantum coherence)
$2I_{1z}I_{2z}, 2I_{1z}I_{2z}$	Unobservable (longitudinal two-spin order)
$-2I_{1z}I_{2y}, -2I_{1y}I_{2z}$	Observable, cross-peak multiplets

of each component operator is orthogonal to the phase of the mixing pulse. The coherence transfer process for spins 1 and 2 is illustrated in Figure 13.57 by non-classical vector diagrams. The observability and appearance of various magnetizations of σ_3 are listed in Table 13.2.

Examination of the observable magnetizations in Table 13.1 in conjunction with the corresponding trigonometric functions appearing in the expression for σ_3 (equation (13.216)) would indicate that those containing the $\cos\left(\pi/2\right)$ factor will reduce to zero. One may note here that the mixing pulse need not necessarily be a 90° pulse as we have been using. Any degree of rotation caused by a RF pulse amounts to a change in energy of the spin magnetization, so the time duration of the mixing pulse could be more or less than that of a $\pi/2$ rotation. For a 90° mixing pulse, the only signals obtained from a cosine Fourier-transformed spectrum originate from the following operator terms shown along with the respective trigonometric arguments.

Diagonal peak, pure in-phase negative dispersion peak

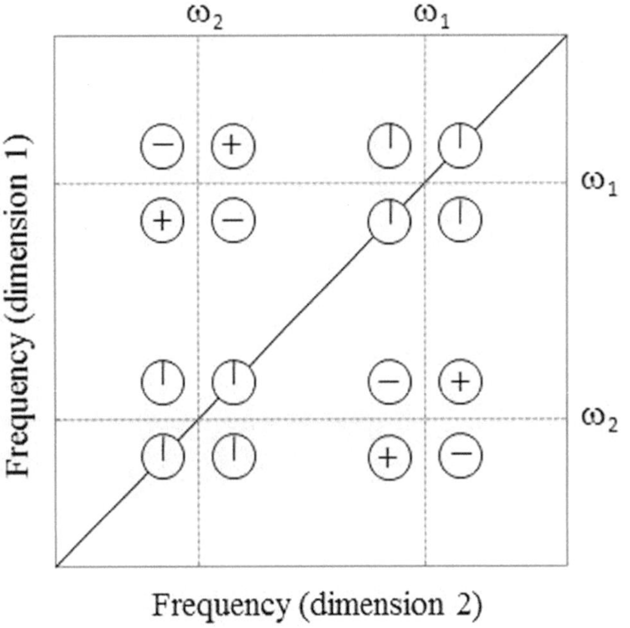

FIGURE 13.58 Schematic of a real Fourier-transformed correlation spectrum (COSY) for a system of two weakly coupled spins 1 and 2. The cross-peak quartet originate from two absorptive antiphase doublets ($2I_{1z}I_{2y}\sin\Omega_1 t_1 \sin\pi J_{12}t_1$ and $2I_{1y}I_{2z}\sin\Omega_2 t_1 \sin\pi J_{12}t_1$), each represented by a pure absorption antiphase lineshape.

$$I_{1x}\sin\left(\Omega_1 t_1\right)\cos\left(\pi J_{12}t_1\right)$$

$$I_{2x}\sin\left(\Omega_2 t_1\right)\cos\left(\pi J_{12}t_1\right)$$

Cross peak, pure antiphase absorption peak

$$-2I_{1z}I_{2y}\sin\left(\Omega_1 t_1\right)\sin\left(\pi J_{12}t_1\right)$$

$$-2I_{1y}I_{2z}\sin\left(\Omega_2 t_1\right)\sin\left(\pi J_{12}t_1\right).$$

The schematic of the 2D spectrum for correlation of two weakly coupled spins 1 and 2 is shown in Figure 13.58, and the simulated 2D lineshapes of cross peak and diagonal multiplets are depicted in Figure 13.59.

13.29 RELAYED CORRELATION SPECTROSCOPY (RELAY COSY)

The cross-peak intensity in the basic COSY spectrum is proportional to the strength of scalar coupling, but a cross peak will not appear if the J-value for a pair of spins in a spin network tends to zero. For example, consider three protons 1, 2, and 3 in a network

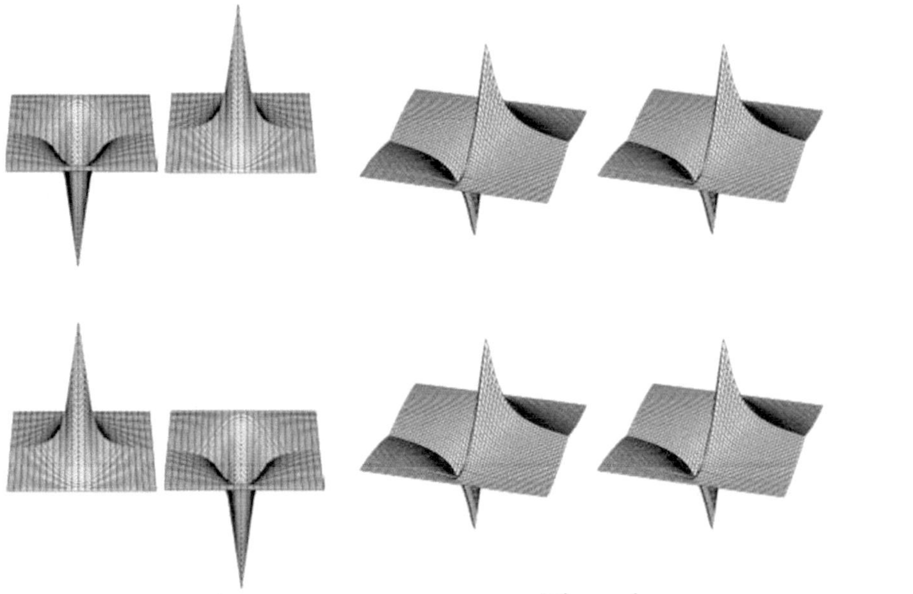

FIGURE 13.59 Simulated 2D lineshapes for pure antiphase absorption doublets (*left*) and pure in-phase negative dispersion (*right*). They are arranged to imitate the appearance of cross (*left*) and diagonal (*right*) multiplets.

such that spin 2 is weakly coupled to both spins 1 and 3, but spins 1 and 3 are not scalar-coupled (or even if coupled, the coupling is negligibly small); so J_{12} and J_{23} exist, but $J_{13} = 0$. In the relay experiment (Eich et al., 1982; Bolton and Bodenhausen, 1982), coherence is transferred from spin 1 to spin 3 via spin 2. The idea is to extend the COSY pulse scheme by introducing two equal mixing times τ which sandwich a 180° refocusing pulse before the third 90° observation pulse arrives (Figure 13.60). Importantly, the rationale for introducing the period τ is to transform the antiphase operator $2I_{1z}I_{2y}$ to the antiphase operator $2I_{2y}I_{3z}$. The 180° pulse halfway through the mixing period only refocuses the chemical shift. The precession has no effect on scalar coupling. In other words, the chemical shift evolution is prevented during the mixing period.

The evolution of operators can be shown without any difficulty. One may start with the I_{1z} operator at σ_0 and work through by following the same evolutions up to σ_3 to recover the $2I_{1z}I_{2y}$ antiphase operator. Now consider both J_{12} and J_{23}, so the evolution of the antiphase operator alone from σ_3 to σ_4 is expressed by

$$-2I_{1z}I_{2y} \xrightarrow{2\pi J_{12}I_{1z}I_{2z}\tau_m + 2\pi J_{23}I_{2z}I_{3z}\tau_m} 2I_{2y}I_{3z}\sin(\pi J_{12}\tau_m)\sin(\pi J_{23}\tau_m). \quad (13.217)$$

A final 90_x reading pulse brings about coherence transfer between spins 2 and 3. The antiphase term immediately after applying the final 90° pulse ($t_2 = 0$) is

$$-I_{2z}I_{3y}\sin(\Omega_1 t_1)\sin(\pi J_{12}t_1)\sin(\pi J_{12}\tau_m)\sin(\pi J_{23}\tau_m)$$

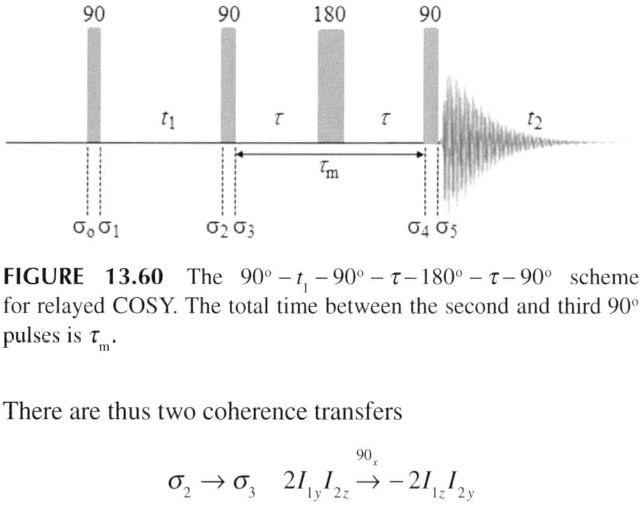

FIGURE 13.60 The $90° - t_1 - 90° - \tau - 180° - \tau - 90°$ scheme for relayed COSY. The total time between the second and third 90° pulses is τ_m.

There are thus two coherence transfers

$$\sigma_2 \to \sigma_3 \quad 2I_{1y}I_{2z} \xrightarrow{90_x} -2I_{1z}I_{2y}$$

$$\sigma_4 \to \sigma_5 \quad 2I_{2y}I_{3z} \xrightarrow{90_x} -2I_{2z}I_{3y}. \quad (13.218)$$

It should not be taken that the usual COSY cross peaks will disappear from a relayed COSY spectrum due to the relayed transfer of the antiphase operator $-2I_{1z}I_{2y} \to 2I_{2y}I_{3z}$ during the mixing time. They will not disappear, because the transfer is not entirely complete (Figure 13.61). When $J_{13} = 0$ Hz, the coherence transfer efficiency for an AMX type of system is given by

$$f = \sin(2\pi J_{12}\tau)\sin(2\pi J_{23}\tau). \quad (13.219)$$

The value of τ is suitably adjusted to obtain adequate transfer of magnetization from spin 1 to 3.

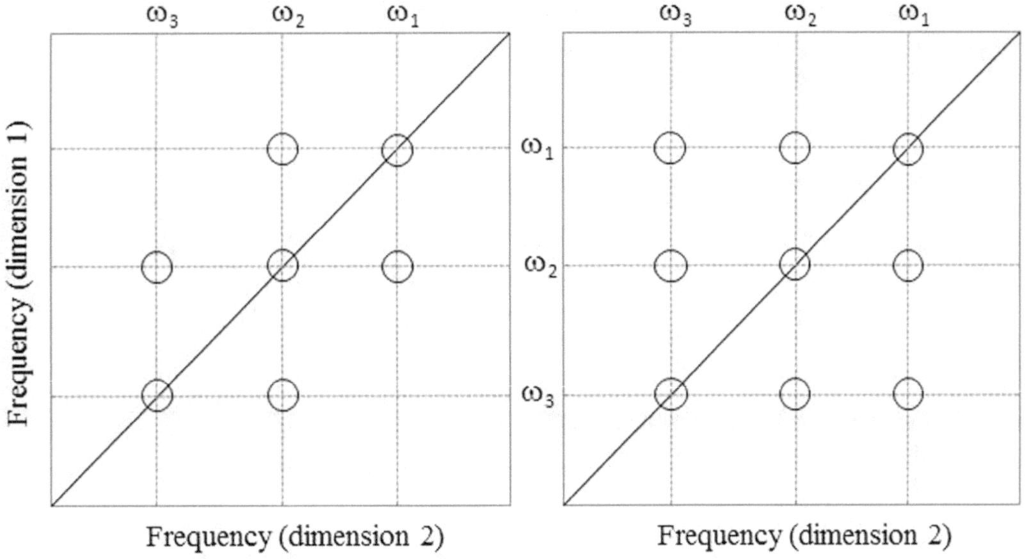

FIGURE 13.61 Schematic contour diagram of COSY (*left*) and relayed COSY (*right*) spectra for an AMX spin system.

13.30 TOTAL CORRELATION SPECTROSCOPY (TOCSY)

Coherence transfer can be achieved up to several spins in a spin network – at least five bonds away, not just within three spins as some organic chemists take for – by adopting a different transfer strategy which rests on the principle that all eigenmodes (collective state functions) can mix and transfer coherence within them, provided the spin system is driven by continuous or periodic perturbations (Figure 13.62).

The experiment starts as usual with a 90° pulse following which chemical shift and scalar coupling evolve during t_1. At the end of the evolution, a long trim pulse disperses (defocuses) those magnetizations that are not in phase with the x-axis. The spin system now enters a regime of mixing time and TOCSY transfer, where different spins are locked by a strong RF field to facilitate magnetization transfer from one to another spin that have the same transition energy. Having the same transition energy allows for cross polarization of the homonuclear system. The most efficient spin-locking in TOCSY refers to an integral number of repetitions of a pulse train called MLEV-17 that consists of a MLEV-16 composite pulse sequence followed by a 180_x pulse. The elements of the MLEV-16 pulses are defined in Figure 13.62. Repetitive MLEV-17 is followed by another long trim pulse to defocus the magnetizations that are not parallel to the x -axis.

Such a mixing sequence, called isotropic sequence, applied in the form of periodic perturbations creates a scalar-coupling average Hamiltonian $\bar{\mathcal{H}}$ as explained in Box 13.1. Under the spin-locked condition, $\bar{\mathcal{H}}$ is an isotropic strong-coupling Hamiltonian

$$\bar{\mathcal{H}}_{\text{iso}} = \sum_{k<l} 2\pi J_{kl} \mathbf{I}_k \mathbf{I}_l \qquad (13.220)$$

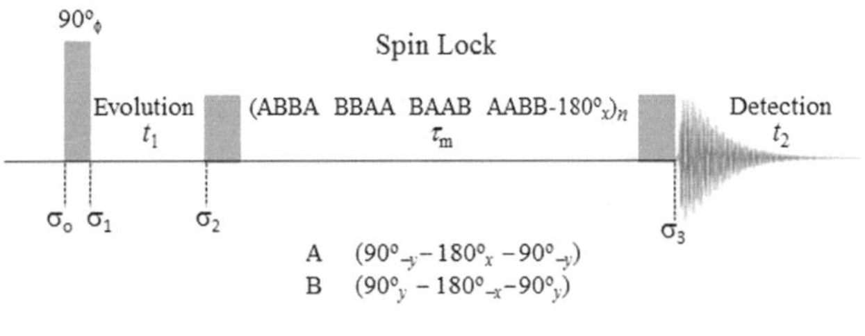

FIGURE 13.62 Pulse sequence for a TOCSY experiment.

where I indicates that the operator contains both secular and nonsecular components (Braunschweiler and Ernst, 1983). In isotropic Hamiltonian all the Cartesian components of the angular momentum operator behave identically. For several spins in the network, we may write the Hamiltonian as

$$\bar{\mathcal{H}}_{\text{iso}} = \sum_{k<l} 2\pi J_{kl} \left(I_{kx} I_{lx} + I_{ky} I_{ly} + I_{kz} I_{lz} \right). \qquad (13.221)$$

It is a property of strongly coupled spin systems that the individual spin operators, be they single spin operators or product operators, mix and combine linearly to produce a set of strong-coupling eigenfunctions, also called eigenmodes or collective eigenfunctions. The evolution of these eigenmodes under the influence of $\bar{\mathcal{H}}_{\text{iso}}$ leads to multiple relayed coherence transfer in the coupled spin network. To clarify further, consider a system of two spins 1 and 2. The evolution of any of the three Cartesian components of one of the two spins can be written as (see Braunschweiler and Ernst, 1983)

$$I_{1\alpha} \xrightarrow{\bar{\mathcal{H}}_{\text{iso}} \tau_{\text{m}}} \frac{1}{2} I_{1\alpha} \left[1 + \cos\left(2\pi J_{12} \tau_{\text{m}} \right) \right]$$
$$+ \frac{1}{2} I_{2\alpha} \left[1 - \cos\left(2\pi J_{12} \tau_{\text{m}} \right) \right] + \left(I_{1\beta} I_{2\gamma} - I_{1\gamma} I_{2\beta} \right) \sin\left(2\pi J_{12} \tau_{\text{m}} \right), \qquad (13.222)$$

where $\alpha, \beta, \gamma = x, y, z,$ and the last term corresponds to zero-quantum coherence. Similarly, if spins 2 and 3 are coupled, the expression for the evolution of the spin system would turn out to

$$I_{2\alpha} \xrightarrow{\bar{\mathcal{H}}_{\text{iso}}\tau_{\text{m}}} \frac{1}{2} I_{2\alpha}\left[1 + \cos\left(2\pi J_{23}\tau_{\text{m}}\right)\right] + \frac{1}{2} I_{3\alpha}\left[1 - \cos\left(2\pi J_{12}\tau_{\text{m}}\right)\right] + \left(I_{2\beta}I_{3\gamma} - I_{2\gamma}I_{3\beta}\right)\sin\left(2\pi J_{12}\tau_{\text{m}}\right), \quad (13.223)$$

and so on.

The evolution of $I_{1\alpha}$ through $I_{1\alpha}$ and $I_{2\alpha}$ (equation (13.222)) can be determined by calculating the expectation value

$$\langle I_{1\alpha} \rangle = tr\left(\sigma I_{1\alpha}\right), \quad (13.224)$$

but an easier way to see the evolution is to express this same equation in the form

$$I_{1\alpha} \xrightarrow{\bar{\mathcal{H}}_{\text{iso}}\tau_{\text{m}}} I_{1\alpha}\cos^2\left(2\pi J_{12}\frac{\tau_{\text{m}}}{2}\right) + I_{2\alpha}\sin^2\left(2\pi J_{12}\frac{\tau_{\text{m}}}{2}\right), \quad (13.225)$$

where we have neglected the dispersive antiphase components $(I_{1\beta}I_{2\gamma} - I_{1\gamma}I_{2\beta})$, and used the identities

$$1 + \cos\theta = 2\cos^2\left(\frac{\theta}{2}\right)$$

$$1 - \cos\theta = 2\sin^2\left(\frac{\theta}{2}\right).$$

Clearly, the I_α coherence oscillates between I_1 and I_2 with τ_{m} (Figure 13.63). Identical oscillations are observed for any pair of coupled spins. This result suggests that the intensities of both diagonal and cross peaks will oscillate with τ_{m}, the intensities of cross peaks reach a maximum when those of diagonals show a minimum and vice versa. The result also suggests that magnetization transfer in a network of spins follows

$$I_{1x} \rightarrow I_{2x} \rightarrow I_{3x} \rightarrow \cdots I_{nx} \quad (13.226)$$

However, coherence transfer in extended network of spins occurs by many frequency components, and the relative intensities of the magnetizations varies with τ_{m} in a complex manner. The magnetization transfer is limited to directly coupled protons when τ_{m} is short $(< 40\ \text{ms})$, but is relayed to several spins in the network at longer τ_{m}, up to ~100 ms.

An important property of coherence transfer under the influence of $\bar{\mathcal{H}}_{\text{iso}}$ is the invariance of the coherence phase during transfer. Equations (13.222), (13.223), and (13.224) indicate that TOCSY transfers in-phase coherence to in-phase coherence, yielding in-phase cross peaks. Signals from antiphase coherence components can be eliminated by adding signals obtained at a few different τ_{m}. Since both cross peaks and diagonals are in-phase, they can be phased to pure 2D absorption lineshape.

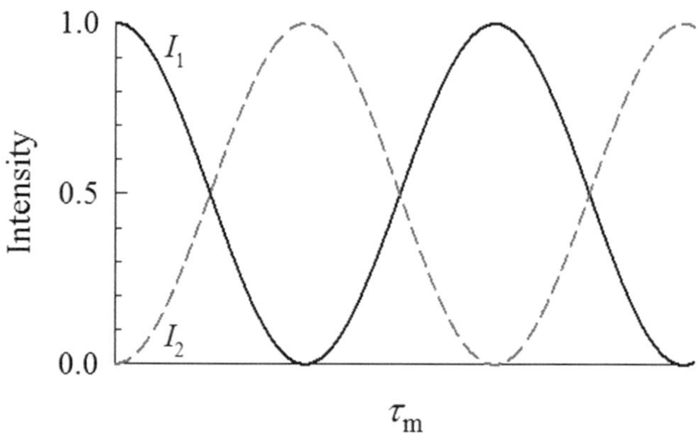

FIGURE 13.63 Oscillation of coherence of $I_\pm$ between I_1 and I_2 in the TOCSY mixing time.

13.31 2D NUCLEAR OVERHAUSER ENHANCEMENT SPECTROSCOPY (NOESY)

Magnetization transfer through cross-relaxation between two spins leads to intensity changes, which is summarily the nuclear Overhauser effect (NOE) discussed in detail earlier. NOE can be studied in a conventional manner by a somewhat laborious way in which a signal in a 1D spectrum is selectively saturated and the effect of the saturation on the intensity of any other signal(s) is examined in a spectrum taken immediately after saturation. The experiment does provide a fair amount of information regarding NOE build-up, the distance between the dipole-coupled spins, and relayed Overhauser effect, if any. The 2D version of transient NOE spectroscopy (NOESY) offers the advantage of mapping out all the transfer pathways, more often assisting in the sequential resonance assignment and structural pattern recognition in a short time.

The homonuclear NOESY pulse sequence (Figure 13.64) employs a mixing duration τ_{m} to allow for the NOE build-up. For a system of two spins 1 and 2, the form of density operators σ_0, σ_1, and σ_2 up to the end of the evolution period t_1 is the same as for the COSY experiment described before. But things take a different turn after the end of the evolution period t_1. Immediately after the second 90° pulse, which has the same RF phase as the first 90° pulse does, the density operator can be readily calculated to obtain the following expression. It is important to be familiar with the expression.

$$\sigma_3\left(t_1, \tau_{\text{m}} = 0\right) = \left[-I_{1z}\cos\left(\Omega_1 t_1\right) + I_{1x}\sin\left(\Omega_1 t_1\right) - I_{2z}\cos\left(\Omega_2 t_1\right) + I_{2x}\sin\left(\Omega_1 t_1\right)\right]\cos\left(\pi J_{12} t_1\right)$$

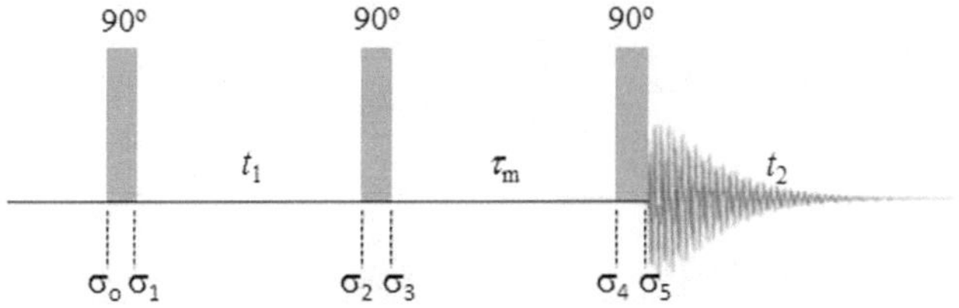

FIGURE 13.64 The skeletal homonuclear 2D NOESY pulse scheme.

$$
\begin{aligned}
&+\frac{1}{2}\left[\left(2I_{1x}I_{2y}+2I_{1y}I_{2x}\right)-\left(2I_{1y}I_{2x}-2I_{1x}I_{2y}\right)\right] \\
&\quad\cos\left(\Omega_1 t_1\right)\sin\left(\pi J_{12}t_1\right) \\
&+\frac{1}{2}\left[\left(2I_{1x}I_{2y}+2I_{1y}I_{2x}\right)+\left(2I_{1y}I_{2x}-2I_{1x}I_{2y}\right)\right] \\
&\quad\cos\left(\Omega_2 t_1\right)\sin\left(\pi J_{12}t_1\right) \\
&+2I_{1z}I_{2y}\sin\left(\Omega_1 t_1\right)\sin\left(\pi J_{12}t_1\right) \\
&+2I_{1y}I_{2z}\sin\left(\Omega_2 t_1\right)\sin\left(\pi J_{12}t_1\right)
\end{aligned}
\tag{13.227}
$$

Notice the terms in the first line – these are longitudinal and in-phase terms corresponding to each spin. Lines 2 and 3 contain linear superpositions of two-spin coherence operators, representing zero- and double-quantum coherences. Lines 4 and 5, each contains a coherence antiphase operator, antiphase with respect to the longitudinal magnetization of the coupled spin.

Because the objective of the experiment is to observe cross-relaxation and chemical exchange, both of which require longitudinal magnetization, it will be advantageous to select only the relevant terms and get rid of the burdensome ones. The selection of I_{1z} and I_{2z} operators are achieved by destroying the unwanted magnetization using an inhomogeneous magnetic field or RF phase cycling (see below), but the zero-quantum coherence term, $2I_{1y}I_{2x}-2I_{1x}I_{2y}$, stays on because it responds to the RF phase of the second pulse in the same manner as does the longitudinal magnetization. The indifference of the zero-quantum term to phase cycling, and hence its persistence, is often called zero-quantum interference. We then have the following effective σ_3 that undergoes transformation during the mixing period τ_m

$$
\begin{aligned}
\sigma_3\left(t_1,\tau_m\right) = &-\frac{1}{2}\left[2I_{1z}\cos\left(\Omega_1 t_1\right)+I_{2z}\cos\left(\Omega_2 t_1\right)\right]\cos\left(\pi J_{12}t_1\right) \\
&+\frac{1}{2}\left(2I_{1y}I_{2x}-2I_{1x}I_{2y}\right) \\
&\left[\cos\left(\Omega_2 t_1\right)-\cos\left(\Omega_1 t_1\right)\right]\sin\left(\pi J_{12}t_1\right).
\end{aligned}
\tag{13.228}
$$

The precession of the operator terms in the second line evolves during τ_m as the difference of precession of the individual spin magnetizations, i.e.,

$$
\begin{aligned}
&\frac{1}{2}\left(2I_{1y}I_{2x}-2I_{1x}I_{2y}\right)\xrightarrow{\left(\Omega_1 I_{1z}+\Omega_2 I_{2z}\right)\tau_m}\frac{1}{2}\left(2I_{1y}I_{2x}-2I_{1x}I_{2y}\right) \\
&\cos\left(\Omega_1-\Omega_2\right)\tau_m+\frac{1}{2}\left(2I_{1y}I_{2x}+2I_{1x}I_{2y}\right)\sin\left(\Omega_1-\Omega_2\right)\tau_m,
\end{aligned}
\tag{13.229}
$$

where the cos- and sin-modulated terms correspond to y- and x-components, respectively, of zero-quantum coherence. The longitudinal terms appearing in line 1 of equation (13.228) evolve by mutual migration of the respective magnetization to each other's site due to cross-relaxation and exchange effects. The density operator after the evolution through the period τ_m is

$$
\begin{aligned}
\sigma_4 = &\left[a_{11}I_{1z}\cos\left(\Omega_1 t_1\right)+a_{21}I_{1z}\cos\left(\Omega_2 t_1\right)\right. \\
&\left.+a_{12}I_{2z}\cos\left(\Omega_2 t_1\right)+a_{22}I_{2z}\cos\left(\Omega_2 t_1\right)\right]\cos\left(\pi J_{12}t_1\right) \\
&+\left(I_{1y}I_{2x}-2I_{1x}I_{2y}\right)\cos\left(\Omega_1-\Omega_2\right)\tau_m \\
&+\left(I_{1x}I_{2x}-2I_{1y}I_{2y}\right)\sin\left(\Omega_1-\Omega_2\right)\tau_m,
\end{aligned}
\tag{13.230}
$$

in which the mixing of the longitudinal magnetizations are characterized by the mixing coefficients a_{11}, a_{12}, a_{21}, and a_{22}. Since the longitudinal magnetizations are expected to decay during τ_m by longitudinal (spin-lattice) relaxation, the coefficients are determined by both T_1 and τ_m. If the exchange of the two sites is symmetric with a rate constant k ($k_+ = k_-$), then

$$
\begin{aligned}
a_{11} = a_{22} = 0.5e^{-\frac{\tau_m}{T_1}}\left(1+e^{-2k\tau_m}\right) \\
a_{12} = a_{21} = 0.5e^{-\frac{\tau_m}{T_1}}\left(1-e^{-2k\tau_m}\right).
\end{aligned}
\tag{13.231}
$$

The dependence of the coefficients on τ_m is shown in Figure 13.65. The initial rise of the $a_{12} = a_{21}$ curve is a measure of the first-order NOE build-up rate.

The density operator at the end of the mixing period is transformed by the third 90° pulse into the detectable magnetization. Note that the 90° pulse will convert only the cosine term of line 2 in the expression for σ_4 (equation (13.230)). Consequently, we get the following expression for $\sigma_5(t=0)$, which forms the basis for the detected NOESY signal

$$
\begin{aligned}
\sigma_5 = &\left[a_{11}I_{1y}\cos\left(\Omega_1 t_1\right)+a_{21}I_{1y}\cos\left(\Omega_2 t_1\right)\right. \\
&\left.+a_{12}I_{2y}\cos\left(\Omega_1 t_1\right)+a_{22}I_{2y}\cos\left(\Omega_2 t_1\right)\right]\cos\left(\pi J_{12}t_1\right) \\
&+\left(I_{1z}I_{2x}-2I_{1x}I_{2z}\right)\cos\left(\Omega_1-\Omega_2\right)\tau_m\sin\left(\pi J_{12}t_1\right)
\end{aligned}
\tag{13.232}
$$

We see that the operators having the coefficients a_{11} and a_{22} both continue to precess at their respective frequencies producing the diagonals, but those two having a_{12} and a_{21} precess at the frequency of the respective coupled spin, thus giving

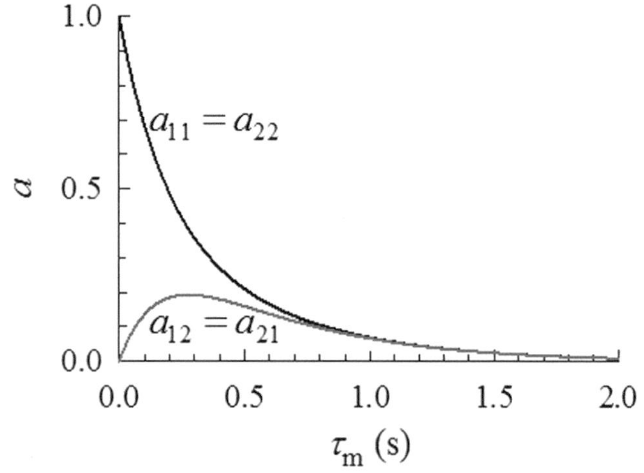

FIGURE 13.65 Variation of the mixing coefficients with the mixing time τ_m. Both coefficients are biexponential functions, the lower one (in red) corresponds to cross peaks. The simulation assumes $T_1 = 0.5$ s, and $k = 2$ s^{-1}.

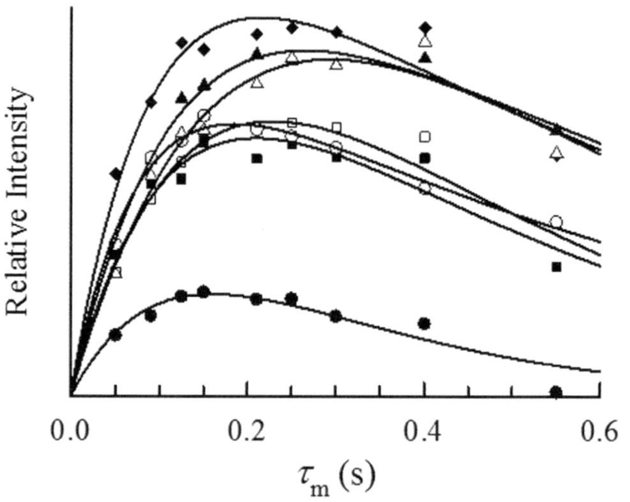

FIGURE 13.66 NOE build-up curves for a few unassigned peaks of the protein lysozyme, pH 4.2, 295 K. Iterated values of T_1 for these protons are in the range 0.12 to 0.67 s. (Courtesy: Kirthi Joshi.)

rise to cross peaks. Both diagonals and cross peaks will be in-phase because the operators I_{1y} and I_{2y} yield absorptive in-phase signal along ω_2. But, since the trigonometric functions $\cos(\Omega_1 t_1)\cos(\pi J_{12} t_1)$ and $\cos(\Omega_2 t_1)\cos(\pi J_{12} t_1)$ also yield in-phase signals along ω_1, the peaks in the NOESY spectrum appear as in-phase absorption signals.

The NOESY spectrum is however riddled by J-cross peaks. One may notice that the $\sigma_4 \to \sigma_5$ transformation describes the conversion of the zero-quantum coherence of σ_4 ($I_{1y}I_{2x} - 2I_{1x}I_{2y}$) to the observable single-quantum coherence of σ_5 ($I_{1z}I_{2x} - 2I_{1x}I_{2z}$). As a rule, both $I_{1z}I_{2x}$ and $I_{1x}I_{2z}$ produce dispersive antiphase doublet along ω_2. Also, $\cos(\Omega_1 - \Omega_2)\,\tau_m \sin(\pi J_{12} t_1)$ is associated with dispersive antiphase doublet. So, the single-quantum term gives rise to antiphase diagonal and cross peaks as observed in the COSY spectrum. This should happen, because these single-quantum terms originate from zero-quantum interference existing in σ_4 by coherence transfer under the influence of the third 90° pulse. These peaks in the NOESY spectrum are obviously not NOE peaks, but J-cross peaks.

It is desirable at times to eliminate the J-cross peaks from the NOESY spectrum. One way to suppress J-cross peaks is to run several NOESY spectra at different values of τ_m and add the signals. This strategy does not suppress J-cross peaks appearing near the diagonal. Another approach uses a fixed τ_m but inserts a 180° pulse during mixing. The insertion of the 180° pulse at different time points along the length of τ_m largely removes J-cross peaks.

The NOE build-up with shorter values of τ_m reaches a plateau and decreases thereafter (Figure 13.66). The decreasing exponential is due to the spin relaxation effect, but the initial build-up of NOE intensity is related to intramolecular and global tumbling dynamics of the molecule. The build-up provides a good estimate of the cross-relaxation rate constant, R_c, which can be approximately correlated with the distance between the two cross-relaxing spins. In other words, the cross-peak intensity at small τ_m is proportional to R_c. There are, of course, contributions from the magnetization relaxation and possible spin diffusion to other nuclei. Second-order NOE between spins 1 and 2, which means NOE mediated by cross-relaxation of an intervening spin k ($1 \to k \to 2$), can also influence the cross-peak intensity. The second-order effect is ideally manifested by a lag phase to the NOE build-up. Nevertheless, the cross-peak intensity developed with short τ_m can be used to approximate r_{12}, the spatial distance between the two spins

$$I_{12}(\tau_m) \sim \frac{\tau_c \tau_m}{r_{12}^{\;6}} \tag{13.233}$$

The correlation time τ_c can be determined by several methods, including dynamic light scattering, hydrodynamics, field-gradient NMR, and spin relaxation.

13.32 PURE EXCHANGE SPECTROSCOPY (EXSY)

The nuclear cross-relaxation effectively has two parts – longitudinal and transverse cross-relaxations. The NOESY method discussed above describes 'incoherent transfer' of longitudinal magnetization in the laboratory frame. The transfer of transverse magnetization, which also is an incoherent transfer process, is studied in the rotating frame, although it is necessary to eliminate the 'coherent transfer' processes in this case. Now consider two cross-relaxing spins, I_1 and I_2. The effective cross-relaxation rate constant, $R_c \approx \sigma_{12}^{(\mathrm{eff})}$, in the limit of negligible scalar coupling of the two spins is (see Griesinger and Ernst, 1987)

$$\sigma_{12}^{(\mathrm{eff})} = \cos\theta_1 \cos\theta_2 \sigma_{12}^{(l)} + \sin\theta_1 \sin\theta_2 \sigma_{12}^{(t)} \tag{13.234}$$

where θ_1 and θ_2 describe the effective tilt angles of the respective spin magnetization vectors with reference to the z-axis, $\sigma_{12}^{(l)}$ is the longitudinal cross-relaxation rate, and $\sigma_{12}^{(t)}$ is the transverse relaxation rate. Importantly, the signs of $\sigma_{12}^{(l)}$ and $\sigma_{12}^{(t)}$ should be opposite, which originates from the difference in the reference frame used for the measurement of cross-relaxation – longitudinal (laboratory frame) and transverse (rotating frame). This specification of the sign of $\sigma_{12}^{(l)}$ and $\sigma_{12}^{(t)}$ holds for large molecules in particular, for which $\omega_0\tau_c \gg 1$. The significance of the opposite signs of $\sigma_{12}^{(l)}$ and $\sigma_{12}^{(t)}$ in the context of equation (13.234) should be clear; $\sigma_{12}^{(\text{eff})}$ tends to zero if the longitudinal and transverse relaxation effects are balanced. When $\sigma_{12}^{(\text{eff})} \to 0$ does occur, the signals remaining in the NOESY spectrum will be those due to chemical exchange alone – they are called pure exchange peaks.

EXSY pulse schemes reported with these considerations filter out all signals but the pure exchange ones (Fejzo et al., 1991; Rao and Bhuyan, 2007). The basic idea is to suppress both longitudinal and transverse cross-relaxations in the mixing period by inserting pulses such that the time duration the magnetization spends along the longitudinal axis is half the time it spends along transverse axes, $\tau^{(l)} = 2\tau^{(t)}$. The time $\tau^{(t)}$ should be short enough so as to render dephasing of the transverse magnetization negligible. Several pulse schemes based on these arguments are possible, and the best performing one is shown in Figure 13.67, in which $\tau^{(l)} = 2\tau^{(90)}$, the time $\tau^{(90)}$ being the 90°-pulse time. Note the reflection symmetry of the 90° mixing pulses. The mixing sequence symmetry with respect to both phase and time reversal compensates for resonance offset and possible inhomogeneity in the RF field. The unit sequence, which effectively covers the total time of six 90° pulses, is cycled n times to wrap the entire duration of the NOESY mixing time τ_m. Practically, for a 8 μs 90° pulse, the unit mixing pulse scheme is cycled 2,048 times to wrap a typical mixing time τ_m of 100 ms as employed for a usual NOESY experiment with a large molecule.

The spectral lines in the 2D exchange spectrum broaden out as the exchange rate increases, and coalesce at a sufficiently fast exchange. A still-faster exchange results in exchange-narrowing. The rate k_{AB} of a symmetric exchange between sites A and B, exemplified by aromatic ring flips for example (Figure 13.68), can be conveniently determined from

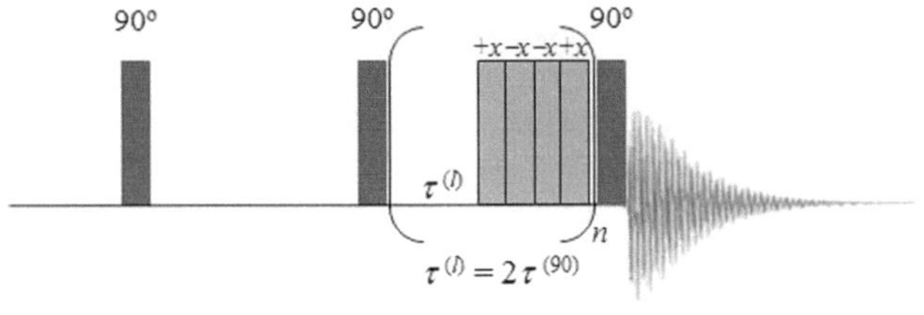

FIGURE 13.67 A performing EXSY pulse scheme. The four 90° pulses in the unit of the pulse train that fills the mixing time of the NOESY sequence are symmetric.

the diagonal and cross peak intensities if spectra at variable mixing time are recorded

$$\frac{V_c}{V_c + V_d} = \frac{1}{2}\left(1 - e^{-2k_{AB}\tau_m}\right), \tag{13.235}$$

where V_c and V_d are volumes of cross- and diagonal peaks, respectively. Figure 13.68 also shows an example of pure exchange peaks obtained from EXSY.

13.33 PHASE CYCLING, SPURIOUS SIGNALS, AND COHERENCE TRANSFER

The phase cycling operation in multipulse NMR experiments refers to selection of phases of RF pulses as well as the signal receiver in such a way that each cycle of a block of N acquisitions may have $\frac{2\pi}{N}$ steps, where the minimum value of N is 2. A world of perfect RF pulses would have ameliorated the need of phase cycling, but is not a reality. The necessity of phase cycling is mainly rooted in imperfections of RF pulses and quadrature detection electronics, spin relaxation during evolution, and the requirement of the selection of coherence transfer pathways in two- and multidimensional spectroscopy. In essence, the undesired magnetization components are filtered out by phase cycling. Let us briefly consider the main causative factors that make the use of phase cycling imperative.

1. *Imperfect RF pulses.* Pulse imperfections arise from both spatial inhomogeneity of the B_1 field and uncalibrated angle of magnetization rotation, and both of these occur inadvertently. Some degree of imperfection persists notwithstanding the care the magnet manufacturer and the experimenter exercise. The consequence of pulse imperfection is, however, the misleading appearance of spurious signals. The earliest example of this was provided by the emergence of the so called 'ghost signals' in the spin-echo experiment (Bodenhausen et al., 1977). If the 180° refocusing pulse is imperfect, a part of the total magnetization remains unfocused. Imagine the NMR sample being divided into n number of volume elements. The spatial inhomogeneity of the RF field results in slightly different B_1 sensed by different elements. In the context of the refocusing pulse, the time taken to produce the spin echo will not be uniform across the sample. Artifacts also arise from inaccurate flip angles of either or both of 90° and 180° pulses. If the initial 90° pulse deviates from the desired rotation, a part of the magnetization remains as the residual z-component, which will also be flipped by the refocusing pulse if it already deviates from the intended 180° rotation.

These artifacts are eliminated by a four-step reversed phase cycle called EXORCYCLE (Bodenhausen

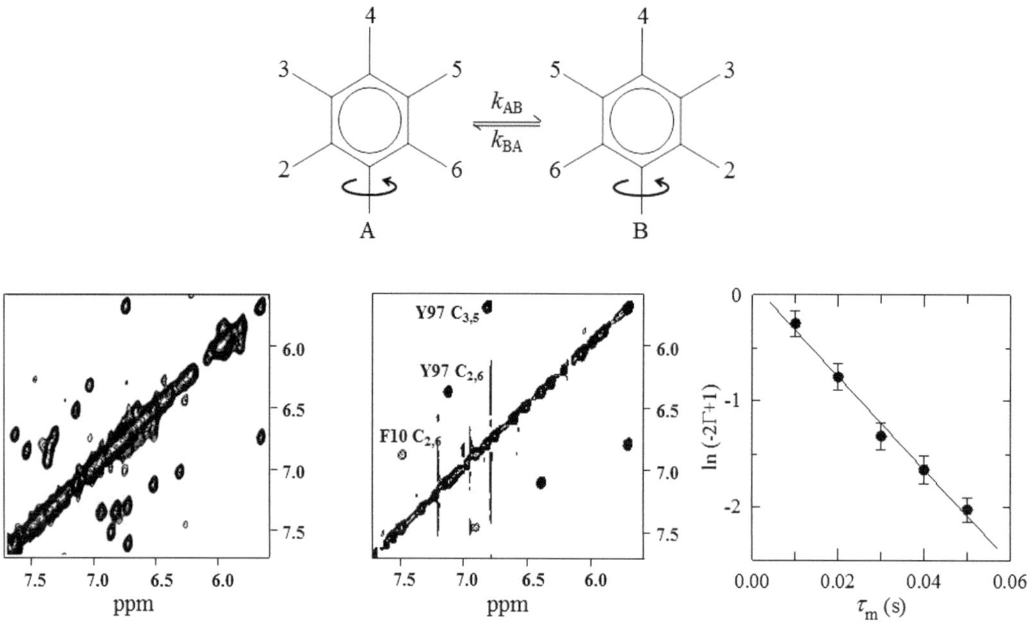

FIGURE 13.68 Chemical exchange. *Top*, flipping motion of phenylalanine ring, the rate constant being k_{AB}. *Bottom*, a region of a NOESY spectrum (*left*) and the corresponding region of an EXSY spectrum (*middle*) of the protein cytochrome *c* to show pure chemical exchange peaks in the latter spectrum. The cross peaks in the EXSY spectrum are identified with the aromatic ring protons of residues F10 and Y97. The exchange rate constant k_{AB} according to equation (13.235) is determined by analyzing a set of spectra of variable mixing time (*right*). Results reproduced from the author's work.

et al., 1977) produced in Figure 13.69. The rationale for phase-shifting the 180° pulse by $\frac{2\pi}{4}$ in each step is based on the idea that the spin echo is preserved but the sign of the other signals is reversed, leading to their cancellation. The shift in the phase of the signal due to the shift in pulse phases is compensated by a shift of the receiver phase.

2. *Longitudinal relaxation during evolution.* The recovery of the z-magnetization in the t_1 period contributes a term to the density operator, which is carried forward to the detection period. The fraction of the magnetization that recovers by T_1 does not evolve during t_1, and hence produces signal on the ω_2 axis registering $\omega_1 = 0$. These undesirable signals are called axial peaks, which are suppressed by phase cycling. A four-step phase cycle suitable for the COSY experiment is shown in Figure 13.70.
3. *Coherence transfer pathway.* A fundamental concept in pulsed NMR is the coherence order of the density operator along a sequence of pulses. The coherence order p can be considered a quantum number that determines the level of coherence of the density operator. The quantum number p specifies a transition in an ensemble of spins such that the two eigenstates differ by p. For a system of two spin-half nuclei ($N = 2$), p can take on an eigenvalue from the set $-2, -1, 0, +1, +2$. Similarly, p can have values from -3 to $+3$ through zero when $N = 3$. Given a system of N spins, each having the same spin quantum number, the coherence order p is allowed to have only integer values from $-N$ to $+N$. It is profound to say that 'the density operator is quantized' in terms of the coherence order

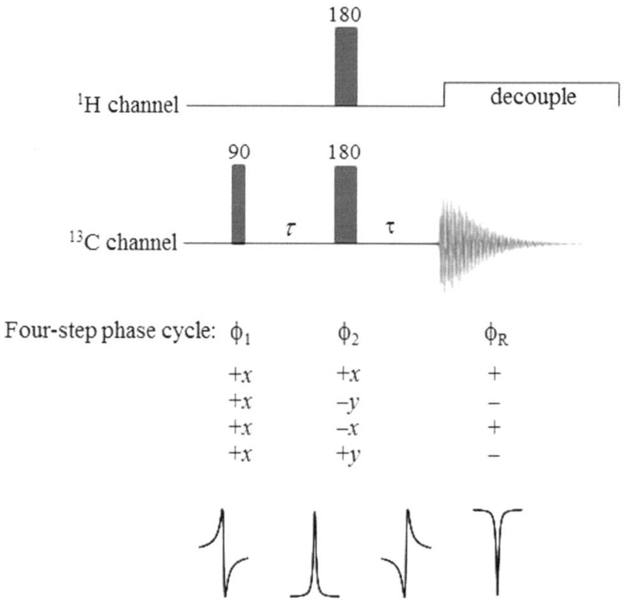

FIGURE 13.69 The EXORCYCLE phase cycling for observation of ^{13}C echo modulation of J-coupled CH spins. The symbols ϕ_1 and ϕ_2 refer to transmitter phases set for 90° and 180° pulses, respectively. The phase ϕ_R is the receiver phase. Shown at the bottom are the lineshapes of the spurious signals – from *left* to *right* are consecutive lineshapes from steps 1 to 4.

$$\sigma(t) = \sum_{p=-2L}^{+2L} \sigma^p(t), \tag{13.236}$$

where the σ^p term is given by (Ernst et al., 1988)

$$\sigma^p = \sum_{a,b} \sigma_{ab} |a\rangle\langle b|. \tag{13.237}$$

The coherence order of the density operator does not change during free precession, but a RF pulse induces a quantized rotational transition. The rotation of σ^p by an angle ϕ measured with respect to the z-axis in the rotating frame can be understood by invoking the rotation superoperator $\hat{\hat{R}}$ given by (Bodenhausen et al., 1984)

$$\hat{\hat{R}} = e^{-i\phi F_z}$$

$$e^{-i\phi F_z} \sigma^p e^{i\phi F_z} = \sigma^p e^{-ip\phi}. \tag{13.238}$$

The quantity F_z is the z-component of the total spin angular momentum taken as a sum of spin angular momenta of all spins k in the system considered

$$F_z = \sum_{k=1}^{N} I_k. \tag{13.239}$$

The expression in equation (13.238) shows that the phase of the density operator is altered to an extent proportional to both p and ϕ. This means that if the rotation due to a RF pulse changes the order of a coherence by Δp, then the coherence must also undergo a phase shift equal to $\phi \Delta p$.

13.34 COHERENCE TRANSFER PATHWAYS

The idea that a change in the coherence order is accompanied by a shift in the phase of the coherence forms the basis of designing phase cycles so as to select a specific coherence transfer pathway. Since this area of NMR spectroscopy may appear confusing at times, it is useful to introduce qualitatively a few basic concepts by taking an arbitrarily defined sequence of two 90° pulses (Figure 13.71).

1. Before an experiment begins all spins are in thermal equilibrium, so $p = 0$. The first 90° pulse creates a transverse magnetization, changing the value of the quantum number p from 0 to ±1. This change in the coherence order, $\Delta p = \pm 1$, is called single-quantum excitation. The first pulse always produces single-quantum excitation, which is shown in the figure immediately below the pulse scheme by the oblique lines connecting $p = 0$ and $p = \pm 1$. The phase shift of the magnetization at the end of the first pulse is $\Delta p_1 \phi_1$, in which we have included the subscript '1' only to indicate that the result is obtained after the first pulse.
2. The phase shift of the coherence, $\Delta p_1 \phi_1$, is not altered during the period of precession of the magnetization. The second 90° pulse changes the coherence order from $p = +1$ to $p = -1$ yielding $\Delta p_2 = -2$, and hence a coherence shift $\Delta p_2 \phi_2$, where ϕ_2 is the phase of the RF pulse. Changing the value of p to −1 is required only for the receiver to detect the signal. Although the receiver can detect both +1 and −1 senses of p, it is a convention to select −1. Note that +1 and −1 correspond to the two possible senses of rotation of a magnetization in the rotating frame.
3. The effective coherence phase shift available for detection is the sum of the phase shifts produced by the first and the second pulse. If there were n pulses, then the total phase shift would add up, $\sum_n \Delta p_n \phi_n$.
4. There also is the concept of restricting the coherence order. For example, obtaining $\Delta p_2 = -2$ is termed restricting the coherence transfer via the $p = 0 \rightarrow +1 \rightarrow -1$ pathway by phase cycling shown at the lower part of the figure. For each step of the phase cycle the magnetization can be accurately tracked at different stages of the pulse scheme. The third column of the lower part of the figure provides the additive phase of the magnetization vector, and the fourth column shows the phase of the receiver ϕ_R.

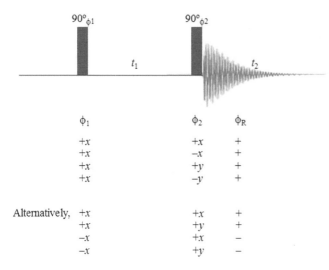

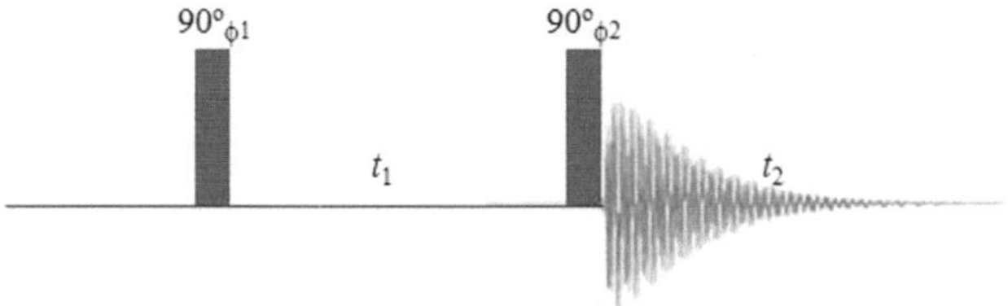

	ϕ_1	ϕ_2	ϕ_R
	+x	+x	+
	+x	−x	+
	+x	+y	+
	+x	−y	+
Alternatively,	+x	+x	+
	+x	+y	+
	−x	+x	−
	−x	+y	−

FIGURE 13.70 A four-step phase cycle for COSY. The number of scans must be set equal to nm, where n is an integer and m is the number of steps.

These concepts are useful to select Δp. For an example, consider the NOESY pulse scheme where the final pulse produces $\Delta p = -1$. We do not need to cycle the first pulse because it always generates $\Delta p = \pm 1$. Nor do we need to select the Δp after the second pulse, since this pulse generates z-magnetization alone. The third pulse transforms the coherence from $p = 0$ to $p = -1$. The coherence order $\Delta p = -1$ can be selected by employing the four-step cycle (Figure 13.72). Elaborate phase cycles with more steps are also possible, any of which must be based on the filter method adopted. Phase cycling is still a growing area of study.

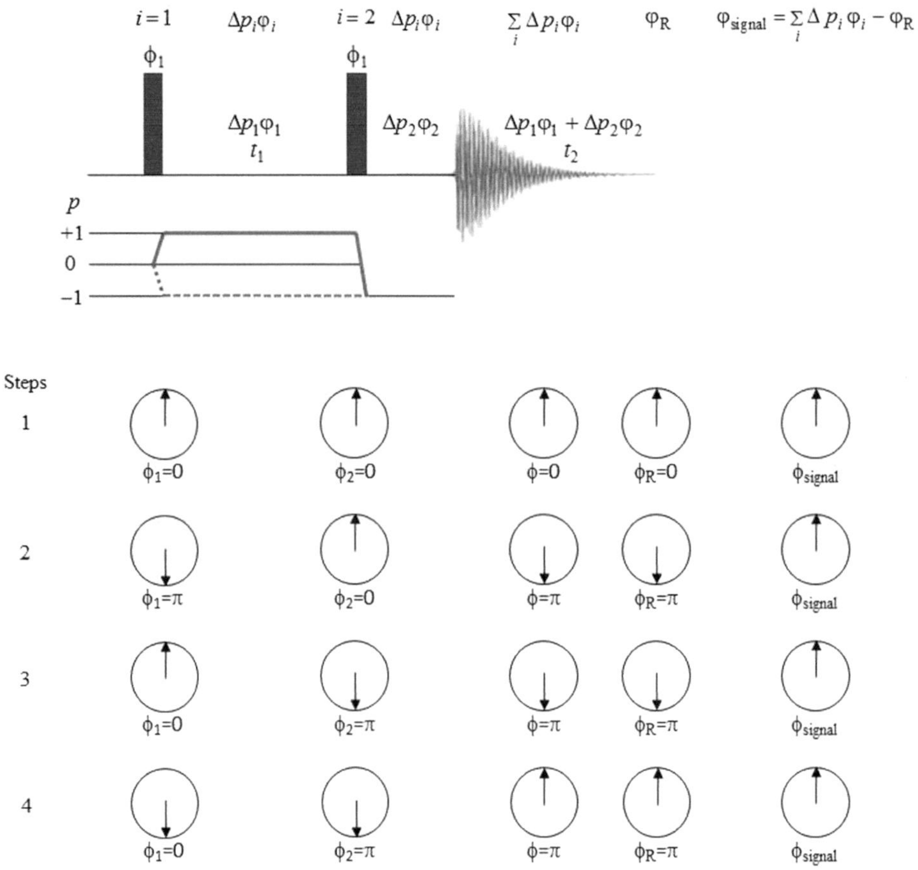

FIGURE 13.71 Coherence transfer, phase cycling, and magnetization tracking for a two-pulse scheme. The arrows indicate a magnetization in the rotating frame, and each phase may be associated with a transverse axis. For example, the phases in steps 1 to 4 may be written as +*x*+*x*+*x*+*x*, −*x*+*x*−*x*−*x*, +*x*−*x*−*x*−*x*, and −*x*−*x*+*x*+*x*, respectively.

13.35 MAGNETIC FIELD GRADIENT PULSE

Phase cycling would seem burdensome considering the complexity of designing a suitable cycle sequence, especially for higher-order multipulse experiments. The time consumed in running the cycles is also a concern. Phase cycling can be substituted by pulsed field gradient (PFG) to disperse the undesired transverse magnetizations whenever necessary (Barker and Freeman, 1985). The idea is to remove phase anomalies by defocusing the residual transverse magnetization by a PFG before applying a RF pulse. Similarly, a PFG can be applied immediately before signal acquisition. Figure 13.73 illustrates the simplest field gradient application in homonuclear correlation (COSY) and pure exchange (EXSY) experiments. Usually, two identical gradient recovery times of tens of microseconds flank each gradient pulse. The effect of the recovery pulses on the field-frequency lock signal is manifested in a sharp drop during the gradient and recovery thereafter. This can be observed in real time in the computer screen of the spectrometer console.

In solution NMR the magnetic field gradient is linear along the z-axis, and is generated by using shielded gradient coils that can produce a field strength $dg/dz \sim 50\,\mathrm{G\ cm^{-1}}$ at the peak current through the coils, although such steep gradient is rarely used. The recovery time of steeper gradient is longer. Figure 13.74 is a cartoon depiction of gradient-induced dispersion and refocusing of transverse magnetization. Clearly, the effective field is the sum of the static field B_0 and the gradient field g. Since g varies along the *z*-axis, we have

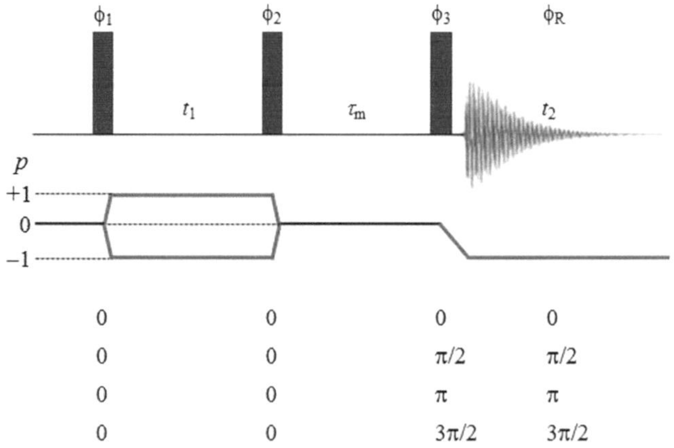

FIGURE 13.72 NOESY coherence transfer pathway, and the simplest four-step phase cycle.

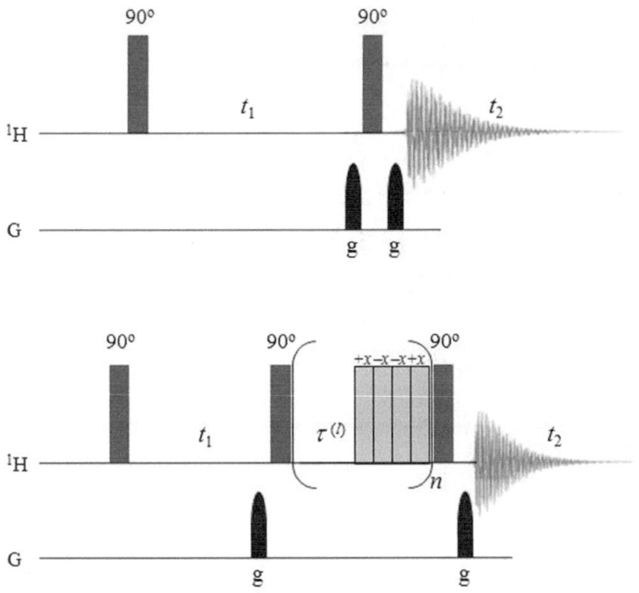

FIGURE 13.73 The use of PFG for coherence pathway selection. A dedicated channel G is used to synthesize the gradient (g). Shown at the *top* is a homonuclear COSY pulse sequence, and at the *bottom* is a homonuclear EXSY scheme.

$$B_z = B_o + g_z,$$

and therefore, the magnetization precession varies linearly along z

$$\omega_z = \omega_0 + \gamma g_z,$$

which essentially dephases the transverse magnetization. The dispersion induced by the field gradient can be reversed or refocused by reversing the gradient, i.e., a gradient of identical strength but opposite in polarity. Reversal of the defocused magnetization is unwanted for application of PFG to eliminate phase anomalies. If the PFG is not reversed, the molecular self-diffusion eventually leads to the loss of phase memory, thus irreversibly dephasing the unwanted transverse magnetization, and hence removing phase irregularities.

The field gradient required to disperse the transverse components of the magnetization (coherence) is related to the self-diffusion coefficient of the molecule by

$$\gamma g_z = \sqrt{\frac{3}{D\delta^2 t}}, \tag{13.240}$$

where D is the diffusion coefficient, δ is the duration of the PFG, and t is the waiting time after the gradient pulse. This relation is central to the application of PFG in studies of a plethora of problems. Since γ is given in the unit of rad G^{-1} s^{-1} and g_z is in G cm^{-1}, $\gamma g_z / 2\pi$ is in Hz cm^{-1}. The dependence of $\left(\gamma g_z\right)^2$ on D^{-1} suggests that the gradient field strength required to annihilate the transverse magnetization fundamentally depends on the molecular mass. Stating otherwise,

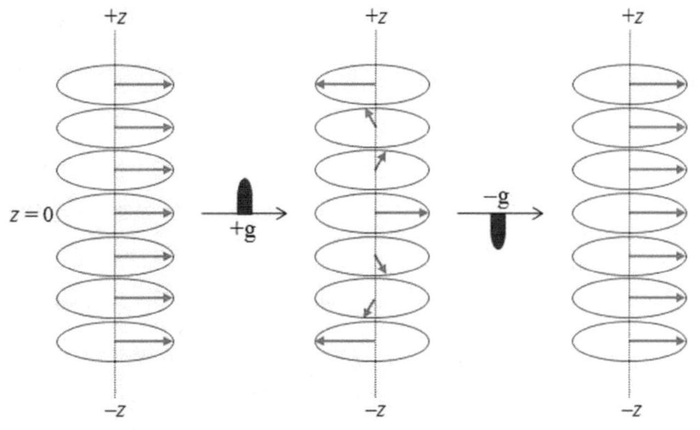

FIGURE 13.74 Dispersion and refocusing of magnetization by positive and negative gradients, respectively. The two gradients are identical except for the reverse polarity.

the larger the strength of the PFG, the lesser the likelihood of refocusing, and if not refocused, the signal will vanish with the strength of the gradient (Figure 13.75).

The argument above forms the basis of the NMR determination of the coefficient of self-diffusion of molecules, originally suggested by Stejskal and Tanner (1965). In brief, the extent of refocusing given by the echo amplitude $A(2\tau)$ is related to the diffusion coefficient D by

$$A(2\tau) = A(0)e^{\left[-(\gamma\delta g_z)\left(\Delta - \frac{\delta}{3}\right)D\right]} \tag{13.241}$$

in which τ represents the time between the first 90° pulse and the 180° refocusing pulse such that the echo occurs at a time $t = 2\tau$. The parameter δ is the duration of a PFG pulse, and Δ is the interval between PFG pulses. The recovery time of a gradient pulse is proportional to the duration of its application. As an example of such measurements, Figure 13.75 shows the attenuation of the signal amplitude with increasing strength of the field gradient. A discussion on further utilities of PFG here will be protracting, but the introduction provided hints at the significance and importance of field gradients.

13.36 HETERONUCLEAR CORRELATION SPECTROSCOPY

Correlation generally means correlating chemical shifts of two scalar-coupled spins. In correlation experiments, the γ-values of the coupled spins play an important role. In the homonuclear ^{1}H COSY for example, J-correlations of ^{1}H are observed, but γ is the same from one ^{1}H to another. This obviously would not be the case if one wished to establish correlation of heteronuclear spins. In heteronuclear 2D correlation, the chemical shift of ^{1}H is correlated with that of a less sensitive nucleus, ^{13}C or ^{15}N for example. The reference of sensitivity to two heteronuclear spins points to two factors – their respective gyromagnetic ratios γ, and the respective Boltzmann distribution of population in the low- (α) and high- (β) energy states.

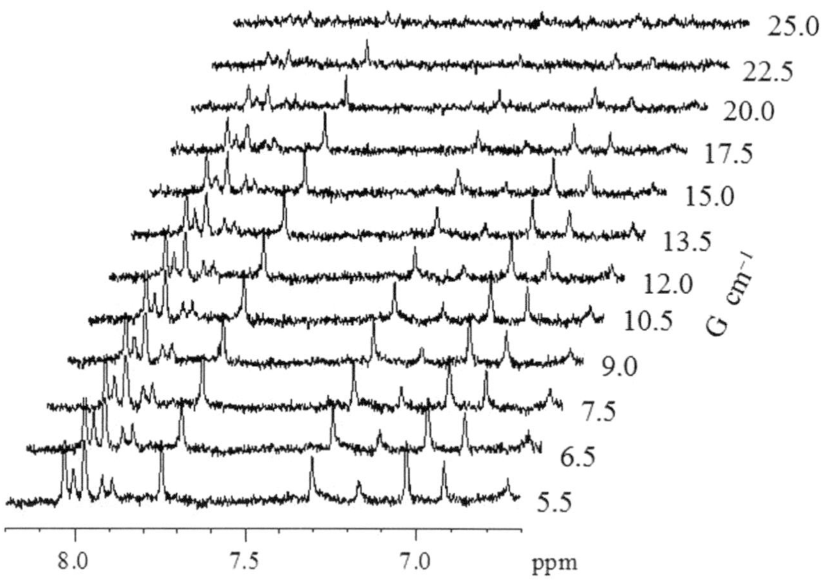

FIGURE 13.75 A region of the 1D spectrum of the protein myoglobin to show the disappearance of refocused signals with the PFG strength. At sufficiently high PFG strength, the spectrum appears bare due to the loss of detectable transverse magnetization.

If two spins of different γ are pulsed simultaneously to produce polarization transfer from the spin with higher to that with lower γ, the signal enhancement for the latter spin will depend on which spin we choose to detect the signal with. Consider the popular amide group (–NH) consisting of ^{1}H and ^{15}N spins with which we wish to perform a heteronuclear correlation experiment. Recall that the use of the INEPT pulse sequence (Figure 13.46) allows for recording the signal in the ^{15}N channel, and the ^{15}N signal enhancement as such will be $\gamma_{1_H} / \gamma_{15_N}$. But the factor of spin-lattice relaxation (T_1) which determines the signal-averaging time comes to the fore. Of all nuclei, ^{1}H has the shortest relaxation time; ^{13}C and ^{15}N relax rather slow in comparison. If the signal is detected using the ^{1}H channel, only the T_1 of ^{1}H will determine the recycling delay during the signal acquisition; the relaxation times of ^{13}C or ^{15}N would not matter. In fact, polarization transfer detected via ^{1}H acquisition offers an advantage, because in a given time many ^{1}H-detected scans can be signal-averaged compared with fewer for the coupled heteronuclear signal acquisition. Thus, in a heteronuclear correlation experiment where ^{1}H is used for signal acquisition, the signal enhancement factorizes to the Boltzmann population difference and the faster proton relaxation. The overall enhancement is given by $\left(\gamma_I / \gamma_S\right)^{5/2}$ which is substantial; for the –NH correlation, the ^{15}N enhancement would be ~300-fold.

A heteronuclear correlation experiment based on these ideas is called heteronuclear single-quantum coherence (HSQC) initially reported by Bodenhausen and Ruben (1980). The HSQC pulse scheme (Figure 13.76) consists of an INEPT sequence to enable polarization transfer from ^{1}H to a heteronucleus, ^{13}C or ^{15}N. The heteronucleus is frequency-labeled under free precession in the evolution period t_1, in the middle of which a 180° refocusing pulse on the ^{1}H channel decouples the two spins. At the end of t_1, the heteronuclear antiphase signal is converted to ^{1}H in-phase magnetization by implementing a refocused INEPT pulse, and the signal is detected during t_2. As discussed under the INEPT and refocused INEPT sections earlier, the delays τ in the HSQC sequence are set to $1/\left(4J_{IS}\right)$. Typical values of $^1J_{15_{NH}}$ and $^1J_{13_{CH}}$ are 90 and 140 Hz, respectively.

The transformation of the density operator through the HSQC sequence can also be worked out. With reference to the scheme in Figure 13.76, we have

$$\sigma_0 = I_z$$

$$I_z \xrightarrow{90_x} I_z \cos\left(\frac{\pi}{2}\right) - I_y \sin\left(\frac{\pi}{2}\right)$$

$$\sigma_1 = -I_y \sin\left(\frac{\pi}{2}\right). \qquad (13.242)$$

There is no chemical shift evolution for I_y during the ensuing $2\times\tau$ periods (Figure 13.76); the reason is that the '$-\tau-180_x^{(I)}-\tau$' segment corresponds to a spin-echo scheme where the $180_x^{(I)}$ pulse only refocuses the spins. Therefore, we need to calculate the scalar coupling evolution alone. One other point is that although the $180_x^{(S)}$ pulse is simultaneous with the $180_x^{(I)}$ pulse, they are taken as consecutive for the sake of calculations. Let us do the scalar coupling in a consecutive mode

$$-I_y \xrightarrow{2\pi J_{IS}\tau I_z S_z} -I_y \cos\left(\pi J_{IS}\tau\right) + 2I_x S_z \sin\left(\pi J_{IS}\tau\right)$$

$$\xrightarrow{180_x^{(I)}} I_y \cos\left(\pi J_{IS}\tau\right) + 2I_x S_z \sin\left(\pi J_{IS}\tau\right)$$

$$\xrightarrow{180_x^{(S)}} I_y \cos\left(\pi J_{IS}\tau\right) - 2I_x S_z \sin\left(\pi J_{IS}\tau\right)$$

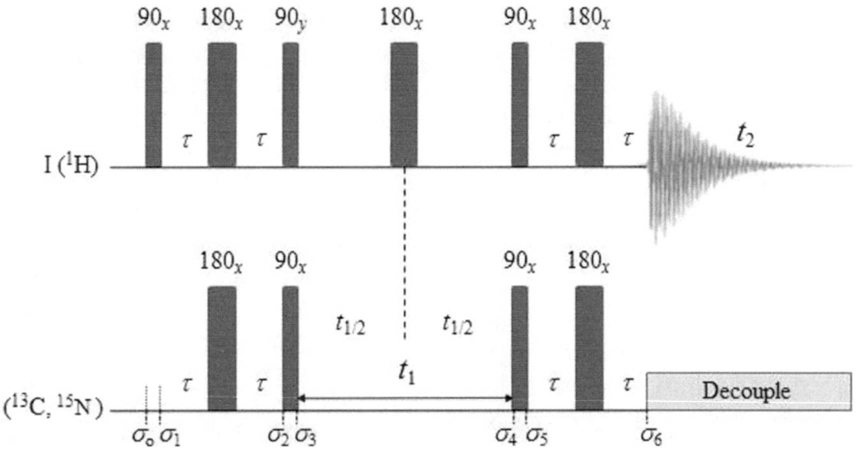

FIGURE 13.76 The HSQC pulse scheme.

$$\xrightarrow{2\pi J_{IS}\tau I_z S_z} \cos(\pi J_{IS}\tau)\left[I_y\cos(\pi J_{IS}\tau) - 2I_xS_z\sin(\pi J_{IS}\tau)\right]$$
$$-\sin(\pi J_{IS}\tau)\left[2I_xS_z\cos(\pi J_{IS}\tau) + I_y\sin(\pi J_{IS}\tau)\right]$$
$$= I_y\cos(2\pi J_{IS}\tau) - 2I_xS_z\sin(2\pi J_{IS}\tau). \quad (13.243)$$

The final expression is made using the identities $\cos(2\theta) = \cos^2\theta - \sin^2\theta$ and $\sin(2\theta) = 2\sin\theta\cos\theta$. Since $\tau = 1/(4J_{IS})$, only the second term is carried forward as

$$\sigma_2 = -2I_xS_z. \quad (13.244)$$

Next in the pulse sequence, the simultaneous application of $90_y^{(I)}$ and $90_x^{(S)}$ pulses transfer the single-quantum coherence as

$$-2I_xS_z \rightarrow -2I_zS_y$$

to yield the next density operator in the sequence

$$\sigma_3 = -2I_zS_y. \quad (13.245)$$

The magnetization now enters the evolution period t_1. The chemical shift evolution of the I-spin should not occur because the magnetization component of I in σ_3 is the longitudinal magnetization I_z. To insist on this, we can write trivially

$$-2I_zS_y \xrightarrow{\Omega_I t_1 I_z} -2I_zS_y. \quad (13.246)$$

Then occurs frequency labeling. Consider the scalar-coupling evolution during the first half ($t_{1/2}$) of the evolution which is expressed by

$$-2I_zS_y \xrightarrow{2\pi J_{IS}t_{1/2}I_zS_z} -2I_zS_y\cos(\pi J_{IS}t_{1/2}) + S_x\sin(\pi J_{IS}t_{1/2}). \quad (13.247)$$

The influence of the refocusing π-pulse selectively on the I-spin after the $t_{1/2}$ period is to change the sign of the single-quantum antiphase term

$$-2I_zS_y\cos(\pi J_{IS}t_{1/2}) + S_x\sin(\pi J_{IS}t_{1/2}) \xrightarrow{180_x^{(I)}}$$
$$2I_zS_y\cos(\pi J_{IS}t_{1/2}) + S_x\sin(\pi J_{IS}t_{1/2}). \quad (13.248)$$

In the second half of the evolution period, the scalar coupling evolves as

$$2I_zS_y\cos(\pi J_{IS}t_{1/2}) + S_x\sin(\pi J_{IS}t_{1/2}) \xrightarrow{2\pi J_{IS}t_{1/2}I_zS_z}$$
$$\left[2I_zS_y\cos(\pi J_{IS}t_{1/2}) - S_x\sin(\pi J_{IS}t_{1/2})\right]\cos(\pi J_{IS}t_{1/2}) +$$
$$\left[2I_zS_y\sin(\pi J_{IS}t_{1/2}) + S_x\cos(\pi J_{IS}t_{1/2})\right]\sin(\pi J_{IS}t_{1/2}), \quad (13.249)$$

which, after invocation of trigonometric identities, yields $2I_zS_y$.

The result that the magnetization does not evolve by J-coupling, but merely changes the sign of the magnetization from $-2I_zS_y$ to $2I_zS_y$, is the consequence of using the refocusing pulse on spin I halfway through the t_1 period. However, the heteronuclear spin S is spared by the refocusing pulse, which suggests that the evolution of the density operator from σ_3 to σ_4 should account for the precession of S_y alone. The $2I_zS_y$ term also has to be transformed under the influence of chemical shift during t_1. At the end of the evolution period, we obtain the following expression for the density operator

$$\sigma_4 = 2I_zS_y \xrightarrow{\Omega_S t_1 S_z} 2I_zS_y\cos(\Omega_S t_1) - 2I_zS_x\sin(\Omega_S t_1). \quad (13.250)$$

The subsequent transformation of the density operator occurs under the action of the refocused INEPT (also called reverse INEPT) pulses. The $90_x^{(I)}$ and $90_x^{(S)}$ pulses applied simultaneously transform the antiphase magnetizations to yield

$$\sigma_5 = 2I_zS_y\cos(\Omega_S t_1) - 2I_zS_x\sin(\Omega_S t_1) \xrightarrow{90_x}$$
$$2I_yS_z\cos(\Omega_S t_1) - 2I_yS_x\sin(\Omega_S t_1), \quad (13.251)$$

in which the $2I_yS_x$ magnetization represents a double-quantum coherence, which is not observable and hence need not be carried forward.

The product operator then evolves under the $-\tau-180_x-\tau$ segment of the pulse scheme (Figure 13.76) where the 180_x pulse is applied simultaneously on both I and S

$$\sigma_6 = 2I_yS_z\cos\left(\Omega_S t_1\right)\xrightarrow{-\tau-180_x-\tau} I_x\cos\left(\Omega_S t_1\right). \quad (13.252)$$

The signal is acquired only to detect I while the spin S is decoupled. The magnetization I_x produces an absorptive doublet of spin I centered at Ω_I which is modulated by the cosine term containing the frequency offset Ω_S. This produces cross peaks $\left(F_1, F_2\right) = \left(\Omega_I, \Omega_S\right)$.

As mentioned already, HSQC is highly sensitive and less time-consuming – as little as a few minutes, due to the reliance on ^{1}H spin-lattice relaxation time (Figure 13.77). Further, HSQC acts as a gateway to heteronuclear 3D spectroscopy discussed below.

13.37 3D NMR

The possibility of 3D spectroscopy was perhaps realized soon after the advent of Fourier spectroscopy in the 1960s, but the requirement of enormous data storage space, computer memory, and display software delayed the implementation of these experiments to the late 1980s. A range of 3D pulse schemes are now available in the repository to acquire both homo- and heteronuclear 3D spectra, even though the time needed to acquire a traditional-plane heteronuclear 3D spectrum of a large molecule may still run into several days, notwithstanding the use of recent spectrometer hardware. The requirement of narrow spectral width for homonuclear experiments involving ^{1}H reduces the acquisition time at no expense of resolution, but this advantage is of little use in the realm of large-molecule spectroscopy that invariably uses homo- and heteronuclear pulse schemes in conjunction. Thus, 3D NMR is expensive, but the expense is compensated by the resolution achieved and the information content in the spectra.

The basic idea of 3D spectroscopy is easy to see from Figure 13.78 where the acquisition time for the 1D experiment is labeled with t_1 only to be uniform with what follows. The number of evolution periods determines the dimensionality – no evolution is the commonplace 1D NMR, one evolution period (t_1) produces a 2D spectrum, two evolution periods (t_1, t_2) give rise to a 3D spectrum, and so on. The evolution time and hence the occurrence of coherence transfer provides a conservative description of multidimensional NMR, which means all evolution periods must evoke magnetization transfer. An 'accordion' type of spectroscopy, in which a number of 2D spectra are recorded with increments of delay τ between the spectra, should not be called 3D spectroscopy even though the signal $S\left(t_1, \tau, t_2\right)$can be Fourier transformed to obtain a third dimension $S\left(\omega_1, \omega, \omega_2\right)$. Because the delay τ is not associated with coherence transfer, accordion spectroscopy does not belong to the class of 3D.

The two evolution periods t_1 and t_2 in a 3D pulse scheme are created by combining two 2D pulse schemes. Figure 13.79 depicts the combination of two homonuclear COSY schemes to obtain a 3D COSY–COSY pulse sequence in which the first COSY scheme, excluding the detection period, acts as the preparation period for the second COSY experiment. Similarly, 3D COSY–TOCSY, NOESY–TOCSY, NOESY–NOESY, EXSY–EXSY, and a whole lot of these experiments can be carried out. The combination of homo- and heteronuclear 2D schemes provide 3D schemes such as NOESY–HSQC and TOCSY–HSQC. A typical 4D experiment where two heteronuclear 2D experiments are combined with one homonuclear 2D experiment is [^{1}H–^{13}C]

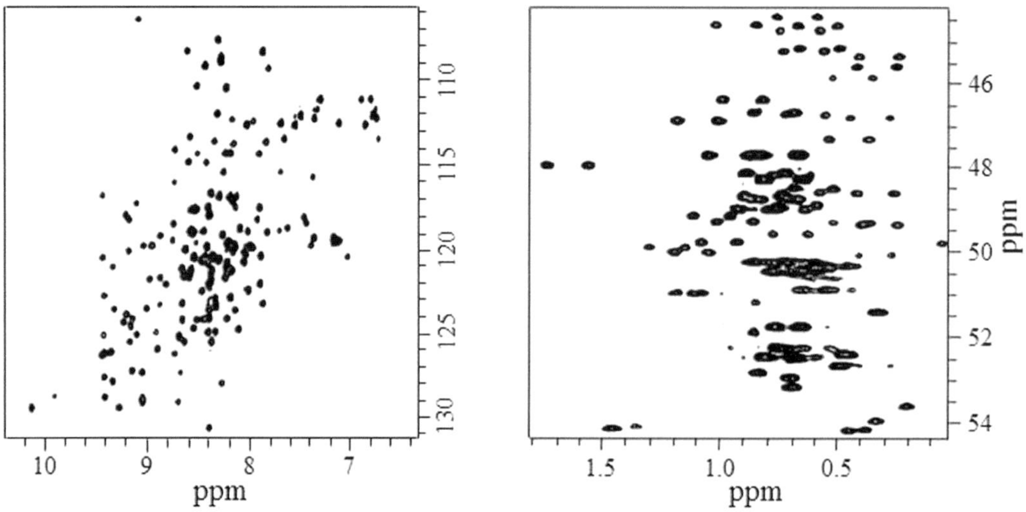

FIGURE 13.77 Regions of unassigned HSQC spectra of an intrinsically disordered plant phloem protein called *At*PP16-1; *left*, [^{1}H–^{15}N] spectrum, pH 4.1, 23°C; *right*, [^{1}H–^{13}C] spectrum, pH 4.1, 23°C.

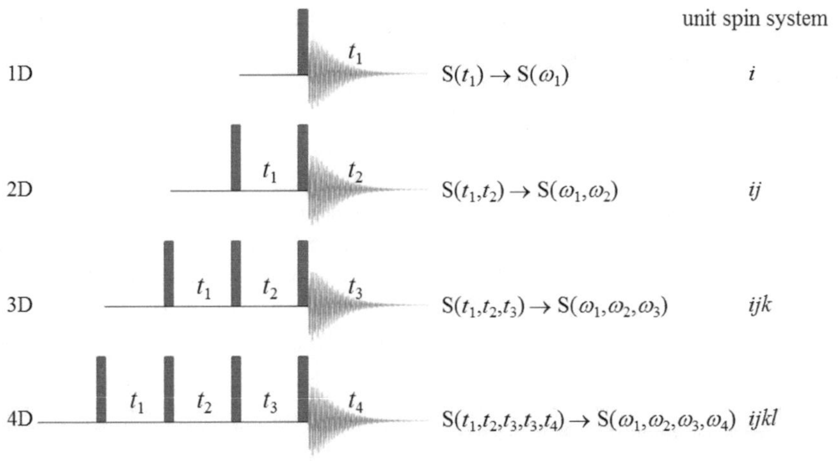

FIGURE 13.78 Correspondence of the number of dimensions (frequency axes) and the evolution periods in Fourier spectroscopy. The number of spins within which magnetization transfer occurs is equal to the number of dimensions.

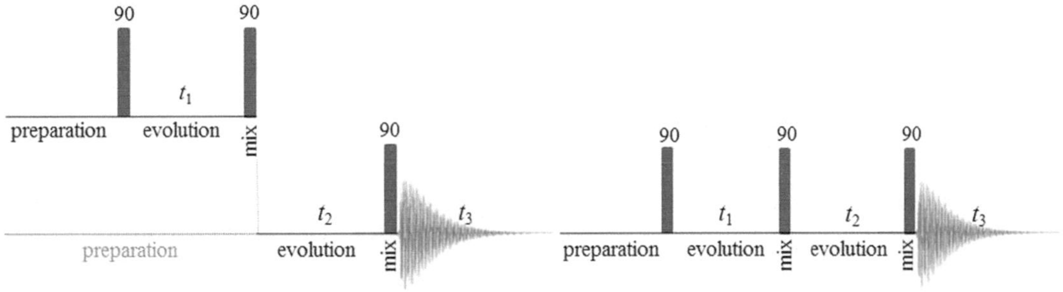

FIGURE 13.79 Sequential combination of two homonuclear 2D COSY schemes (*left*) to obtain a homonuclear 3D COSY–COSY sequence (*right*).

HMQC–NOESY–[^{1}H–^{15}N]HSQC, in which HMQC is heteronuclear multiple quantum coherence.

The three frequencies of a spectral peak $(\omega_1, \omega_2, \omega_3)$ must also mean the requirement of three spins *ijk*, homo- or heteronuclear, among which magnetization transfer can take place during the periods t_1 and t_2. If the spins *i*, *j*, and *k* are scalar-coupled, then coherence transfer within them establishes correlations of the three spins and we obtain a 3D correlation spectrum. In the case where two of the three spins are dipole-coupled, say *i* and *j* leaving the other two *j* and *k* as scalar-coupled, the transfer of longitudinal magnetization (incoherent) from *i* to *j* occurring in t_1 can be transferred to spin *k* in t_2, and the resultant 3D spectrum is an exchange-correlation spectrum. If the spins *i*, *j*, and *k* are only dipole-coupled, say in *i* and *j*, and *j* and *k* pairs for instance, then incoherent transfer of longitudinal magnetization during both t_1 and t_2 periods will give rise to a 3D cross-relaxation spectrum.

13.37.1 Dissection of a 3D Spectrum

We have learned that Fourier transformation of the 2D time domain signal $S(t_1, t_2)$ with respect to the acquisition time variable t_2 alone yields a series of 1D spectra $S(t_1, \omega_2)$ corresponding to different values of the time increment t_1. The columns of the $[t_1, \omega_2]$ matrix show sinusoidal variation in the signal intensity as a function of t_1. These 1D spectra $S(t_1, \omega_2)$ can be thought as 1D projections of the 2D spectrum $S(\omega_1, \omega_2)$. By the same analogy, the time-domain signal of a 3D spectrum $S(t_1, t_2, t_3)$ can be Fourier transformed with respect to the time variables t_3 and t_1 sequentially to obtain a set of 2D spectra $S(\omega_1, t_2, \omega_2)$ corresponding to the incremented duration of the evolution period t_2. For each t_2 we have a 2D spectrum (a plane) and the intensity of each of the cross peaks varies sinusoidally with time-incremented (t_1) planes. Instructively, the intensity of a 2D cross peak varies with t_2 in partial analogy of the way the $\mathbf{E}_o$ vector of a plane monochromatic wave varies with time (see Figure 1.5). Note the negative intensity of 2D cross peaks in the second half of the sine wave. This outcome is obviously the result of excluding the t_2 times from Fourier transformation. Figure 13.80 shows the variation of the 2D cross peaks with incremented t_2. In this sense, these spectra are 2D projections of the 3D spectrum. Finally, Fourier transformation in the t_2 dimension provides the frequency axis F_2, which represents the chemical shift of ^{15}N. The 3D frequency-domain spectrum can be generated by Fourier transformation of the 3D matrix $[t_2, \omega_1, \omega_3]$ with respect to the sinusoidal modulations represented by columns t_2. The diagonal of the 3D frequency-domain spectrum runs between two diagonally opposite vertices of a cube or a cuboid formed by the stack of the 2D planes. A projected 2D spectrum $S(\omega_1, \omega_3)$ can be defined if ω_2 is specified.

13.37.2 NOESY–[1H–15N]HSQC

As an example of a 3D double-resonance spectrum, let us briefly discuss the NOESY–[^{1}H–^{15}N]HSQC experiment that has gained enormous popularity in studies of large molecules like proteins. The experiment is not only simple and sensitive, but also provides an easy approach for resonance assignment of side-chain protons. The pulse scheme (Figure 13.81) is designed to carry out a ^{1}H homonuclear experiment, during the evolution period (t_1) of which the ^{15}N spins are decoupled. After the NOESY mixing period τ_m, the density operator is converted to observable magnetization by a 90° ^{1}H pulse, thus concluding the NOESY part of the experiment. The proton spins are already frequency-labeled and cross-relaxed. The HSQC part of the scheme starts out with simultaneous 180° pulses on ^{1}H and ^{15}N spins to induce polarization transfer from the former to the latter (INEPT). Then simultaneous 90° pulses on both of these nuclei create transverse magnetization of the ^{15}N spins by a single-quantum coherence transfer (HSQC) process. The magnetization now evolves in the time period t_2, at the completion of which a reverse INEPT sequence transfers the polarization to ^{1}H for detection. This short qualitative description is schematized in Figure 13.81 in which the 2D details of the individual experiments are already familiar to us.

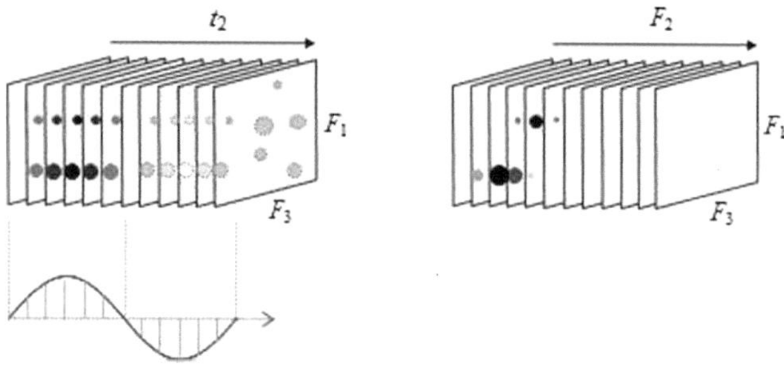

FIGURE 13.80 *Left*, 2D spectral planes of a 3D spectrum obtained by Fourier transforming $S(t_1,t_2,t_3)$ to $S(\omega_1,t_2,\omega_2)$. The variation of signal intensities are shown for one cycle of a sine wave. *Right*, intense peaks in fewer planes result after complete Fourier transformation.

The analysis of the data may be viewed in perspective of what has been discussed in the preceding. The same sinusoidal variation of peak intensity will be observed when Fourier transformed by excluding t_2. We then Fourier transform in the t_2 dimension to obtain the frequency axis F_2, which represents the chemical shift of ^{15}N. The final processed spectrum is analyzed by taking F_1F_3 slices at different F_2 frequencies corresponding to chemical shifts of different ^{15}N spins. In practice, the slices may be placed next to each other in some order of convenience. In protein NMR, the slices are arranged in stretches of contiguous amino acids (Figure 13.82). If the ^{15}N resonance assignments are available already, by a 2D [^{1}H–^{15}N]HSQC for example, the ^{1}H resonances contained in a slice can be assigned to the side-chain and C$^\alpha$ protons. In fact, the 3D NOESY–[^{1}H–^{15}N]HSQC spectrum is of great help for side-chain spin assignments. The assignments can also be verified by a 3D NOESY–[^{1}H–^{13}C]HSQC experiment which runs with the same pulse sequence as shown in Figure 13.81, where $\tau\left(=1/4J_{^1\mathrm{H}^{13}\mathrm{C}}\right)$ is now set according to the appropriate coupling constant (~140 Hz for one-bond J-coupling).

13.37.3 Triple-Resonance 3D Spectroscopy

These experiments developed some 30 years ago to study proteins integrate a heteronuclear 2D experiment with an indirect-detection 2D experiment (Kay et al., 1990). The experiments use one- and two-bond scalar coupling interactions to correlate three different spins, typically ^{1}H, ^{13}C, and ^{15}N of the protein backbone (Figure 13.83). There is a battery of such specialized experiments whose discussion here may seem out of place. A brief qualitative discussion of a prototype is given below.

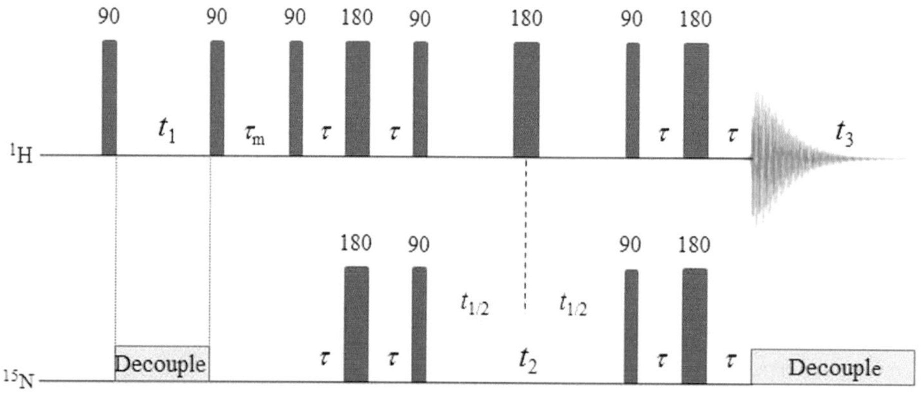

FIGURE 13.81 Pulse scheme for 3D NOESY–[^{1}H–^{15}N]HSQC shown without phase cycles.

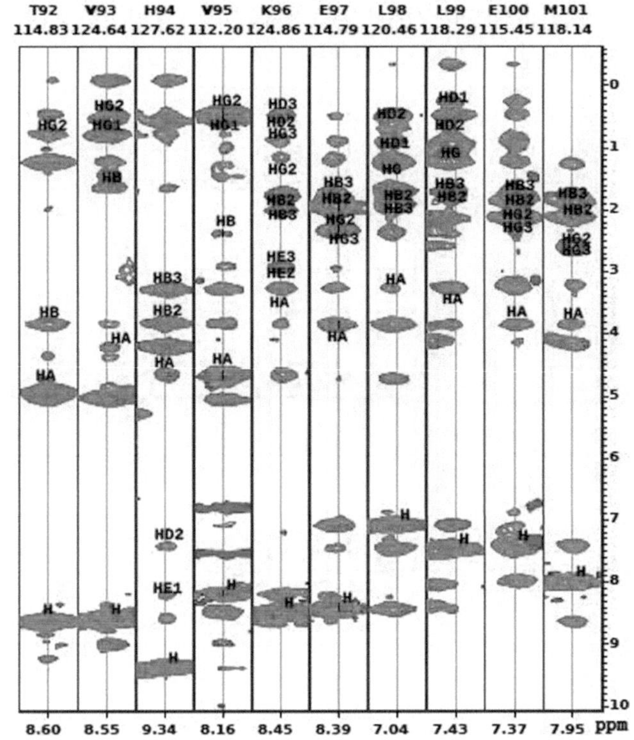

FIGURE 13.82 NOESY–[^{1}H–^{15}N]HSQC spectrum of *At*PP16-1, a plant phloem protein. Representative slices of $F_1(^1\mathrm{H})F_2(^1\mathrm{H}^\mathrm{N})$ at the indicated $F_3(^{15}\mathrm{N})$ chemical shifts are shown along with peak assignment labels. Amide protons ^{1}H$^\mathrm{N}$ are labeled simply using H, main-chain α-protons are indicated with the letter A, and side-chain β, γ, δ, and ε protons are indicated with B, C, D, and E, respectively.

An experiment called HNCA, dubbed for HNC$^\alpha$, relies upon both one- and two-bond scalar-couplings in the protein backbone to establish *J*-correlations between ^{1}H$^\mathrm{N}$, ^{15}N, and ^{13}C$^\alpha$. Consider a segment of the backbone containing ^{13}C$^\alpha$ spins of the ith and $i-1^{\mathrm{th}}$ residues along with the HNCA pulse scheme (Figure 13.84).

Let us represent the spins $^1\mathrm{H}_i^\mathrm{N}$, $^{15}\mathrm{N}_i$, $^{13}\mathrm{C}_i^\alpha$, and $^{13}\mathrm{C}_{i-1}^\alpha$ by I_1, I_2, I_3, and I_4, respectively. It is interesting that the effect of the 90° pulse operators immediately before and after the evolution periods t_1 and t_2 is to change the spin coherence. The effects of the pulses on the appropriate operators appearing along the sequence, and evolution of the two-spin antiphase operator during t_1 and the three-spin coherence operator during t_2 can be worked out with a little effort. The INEPT scheme affects magnetization transfer from $^1\mathrm{H}_i^\mathrm{N}$ to $^{15}\mathrm{N}_i$ to create antiphase operator of these two spins. The evolution of the $^{15}\mathrm{N}_i$ chemical shift occurs during the period t_1, in the middle of which refocusing pulses on ^{1}H$^\mathrm{N}$ and ^{13}C$^\alpha$ decouple them from the ^{15}N spin. At the end of t_1, a further delay δ $(= n / J_{\mathrm{NH}}$, where n=1, 2, 3, ⋯) is allowed to have ^{15}N magnetization antiphase with respect to the ^{13}C$^\alpha$ magnetization. Then the pulse operator corresponding to simultaneous 90° pulses on ^{1}H$^\mathrm{N}$ and ^{13}C$^\alpha$ yields coherence (operator) of all three spins ^{1}H$^\mathrm{N}$, ^{15}N, and ^{13}C$^\alpha$, also called three-spin coherence. The chemical shift of ^{13}C$^\alpha$ evolves during the evolution period t_2. The simultaneous refocusing pulses on ^{1}H$^\mathrm{N}$ and ^{15}N halfway through the t_2 period serves to obtain the evolution of the chemical shift of only ^{13}C$^\alpha$. Since the three-spin coherence has already been established, the chemical shift evolution of ^{13}C$^\alpha$ determines the evolution of the three-spin coherence. At the end of the t_2 period the three-spin coherence is destroyed by 90° pulses on ^{1}H$^\mathrm{N}$ and ^{13}C$^\alpha$. Finally, the magnetization is transferred to the ^{1}H$^\mathrm{N}$ spin by the refocusing INEPT for detection by the ^{1}H channel of the receiver.

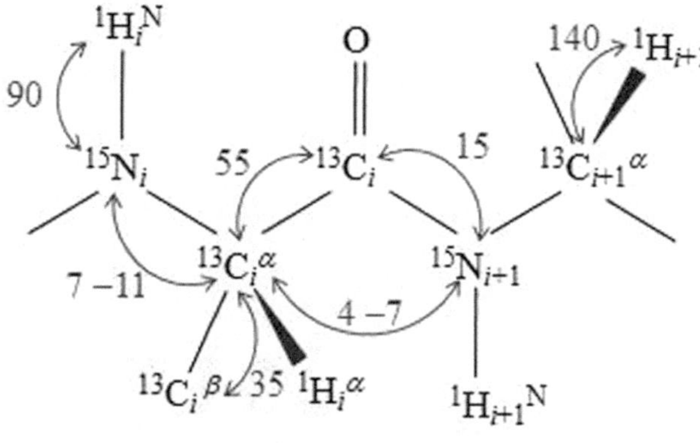

FIGURE 13.83 Protein backbone spins and *J*-couplings (numbers in Hz shown alongside the curved arrows) across one- and two-bonds. Since the experiments are invariably carried out on isotope-labeled samples, the carbon and nitrogen spins are shown with the isotope label. The subscript i and $i+1$ denote sequential amino acid residues i and $i+1$.

The one-bond coherence transfer $^1\mathrm{H}_i^\mathrm{N} \rightarrow {}^{15}\mathrm{N}_i \rightarrow {}^{13}\mathrm{C}_i^\alpha$ and the reverse $^{13}\mathrm{C}_i^\alpha \rightarrow {}^{15}\mathrm{N}_i \rightarrow {}^1\mathrm{H}_i^\mathrm{N}$ establish correlations of the three spins of the same residue i (intraresidue correlations). However, correlations also arise due to two-bond scalar coupling between $^{15}\mathrm{N}_i$ and $^{13}\mathrm{C}_{i-1}^\alpha$, in which case we have

$$^1\mathrm{H}_i^\mathrm{N} \xrightarrow{\text{one-bond transfer}} {}^{15}\mathrm{N}_i \xrightarrow{\text{two-bond transfer}} {}^{13}\mathrm{C}_{i-1}^\alpha$$

and the reverse $^{13}\mathrm{C}_{i-1}^\alpha \rightarrow {}^{15}\mathrm{N}_i \rightarrow {}^1\mathrm{H}_i^\mathrm{N}$. Since $^2J_{\mathrm{NC}^\alpha} \le {}^1J_{\mathrm{NC}^\alpha}$, the correlation of $^1\mathrm{H}_i^\mathrm{N}$ and $^{15}\mathrm{N}_i$ with $^{13}\mathrm{C}_{i-1}^\alpha$ is expected to be weaker. Nevertheless, this interresidue correlation is generally detectable and it provides an elegant approach to sequentially assign ^{13}C$^\alpha$ resonances along the protein backbone.

As an example of sequential assignment, Figure 13.85 shows sequentially arranged strips of ^{13}C$^\alpha$ chemical shift of residues W59 to G67 of the protein *At*pp16-1. A strip corresponding to the ith residue aligns its ^{13}C$^\alpha$ peak with that of the preceding residue ($^{13}\mathrm{C}_{i-1}^\alpha$), which in turn connects with the resonance of the residue preceding ($^{13}\mathrm{C}_{i-2}^\alpha$), and so on. Note that CA in all strips refers to the ^{13}C$^\alpha$ resonance corresponding to that strip. In the analysis shown, the resonance of the residue 64 is missing, but G67 is sequentially connected to G65, and F63 is connected to W59 through A61 and R60.

Other common triple resonance 3D experiments include HNCO, HN(CO)CA, HN(CA)CO, HN(CA)HA, HN(COCA)HA, HNCACB, and CBCANH. Often more than one of these experiments are performed to obtain assignments for a

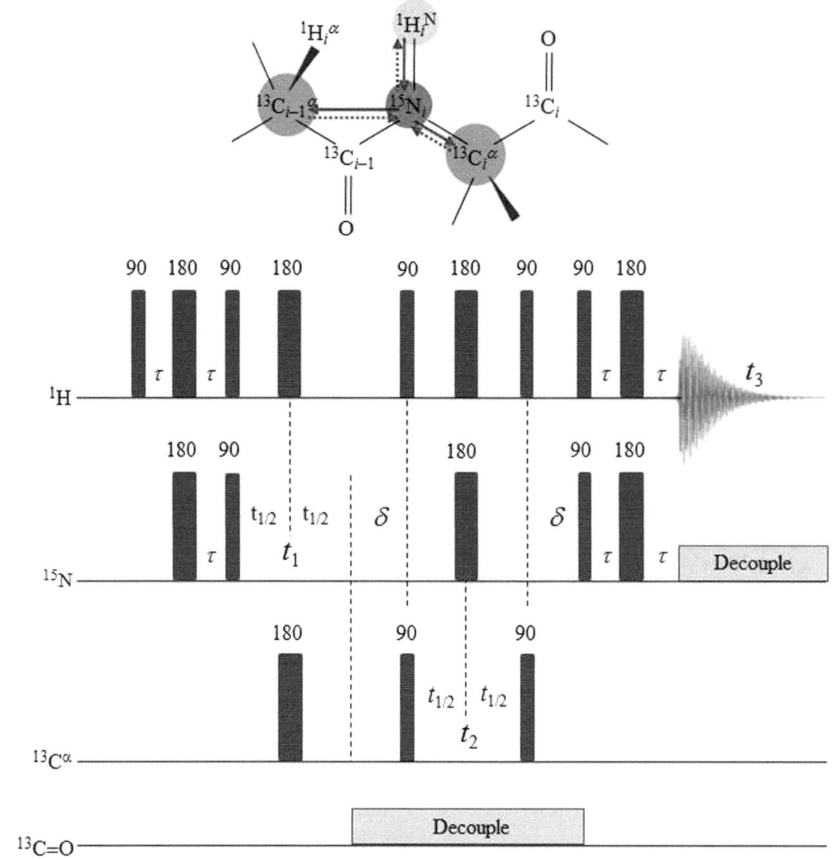

FIGURE 13.84 *Top*, two relevant amino acid residue groups showing intraresidue (residue *i*) coherence transfer across one bond from $^1H_i^N$ to $^{15}N_i$, and then to $^{13}C_i^{\alpha}$ and the reversal of both transfers, and interresidue (residues *i* and *i*–1) transfer across one bond (from $^1H_i^N$ to $^{15}N_i$) and two bonds (from $^{15}N_i$ to $^{13}C_{i-1}^{\alpha}$), and their reversals. The reversals of the one-bond coherence transfer are indicated by broken arrows. *Bottom*, the HNCA pulse scheme originally given in Kay et al. (1990).

reasonably large set of residues in order to calculate molecular structure.

13.38 CALCULATION OF 3D MOLECULAR STRUCTURE

If the resonances of a reasonable set of spins in a molecule are assigned by through-bond (coherent) experiments and pairwise distances between them are determined by through-space (incoherent) Overhauser experiments, one can place the atoms in a Cartesian frame according to pairwise interatomic distances. The lowest energy conformation of the molecule with these distance constraints is thermodynamically the most stable. Although the principle of distance-geometry structure calculation may appear straightforward and algorithm-based, the practical difficulties are too many and they multiply with the size of the molecule. For example, many resonances stay unassigned, mainly due to severe resonance overlaps, and the distances between these spins remain unknown. One should also acknowledge that the problem of this magnitude arises little in the case of small organic molecules whose structures are routinely determined by fewer stereo-specific resonance assignments, coupling constants, and interatomic distance constraints by moderate level of NMR experiments. Of course, calculation of small molecular structure is largely assisted by elemental analysis, mass data, and IR spectra. In a similar vein, structure of large molecules, as large as ~10 kDa can be determined by 2D homonuclear methods without the need of isotope enrichment; such exercises are protracting, however. The advent of 3D NMR methods and concomitant development of various software along with already existing molecular biology protocols to isotopically enrich proteins have greatly influenced the area of NMR structure determination, especially in the solution state. Numerous activities in the past 30 years or so have been devoted to finding approaches toward efficient ways of structure determination. For example, structure calculation with torsion angle constraints instead of distance geometry has been worked out. These activities are interdisciplinary in nature, and an exhaustive discussion of them is outside the scope here.

In brief, the peak volumes of assigned resonances extracted from 3D NOESY–HSQC spectra provide distance constraints and structural information. The chemical shifts of backbone atoms can be used to obtain restraints

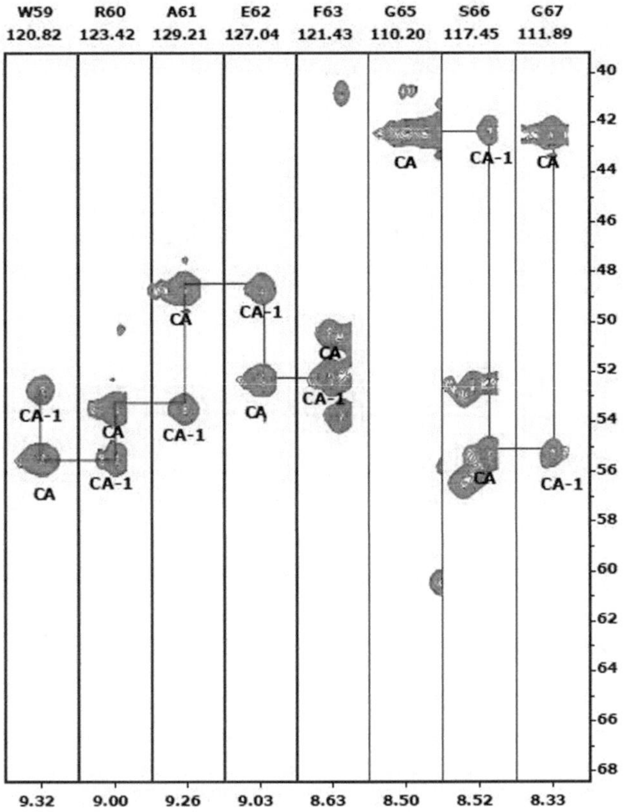

FIGURE 13.85 Representative $F_1(^{13}C^{\alpha})F_3(^{1}H^{N})$ slices at the indicated $F_2(^{15}N)$ chemical shifts from the HNCA spectrum of the protein *At*PP16-1.

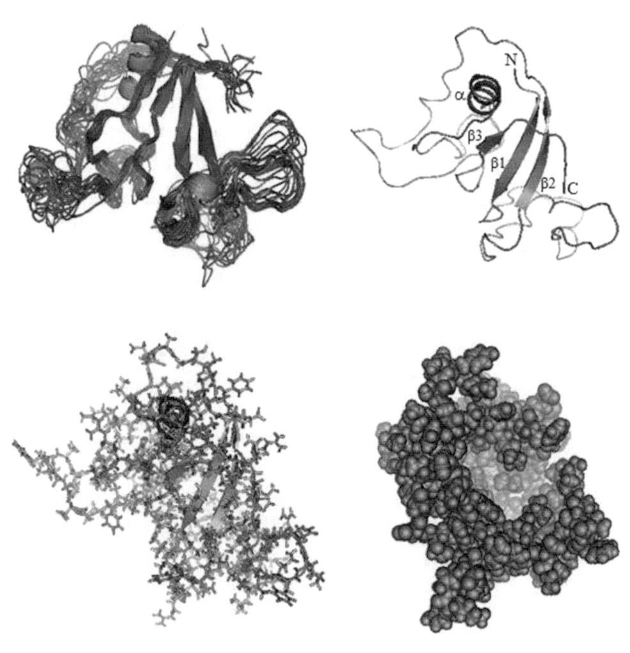

FIGURE 13.86 *Top left*, 20 best calculated structures of *At*PP16-1 superimposed, *right*, the best of the 20 structures (secondary structure elements labeled) selected by statistical criteria. *Bottom left*, a ball-and-stick model of the best structure, *right*, a space-filled model showing spatial irregularity and a moderately large cavity. Data are from the author's laboratory.

on backbone dihedral angles ϕ and ψ. This is called the torsion angle approach and is largely assisted by software packages, including CYANA, for which the torsion angle constraints, the amino acid sequence of the protein, and the NOESY–HSQC peak list are requisites. The idea is to vary the torsion angles alone using a large and variable target function so as to generate an ensemble of conformers, often in hundreds. Of these, the best structures are those that are lowest in energy and which violate the torsion constraints minimally. The range of the best structures can be narrowed down by examining the relative structural statistics of best structures in the ensemble. To reiterate, structure calculation is hugely aided by algorithms and software packages. As an example of structure calculation, Figure 13.86 shows an ensemble of the best 20 of some 200 computed structures. The global displacement and main-chain flexibility of these 20 structures are evident from the figure. At the end of strict evaluation of these structures, one pronounces that the structure consists of a single helix and three β-strands. An overwhelming region of the molecule appears structurally random coil-like, and hence structurally disordered. Such proteins, called intrinsically disordered proteins, account for about a-half of the total cellular proteins.

There is much more to NMR spectroscopy and its applications in molecular science. Exhaustive theoretical treatments of different topics are also available; all that are out of scope of this monograph. One is confident nonetheless that there is little that NMR cannot do.

Box 13.1 The Average Hamiltonian

Suppose a system is evolving in the time interval (t_1, t_2) under the influence of a time-dependent Hamiltonian. It is convenient to define an average Hamiltonian corresponding to the time interval $t_2 - t_1 = \Delta t$. Since the Hamiltonian works all through the interval, we denote it as $\mathcal{H}(t_1, t_2)$ only to indicate the time limits of the operator validity; it should, otherwise, be an average time-independent function. We show below that such a time-independent average Hamiltonian working within a specified time interval can be constructed from a periodic time-dependent operator $\mathcal{H}(t)$.

Say the system is driven by a time-periodic Hamiltonian as shown in Figure 13.87, such that the driving Hamiltonian and the time-dependent Schrödinger equation are given by

$$\mathcal{H}(t) = \mathcal{H}^0 + \mathcal{H}(t+T)$$

$$i\hbar\frac{\partial}{\partial t}|\Psi(t)\rangle = \mathcal{H}(t)\Psi(t),$$

where $\mathcal{H}^0$ is the time-independent (static) part of the Hamiltonian, and $T = 2\pi/\omega$ is the period. Generally, we are used to solve a time-dependent Schrödinger

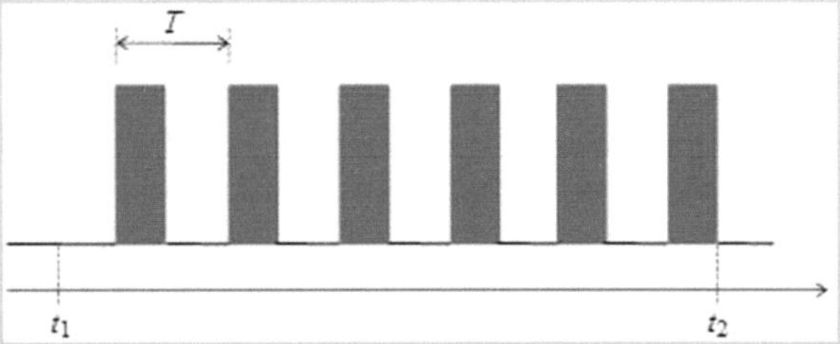

FIGURE 13.87 A sequence of periodic and uniform time-dependent Hamiltonian within the time interval $t_2 - t_1 = \Delta t$, appearing with a period T.

equation with some specified conditions in mind, the state function at time $t = 0$, $|\Psi(0)\rangle$, for example. But, to obtain all solutions of $|\Psi(t)\rangle$ a unitary time-evolution operator $U(t,0)$ is introduced. Accordingly, the propagation of our system to some longer time within the stipulated time $\Delta t = t_2 - t_1$ will be given by

$$|\Psi(t)\rangle = U(t,t_1)|\Psi(t_1)\rangle. \qquad (13.B1.1)$$

Because the driving Hamiltonian is time-periodic, the unitary evolution operator is

$$U(t+T,t_1) = U(t,t_1)U(T,t_1), \qquad (13.B1.2)$$

implying that the operator $U(t,t_1)$ in the interval $t_1 \le t \le T$ provides for expressing the operator for any or all the times $t \ge t_1$. If we write the unitary operator as a one-period propagation operator

$$U(T,t_1) = e^{-\frac{i}{\hbar}\mathcal{H}T},$$

in which $\mathcal{H}$ is Hermitian, then the exponential containing this Hermitian operator facilitates the introduction of another unitary operator $P(t)$ such that

$$P(t) = U(t,t_1)e^{-\frac{i}{\hbar}\mathcal{H}T}, \qquad (13.B1.3)$$

and

$$P(t+T) = U(t+T,t_1)e^{-\frac{i}{\hbar}\mathcal{H}(t+T)}$$

$$= U(t,t_1)\left\{U(T,t_1)e^{-\frac{i}{\hbar}\mathcal{H}T}\right\}e^{-\frac{i}{\hbar}\mathcal{H}t}. \qquad (13.B1.4)$$

Comparison of equations (13.B1.3) and (13.B1.4) suggests that $P(t) = P(t+T)$, implying time-periodicity of $P(t)$. This implication also allows one to write

$$U(t,t_1) = P(t)e^{-\frac{i}{\hbar}\mathcal{H}t} \qquad (13.B1.5)$$

showing a very useful concept that 'the product of unitary functions is a unitary function'.

The point here is that equation (13.B1.5) can be written in the form of an 'average Hamiltonian' $\bar{\mathcal{H}}(\Delta t)$

$$U(\Delta t) = e^{-\frac{i}{\hbar}\bar{\mathcal{H}}(\Delta t)\Delta t}. \qquad (13.B1.6)$$

The eigenvalues of $U(\Delta t)$ are written as a set $\left\{e^{-\frac{i}{\hbar}E_n t}\right\}$ along with the eigenstates to give the spectrum as

$$U(\Delta t,t_1) = \sum_n |n\rangle e^{-\frac{i}{\hbar}E_n t}\langle n|.$$

The relations above also yield the equality

$$e^{-\frac{i}{\hbar}\bar{\mathcal{H}}t|n\rangle} = e^{-\frac{i}{\hbar}E_n t}.$$

PROBLEMS

13.1 Higher static field B_0 provides higher intensity and better resolution of NMR spectra. To this boon comes certain disadvantages. List the undesirable effects brought about by high-field Zeeman interaction and justify the entries in the list.

13.2 The table below gives relevant information about Group 2 elements, except radium, whose nuclei are not preferred for NMR. Nevertheless, calculate the frequencies at which each of them will resonate when they are placed in 2, 6, 10, 14, and 18 T magnetic fields. Draw a graph of magnetic field dependence of the resonance frequency for each nucleus. How do the slopes for different nuclei compare as one moves down in the group from Be to Ba?

Elements	Isotope	Spin (s)	Gyromagnetic ratio, γ(rad T^{-1} s^{-1})
Be	^{9}Be	3/2	-3.759×10^{7}
Mg	^{25}Mg	5/2	-1.639×10^{7}
Ca	^{43}Ca	7/2	-1.802×10^{7}
Sr	^{87}Sr	9/2	-1.163×10^{7}
Ba	^{135}Ba	3/2	$+2.671\times10^{7}$

13.3 Draw stick diagrams of the ^{1}H spectrum of glycerol at pH 2 and 7. Discuss relative intensities and splitting patterns.

13.4 The concept of rotating frame is fundamentally important in NMR. Following are a few questions on the rotating frame idea that may be answered qualitatively.

(a) What will happen if the RF frequency ω_{rf}, also called carrier frequency, is matched absolutely with the Larmor frequency ω_0? Is some offset, meaning $\Omega = |\omega_{rf} - \omega_0|$, necessary?

(b) Why is Ω called chemical shift?
(c) Suppose $\Omega = 0$ for all nuclei, say ^1H, in the molecule. Does one still expect to see chemical shift dispersion in the spectrum?
(d) The carrier frequency is usually placed in the middle of the spectral width. Why?
(e) Often the true RF field in the rotating frame is not the one corresponding to ω_{rf}, but a modified effective field called ω_{eff}. How does the effective field influence the motion of a magnetization vector in the rotating frame?
(f) Is the effective field identical throughout the sample volume, or varying from one element to another? Justify.

13.5 Discuss the terms absorption and dispersion, each positive and negative.

13.6 Consider treating the proton spin using the formalism of two-level system for which the two eigenstates $|\alpha\rangle$ and $|\beta\rangle$ are defined in the presence of an external magnetic field $\mathbf{B}_0$. Use a $\mathbf{B}_1$ field due to RF to carry out an on-resonance transition. The interaction Hamiltonian is simply

$$\mathcal{H} = -\boldsymbol{\mu} \cdot \mathbf{B}_1$$

where $\mu = \gamma I$.
(a) Show that the NMR analogue of Rabi frequency is

$$\Omega_{NMR} = \frac{\mathbf{B}_1(t) \cdot \boldsymbol{\mu}_{\alpha\beta}}{\hbar}.$$

(b) Check that Ω_{NMR} has the dimension of Hz.

13.7 What are the main sources of chemical shielding for the alpha-proton and the amide hydrogen in the dipeptide Glycine–Glycine? Assume there is no ring nearby.

13.8 Describe how bond hybridization and dihedral angles affect the coupling constant.

13.9 Consider two ^1H nuclei 5 Å apart. The line joining the two nuclei is at 60° from the z-axis of the laboratory frame. Calculate the energy of interaction of the two nuclei.

13.10 In the description of dipolar relaxation the term 'transition rate' W is used. This is not the rate used in chemical kinetics. Some books and literature call W just probability, which is not quite right. With reference to the single-quantum pathway for two dipolar-coupled spins $|\alpha\beta\rangle \rightleftharpoons |\beta\beta\rangle$ derive an expression to provide the exact meaning of W.

13.11 The single-quantum transition rate W_1 for two dipolar-coupled spins k and l is

$$W_1 \approx 0.74 b_{kl}{}^2 J(\omega_k)$$

where ω_k is the relaxation frequency, and the dipolar interaction energy is

$$b_{kl} = \frac{\mu_0 \gamma_k \gamma_l \hbar}{4\pi r^3}$$

with $\mu_0 = 4\pi \times 10^{-7}$ kg m s^{-2} A^{-2} (A is area). Now, let k be a ^1H nucleus (26.752×10^7 rad T^{-1} s^{-1}) and l are the following

l	γ(rad T^{-1} s^{-1})
^{13}C	$+6.728\times10^7$
^{15}N	-2.712×10^7
^{19}F	$+25.181\times10^7$
^{31}P	$+10.841\times10^7$

(a) Draw a graph of W_1 with γ at $J(\omega_k) = 0.5$ for each kl pair. What do the graph properties indicate?
(b) If one carried out polarization experiments, where cross-polarization from k to one of the l nuclei is achieved, in what order the sensitivity of the latter nuclei will appear?

13.12 Let there be four homonuclear spins k, l, m, n within ~4 Å of each other's proximity, and let them cross relax.
(a) Note that the mere condition of being in close proximity does not qualify a set of nuclei to cross relax. Why so?
(b) Discuss how and why the spin will cross relax or not.
(c) Write down the Solomon equation for the network of the four spins in matrix format.
(d) Reduce the problem to the level of three spins k, l, m and solve for the eigenvalues (determinant) and eigenvectors. One may choose to use Kramer's rule to analytically solve for the eigenvectors.

13.13 Discuss critically the difference between NOESY and ROESY.

13.14 How is transient NOE different from steady-state NOE? Suppose one wishes to study bond vector dynamics in a molecule. Which of transient and steady-state NOE is useful? Justify the answer.

13.15 A flipping ring can also sway by virtue of its location in a molecule. Such librations of a segment, say to the right and then to the left, are distributed according to Boltzmann distribution. Let the resident segment of the flipping ring librate. Predict the effect of libration on the exchange peaks of the ring. Draw a 1D stick spectrum and schematic contours of the 2D exchange spectrum.

13.16 Write down the expectation value (signal) of a NOESY scheme in density matrix representation.

13.17 Take just one spin-half nucleus having the eigenstates $|\alpha\rangle$ and $|\beta\rangle$. Use the eigenvalue equations

$$I_z|\alpha\rangle = \frac{\hbar}{2}|\alpha\rangle$$

$$I_z|\beta\rangle = -\frac{\hbar}{2}|\beta\rangle$$

to derive the Pauli matrices. Note that $\hbar$ appears because of energy.

13.18 Consider three coupled spins I_1, I_2, and I_3 in a Cartesian basis. Show the matrices corresponding to the operators $I_{1x}I_{2x}I_{3x}$, $I_{1x}I_{2y}I_{3x}$, and $I_{1y}I_{2x}I_{3x}$.

13.19 Show the transformations of spin product operators of just one spin I in some arbitrary schemes where all pulses are along $+x$. The schemes need not be rational.

13.20 Draw schematic contours of a COSY spectrum of 2(methyl-cyclohexyl)4,6-dimethyl phenol.

13.21 Consider recording four TOCSY spectra of glutamic acid using 40, 80, 120, and 160 ms mixing times. What differences one expects to see amongst these four spectra. Show contour schematic of the spectra.

BIBLIOGRAPHY

Abragam, A. (1961) *Principles of Nuclear Magnetism*, Clarendon Press.

Aue, W. P., E. Bartholdi, and R. R. Ernst (1976) *J. Chem. Phys.* 64, 2229.

Avbelj, F., D. Kocjan, and R. L. Baldwin (2004) *Proc. Natl. Acad. Sci. USA* 101, 17394.

Barker, P. and R. Freeman (1985) *J. Magn. Reson.* 64, 334.

Bhuyan, A. K. and J. B. Udgaonkar (1999) *Biochemistry* 38, 9158.

Bloembergen, N., E. M. Purcell, and R. V. Pound (1948) *Phys. Rev.* 73, 679.

Bodenhausen, G., R. Freeman, and D. L. Turner (1977) *J. Magn. Reson.* 27, 511.

Bodenhausen, G., H. Kogler, and R. R. Ernst (1984) *J. Magn. Reson.* 58, 370.

Bodenhausen, G. and D. J. Ruben (1980) *Chem. Phys. Lett.* 69, 185.

Bolton, P. H. and G. Bodenhausen (1982) *Chem. Phys. Lett.* 89, 139.

Braunschweiler, L. and R. R. Ernst (1983) *J. Magn. Reson.* 53, 521.

Buckingham, A. D. (1960) *Can. J. Chem.* 38, 300.

Eich, G., G. Bodenhausen, and R. R. Ernst (1982) *J. Am. Chem. Soc.* 104, 3731.

Ernst, R. R. (1966) *Adv. Magn. Reson.* 2, 1.

Ernst, R. R., G. Bodenhausen, and A. Wokaun (1988) *Principles of Nuclear Magnetic Resonance in One and Two Dimensions*, Oxford University Press.

Ernst, R. R. and W. A. Anderson (1966) *Rev. Sci. Instrum.* 37, 93.

Fejzo, J., W. W. Westler, S. Macura, and J. L. Markley (1991) *J. Magn. Reson.* 92, 20.

Griesinger, C. and R. R. Ernst (1987) *J. Magn. Reson.* 75, 261.

Hahn, E. L. (1950) *Phys. Rev.* 80, 580.

Kay, L. E., M. Ikura, R. Tschudin, and A. Bax (1990) *J. Magn. Reson.* 89, 496.

Macura, S., Y. Huang, D. Suter, and R. R. Ernst (1981) *J. Magn. Reson.* 43, 259.

Marshall, T. W. and J. A. Pople (1958) *Mol. Phys.* 1, 199.

McConnell, H. M. (1958) *J. Chem. Phys.* 28, 430.

Noggle, J. H. and Schirmer (1971) *The Nuclear Overhauser Effect: Chemical Applications*, Academic Press.

Ramsey, N. F. (1950) *Phys. Rev.* 78, 699.

Rao, D. K. and A. K. Bhuyan (2007) *J. Biomol. NMR* 39, 187.

Resnick, R. and D. Halliday (1966) *Physics*, *Part II*, John Wiley & Sons.

Solomon, I. (1955) *Phys. Rev.* 99, 559.

Sorensen, O. W., G. W. Eich, M. H. Levitt, G. Bodenhausen, and R. R. Ernst (1983) *Prog. NMR Spectrosc.* 16, 163.

Stejskal, E. O. and J. E. Tanner (1965) *J. Chem. Phys.* 42, 288.

Thibaudeau, C., J. Plavec, and J. Chattopadhyaya (1998) *J. Org. Chem.* 63, 4967.

Wüthrich, K. (1976) *NMR in Biological Research: Peptides and Proteins*, North-Holland and American Elsevier.

Index